Into Statistics

Springer
Berlin
Heidelberg
New York
Barcelona
Budapest
Hong Kong
London
Milan
Paris
Santa Clara
Singapore
Tokyo

Into Statistics

A guide to understanding statistical concepts in engineering and the sciences

Peter J Smith

Springer

Peter J Smith
Department of Statistics
and Operations Research
Royal Melbourne Institute of Technology
360 Swanston Street
Melbourne, Victoria 3001
Australia

First published by Thomas Nelson Australia
under ISBN 0-17-008710-7

ISBN 981-3083-20-4

Printed in Australia.

SPIN 10570073 5 4 3 2 1 0

Contents

Preface

About the book

Into Statistics is moulded for the Australasian environment with examples and data sets mainly from the Australasian environment. In this sense the text is locally data driven. The data sets are supported in a framework of explanation and worked examples. There are few recipes. Formulae that are served as important are justified at the same time that they are used. For this, some of the formality of the reference text philosophy is maintained. There are theorems, there are definitions and there are proofs. The theorems and definitions are put precisely, the proofs are put simply. Overriding the text is an emphasis on understanding . . . *A Guide to Understanding Statistical Concepts in Engineering and the Sciences.*

Data sets

Data are presented from many sources: *real*, from carefully referenced publications; *real*, drawn from experiences at the consulting interface with users of statistics (possibly with changed references to ensure client confidentiality); *realistic*, to highlight educational points that may otherwise be smothered in the complexity of real data.

Data have been chosen to attract a wide audience. Examples are not all hard core mathematics/science/engineering. We believe that students hold interests beyond their areas of specialised study. Many are interested in the odds at a roulette table at Burswood Island, in scuba diving at the Abrolhos Islands, in Labor's odds in a federal election, in the height of the surf at Bells Beach, in the profile of the AFL's Collingwood football team, in Melbourne/California air pollution comparisons.

Users

In an ideal setting, this text could be used in statistics courses for science and engineering with co-requisite calculus. The text is designed for first encounters with probability and statistics. In many tertiary institutions, these occur from the second year of science and engineering study at a time when students need statistical ideas for investigations of their own.

First courses are either 'first and only', in which case the wide variety of local data, examples and worked examples helps to motivate the student; or 'first of many' in which case the more formal aspects of the text provide an appropriate foundation for future courses. For example, this text has been used in the introductory course for the Graduate Diploma in Statistics/Operations Research at RMIT in Melbourne.

Teaching philosophy

The ideas presented in this text accumulate from nearly two decades of tertiary teaching. The impression gained from this experience is that the student of statistics must be able to *do*; and having done, must be able to *understand* what has been done. This is preparation for the next time when the statistical task at hand is slightly different, the problem slightly changed, the computer output altered.

Computers are now widespread in teaching statistics. Fortunately, we are being taught how to *do* with ease: how to handle computer files; how to produce output easily. It is vital that we are prepared for an intelligent use of computer output: when a residual plot looks like *this*, what should we do? Should we transform the response? Should the model be changed?

For this reason there is little computer output in the text. The emphasis is on the interpretation and presentation of results rather than computer familiarity. If you have a favourite computer package, then *use it*. We have used MINITAB for examples and solutions. Such packages have enormous benefit in making concepts such as the Central Limit Theorem instantly unfold.

Chapter by chapter

The topic flow is standard. There is a greater emphasis on EDA techniques (Chapter 1) than in many texts. In particular, in readiness for computer-driven data analysis, an entire section on the role of power transformations has been included in the first chapter. Analysis of centre, spread and skewness of data occupies most of the first 100 pages of the book. It was quite intentional to expand the importance of this area of study before the simple probability discussions of Chapter 3.

The concept of independent random variables has been introduced early in the text (Chapter 5). This allows for an understanding of the binominal distribution as a sum of independent 0–1 random variables. In fact, many of the distributions introduced in Chapters 6 and 7 are built from sums of independent random variables, even members of the Poisson and Gamma families.

What happens to such sums when independence fails is discussed in Chapter 8—this provides an essential basis for understanding sampling without replacement. For example, the hypergeometric distribution is recognised, not through a complicated combinatorial formula, but rather as a sum of dependent 0–1 random variables in a without–replacement random sample.

The standard topics for statistical inference follow from Chapter 10. These include the careful construction and interpretation of confidence intervals (Chapter 10) and an easily–followed description of significance testing (Chapter 11 and 12). Significance tests from the same pivot are presented in a unified manner; this avoids the props of endless formula sheets. There is full discussion on the use of p-values which are prevalent in much published scientific work.

Linear regression is given extensive treatment to include regression diagnostics, points of influence and residual plots.

Short courses may choose to place less emphasis on specialised topics from Chapters 5–10.

Acknowledgments

I owe a great debt to past colleagues (or rather mentors) who helped shape my attitudes to the teaching of statistics: Beryl Hume, Robin Milne, Terry Speed, Robert Berk.

I owe a great debt to all my current colleagues and students for their enthusiasm and patience. Many went so far as to assist with data sets, references and comments on early drafts of the manuscript: Ian Cox, Dan Kildea, David Stewart, Raj Vasudeva, Barry McDonald, Santosh Kumar, Panlop Zeephongsekul, Ian Grundy, Kaye Marion and Ken Mavin. I particularly thank: Ian Peckham for surveying the manuscript for mathematical errors and for working solutions; Gary Fitz-Gerald and Jiami Zhang for their stylish typesetting; the publishers and reviewers for their criticisms, comments and suggestions. To all, my sincere thanks.

I dedicate this work to Mary Allison, Simon Smith and Lily Smith.

PJ S

CHAPTER ONE

On Sorting Data

In this chapter you will find

1.1 UNDERSTANDING DIFFERENT TYPES OF DATA

On Monday a commuter caught the 8.05 am train at Darling Station and arrived in Melbourne at Flinders Street Station 22 minutes later. On Tuesday the commuter caught the morning express at Darling, reducing the commuting time to 15 minutes. The commuting times for the next 20 weekdays were 24, 20, 22, 22, 25, 26, 23, 24, 21, 21, 21, 25, 21, 20 minutes, including 15, 14, 15, 15, 21, 16 minutes via the express. How long does the commuter take to get to work? Certainly the above record of commuting times helps us to reply; for example, we may quote the most frequently occurring recorded time, or, calculate an 'average' of the times listed. We shall see that such a record assists in answering much more searching questions such as: Is the commuting time significantly reduced by catching the express? Seeking answers to such statistical questions is often at the core of our endeavours.

Quantitative data

Frequently, through media reports, our own research efforts or the reports of other researchers, we encounter facts, measurements, or information called **data**. This data may reach us in various forms since there are many forms that facts, measurements and information may take. Typically, data consists of observed measurements (say lengths from ruler, callipers, ...), observed counts (say number of cars crossing the Sydney harbour bridge per hour, ...), or, observed labels (say eyecolour of twins, the trainers of Melbourne Cup winners, ...). For example, a market research firm may use telephone interviews to obtain data on product preference. The interviewer may ask over the phone: Which of the television channels 2, 7, or 9 is your most preferred station? The information provided by such a question not only provides a record of who watches what, but also may indicate to potential advertisers the audience and market share of the TV stations.

Example 1.1

How long does it take to get to work?

Commuting times, in minutes, from Darling Station to Flinders Street Station over 22 weekdays. (†denotes time for journey via the express.)

22 15† 24 20 22 22 15† 25 26
23 14† 15† 24 21 15† 21† 21 21
25 21 20 16†

In obtaining data we are observing the value of a **characteristic** or **variable**. For the commuter the variable of interest is 'commuting time' ; it is a variable since its value changes depending on when, how, or with what we observe it. The observations or realised values of this variable collectively form the data set listed in Example 1.1. Since this data set consists of numbers, 'commuting time' is a **quantitative variable**. Similarly, the quantitative variables 'weight' , 'height' or even 'emissions per second from a radioactive source' all give rise to **quantitative data**.

Quantitative data may be obtained through different scales of measurement with varying degrees of complexity. For example, weights in kg are measured on the **ratio scale**. This level of measurement is characterised by a constant difference between each successive unit on the scale, and the fact that there is a zero value located in the scale which has a natural physical interpretation. Suppose that we first think about constant differences. The difference in weight between an 86 kg person and an 85 kg person is the same as the difference in weight between a 60 kg person and a 61 kg person. There is more. The zero in the scale ensures that the weights may be compared via ratios (as the name of the scale suggests). That is, a 90 kg person is 1.5 times as heavy as a 60 kg person. Try to think about whether the same scale would apply if the weights were measured, say, on the moon (where gravity is one-sixth as formidable). Is the scale still ratio? What if we create a new quantitative variable 'natural logarithm of weight'? Numerically, the quantities $\log_e 86 - \log_e 85$ and $\log_e 61 - \log_e 60$ are no longer the same.

Example 1.2

A census of Australian jails

Prison profile: summary of prisoners by 'age' and 'sex' (Australia, 30 June 1983)

(*Source*: Cameron, R.J. (1985). *Year Book Australia 1985.*)

Under 16 years	1
16 years	14
17 years	97
18 years	376
19 years	575
20–24 years	2958
25–29 years	2347
30–34 years	1503
35–39 years	998
40–44 years	613
45–49 years	299
50–54 years	195
55–59 years	140
60–64 years	47
65 years and over	21
Unknown	12
Total persons	10196
Total males	9797
Total females	399

A measurement scale which possesses constant interval size but lacks a fixed origin is termed an **interval scale**. Temperature scales in degrees Celsius and degrees Fahrenheit are the standard examples. Clearly, on the Celsius scale, the difference between 20 degrees and 10 degrees is 10 degrees, as is the difference between 30 degrees and 20 degrees. However, without an absolute zero on the Celsius scale it is not meaningful to say that 20 degrees is twice as hot as 10 degrees. (Note that temperatures in degrees Kelvin are measured on a ratio scale.)

Discrete and continuous data

Discussion of scales of measurement relates to how we may handle or interpret the observed values of a variable. Quantitative variables are of two distinct types classified according to the possible values the variable may assume. A quantitative variable is said to be a **continuous variable** (and yields continuous data) if it may assume, either practically or conceivably, any value within a given range of values. For example, data from volume measurements is continuous—the actual fluid contents of a 2 litre container of water could theoretically take any value in the interval 1.9 litres to 2.1 litres. We may write this mathematically using **interval notation**: the interval of values occurring between 1.9 and 2.1 inclusive is written $[1.9, 2.1]$. The square bracket indicates that the endpoints are included in the interval. A round bracket may be used to indicate endpoint exclusion: $[1.9, 2.1)$ denotes values between 1.9 and 2.1 but excluding 2.1.

Discrete data arises from **discrete variables** which may assume only specified values. Such data frequently comes from a counting process. Indeed, 10 people may be counted entering a Coles supermarket in a specified 5 minute time period, but 10.5 people could never be counted walking through the supermarket doors. Thus the variable 'number of people entering a Coles supermarket in a 5 minute period' may assume only consecutive integer values 0, 1, 2, 3, ...

Often the limits of our measuring instruments tend to **discretise** continuous data; that is, make continuous data *appear* discrete. Under the current limits of electronic timekeeping devices, the only possible observed times for a 100 metre Olympic sprint are specific values 0.001 second apart in an interval near the world record. Even so, such data are still regarded as continuous.

Qualitative data

So far our discussion has involved data sets consisting of numbers. Other types of data also frequently arise. The values of a **qualitative variable** are labels, titles or categorical listings. Such values are termed **levels**. For example, the qualitative variable 'occupational status' may have levels: employer, self-employed, employee, helper, looking for first job, other unemployed, not in labour force. The variable 'occupational status' is measured on the **nominal scale**. This scale is used when the levels act simply as labels for categorical membership.

In the case of gradings (poor, good, excellent) of a wine-tasting panel, measurement is more sophisticated. This is not because the panel consists of connoisseurs (although it may well do), but rather, the levels show an inherent order; 'excellent' is better than 'good' which is better than 'poor'. Here the panel is measuring wine quality on an **ordinal scale**.

Notice that the scales of measurement discussed here form a hierarchy. The ordinal scale possesses all attributes of the nominal scale plus more: an ordering. Similarly, the interval scale may be regarded as more sophisticated than the ordinal scale with the ratio scale affording the highest level of measurement.

Why do we need data? From sample to population

Demography is the study of populations of people. Certainly many statistical techniques are used in this study. However, our use of the term 'population' is wider than its usual demographic meaning. A **target population** consists of all possible observations on a variable which holds our interest in a particular study.

The word 'target' (which is often omitted for brevity) helps to remind us of the difficulty in defining, pinning down or getting at a listing of population members. For example, when studying tropical diseases of goldfish, the fish themselves are difficult to locate since lists of owners do not exist. A complete listing of the members of a finite population is termed a **census**. For some populations such a list is relatively easy to compose. The counts in the various age categories in Example 1.2 will have been compiled from the files on prisoners at institutions around Australia. Here a census is possible.

Often the variable of interest may be a continuous one. Then the population is hypothetical with uncountably many members and a census is out of the question. For example, consider an engineering firm planning to build a bridge across the Murray River. Then, at a particular crossing, there are uncountably many possible depth measurements. Just try to count them all!

The age profile of the entire prison population in Example 1.2 consists of a relatively small 10 196 observations; this information is readily stored in a computer. However, for much larger populations counting becomes difficult. After all, a census of the people of Australia is conducted only about twice per decade. When time, cost and accessibility factors make complete knowledge of a population prohibitive, it becomes necessary to extract a representative sample from the population. Put simply: a **sample** is a subcollection of the population. The idea is to use this sample to try to predict qualities of interest in the whole population. Usually, the number of items in the population that we sample is very much smaller than the totality of items in the population. To be of use, a sample must reflect the nature of the population.

Generally our data sets result from a sampling procedure. Even in recording the data set of commuting times in Example 1.1, we are sampling values from the theoretical population of all possible commuting times. By displaying, studying and analysing the sample we may glean information about the population from which the sample came.

This process, called **statistical inference**, is discussed in the later chapters of this book. A simple example: in a general election, a candidate may concede defeat a few hours after the polls close if his/her electoral support trails badly even with only 50% of the vote counted, provided (and this is the important aspect) the 50% is a *representative* sample of the constituents in the electorate. But note that pundits, also known as psephologists, often make incorrect inferences about election swings simply because the first votes counted on election night are not necessarily representative of the voting patterns to emerge later across the electorate.

In order to proceed with the analysis of data sets, we need an algebraic way of representing them. If we are observing the outcomes of a discrete variable X, we write the general data set, consisting of n observations on X, as $x_1, x_2, x_3, \ldots, x_n$.

The subscript notation identifies the recorded position of the observation in the data set. In general, throughout this text, we use upper case letters to denote the variables themselves and lower case letters to denote the values that they may assume. This means that X may denote a variable and x an observation of X.

Rounding data: sensible accuracy

We often have choice in the degree of accuracy with which quantitative data are reported. This especially applies to the results of calculations based on our data. It does not mean that we should make data misleading; it means that many numbers carry the weight of extra digits that disguise a quick reading of the magnitude of the number. This applies, not just to data sets, but to quantitative numbers of all kinds. For example, an advertised '\$100 000 salary package' (often abbreviated further as: 'salary package to 100K') conveys, for advertising purposes, at least as much as 'salary package of \$99 236.98 offered'. The extra digits tell us nothing. The same applies to the results of our statistical calculations—we rarely need all 8 digits copied from a calculator display.

Intuitively, the **rounding of a number** x is based on the following idea: assign x to its nearest neighbour in units of the desired order of magnitude. Such units of magnitude are often powers of 10. Thus \$99 236.98 is \$100 000 when rounded to the nearest \$10 000. (\$99 236.98 is much closer to \$100 000 than it is to \$90 000 when considering units of \$10 000.) Further, \$99 236.98 is \$99 237 when rounded to the nearest dollar. (\$99 236.98 is much closer to \$99 237 than it is to \$99 236 when considering units of \$1.)

Practically, discard unwanted trailing digits, and, increase by one the last digit retained if the first digit dropped is 5 or more. Thus, \$99 236.98 is \$99 237.0 when rounded to 1 decimal place: 8 is discarded and the last retained digit, 9, is increased by 1 (causing an increase in the units digit). We use the notation 'dp' in this text to denote rounding to decimal places. This form of rounding is very commonly used in engineering and the sciences. We offer this advice on accuracy:

- Report the results of calculations to 1 dp more than the data.
- Do not round during a calculation.

Another way of reporting numbers is through **scientific notation**. Most calculators use this method of displaying the order of magnitude of a number as a power of 10. To convert a number x into scientific notation, we determine $a \in [0, 10)$ and an integer k so that $x = a \times 10^k$. Thus 99 236.98 = 9.923 698 $\times 10^4$ and 0.000 992 369 8 = 9.923 698 $\times 10^{-4}$.

In scientific notation form, some of the decimal places in the number a may seem superfluous. If we round the number a in $x = a \times 10^k$ to have r digits, then the new value of x so obtained is said to have $\boldsymbol{r}$ **significant figures**. Thus 99 236.98 = 9.923 698 $\times 10^4$ and is 9.924 $\times 10^4$ = 99 240 to 4 significant figures. 0.000 992 369 8 = 9.923 698 $\times 10^{-4}$ and is 9.92 $\times 10^{-4}$ = 0.000992 to 3 significant figures. We use the notation 'sf' in this text to denote rounding to significant figures.

Example 1.3

Reporting the size of States and Territories

The Age, Melbourne, 22 February 1989, reported data on the relative size, in square kilometres, of Australia's State/Territory components. We also report the resident population as estimated (counted by census) at 30 June 1989. (*Source*: Castles, I. (1990). *Year Book Australia 1990*.)

State/Territory	Area (km^2)	Population
Western Australia	2 525 500	1 591 077
Queensland	1 727 200	2 830 198
Northern Territory	1 346 200	156 147
South Australia	984 000	1 423 337
New South Wales	801 600	5 761 919
Victoria	227 600	4 315 170
Tasmania	67 800	450 956
Australian Capital Territory	2 400	277 926

Clearly, the two sets of data are reported with different degrees of accuracy. The area data is apparently reported to the nearest 100 km^2. This gives an indication that each data point is an estimate of the true area. The census data is reported to the nearest count—the context of the report emphasises that these are estimates as no census could hope to deterministically count *every* resident once and *once only*.

The question now is: how densely packed are we across the States/Territories? In the table below we show the population densities (persons km^{-2}) obtained by dividing the population for each State by its area. Notice the extremely high density for the ACT. The three numerical columns of the table represent the calculation: from the pocket calculator, rounded to 1 decimal place, rounded to 2 significant figures. This is done to illustrate how the methods of rounding compare. For these numbers, the 1 dp rounding seems most suitable for making interstate comparisons easy.

State/Territory	Population density (persons km^{-2}) from calculator	to 1 dp	to 2 sf
Western Australia	0.6300048	0.6	0.63
Queensland	1.6386047	1.6	1.6
Northern Territory	0.1159909	0.1	0.12
South Australia	1.4464807	1.4	1.4
New South Wales	7.1880227	7.2	7.2
Victoria	18.959446	19.0	19
Tasmania	6.6512684	6.7	6.7
Australian Capital Territory	115.8025	115.8	120

EXERCISES

1.1 The current records of 10 patients visiting the outpatient clinic of a Perth hospital gave the following information.

Year of birth	Blood type	Religion	Blood pressure
1954	A	Catholic	low
1938	B	Lutheran	normal
1972	A	unrecorded	normal
1941	A	Adventist	high
1942	A	unrecorded	normal
1923	AB	Anglican	normal
1957	O	Anglican	normal
1944	B	Presbyterian	high
1966	A	unrecorded	low
1918	O	Catholic	normal

Classify the four variables according to their scale of measurement:

nominal, ordinal, interval or ratio.

1.2 For each of the following variables, indicate the scale of measurement and state whether each variable is discrete or continuous:

(a) Time of day on the 24 hour army clock (where, for example, 9.35 a.m. is represented by 0935).

(b) Shoe size (with values 5, 5.5, 6, 6.5, 7, ...).

(c) Height of a mountain above sea level (in metres).

(d) Temperature (in degrees Kelvin).

(e) Size classification of hen eggs (with values 43, 45, 55, 65, ... grams).

1.3 Would the following variable be considered discrete or continuous? 'The ratio of defectives to non-defectives in a batch of 10 manufactured items which are not all defective.' Explain.

1.4 When individuals become involved in an exercise program, many gymnasiums offer 'fitness assessments' as part of their service. One aspect of such an assessment is the determination of 'percentage of body-fat' using a skinfold measure; callipers are used to measure the thickness of a pinch of skin (and underlying fat). The results combined from several body locations are averaged and then may be gauged against the following ready reckoner:

How you rate	Percentage body fat(%) Women	Men
Low fat	14.0–16.9	7.0– 9.9
Ideal	17.0–19.9	10.0–12.9
Average	20.0–23.9	13.0–16.9
Above average	24.0–26.9	17.0–19.9
High fat	27.0–29.9	20.0–24.9
Obese	$\geq$ 30.0	$\geq$ 25.0

(a) Which scale is used to measure the 'How you rate' variable?

(b) Which scale is used to measure the 'Percentage body fat' variable?

(c) There is no apparent category for an individual measured, in the above manner, with 19.95% body fat. What does this say about the accuracy of percentage body fat measurements?

(d) Is it likely that an individual with a percentage body fat assessment of 8.5% is a man? Explain.

(e) Consider the following measurements of percent body fat in men: 22, 12, 12, 16, 13.5, 15, 16, 22, 22.5, 20, 17, 11, 19, 19.5, 9, 21, 16, 16.5, 18.5, 10.

What proportion of these men had an 'Ideal' level of body fat? Would this measure be useful to predict the true proportion of men in society at large with an 'Ideal' level of body fat?

(f) Turn the quantitative data in (e) into observations on the qualitative variable 'How you rate'.

1.5 The following data show AFL (Australian Football League) career games and goals for three players from the Geelong Club.

	Games	Goals
Ablett	146	514
Bairstow	89	107
Barnes	12	12

(*Source:* *The Sunday Age* Football Magazine, Season 1992.)

Determine, to an appropriate number of decimal places, the value of the ratio

$$\frac{\text{Goals}}{\text{Games}}$$

for each player.

1.6 For the data summarised in the 'Age' census of Australian jails in Example 1.2 determine answers, where possible, to the following questions:

(a) What percentage of the prison population was women?

(b) What percentage of the prison population was under 20 years?

(c) What percentage of the prison population was women under 20 years?

(d) What percentage of the prison population was under 22 years?

1.7 Round:

21 591.561 to 2 sf;
21 960.729 to 3 sf;
22 001.772 to 4 sf.

What is noticeable about this result?

(The rounded scientific notation form 2.2×10^4 indicates that two figures are significant. Similarly, the rounded scientific notation form 2.20×10^4 indicates that three figures are significant. However, when multiplied out, both forms are written 22 000, and the possible significance of some or all of the 0 digits is hidden.)

1.8 The following data give the chemical composition of quartz-kaolinite rock by laboratory determination.

	% Weight
SiO_2	71.91
TiO_2	0.58
Al_2O_3	14.38
Fe_2O_3	3.56
MnO	0.06
MgO	1.24
CaO	2.81
Na_2O	4.98
K_2O	0.26
P	0.18
S	0.04

(*Source*: Letich, E. C. and Feenan, J.P. (1989). Quartz-kaolinite rocks from northeastern New South Wales: Distal products of a pluton-driven hydrothermal system. *Australian Journal of Earth Sciences*, 36, 275-81.)

Round these determinations to 1 decimal place.

1.9 February 29th is an extra day occurring in 'leap years' to catch up for lost time: a 365 day calendar is short of the 365.242199 days that it takes the earth to circle the sun.

Assume that the solar year consists of 365.242199 days.

(a) Round 365.242199 to:

i. 4 decimal places;

ii. 4 significant figures.

(b) What is the absolute difference (in days) between the number of days in four solar years and the total number of calendar days in the years 1991, 1992, 1993 and 1994? Give your answer in scientific notation.

1.10 The data of Example 1.3 are given below.

State/Territory	Area(km^2)	Population
Western Australia	2 525 500	1 591 077
Queensland	1 727 200	2 830 198
Northern Territory	1 346 200	156 147
South Australia	984 000	1 423 337
New South Wales	801 600	5 761 919
Victoria	227 600	4 315 170
Tasmania	67 800	450 956
ACT	2 400	277 926

For this data:

(a) express the area of each State to 2 sf;

(b) round the population of each State to the nearest 1000.

1.11 Consider the following definition of **rounding**:

x is rounded to y to the nearest r if there exists an integer a such that y is the endpoint of the interval

$$ar \leq x \leq (a+1)r$$

closest to x.

Use this definition to round 99 236.98 to the nearest:

(a) 500; (b) 50; (c) 10; (d) 1.

1.2 STEM-AND-LEAF PLOTS: SHOWING OFF THE DATA

There are many ways to represent data visually. In particular, in this section we consider the construction of **stem-and-leaf plots** (or **S/L plots**), a relatively recent invention of J.W. Tukey (1977) now having universal acceptance, not only by other authors, but also as output from statistical computer packages.

In the construction of S/L plots our prime purpose is to display data and suffer only a minimal loss of information. That is, the collective size (or magnitude) and shape (or proximity) of the numbers in the data set should be visible at a glance. Using this visual data summary it is also possible to highlight qualitative aspects of a data set that would otherwise be discarded.

S/L construction

The following technique is for 'looking closely at' quantitative data. Begin by examining the magnitude of the data; look carefully at the numbers and ask: Are they of a similar size? Are there physically meaningless digits trailing the decimal point? What aspects of each data point make it different from other data points in the data set? What is the smallest (largest) data point?

Specifically, for each data point identify a **stem** to categorise that data point relative to others, and a **leaf**, often in the form of a trailing digit, to maintain the identity of the data point within the category that it has been placed. In the simplest case, the division between stem and leaf happens at the decimal point. For a data point like 20.2 (extracted from a data set of maximum daily temperatures in Geelong) the division may fall as

stem	leaf
20	2

given that the temperatures generally vary from 10 to 40 degrees, and one decimal place would generally be a sensible accuracy for scientific reporting of such temperatures. (Media such as television report maximum daily temperatures as whole numbers.)

Different data points with the **same stem** are placed on the **same line** in the plot; they are identified by **separate leaves**. A **key** and a **title** should accompany all S/L plots. This is designed to provide enough information for immediate visual interpretation and to give a sense of scale to the data presented.

Example 1.4

How hot does it get during the day?

Maximum daily temperatures in Geelong and environs. A random sample of $n = 70$ temperatures recorded both at city and rural locations during the autumn and summer of 1983. Temperatures (to 1 dp) in degrees Celsius:

18.4	18.4	27.9	24.7	18.4	15.1	28.2	29.6
22.5	12.1	14.8	19.5	26.2	20.2	20.3	32.9
16.9	16.1	26.6	33.7	17.4	11.3	22.4	27.5
15.9	14.3	31.7	26.2	19.1	23.4	16.3	25.2
14.7	27.5	17.3	17.2	24.7	37.5	38.4	27.1
31.2	25.1	23.3	38.7†	15.3	29.4	25.7	24.6
32.1	17.2	35.3	13.9	25.3	21.6	31.5	30.3
15.8	21.1	21.3	21.5	28.8	22.1	23.2	25.2
28.2	23.2	13.0	26.8	24.4	33.8		

Geelong temperatures

Scale: stems—°C; leaves—10^{-1} °C

20 | 2 represents 20.2 degrees Celsius

S/L Plot 1		S/L Plot 2	
11	3	11	3
12	1	12	1
13	9 0	13	0 9
14	3 7 8	14	3 7 8
15	1 3 8 9	15	1 3 8 9
16	3 9 1	16	1 3 9
17	2 4 2 3	17	2 2 3 4
18	4 4 4	18	4 4 4
19	1 5	19	1 5
20	3 2	20	2 3
21	6 5 1 3	21	1 3 5 6
22	4 1 5	22	1 4 5
23	4 2 2 3	23	2 2 3 4
24	7 6 4 7	24	4 6 7 7
25	2 7 3 2 1	25	1 2 2 3 7
26	2 8 2 6	26	2 2 6 8
27	1 5 5 9	27	1 5 5 9
28	2 8 2	28	2 2 8
29	4 6	29	4 6
30	3	30	3
31	5 2 7	31	2 5 7
32	9 1	32	1 9
33	7 8	33	7 8
34		34	
35	3	35	3
36		36	
37	5	37	5
38	4 7†	38	4 7†

†denotes Ash Wednesday temperature

By attending to each of the data points one at a time, S/L Plot 1 in Example 1.4 is rapidly assembled. Once we decide that each degree will comprise a stem, then the decimal places occupy leaf positions. This provides us with the **key**. The plot is then completed with a bold **title**.

Also given in Example 1.4 is the corresponding plot (labelled S/L Plot 2) where the leaves have all been ordered from smallest to largest. This is all easy preparation for later calculation of quantities (such as medians, quartiles, percentiles) which depend essentially on the natural ordering in the data—we will fully discuss such calculations later in this chapter. For now, simply notice that by ordering the leaves from smallest to largest on each stem we have ordered the entire data set from smallest to largest since the stems themselves had an in-built ordering. Also in S/L Plot 2 we show how to retain qualitative information in the plot. This is achieved by using a label (dagger in this example) associated with certain leaves in the plot; the meaning is defined in a key beneath the plot. Put simply: the more information conveyed in the plot the better. This is very useful for maintaining information on unusual observations in the plot.

In constructing a S/L plot it is important to assign to each leaf the same sized space as every other leaf. Then on a given stem, the total space taken up by all the leaves is directly proportional to the number of data points represented on that stem. Therefore, by directly comparing different stems within the plot, the *shape* of the data set becomes apparent. We now define some terms to describe the shape of data.

The extremes (high or low) of a distribution of data are termed **tails** and a data set exhibits **skewness** if the tails are of differing length. If a S/L plot is rotated anti-clockwise 90 degrees then the left tail corresponds to low data values and the right tail to high data values. This rotation would correspond to writing the stems on a horizontal axis. Then, if the left tail is longer than the right, the resulting plot is **left-skewed**; if the right tail is longer than the left, the plot is **right-skewed**. The plot is **symmetric** about its centre if the tails are of similar length. When distributions of data are skewed, then so are the S/L plots representing them.

How do we choose the right number of stems? There are no abiding rules; selecting too few stems results in a loss of information—too many in overkill and aspects of the shape of the data are lost in the resulting jumble. When forced to put a number to it, we quote Dixon and Kronmal (1965) with the suggestion

$$k = \text{number of stems} = [10 \log_{10} n]\,,$$

where $[x]$ denotes the **integer part** of the real number x. Hoaglin, Mosteller and Tukey (1983) report that this may provide a convenient choice for the number of stems when n is large (say for more than 50 data points).

If we compress the number of stems to a multiple of 5, then the following notation is often of assistance in describing the stems:

* for digit leaves 0 or 1
t for digit leaves 2 or 3
f for digit leaves 4 or 5
s for digit leaves 6 or 7
. for digit leaves 8 or 9

Example 1.5

Compressing the stems

Example 1.4 revisited. Maximum daily temperatures in Geelong and environs. A random sample of $n = 70$ temperatures recorded at city and rural locations during the autumn and summer of 1983. Temperatures have been truncated to degrees Celsius (and not a lot of useful information is lost in this process).

Suppose that we calculate the value of k with a view to compressing the number of stems in Example 1.4. In this case: $n = 70$, and $k = [10 \log_{10} 70] = 18$.

This suggests that it may be worth choosing fewer stems to represent the data in Example 1.4. The profile of the data is still apparent when we compress the data on 15 stems (approximately half the number used previously). The stems now correspond to temperature intervals of width two degrees; the symbol * denotes a stem for leaf digits 0 and 1, etc. Some may feel that the display has now become too restricted in presentation!

Geelong temperatures

Scale: stems—10 °C; leaves—°C

1t | 3 represents 13 degrees Celsius

```
1* | 1
 t | 2 3 3
 f | 4 4 4 5 5 5 5
 s | 6 6 6 7 7 7 7
 . | 8 8 8 9 9
2* | 0 0 1 1 1 1
 t | 2 2 2 3 3 3 3
 f | 4 4 4 4 5 5 5 5 5
 s | 6 6 6 6 7 7 7 7
 . | 8 8 8 9 9
3* | 0 1 1 1
 t | 2 2 3 3
 f | 5
 s | 7
 . | 8 8†
```

†*denotes Ash Wednesday temperature*

Using such (descriptive to the letter) notation, 10 lines of stems in a plot may be compressed down to 5 lines with the assistance of truncation.

The opposite of the situation just described may also occur; that is, a S/L plot may have too few stems resulting in a formless profile of the data. In this case a given stem may need to be split into two or even five pieces. The notation for a five-piece split is described as above: *, t, f, s, . The corresponding notation for a two-piece split is an abbreviation of this: *, and . so that here we use

* for digit leaves 0,1,2,3,4
. for digit leaves 5,6,7,8,9.

A plot so labelled is called a **split-stem plot**.

After some practice at S/L plot construction, it soon becomes apparent that some numbers are easier to represent with stems and leaves than others. Numbers with many digits may need to be chopped (or **truncated**) before they break readily into a stem portion and a leaf portion. This is often preferable to rounding data (up or down), since the process of rounding may cause a data point to be assigned to an adjacent stem.

Example 1.6

Making a meal stretch further

Combined weights (g) of the ham/crepe/sauce/tomato/onion components from a sample of in-flight meals served by Ansett Airlines:

189.1	188.4	205.2	188.4	176.3
199.1	188.1	181.9	189.8	182.5
198.1	197.9	197.4	198.3	211.6
186.5	188.2	198.1	202.4	203.6
177.1	171.9	172.3	183.6	183.7
191.6	204.2	204.5	183.2	

To construct a S/L plot for this data, we notice that the trailing digits do not contribute to the overall data profile. If these trailing digits are chopped, then a first attempt at a S/L plot of the remaining integers may be:

The weight of airline meals
Scale: stems—10 g; leaves—g
17 | 6 represents 176 grams

17	6 7 1 2
18	9 8 8 8 1 9 2 6 8 3 3 3
19	9 8 7 7 8 8 1
20	5 2 3 4 4
21	1

To stretch this plot out, the * and . notation can be used to assign two lines to each stem. This results in the stretched (preferable) plot:

The weight of airline meals
Scale: stems—10 g; leaves—g
17. | 6 represents 176 grams

17*	1 2
.	6 7
18*	1 2 3 3 3
.	6 8 8 8 8 9 9
19*	1
.	7 7 8 8 8 9
20*	2 3 4 4
.	5
21*	1

The stretched plot shows the data dividing into two groups. This aspect is worth reporting and it was not evident in the original plot.

Back-to-back plots

The plotting technique for S/L plots is well suited to comparing two data sets. Using a central stem, the leaves from one data set may trail to the left of this stem and the leaves from the second data set may trail to the right of the same stem. The shapes so traced out by the two sets of leaves immediately become visually comparable.

Example 1.7
Currency movements back-to-back
Exchange rates for the New Zealand dollar recorded during the month of January for both 1989 and 1990.

\$A100 in terms of \$NZ
Scale: stems—\$NZ; leaves—10^{-1}\$NZ
| 128 | 3 represents \$NZ128.3

January 1989		January 1990
	126	
4 2 6 0 0 8 2 4 5 9 0 2	127	
3 5 1 3 3 0 7 3 6 3	128	0 2 3
2 4 6	129	8
	130	7 8 9 2
	131	2 0 0 9 4
	132	4 6 5 7 6 0 3 0 2 2
	133	3 9
	134	

What can we tell immediately from this plot? The exchange rate tends to be centred on different stems in different years. The rate appears to be more variable in 1990, where the leaves are spread across a larger number of stems. The January 1990 data is left-skewed. The notions of 'centre', 'spread' and 'skewness' are very useful in describing the shape of S/L plots.

EXERCISES

1.12 A one metre steel beam, placed horizontally with ends supported, had a fixed weight placed at its centre. The following S/L plot gives the deflection. The data were recorded in the plot without any rounding or chopping.

Deflection of a steel beam
Scale: stems—10^{-2}mm; leaves—10^{-3}mm
8 | 3 represents 0.083 mm

1	
2	5 7 8 0 5 3
3	6 2 2 1 9
4	3 6 5 7 1
5	2 9 9 9 0 1 2 3 1 1 1
6	4 4 4 1 0 8
7	4 9 2 1
8	3 9 0 5 5
9	
10	
11	
12	
13	
14	
15	
16	4

(a) Write down the smallest and largest observed deflections.

(b) Rewrite the S/L plot so that the leaves are sorted within a stem.

(c) How many deflections were measured?

(d) What percentage of the measured deflections exceeded 0.065 mm?

(e) Is there a possible explanation for the data point that is separated from the rest of the plot?

1.13 The S/L plot gives transformed reliabilities for components from a failure analysis profile.

Failure analysis profile
Scale: stems—10^{-2}units; leaves—10^{-3}units
−12. |8 represents −0.128 units

−14*	4
−13.	
*	1 0 0
−12.	8 5 5
*	3 1
−11.	9 9 8 7 6 5
*	1 1 0
−10.	9 9 8 7 7 6 6 5 5 5 5 5
*	3 2 0
−9.	9 9 6
*	1 1 0 0
−8.	9 9 8 7 6 5
*	3 3 2 1 1 1
−7.	9 9 9 8 6 5 5 5
*	4 3 2 2 1
−6.	7 6 6 5
*	3 1 1 1
−5.	9 5
*	4

(*Source*: Mauri, P. (1989). A computer system for the analysis of integrated circuit reliability. *Quality and Reliability Engineering International*, 5, 273-82.)

(a) Determine the difference between the largest and smallest data values as recorded in the S/L plot presentation.

(b) Rewrite the S/L plot with stems (in units) $-14, -13, -12, -11, \ldots, -5$.

(c) Have the leaves been sorted so that the data depicted are sorted from smallest to largest?

1.14 Consider the following data extracted from Example 1.7:

Exchange rates for the New Zealand dollar recorded during the month of January 1990.

$A100 in terms of $NZ
Scale: stems—$NZ; leaves—$10^{-1}$$NZ
128 | 3 represents $NZ128.3

127	
128	0 2 3
129	8
130	7 8 9 2
131	2 0 0 9 4
132	4 6 5 7 6 0 3 0 2 2
133	3 9
134	

(a) Write down the original data set by using only the information contained in this plot. Give the data in dollars truncated after 1 decimal place. [This question is designed to illustrate the idea that the original data is essentially retrievable from a S/L plot.]

(b) Explain the meaning of the following statement:

'The January 1990 exchange rate data is left-skewed.'

(c) Using only the plot as a guide, try to explain how the exchange rate fluctuated during the month of January. Can you infer from the plot the exchange rate for a specific day?

1.15 In Exercise 1.14, construct a S/L plot based on the data which shows the exchange rate operating in reverse—giving $NZ100 in terms of $A. How are the shapes of the two plots related?

1.16 Bill lengths (in mm), for various species of hummingbird, are given below. Of all the nectar-feeding (nectarivorous) birds including sunbirds, hummingbirds and honey-eaters, the greatest range in bill length is found between species of hummingbird.

11.9	15.0	16.2	15.7	20.0
14.4	14.5	20.2	19.2	19.4
16.0	17.6	20.0	19.9	17.8
19.9	23.1	19.7	20.0	23.7
25.0	18.0	23.3	22.5	21.2
20.4	22.5	17.0	22.3	20.7
20.9	23.8	26.0	23.9	26.6
24.6	28.8	24.4	20.0	20.0
15.3	17.8	10.5	21.3	17.1
12.0	18.7	20.4	16.3	39.7†
15.0	18.1	19.0	17.9	92.2††

†*species: Androdon aequatorialis*

††*species: Ensifera ensifera*

(*Source*: Paton, D.C. and Collins, B.G. (1989). Bills and tongues of nectar-feeding birds: A review of morphology, function and performance, with intercontinental comparisons. *Australian Journal of Ecology*, 14, 473-506.)

Construct a S/L plot for these data.

Is there evidence in the S/L plot of 'short-billed' and 'long-billed' hummingbirds? Explain briefly.

1.17 Write down the data, truncated after 1 decimal place and then sorted from smallest to largest, that is represented in this S/L plot:

Glucose concentration (mg L^{-1})
Scale: stems—units; leaves—10^{-1} units

2*	0 1
t	2 2 3 3
f	5 5 5
s	6 7 7
.	8 9
3*	1
t	2 3
f	5

What would a S/L plot look like for the specification:

Glucose concentration (g L^{-1})
Scale: stems—10^{-3} units;
leaves—10^{-4} units?

1.18 Listed below are telephone numbers randomly selected from the 1990 Melbourne White Pages (telephone directory):

4571361	6874717	4662069	7362448
6541733	7981636	8171803	4804416
5515948	5846293	7441944	8668286
4803564	4973871	8270772	5606953
5966686	7441737	6909866	4358609
8075614	6875711	7007116	4351531
8442313	7296561	4280360	5601416
5516582	8074889	7982463	7221540
5600121	8576240	3873544	5784993
7260645	7002080	6464134	3260457
4270944	3114125	3660272	8787111
7915280	8793857	3068366	8701401
3801831	8220710	6891592	7259296
7002236	7874140	7528077	

(a) The first digits of a phone number refer to a particular locality or suburb. Construct a stem-and-leaf plot showing the first two digits of the above phone numbers.

(b) Construct a stem-and-leaf plot of the *last* two digits of each phone number. Does a plot with the same shape (as in (a)) emerge?

(c) What would the stem-and-leaf plot look like for data comprising 1000 two digit numbers where each digit was completely randomly selected?

1.19 Consider the following data set from a laboratory investigation into vaginal cancer in female rats.

Days until the development of vaginal cancer in rats				
142	156	163	198	204†
205	232	233	233	233
233	239	240	261	280
296	296	323	344†	

† *Censored*

(*Source*: Adams, G. and Watson, R. (1989). The analysis of failure time data. *The Australian Journal of Statistics*, 31, 3.)

The **censored data** points are due to limits in our ability to determine the true survival time from the specified cancer: that is, the subject (rat) may have escaped, died from other causes, or the study terminated with the subject still alive.

Construct a S/L plot for these data. It is essential that the qualitative information regarding censoring be retained in the plot.

1.20 The retail price (in dollars) of 1 kg packets of Rinso washing powder sampled in supermarkets, delis and after-hours food stores (Food Plus, Seven Eleven) in metropolitan Melbourne gave the following data:

3.22	3.25	3.20	3.27	3.39	3.34
2.99	3.24	3.24	3.43	3.21	3.19
3.33	3.21	3.22	3.21	3.21	3.21
3.21	3.24	3.22	4.12	4.44	3.29
3.27	3.24	3.22	3.12	3.26	3.30
4.47	3.28	3.32	3.32	3.32	3.21
3.45	4.62				

Construct a stem-and-leaf plot for these data and try to explain the presence of unusually high values and/or unusually low values in the plot.

1.21 The following data (pH values) shows the acidity of 35 samples of rainfall collected at sites in metropolitan Sydney, Newcastle and Wollongong during 1989. [The pH, Y, of a solution is given by $Y = -\log_{10} X$, where X is the concentration of hydrogen ions in moles per litre.] This means that a pH of 7 is neutral; lower values are more acidic.

4.38	4.52	4.60	4.09	4.55	4.42
4.69	4.21	4.57	5.62	4.13	5.30
4.13	4.30	4.38	4.45	5.67	4.39
4.26	4.26	4.40	5.78	4.74	5.09
4.13	5.56	4.85	4.66	4.29	4.61
3.98	4.46	4.52	4.42	5.95	

Construct a S/L plot for the pH values and describe the distribution that results. Is it reasonable to try and find a single 'typical' value of pH for these east coast cities?

1.22 Two dispensing nozzles in a factory fill 500 g Vegemite containers automatically. As a check on the accuracy of the product label specification, random samples of containers filled by each of the two nozzles had their contents measured accurately with the following results (kg):

Nozzle 1	Nozzle 2
0.46	0.47
0.45	0.46
0.51	0.52
0.56	0.58
0.40	0.49
0.42	0.54
0.58	0.44
0.60	0.40
0.44	0.51
0.41	0.50
0.49	0.55

Construct a back-to-back S/L plot of the data and comment on what is revealed in the plot.

1.23 The following back-to-back S/L plot shows the time from purchase until failure of several digital watches classified by country of manufacture (or source of importation into Australia).

Failure time of digital watches
Scale: stems—10^2 days; leaves—10 days
|3*| 2 represents 320–329 days

Source 1		Source 2
	0*	2 2 3 4 4
	.	6
	1*	1 1 2
5 6	.	5 6 6 9 7
1 3 3 4	2*	3 4 4
5 5 7 7	.	
1 3 4 4 4	3*	2
6 7	.	

(a) What is the shortest survival time (or time taken from purchase to failure) observed for digital watches from each source?

(b) Is there any evidence that the data have been truncated?

(c) Which source provides watches that are, in general, most reliable? In fact, *how* is reliability measured in this example?

(d) If the importers were to offer a guarantee period during which time they would anticipate replacing at most 25% of Source 2 watches, determine from the plot an estimate of how long the guarantee period should be.

1.24 Hoaglin, Mosteller and Tukey (1983), suggest that when n is small (say, less than 50), a useful criterion for selecting a suitable number of stems in a S/L plot is

$$k = [2\sqrt{n}],$$

where n is the number of observations, k is the number of stems and $[x]$ denotes the integer part of the real number x. For $n=$ 10, 20, 30, 40, ...,100, compare this value of k with that suggested in this section; namely $[10 \log_{10} n]$.

1.3 TRANSFORMATIONS: NEW SCALES FOR OLD DATA

Quantitative variables such as 'speed' , 'temperature' and 'weight' are often measured on different scales in different parts of the world. Daily temperatures in weather reports from the United States of America are commonly reported in degrees Fahrenheit; in Australia the Celsius scale is used. For weights, Australians talk of kilograms (kg) and Americans talk of pounds (lb). If we wish to compare two such scales, or change one scale into another, then we say that we *transform* one scale (or variable) into another. Chambers *Twentieth Century Dictionary* defines:

> **transform**, *trans-förm, v.t.* to change the shape of: to change to another form, appearance, substance, character: to change the form of (an algebraic expression or geometrical figure).

Linear transformations: changing origin and scale

Transformations are used for the purpose that the name suggests; that is, to change. If we have the true weights, X, in *grams*, of twenty (supposedly) 500 g Weetbix packets from the supermarket shelf, then to change these weights to corresponding kilogram measurements, Y, we multiply by 0.001. This means that X and Y are related by $Y = 0.001X$.

Example 1.8
From Celsius to Fahrenheit
Maximum daily temperatures in Parramatta sampled during a 12 month period. 20 observations recorded in both Celsius and Fahrenheit.

Celsius					Fahrenheit				
20	17	26	18	20	68.0	62.6	78.8	64.4	68.0
22	16	27	18	29	71.6	60.8	80.6	64.4	84.2
30	17	20	18	17	86.0	62.6	68.0	64.4	62.6
17	21	23	17	19	62.6	69.8	73.4	62.6	66.2

The comparable Fahrenheit and Celsius temperatures are related in the following way: if a temperature is recorded in Celsius as $°C$, then on the Fahrenheit scale it is $°F$, where

$$F = 1.8C + 32.$$

Using this formula, repeated substitution of the observed Celsius temperatures into the right-side of the equation yields the corresponding Fahrenheit temperatures (also listed in this example).

These examples show the equations of simple kinds of transformations of a type that we call **linear transformations**. We give the more general framework in the following definition.

Definition 1.1 *For quantitative variables X and Y we say that Y is a* **linear transformation** *of X if*

$$Y = a + bX$$

for supplied constants a and b.

□

The linear transformation is termed **non-trivial** if b is non-zero. If $b = 0$, we are reduced to the trivial situation where Y is always constant (and equal to a). This is of little interest, so it is assumed in the following discussion that b is non-zero. To examine algebraically how a linear transformation affects an entire data set, let $y_1, y_2, y_3, \ldots, y_n$ denote the results of applying the transformation $Y = a + bX$ successively to the given data set $x_1, x_2, x_3, \ldots, x_n$; then the observed x data values and the 'derived' y data values are related by

$$y_i = a + bx_i, \qquad i = 1, 2, 3, \ldots, n.$$

In Example 1.8, the identifications $n = 20$, $a = 32$ and $b = 1.8$ are readily noted; Celsius temperature is the X variable and Fahrenheit temperature is the Y variable.

Several special cases of linear transformations arise in practice. When $b = 1$ and a is non-zero, the linear transformation $Y = a + X$ represents a **change of origin**. (If $a = 0$, the transformation $Y = X$ is called the **identity**.) When $b \neq 1$, the linear transformation $Y = bX$ represents a **change of scale**. A **change of origin and scale** is accomplished with the general linear equation $Y = a + bX$ for suitable constants a and b. Here, multiplication of X by b changes the scale of X to a new variable; Y is created from this new variable by the change of origin when the constant a is added. This is the reason that linear transformations are sometimes referred to as *changes of origin and scale*.

Example 1.9

Scuba diving at the Abrolhos Islands off the coast of Western Australia

Deep sea diving, quite literally, involves a natural change in origin from sea level to sea bed. The reef setting of the Abrolhos Islands offers a wide choice of relatively shallow dive sites. In this example we consider the equations of the transformations rather than realised data. Suppose that we define the following variables and constants:

$$\begin{aligned} \text{Water depth at dive site (km):} &\quad 0.3 \\ \text{Altitude of diver above sea bed (km):} &\quad X \\ \text{Altitude of diver above sea level (km):} &\quad Y \\ \text{Altitude of diver above sea level (m):} &\quad Z \end{aligned}$$

Before the dive commences (and the diver is still in the boat), X, Y and Z are all positive. Once the dive commences and the diver is underwater, X remains positive, but Y and Z become negative.

In terms of these variables, $Y = -0.3 + X$ represents a *change in origin* from the sea bed to sea level. The equation $Z = 1000Y$ represents a *change in scale*; the depth of the diver is changed to the more appropriate measurement in metres. Both these combine in the single equation $Z = -300 + 1000X$ representing a *change in origin and scale.*

There are many situations where it may be useful to perform a scale change to remove the presence of a decimal point. This is illustrated by the change from Y to Z in Example 1.9. However, of overriding importance is the possibility of bending and improving the shape of data by applying a suitably chosen transformation. This aspect will be discussed in relation to many different types of transformations (and not just linear ones). The 'shape' of data refers to how spread out or bunched up it is. These characteristics may not be readily apparent simply by looking at the numbers, but they become very apparent after the data has been plotted in a S/L plot.

Example 1.10

Coding

The following two data sets are related by a change of origin and scale:

i	1	2	3	4	5	6	7
x_i	−7.79	14.03	3.12	−18.70	−7.79	79.49	3.12
y_i	−1	1	0	−2	−1	7	0

In this example, the spread of the X values is difficult to take in at a glance. On the other hand, it is easy to see that the corresponding Y values are clustered about 0 with one observation being 'atypical' with value 7. In this example, how can the equation of the linear transformation that sends X to Y be identified? The equation must be of the form $Y = a + bX$ for some (as yet) undisclosed values of a and b. All 7 pairs of data must satisfy this equation, so, choosing the (x, y) pairs (3.12, 0) and (14.03, 1) for substitution we get:

$$0 = a + 3.12b$$
$$1 = a + 14.03b$$

This pair of simultaneous equations has solution $a = -0.285976$ and $b = 0.091659$ and the linear transformation may be written $Y = -0.2860 + 0.0917X$.

As the name suggests, linear transformations are geometrically represented by straight lines. Graphical methods are sometimes useful to illustrate the effect of the linear transformation $Y = a + bX$. Careful sketching of the line in the $x - y$ plane can provide a ready reckoner for data conversion from X to Y. This is done by first plotting two points that satisfy the equation and then connecting these points with the unique straight line passing through them. From the discussion at the beginning of this section, we can see that the effect of the linear transformation $Y = a + bX$ is to create $y_0 = a + bx_0$ from an observed value $X = x_0$.

Geometrically, this means that a vertical line is drawn through the point x_0 on the horizontal axis to intersect the line; from where y_0 is located by tracing horizontally across to the y-axis. To follow this procedure literally would invite error caused by imprecise drawing. However the geometry should assist in understanding the relationship between X and Y.

The geometry of a linear transformation

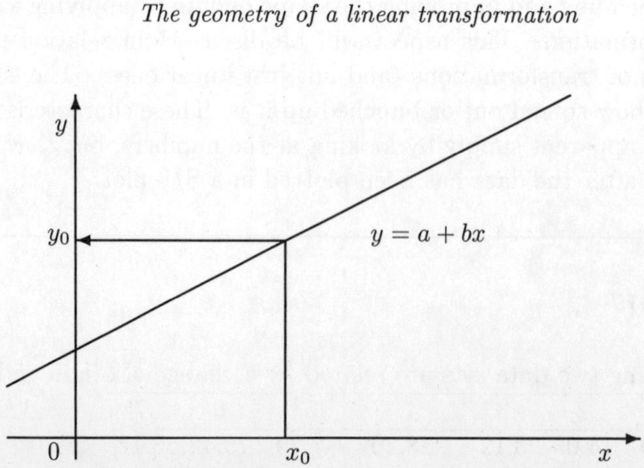

Non-linear transformations: the shape changers

Linear transformations can do little to adjust for distortions in the shape of a data set. There are many data sets that possess unusually large values (or even unusually small values) compared with the majority of values in the data set. These unusual values could easily be simply transcribing mistakes in recording the data on our page (computer screen). At the very least, such points should be highlighted, then checked. When possible, recording mistakes should be rectified or removed from the data set.

Example 1.11
Calcium and cold apples
Determinations of the calcium content (ppm) of batches of Jonathan apples after 3 months cold storage:

224	186	211	262	241	252	212	243
197	192	253	209	237	222	198	2231
218	229	216	201	214	236	213	199

One data point in this array appears to be an 'outlier'. It is very likely to be a mistake in either original measurement or in transcribing the original measurements to the recorded data set listed here.

Later in this chapter is a simple computational guide to the *detection* of 'outliers'. [See Definition 1.8.] Example 1.11 provides a data set with such an observation which is likely to be a recording error. On the other hand, the data set in the following example on the survival time of lightbulbs contains some very small (and possibly very large) atypical values that we would definitely want to keep. A non-linear transformation may be required to compress the time scale to make seemingly disparate values more comparable. A suitable candidate (as we shall see) is the logarithmic transformation $y = \log_e x$.

Example 1.12
Time to lights out
The National Gallery in Canberra houses the country's art treasures in large well-lit exhibition spaces. The spotlights that illuminate individual paintings are installed at ceiling level high above the floor. This makes changing a lightbulb difficult; naturally the gallery staff are very interested in how long a new bulb will last (especially if a cherry-picker is needed for a changeover). The following data give the survival times (days) for

a particular brand of bulb in an industrial life test. (†denotes lifetime for a bulb that has already failed.)

107	94	21	233	7	245	57
189	42	125	12	437	0†	3
211	103	2	109	677	22	283
12	62	32	759	1	177	160
29	97	5	3	552	83	1

Survival times for lightbulbs
Scale: stems—10^2days; leaves—10 days
2. | 8 represents 280–289 days

```
0* | 0†0 0 0 0 0 0 0 0 1 1 2 2 2 3 4
 . | 5 6 8 9 9
1* | 0 0 0 2
 . | 6 7 8
2* | 1 3 4
 . | 8
3* |
 . |
4* | 3
 . |
5* |
 . | 5
6* |
 . | 7
7* |
 . | 5
```

†denotes failed prior to installation

All observations in a data set must be strictly positive before transforming by logarithms since the logarithm is not defined for negative or zero quantities. This problem is overcome by adding a positive amount (called a **start**) to each data point before transformation. This corresponds to a change to a positive origin. With this in mind we note that the effect of the 0 is removed by adding a start of 1. Then, on taking natural logarithms, the transformed data set becomes:

4.68	4.55	3.09	5.46	2.08	5.51	4.06
5.25	3.76	4.84	2.56	6.08	0.0†	1.39
5.36	4.64	1.10	4.70	6.52	3.14	5.65
2.56	4.14	3.50	6.63	0.69	5.18	5.08
3.40	4.58	1.79	1.39	6.32	4.43	0.69

Log survival times for lightbulbs
Scale: stems—units; leaves—10^{-1}units
6 | 6 represents 6.6 units

```
0 | 0†6 6
1 | 1 3 3 7
2 | 0 5 5
3 | 0 1 4 5 7
4 | 0 1 4 5 5 6 6 7 8
5 | 0 1 2 3 4 5 6
6 | 0 3 5 6
7 |
```

†denotes failed prior to installation

in corresponding order with entries given to 2 decimal places. As can be seen from the accompanying S/L plot the shape of this data set has been improved considerably; there are no more unwieldy numbers and yet information about whether survival times were very long or very short has not been sacrificed. Perhaps survival times should naturally be measured on a logarithmic scale. Perhaps the transformation has overcompensated because now the S/L plot tends to be left-skewed (whereas the data was originally right-skewed). In the next section we discuss other transformations some of which produce similar effects to the logarithm.

We will generally use natural logarithms, but logarithms expressed in other bases may be just as readily used. Recall the relationship

$$\log_e x = (\log_e 10)(\log_{10} x)$$

which lets us move between natural logarithms and base 10 logarithms. Therefore, the two types of logarithms are related by a *scale change* since $\log_e 10$ is simply a constant.

The power family

There are certainly other transformations which compress (or expand) data. If a stretching out is required, then taking the squares or even cubes of each data point may suffice. For contraction, square roots and cube roots will bring big numbers down. These operations have one thing in common; they all involve taking powers. This idea is formalised in Definition 1.2 where the powers are classified by using an index κ so that every new value of κ gives a new transformation. For example:

$\kappa = 2$	gives the square transformation
$\kappa = 1$	gives the linear transformation
$\kappa = 0.5$	gives the square root transformation
$\kappa = 0$	gives the logarithmic transformation
$\kappa = -1$	gives the inverse transformation.

This covers the range of values of κ that are most commonly used. The transformations are defined only on positive data so that a start may be required if there are some negative values in the data set.

Definition 1.2 *The* **power family** *of transformations is a collection of transformations of the form*

$$y = \begin{cases} a + bx^{\kappa} & \text{for } \kappa \neq 0; \\ a + b\log_e x & \text{for } \kappa = 0. \end{cases}$$

a and b are supplied constants and κ is a real number called the **index** *of the transformation and $x > 0$.*

□

There is a hierarchy involved which is described in Tukey (1977) as the **ladder of power transformations**. Powers of κ above $\kappa = 1$ have the effect of increasing large data values, while powers of κ below $\kappa = 1$ have the effect of compressing large data values. The extent of the increase (compression) can be assessed by drawing graphs of these transformations (which is discussed in the exercises at the end of this section) and such graphs also help to explain why the logarithmic transformation appears in the centre of the ladder at $\kappa = 0$. After all, the transformation $y = x^0$ would be of little use, sending all data to the same constant 1. The transformation $y = \log_e x$ is the appropriate alternative, its graph lying mid-way between the graphs of $y = a + bx$ and $y = c + dx^{-1}$, for appropriate choices of constants a, b, c and d. Notice that the inverse transformation (for which $\kappa = -1$) will have the effect of *reversing* the order of the data values unless b is chosen (from Definition 1.2) to be negative.

This raises the question: How do we choose a power transformation? The choice may be difficult because three quantities—namely a, b and κ—need to be decided on. a and b relate to origin and scale. The data analyst may essentially choose κ by trial and error but is usually armed with inside information. This information is precisely that gathered by plotting, drawing or graphing the data set in the manner described in the first two chapters of this book. Once a data set is turned into a picture, then its shape is obvious and the extent to which its shape may need improvement is also clarified; for example, do large data values need to be stretched out or shrunk down? When working with a data set on a computer, effective transformation and storage of both the new and old data sets is readily accomplished at the keyboard. Plots can be made for visual comparison. Such facilities are now standard for many computer packages. We stress the need to use such facilities and leave the interested student to 'unwrap' the packages for themselves.

EXERCISES

1.25 In Example 1.8, the formula

$$F = 1.8C + 32$$

showed how Fahrenheit temperatures may be obtained from Celsius temperatures.

(a) Rearrange this formula so that C may be determined from F. That is, find the value of constants a and b so that $C = a + bF$.

(b) Use the formula constructed in (a) to determine the Celsius temperature corresponding to 100 degrees Fahrenheit (to 1 dp).

1.26 Suppose that rectangular metred car parking spaces parallel to the curb in William St are all 5.5 metres long with white line boundaries painted on the street.

Space 1	Space 2	Space 3	...

(a) If a car of length 3.5 metres parks in Space 1 with its front X metres from the front of the space, express the distance, Y, from the back of the car to the back of its parking space as a linear transformation of X.

(b) If the car that parks in Space 2 intrudes into the Space 1 park by 0.4 m, express the distance between the parked cars as a linear transformation of X.

1.27 A supermarket chain is promoting a scheme where items may be purchased using a credit card and the user is billed monthly. On a total monthly purchase of X dollars, the supermarket charges a 1% levy for the administration and handling of the account and this expense is included in the monthly amount billed. However, as an inducement to customers, if the monthly amount billed is paid by the due date, a 1% discount is allowed. Write down the linear transformation which expresses Y, the net income to the supermarket chain for a monthly account paid on time, in terms of X. Use this transformation to determine if it is cheaper to pay cash or to use the credit card facility on offer, with prompt monthly repayments.

1.28 The acidity in rainfall in many Australian cities causes corrosion to the facades of buildings. In particular, it reduces the life of many new high-rise office blocks. The outside cladding on such a 5 year old building was originally 0.025 cm thick. Ten core samples taken from the facade showed depths left uncorroded, X, in cm, of

0.024	0.017
0.022	0.021
0.019	0.009
0.006	0.022
0.017	0.022

(a) Transform these data to give the depth of corrosion, Y, in mm.

(b) Determine the values of a and b so that $Y = a + bX$.

1.29 A bookshop sells X books, out of 100 copies ordered, at \$45 each. The remainder are finally sold at 20% discount (apart from 3 copies which became shop-soiled and are returned to the supplier for complete refund). Express the total income, Y (dollars), to the bookshop as a linear transformation of X. That is, determine the values of a and b so that

$$Y = a + bX.$$

The bookshop pays \$20 per copy. Show that the total profit, P (dollars), can be written as the change of origin

$$P = Y - 1940.$$

Hence express P in terms of X.

1.30 The data given below show the annual frequency with which the safety 0.12 ppm level of ozone was breached during 1986 in a comparison of Melbourne with various Californian cities.

[0.12 ppm, termed a 1-hour acceptable level, is breached if average ozone readings exceed the level in the course of an hour of monitoring. Ozone is the main constituent of photochemical oxidant.]

City	(X) **Frequency 0.12 ppm level exceeded**
San Francisco	5
Los Angeles - Central	48
Los Angeles - West	30
Pasadena	110
Glendora-Laurel	147
Long Beach	10
San Diego	42
Burbank	93
Palm Springs	54
Sacramento	9
Santa Barbara	1
San Jose	1
Anaheim	28
Melbourne	4

(*Source*: Morgan, P.A. (1988). *Air Quality in Victoria*. EPA Scientific Services Publication.)

Demonstrate that the transformations $y = \log_{10}(x)$ and $y = \sqrt[3]{x}$ produce similar effects on this data set: do this by constructing S/L plots for the data under each of the transformations. What has been the overall effect of these transformations on the data set? Has skewness increased or decreased?

1.31 Which of the given functions are members of the power family of transformations?

(a) $y = (1-x)^{-1}$

(b) $y = 1 - x^{-1}$

(c) $y = 2^x$

(d) $y = x^2$.

1.32 The following list is of functions $y = f(x)$ where f is a member of the power family of transformations. Identify the index κ of each transformation and determine the values of a and b as given in Definition 1.2.

Function	κ	a	b
$1 - \frac{2}{x^2}$			
$\sqrt[3]{x}$			
$\log_e 2x$			
$(1+x)(1-x)$			

1.33 Show that for any base a,

$$y = \log_a x$$

is a member of the power family of transformations (of Definition 1.2).

1.34 Consider the curves

$$y_q = \begin{cases} \frac{x^q - 1}{q} & \text{for } q > 0; \\ \log_e x & \text{for } q = 0, \end{cases}$$

indexed by q for $x > 0$. (These are the Box-Cox Transformations considered in Box and Cox (1964).)

(a) Show that these transformations are members of the power family of transformations. (That is, determine the values of a and b from Definition 1.2.)

(b) Sketch these transformations for $x > 0$ and $q = -2, -1, 0, 1$ and 2. Notice the natural hierarchy in these sketches; in particular, the position of the logarithm.

(c) Show that $y = -1+x$ is the common tangent to these curves at $x = 1$.

1.35 In the ladder of power transformations, the role of the logarithm is suitably filled by the power $p = 0$. In support of this idea, prove that

$$\lim_{p \to 0} \frac{x^p - 1}{p} = \log_e x.$$

1.36 A manufacturer of steel ballbearings is investigating the quality of its 1.5 cm diameter product in response to complaints by workers on a car assembly line that the ballbearings were too *variable* in diameter (and therefore often did not fit the purpose intended). Weights, X (g), of a sample of 52 ballbearings were recorded as:

4.1	4.5	4.6	4.7	4.7	4.8	4.8
4.8	4.8	4.9	4.9	5.0	5.0	5.0
5.0	5.0	5.0	5.1	5.1	5.1	5.1
5.2	5.2	5.2	5.3	5.3	5.3	5.3
5.3	5.3	5.4	5.4	5.4	5.4	5.5
5.6	5.6	5.8	5.8	6.0	6.0	6.1
6.2	6.2	6.4	6.6	6.7	6.9	7.2
7.4	7.6	7.6				

(a) Construct a S/L plot for these data.

(b) We will assume that the steel used in production has a constant density of 10 g cm^{-3}. Show that if $x = f(d)$ is the weight of a ballbearing with diameter d, then

$$x = f(d) = \frac{5}{3}\pi d^3,$$

if the ballbearings are assumed to be spherical. [The volume of a sphere of radius r is $\frac{4}{3}\pi r^3$.]

(c) The plot in (a) shows a highly skewed distribution. One explanation could be that a small change in the diameter of a spherical ballbearing causes a larger change in the volume (and hence in x). Investigate this effect by simplifying

$$f(d + \Delta d) - f(d).$$

Determine the size of this discrepancy in the case that $d = 1.5$ and $\Delta d = 0.05$.

(d) Find the inverse transformation

$$d = f^{-1}(x).$$

Show that this transformation is a member of the power family of transformations.

(e) Show that a S/L plot of the values

$$f^{-1}(x_i),\ i = 1, 2, 3, \ldots, 52,$$

produces a S/L plot of greater symmetry than that of (a). [Intuitively, the variability in the process is best measured by the diameters of the ballbearings and not their weights.]

1.4 RANK

Data often reaches the data analyst in a completely unsorted form. Observations are often written down serially, in the order of occurrence. This serial recording is how data sets are built up. Any investigation of the size, or magnitude, of the data will require knowing which are the smallest data points, which ones have a magnitude that we expect to see, and which data points seem unusually large. In short, we need to sort the data from smallest to largest. This is one of the first tasks of the data analyst; but, it is not a separate task. An automatic ranking will occur naturally at the same time that we construct a stem-and-leaf plot. It is prudent to take advantage of this.

Ranked data

We use the notation

$$x_1, x_2, x_3, \ldots, x_n,$$

to denote the observations in a data set of size n. Here x_1 is the first recorded observation, x_2 is the second recorded observation, and so on. We reserve the notation

$$x_{(1)}, x_{(2)}, x_{(3)}, \ldots, x_{(n)},$$

for the same data set that has been sorted or ranked from smallest to largest. Such data is called **ranked data**. The notation implies that

$$x_{(1)} \leq x_{(2)} \leq x_{(3)} \leq \ldots \leq x_{(n)},$$

and $x_{(i)}$ is called the ***i*th order statistic**.

Definition 1.3 *Suppose that a data set $x_1, x_2, x_3, \ldots, x_n$ has no tied values. Then, for any $k = 1, 2, 3, \ldots, n$, a data point x_k is said to have* **rank *i*** *if i is the index so that $x_k = x_{(i)}$.*

□

In the case of tied data, the common rank assigned to each data point is the average of the ranks that would be assigned if the data points had been slightly different. As an illustration:
If $x_1 = 2$, $x_2 = -5$, $x_3 = -1$, $x_4 = 3$ and $x_5 = -2$, then obviously $x_{(1)} = -5$, $x_{(2)} = -2$, $x_{(3)} = -1$, $x_{(4)} = 2$ and $x_{(5)} = 3$. We could then summarise the ranks for the original data (x values):

data	x_1	x_2	x_3	x_4	x_5
rank	4	1	3	5	2

When data is presented in a S/L plot then a complete ordering of the data, as above, from smallest to largest, is achieved when the leaves on each stem are sorted from smallest to largest. (See Example 1.4 where S/L Plot 1 presents the original data and S/L Plot 2 presents the same data with sorted leaves on each stem.) Very large data sets will require a computer algorithm for sorting!

Example 1.13

Sailmakers scrutinise the canvas

At the time of the America's Cup in Perth in the summer of 1986 a lot of canvas was being sewn into sailcloth. When canvas is used for such an important function as a mainsail, any imperfections in the cloth could lead to tearing under stress (with disastrous consequences).

Regular quality control checks on four canvas suppliers gave the following observed data of faults per 100 metres of canvas:

Supplier	Observed faults per 100 metres								
A	0	2	1	0	0	1	3	0	0
B	1	5	2	1	0	7	9	5	3
C	3	5	10	4	2	12	6	3	15
D	6	38	5	9	7	7	2	13	10

The data is simply ranked by placing it in a S/L plot. (We will do this without regard to supplier so that we obtain a reasonably large data set to work with.)

Flaws in canvas per 100 metres

Scale: stems—tens; leaves—units

1*| 0 denotes 10 flaws

```
0* | 0 0 0 0 0 0 1 1 1 1 2 2 2 2 3 3 3 3 4
 . | 5 5 5 5 6 6 7 7 7 9 9
1* | 0 0 2 3
 . | 5
2* |
 . |
3* |
 . | 8
```

This S/L plot has been written for convenience with sorted leaves. Since $n = 36$, the rank of the median is $(\frac{1}{2})36 + \frac{1}{2} = 18.5$. This places the median of the entire data set halfway between the 18th and 19th largest observations. That is,

$$\tilde{x} = \frac{x_{(18)} + x_{(19)}}{2} = \frac{3+4}{2} = 3.5.$$

If we now return to the original data and determine the median of the imperfections from Supplier B, then $n = 9$ and the rank of the median is $(\frac{1}{2})9 + \frac{1}{2} = 5$. This places the median at the 5th largest observation in the data set for Supplier B. This data value is 3.

Since there are 9 data points for each supplier, the median for each supplier corresponds to the 5th largest observation for that supplier. For Supplier D, we find that $\tilde{x} = 7$. This assessment of the centre of the distribution for Supplier D is not affected by the extremely large observation of 38 faults per 100 metres.

The median: going for the centre

One advantage of dealing with ranked data is that it becomes easy to locate the **median**, a measure of central location of the data. It is a value which divides the data set in half; that is, 50% of the data lies at or below the median, and 50% of the data lies at or above the median. This means that there may be more than one median, so that in order to be able to talk of **the** median we need to adopt the strategy (or convention) of the following definition.

Definition 1.4 *The* **median**, $\tilde{x}$, *of the ranked data set* $x_{(1)} \le x_{(2)} \le x_{(3)} \le \ldots \le x_{(n)}$, *is given by*

$$\tilde{x} = \begin{cases} x_{(\frac{n+1}{2})} & \text{if } n \text{ is odd;} \\ \frac{x_{(\frac{n}{2})} + x_{(\frac{n}{2}+1)}}{2} & \text{if } n \text{ is even.} \end{cases}$$

□

This strategy is: if the median falls between two data points, let's say that it falls *halfway* between. There is a sensible reason for this. After all, the median is that data value with rank $\frac{n}{2} + \frac{1}{2}$, which is as it should be, that is, the unique middle value, if $\frac{n}{2} + \frac{1}{2}$ is an integer (which occurs if n is odd). But if $\frac{n}{2} + \frac{1}{2}$ is not an integer, the $+\frac{1}{2}$ means that $\frac{n}{2} + \frac{1}{2}$ is a **mid-rank**, or halfway between the two adjacent ranks $\frac{n}{2}$ and $\frac{n}{2} + 1$. We then call on the average of the data points carrying these adjacent ranks to be the median. The algebra tends to hide the simplicity of the idea but the definition is essential. Example 1.13 illustrates the simplicity. The same example illustrates how the median is fortified against the effect of extremely large and/or extremely small data points. The size (magnitude) of such points is not used in the median calculation.

From order statistics to percentiles

We have seen that a median splits a data set in half. The formal definition of the median requires that at least 50% of the data set is $\leq \tilde{x}$ *and* at least 50% of the data set is $\geq \tilde{x}$. The median then belongs in the upper half of the data *and* the lower half of the data. To achieve a sense of symmetry, so that both halves of the distribution are treated equally, when the median falls on a data point we imagine that point chopped into two pieces; one piece is assigned to the lower half of the data set and the other piece is assigned to the upper half of the data set. This process may be generalised; we can imagine the order statistics dissected in the same way. For example, if $n = 10$, the first order statistic $x_{(1)}$ has at least $\frac{(1-\frac{1}{2})}{10}100 = 5\%$ of the distribution at or below it and at least 95% of the distribution at or above it. The second order statistic $x_{(2)}$ has at least $\frac{(2-\frac{1}{2})}{10}100 = 15\%$ of the distribution at or below, and at least 85% of the distribution at or above, and so on. In each case the $-\frac{1}{2}$ in the calculation is the order statistic 'taking sides'. That is, the $-\frac{1}{2}$ ensures that the order statistic lies in both the part of the distribution that is 'at or above', and the part of the distribution that is 'at or below'. Finally, the largest value $x_{(10)}$ has at least $\frac{(10-\frac{1}{2})}{10}100 = 95\%$ of the distribution at or below it, and at least 5% of the distribution at or above it. What we have done is turned the order statistics into **percentiles**.

Definition 1.5 *For* $0 < p < 1$*, a* **100*p*th percentile** *of the data set is any value* x_p *such that at least* $100p\%$ *of the data lies at or below* x_p *and at least* $100(1-p)\%$ *of the data lies at or above* x_p*.*

□

This definition lets us see exactly what percentiles our ranked data points correspond to. Consider the *i*th largest data point $x_{(i)}$. Data points with ranks $1, 2, 3, \ldots, i-1, i$, all lie at, or below, $x_{(i)}$. Hence the percentage of data points $\leq x_{(i)}$ is $100(i/n)$ which certainly exceeds $100(i - \frac{1}{2})/n$. On the other side of our order statistic, data points with ranks $i, i+1, i+2, \ldots, n-1, n$, all lie at, or above, $x_{(i)}$. Therefore the percentage of data points $\geq x_{(i)}$ is $100[n-(i-1)]/n = 100[n-i+1]/n$, which exceeds $100 - 100(i-\frac{1}{2})/n$. We have reasoned through the proof of the following theorem. We refer also to Definition 1.5 for an understanding of percentile.

Theorem 1.1 *For the ranked data set* $x_{(1)} \leq x_{(2)} \leq x_{(3)} \leq \cdots \leq x_{(n)}$,

$$x_{(i)} \text{ is the } 100p\text{th percentile where } p = \frac{(i-\frac{1}{2})}{n}.$$

■

Note that other authors may present slightly different approaches to the sample percentiles. A good discussion is given in Hoaglin, Mosteller and Tukey (1983). For the example $n = 10$ we list the order statistics and the percentiles corresponding to them:

ordered data	$x_{(1)}$	$x_{(2)}$	$x_{(3)}$	$x_{(4)}$	$x_{(5)}$	$x_{(6)}$	$x_{(7)}$	$x_{(8)}$	$x_{(9)}$	$x_{(10)}$
percentile	5	15	25	35	45	55	65	75	85	95

From percentiles to order statistics

We now pose the questions of the previous section in reverse. That is, given a specific value of p, how can the 100pth percentile, x_p, be determined in terms of the observed data (or order statistics)? For $n = 10$, using the above table in reverse gives the order statistics corresponding to percentiles 5, 15, 25, 35, ... But, what of the 10th percentile, or the 23rd percentile? Or the pth percentile? The answer comes from using Theorem 1.1 in reverse. If $x_{(i)}$ is the 100pth percentile where $p = (i - \frac{1}{2})/n$, then, solving this last equation for i gives

$$i = np + \tfrac{1}{2}.$$

When this value of i is treated as a rank, the data point(s) to which it refers must be located. This is accomplished very simply by rounding i to the nearest integer (rank) and then looking at the data point(s) with this rank to identify x_p. Notice that if i is a mid-rank then there will be two such nearest integer ranks. In this case we take x_p to be the average of the two data points with these nearest integer ranks. (See, for example, the definition of the median.) This ensures that the correct proportions lie 'at or above' and 'at or below' x_p. Essentially we have proved Theorem 1.2 below; this theorem formally sets out these discussions about how to find x_p for *any* specified value of p. It provides a rule which works for all data sets. It provides a rule which satisfies our intuition.

So that this 'percentiles theorem' can be expressed in a form which suitably generalises our previous definition of median, we introduce the following: The notation $< i >$ will be used to denote the integer(s) nearest to a real number i, or the **nearest integer** to i. So, for example: $< 1.0 >= 1$; $< 1.1 >= 1$; $< 1.49 >= 1$; $< 1.5 >= 1$, or 2; $< 1.500001 >= 2$ and $< 1.9 >= 2$.

Theorem 1.2 *Given the ranked data set $x_{(1)} \leq x_{(2)} \leq x_{(3)} \leq \cdots \leq x_{(n)}$, a 100pth percentile, x_p, for $0 < p < 1$, is given by*

$$x_p = \begin{cases} x_{\left(<np+\frac{1}{2}>\right)} & \text{if } np \text{ is not an integer;} \\ \frac{x_{(np)}+x_{(np+1)}}{2} & \text{if } np \text{ is an integer,} \end{cases}$$

where $< r >$ *is the nearest integer to* r. *The quantity* $i = np + \frac{1}{2}$ *is called the* **rank of x_p**.

■

Notice that if n is odd and $p = 0.5$, then

$$< \tfrac{1}{2}n + \tfrac{1}{2} > = \tfrac{1}{2}n + \tfrac{1}{2},$$

which is the rank of the median. Therefore, it is clear that with $p = 0.5$, the formula in the theorem reduces to the previous definition of the median. That is, the median is the 50th percentile.

The **quartiles** are also calculated from sorted data. Together with the median they split the data set into four equal pieces.

Definition 1.6 *The* **lower quartile** *is* $x_{0.25}$*; the value 'at or below' which 25% of the data lies, and 'at or above' which 75% of the data lies. Similarly, the* **upper quartile** *is* $x_{0.75}$*; the value 'at or below' which 75% of the data lies, and 'at or above' which 25% of the data lies.*

□

Example 1.14

Finding percentiles

Simple numerical examples show Theorem 1.2 put to work to produce the median, quartiles and the 81st percentile! Consider the data set

$$2.7 \quad 4.7 \quad -0.5 \quad 0.9 \quad 2.9 \quad 1.1 \quad 4.4 \quad 11.0 \quad 0.4 \quad 1.6 \quad 3.7 \quad -1.2$$

consisting of 12 data points (the weight gains, in kilograms, of 12 individuals on a low-fat diet).

Worked example:

For both the entire data set and also the restricted data set of the seven 'most successful' dieters, determine the median, quartiles and 81st percentile.

n = 12

i:	1	2	3	4	5	6	7	8	9	10	11	12
$x_{(i)}$:	−1.2	−0.5	0.4	0.9	1.1	1.6	2.7	2.9	3.7	4.4	4.7	11.0

Calculations

p	np	$np + \frac{1}{2}$	x_p		
0.25	3	3.5	lower quartile	$= \frac{x_{(3)}+x_{(4)}}{2}$	$= 0.65$
0.50	6	6.5	median	$= \frac{x_{(6)}+x_{(7)}}{2}$	$= 2.15$
0.75	9	9.5	upper quartile	$= \frac{x_{(9)}+x_{(10)}}{2}$	$= 4.05$
0.81	9.72	10.22	81st percentile	$= x_{(10)}$	$= 4.4$

The same considerations applied to the seven smallest data values gives:

n = 7							
i :	1	2	3	4	5	6	7
$x_{(i)}$:	−1.2	−0.5	0.4	0.9	1.1	1.6	2.7

Calculations

p	np	$np + \frac{1}{2}$	x_p		
0.25	1.75	2.25	lower quartile	$= x_{(2)}$	$= -0.5$
0.50	3.50	4	median	$= x_{(4)}$	$= 0.9$
0.75	5.25	5.75	upper quartile	$= x_{(6)}$	$= 1.6$
0.81	5.67	6.17	81st percentile	$= x_{(6)}$	$= 1.6$

We have shown percentile calculations for very small data sets simply for the purpose of illustrating the technique. In general, a large number of data points are needed before there is much gain in splitting a data set into more than quarters.

EXERCISES

1.37 Regular quality control checks on canvas Supplier C (see Example 1.13) gave the following observed data of faults per 100 metres of canvas:

3 5 10 4 2 12 6 3 15

Determine the median and quartiles for these data.

1.38 The following are cooking times (minutes) for certain vegetables.

Vegetable	Cooking time (mins)
Carrots (frozen)	9
Cabbage (small quarter)	12
Cauliflower (small quarter)	12
Celery (fresh)	8
Corn on cob (frozen)	5
Mushrooms (button)	10
Onions (small)	6
Peas (fresh)	6
Potatoes (small whole)	14
Artichokes	20
Asparagus	8
Beans (fresh)	7
Brussel sprouts	13
Carrots (chopped)	6

Determine the median of these data. Check that your value is indeed the median by verifying that at least 50% of the data is (at or) less than your chosen value and at least 50% of the data is (at or) greater than your chosen value.

1.39 Explain why the median is not a useful measure of the centre of this data set.

8.9	0.1	8.1	1.0	2.1	9.3	0.3
8.0	9.2	9.8	9.6	1.1	9.7	1.8
1.5	9.6	1.6	0.2			

1.40 The following are sector flight distances (km) for Air New Zealand flights operating out of Christchurch:

Christchurch to:	km
Brisbane	2494
Hobart	2025
Melbourne	2412
Norfolk Island	1674
Perth	5053
Sydney	2124

(*Source*: *Pacific Way*, July 1991.)

Determine the rank of the Melbourne flight sector distance.

1.41 If the flight sector distance data (km) of Exercise 1.40 is summarised in the order listed by $x_1 = 2494, x_2 = 2025, \ldots, x_6 = 2124$, determine the value of $x_{(5)}$ and give the rank of the median of the data set.

1.42 Haemodynamic (blood flow) test results for 15 coronary artery disease patients were obtained at low exercise levels on a calibrated bicycle ergometer.

The data show the observed percentages of age-related peak heart rate achieved with a Thallium-201 injection.

64	84	78	70	60
84	74	85	74	78
79	63	82	82	82

(*Source*: Verzijlbergen, J.F. et al (1991). Myocardial perfusion imaging with dipyridamole and exercise Thallium-201. *Journal of Nuclear Medicine*, 32, 2071-8.)

(a) Determine the rank of each data point.

(b) Determine the upper quartile for these data.

1.43 Complete the following chart to establish the percentiles for 6 ranked data points:

Ranked data	$x_{(1)}$	$x_{(2)}$	$\cdots$	$x_{(5)}$	$x_{(6)}$
Percentile					

1.44 For the cooking time data in Exercise 1.38 copy and complete the chart:

p	np	$np + \frac{1}{2}$	x_p	
0.25			lower quartile	=
0.50			median	=
0.75			upper quartile	=
0.80			80th percentile	=

1.45 Prefabricated concrete slabs for use in building are reinforced with steel rods for strength. When a shear force is applied to a rod, the yield strength, X (MPa), is the point at which the steel rod loses its elasticity and suffers deformity. For reinforcing bars the minimum Australian Standard is 410 MPa. The following observations on X were sampled from a production line:

440	425	413	435	439	411
406	431	417	434	432	428
427	432	450	442	433	435
420	426	410	441	419	446

(a) Construct a S/L plot (with ordered leaves) for the data.

(b) By counting in the S/L plot, determine the rank of the observation 410, which is exactly at the minimum Australian Standard.

(c) What percentage of the data meets the requirements of the minimum Australian Standard?

1.46 (a) What is the effect of the linear transformation $Y = X - 410$ on the yield strength data in Exercise 1.45? Can a physical meaning be attached to the variable Y?

(b) How does a S/L plot of the Y values differ from the S/L plot for the X values? Explain.

1.47 The following data give the 'contact resistance', in ohms.mm, when rapid isothermal processing is used in an annealing process.

90×10^{-3}	280×10^{-3}
100×10^{-3}	85×10^{-3}
90×10^{-3}	90×10^{-3}
85×10^{-3}	75×10^{-3}
80×10^{-3}	110×10^{-3}
90×10^{-3}	55×10^{-3}
100×10^{-3}	95×10^{-3}
75×10^{-3}	95×10^{-3}
90×10^{-3}	100×10^{-3}
80×10^{-3}	90×10^{-3}
80×10^{-3}	95×10^{-3}
85×10^{-3}	150×10^{-3}
85×10^{-3}	130×10^{-3}

(*Source*: Pearton, S.J. et al (1992). Rapid Isothermal Processing for Fabrication of GaAs-Based Electronic Devices. *IEEE Transactions on Electron Devices*, 39, 154-8.)

Determine the median contact resistance.

Identify the location of the median in a S/L plot of these data.

1.48 Daily sales for a small retail establishment (specialising in Speedo swimwear) for the months of June and September are given in the S/L plot below:

Daily retail sales ($)
Scale: stems—10$; leaves—$
| 9 | 1 denotes $91

June		September
0	5	
5 3 3 2 2	6	0
9 7 7 7 5 3 3 0	7	1 6
7 3 3 2 2 2 1	8	2 7
4 3 3 0 0	9	1 1 4 4 5
5 4 1 1	10	0 0 0 0 1
	11	0 1 2 2 7
	12	2 5 6 9
	13	1 3 5 6 9 9
	14	

Compute the median and quartiles for each month. Discuss the skewness of the two plots.

1.49 Determine the 80th percentile for the September retail sales in Exercise 1.48.

1.50 Tukey (1977) defines **fourth** as a measure closely related to the concept of quartile. A **lower fourth** is that value with rank

$$\frac{[\text{rank of median}] + 1}{2},$$

where $[x]$ denotes the integer part of x.

Show that for $n = 11$ data points the lower fourth and the lower quartile may give different values. (For large data sets, any discrepancy is negligible.)

1.5 BOX-AND-WHISKER PLOTS

5-point number summaries

Often the information collected from 'ranking' calculations (such as those above) is displayed in a simple diagram called a **5-point number summary**. The quantities summarised are $x_{(1)}$, $x_{0.25}$, $\tilde{x}$, $x_{0.75}$ and $x_{(n)}$. These are the main **summary statistics** based on ranked data and they occupy places in the number summary as indicated:

n	data label	
	$\tilde{x}$	
$x_{0.25}$		$x_{0.75}$
$x_{(1)}$		$x_{(n)}$

It is noticeable in this display that the summary statistics are displayed at various levels; the median occupies a level in the display above the quartiles (which share the same level); the quartiles occupy the level above the extremes. This idea is certainly related to that of rank. In general, for a data set with n distinct elements,

> $x_{(i)}$ has **upward rank** i and **downward rank** $n - (i - 1) = n - i + 1$.

The upward rank of the lower quartile is the same as the downward rank of the upper quartile. These ranks tell us how far $x_{(i)}$ is from each end of the data set.

> The **depth** of a data point is the *smaller* of its upward rank and its downward rank.

Each line in a 5-point number summary lies at the same depth; these values are sometimes written on the left side of the display. Similarly, depths are often written against the left-side of a S/L plot in the form of the 'greatest depth' for all data points on the same stem.

Box-and-whisker plots

5-point number summaries do not provide an immediate impression of the *shape* of the data; they provide a list of the values of the main statistics based on ranks. In a **box-and-whisker plot** (we shall abbreviate the title to **B/W plot**), this information is set out in a format that indicates

- the location of the centre of the data,
- how spread out the data is, and
- whether there seem to be any unusually large or unusually small data values.

Example 1.15

Collingwood players tipping the scales

In Australian Rules Football, the weight of a player often depends on the position they play within the team. Not all players are super-heavyweights, as the following data on players in the Collingwood Football Club shows. Collingwood played Essendon at the MCG on 5 May 1990 to a crowd of 63 000 people.

(*Source*: *The Football Record*, 4–6 May 1990.)

Player	Weight (kg)
Alexander	72.0
Banks	81.0
Barwick	85.0
Bayliss	90.0
Brown	80.0
Christian	93.0
Croall	76.0
Crosisca	85.5
Daicos	84.0
Fellowes	106.0
Francis	68.0
Gayfer	82.5
Gladman	80.5
Hawke	76.0
Hecker	76.0
Hrysoulakis	73.0
Keays	82.5
Kelly	90.0
Kerrison	82.5
Manson	95.0
McGuane	82.5
McKeown	82.5
Millane	89.0
Monkhorst	102.0
Morwood	83.0
Mrakov	91.0
Richardson	76.0
Russell	80.0
Saunders	79.5
Shaw	79.5
Shephard	73.0
Starcevich	89.0
Styles	75.0
Tarpey	89.0
Taylor, B	101.0
Taylor, S	80.0
Trantor	75.0
Tuddenham	85.0
Turner	80.5
Wrensted	85.0
Wright	79.0

Weight (kg) of the Collingwood players

Scale: stems—tens; leaves—units

h represents 0.5 kg

s	
.	8
7*	
t	2 3 3
f	5 5
s	6 6 6 6
.	9 9h 9h
8*	0 0 0 0h 0h 1
t	2h 2h 2h 2h 2h 3
f	4 5 5 5 5h
s	
.	9 9 9
9*	0 0 1
t	3
f	5
s	
.	
10*	1
t	2
f	
s	6

Worked example:
Construct a 5-point number summary for these data.
The required percentiles for the 41 data points are determined as follows:

p	np	$np+\frac{1}{2}$	x_p		
0.25	10.25	10.75	lower quartile	$= x_{(11)}$	$= 79$
0.50	20.5	21	median	$= x_{(21)}$	$= 82.5$
0.75	30.75	31.25	upper quartile	$= x_{(31)}$	$= 89$

$n = 41$	Football players' weights	
	82.5	
11	79	89
1	68	106

The quartiles have depth 11 as indicated in the 5-point number summary.

By definition, $x_{(1)}$ and $x_{(n)}$ are useful indicators of extreme values in a data set. These give unusual values when there are gross errors in recording the data and this is why they are recorded in 5-point number summaries. In fact the order statistics with high ranks and the order statistics with low ranks are all useful for detecting outliers. We have seen the median as a useful measure of the location of the centre of a data set. In Example 1.15, the median not only locates the centre of the data set but also the most frequently occurring observation (in this case 82.5). The separation of the quartiles becomes our first measure of how spread out the data is.

Definition 1.7 *For the ranked data set $x_{(1)} \le x_{(2)} \le x_{(3)} \le \ldots \le x_{(n)}$, the* **support** *of the data set is the interval $[x_{(1)}, x_{(n)}]$ containing the data. The* **range** *of the data, denoted by R, is the length of this interval and so is given by $R = x_{(n)} - x_{(1)}$. The* **interquartile range,** *or* **quartile spread,** *q_s, is given by $q_s = x_{0.75} - x_{0.25}$.*

□

The range, as a measure of spread, must be treated with care, since its value is directly affected by observations with extreme ranks. The q_s is not as badly affected in this way and may be used to assess whether a data set contains potential outliers. Tukey (1977) indicated how to make such an assessment: 'fence off' a multiple of q_s and then detect whether any data lies trapped 'inside'.

Definition 1.8 *For a ranked data set, any points that lie* outside *the interval*

$$[x_{0.25} - \tfrac{3}{2}q_s,\ x_{0.75} + \tfrac{3}{2}q_s]$$

are called **mild outliers**. *The endpoints of this interval are called* **inner fences**. *Further, any points in the data set that also lie* outside *the larger interval*

$$[x_{0.25} - 3q_s, \, x_{0.75} + 3q_s]$$

are classified as **extreme outliers**. *The endpoints of this interval are called* **outer fences**.

□

This definition gives us a rule-of-thumb for *flagging* outlier values. This may assist in identifying mistakes in recording the data; such mistakes should be rectified. However, outlier values should *never* be deleted from a data set automatically.

We shall use the following procedures for constructing B/W plots with the building blocks: $\tilde{x}$, $x_{0.25}$, $x_{0.75}$, q_s and the inner and outer fences:

1. Determine and position the median and the quartiles on an appropriately labelled horizontal axis.
2. Directly above the quartiles, above the axis, construct a horizontal box of length q_s. Divide the box at the median.
3. Extend from the central outer edges of the box, from the quartiles towards the extremes, horizontal lines (called *whiskers*) to the most extreme data points still within the inner fences.
4. Identify potential outliers beyond the whiskers as either mild (with an open circle), or extreme (with a filled-in circle).
5. If there are several data sets to compare, repeat this procedure to obtain a 'stack' of plots lying horizontally one above the other, all measured against the *same* scale.

The process of constructing a B/W plot is very easy. The given recipe, once followed, is logical and efficient. Practise it whenever you see a data set. This is sound advice even though many software packages now give B/W plots as output options. Such output will save on arithmetic, but it is important to be able to interpret the output: What does the shape tell us? When distributions of data are skewed, then so are the B/W plots representing them. The skewness shows up in long boxes and long whiskers.

Example 1.16

A B/W plot for the Collingwood players: Example 1.15 revisited

For this data set we have already computed a 5-point number summary. Therefore, following the B/W plot scheme,

$$q_s = x_{0.75} - x_{0.25} = x_{(31)} - x_{(11)} = 89 - 79 = 10.$$

The inner fences are at the endpoints of the interval

$$[x_{0.25} - \tfrac{3}{2}q_s,\ x_{0.75} + \tfrac{3}{2}q_s] = [79 - (1.5)(10),\ 89 + (1.5)(10)] = [64, 104].$$

The whiskers extend to the most extreme data points in this interval, that is to 68 below and to 102 above. Since 106 lies beyond this inner fence, it may be considered as a mild outlier (circled in the plot below). The outer fences are the endpoints of the interval

$$[x_{0.25} - 3q_s,\ x_{0.75} + 3q_s] = [79 - (3)(10),\ 89 + (3)(10)] = [49, 119].$$

Since none in the team tips the scales at these weights, there are no extreme outliers (i.e. data points beyond the outer fences). This gives the following B/W plot:

Weights of the Collingwood players

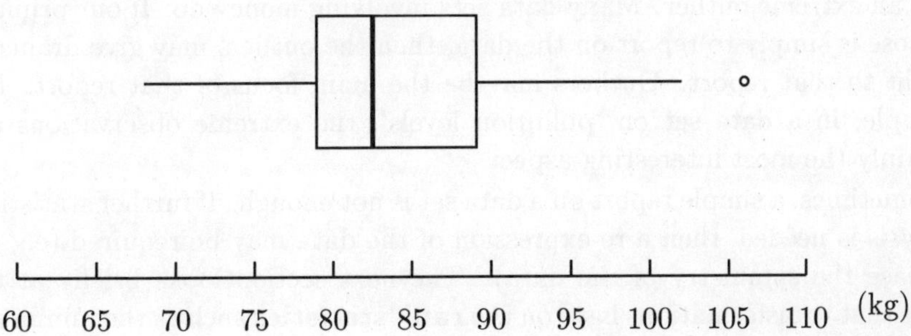

The 'stack' of B/W plots below were constructed from the above sequence of instructions. Imagine them representing the plots for three different data sets. For simple comparison, the plots are drawn against the same horizontal axis according to item 5 above. Notice that, because of the way that these diagrams are constructed, the boxed areas in each plot represent the same *proportion* (i.e. 25%) of observations from the data set being depicted. This is because 25% of the data lies between the median and a given quartile. But, and this is important, the boxes are of different lengths: If a quartile box is long, it means that the data it represents are thinly spread out; if a quartile box is very short, it means that the represented data are densely packed (in the corresponding area on the horizontal scale). Simple ideas that may be immediately 'read' from these plots: all plots show a median $\widetilde{x} = 4$; Plot C shows an interquartile range $q_s = 3$; in Plot A, 25% of the data depicted lies between 4 and 4.5; Plot A is left-skewed, Plot B is symmetric and Plot C is right-skewed.

A comparison of B/W plots of different shapes

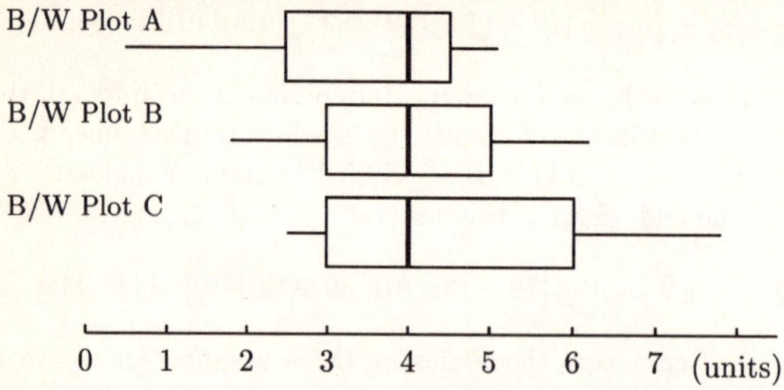

Having discussed the shape and skewness of B/W plots, the next question is: What do the outliers tell us? In Example 1.15 there were no extreme outliers in the data set (represented by shaded circles beyond the whiskers). This is to be expected; almost by definition, a football player is unlikely to be either extraordinarily light or extraordinarily heavy. Example 1.17 on the export revenue for some of Australia's major manufacturing exporters, *does* have an extreme outlier. Many data sets involving money do. If our primary purpose is simply to report on the data, then the outliers may give dramatic weight to that report. Outliers may be the main focus of that report. For example, in a data set on 'pollution levels', the extreme observations are certainly the most interesting aspect.

Sometimes, a simple report on a data set is not enough. If further statistical analysis is needed, then a re-expression of the data may be required (e.g. to increase the symmetry of the data). The next section looks briefly at the effect that transformations have on the **rank statistics** such as the minimum, maximum, median, quartiles, percentiles and on the B/W plots themselves which are built from these statistics.

Change of scale/change of rank?

A simple question which sometimes arises is: If data are transformed, what happens to the ranks? Do they transform in the same manner? Sometimes!

Some transformations that have been discussed are *order-preserving* and some are *order-reversing*. In fact a transformation can only be useful to us if it acts in a consistent way with the data, that is, it is always order-preserving or is always order-reversing. Suppose that X is the variable that we know about or have data for. In a simple case we may have the following observations on $X : -2, 3, 8$. These are in order from smallest to largest. This order doesn't change if the data is multiplied by a *positive* constant or is added to an arbitrary constant. However, if the data is multiplied by -1 (as an example of multiplying by a *negative* constant) then the order reverses: $-X : -8, -3, 2$. If we think in a more general framework, for constants a and b the linear transformation $y = ax + b$ is order-preserving if $a > 0$, and order-reversing if $a < 0$. Geometrically, the difference is in whether or not the graph of the line $y = ax + b$ has positive or negative slope. If we use the p subscript to denote percentiles (as given in Definition 1.5) then, if $a > 0$, $y_p = ax_p + b$ and so the pth largest x observation is transformed to the pth largest y observation. If, on the other hand $a < 0$, then $y_p = ax_{1-p} + b$, so that the $(1 - p)$th largest x observation is transformed to the pth largest y observation. In such a case, the upper quartile is transformed to the lower quartile (and vice versa).

The sign of the slope or gradient is also the deciding factor in whether ranks change when data is re-expressed with a power transformation. In a simple case where we take $a = 0$ and $b = 1$ and $\kappa \geq 0$ in

$$y = \begin{cases} a + bx^{\kappa} & \text{for } \kappa \neq 0; \\ a + b\log_e x & \text{for } \kappa = 0, \end{cases}$$

from Definition 1.2, we are left with increasing functions of x which preserve the order in ranked data. In the case $\kappa = 0$, the transformation becomes $y = \log_e x$ which is order-preserving.

Example 1.17

Exporters worth millions

The following list of who's who in the competitive world of manufacturing exporters during 1989 was published in *The Age*, Melbourne, 12 July 1990.

Company	**Exports ($m, 1989)**
BHP Steel International	900
GM Holden	248
IBM	180
Kodak	177
Hawker de Havilland	125
FH Faulding	120
Kraft	110
Petersville Sleigh	105
ICI	100
Nucleus	100
Chemplex	65
Ford Motor Co	59
Aerospace Technologies	57
ANI	56
Varian Techtron	54
Burns Philp	50
AMI Toyota	48
United Milk of Tas.	42
Tubemakers	31
ROH Industries	30
BTR Nylex	30
Labtam	28
Containers	27

The data set is a list of the top 23 companies ranked according to how many millions of dollars of manufactured goods they exported during 1989. The B/W plot maintains the labels for companies that are potential outliers. Certainly, in this data set, these so-called outliers are a main feature of the data set.

Australia's top manufacturing exporters

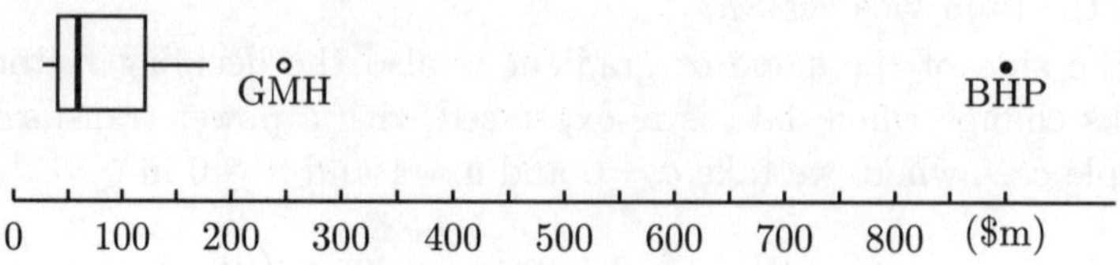

Australia's top manufacturing exporters measured on a log scale

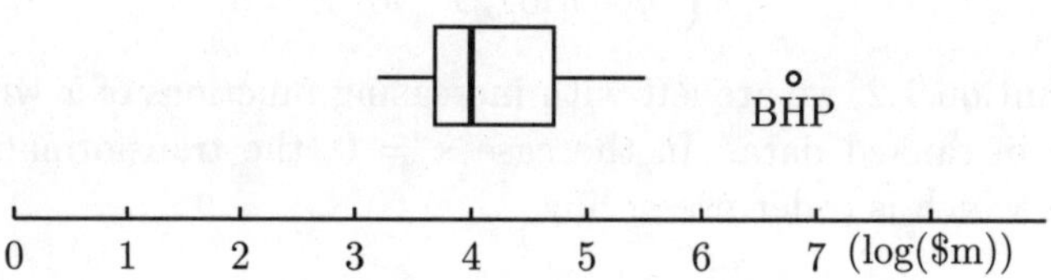

The effect of the logarithm transformation on the B/W plot, in particular, is illustrated in Example 1.17. The long whisker in the upper tail of the plot is shortened by taking logarithms.

EXERCISES

1.51 Suppose that data showing 'Dam Capacity' for Melbourne's water supply is as follows:

Dam	Total capacity (ML)	Percentage full
Yan Yean	30 000	82
Greenvale	27 000	94
Sugarloaf	96 000	95
Maroondah	22 000	99
O'Shannassy	3 000	41
Upper Yarra	200 000	74
Silvan	40 000	90
Cardinia	287 000	89
Thomson	1 068 000	97

(*Source*: *The Age*, Melbourne, 2 February 1992.)

Determine the *upward rank*, *downward rank* and *depth* for each data point comprising the total capacity data.

1.52 In engineering applications, the lifetime (or reliability) of electronic components is of interest; in medical applications, the lifetime (or remission) of cancer patients is of interest. But what of the useful life of Prime Ministers?

The following data show the Australian Commonwealth Government Ministries of this century. Consecutive terms of office won by the same Prime Minister have been added to give total time in office.

Prime Minister	Term in office
Barton	1 Jan 1901–24 Sept 1903
Deakin 1	24 Sept 1903–27 Apr 1904
Watson	27 Apr 1904–17 Aug 1904
Reid-McLean	18 Aug 1904–5 Jul 1905
Deakin 2	5 Jul 1905–13 Nov 1908
Fisher 1	13 Nov 1908–2 Jun 1909
Deakin 3	2 Jun 1909–29 Apr 1910
Fisher 2	29 Apr 1910–24 Jun 1913
Cook	24 Jun 1913–17 Sept 1914
Fisher 3	17 Sept 1914–27 Oct 1915
Hughes	27 Oct 1915–9 Feb 1923
Bruce-Page	9 Feb 1923–22 Oct 1929
Scullin	22 Oct 1929–6 Jan 1932
Lyons	6 Jan 1932–7 Apr 1939
Page	7 Apr 1939–26 Apr 1939
Menzies 1	26 Apr 1939–29 Aug 1941
Fadden	29 Aug 1941–7 Oct 1941
Curtin	7 Oct 1941–6 July 1945
Forde	6 July 1945–13 July 1945
Chifley	13 July 1945–19 Dec 1949
Menzies 2	19 Dec 1949–26 Jan 1966
Holt	26 Jan 1966–19 Dec 1967
McEwen	19 Dec 1967–10 Jan 1968
Gorton	10 Jan 1968–10 Mar 1971
McMahon	10 Mar 1971–5 Dec 1972
Whitlam	5 Dec 1972–11 Nov 1975
Fraser	11 Nov 1975–11 Mar 1983
Hawke	11 Mar 1983–19 Dec 1991

(*Source*: Cameron, R.J. (1985). *Year Book Australia 1985.*)

(a) Construct a data set showing the number of consecutive months in office (including commencing and terminating months) for each Prime Minister.

(b) Record the results in a B/W plot. Discuss the presence of outliers.

1.53 Suppose that a data set consists of $n = 10$ distinct observations which form a highly skewed B/W plot. Is it possible to add another observation to the data set so that the B/W plot remains completely unchanged?

1.54 Referring to Exercise 1.51, construct data showing the *available* water capacity at each dam site. Summarise your findings in a 5-point number summary.

1.55 The computer package MINITAB produced the S/L plot below which records the mass of precipitate (mg) from a platinum solution.

```
Stem-and-leaf of C1   N  = 21
Leaf Unit = 0.010

  4  102 2233
  6  102 45
  7  102 7
 (4) 102 8899
 10  103 01
  8  103 3333
  4  103 4
  3  103 6
  2  103 88
```

The maximum depths are recorded on the left-side of this display; stems 102, 103 in the centre; single digit leaves on the right-side. (4) denotes the frequency of data on the stem where the median occurs.

21 data points are depicted without rounding or chopping.

(a) Use the information provided by the depths to determine the median, quartiles and inner fences.

(b) Are there any outliers lying beyond the inner fences?

1.56 The Pavement Research Group at RMIT is investigating the durability of brick pavers subjected to high volume pedestrian traffic. Under experimental conditions, holes are gouged in fixed pavers subjected to the vibrating head of a tip-hardened nail punch. The volume of the hole (as assessed by liquid displacement) is a measure of the durability of the paver.

20 pavers gave the following liquid displacements (mL $\times 10^{-3}$):

401	408	397	406	437	418	411
406	406	409	422	405	405	405
403	400	409	407	404	399	

(a) Construct a 5-point number summary for these data.

(b) Determine the inner fences for this data set. Hence give a detailed account of possible outliers.

(c) Construct a B/W plot and comment on the results.

1.57 Construct a B/W plot for the precipitate data from the S/L plot generated by MINITAB in Exercise 1.55.

1.58 Let R denote the range of the 10 data points $x_1, x_2, x_3, \ldots, x_{10}$. Is the equation

$$R = x_{10} - x_1$$

always true/ sometimes true/ never true? Select *one* and explain your choice.

1.59 A company produces wood-grained panels to be used for the lining of interior walls. The production line is constantly monitored to ensure that a quality product is manufactured to a specified standard. The following data give the widths observed (in metres) for 50 panels randomly selected from the production line in a particular week.

0.980	0.997	1.030	1.016	0.954
1.007	0.982	0.995	1.013	0.960
1.050	0.972	1.022	1.020	1.038
0.970	1.010	0.944	1.004	0.990
0.996	0.976	1.012	1.060	1.008
0.964	1.028	0.966	1.042	1.024
1.026	1.006	1.034	0.974	1.018
0.995	0.995	0.992	1.002	1.001
1.000	1.000	0.998	1.014	0.991
0.983	1.011	0.984	1.032	0.988

Construct a B/W plot for these data. Ensure that your plot has a title and a clearly defined scale.

1.60 The data recorded below show a comparison of the second highest ozone values (ppm) for United States cities with populations greater than 2 million, with Melbourne, for 1982 and 1983.

City	1982	1983
New York	0.17	0.19
Los Angeles	0.32	0.37
Chicago	0.12	0.17
Philadelphia	0.18	0.19
Detroit	0.16	0.14
San Francisco	0.14	0.17
Washington DC	0.15	0.17
Dallas-Fort Worth	0.17	0.16
Houston	0.21	0.28
Boston	0.16	0.18
St Louis	0.16	0.18
Pittsburgh	0.14	0.14
Baltimore	0.14	0.19
Minneapolis	0.10	0.13
Atlanta	0.14	0.17
Melbourne	0.22	0.20

(*Source*: Morgan, P.A. (1988). *Air Quality in Victoria*. EPA Scientific Services Publication.)

(a) Construct a back-to-back S/L plot for the data. Comment on what is revealed in the plot.

(b) Discuss the presence of outliers.

(c) Draw comparative B/W plots for each year. Has there been an obvious change in extreme ozone concentrations?

1.61 In an experiment to measure the effect of high ozone concentration on tissue growth, a group of 25 mature rats, called the **ozone group**, were kept in an environment containing ozone for seven days and their weight gains recorded in grams. The data are given in the S/L plot below:

Weight gains
Scale: stems—g; leaves—10^{-1}g
1* | 2 represents 1.2 g

Stem	Leaves
−2*	0
−1.	5
*	2
−0.	9 5
*	2
0*	2 3
.	6 6 7 8
1*	0 2 4 4
.	5 5 7
2*	0
.	8
3*	
.	9
4*	0 4
.	
5*	4

(a) Construct a box-and-whisker plot for these data. Indicate in your plot the presence of any mild or extreme outliers. Ensure that your plot has a title and a clearly defined scale.

(b) Another group of 25 mature rats, called the **control group**, were kept in an ozone-free environment for a similar time and their weight gains (in grams) were recorded as:

−0.4	1.3	1.5	1.7	1.7	1.8
1.9	2.1	2.1	2.1	2.2	2.2
2.5	2.6	2.6	2.6	2.6	2.7
2.8	2.9	3.8	4.1	1.9	2.4
2.7					

Using the same scale as in (a), construct a B/W plot for this data. Compare and contrast the **ozone** and **control** groups.

1.62 The company in Exercise 1.59 divides its production of wood-grained panels into three different quality groups. Ten samples of 100 panels each were taken from each group and the number of imperfections per 100 panels were counted.

Quality group	Observations				
I	0	2	1	1	0
	1	0	2	3	0
II	4	1	7	1	4
	0	2	5	6	9
III	6	7	32	4	14
	10	10	15	11	9

Use these data to answer the following questions:

(a) On the same scale, construct 3 box-and-whisker plots to compare the 3 quality groups. Construct your plots horizontally, one above the other. Comment very briefly on what is revealed in the plots; your comments should address the terms *centre*, *skewness* and *outliers.*

(b) Let the random variable Y denote the number of imperfections per 100 panels. The transformation

$$W = \sqrt{Y}$$

is called a *variance stabilising transformation.* Calculate the effect of this transformation on the interquartile range for each of the 3 quality groups.

1.63 Consider the following summary measures:

median, lower quartile, range, interquartile range, minimum.

Which of these remain unaffected by the linear transformation
$y = -x - 10$?

1.64 The following results are raw weight loss test data (grams) from a series of laboratory abrasion tests on four materials.

Material G	Material H
35.3	41.4
86.7	18.9
72.4	44.8
74.6	34.1
85.1	37.7
79.3	40.0
80.5	27.8
69.7	25.7

Material Y1	Material Y2
13.4	5.7
19.9	5.2
18.0	6.4
18.6	4.5
17.9	11.4
15.4	11.1
17.1	10.7
15.8	5.0

(*Source*: Sare, I.R. and Constantine, A.G. (1991). Design and Analysis of Jaw Crusher Gouging and Abrasion Tests. *Journal of Testing and Evaluation*, 19, 115-22.)

Construct four B/W plots to compare the abrasion test results for the four materials. Discuss what is revealed in the plots.

Repeat the B/W plot construction after the data have been transformed by logarithms. Has the symmetry of the plots increased?

1.65 Construct a 5-point number summary for the data:

40	25	13	35
39	11	41	06
31	17	34	32
28	19	27	32
50	42	33	35
46	20	26	10

Hence use transformations to construct a 5-point number summary for the data

−14.40	−14.25	−14.13	−14.35
−14.39	−14.11	−14.41	−14.06
−14.31	−14.17	−14.34	−14.32
−14.28	−14.19	−14.27	−14.32
−14.50	−14.42	−14.33	−14.35
−14.46	−14.20	−14.26	−14.10

CHAPTER TWO

The Size/Shape/Spread of Data

In this chapter you will find

2.1 BAR CHARTS AND PARETO DIAGRAMS

How do graphs and diagrams help when dealing with numbers? Simply: the displaying of the most important aspects of our data set is the first step in finding out what the data are really like and communicating that discovery to others.

In Chapter 1, we used S/L plots and B/W plots to display the shape of ranked data. Both plots provide insight into where quantitative data is centred and how spread out it is about the centre. In this chapter we add histograms to our graphics options. Consideration will also be given to versions of histograms appropriate to the display of qualitative data.

It should always be remembered that in data analysis the idea of a graphical display of the data serves several purposes. The idea of a graph is familiar from various media sources so the potential audience is large ... better communication! Also, mistakes in recording the data (the kind of *atypical observations* that we do not want) tend to become very noticeable. Further, diagrams and pictures help us model what we would expect future observations to be like. For the record, pictorial information should be clear, informative and easily interpreted. In a graph, misuse of scale and origin placement leads the eye to take in incorrect information. To avoid this, axes should be given titles with labels indicating the scale and where the origin is located. We will first consider displays of qualitative data.

Frequencies and relative frequencies

Intuitively the word 'frequency' refers to how often something happens. We shall use the qualitative variable 'intended party preference in the next federal election ... ' to illustrate the ideas. Suppose that this variable has four levels: 'Labor', 'Liberal', 'Other' and 'Undecided'. (Are we measuring here on a nominal scale?) For each level we record the **frequency** or the number of times that level is observed in the course of data collection. (How many people will vote Labor? Liberal? etc.)

It is worth stopping momentarily and thinking about how you would record these counts accurately 'out in the field' at the point where the data is collected. A simple answer is to use the **tally method**. This quick and simple method recalls our first days of learning to count. Simply, the counts, as they are observed, are struck off in groups of 5. Thus, a tally sheet recording 23 cars in the southbound lane through an intersection in a 1 minute period may have an entry looking like

~~////~~ ~~////~~ ~~////~~ ~~////~~ ///

the marks being made as the cars were seen. Returning to the 'how will you vote?' example: how do the scrutineers at a polling booth accurately tally the votes? The emphasis is on 'accurately'.

Given that the frequency is the number of times that the level of the variable is observed, then the **relative frequency** is the proportion of times that the level is observed. This, for each level, is the ratio of the frequency to the total number of observations in the data set and is sometimes, for convenience, expressed as a percentage. (What is the percentage of electors in Curtin in Example 2.1 who would vote for Liberal? Are undecided?)

Example 2.1

How will you vote ?

Frequency distribution of intended voting patterns in the seat of Curtin, Western Australia, prior to the July 1987 federal election.

In the weeks leading to a parliamentary election, the media focus on the results of polls which sample electors' intended voting preference. This helps to detect swings in voting patterns and successes in campaign performance. Note that many people decide only on polling day which candidate they will favour with their vote; often decisions are made at the last minute in the polling booth. 100 responses were recorded as:

Liberal	Liberal	Liberal	Labor	Labor
Labor	Liberal	Labor	Undecided	Labor
Labor	Labor	Liberal	Other	Undecided
Labor	Liberal	Liberal	Labor	Labor
Labor	Labor	Liberal	Undecided	Labor
Labor	Labor	Labor	Liberal	Liberal
Undecided	Liberal	Liberal	Undecided	Labor
Labor	Liberal	Liberal	Labor	Liberal
Labor	Labor	Liberal	Other	Labor
Liberal	Liberal	Liberal	Undecided	Labor
Liberal	Liberal	Labor	Other	Labor
Liberal	Liberal	Liberal	Undecided	Liberal
Labor	Liberal	Liberal	Other	Labor
Liberal	Labor	Liberal	Liberal	Labor
Undecided	Liberal	Liberal	Undecided	Liberal
Labor	Liberal	Liberal	Liberal	Labor
Labor	Undecided	Liberal	Liberal	Liberal
Liberal	Liberal	Liberal	Liberal	Labor
Labor	Liberal	Labor	Labor	Labor
Labor	Liberal	Liberal	Undecided	Labor

These qualitative observations may be accurately counted using data tallies

Liberal	~~////~~ ~~////~~ ~~////~~ ~~////~~ ~~////~~ ~~////~~ ~~////~~ ~~////~~ ~~////~~ /
Labor	~~////~~ ~~////~~ ~~////~~ ~~////~~ ~~////~~ ~~////~~ ~~////~~ ////
Undecided	~~////~~ ~~////~~ /
Other	////

then appropriately summarised in the (relative) frequency distribution:

Party	Frequency	Relative frequency
Liberal	46	0.46
Labor	39	0.39
Undecided	11	0.11
Other	4	0.04
Total	100	1.00

It should always be remembered that this (relative) frequency distribution is a *summary* of the original 100 observations. When working with such summaries it is easy to forget what the original data look like (in this case a large list of 100 labels).

For general notation: Suppose that a qualitative variable has k different levels and that we record n observations on this qualitative variable. For each $i = 1, 2, 3, \ldots, k$, we will use the symbol f_i to denote the frequency for level i. This means that the frequencies for the k levels of the qualitative variable are given by $f_1, f_2, f_3, \ldots, f_k$ and clearly these frequencies add to the total sample size: $f_1 + f_2 + f_3 + \ldots + f_k = n$. The relative frequency r_i, for level i, is given by $r_i = \frac{f_i}{n}$. Therefore, since the frequencies sum to the total number of observations, the relative frequencies theoretically sum to 1. (Rounding error in reporting relative frequencies may cause slight discrepancies.)

All this information, when collected together, gives the 'distribution' of the data as defined in Definition 2.1.

Definition 2.1 *A tabular listing of the levels of the qualitative variable together with the (relative) frequencies associated with each level is called a* **(relative) frequency distribution for qualitative data**. □

The structure of the (relative) frequency distribution is:

Variable	**Frequency** f_i	**Relative frequency** $r_i = \frac{f_i}{n}$
Level 1	f_1	r_1
Level 2	f_2	r_2
Level 3	f_3	r_3
⋮	⋮	⋮
Level k	f_k	r_k
Total	n	1.00

In Example 2.1, the qualitative variable 'Intended party preference in the next federal election ...', or, 'Party', has $k = 4$ levels with $n = 100$ observations. The columns of the table with headings 'Party' and 'Frequency' constitute a frequency distribution, while the columns with headings 'Party' and 'Relative frequency' comprise a relative frequency distribution.

Bar charts

A graph of nominal or ordinal data with (relative) frequency labelling the vertical axis and level on the horizontal axis is called a **bar chart**. The prototype has bars of equal width with heights proportional to the (relative) frequency; often a space separates each bar. Such spaces are useful to visually separate out the levels depicted on the horizontal axis. Example 2.2 illustrates the nature of a bar chart. There are many variations to this prototype. The bars may be conglomerates of stacked coins (perhaps representing discretised units of Australia's current account deficit), or even accumulations of stick figures (where each figure may represent a count of 20 000 people). These are broadly termed **pictograms**. The **pie chart**, as the name suggests, consists of a segmented circle. The segments, or pieces of the pie, have size proportional to the relative frequency of the category depicted. Such pictures certainly have visual appeal, but the actual sizes of the depicted pie pieces are not numerically comparable, unless pieces are overwritten with relative frequencies. Notice that bar charts are the appropriate graphical treatment for qualitative data measured on a nominal or ordinal scale. In essence, this same type of graphical treatment may be used for displaying interval and ratio data; adjacent bars are drawn together, giving a smooth transition between divisions on the horizontal axis appropriate to the measurement of quantitative data.

Example 2.2
Voting revisited

Bar chart of intended voting patterns of $n = 100$ electors in the seat of Curtin, Western Australia, prior to the July 1987 federal election.

The levels of the variable 'intended party preference in the next federal election ...' have been labelled from 1 to 4 for easy identification. This gives, for example, $f_1 = 46$ and $r_1 = 0.46$ for the frequency and relative frequency of those intending to vote Liberal in the next federal election. Notice how these measurements translate into the height of the bars in the bar chart.

i	**Party**	**Frequency** f_i	**Relative frequency** $r_i = \frac{f_i}{n}$
1	Liberal	46	0.46
2	Labor	39	0.39
3	Undecided	11	0.11
4	Other	4	0.04
	Total	100	1.00

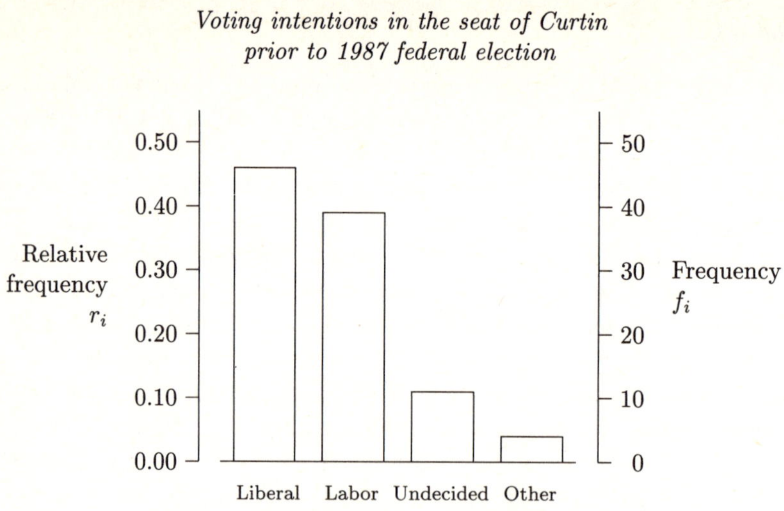

Pareto diagrams

In their simplest form, Pareto diagrams are bar charts for ordinal data. McConnell (1986) reports that Pareto diagrams (sometimes called Pareto charts) are named after the 19th century Italian Vilfredo Pareto who, on studying the distribution of wealth in Europe, concluded that 80% of the wealth belonged to 20% of the people. The idea of Pareto diagrams is to isolate such major discrepancies. They are generally associated with a special kind of ordinal data for ordinal variables whose levels are ordered from *worst* to *best*; they are usually applied in situations where the worst category needs to be identified.

These situations occur in industrial quality control, for example, where an observed lack of quality in a product produced on a production line may be traced back to the levels Reason 1, Reason 2, ..., Reason k of the qualitative variable 'Cause'. Constructing a bar chart for these categories immediately visually identifies the worst cause (which is Reason 1 in our notation). A simple, easily read description of the variety of uses for Pareto diagrams is given in McConnell (1986).

Once the worst cause has been identified, steps may be taken to rectify the problem. Then, a new diagram may be constructed, hopefully with a new ordering associated with the levels of the qualitative variable 'Cause'. By comparing the two diagrams, the reduced effect of 'Reason 1' can be observed directly. For example, in the manufacture of flour in a wheat mill, the worst reasons affecting the quality of production may be identified as:

Reason 1: Water absorption
Reason 2: Starch damage
Reason 3: Poor wheat cleaning
Reason 4: Sieve alignment

Pareto diagram
of product quality violations

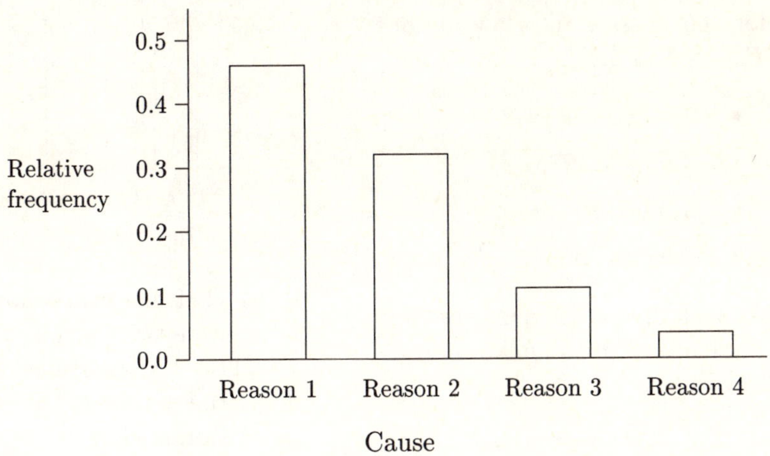

Samples taken from production runs are assigned to one of these four categories and then a Pareto diagram may be constructed from the frequencies. Once the degree of water absorption has been controlled, a new Pareto diagram would show water absorption occurring at a lower level.

EXERCISES

2.1 The *Matrimonial Causes Act 1959-66* allowed divorces to be filed on one (or more) of 14 grounds. It came into operation in 1961. The *Family Law Act*, a piece of more recent legislation, radically reduced divorce grounds to one: irretrievable marriage breakdown. A compilation of recent divorce statistics is given in the bar chart.

Number of divorces
1972-1982

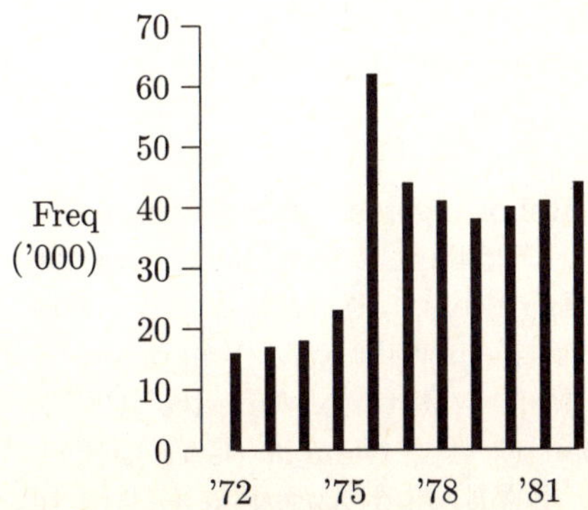

(a) The horizontal axis measures *time*, a continuous variable. Why, when this data set is graphed, is it appropriate to use a bar chart?

(b) Based on information contained in the bar chart, which of the following dates is the most likely for the passing into law of the *Family Law Act*?

- 31 December 1972
- 5 January 1976
- 9 June 1979
- 1 August 1981

(c) Annual divorce frequencies are summarised in the following table:

Year	Number of divorces granted
1979	37 854
1980	39 258
1981	41 412
1982	44 088
1983	43 525
1984	43 012

(*Source*: Cameron, R.J. (1985). *Year Book Australia 1985*.)

Construct a relative frequency distribution for this data.

2.2 The following bar chart appeared in an advertisement for The Institution of Engineers, Australia. It compares the population of engineers in the United States of America (USA), Japan (JAP), Canada (CAN), Sweden (SWE) and Australia (AUS) in 1985.

Professional engineers per 10 000 of population

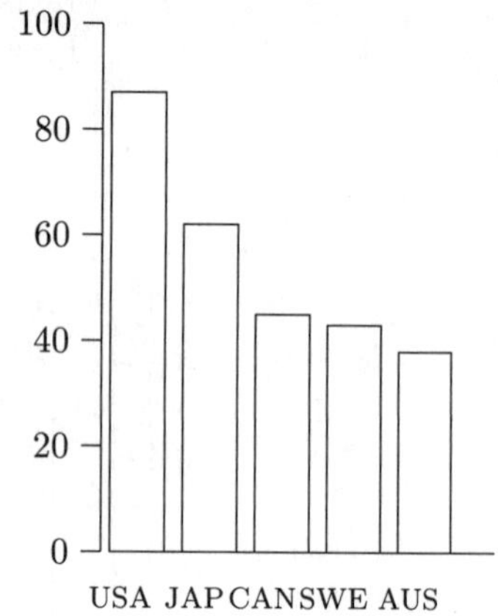

(*Source*: *The Australian*, 7 November 1990.)

(a) 'This bar chart shows that there were twice as many engineers in the United States of America as there were in Australia in 1985.' Comment.

(b) Assuming that the Australian population was approximately 16 million in 1985, estimate the number of engineers in Australia in 1985.

2.3 The following frequency distribution given by Cameron (1985) in *Yearbook Australia 1985* shows traffic deaths in Australia across all States and Territories in 1978 and 1983.

Traffic deaths Australia	1978	1983
NSW	1384	966
Vic.	869	664
Qld.	612	510
SA	291	266
WA	345	203
Tas.	106	70
NT	68	48
ACT	30	28

(*Source*: Cameron, R.J. (1985). *Year Book Australia 1985*.)

(a) What qualitative variable is being measured to obtain the data for each year? What are the levels of this variable?

(b) What information does this particular frequency distribution tell us at a glance?

(c) How many traffic deaths were there in Australia in 1983?

(d) Construct a bar chart for the 1978 data. On this bar chart, superimpose and shade the rectangles for a bar chart of the 1983 data. Discuss what is revealed by the comparison of these charts.

2.4 In answer to the question:

Should Australia become a republic?

1000 voters telephoned nationally aligned themselves against specific alternatives as:

	Total %	Men %	Women %
Remain in Commonwealth Queen head of state	53	47	59
Remain in Commonwealth as a republic	32	35	30
Become a republic outside Commonwealth	13	16	9
Don't know	2	2	2

(*Source*: *The Age*, Melbourne, 9 May 1988.)

(a) What qualitative variable(s) is/are being measured in this data set?

(b) On what scale is this variable measured? (nominal/ordinal/interval/ratio)

(c) Construct a bar chart showing the distribution for all respondents.

(d) What additional information is necessary before a bar chart showing frequencies may be constructed for each of 'Men' and 'Women'?

(e) Give a relative frequency distribution for the response of women to the republic question. Construct bar charts to compare and contrast the responses of men and women.

[*Note*: By 1992, a similar survey indicated that the percentage in favour of republican status had increased to more than 50%.]

2.5 Many major accidents on the Hume Highway (between Sydney and Melbourne) have been attributed to speeding. Data recorded using radar guns (the accuracy of which we do not question here) was categorised as: motorist speeding (S), motorist not speeding (N). Speeding motorists were not apprehended. Instead, the following data show whether or not a motorist was speeding and the recorded time (hrs:mins:secs) as the vehicle passed a hidden police observation post near Wodonga.

03:36:25	N	04:03:28	S	04:47:22	N
03:36:30	N	04:04:02	S	04:47:27	N
03:36:35	N	04:13:17	S	04:47:52	N
03:38:03	N	04:14:24	S	04:49:37	S
03:39:29	S	04:17:23	S	04:49:50	S
03:44:01	N	04:19:43	S	04:52:32	S
03:44:06	N	04:22:24	S	04:53:07	S
03:44:12	N	04:26:21	N	04:57:22	S
03:44:18	N	04:27:28	N	04:57:32	S
03:44:25	N	04:33:16	S	04:57:45	S
03:44:36	N	04:42:17	S	04:59:53	S
03:53:11	S	04:44:48	S	05:00:19	S
03:59:08	S	04:47:08	S	05:04:15	S

(a) Summarise the qualitative data in a relative frequency distribution and construct a bar chart.

(b) Try to explain the pattern of 'N' and 'S' symbols in this data set. For example, what could cause there to be a series of 'N' symbols in a row?

2.6 In the production of aluminium ingots, quality checks are made on the production line to determine the extent of metallurgical defects. In each case a defective ingot, sampled from the production line, is classified according to its most outstanding defect. Construct a Pareto diagram from the following frequency distribution of defects:

Defect	**Frequency**
Shrinkage cavities	22
High porosity	31
Blistering	40
Inclusions	17
Other defects	4

2.2 HISTOGRAMS: BUILDING WITH BLOCKS

The levels of a qualitative variable create natural categories for counting the frequency of observations. In the case of quantitative data, such categories need to be constructed, usually in the form of contiguous intervals covering the support of the data set. The methods described here are useful for summarising large sets of continuous data. Once intervals have been determined, the data from the original data set may be placed into the intervals. It is easy to make copy-over errors, so the tally method is recommended for allocating data to intervals.

The timber data of Example 2.3 give a useful illustration: the tally method is used to construct 10 equal width intervals from the raw data. Since the variable 'timber density' is *continuous*, care must be exercised in defining the interval endpoints; each data point must be uniquely assigned to an interval. (Recall that for continuous data, any value is possible within a given range.) Perhaps the simplest method (used here) of ensuring unique data assignment is to use **from** ... **to under** ... notation. In relation to the first interval listed in the distribution of Example 2.3, all the notations

from 25 to under 30
25–under 30
$25\text{–}30^-$
$[25, 30)$

used by various authors have precisely the same coverage; included are all observations greater than or equal to 25 but strictly less than 30. The other 9 intervals in Example 2.3 follow accordingly, each having the same width of 5 units.

Extending the notation used previously for bar charts: Suppose that for n pieces of quantitative data we have created k intervals (for $k < n$) $[e_0, e_1)$, $[e_1, e_2), [e_2, e_3), \ldots, [e_{k-1}, e_k)$, where $e_0, e_1, e_2, \ldots, e_k$ are the **class boundaries** or **interval endpoints**. Then, for each interval, we can make data tallies (as we would for qualitative data) and from this, find the frequency, or number of data points occurring in each interval. For each $i = 1, 2, 3, \ldots, k$, we will use the notation f_i to denote the frequency and $r_i = \frac{f_i}{n}$ to denote the relative frequency for the ith interval $[e_{i-1}, e_i)$. Then the frequencies for the k intervals are given by $f_1, f_2, f_3, \ldots, f_k$ and clearly sum to n, the total number of observations. Therefore, the relative frequencies theoretically sum to 1. (This will be shown algebraically in Section 2.3.)

Definition 2.2 *A* **grouped (relative) frequency distribution for quantitative data** *is a tabular listing of non-overlapping class intervals together with the (relative) frequency of occurrence in each interval. An interval with the highest (relative) frequency is called a* **modal class**.

□

The structure of a (relative) frequency distribution is:

Interval	**Frequency** f_i	**Relative frequency** $r_i = \frac{f_i}{n}$
$[e_0, e_1)$	f_1	r_1
$[e_1, e_2)$	f_2	r_2
$[e_2, e_3)$	f_3	r_3
$\vdots$	$\vdots$	$\vdots$
$[e_{k-1}, e_k)$	f_k	r_k
Total	n	1.00

Notice that the process of constructing a frequency distribution is not reversible (unless we keep a copy of the original data) since the identity of an original data point is gone once it has been allocated to an interval. In Example 2.3 it is not possible to reconstruct the original data of timber densities by using the frequency distribution alone. Intervals don't need to be the same width. In fact, intervals of varying width may be used to emphasise asymmetrical qualities in some types of data sets. For example, it would seem inappropriate to report the frequencies of income data in narrow intervals of equal width; the top end of the scale may produce some empty intervals.

Relative frequency histograms

A relative frequency histogram is an appropriate graphical display for a relative frequency distribution of quantitative data. Both frequency histograms and relative frequency histograms can be presented as the same graph. The two differ by a *change of scale* (or *relabelling*) of the vertical axis. This may be seen via the linear transformation

$$r = (\frac{1}{n})f,$$

where r represents relative frequency, f represents frequency and n is fixed (as the total number of observations).

The construction given in Example 2.3 should be studied carefully. The relative frequency histogram is constructed from the relative frequency distribution by representing the 10 intervals on the horizontal axis. Rectangles are constructed on these intervals so that the *area* of the rectangle represents the relative frequency. It follows (from our knowledge of how to determine the area of rectangles) that, for each interval:

$$\boxed{\text{height of rectangle} = \frac{\text{relative frequency}}{\text{interval width}}}$$

The left vertical axis is then labelled **relative frequency per unit density**. The (relative) frequencies associated with each interval can be usefully bracketed above each rectangle. To create a frequency distribution on the same graph it is convenient to place a vertical scale in the right margin labelled **frequency per unit density**. Of course, when the intervals are of unit width, the heights of the rectangles will correspond exactly to the relative frequencies. Recalling for each $i = 1, 2, 3, \ldots, k$, the notation $[e_{i-1}, e_i)$ for the ith interval, then our methods suggest that the rectangles are constructed with widths $w_i = e_i - e_{i-1}$ and heights $h_i = \frac{r_i}{w_i}$. The left axis is labelled **relative frequency per unit** ..., and the scales on the right and left axes are related by the linear transformation $f = n\ r$, which is a change of scale since n is fixed.

Example 2.3

How dense is timber?

The following observations on the air-dried density for specimens of 50 different varieties of timber (mg cm^{-3}) are part of a larger data set used by Brown and Maritz (1982) to illustrate their work on distribution-free regression methods:

55.2	50.3	42.5	28.2	58.7	39.0	50.2	52.8	61.3	61.5
32.5	42.3	37.1	50.3	31.4	58.3	59.5	30.7	70.8	53.8
40.3	38.3	68.1	63.3	51.3	56.9	36.8	40.3	49.0	53.9
63.2	43.0	54.9	55.3	55.1	57.3	43.0	25.3	46.7	68.9
40.6	28.6	58.6	68.9	42.4	40.1	29.1	60.8	51.7	39.6

(*Source*: Brown, B.M. and Maritz, J.S. (1982). Distribution-free methods in regression. *Australian Journal of Statistics*, 24, 3, 318-31.)

With 10 intervals each of width 5, a (relative) frequency distribution is built from the tallies:

Interval	**Tally**	**Frequency**	**Relative frequency**
25–30⁻	////	4	0.08
30–35⁻	///	3	0.06
35–40⁻	~~////~~	5	0.10
40–45⁻	~~////~~ ////	9	0.18
45–50⁻	//	2	0.04
50–55⁻	~~////~~ ////	9	0.18
55–60⁻	~~////~~ ////	9	0.18
60–65⁻	~~////~~	5	0.10
65–70⁻	///	3	0.06
70–75⁻	/	1	0.02
Total		50	1.00

(Relative) frequency histogram for air-dried density of timber

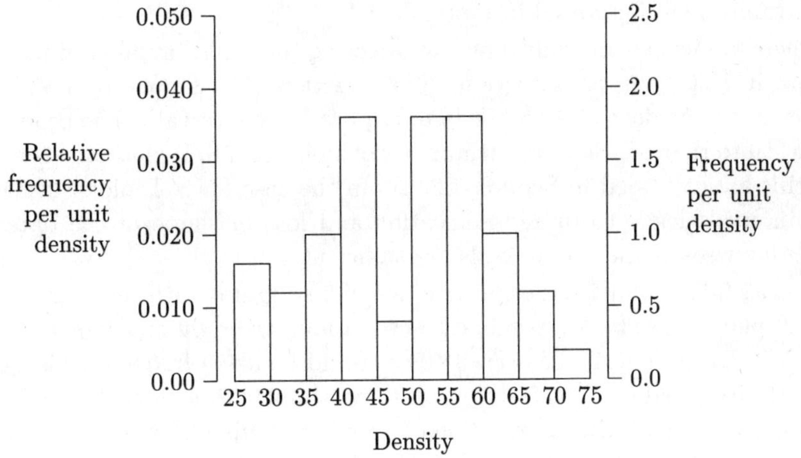

In Example 2.3,

$$h_i = \frac{r_i}{5} = \frac{f_i}{5n} = \frac{f_i}{250}.$$

Therefore, h_1, the height of the first rectangle standing on $[e_0, e_1) = [25, 30)$ is

$$h_1 = \frac{f_1}{250} = \frac{4}{250} = 0.016.$$

Similar calculations follow for the other rectangles.

There are several consequences of this method of histogram construction.

The preference for consideration of relative frequencies rather than frequencies is an introduction to the role of probabilities and probability distributions (see Chapter 4). Probability distributions are sometimes used as a theoretical model to help explain and predict the behaviour of relative frequencies. This also ties in the view of relative frequencies as *areas*; since the areas of the constructed rectangles represent the relative frequencies, the sum of the areas of all these rectangles must be 1. (Relative frequencies sum to 1.) By analogy, in Chapter 4, the area under a theoretical probability curve (or probability density function) is 1.

The notation used to describe the intervals in Example 2.4 differs from previous examples. Since the data consist of whole numbers, the intervals are unambiguously specified by specifying their endpoints to one decimal place (more than the data). There is no doubt where to put each data point. In fact, this method of specifying the class limits gives the reader a guide as to how accurately the (continuous) data have been measured.

The most important feature of the given method of drawing histograms is that it automatically accommodates, without bias, the use of intervals of differing widths on the horizontal axis. This means that rectangles standing on wider intervals are not eye-catching and exaggerated since their areas cover an appropriate fraction of the area of the entire histogram. This aspect is certainly well illustrated in Example 2.4.

There are no specific guidelines for selecting the 'right' number of intervals to use in histogram construction. The selection of k is done by keeping in mind that large data sets are likely to require more intervals. The problem is identical to that of selecting the number of stems in a S/L plot and strategies for this are discussed in Section 1.2. As in the case for S/L plots, using too few intervals leads to oversimplification and loss of shape of the data set. Often between 5 and 15 intervals are sufficient.

It is useful to compare the construction of histograms with the construction of S/L plots. If we took a pen and traced around the right-side fringes of S/L Plot 1 in Example 1.4, the overall effect would be like a frequency histogram turned on its side. This is important; the implication is that a S/L plot provides us with all the information of a frequency histogram and then some! In a S/L plot the data is essentially retrievable. (Try writing out the data in Example 1.4 by using only the information from the stems and leaves of S/L Plot 1.) 'Essentially retrievable' means fully retrievable if the leaves in the S/L plot have not suffered rounding or chopping. This is not so in the case of frequency histograms. Of course, the reason why this works is that, by construction, each leaf in an S/L plot occupies the same sized space as every other leaf. Then, collectively, on a given stem, the total space taken up by all the leaves is proportional to the *frequency* for that stem. Even so, histograms have an advantage over S/L plots: they allow the flexibility of widely differing intervals in their definition. A S/L plot of the inter-arrival times of Example 2.4 would be difficult to collate with stems of fixed width and the data analyst would be drawn towards first transforming the data.

Example 2.4
Inter-arrival times
Listed is a sample of 100 times (in seconds) between the arrivals of customers at a self-service petrol station in the Perth suburb of Victoria Park. The study was commissioned by the petroleum supplier to determine whether a further service station should be established in the area. The data was collected between 3 and 4 p.m. on consecutive weekdays.
Also given are the first steps in constructing a relative frequency histogram by first constructing a relative frequency distribution using the tally method for accuracy.
The listing of time intervals in this example shows a new departure. Because of the relative scarcity of large observations in this data set, it is appropriate to use intervals of differing sizes, using considerably larger widths towards the top of the distribution. The final column in the table of the relative frequency distribution gives a listing of the heights of the rectangles necessary for the construction of the relative frequency histogram. The reader may wish to check these entries. For example, for the interval 0.5–25.5 of width 25 seconds,

$$\text{height} = \frac{\text{relative frequency}}{\text{interval width}} = \frac{0.36}{25} = 0.0144.$$

The remaining heights follow from similar calculations.

103	53	23	8	49	8	6
57	24	4	83	9	6	3
32	12	106	53	15	33	73
52	191	10	1	45	127	81
96	43	42	189	135	376	11
1	67	42	64	145	15	62
61	426	9	30	70	35	34
23	114	87	24	19	193	12
34	28	237	59	79	97	17
118	23	70	139	36	86	2
139	7	47	42	26	23	2
37	101	15	17	18	71	16
79	185	4	2	108	14	51
37	85	11	3	128	81	31
28	271					

Interval	Tally
0.5–25.5	𝍸 𝍸 𝍸 𝍸 𝍸 𝍸 𝍸 /
25.5–50.5	𝍸 𝍸 𝍸 𝍸
50.5–75.5	𝍸 𝍸 ////
75.5–100.5	𝍸 𝍸
100.5–150.5	𝍸 𝍸 //
150.5–200.5	////
200.5–300.5	//
300.5–500.5	//

Interval	Interval width	Frequency	Relative frequency	Interval height
0.5–25.5	25	36	0.36	0.0144
25.5–50.5	25	20	0.20	0.0080
50.5–75.5	25	14	0.14	0.0056
75.5–100.5	25	10	0.10	0.0040
100.5–150.5	50	12	0.12	0.0024
150.5–200.5	50	4	0.04	0.0008
200.5–300.5	100	2	0.02	0.0002
300.5–500.5	200	2	0.02	0.0001
Total		100	1.00	

(Relative) frequency histogram
for the inter-arrival times of cars at
a self-service petrol station

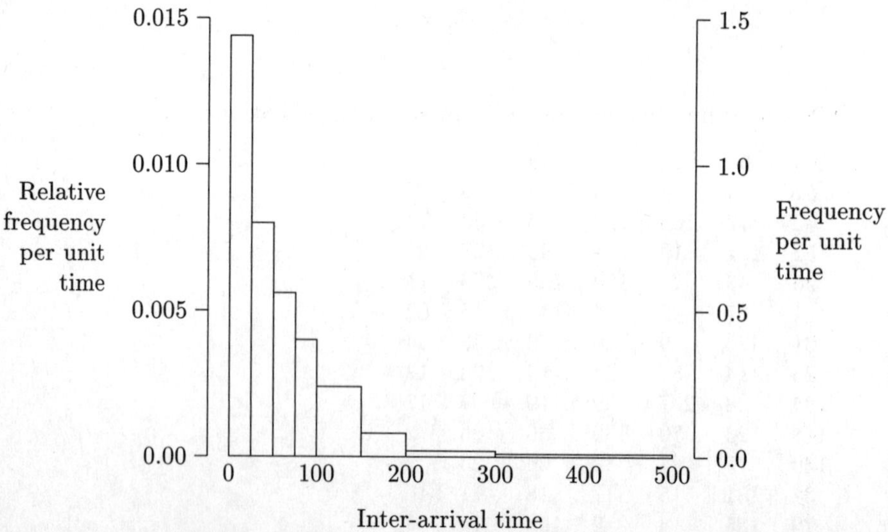

Skewness

An obvious feature of the relative frequency histogram in Example 2.4 is that most of the inter-arrival times (blank periods for the service station's cash register) last less than 100 seconds, with few observations occurring at the top end of the distribution. Returning to the discussion of skewness that began in Section 1.5: the *ends* of a histogram are termed **tails** and a histogram exhibits **skewness** if the tails are of differing length. If the left tail is longer than the right, the histogram is **left-skewed** (such skewness is often termed **negative skewness**); if the right tail is longer than the left, the histogram is **right-skewed** (or shows **positive skewness**) as in the case of the histogram of inter-arrival times. The histogram is **symmetric** if the tails are of similar length (as in the case of the distribution of timber densities).

Bending histograms

Skewed data sets will produce skewed histograms with long visible tails. These tails may be the main interest for the data set; for example, it is the extreme values in air pollution data that may be of interest. On the other hand, many statistical tests require assumptions of more symmetric distributions of data, in which case a skewed distribution would need transforming.

Example 2.5

Inter-arrival times symmetrised by a power transformation

Shown below is a histogram for the distribution of $Y = X^{0.3}$, where X is the inter-arrival time variable of Example 2.4. A glancing comparison with the histogram of the original data prior to transformation (Example 2.4) shows that the data has been bent towards symmetry; the right-skewness however has not been completely removed. An attempt to use the transformation $Y = \log_e X$ resulted in an overcompensation for the skewness (so that the resulting distribution became left-skewed).

The reader may like to experiment with other brands of power transformations. Useful choices include values of κ, the power transformation index, in the range $0 < \kappa < 0.3$, where, from Section 1.2, the role of $\kappa = 0$ is occupied by the logarithm and $\kappa = 0.3$ is given in this example. Since the power transformation (logarithm) with $\kappa = 0$ overcompensates with slight left-skewness, and using the index $\kappa = 0.3$ for the power transformation causes a slight right-skewness to be retained, we need to look in the range $0 < \kappa < 0.3$ for an index which provides transformation to symmetry.

Power transformation $Y = X^{0.3}$ of inter-arrival times (X)

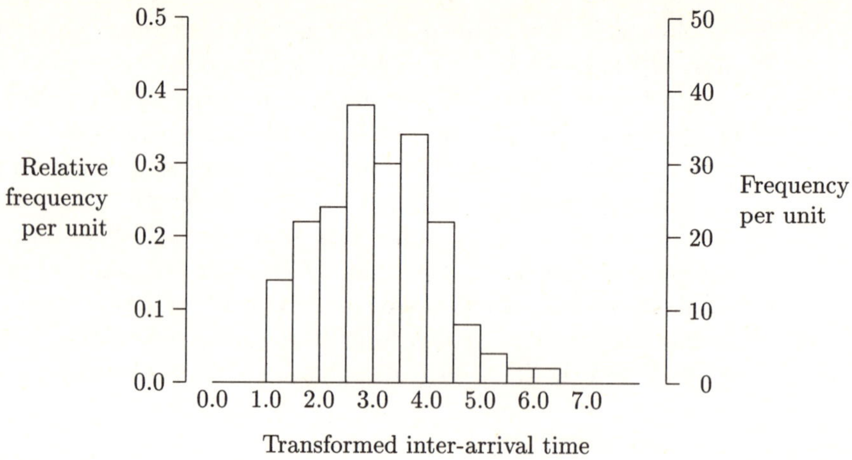

The power transformation $Y = X^{0.3}$ was applied to the histogram in Example 2.4 to produce the histogram in Example 2.5. The comments made in Section 1.5 are all relevant here, so that logarithms are a useful first choice when attempting to symmetrise highly right-skewed distributions. The point is that now, with the aid of a histogram or a S/L plot, the assessment of which transformation to use and the effect that it has on the data can be more easily made.

EXERCISES

2.7 After the first year of operation of the Medicare health system in Australia, 60 respondents reported the number of Medicare claims that they had filed. The results were as follows.

9	10	14	14	12	14	0	2
6	2	3	5	7	3	15	7
4	3	6	4	5	4	7	4
5	9	4	10	8	4	1	3
4	8	6	10	6	12	0	3
12	7	7	3	6	6	4	3
7	7	3	7	6	6	4	10
12	9	14	14				

(a) Use the tally method to construct the following frequency distributions:

Frequency distribution A

Interval	Frequency
0–under 2	
2–under 4	
4–under 6	
6–under 8	
8–under 10	
10–under 12	
12–under 14	
14–under 16	

Frequency distribution B

Interval	Frequency
0–under 3	
3–under 6	
6–under 9	
9–under 12	
12–under 15	
15–under 18	

(b) For each of the frequency distributions A and B determine intervals of highest and lowest frequency.

(c) Describe some of the effects of reducing the number of intervals used in (a) to describe this data set.

2.8 The following are observed weights (kg) of containers of PVC powder with a label weight of 25 kg.

25.43	25.37	25.07	25.01
25.45	25.50	25.40	25.40
24.87	25.38	25.03	25.38
25.13	25.38	24.67	25.08
24.83	25.38	25.22	25.17
24.72	25.18	24.88	24.98
25.15	25.25	25.10	25.07
25.42	25.34	25.08	25.28
25.00	25.36	25.45	25.19
25.14	25.24	25.28	25.83

(a) Construct a relative frequency distribution for these data.

(b) Construct a relative frequency histogram for these data.

(c) How does the shape of the histogram change if the bag label weight of 25 kg is subtracted from each observation?

(d) Construct a suitable S/L plot for these data. Does the S/L plot convey more information than the histogram?

2.9 The diagram shows a relative frequency histogram constructed from 100 daily measurements of ozone concentration (ppm) taken from metropolitan areas around Australia. The data were measured to one decimal place.

Relative frequency histogram for ozone concentration

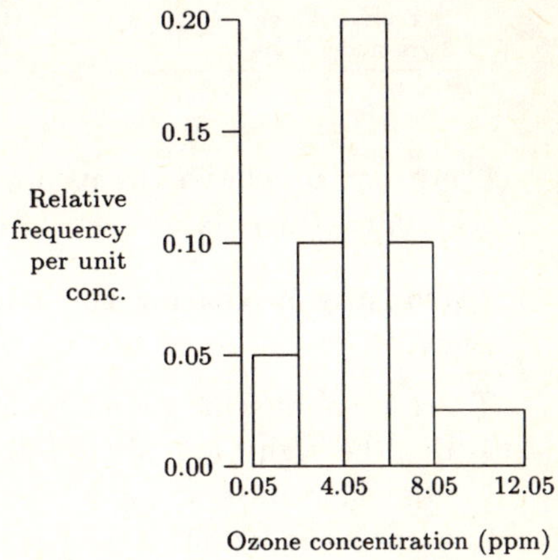

(a) How can you tell from looking at the histogram (only) that the data were measured to 1 decimal place?

(b) Determine the frequency associated with the interval 4.05–6.05. Repeat this for all the intervals depicted in the diagram and hence collect the information in a frequency distribution.

(c) On what proportion of days did the ozone concentration sink below 2 ppm?

2.10 **Age distribution of children in commercial daycare centres.** This example uses intervals of variable width but this time with equal frequencies across the intervals. The example illustrates the pictorial significance for the areas of component rectangles to represent the relative frequency in a relative frequency histogram.

Relative frequency distribution
†$n = 40$. Each interval has frequency 10.

Interval †	Interval width	Relative frequency
0–under 6 mths	0.5	0.25
6 mths–under 1 yr	0.5	0.25
1 yr–under 2 yrs	1.0	0.25
2 yrs–under 4 yrs	2.0	0.25

Construct a relative frequency histogram for these data.

2.11 **Collapsing boxes for the timber data.**

There are no specific guidelines for selecting the 'right' number of intervals to use when constructing a (relative) frequency histogram. Using too few intervals leads to over-simplification and loss of shape of the data set. Notice the particular bi-modal shape of the timber data of Example 2.3. This information may be lost if the number of intervals is greatly reduced. Consider what would happen to the shape of the histogram if we halved the number of intervals used: construct a relative frequency histogram based on 5 intervals each 10 units wide.

Comment on the results.

2.12 Austral Ware ceramic mugs are manufactured in large batches and then kiln fired. After the firing a batch is rejected if it contains any defective mugs due to faults in the glazing. Analysis of 1000 defective batches showed percentages defective as recorded in the following relative frequency histogram.

Relative frequency histogram for defective ceramic mugs

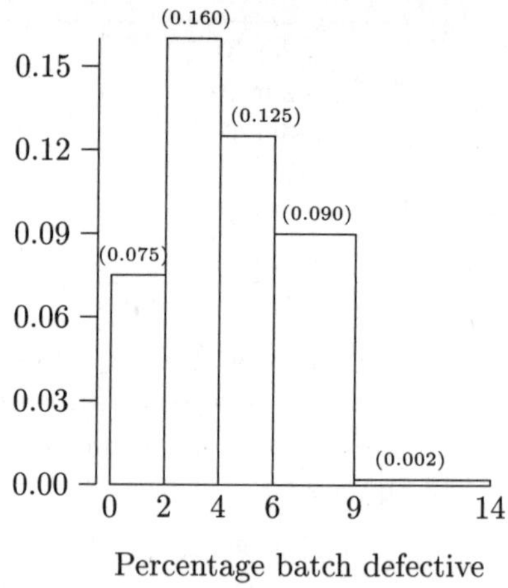

(a) Copy and complete the following description of intervals on the horizontal axis: 0–under 2; 2–under 4 ...

(b) Determine the total area of all the rectangles comprising the histogram. (The relative frequencies per unit percentage are conveniently given bracketed above the intervals.)

(c) Construct a relative frequency distribution.

(d) Explain how to convert the given histogram to a frequency histogram. What will be the total area of all the rectangles in this frequency histogram ?

(e) Obtain from (c) a frequency distribution for the data.

(f) Hence give the number of batches containing less than four percent defective mugs.

2.13 Suppose that a random sample of 100 first offenders from New South Wales prisons gave the following frequency distribution for length of sentence (in years).

Sample A: First offenders	
Interval	**Number**
less than 1.5	35
1.5–under 3	30
3–under 5	15
5–under 10	10
10–under 30	6
life (30+ years)	4

For comparison, the records of 45 New South Wales repeat offenders were also accessed giving, in ascending order, the current sentence lengths (in months/years) listed below:

Sample B: Repeat offenders			
2 months	4 months	6 months	6 months
1 year	1 year	1 year	18 months
18 months	2	2	3 years
3 years	3	3	4
4	4	4	4
4	4	4	5
5	8	8	8
10	10	12	12
12	20	20	25
25	25	25	25
25	life	life	life
life			

(a) Construct relative frequency histograms for both samples to allow for easy comparison. Use an interval of width 20 to cover the 'life' sentence.

(b) Discuss whether there is any evidence to support the claim that repeat offenders are currently serving longer sentences.

2.14 Show that for the data of Example 2.4, the inter-arrival times, X, of cars at a petrol station in Victoria Park, the (power) transformation $Y = \log_e X$ causes a change in the skewness of the data. In fact, the data change under this transformation from being strongly right-skewed to mildly left-skewed.

2.15 Which of the following transformations has the effect of improving the symmetry (reducing the skewness) of the given histogram?

$$Y = X^3$$
$$Y = X^2$$
$$Y = X^{1.5}$$
$$Y = X^{0.5}$$

(a)

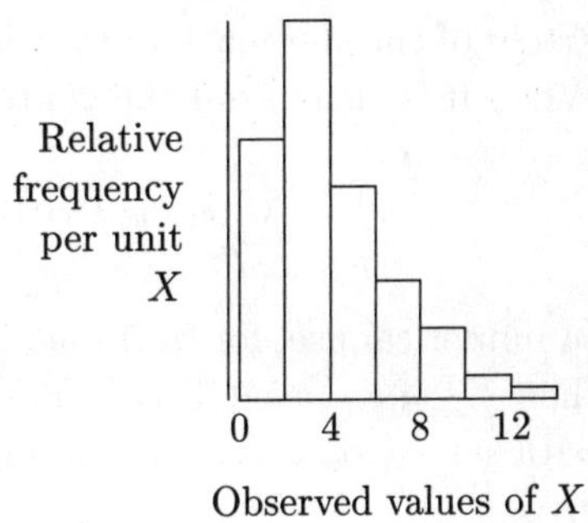

(b)

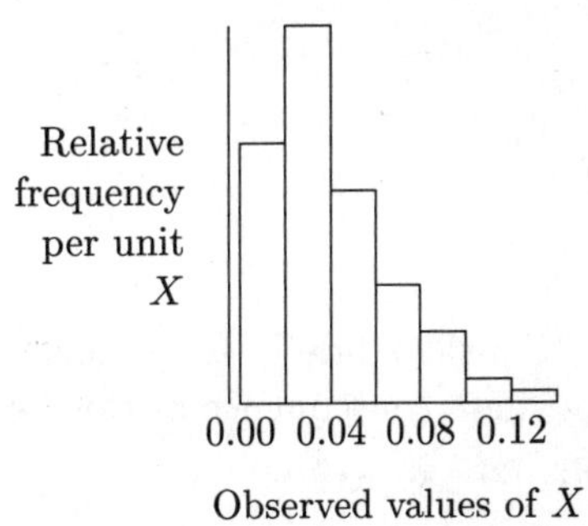

2.3 SUMMATION NOTATION AND CUMULATIVE FREQUENCIES

Extensive use will be made of **summation notation** in different frameworks throughout this text. This notation is now carefully defined and used immediately in the context of adding frequencies and relative frequencies.

Definition 2.3 *Given n quantities $x_1, x_2, x_3, \ldots, x_n$, the* **summation notation** $\sum$ *acts as a summary for their sum:*

$$\sum_{i=1}^{n} x_i = x_1 + x_2 + x_3 + \ldots + x_n.$$

□

The subscripts 1 and n above and below the summation sign indicate the *extent* of the summation; that is, the range of values actually being summed. When it is clear from the context, these limits are often omitted and

$$\sum_{i=1}^{n} x_i \quad \text{is written} \quad \sum_{i} x_i \quad \text{or simply} \quad \sum x_i.$$

In numerical use, for example, if $x_1 = 5, x_2 = 3, x_3 = -4, x_4 = 1$, and $x_5 = 2$, then $\sum_{i=1}^{5} x_i = 5 + 3 + (-4) + 1 + 2 = 7$. Notice that for the very special data set $x_1, x_2, x_3, \ldots, x_n$, where $x_1 = x_2 = x_3 = \ldots = x_n = 1$, Definition 2.3 gives

$$\begin{aligned} \sum_{i=1}^{n} x_i &= \sum_{i=1}^{n} 1 \\ &= 1 + 1 + 1 + \ldots + 1 \\ &= n. \end{aligned}$$

Consider briefly some general applications of Definition 2.3. It is an immediate consequence of this definition that for a constant b and n quantities $x_1, x_2, x_3, \ldots, x_n$

$$\begin{aligned} \sum_{i=1}^{n} bx_i &= bx_1 + bx_2 + \ldots + bx_n \\ &= b(x_1 + x_2 + \ldots + x_n) \\ &= b\sum_{i=1}^{n} x_i, \end{aligned}$$

which gives

$$\sum_{i=1}^{n} bx_i = b\sum_{i=1}^{n} x_i \tag{2.1}$$

so that constants multiply through summation signs. Equation 2.1 is called the **scalar multiplication property of the $\sum$ sign**. For example, if $b = 1.5$ and all the data in a data set are multiplied by 1.5, then Equation 2.1 indicates that the sum of the data is also multiplied by 1.5.

Further, the summation sign allows us to add two data sets $x_1, x_2, x_3, \ldots, x_n$ and $y_1, y_2, \ldots, y_n$ together either by pairs or by adding together the separate sums of each data set:

$$\begin{aligned}\sum_{i=1}^{n}(x_i + y_i) &= (x_1 + y_1) + (x_2 + y_2) + \ldots + (x_n + y_n) \\ &= (x_1 + x_2 + \ldots + x_n) + (y_1 + y_2 + \ldots + y_n) \\ &= \sum_{i=1}^{n} x_i + \sum_{i=1}^{n} y_i.\end{aligned}$$

This may be summarised as

$$\sum_{i=1}^{n}(x_i + y_i) = \sum_{i=1}^{n} x_i + \sum_{i=1}^{n} y_i \tag{2.2}$$

and is called the **distributive property of the $\sum$ sign**.

Adding frequencies

Suppose that a qualitative/quantitative variable has k different levels/intervals and that we record n observations on this variable. As in previous sections, for each $i = 1, 2, 3, \ldots, k$, we will use the notation f_i to denote the frequency for level i/ interval i. This means that the frequencies for the k levels/intervals are given by $f_1, f_2, f_3, \ldots, f_k$ and clearly

$$\sum_{i=1}^{k} f_i = n.$$

The relative frequency r_i, for level i/ interval i, is given by $r_i = \frac{f_i}{n}$. Therefore, since the frequencies sum to the total number of observations, the relative frequencies theoretically sum to 1. That is, since by Equation 2.1 the constant $\frac{1}{n}$ multiplies through the summation sign,

$$\sum_{i=1}^{k} r_i = \sum_{i=1}^{k} \frac{f_i}{n} = \frac{1}{n} \sum_{i=1}^{k} f_i = \frac{1}{n}(n) = 1.$$

(Rounding error in reporting relative frequencies may cause slight discrepancies.)

We can also easily describe cumulative frequencies which are generally used for quantitative data: if f_i is the frequency for the ith interval $[e_{i-1}, e_i)$, then the **cumulative frequency**, c_j, for the jth interval is given by

$$c_j = \sum_{i=1}^{j} f_i, \quad j = 1, 2, 3, \ldots, k,$$

and represents the sum of frequencies over the intervals $[e_0, e_1), [e_1, e_2), \ldots,$ $[e_{j-1}, e_j)$. The cumulative frequencies may then be used to describe a **cumulative grouped frequency distribution** which has the structure:

Interval	**Frequency** f_j	**Cumulative frequency** $c_j = \sum_{i=1}^{j} f_i$
$[e_0, e_1)$	f_1	c_1
$[e_1, e_2)$	f_2	c_2
$[e_2, e_3)$	f_3	c_3
$\vdots$	$\vdots$	$\vdots$
$[e_{k-1}, e_k)$	f_k	$c_k = n$

Example 2.6 gives a cumulative frequency distribution for the drift velocity of electrons. We do not present the original data, but list intervals and observed (cumulative) frequencies in those intervals. This is often how experimental results are reported: seldom is the reader of a published scientific journal given immediate access to the original data. However, all information is not lost and, using the methods that we now give, the sample median (for example) still may be estimated from the frequencies and intervals alone.

When we determine the cumulative frequencies in a frequency distribution, the process is like counting up the leaves on each stem in a S/L plot. To determine the sample median in a S/L plot, we determine the rank of the median, identify the correct stem and then count up the leaves on that stem until the appropriate point is reached. Exactly the same counting process is worked through to estimate the median of a frequency distribution: determine the rank of the median, identify the interval where the median lies, and then count up from the bottom of the median interval the correct number of points until the median is reached. There is an assumption involved in determining the 'correct number of points'. We assume that the data ascribed to a given interval in a given frequency distribution are *spread uniformly across that interval.* This means that each data point lying in the interval $[e_{i-1}, e_i)$ is allocated the fraction $\frac{1}{f_i}$ of the interval width w_i. We phrase the definition that follows in terms of percentiles, so that the commonly used median and quartiles may be extracted as special cases.

Definition 2.4 *For data having (cumulative) grouped frequency distribution*

Interval	Width w_j	Frequency f_j	Cumulative frequency $c_j = \sum_{i=1}^{j} f_i$
$[e_0, e_1)$	w_1	f_1	c_1
$[e_1, e_2)$	w_2	f_2	c_2
$[e_2, e_3)$	w_3	f_3	c_3
$\vdots$	$\vdots$	$\vdots$	$\vdots$
$[e_{k-1}, e_k)$	w_k	f_k	$c_k = n$

an **approximate 100*p*th percentile,** x_p^* *is given by*

$$x_p^* = e_{l-1} + \left((np + \tfrac{1}{2}) - c_{l-1}\right)\frac{w_l}{f_l},$$

where l is the index such that x_p lies somewhere in the lth interval $[e_{l-1}, e_l)$. □

(We use the notation * to differentiate between x_p calculated directly from the data and the same percentile *approximated* from the frequency distribution.)

Example 2.6

The drift velocity of electrons in water vapour at 294 K

These data are taken from a paper by Cheung and Elford (1990) appearing in *The Australian Journal of Physics.* They provide reliable data for the study of cross-sections for electron scattering in water vapour at low energies. The velocities were reported for E/N values between 1 Td and 40 Td. (E is the electric field strength and N the gas number density.) E/N is measured in units: 1 Td $= 10^{-17}$ V cm^2. The data are measured in 10^4 cm s^{-1}, the scatter being observed at various pressures although the data suggests that drift velocity does not depend on pressure.

Interval	Width w_j	Frequency f_j	Cumulative frequency $c_j = \sum_{i=1}^{j} f_i$
$[0, 5)$	5	7	7
$[5, 10)$	5	15	22
$[10, 20)$	10	16	38
$[20, 40)$	20	16	54
$[40, 80)$	40	11	65
$[80, 150)$	70	5	70

(*Source*: Cheung, B. and Elford, M.T. (1990). The drift velocity of electrons in water vapour at low values of E/N. *Australian Journal of Physics*, 43, 755-63.)

Worked example:
Determine the approximate median drift velocity.

The calculations proceed as follows:

- $n = 70$ and, for the median, $p = \frac{1}{2}$.
- The rank of the median is $np + \frac{1}{2} = \frac{70}{2} + \frac{1}{2} = 35.5$.
- This locates the median in the 3rd interval ($l = 3$), since $c_2 = 22$, so that 22 observations appear by the start of the 3rd interval and $c_3 = 38$, so that 38 observations have appeared by the end of the 3rd interval. Therefore, $[e_{l-1}, e_l) = [10,\ 20)$ is the median interval.
- The width of the median interval is 10 units so that each data point occupies a space $\frac{w_l}{f_l} = \frac{10}{16}$.
- $x^*_{0.5} = e_{l-1} + \left(\left(\frac{n}{2} + \frac{1}{2}\right) - c_{l-1}\right)\frac{w_l}{f_l} = 10 + (35.5 - 22)\frac{10}{16} = 18.4375$.
  Hence the estimated sample median is 18.4375×10^4 cm s^{-1}.

EXERCISES

2.16 (a) The diagram shows a square of length of side $x_1 + x_2$ units subdivided into smaller rectangles each with a length of side of x_1 units or x_2 units. By a consideration of areas, show that $(x_1+x_2)^2$ does not equal $x_1^2+x_2^2$.

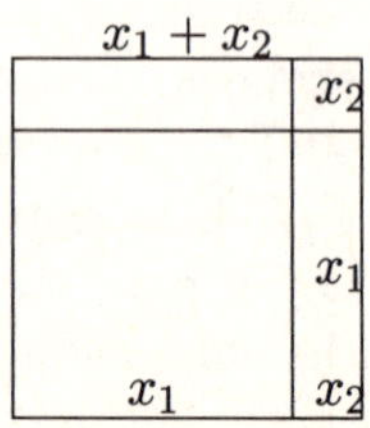

(b) What geometric picture could you use to illustrate the fact that $(\sum_{i=1}^{n} x_i)^2$ and $\sum_{i=1}^{n} x_i^2$ are different? Evaluate each of these quantities for $n = 4$ and $x_1 = 3, x_2 = 7, x_3 = 1, x_4 = 1$.

2.17 If $x_1 = -3, x_2 = 2, x_3 = -\frac{1}{2}, x_4 = 1$ and $x_5 = -1$, determine the value of

$$\sum_{i=1}^{5}(-2x_i + 2).$$

2.18 A market research firm conducted a survey of 100 people of which 36 were men. Part of the recorded information was the gender of each individual. This part produced a data set $x_1, x_2, x_3, \ldots, x_{100}$, in which each observed point is a 0 or a 1 (corresponding to coded indicator values 0 = male, 1 = female). Determine the value of

$$\sum_{i=1}^{100} x_i$$

2.19 Evaluate

$$\sum_{i=10}^{100} 100.$$

2.20 Evaluate

$$\sum_{i=1}^{100} i.$$

2.21 (a) Determine the estimated interquartile range $x^*_{0.75} - x^*_{0.25}$ for the data from Example 2.6 on the drift velocity of electrons. The tabled values have units $\times 10^4$ cm s^{-1}.

Interval	f_i	**Cumulative frequency**
[0, 5)	7	7
[5, 10)	15	22
[10, 20)	16	38
[20, 40)	16	54
[40, 80)	11	65
[80, 150)	5	70

(b) Is it true in general that

$$x^*_{0.75} - x^*_{0.25} = x_{0.75} - x_{0.25}?$$

That is: Does the assumption that data points are spread uniformly in a given interval have any effect on the interquartile range or does the effect cancel out and give the same answer as would be obtained from the original data?

2.22 Determine the value of the constant c so that for the data set $x_1, x_2, x_3, \ldots, x_n$,

$$\sum_{i=1}^{n}(x_i - c) = 0.$$

2.23 (a) Construct a relative frequency histogram for the following aircraft waiting times at Kingsford Smith Airport. Aircraft waiting time, in minutes, for a departing aircraft is defined to be the difference between scheduled departure time and actual departure time.

Interval $[e_{i-1}, e_i)$	**Frequency** f_i
[0, 10)	32
[10, 20)	40
[20, 45)	25
[45, 90)	13
[90, 180)	15

(b) Sketch the function

$$h(x) = \frac{f_i}{n(e_i - e_{i-1})}$$

for $x \in [e_{i-1}, e_i)$, $i = 1, 2, 3, 4, 5$. n as usual denotes the sum of the frequencies and in this example, $e_0 = 0$ and $e_5 = 180$.

[A function such as h with a graph consisting of horizontal line segments is termed a **step function**.]

(c) By calculating the areas of rectangles under the step function h, find

$$\int_0^{90} h(x)dx.$$

2.24 It is difficult to assess the variability of rainfall over a large geographic region. Cameron (1985) indicates that in Australia, regional rainfall may be sparse (in the Simpson Desert) or plentiful (in Tully, Queensland) or a varied mixture (at Whim Creek, WA, 747 mm of rain have been recorded on a single day, but only 4 mm were received during the whole of 1924).

One method for assessing rainfall variability is by the **variability index**, ζ, defined for data given in a frequency distribution by

$$\zeta = \frac{x^*_{0.90} - x^*_{0.10}}{x^*_{0.5}}.$$

(a) Would the value $\zeta = 3$ be regarded as a high or a low value for this variability index? Explain briefly.

(b) Show that ζ can still be calculated approximately by using a relative frequency distribution when the total number of observations is unknown.

(c) Determine the value of ζ for the annual rainfall for the Northern Territory based on the following relative frequency distribution. (The annual rainfall is reported from a large number of stations covering a wide variety of climatic regions.)

Annual rainfall (mm)	**Relative frequency**
[0, 200)	0.155
[200, 300)	0.356
[300, 400)	0.090
[400, 500)	0.066
[500, 600)	0.058
[600, 800)	0.116
[800, 1200)	0.096
Above 1200	0.063

2.25 For the data set $x_1, x_2, x_3, \ldots, x_n$, determine the value of

$$\sum_{i=1}^{n} \left(x_i - x_{(i)}\right),$$

where $x_{(1)} \le x_{(2)} \le \ldots \le x_{(n)}$.

2.4 CENTRE AND SPREAD

We now focus on the centre of a data set and look carefully at how scattered the data points are about that centre. The idea is to capture the concept of centre and the concept of spread by using a single number summary in each case. There are several choices; for example, the median has already been investigated in Chapter 1 as an estimator of the centre of a data set, simply because it splits the ranked data set in half.

Sample means and variances

The *location* of the median is based on relative rankings within the data set rather than how big the data actually are. The *magnitude* of the median is based on the magnitude of a single middle value or the magnitude of a pair of middle values. By comparison, the sample mean provides us with a measure based on the actual size of *all* the data.

Consider first the outcomes $x_1, x_2, x_3, \ldots, x_n$ of a quantitative variable. This is the general notation for a quantitative data set. To mesh with concepts in later chapters, imagine these n outcomes to constitute a *sample* from the target population of all possible outcomes.

Definition 2.5 *For the quantitative data set $x_1, x_2, x_3, \ldots, x_n$, the* **sample mean**, *denoted by $\overline{x}$, is defined by*

$$\overline{x} = \frac{\sum_{i=1}^{n} x_i}{n}.$$

When $n > 1$, the **sample variance**, *denoted by s_x^2, is defined by*

$$s_x^2 = \frac{\sum_{i=1}^{n} x_i^2 - n\overline{x}^2}{n-1}.$$

The **sample standard deviation**, *denoted by s_x, is the positive square root of the sample variance.*

□

The sample mean, variance and standard deviation are relatively well known and may have been previously encountered. The sample mean is often described as an **arithmetic average**, for it represents the fulcrum when a histogram is balanced on a see-saw so as to tip neither one way nor the other. In this context, notice that a child placed at an extreme end of a see-saw will exert plenty of leverage; the same child placed above the fulcrum exerts no net effect.

Example 2.7

How many times have your numbers come up in Tattslotto?

The following chart, published in *The Herald*, Melbourne, 22 August 1990, shows the number of times that each number has been drawn in Tattslotto since drawing began. The chart is listed in the same order as a Tattslotto card, that is, left to right along the first row are the numbers 1–8, left to right along the second row are 9–16, etc. The game currently has 45 numbers (from which the player selects 6); however we list (in smaller type) only numbers 1–40 which were in the original game.

1	2	3	4	5	6	7	8
162	170	146	155	165	153	176	192
9	10	11	12	13	14	15	16
168	156	159	155	179	170	171	156
17	18	19	20	21	22	23	24
161	155	153	202	156	152	174	160
25	26	27	28	29	30	31	32
179	152	151	167	155	152	167	155
33	34	35	36	37	38	39	40
157	181	175	154	159	160	140	154

Worked example:

Determine the sample mean variance and standard deviation for these data.

According to the way the Tattslotto card is structured, $x_1 = 162, x_2 = 170, x_3 = 146, \ldots, x_{40} = 154$. Therefore,

$$\textstyle\sum_{i=1}^{40} x_i = 162 + 170 + 146 + \ldots + 154 = 6504$$
$$\textstyle\sum_{i=1}^{40} x_i^2 = (162)^2 + (170)^2 + (146)^2 + \ldots + (154)^2 = 1063644$$

and simple calculations give

$$\overline{x} = \tfrac{6504}{40} = 162.6$$
$$s_x^2 = \tfrac{1063644-(40)(162.6)^2}{40-1} = 156.24615$$
$$s_x = \sqrt{156.24615} = 12.499 \text{ (becoming 12.5 to 1 dp.)}$$

To follow through the see-saw theme, imagine each point in the data set $x_1, x_2, x_3, \ldots, x_n$ to be a unit point mass sitting against a measured scale.

The distances $x_i - \overline{x}$, which can be positive or negative, from the point mass at x_i to the centre $\overline{x}$, effectively balance. These distances help measure how spread out the data is. When such distances are added, the algebraic sum is zero:

$$\begin{aligned}\textstyle\sum_{i=1}^{n}(x_i - \overline{x}) &= \textstyle\sum_{i=1}^{n} x_i - \sum_{i=1}^{n} \overline{x} \\ &= \textstyle\sum_{i=1}^{n} x_i - n\overline{x} \\ &= n\overline{x} - n\overline{x} \\ &= 0\end{aligned}$$

so that

$$\sum_{i=1}^{n}(x_i - \overline{x}) = 0. \tag{2.3}$$

When such distances are squared, the algebraic sum is very recognisable:

$$\begin{aligned}\textstyle\sum_{i=1}^{n}(x_i - \overline{x})^2 &= \textstyle\sum_{i=1}^{n} [x_i(x_i - \overline{x}) - \overline{x}(x_i - \overline{x})] \\ &= \textstyle\sum_{i=1}^{n} x_i(x_i - \overline{x}) - \overline{x}\sum_{i=1}^{n}(x_i - \overline{x}) \\ &= \textstyle\sum_{i=1}^{n} x_i(x_i - \overline{x}) \\ &= \textstyle\sum_{i=1}^{n} x_i^2 - \overline{x}\sum_{i=1}^{n} x_i \\ &= \textstyle\sum_{i=1}^{n} x_i^2 - n\overline{x}^2\end{aligned}$$

This last expression is the numerator of the sample variance. Therefore, according to Definition 2.5 we may now write the sample variance s_x^2 as

$$s_x^2 = \frac{\sum_{i=1}^{n}(x_i - \overline{x})^2}{n-1}. \tag{2.4}$$

This equation shows s_x^2 explicitly and symmetrically measuring the spread of the data in terms of the **mean deviations** $x_1 - \overline{x}, x_2 - \overline{x}, \ldots, x_n - \overline{x}$. Collectively, the mean deviations constitute a **centred data set** in which the sample mean has been subtracted off each observation.

By Equation 2.3, knowledge of $(n-1)$ observations in a centred data set implies that all n observations are known through

$$(x_n - \overline{x}) = -\sum_{i=1}^{n-1}(x_i - \overline{x}).$$

The importance of only $(n-1)$ of the n observations in a centred data set is formalised by saying that the centred data set has $(n-1)$ **degrees of freedom**. This is reflected in the use of $(n-1)$ as a denominator in the sample variance. Staudte (1990) is one of several authors who advocate n instead of $(n-1)$ in the sample variance denominator. Such discrepancy is negligible provided n is large.

The sample variance is a measure in $(units)^2$ and the sample standard deviation is measured in $(units)$: if length data is in *cm*, then s_x^2 is in cm^2 and s_x is in *cm*. Commonly, the sample standard deviation rather than the sample variance is used as a measure of spread since its units are the same as those of the original data.

The term 'measure of spread' needs further clarification. How *does* s_x measure the spread of data about its centre? The answer lies in the close relationship enjoyed by $\overline{x}$ and s_x through the theorem of Chebychev.

Theorem 2.1 (Chebychev's Inequality)
For any $r > 1$, *at least* $100\left(1 - \frac{1}{r^2}\right)\%$ *of the data set* $x_1, x_2, x_3, \ldots, x_n$ *lies in the interval* $(\overline{x} - rs_x,\ \overline{x} + rs_x)$.

■

The proof of this result is developed in the final exercise of this section. The interpretation of this inequality for various values of r is:

$r = 2$: At least $100(1 - \frac{1}{4})\%$ or 75% of ANY data set lies within 2 standard deviations of its mean.
$r = 3$: At least $100(1 - \frac{1}{9})\%$ or 88.9% of ANY data set lies within 3 standard deviations of its mean.

Example 2.8
How long since your numbers came up in Tattslotto?
The following chart, published in *The Herald*, Melbourne, 22 August 1990, shows the number of weeks since each number was drawn in Tattslotto. The chart is listed as in Example 2.7 but with all 45 numbers that were in the game as of 22 August.

1	2	3	4	5	6	7	8
0	3	5	5	0	21	5	3
9	10	11	12	13	14	15	16
4	1	18	6	3	7	2	1
17	18	19	20	21	22	23	24
7	1	2	1	1	3	5	9
25	26	27	28	29	30	31	32
3	6	5	2	0	9	1	0
33	34	35	36	37	38	39	40
0	4	0	9	0	2	0	6
41	42	43	44	45			
2	2	3	4	4			

Worked example:
Determine the sample mean, variance and standard deviation for these data.

The $k = 11$ distinct observations occur with the following frequencies:

x_i	0	1	2	3	4	5	6	7	9	18	21
f_i	8	6	6	6	4	5	3	2	3	1	1

For these data:

$$\begin{aligned}\textstyle\sum_{i=1}^{11} f_i x_i &= (0)(8) + (1)(6) + (2)(6) + \ldots + (21)(1) = 175\\ \textstyle\sum_{i=1}^{11} f_i x_i^2 &= (8)(0)^2 + (6)(1)^2 + (6)(2)^2 + \ldots + (1)(21)^2 = 1487.\end{aligned}$$

Using these values in the sample mean and sample variance calculations gives:

$$\overline{x} = \frac{\sum_{i=1}^{11} f_i x_i}{45} = \frac{175}{45} = 3.8889,$$

or 3.9 to 1 dp.

$$s_x^2 = \frac{\sum_{i=1}^{11} f_i x_i^2 - n\overline{x}^2}{44} = \frac{1487 - 45(3.8889)^2}{44} = 18.3283.$$

$$s_x = \sqrt{18.3283} = 4.2812 \quad \text{or} \quad 4.3 \quad \text{to} \quad 1 \text{ dp.}$$

For $r = 3$ Chebychev's Interval containing at least $100\left(1 - \frac{1}{9}\right)$ % or 88.9% of the data is

$$(\overline{x} - rs_x,\ \overline{x} + rs_x) = (3.8889 - 3(4.2812),\ 3.8889 + 3(4.2812)) = (-9.0, 16.7)$$

The only data points that lie outside this interval are $x_{10} = 18$ and $x_{11} = 21$ and these comprise only 4.4% of the data, leaving 95.6% of the data inside the interval. This shows that the percentage provided by Chebychev in Theorem 2.1 is conservative.

This indicates that spread from the centre may be measured by a multiple of standard deviations.

In many data sets, a particular observation may occur more than once. Observed data from a discrete variable X, which may assume only certain specified values, will often show this effect. For example, observed values of the variable 'number of incoming phonecalls through the switchboard of a large media organisation (such as the ABC) during a 10 minute period' are likely to be 0, 1, 2, ..., depending on which 10 minute period we consider. Over several 10 minute periods, the data set will be constructed as the values 0, 1, 2, ..., are repeated with certain frequencies.

Suppose that in general the distinct points in a data set are $x_1, x_2, x_3, \ldots, x_k$ and that for each i, the data point x_i occurs with frequency f_i. For calculations involving summation, rather than writing

$$\underbrace{x_i + x_i + \ldots + x_i}_{f_i \ terms}$$

it is easier to write the same quantity as $f_i x_i$, so that the definitions of $\overline{x}$ and s_x^2 in Definition 2.5 become

$$\overline{x} = \frac{\sum_{i=1}^{k} f_i x_i}{n} \tag{2.5}$$

and

$$s_x^2 = \frac{\sum_{i=1}^{k} f_i x_i^2 - n\overline{x}^2}{n-1}. \tag{2.6}$$

Such calculations are illustrated in Example 2.8.

Proportions as sample means

Suppose that in a data set, the only values that occur are $x_1 = 1$ and $x_2 = 0$; the value 1 occurs with frequency 11 and the value 0 occurs with frequency 89. Then, for this data set

$$\overline{x} = \frac{\sum_{i=1}^{2} f_i x_i}{100} = \frac{(11)(1) + (89)(0)}{100} = \frac{11}{100}.$$

This value is the *proportion* of non-zero items in the data set. This proportion is, in value, identical to the sample mean of the observed zeroes and ones in the data set. Proportions and relative frequencies are a special case of means.

By returning to Example 2.1 concerning intended voting patterns in the seat of Curtin, we find the following record:

Liberal	Liberal	Liberal	Labor	Labor
Labor	Liberal	Labor	Undecided	Labor
Labor	Labor	Liberal	Other	Undecided
Labor	Liberal	Liberal	Labor	Labor
Labor	Labor	Liberal	Undecided	Labor
Labor	Labor	Labor	Liberal	Liberal
Undecided	Liberal	Liberal	Undecided	Labor
Labor	Liberal	Liberal	Labor	Liberal
Labor	Labor	Liberal	Other	Labor
Liberal	Liberal	Liberal	Undecided	Labor
Liberal	Liberal	Labor	Other	Labor
Liberal	Liberal	Liberal	Undecided	Liberal
Labor	Liberal	Liberal	Other	Labor
Liberal	Labor	Liberal	Liberal	Labor
Undecided	Liberal	Liberal	Undecided	Liberal
Labor	Liberal	Liberal	Liberal	Labor
Labor	Undecided	Liberal	Liberal	Liberal
Liberal	Liberal	Liberal	Liberal	Labor
Labor	Liberal	Labor	Labor	Labor
Labor	Liberal	Liberal	Undecided	Labor

If we relabel these responses as 1 = *Undecided* and 0 = *All other voters*, then 'Liberal', 'Labor' and 'Other' are all collapsed to 0 and we obtain the data set

x_i	1	0
f_i	11	89

which, from the above discussion, has $\overline{x} = 0.11$. Therefore we have used the idea of a sample mean to describe the relative frequency of occurrence of a level of a qualitative variable.

Mode

A simple measure of the centre of a data set that is easily located is that value which occurs most often. It is a measure that is not used very often!

Definition 2.6 *A most frequently occurring observation in a data set is termed a* **mode** *of the data set. A data set is termed* **bimodal** *if two values occur with the same highest frequency.*

□

As an illustration, return to Example 2.8: 'How long since your numbers came up in Tattslotto?' The mode of this data set is 0. Indeed, 0 occurred with frequency 8, indicating that this week (as is the case each week) 8 numbers were drawn. This 0 value does not address the fact that in this example the observations 1, 2 and 3 each occurred with frequency 6. (For this data set does the mode of 0 appear to be a more descriptive summary of centre than the sample mean $\overline{x} = 3.9$? Does a quotation of the mode answer the question: How long does it take for Tattslotto numbers to recur?)

One conclusion from Definition 2.6 is that modes need not be unique. In a grouped frequency distribution, a modal class is any interval with the highest frequency. (Many authors define the modal class to be any interval that corresponds to a peak in the frequency histogram.) By assuming that in each interval the data are spread uniformly and act through the centre of that interval (imagine a see-saw again) then a mode for a grouped frequency distribution is approximately the midpoint of a modal class. We also invoke this assumption in the next section.

Approximate mean/variance from grouped frequency distributions

We have seen that the identity of original data becomes lost in a grouped frequency distribution. It cannot be retrieved from the distribution. Even so there are approximations $\overline{x}^*$ and s_x^{2*} to the respective sample mean $\overline{x}$ and sample variance s_x^2 that would be found if the original data were known. It may be that n is so large that summarising data in a grouped frequency distribution is the only manageable abbreviation. As in the last section, we operate on the following assumption: Data in an interval $[e_{i-1}, e_i)$ are spread evenly in that interval and the effects are felt at the centre

$$m_i{}^* = \frac{e_{i-1} + e_i}{2}$$

of the interval. The intervals of the grouped frequency distribution

Interval	Frequency
$[e_0, e_1)$	f_1
$[e_1, e_2)$	f_2
$[e_2, e_3)$	f_3
$\vdots$	$\vdots$
$[e_{k-1}, e_k)$	f_k

are replaced by the data set $m_1{}^*, m_2{}^*, m_3{}^*, \ldots, m_k{}^*$ where $m_i{}^*$ occurs with frequency f_i. Equation 2.5 and Equation 2.6 may then be used to give

$$\overline{x}^* = \frac{\sum_{i=1}^{k} f_i m_i{}^*}{n} \tag{2.7}$$

as an approximation to $\overline{x}$ and

$$(s_x^2)^* = \frac{\sum_{i=1}^{k} f_i (m_i{}^*)^2 - n(\overline{x}^*)^2}{n-1} \tag{2.8}$$

as an approximation to s_x^2.

A sample mean/sample variance comparison of two data sets

Whether the intention is simply to summarise sample information, or to go deeper and use such summary for inference and prediction about the population from which the sample came, the fact remains: the sample mean and the sample standard deviation are very basic tools for the data analyst. For a single data set taken in isolation it is intuitively hard to see their appeal (apart from their being mathematically very convenient). They are used most powerfully as *relative measures*, either pitted in ratio one against the other (the coefficient of variation) or used in comparison of one data set against another.

Definition 2.7 *For the data set* $x_1, x_2, x_3, \ldots, x_n$*, the* **sample coefficient of variation**, v_x*, is defined by*

$$v_x = \frac{s_x}{\overline{x}}.$$

□

Since both $\overline{x}$ and s_x are expressed in the same units as the original data, the sample coefficient of variation is a dimensionless quantity. v_x is often expressed as a percentage and may be usefully compared across two data sets. For example: Whales are, on average, much larger than tuna fish. Indeed, they probably have a much larger variability in size than tuna fish. But, *relative to their average size* do they have a larger variability in size than tuna fish?

Example 2.9

Pavement wear: drilling holes in bricks

The following data represent volumes eroded (mL) in two minutes when small rectangular brick pavers are impacted by a metal tip-hardened punch attached to a high frequency vibrating plate. The equipment was designed to simulate high-traffic wear of brick pavements.

The 20 wet paver samples were subjected to drilling shortly after manufacture; 20 different units from the same shipment were dried for a period of one month before drilling. The results are summarised in the following back-to-back S/L plot.

Pavement wear: volume eroded

Scale: stems—10 mL, leaves—mL

| 22 | 1 represents 221 mL

Wet pavers		Dry pavers
W		D
9 7	19	
7 6 4 3	20	
9 7 7 5 4 3 1 1 1 0	21	
4	22	1 1
5 3 0	23	0 0 2 6 9
	24	0 2 3 6 8
	25	1 2 6 6
	26	1 4 6
	27	4

The back-to-back S/L plot shows the shape of these distributions clearly. The distributions are centred in different locations on the volume eroded scale and clearly (perhaps) our assessment of the spread should give different values for the two data sets. The question is: how different?

Calculations give the following values for the sample means and sample variances:

sample mean	$\overline{w} = 213.8$	$\overline{d} = 245.4$
sample variance	$s_w^2 = 110.17$	$s_d^2 = 218.88$
coefficient of variation	$v_w = 0.049$	$v_d = 0.060$

From this we may numerically compare centre and spread for the two data sets. Notice first that the sample variances differ by a factor of almost 2. This gives a numerical value to the relative spread of the two data sets with shapes given in the above S/L plot. *Further*, the sample coefficients of variation for the two data sets are almost the same; the conclusion is that relative variability in the volume eroded in wet pavers and dry pavers is much the same.

Pooling sample means

A final note concerns the dangers of attempting to locate the centre of two data sets combined (or pooled) by taking the 'mean of the means' of the individual data sets. The result will be incorrect if the two data sets have different numbers of elements. To determine the mean of two data sets combined, the total data sum (from both data sets) must be divided by the total sample size (from both data sets). Therefore, if $x_1, x_2, x_3, \ldots, x_n$ has sample mean $\overline{x}$ and $y_1, y_2, y_3, \ldots, y_m$ has sample mean $\overline{y}$, then the sample mean of the combined data set is

$$\frac{\sum_{i=1}^{n} x_i + \sum_{j=1}^{m} y_j}{m+n} = \frac{n\overline{x} + m\overline{y}}{m+n}$$

and this is $\frac{\overline{x}+\overline{y}}{2}$ only when $m = n$.

For example, if the sample mean volume eroded from 10 wet brick pavers was 200 mm^3 and the sample mean volume eroded from 20 dry brick pavers was 245 mm^3, the sample mean volume eroded from all 30 brick pavers is $\frac{(10)(200)+(20)(245)}{30} = 230$ mm^3.

EXERCISES

2.26 Ten determinations of the modulus of elasticity of timber varieties are:

188	238	99	189	213
165	210	224	228	209

(*Source*: Brown, B. M. and Maritz, J. S. (1982). Distribution-free methods in regression. *Australian Journal of Statistics*, 24, 3, 318-31.)

(a) Determine the sample mean of these data.

(b) Determine the sample variance of these data.

(c) Construct a B/W plot of the mean deviations for this data set.

(d) What does the B/W plot indicate about the size of these mean deviations?

(e) For this particular data set, does the largest (in absolute value) mean deviation seem to be inflating the size of the sample standard deviation? Discuss.

2.27 The following are 30 *residuals* (differences between observed data values and predicted values) when the heights of 10 year old sapling pines are estimated as a specific function of girth at 1 metre above ground level.

−0.122	1.450	1.110	0.344
0.002	−0.425	−1.725	−1.034
0.455	−2.789	0.006	0.120
−0.322	−0.022	−0.135	−0.139
−0.275	−0.156	1.463	0.000
−1.025	1.890	−1.004	−0.010
0.274	−0.007	−1.912	4.921
0.067	0.000		

(a) Determine the sample mean $\overline{x}$ and sample variance s_x^2 of this data set.

(b) Residuals have the algebraic property that they sum to zero. Based on your answer in (a), does this occur for this data set?

(c) What percentage of the data set falls in the interval

$$(\overline{x} - 2s_x,\ \overline{x} + 2s_x)?$$

How does this observed result compare with the theoretical result provided by Chebychev's Inequality?

2.28 The following data give rainfall rate (mm h^{-1}), in time sequence, recorded electronically every 2 minutes during the 25 minute duration of a late afternoon thunderstorm over Darwin:

10	130	70	60	60	70
35	10	20	35	10	10

(a) Determine the sample mean and sample variance.

(b) Is important information being lost when we use this sample mean/variance combination to describe rainfall rate during this thunderstorm?

(c) The observation of 130, recorded second in time sequence in this data set, seems unusually high. Would you regard this value as an outlier? Why?

(d) By what percentage is the sample variance reduced if the observation of 130 is (hypothetically) removed from this data set?

2.29 The following prices for shorn lambs at the Ballarat sales were reported in *Stock and Land* on 6 December 1990.

Quantity	Buyer	\$/hd
42	D Chester, Stawell	\$20.10
120	C Hussey, Cressy	\$19.60
77	K I Hall, Stawell	\$19.00
98	F C & G L Seers & Sons, Glengower	\$18.29

Determine the average shorn lamb price. How many shorn lambs were sold for above average price?

2.30 The following data give the height of the waterline (in units of 0.2 m) with respect to a marker-post standing in the Swan river:

−2 6 4 2 0 −6 3

(a) Give the measurement scale used for these data. Choose from

nominal/ordinal/interval/ratio.

(b) Give the distance, in metres, between the river's highest observed level and the river's lowest observed level as recorded at this marker-post.

(c) Determine the sample mean, the sample variance, and the sample standard deviation for these data.

2.31 For 5 data values, x_1, x_2, x_3, x_4, x_5, the following calculations were made:

$$\sum_{i=1}^{5} x_i = 20, \qquad \sum_{i=1}^{5} x_i^2 = 116.$$

Determine the values of $\overline{x}$ and s_x.

2.32 After the first year of operation of the Medicare health system in Australia, 60 respondents reported the number of Medicare claims that they had filed. The results were as follows.

9	10	14	14	12	14	0	2
6	2	3	5	7	3	15	7
4	3	6	4	5	4	7	4
5	9	4	10	8	4	1	3
4	8	6	10	6	12	0	3
12	7	7	3	6	6	4	3
7	7	3	7	6	6	4	10
12	9	14	14				

Determine the sample mean and the sample variance of these data.

2.33 Write down a data set with 10 elements which are either 0 or 1 so that:

(a) $\overline{x}$ is as large as possible;

(b) s_x^2 is as large as possible;

(c) the data set has a single mode which is as large as possible;

(d) the data set is bimodal.

2.34 (a) Show that for the data set $x_1, x_2, \ldots, x_n$, if $s_x^2 = 0$ then $x_1 = x_2 = \ldots = x_n$.

(b) Show that if $x_1 = x_2 = \ldots = x_n$, then $s_x^2 = 0$.

2.35 Suppose that x_1, x_2 represents a data set of size $n = 2$ observations. Show that the sample variance s_x^2 may be written in the form

$$s_x^2 = \tfrac{1}{2}(x_1 - x_2)^2.$$

2.36 Determine the approximate mean and standard deviation for the data from Example 2.6 on the drift velocity of electrons. The tabled values have units $\times 10^4$ cm s^{-1}.

Interval	Frequency
[0, 5)	7
[5, 10)	15
[10, 20)	16
[20, 40)	16
[40, 80)	11
[80, 150)	5

2.37 For the data set $x_1, x_2, x_3, \ldots, x_n$:

(a) Explain in words the difference between the quantities $\overline{x^2}$ and $\overline{x}^2$.

(b) Show that the sample variance s_x^2 may be expressed in the form

$$s_x^2 = \frac{n}{n-1}\left(\overline{x^2} - \overline{x}^2\right).$$

This is the version of s_x^2 used in Staudte (1990).

(c) What are the required conditions on $x_1, x_2, x_3, \ldots, x_n$ for $\overline{x^2} = \overline{x}^2$?

2.38 Let $x_1, x_2, x_3, \ldots, x_n$ be n data values with mean $\overline{x}$.

(a) Draw a sketch of the quadratic function $f(a)$ of a where $f(a) = (a - \overline{x})^2$.

(b) Prove that, for any value of a,

$$\sum_{i=1}^{n}(x_i-a)^2=\sum_{i=1}^{n}(x_i-\overline{x})^2+n(a-\overline{x})^2.$$

(c) Use this result to determine which of the following is smallest:

- $\sum_{i=1}^{n}(x_i-2\overline{x})^2$
- $\sum_{i=1}^{n}(x_i-\overline{x})^2$
- $\sum_{i=1}^{n}x_i^2$
- $\sum_{i=1}^{n}(x_i+\overline{x})^2$
- $\sum_{i=1}^{n}(x_i+2\overline{x})^2$.

2.39 For the data set $x_1, x_2, x_3, \ldots, x_n$ with sample mean $\overline{x}$ and sample standard deviation s_x, show that the difference between the ith observation and the jth observation, for any $i, j = 1, 2, 3, \ldots, n$, may be expressed as

$$x_i - x_j = (x_i-\overline{x})-(x_j-\overline{x}).$$

Hence, by squaring and adding together such differences, show that

$$s_x^2=\frac{\sum_{i=1}^{n}\sum_{j=1}^{n}(x_i-x_j)^2}{2n(n-1)}.$$

2.40 Example 1.7 in Chapter 1 gave the exchange rates for the New Zealand dollar recorded during the month of January for both 1989 and 1990. Compare the two data sets using sample means, variances and coefficients of variation. In which year was the exchange rate relatively the least variable?

2.41 A machine cuts metal templates from thin sheets of galvanised iron. The template shapes cut out are perfect equilateral triangles, but with a length of side that is variable. Measurements of side lengths of a sample of n templates produced the observed lengths $x_1, x_2, x_3, \ldots, x_n$.

(a) Show that the average *area* of the n templates in the sample is

$$\frac{\sqrt{3}\left(\overline{x^2}\right)}{4}.$$

(b) Show that the average area in (a) may be expressed in terms of $\overline{x}$, s_x and n.

2.42 Recall from Chapter 1 that the **range**, R, of the data $x_1, x_2, x_3, \ldots, x_n$ is given by $R = x_{(n)} - x_{(1)}$, where $x_{(n)}$ is the largest observation and $x_{(1)}$ is the smallest observation in the data set. Decide, with reasons, whether the following statement is true or false:

For some data sets the standard deviation may be strictly larger than the range.

2.43 Consider the data $x_1, x_2, x_3, \ldots, x_n$. For a given value of r, suppose that ℓ elements of this data set belong to the set

$$\nabla=\{x_i : |x_i-\overline{x}|\geq rs_x\}.$$

Show that

$$\sum_{i:x_i\in\nabla}(x_i-\overline{x})^2\geq \ell r^2 s_x^2.$$

Hence prove Chebychev's Inequality:

$$\frac{\ell}{n}<\frac{1}{r^2}.$$

2.5 DETECTING SKEWNESS

Discussion of the skewness of a data set has been in the context of 'long tails' of stem-and-leaf plots (in Section 1.2), of box-and-whisker plots (in Section 1.5) and of histograms (in Section 2.2): in all cases the term 'right-skewed' indicates a long tail towards the right. These are visible signs of skewness—a direction measure (right or left) indicating where the extreme values are. In this section, we try to clarify our intuition and give a comparative assessment of skewness with means and medians, and finally give a numerical measure based on the magnitudes of the data. Many statistical computer packages provide numerical measures for skewness. Such measures are useful to *assist* in the interpretation of the shape of a data set. Primarily, shape is made visible through stem-and-leaf plots, box-and-whisker plots and histograms.

Comparing sample means and medians

The sample mean is affected by extremes in a data set but the median is not. The reason is simple. The median splits a ranked data set into two halves, regardless of the actual magnitude of the data in each half. However, the mean involves these magnitudes in its calculation; the presence of large data values tends to increase the sample mean compared to the median (and, similarly, the presence of many small data values reduces the sample mean compared to the median).

Suppose that a data set $x_1, x_2, x_3, \ldots, x_n$ has sample median $\tilde{x}$ and sample mean $\overline{x}$.

If

$\overline{x} > \tilde{x}$, then the data set is generally right-skewed (positively skewed);

$\overline{x} < \tilde{x}$, then the data set is generally left-skewed (negatively skewed);

$\overline{x} \approx \tilde{x}$, then the data set is symmetrically distributed about its centre.

For a simple example, suppose that $-2, -2, -2, 0, 1, 1, 2, 3, 10, 89$ represents a data set with $n = 10$ observations. Simple calculations give $\overline{x} = 10$ and $\tilde{x} = 1$. Clearly $\overline{x}$ has been inflated past the 90th percentile. This does not indicate the 'centre' of the data set. If the relatively large observation of 89 is excluded from the data set, then the sample mean plummets to approximately to 1.2 and the median remains at 1.

A measure that assesses the *degree of skewness* of a data set by utilising the discrepancy between the sample mean and sample median is *Pearson's coefficient of skewness*, g_2, defined by:

A measure that assesses the *degree of skewness* of a data set by utilising the discrepancy between the sample mean and sample median is *Pearson's coefficient of skewness*, g_2, defined by:

Definition 2.8 *For a data set* $x_1, x_2, x_3, \ldots, x_x n$, *with sample mean* $\overline{x}$, *sample median* $\widetilde{x}$ *and sample standard deviation* s_x, *the* **Pearson coefficient of skewness,** g_2, *is given by*

$$g_2 = \frac{3(\overline{x} - \widetilde{x})}{s_x}.$$

□

Notice that the Pearson coefficient is independent of the scale of the data since $\overline{x}$, $\widetilde{x}$ and s_x all carry the same unit of measurement as the data. Clearly, $g_2 > 0$ for right-skewed data sets, $g_2 < 0$ for left-skewed data sets and $g_2 = 0$ when $\overline{x} = \widetilde{x}$.

A question often raised is: When should we use sample means and when should we use sample medians in data analysis? A simple reply would suggest that medians more accurately reflect the centre of skewed distributions. In this case, for simple data summary, the median is ideal. However, the sample median is not commonly used for statistical inference, since the mathematical theory concerning the behaviour of medians from one sample to another is quite difficult (compared with the same theory for sample means given in later chapters of this book). Of course, an obvious way around this difficulty is to transform skewed data sets towards symmetry (and we have already seen cases of this). Power transformations are very useful in this sense. If a data set is fairly symmetric, then the sample mean and median tend to coincide.

Sample moments about the mean

Given the data set $x_1, x_2, x_3, \ldots, x_n$, then according to Definition 2.5 the simple sums

$$\sum_{i=1}^{n} x_i \quad \text{and} \quad \sum_{i=1}^{n} x_i^2$$

are the first quantities to calculate in order to determine the sample mean $\overline{x}$ and the sample variance s_x^2. Sums of higher powers of the data can also give some insight into the shape of the data. We will look, in particular, at the sums: $\sum_{i=1}^{n} x_i^3$ and $\sum_{i=1}^{n} x_i^4$. Averages of such powers of data are termed **sample moments.**

Definition 2.9 *The* r*th* **sample moment about the sample mean** *for a data set* $x_1, x_2, x_3, \ldots, x_n$ *is given by*

$$m_r = \frac{\sum_{i=1}^{n} (x_i - \overline{x})^r}{n},$$

where r is a positive integer.

□

The m_r are defined on the **centred** data set; that is, the sample mean $\overline{x}$ is subtracted from each data point before powers and then sums are determined. A list will show the uses of these moments:

m_1 refers to central location;
m_2 refers to spread about the centre;
m_3 refers to skewness about the centre;
m_4 refers to how much data is massed at the centre.

Writing the moments in terms of the data we see that

$$m_1 = \frac{\sum_{i=1}^{n}(x_i - \overline{x})}{n} = 0$$

by Equation 2.3. This shows that the centred data has mean 0.

$$m_2 = \frac{\sum_{i=1}^{n}(x_i - \overline{x})^2}{n} = \frac{(n-1)s_x^2}{n}$$

by Equation 2.4. This indicates that the second moment is a multiple of the sample variance and is therefore a measure of how spread out the data are. m_2 behaves like s_x^2 when n is large and obviously results from using an n in the denominator of sample variance instead of $(n-1)$. Further, the 3rd and 4th sample moments are used in Definition 2.10.

Definition 2.10 *The* **sample coefficient of skewness,** g_3, *is given by*

$$g_3 = \frac{m_3}{(\sqrt{m_2})^3}$$

and the **sample coefficient of kurtosis,** g_4, *is given by*

$$g_4 = \frac{m_4}{(\sqrt{m_2})^4}.$$

□

These definitions result from ideas presented in Snedecor and Cochran (1967). In general, $m_3 > 0$ if large data values are present far above $\overline{x}$, for then $(x_i - \overline{x})^3$ is large and positive. Such large values are found in the long tails of right-skewed histograms. (The downward pull of low data values well below $\overline{x}$ account for $m_3 < 0$.) When $m_3 = 0$ the data values balance symmetrically about the mean.

Notice that g_3 and g_4 are essentially standardised versions of m_3 and m_4 respectively. Dividing in each case by the appropriate power of $\sqrt{m_2}$ ensures that the coefficients of skewness and kurtosis do not depend on the units that the data are measured in. For example, a data set of lengths will always have the same value of g_3 irrespective of whether the lengths are in millimetres, centimetres or kilometres. The theoretical aspects of **kurtosis** for which m_4 provides some insight will later be discussed in relation to the normal probability distribution.

The calculations for m_3 and m_4 are tedious, although simply implemented on a computer. They are included here to give an impression of how the magnitude of the data affects the overall shape that a data set assumes. Hand calculations are assisted by the following relationships between the rth sample moment m_r and the averages

$$\overline{x^r} = \frac{\sum_{i=1}^n x_i^r}{n}.$$

(These averages are sometimes called the sample moments about zero.) Simple algebraic manipulations with summation signs then give

$$\begin{aligned} m_1 &= 0 \\ m_2 &= \overline{x^2} - \overline{x}^2 \\ m_3 &= \overline{x^3} - 3\overline{x}(\overline{x^2}) + 2\overline{x}^3. \end{aligned}$$

Example 2.10

How skewed is the flawed canvas data?

We return to Example 1.13 and give the data set for quality control checks on four canvas suppliers at the time of the America's Cup in Perth. Let the variable X denote the number of flaws per 100 metres of canvas.

Supplier	Observed faults per 100 metres								
A	0	2	1	0	0	1	3	0	0
B	1	5	2	1	0	7	9	5	3
C	3	5	10	4	2	12	6	3	15
D	6	38	5	9	7	7	2	13	10

		A	B	C	D
Sample mean	$\overline{x}$	0.78	3.67	6.67	10.78
Sample variance	s_x^2	1.19	9.25	21.00	113.94
Coefficient of variation	v_x	1.40	0.83	0.68	0.99
Skewness (Pearson coefficient)	g_2	2.14	0.66	1.09	1.06
Skewness (sample coefficient)	g_3	1.06	0.48	0.75	2.08
Kurtosis (sample coefficient)	g_4	2.80	2.02	2.12	6.03

From these results it is noticeable that the coefficients of skewness increase as the data sets become more right-skewed. The sample coefficient of skewness g_3 is particularly influenced by the large observation of 38 for Supplier D. This observation has a greater effect on the sample coefficient of skewness than it does on the Pearson coefficient of skewness (where the effect of extreme values is softened by the $\bar{x} - \tilde{x}$ calculation). With regard to variability, the coefficient of variation is highest for Supplier A, indicating that this supplier's output is relatively more variable than the output from each of Suppliers B, C and D.

Comparing the interquartile range and the sample standard deviation

The choices involved here are concerning measures of spread. For the data analytic methods based on ranks discussed in Chapter 1, the interquartile range, $q_s = x_{0.75} - x_{0.25}$, has wide acceptance, giving the length of the box in a B/W plot. It is impervious to the extremes of the distribution, as are medians. It is used in tandem with the sample median in the same way as the sample mean/sample variance combination has been used in this chapter. Following Hamilton (1990) we introduce the pseudo-standard deviation to provide an appropriately scaled version of q_s; the scaling is relative to the interquartile range of the theoretical standard normal curve studied in Chapter 4. Although we do not want to pre-empt such study, we mention here that the standard normal curve has interquartile range

$$\Phi^{-1}(0.75) - \Phi^{-1}(0.25) = 0.675 - (-0.675) = 1.35,$$

where

$$\Phi(t) = \int_{-\infty}^{t} \frac{1}{\sqrt{2\pi}} e^{-\frac{1}{2}u^2}\, du.$$

This theoretical interquartile range of 1.35 for the standard normal curve is used directly as the scale factor in the definition of pseudo-standard deviation.

Definition 2.11 *The* **pseudo-standard deviation**, p_s *is defined by*

$$p_s = \frac{q_s}{1.35}, \quad \text{where } q_s = x_{0.75} - x_{0.25} \text{ is the interquartile range of the data.}$$

□

p_s is most sensibly applied to *symmetric* data sets to detect whether their symmetry corresponds to that of the theoretical normal probability curve. p_s is, therefore, a measure of kurtosis that is easier to calculate than g_4.

Suppose that a **symmetric** data set $x_1, x_2, x_3, \ldots, x_n$ has standard deviation s_x and pseudo-standard deviation p_s.

> If
> $p_s > s_x$, then the symmetric data set is generally light-tailed;
> $p_s < s_x$, then the symmetric data set is generally heavy-tailed;
> $p_s \approx s_x$, then the symmetric data set is approximately normally distributed.

The ideas 'heavy-tailed' and 'light-tailed' really refer to how many data lie in the extremes (or tails) of the distribution compared with the centre of the distribution. The standard normal distribution (of Chapter 4) provides the basis of comparison and we will return to this discussion in that chapter.

EXERCISES

2.44 The data from canvas Supplier A in Example 2.10

0 2 1 0 0 1 3 0 0,

gave the value $g_2 = 2.14$ for the Pearson coefficient of skewness. This indicates that the data is right-skewed.

Suppose that canvas Supplier A supplied a slightly poorer quality product, so that in the above data set one of the 0's is changed to a 1. With this minor reduction in quality the data set becomes

0 2 1 1 0 1 3 0 0.

(a) Construct a B/W plot for the original data set. Does this show that the original data set is right-skewed?

(b) Show that now, for the revised data set, $\overline{x} < \widetilde{x}$ so that g_2 is negative.

(c) Determine the magnitude of g_2.

(d) Construct a B/W plot for the revised data set. Is the distribution still right-skewed?

(This problem indicates that caution is required in determining numerical measures of skewness. Plots showing long tails are an effective indicator of skewness.)

2.45 Daily sales ($) for a small retail establishment for the months of April, June and September are given in the S/L plots below:

Daily retail sales
Scale: stems—$100; leaves—$10
4t| 2 represents $420

APRIL

4t	2 2 2 2 2 3 3 3 3 3
f	
s	
.	
5*	0 0 0 0 0 0 0 0 1 1
t	
f	4 4 5 5
s	6
.	
6*	0 0
t	2
f	5
s	7
.	
7*	
t	

JUNE

4t	
f	
s	
.	
5*	0
t	2 2 3 3 3
f	4 4 5 5 5 5
s	6 6 7 7 7 7
.	8 8 9 9 9 9
6*	0 0 1 1 1
t	2
f	
s	
.	
7*	
t	

SEPTEMBER

4t	
f	
s	
.	
5*	
t	
f	4
s	6 6
.	8 8
6*	0 0 0 1
t	2 2 2 3
f	4 4 5 5 5
s	6 6 7 7 7 7 7
.	8 8 8
7*	0
t	2

Without performing calculations, determine which month showed the smallest

(a) $\overline{x}$
(b) s_x
(c) $\widetilde{x}$
(d) q_s
(e) p_s
(f) $|\widetilde{x} - \overline{x}|$
(g) $\sum_{i=1}^{30} x_{(i)}$.

Give a brief justification for each answer.

2.46 For the daily retail sales data of Exercise 2.45:

(a) Discuss the skewness of these three monthly distributions. Support your answer with B/W plots for each month.

(b) Determine the Pearson coefficient of skewness for the month of June.

(c) Determine the pseudo-standard deviation for the month of June. Is the distribution of June daily sales 'heavy-tailed' or 'light-tailed'?

2.47 (a) Show that the symmetric data set

$$-3 \quad -2 \quad -1 \quad 0 \quad 1 \quad 2 \quad 3$$

has 0 coefficient of skewness.

(b) Determine m_3 for the data set

$$-2.9 \quad -1.9 \quad -0.9 \quad 0.1 \quad 1.1 \quad 2.1 \quad 3.1$$

(c) Determine m_4 for the data set in (b).

2.48 Show that $m_2 = \overline{x^2} - \overline{x}^2$ for any data set $x_1, x_2, \ldots, x_n$. Explain in words what this equation means.

2.49 Consider the following B/W plot:

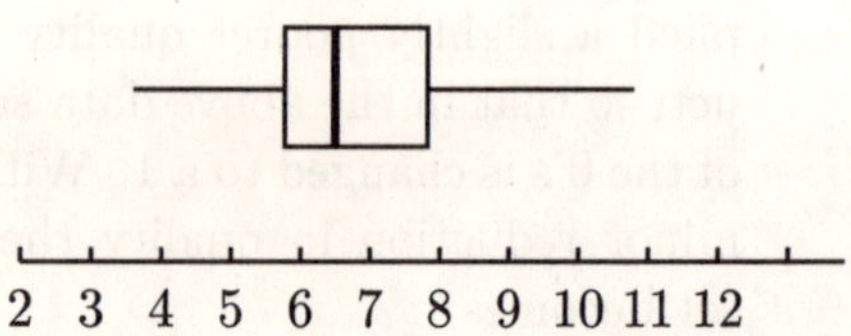

(a) The plot is constructed from data points $x_1, x_2, x_3, \ldots, x_{10}$. Explain briefly why you would expect the value of $\sum_{i=1}^{10} x_i$ to be approximately 70.

(b) Use the plot to determine p_s for the data.

2.50 Show that for any data set $x_1, x_2, \ldots, x_n$,

$$m_3 = \overline{x^3} - 3\overline{x}(\overline{x^2}) + 2\overline{x}^3.$$

2.6 SUMMARY STATISTICS TRANSFORMED

The non-graphical methods described so far in this text for describing the shape of a data set have involved finding a single number, or *measure*, to quantify such concepts as centre, spread, skewness. Measures such as $\overline{x}$, s_x^2 and $\widetilde{x}$ may each be regarded as an extremely brief summary of the data set $x_1, x_2, x_3, \ldots, x_n$. They are defined in terms of the data. Collectively, single number data summaries are called **summary statistics**. Since data sets are often transformed either to improve their shape or to make them less variable, it is important to understand how summary statistics behave under transformation.

The effect of linear transformations on the sample mean and sample variance

Suppose that a bookstore with a pricing policy of 50% mark-up on the cost price, X, of a book incurs a nominal fixed cost of \$3 (per book) for the storing, shelving and displaying of each book. Based on this information, the total profit, Y, when a book originally costing X is sold is given by the linear transformation $Y = -3 + 0.5X$. These ideas on linear transformations were introduced in Chapter 1. Now we wish to apply this transformation to means and standard deviations.

If $x_1, x_2, x_3, \ldots, x_{100}$ are a sample of 100 cost prices giving a sample mean $\overline{x} = \$56$ and a sample standard deviation $s_x = \$4$, what are the mean and standard deviation in profit when all 100 books are eventually sold? This is simple to evaluate even without knowledge of the cost price of each individual book in the sample. The summary statistics such as sample mean, sample standard deviation and sample variance may be transformed directly.

Consider the most general case: transformation $Y = a + bX$ of the variable X produces a new variable Y. When applied to the sample of n realised values $x_1, x_2, x_3, \ldots, x_n$ of X, the transformation produces the n realised values $y_i = a + bx_i$, $i = 1, 2, \ldots, n$, comprising the data set $y_1, y_2, \ldots, y_n$.

Theorem 2.2 *Given the data set $x_1, x_2, x_3, \ldots, x_n$ with sample mean $\overline{x}$ and sample variance s_x^2, then for constants a and b, the transformed values $y_i = a + bx_i$, $i = 1, 2, \ldots, n$, have sample mean $\overline{y} = a + b\overline{x}$, sample variance $s_y^2 = b^2 s_x^2$ and sample standard deviation $s_y = |b| s_x$.*

Proof: When the transformed data is added,

$$\sum_{i=1}^{n} y_i = \sum_{i=1}^{n}(a + bx_i) = \sum_{i=1}^{n} a + b\sum_{i=1}^{n} x_i = na + bn\overline{x}.$$

Therefore, when these equations are divided by n,

$$\overline{y} = a + b\overline{x}. \tag{2.9}$$

The sum of squares of the centred transformed data is

$$\sum_{i=1}^{n}(y_i - \overline{y})^2 = \sum_{i=1}^{n}([a + bx_i] - [a + b\overline{x}])^2 = \sum_{i=1}^{n}(bx_i - b\overline{x})^2 = b^2\sum_{i=1}^{n}(x_i - \overline{x})^2.$$

On dividing by $(n - 1)$ we obtain

$$s_y^2 = b^2 s_x^2. \tag{2.10}$$

The result for the standard deviation comes from taking the positive square root of each side of Equation 2.10 and noting that $\sqrt{b^2} = |b|$, the absolute value of b.

■

If we return to our bookstore example introduced at the beginning of this section, then Theorem 2.2 with $a = -3$ and $b = 0.5$ gives for $\overline{x} = \$56$ and $s_x = \$4$:

$$\overline{y} = -3 + (0.5)(56) = \$25;$$
$$s_y^2 = (0.5)^2 4^2 = (\$2)^2;$$
$$s_y = \$2.$$

Certain choices of a and b are most effective. For example, to produce a **standardisation** of the data set $x_1, x_2, x_3, \ldots, x_n$, we subtract $\overline{x}$ from each observation and then divide the result by s_x. Equivalently, take

$$a = -\frac{\overline{x}}{s_x} \quad \text{and} \quad b = \frac{1}{s_x};$$

then the effect of the linear transformation on each data point is

$$y_i = \left(-\frac{\overline{x}}{s_x}\right) + \left(\frac{1}{s_x}\right)x_i = \frac{x_i - \overline{x}}{s_x}.$$

Applying Theorem 2.2 we obtain:

$$\overline{y} = a + b\overline{x} = \left(-\frac{\overline{x}}{s_x}\right) + \left(\frac{1}{s_x}\right)\overline{x} = 0;$$

$$s_y^2 = b^2 s_x^2 = \left(\frac{1}{s_x}\right)^2 s_x^2 = 1.$$

This establishes Theorem 2.3.

Theorem 2.3 *Suppose that the data set $x_1, x_2, x_3, \ldots, x_n$ has sample mean $\overline{x}$ and sample standard deviation s_x. Then the standardised data set*

$$y_i = \frac{x_i - \overline{x}}{s_x} \qquad i = 1, 2, \ldots, n$$

has sample mean $\overline{y} = 0$ and sample standard deviation $s_y = 1$. ■

Some results for non-linear transformations

We have seen examples of data sets where the logarithm transformation has effectively improved the symmetry of the data set. Suppose that the data set $x_1, x_2, x_3, \ldots, x_n$ consisting of positive numbers has been transformed by logarithms to obtain the new data set $y_1, y_2, y_3, \ldots, y_n$, where, for each i, $y_i = \log_e x_i$. The mean of the transformed data has a very special form:

$$\begin{aligned}
\overline{y} &= \frac{\sum_{i=1}^{n} \log_e x_i}{n} \\
&= \frac{1}{n} \log_e (x_1 x_2 x_3 \ldots x_n) \\
&= \log_e (\sqrt[n]{x_1 x_2 x_3 \ldots x_n}) \\
&= \log_e \overline{x}_g
\end{aligned}$$

where $\overline{x}_g = \sqrt[n]{x_1 x_2 x_3 \ldots x_n}$ is the **geometric mean** of $x_1, x_2, x_3, \ldots, x_n$. The conclusion, therefore is as follows: If the sample mean is a suitable measure of the centre of a data set that has been transformed by logarithms, then the geometric mean is a suitable measure of the centre of the original data set.

This is a general property of the ladder of power transformations of Section 1.3. Suppose that

$$f_p(x) = \begin{cases} x^p & \text{if } p \neq 0; \\ \log_e x & \text{if } p = 0, \end{cases}$$

represents the power transformation of raising to the power p. The ladder is formed by:

Function	p
reciprocal	−1
logarithm	0
identity	1
square	2

and, as discussed in Section 1.3, the position of $p = 0$ is occupied by the logarithm.

Suppose that f_p is applied to each point in our data set $x_1, x_2, x_3, \ldots, x_n$ in turn, creating a new data set $y_1, y_2, y_3, \ldots, y_n$, where $y_i = f_p(x_i)$. Then, if the transformed data set $y_1, y_2, y_3, \ldots, y_n$ is reasonably symmetric, so that $\overline{y}$ is a suitable measure of its centre, then $f_p^{-1}(\overline{y})$ is a suitable measure of the centre of the original data set. (For some values of p, such as $p = 2$, the inverse power transformation is defined only for positive data.) These ideas invite the following definitions:

Definition 2.12 *Given a data set $x_1, x_2, x_3, \ldots, x_n$ and a power transformation*

$$y = f_p(x) = \begin{cases} x^p & \text{if } p \neq 0; \\ \log_e x & \text{if } p = 0, \end{cases}$$

for $p = -1, 0, 1, 2$, applied to that data set, then:

the **harmonic mean**	*is given by*	$\overline{x}_h = f_{-1}^{-1}(\overline{y})$;
the **geometric mean**	*is given by*	$\overline{x}_g = f_0^{-1}(\overline{y})$;
the **arithmetic mean**	*is given by*	$\overline{x} = f_1^{-1}(\overline{y})$;
the **root mean square**	*is given by*	$\overline{x}_r = f_2^{-1}(\overline{y})$,

where f_p^{-1} is the inverse function of f_p.

□

For example, the harmonic mean, $\overline{x}_h$ of the data $x_1, x_2, x_3, \ldots, x_n$ is given by the power transformation with $p = -1$ (the reciprocal transformation) so that

$$\begin{aligned} \overline{x}_h &= f_{-1}^{-1}(\overline{y}) \\ &= (\overline{y})^{-1} \\ &= \left(\frac{y_1 + y_2 + \ldots + y_n}{n}\right)^{-1} \\ &= \left(\frac{\frac{1}{x_1} + \frac{1}{x_2} + \ldots + \frac{1}{x_n}}{n}\right)^{-1}. \end{aligned}$$

For discussion on transformations of summary statistics based on ranks, see Section 1.5.

EXERCISES

2.51 Suppose that 20 measurements in millimetres give a mean length of 0.19 mm and a variance of 0.04 mm^2. The measurements were then converted to metres. Determine the standard deviation in metres.

2.52 If some temperature data measured in degrees Fahrenheit have a mean of 68.9 °F and a standard deviation of 4.8 °F, determine the mean and standard deviation when the data are converted to degrees Celsius. Note that

$$C = \frac{(F - 32)}{1.8},$$

where C denotes the temperature in degrees Celsius and F denotes the temperature in degrees Fahrenheit.

2.53 A large number of measurements of weekly rainfall (in cm) from the La Trobe Valley region had a sample variance of 9 cm^2. The Bureau of Meteorology wished to compare this result with historical records (in which rainfall was measured in inches). Determine the sample standard deviation of these measurements had they been reported in inches.

(1 inch = 2.54 cm.)

2.54 Measuring the distance between two laser beams assisted in the rehabilitation plumbing of the Man and Supply Shaft at Mt Isa Mine. The following data (in metres) were obtained.

0.930	0.930	0.935	0.938
0.945	0.945	0.950	0.945
0.945	0.940	0.940	0.935
0.930	0.930	0.930	0.925
0.925	0.915	0.915	0.910
0.910	0.900	0.900	

(*Source*: Broad, B. J. (1986). Shaft plumbing techniques. *The Australian Surveyor*, 33, 4, 292-304.)

(a) Determine the sample mean and sample variance of these data by *first* employing a change of origin and scale. Explain clearly the linear transformation that you have used.

(b) Discuss the effect that the linear transformation has on the range of the data.

2.55 Based on the counts of locust larvae in a square metre of topsoil from six neighbouring locations, the following estimates were generated for the size of the locust plague once the larvae hatch.

25 319 500	25 851 030
25 406 670	25 322 700
25 344 890	25 006 090

(a) It is difficult to grasp the magnitude of these data points because of the large number of digits involved.

Re-express the data to 3 sf.

(b) Demonstrate how to use the linear transformation

$$y = \frac{x - 25\,000\,000}{100\,000}$$

on the rounded data to determine the sample mean and sample variance in estimated population size.

2.56 A sample of 100 trees from a stand of jarrah near Bunbury (Western Australia) gave a chest-level mean tree diameter of 1.4 metres with a standard deviation of 0.4 metres.

If an estimate of the volume, v, of timber (in cubic metres) that can be milled from a tree of diameter d metres is

$$v = \frac{8}{27}d^2,$$

determine the sample mean volume of available timber per tree.

Assuming that this sample is representative of the entire stand, estimate the volume of timber that may be milled from the stand of 5000 trees.

2.57 What proportion of the data set

$$\frac{x_i - \overline{x}}{s_x}, \quad i = 1, 2, \ldots, n,$$

is guaranteed to lie in the interval (−2, 2)? Explain briefly.

2.58 Suppose that data sets $u_1, u_2, \ldots, u_n$ and $x_1, x_2, \ldots, x_n$ are related by

$$u_i = k(c - x_i), \quad i = 1, 2, \ldots, n,$$

where k and c are constants. Derive formulae relating:

(a) $\overline{u}$ and $\overline{x}$;

(b) s_u^2 and s_x^2;

(c) s_u and s_x.

2.59 In the Adelaide Grand Prix, suppose that the winning driver completed the ith lap of the circuit at an average speed of v_i km hr^{-1}. Show that in an n lap race his average speed is

$$\left(\frac{\frac{1}{v_1} + \frac{1}{v_2} + \ldots + \frac{1}{v_n}}{n}\right)^{-1}$$

in km hr^{-1}. Harmonic means occur in surprising places!

2.60 Suppose that data sets $u_1, u_2, \ldots, u_n$ and $x_1, x_2, \ldots, x_n$ are related by

$$u_{(i)} = k(c - x_{(n-i+1)}), \quad i = 1, 2, \ldots, n$$

where k and c are positive constants and, in the usual notation, $u_{(1)} \leq u_{(2)} \leq \cdots \leq u_{(n)}$ and $x_{(1)} \leq x_{(2)} \leq \cdots \leq x_{(n)}$. Derive formulae relating:

(a) $\overline{u}$ and $\overline{x}$;

(b) s_u^2 and s_x^2;

(c) s_u and s_x.

2.61 Determine the geometric mean and the harmonic mean of

$$10, 10^2, 10^3, \ldots, 10^{10}.$$

2.62 (a) Consider a linear transformation

$$y_i = ax_i + b, \quad i = 1, 2, 3, \ldots, n,$$

of a data set $x_1, x_2, x_3, \ldots, x_n$ into a data set $y_1, y_2, y_3, \ldots, y_n$, where a and b are constants.

i. If m_j denotes the jth moment of $x_1, x_2, x_3, \ldots, x_n$ (about the sample mean) and m_j' denotes the jth moment of $y_1, y_2, y_3, \cdots, y_n$ (about the sample mean), show that

$$m_j' = a^j m_j.$$

ii. Show that $x_1, x_2, x_3, \ldots, x_n$ and $y_1, y_2, y_3, \ldots, y_n$ have the same skewness and kurtosis coefficients g_3 and g_4.

iii. Explain why this shows that a data set of length measurements in mm has the same skewness and kurtosis coefficients when each data point is re-expressed in cm.

(b)

Recess gauging (mm)	Frequency
0.052	8
0.053	15
0.054	14
0.055	13

Fifty measurements were taken on the recess gauging of screws used in the assembly of windscreen washers.

i. Determine the mean and standard deviation of this data.

ii. By choosing a suitable linear transformation, find the skewness and kurtosis coefficients for the recess gauging data.

CHAPTER THREE

Assessing Probabilities

In this chapter you will find

3.1 HISTORY: WHERE HAS PROBABILITY COME FROM?

In this chapter we step back from the engrossing activity of data summary, graphics and display. This step is to prepare the groundwork for putting data analysis to active use: we will want to use our data to probe and question the hypotheses about, and models for, the population from which the data came. This is an active role for data analysis that goes beyond the use of data summaries (B/W plots, S/L plots, histograms, Pareto diagrams, sample means, sample medians, sample variances, ...) as an end in themselves.

How do we probe and question hypotheses about populations? We get data from the population. We analyse the data. We summarise it. We display it. We try to extract from it all the information that it contains. Then we ask: If our theories about this population are correct, how *likely* is it that we would observe the data that we find ourselves with? How *likely* is it to observe a sample mean as large as this if we believe the population mean to be ...? How *likely* is it to observe this sized spread of data if we believe that the spot-welding robot arm on the factory production line is set to this accuracy and these specifications?

How *likely*?

This emphasised word is difficult to understand. We are assisted by the theory of probability. This elegant theory had its origins over 300 years ago in the 17th century gaming houses of Europe. The wealthy nobility, who gambled for leisure, or by profession, needed assistance for examining the outcomes in games of dice. In doing this applied consultancy work, Pascal and Fermat, two of the greatest thinkers of the age, developed the classical theory of probability.

The classical definition of probability

This, historically the first version of probability, is still in constant use today in situations of counting simple 'equally likely' outcomes. With reference to (and for many probabilists, reverence for) the subject's historical roots in gambling, we begin with a die (the singular of dice). Suppose that we toss the die. The outcome cannot be determined in advance and therein lies the problem. We know only that, for the resulting uppermost face, the possible outcomes are:

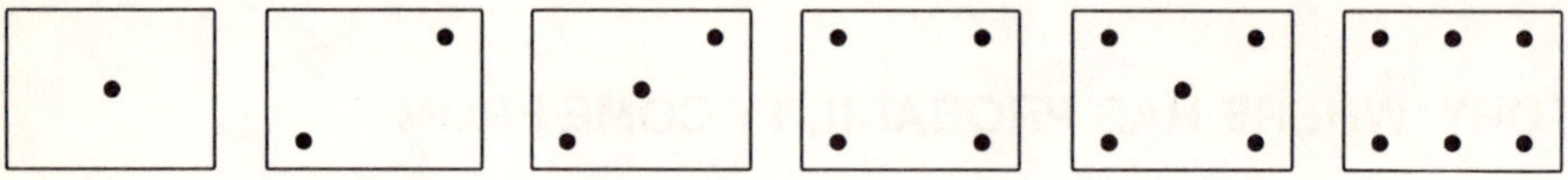

These outcomes may be labelled $1, 2, 3, 4, 5$ and 6. Tossing a die is an example of a probabilistic experiment.

Definition 3.1 *An* **experiment** *is any activity which generates observations. The* **sample space**, *denoted by S, is the set of all possible outcomes of the experiment. S is termed* **discrete** *if it has countably many (including finitely many) outcomes. S is termed* **continuous** *if it contains an interval of real numbers (and therefore has uncountably many outcomes).*

□

The term 'set' is used here in the set-theoretic sense (where all items in the set are distinct) rather than in the data set sense (which allows for repetitions of items). In the die example, $S = \{1, 2, 3, 4, 5, 6\}$ is the *set* of possible outcomes that records complete information on the outcomes of the experiment: 'tossing a die'. It is a discrete sample space with a finite number of elements. Notice that the sample space is completely dependent on the aim of the experiment. If, when the die is tossed, our aim is to know only if a six occurs, then we may write the sample space for *this* experiment as $\mathcal{S} = \{s, n\}$ where $s =$ 'six', and $n =$ 'non-six', since this is a complete record of the information that we require in this case. If the aim of our experiment is to toss the die *until* the first six appears then the sample space takes the form

$$\mathcal{S} = \{s, ns, nns, nnns, nnnns, nnnnns, \ldots\}$$

where, as before, $s =$ 'six', and $n =$ 'non-six'. This discrete sample space has countably (infinitely) many outcomes; in theory, we could continue tossing the die indefinitely waiting for a 'six'. If we consider dice from the point of view of manufacture, then the outcome x for the experiment where we determine the weight of a die may be any value between, say, 0 g and 10 g. Continuous sample spaces are characterised by such interval structure. For this experiment we would write $\mathcal{S} = [0, 10]$. In this section we consider discrete sample spaces. Continuous sample spaces will be considered briefly in Section 3.2.

Suppose we believe that when a die is tossed none of the possible outcomes in $\mathcal{S} = \{1, 2, 3, 4, 5, 6\}$ is 'more likely' than any other. That is, we believe that the dice are not *loaded* towards the occurrence of any particular outcome, although in reality the dice may be loaded and we do not know it, or we choose to ignore it. Our belief and the fact of reality can, indeed, differ. We regard an **event** as a specified potential outcome for an experiment. More precisely, using ideas from set theory:

Definition 3.2 *A is an* **event** *if it is a subset of the sample space* $\mathcal{S}$. *If the experimental outcome is* $x \in \mathcal{S}$, *then the event A is said to* **occur** *if* $x \in A$ *and x is said to be* **favourable to the outcome** *of A.* n_A *is the number of points favourable to the outcome of A.*

□

Algebraically, we denote events by upper case letters, for example, A. The symbol for subset is '$\subset$' and so we write $A \subset \mathcal{S}$ to indicate that A is 'contained in' $\mathcal{S}$.

According to the classical definition, the probability of the event, A, that 'a die will show an odd number uppermost after being tossed high in the air', is determined as follows: Three points, faces 1, 3 and 5, are favourable to our event A occurring. That is, $A = \{1, 3, 5\}$. There are six 'equally likely' faces, one of which must occur. That is, $\mathcal{S} = \{1, 2, 3, 4, 5, 6\}$. Therefore, the probability of the event 'odds' is $\frac{3}{6} = \frac{1}{2}$. This works. It has done for 300 years and will continue to do so. It is a probability assessment that agrees with our experience of reality. Put formally:

Definition 3.3 (Classical definition) *If a sample space $\mathcal{S}$ contains $n_{\mathcal{S}}$ equally likely points, of which n_A are favourable to the event A occurring, then the probability of the event A is given by*

$$\frac{n_A}{n_{\mathcal{S}}}.$$

□

We have said that this definition works; it is, however easy to get it wrong! Definition 3.3 is logically flawed: indeed, probabilities are defined in terms of 'equally likely' outcomes, so that probability is being defined in terms of itself. The concept of 'equally likely' is a symmetry assumption that is gleaned from our experience of the world; an experience that gives us access to the results of tossing a 'fair' die repeatedly, either in our own experiments, or in the experiments of others. This was the main difficulty facing the early probabilists, who perhaps did not link experimental verification to the classical definition's probability assessments. Example 3.1 illustrates the point.

Example 3.1
Playing two-up at the Burswood Island Casino

The fabled game of two-up is traditionally played on the ground around a circle of dust somewhere down a bush track near Kalgoorlie. The polished casino version retains some of the hallmarks of the game. The spinner, chosen from among the circle of bettors, tosses a pair of *identical* coins so as to land within the circle. The events of interest are:

heads: where both coins upturn heads;
tails: where both coins upturn tails;
odds: where one coin gives heads and the other tails.

The sample spaces are more difficult to ascertain; remember that the coins are visually *indistinguishable*. There are two possibilities:

$$\begin{aligned}\mathcal{S}_1 &= \{hh, ht, tt\}\\ \mathcal{S}_2 &= \{hh, ht, th, tt\},\end{aligned}$$

where t denotes 'tails' and h denotes 'heads'. In $\mathcal{S}_1$, no allowance is made for the combinations ht and th to be different; they both carry the same label ht meaning that one coin upturns 'heads' and the other coin upturns 'tails'. In $\mathcal{S}_2$, the combinations ht and th are kept separate.
There is a difficulty choosing between these possibilities $\mathcal{S}_1$ and $\mathcal{S}_2$: Exactly how many equally likely outcomes are there? Are the sample points th and ht really different if the two coins are, and again we emphasise, *completely indistinguishable*?

A careful, logical examination of the contents of $\mathcal{S}_1$ and $\mathcal{S}_2$ will not provide the answer. We only come to know that, from real-world experience in tossing coins (see Example 3.2), identical coins behave as if they were different. This experience (and just ask a two-up player) indicates that the event 'odds' occurs in reality approximately twice as often as the events 'heads' or 'tails'. Therefore, if we want our probability assessments to agree with reality (and for betting purposes this would be wise), the correct sample space is $\mathcal{S}_2$. The correct probabilities are: $\frac{1}{4}, \frac{1}{4}$ and $\frac{1}{2}$ for the events 'heads', 'tails' and 'odds' respectively.

Notice that the difficulty arose because the coins were identical. If they were different in some way so that they could be labelled with subscripts 1 and 2, then the correct sample space would be

$$\mathcal{S}_3 = \{h_1h_2, h_1t_2, t_1h_2, t_1t_2\}.$$

Finally, the 'equally likely' symmetry assumption is necessary for the classical definition to operate. If, before our die experiment we take the die and hit it with a hammer so that it becomes deformed, then we have good reason to assume that some of the faces will be more likely to occur than others. The classical definition can no longer help us. If someone had hit the die with a hammer, without our knowing, then the innocent symmetry assumption would mislead us in our probability assessments. After the hammer incident we should toss the die many times to ascertain its nature. This leads us to a relative frequency approach for probability.

Limiting relative frequencies

This approach to defining probability is an attempt to overcome the problem of defining sample spaces with 'equally likely' points. The approach enables us to make probability assessments about events arising from experiments which can be repeated indefinitely under similar conditions. If an event A occurs k_n times in n repetitions of a simple experiment, then, according to the relative frequency interpretation of probability, the probability of the event A is the limiting value of the relative frequency $\frac{k_n}{n}$ as n becomes large. There is an implicit assumption that the relative frequencies stabilise as n increases; that is, the relative frequencies 'settle down' the more the experiment is repeated.

Example 3.2, in a continuation of Example 3.1, illustrates how relative frequencies may fluctuate along a sequence of 500 tosses of a pair of identical coins.

Subjective probability: a matter of opinion

The relative frequency interpretation of probability, essentially based on scientific evidence, will give results in agreement with reality for experiments that can be repeated under similar conditions arbitrarily often. However, many of our quite natural questions of uncertainty are concerned with events as outcomes from experiments that cannot be repeated. For example, what is the probability that Sydney's Centrepoint Tower will collapse in high velocity winds? Probability assessments in such situations must be determined by our *degree of belief* in the outcome concerned. In this sense probability becomes 'person-based'; two different people, with different knowledge backgrounds, are able to assign different probabilities to the same event. Even weather forecasters on different television channels have, on occasion, offered us different assessments for the chance of rain. Such person-based assessments are valid, but ... we must be prepared to back up our assessment with a bet. This is to keep us honest. We need notation for our losing the bet; that is, when the event A does not occur.

Definition 3.4 *Given an event $A \subset \mathcal{S}$, the* **complementary event**, $\overline{A}$, *consists of all elements in $\mathcal{S}$ but not in A and represents the event of A not occurring. The* **null event**, $\emptyset$, *is defined to be the complement of the sample space:* $\emptyset = \overline{\mathcal{S}}$.

□

In tossing a fair die, $\mathcal{S} = \{1, 2, 3, 4, 5, 6\}$ and the event $A = \{6\}$ of throwing a six (to reduce arguments as to which player starts a game of scrabble) has complementary event $\overline{A} = \{1, 2, 3, 4, 5\}$. The null event cannot occur since it has no elements.

To place a bet on event A, one finds another person (perhaps a professional bookmaker) who backs $\overline{A}$. Then the subjective probabilities come into use. If you are willing to risk losing \$$x$ if $\overline{A}$ occurs so that you may win \$$y$ if A occurs, then your subjective probability assessment for A occurring, in a *fair bet*, is $\frac{x}{x+y}$. In terminology of the racetrack, your **odds are** $y : x$ **against** A. Odds are offered by bookmakers. It is up to you to decide if they agree with your subjective judgment. In a fair bet it makes no difference which side of the bet you take; if A has odds $y : x$ against, then $\overline{A}$ has odds $x : y$ against.

Example 3.2

Detecting odds with relative frequencies

In this continuation of Example 3.1 we tossed two identical coins 500 times under the same conditions. Given first are the results for the first 50 tosses to illustrate the observed pattern of the events labelled $H = \{hh\}$, $T = \{tt\}$, $O = \{th, ht\}$, all of which are subsets of the correct sample space $\mathcal{S}_2$. The symbols are written in the time sequence (reading left to right one row at a time) in which they occurred.

$T\ O\ T\ H\ T\ H\ O\ O\ H\ H$
$T\ H\ T\ H\ H\ O\ H\ H\ O\ T$
$O\ H\ T\ O\ O\ O\ O\ O\ O\ T$
$O\ O\ O\ O\ O\ H\ O\ O\ O\ O$
$H\ O\ O\ O\ O\ O\ O\ T\ O\ H$

With regard to these first 50 results notice that in the Burswood Island game of two-up, players may bet only on H or T. When the spinner throws 'odds', all bets remain frozen and the spinner rethrows. If 'odds' appear on 5 consecutive throws, the casino takes all bets placed. Based on the above data, the 'house' would be pleased with the efforts of the spinner. The 'house' doesn't always win: the next occasion that a string of 5 'odds' occurred was starting from toss number 414.

The relative frequencies for the three events are given in the accompanying table for various n.

n	T	H	O
10	0.300	0.400	0.300
50	0.180	0.260	0.560
100	0.220	0.240	0.540
200	0.230	0.275	0.495
500	0.245	0.270	0.485

Certainly it appears that 'odds' is twice as likely as 'heads' or 'tails'. There seems to be evidence that the coins are 'fair' and that $\mathcal{S}_2$ is the correct sample space.

Some results from set theory

Two sets (and therefore two events) A and B satisfy the equation $A = B$ *if and only if* $A \subseteq B$ *and* $B \subseteq A$. Therefore, two events A and B are identical if a typical outcome from A is favourable to the outcome of B and vice versa: a typical outcome of B is favourable to the outcome of A. Further definitions are required before introducing the probability axioms in the next section.

Definition 3.5 *If events A and B are subsets of a sample space $\mathcal{S}$, we use the notation: $A \cap B$ to indicate the event that both A and B occur; $A \cup B$ to indicate that either A occurs or B occurs or* **both** *A and B occur.*

□

$A \cap B$ is read 'A and B' or 'A intersect B'; $A \cup B$ is read 'A or B' or 'A union B'.

These set-theoretic ideas may be represented pictorially by a **Venn diagram** where $\mathcal{S}$ is represented by a rectangle and the events of interest in $\mathcal{S}$ may be represented by circles lying within the rectangle. Thus:

$A \cap B$

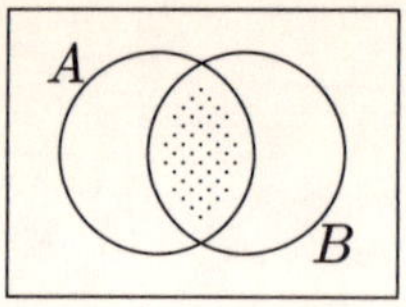

$A \cup B$

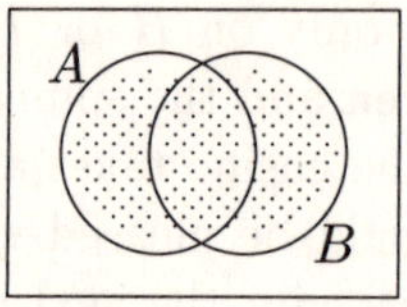

$\overline{A}$

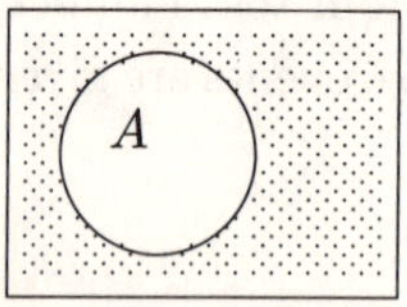

Definition 3.6 *Events A and B in a sample space $\mathcal{S}$ are* **mutually exclusive** *if $A \cap B = \emptyset$.*

□

This definition indicates that mutually exclusive events cannot happen at the same time. The shapes representing A and B in the Venn diagram do not overlap. To simply illustrate 'union' and 'intersection' we will construct the sample space for tossing two fair dice. Individually each die has sample space $A = \{1, 2, 3, 4, 5, 6\}$; however, when the outcomes for the two dice are to be considered together then the 36 ordered pairs (Die 1 outcome, Die 2 outcome) may be regarded as equally likely experimental outcomes comprising a sample space $\mathcal{S}$.

(1,1)	(1,2)	(1,3)	(1,4)	(1,5)	(1,6)
(2,1)	(2,2)	(2,3)	(2,4)	(2,5)	(2,6)
(3,1)	(3,2)	(3,3)	(3,4)	(3,5)	(3,6)
(4,1)	(4,2)	(4,3)	(4,4)	(4,5)	(4,6)
(5,1)	(5,2)	(5,3)	(5,4)	(5,5)	(5,6)
(6,1)	(6,2)	(6,3)	(6,4)	(6,5)	(6,6)

In relation to the events

A: 'the sum of the two dice is six'
B: 'the dice show the same number'
C: 'the sum of the two dice is odd',

we can identify the subset structure of $\mathcal{S}$ corresponding to each of these events:

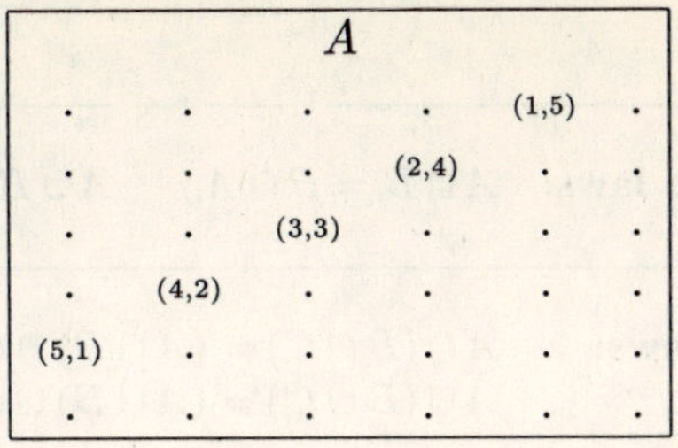

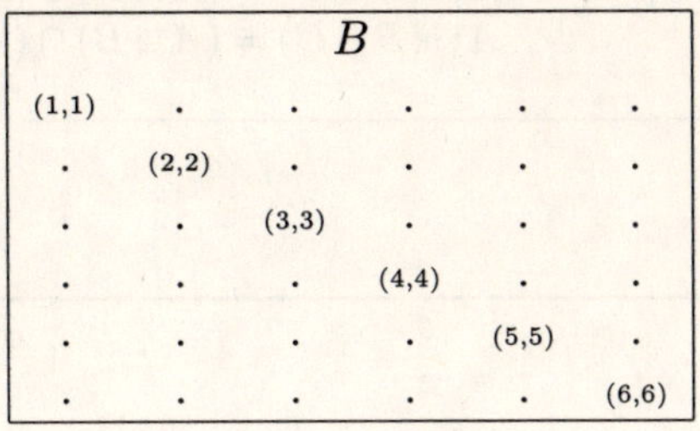

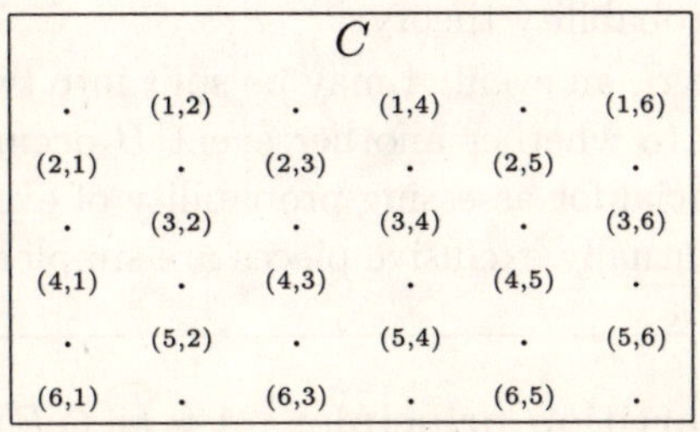

Hence, A and C are mutually exclusive since $A \cap C = \emptyset$. Also, $A \cap B = \{(3,3)\}$, since $(3,3)$ is the only outcome in common to both A and B. To observe all elementary outcomes in $A \cup B$ we may write

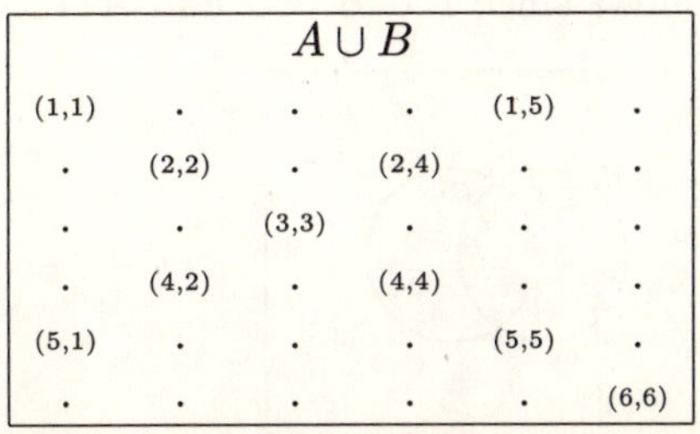

The following basic laws concerning events carry over directly from set theory: they are summarised here for convenience.

Commutative laws:	$A \cap B = B \cap A, \qquad A \cup B = B \cup A$
Associative laws:	$A \cap (B \cap C) = (A \cap B) \cap C = A \cap B \cap C,$ $A \cup (B \cup C) = (A \cup B) \cup C = A \cup B \cup C$
Distributive laws:	$A \cap (B \cup C) = (A \cap B) \cup (A \cap C),$ $A \cup (B \cap C) = (A \cup B) \cap (A \cup C)$

Example 3.3

The partition principle

This example highlights one of the most important relationships between events in basic probability theory:

In a sample space $\mathcal{S}$, an event A may be split into two mutually exclusive pieces, according to whether another event B occurs or does not occur. This becomes crucial for assessing probability of event A when the probabilities of the mutually exclusive pieces are simpler to establish.

The partition principle: $A = (A \cap B) \cup (A \cap \overline{B})$

This result is established quite simply from the Distributive law:

$$A \cap (B \cup C) = (A \cap B) \cup (A \cap C).$$

For if we put $C = \overline{B}$, then $A \cap (B \cup \overline{B}) = (A \cap B) \cup (A \cap \overline{B})$.
The result now follows since $B \cup \overline{B} = \mathcal{S}$ and $A \cap \mathcal{S} = A$.

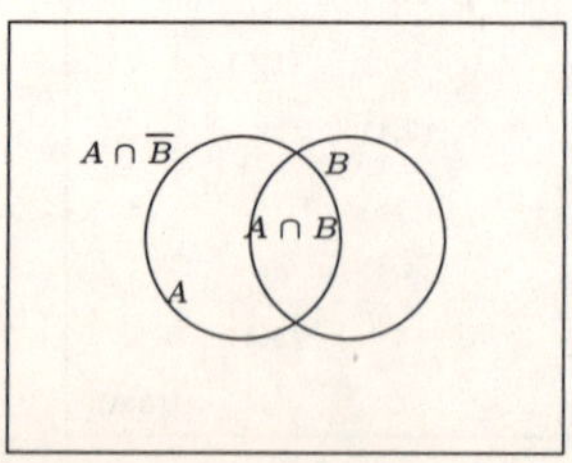

The complements of union and intersection comprise **De Morgan's Laws**:

$$\overline{(A \cup B)} = \overline{A} \cap \overline{B} \tag{3.1}$$
$$\overline{(A \cap B)} = \overline{A} \cup \overline{B} \tag{3.2}$$

These results may be *proved* by using the standard 'containment methods' of set theory. These results may be *verified* by examining intersecting regions of Venn diagrams.

Example 3.4

Random breath testing in Parramatta Road

The following array of data is from the first 50 motorists stopped in Parramatta Road and given a breath test. Motorists with blood alcohol levels exceeding 0.05 mg L^{-1} were classified as 'driving under the influence of alcohol'.

	Blood alcohol level	
	≥ 0.05	< 0.05
Men	6	26
Women	2	16

Suppose that we randomly select a person from this police file and note their gender and blood alcohol level. In this case $\mathcal{S}= \{$ Person 1, Person 2, ..., Person 50 $\}$.

Given the events:

A : 'selected person has a blood alcohol level ≥ 0.05'
B : 'selected person is female'

then n_A, the number of outcomes favourable to the event A, is 8. Similarly it is easy to establish that:

$$\begin{aligned}
n_B &= 16 + 2 = 18 \\
n_{A \cap B} &= 2 \\
n_{A \cup B} &= 2 + 6 + 16 = 24 \\
n_{\overline{A}} &= 50 - 8 = 42 \\
n_{\overline{A \cup B}} &= 50 - 24 = 26.
\end{aligned}$$

Further, $\overline{B}$ is the event 'selected person is male' and $\overline{A}$ is the event 'selected person has a blood alcohol level < 0.05'. Therefore $n_{\overline{A} \cap \overline{B}} = 26$, the number of men with low blood alcohol level (< 0.05).

Since $n_{\mathcal{S}} = 50$, then, according to the classical definition of probability (Definition 3.3), the probabilities of the events $A, B, A \cap B, A \cup B, \overline{A}, \overline{A \cup B}$

and $\overline{A} \cap \overline{B}$ are $\frac{8}{50}$, $\frac{18}{50}$, $\frac{2}{50}$, $\frac{24}{50}$, $\frac{42}{50}$, $\frac{26}{50}$ and $\frac{26}{50}$ respectively. This numerically verifies Equation 3.1: $\overline{A \cup B} = \overline{A} \cap \overline{B}$.

Other relationships between the algebraic labels for events and what the events actually describe:

$\overline{A} \cap B$: 'selected person is female with blood alcohol level < 0.05'
$A \cap \overline{B}$: 'selected person is male with blood alcohol level ≥ 0.05.

EXERCISES

3.1 Consider the following industrial inspection plan:

Select four cables at random from the lot. Test each for breaking strength. Accept the entire lot if all four pieces break above 500 kg.

Describe the probabilistic experiment involved in this inspection plan. Give the sample space for the experiment.

3.2 The proofreader of a textbook of 450 pages reports that:

18 pages contain spelling mistakes and 13 pages have arithmetical mistakes, including 3 pages showing both kinds of errors.

(a) Illustrate the proofreader's report by means of a Venn diagram.

(b) Are the two types of page errors mutually exclusive?

(c) If a page is randomly selected from the book, determine the probability that it is free of mistakes.

3.3 The children in three-child families may be represented by the sample space $\mathcal{S}$ with elements written in birth order 1st born, 2nd born, 3rd born.

bbb bbg bgb bgg
gbb gbg ggb ggg

(a) Assume that the eight points in $\mathcal{S}$ are equally likely so that the classical definition may be used to evaluate probabilities.

Write down the events

A : 'the youngest child is a girl'
B : 'there is one boy'
C : 'no boy has an elder sister'
D : 'there is at least one boy'

as explicit subsets of $\mathcal{S}$.

Determine the probability of

- A
- $C \cap D$
- $A \cup B$.

(b) Assume now that the birth rate for boys is actually higher than it is for girls so that not all points in $\mathcal{S}$ should be regarded as equally likely. Discuss the statement: 'Three-child families are more likely to have three boys than two boys because boys are more likely'.

3.4 Consider the following sample space for tossing two fair dice. The 36 points

(Die 1 outcome, Die 2 outcome)

are to be regarded as equally likely.

(1,1)	(1,2)	(1,3)	(1,4)	(1,5)	(1,6)
(2,1)	(2,2)	(2,3)	(2,4)	(2,5)	(2,6)
(3,1)	(3,2)	(3,3)	(3,4)	(3,5)	(3,6)
(4,1)	(4,2)	(4,3)	(4,4)	(4,5)	(4,6)
(5,1)	(5,2)	(5,3)	(5,4)	(5,5)	(5,6)
(6,1)	(6,2)	(6,3)	(6,4)	(6,5)	(6,6)

(a) Use this sample space to determine the probability that

i. the outcome for Die 1 strictly exceeds the outcome for Die 2;

ii. the sum of the two outcomes is 7.

(b) Identify the events of (a) as subsets of the given sample space. (Essentially, this illustrates the events in the form of a Venn diagram.)

(c) Use the events of (a) to illustrate how the partition principle operates. (Refer to Example 3.3.)

(d) Determine the probability that when two fair dice are tossed they will show different outcomes.

3.5 The following data give the number of deaths in 1983 in Australia per 1000 of population of the same age group and sex. This is called the *Age-Specific Death Rate.*

Age group (years)	**Males**	**Females**
Under 1	10.5	8.8
1-4	0.6	0.5
5-9	0.3	0.2
10-14	0.3	0.2
15-19	1.1	0.4
20-24	1.5	0.5
25-29	1.4	0.5
30-34	1.2	0.6
35-39	1.4	0.9
40-44	2.3	1.3
45-49	3.9	2.3
50-54	7.2	4.0
55-59	12.3	6.1
60-64	19.0	9.6
65-69	30.6	15.0
70-74	47.9	25.2
75-79	77.3	42.6
80-84	116.8	74.3
Over 84	204.5	156.0

(*Source*: Cameron, R.J. (1985). *Year Book Australia 1985*)

These data show that in Australia women generally live longer than men. Discuss.

3.6 Items sampled on a production line may be classified as defective (D) or non-defective (N). List elements in the sample space showing all possible outcomes if the sampling process terminates:

(a) after five items have been sampled;

(b) after three defectives in a row have been observed or five items have been sampled;

(c) when the first defective is observed.

3.7 The Royal Melbourne Institute of Technology has a building with 13 floors all serviced by a lift. The lift (certainly not an express lift) stops at all floors to the 13th on the way up and all floors to the 1st on the way down. If I call for the lift from the Electrical Engineering Department on the 8th floor with the intention of leaving the building, use the classical definition of probability to argue that the lift, when it arrives, is more likely to be 'going up' than 'going down'. Give probabilities for each of these events. (Assume that the lift is not already waiting when called.)

Try to generalise the situation to a building with n floors and we are trying to leave from floor $m < n$.

3.8 A section of Melbourne's Central Business District is represented in the diagram below. The depicted area is that flanked by Swanston St, Flinders St, William St and La Trobe St.

Melbourne Central Business District

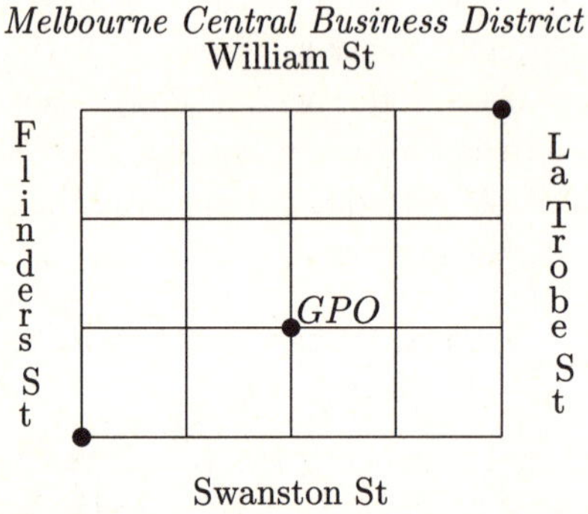

A pedestrian walks from Flinders St Station at the corner of Flinders St and Swanston St to the Flagstaff Gardens at the corner of William St and La Trobe St, always walking either directly towards La Trobe St or directly towards William St.

(a) How many different paths are available?

(b) If all the paths are equally likely, what is the probability that the pedestrian walks via the GPO?

3.9 Let A_1, A_2 and A_3 be three arbitrary events in a sample space $\mathcal{S}$. Then, for example, the set-theoretic expression for the event that of A_1, A_2 and A_3, only A_1 occurs is $A_1 \cap \overline{A_2} \cap \overline{A_3}$.

(a) Give a set-theoretic expression for the event that at most one of the events A_1, A_2, A_3, occurs.

(b) Is it true that $A_1 \cup A_2 \cup A_3 = \mathcal{S}$? Explain briefly.

3.2 SETTING TO WORK WITH THE PROBABILITY AXIOMS

In 1933, the Russian mathematician Kolmogorov was able to present a unified approach to probability assessment founded on three basic assumptions, or axioms. The truth of these axioms is not questioned but, rather, a mathematical theory of probability is built from their interplay through reasoning and logical deduction. The three versions of probability that introduced this chapter (classical, relative frequency and subjective probability) may be derived as special cases of this theory.

Definition 3.7 (Axiomatic definition of probability) *For every event* $A \subseteq \mathcal{S}$, *a* **probability function** P *assigns a real number,* $P(A)$, *called the* **probability of the event** A *so that*

Axiom 1: $P(\mathcal{S}) = 1$.

Axiom 2: $P(A) \geq 0$.

Axiom 3: *If* A_1 *and* A_2 *are mutually exclusive events in* $\mathcal{S}$ *then*

$$P(A \cup B) = P(A) + P(B).$$

If $A_1, A_2, A_3, \ldots,$ *are a sequence of mutually exclusive events in* $\mathcal{S}$ *then*

$$P(\bigcup_{i=1}^{\infty} A_i) = \sum_{i=1}^{\infty} P(A_i).$$

□

The first axiom states that the sample space $\mathcal{S}$ is a **certain event**. This is reasonable since $\mathcal{S}$ is the set of all possible outcomes, so certainly one of them, and hence $\mathcal{S}$, occurs. The second axiom places in position, on the real number axis, the lower end of the scale by which we will compare the relative sizes of probabilities—they are to be non-negative. The final axiom tells us how to add up probabilities for disjoint events (non-intersecting sets) in the sample space.

This set of axioms is not unique. Others may be substituted that will generate the same mathematical theory as that generated by Axioms 1–3; however the given set generates the theory *simply.*

Assigning probabilities to discrete sample spaces

We should be very clear about what the axioms tell us and what they do not tell us. They tell us how to combine probabilities already assigned to events in $\mathcal{S}$. They do not tell us how the probabilities are initially assigned to the events in $\mathcal{S}$. In a discrete sample space, the assignment of probabilities is done one simple event at a time.

Definition 3.8 *Given a discrete sample space* $\mathcal{S} = \{e_1, e_2, e_3, \ldots\}$, *the events* $\{e_i\}$ *consisting of one outcome are termed* **simple events**; *events consisting of more than one outcome are termed* **compound events**.

□

Probabilities must be assigned to each of these simple events, by using classical, relative frequency or subjective probability assessments. Then, the probabilities for all compound events in $\mathcal{S}$ follow by Axiom 3 since each compound event is a disjoint union of simple events. In particular, if $\mathcal{S} = \{e_1, e_2, e_3, \ldots\}$, then since, by Axiom 1, $P(\mathcal{S}) = 1$, we may quite arbitrarily assign positive probabilities to the simple events $\{e_1\}, \{e_2\}, \{e_3\}, \ldots$, provided that, by Axiom 3,

$$P(\{e_1\}) + P(\{e_2\}) + P(\{e_3\}) + \ldots = 1.$$

For example $\mathcal{S} = \{hh, ht, th, tt\}$, represents the sample space of equally likely outcomes when a pair of identical coins is tossed. Therefore, according to the classical definition of probability, we would assign the probabilities $P(\{hh\}) = \frac{1}{4}$, $P(\{ht\}) = \frac{1}{4}$, $P(\{th\}) = \frac{1}{4}$, $P(\{tt\}) = \frac{1}{4}$. Then, by Axiom 3,

$$P(\text{odds}) = P(\{ht\} \cup \{th\}) = P(\{ht\}) + P(\{th\}) = \frac{1}{4} + \frac{1}{4} = \frac{1}{2}.$$

Assigning probabilities in continuous sample spaces

Recall from Definition 3.1 that a continuous sample space contains an interval of real numbers. Because there are uncountably many points in such an interval, it makes no sense to count them one at a time, since between any two there are infinitely many more. That is, we cannot assign probabilities to elementary outcomes—single points—in a continuous sample space $\mathcal{S}$. Any possible point-wise assignment of probabilities, no matter how minuscule, would cause the sum of probabilities for all of $\mathcal{S}$ to exceed 1. This would be in violation of Axiom 1.

Instead, we must consider intervals. In fact, in a continuous sample space $\mathcal{S}$, an **event** is any subset of $\mathcal{S}$ that may be constructed from a countable number of unions, intersections and complements of intervals in $\mathcal{S}$. Specifically then, intervals are events. Suppose that $\mathcal{S}$ is an interval. If we assign

$$P(A) = \text{length of the interval } A \text{ relative to the length of } \mathcal{S},$$

then P is a probability function which satisfies Kolmogorov's axioms in Definition 3.7. In particular, if $\mathcal{S} = [0, 1]$, and $A = [a, b] \subseteq \mathcal{S}$, then $P(A) = b - a$. If we take on board a very general concept of 'interval'—that is, an interval of a continuum—then probabilities in one dimension are relative lengths; in two dimensions, relative areas; in three dimensions, relative volumes. We look at such volumes in Example 3.5.

Example 3.5

Packing a hollow cube with spheres: an insight into crystal structure

A crystal lattice consists of a 3-dimensional grid of atomic particles held together by interconnecting forces. If we simply model the particles as small spheres, then the close packing that we require to hold the 'crystal' together may be simulated by throwing these small spheres into a hollow cube.

A cube of side d mm is filled with balls of radius $\frac{d}{2n}$ mm. Notice that, when the cube is full it will contain n^3 balls. If a point is randomly selected from within the cube, the probability that it lies within one of the balls is

$$\begin{aligned}\frac{\text{total ball volume}}{\text{cube volume}} &= \frac{(\text{number of balls})(\text{volume of each ball})}{\text{cube volume}} \\ &= \frac{n^3\frac{4}{3}\pi\left(\frac{d}{2n}\right)^3}{d^3} \\ &= \frac{\pi}{6}\end{aligned}$$

Since, to 2 decimal places $\frac{\pi}{6}$ is 0.52, this result shows that the close packing of spheres of the same size will fill only 52% of 3-dimensional space.

Simple consequences of the axioms

The axioms provide the foundation on which the mathematical theory of probability is built. The following results are basic to probability calculations in practical situations.

Theorem 3.1 *For any $A \subseteq \mathcal{S}$, the probability of the complement $\overline{A}$ is given by*

$$P(\overline{A}) = 1 - P(A).$$

Proof: This follows by Axiom 3, since $\mathcal{S} = A \cup \overline{A}$ is the disjoint union of two sets, $1 = P(\mathcal{S}) = P(A) + P(\overline{A})$.

■

Theorem 3.1 applied to $A = \mathcal{S}$ gives $P(\overline{\mathcal{S}}) = 1 - P(\mathcal{S}) = 0$. But $\overline{\mathcal{S}} = \emptyset$; hence, we have proved Theorem 3.2.

Theorem 3.2 $P(\emptyset) = 0$.

■

The next in our series of theorems allows us to conclude that probabilities lie between 0 and 1.

Theorem 3.3 *For subsets A and B of a sample space $\mathcal{S}$, if $B \subseteq A$ then $P(B) \leq P(A)$.*

Proof: By the 'partition principle' emphasised in Example 3.3, we may immediately write the larger event A as the sum of two mutually exclusive pieces according to whether or not B occurs.

$$\begin{aligned} A &= (A \cap B) \cup (A \cap \overline{B}) \\ &= B \cup (A \cap \overline{B}), \end{aligned}$$

since $B \subseteq A$ implies that $A \cap B = B$. Therefore, by Axiom 3, $P(A) = P(B) + P(A \cap \overline{B})$ and the result is now immediate since $P(A \cap \overline{B}) \geq 0$.

■

The following very widely used result shows a simple method for evaluating the probability of the union $A \cup B$ in terms of the events A, B and $A \cap B$. The logic is as follows: To cover the area of $A \cup B$ in a Venn diagram, by adding together the areas for A and B, the intersecting area $A \cap B$ is counted twice, and so, in the Addition Law, is once subtracted.

Theorem 3.4 (The Addition Law) *For subsets A and B of a sample space $\mathcal{S}$,*

$$P(A \cup B) = P(A) + P(B) - P(A \cap B).$$

Proof: The events $A \cap B$, $A \cap \overline{B}$ and $B \cap \overline{A}$ form a disjoint union of $A \cup B$. Therefore, by Axiom 3 and the 'partition principle',

$$\begin{aligned} P(A \cup B) &= P(A \cap B) + P(A \cap \overline{B}) + P(B \cap \overline{A}) \\ &= P(A \cap B) + P(A \cap \overline{B}) + P(A \cap B) + P(B \cap \overline{A}) - P(A \cap B) \\ &= [P(A \cap B) + P(A \cap \overline{B})] + [P(A \cap B) + P(B \cap \overline{A})] - P(A \cap B) \\ &= [P(A)] + [P(B)] - P(A \cap B). \end{aligned}$$

■

Example 3.6

The hunt for oil

It takes only one good oil strike to offset the cost of several test drills.

Worked example:

An oil company believes that for a given locality, there is a 25% chance that a given test drill will locate oil and that if two drilling locations are tried, then there is only a 5% chance that oil will be found at *both* sites. Determine the probability of a successful oil strike.

Let E_i denote the event that oil is found at test site i $(i = 1, 2)$. Then $P(E_1) = 0.25 = P(E_2)$. By Theorem 3.4,

$$\begin{aligned} P(\text{successful oil strike}) &= P(E_1 \cup E_2) \\ &= P(E_1) + P(E_2) - P(E_1 \cap E_2) \\ &= 0.25 + 0.25 - 0.05 \\ &= 0.45. \end{aligned}$$

Example 3.7

Choosing a random number from [0,1]

The concept of choosing a random number between 0 and 1 is basic to the simulated construction of other random quantities.

Worked example:

If a number is randomly chosen from the interval [0,1], determine the probability that the number has a 2 in:

1. **the first decimal place;**
2. **the second decimal place;**
3. **both the first two decimal places;**
4. **either the first or the second decimal place.**

For 1, in order for the randomly chosen number (let us denote it by x) to have a 2 in the first decimal place of its decimal expansion we must have $0.2 \le x < 0.3$. If E_1 denotes the event $\{$ 2 in first decimal place $\}$, then

$$\begin{aligned} P(E_1) &= \text{length of the interval } [0.2, 0.3) \text{ relative to the length of } [0,1] \\ &= \frac{0.3 - 0.2}{1 - 0} \\ &= 0.1 \end{aligned}$$

For 2, in order for a decimal expansion to have a 2 in the second decimal place, we seek a number from $\mathcal{S} = [0, 1]$ in any of the 10 mutually exclusive intervals

$$[0.02, 0.03),\ [0.12, 0.13),\ [0.22, 0.23),\ \ldots,\ [0.92, 0.93).$$

If E_2 denotes the event { 2 in second decimal place }, then, by Axiom 3,

$$\begin{aligned}
P(E_2) &= \text{total length of intervals relative to the length of } [0,1]. \\
&= P\left([0.02, 0.03) \cup [0.12, 0.13) \cup \ldots \cup [0.92, 0.93)\right) \\
&= P[0.02, 0.03) + P[0.12, 0.13) + \ldots + P[0.92, 0.93) \\
&= 0.01 + 0.01 + \ldots + 0.01 \\
&= 0.1.
\end{aligned}$$

For 3,
$P(2 \text{ in first two decimal places}) = P(E_1 \cap E_2) = P[0.22, 0.23) = 0.01.$
Finally,

$$\begin{aligned}
P(2 \text{ in first or second decimal place}) &= P(E_1 \cup E_2) \\
&= P(E_1) + P(E_2) - P(E_1 \cap E_2) \\
&= 0.1 + 0.1 - 0.01 \\
&= 0.19.
\end{aligned}$$

EXERCISES

3.10 The following data on *car pooling* were the result of a large survey of respondents in 1981. The survey was taken at a time of great interest in car pooling; a time of rising petrol prices and decreasing patronage of public transport.

	Interest in car pooling			
	N	S	A	W
Drivers				
Male	11	52	26	68
Female	2	6	1	9
Non-drivers				
Male	15	51	1	24
Female	9	13	0	7

Key: N—never
S—strike days only
A—alternate week days only
W—all week days

(*Source*: Raper, R.L. and Kumar, S. (1981). Car pooling—A feasibility study in an Australian environment. *Scientific Management of Transport Systems.*)

(a) A respondent is randomly selected from these survey files. Determine the probability that the selected person is:

i. male;

ii. a non-driver;

iii. a female non-driver;

iv. a person not interested in car pooling;

v. a female driver who wishes to car pool on all weekdays;

vi. a female who does not wish to car pool on all weekdays;

vii. a female interested in some form of car pooling.

(b) Use the data to estimate the following probabilities: (M—male, D—driver)

i. $P(M \cup D)$;

ii. $P(M \cap D)$;

iii. $P(M \cap \overline{D})$;

iv. $P(\overline{M} \cap \overline{D})$;

v. $P(M \cap D \cap A)$.

Express each of these probability statements in words.

3.11 Six laboratory mice are injected with an experimental drug being tested for carcinogenic side-effects. Previous results indicate that the probability that one or more of the mice will die over a two week period is 0.75, while the probability that five or less mice die over the same period is 0.35.

(a) Sketch the events:

A = { one or more mice die in two week period }

B = { five or less mice die in two week period }

in a Venn diagram.

(b) Write down the values of $P(A)$ and $P(B)$.

(c) Explain how to use Axiom 1 to determine $P(A \cup B)$.

(d) Determine $P(\overline{A})$.

(e) Are $\overline{A}$ and $\overline{B}$ mutually exclusive? Show how to use Axiom 3 to determine $P(\overline{A} \cup \overline{B})$.

(f) Determine the probability that, over the two week period:

i. six mice will die;

ii. no mice will die;

iii. either 1, 2, 3, 4 or 5 mice will die.

3.12 Determine the probability that a card, drawn at random from a pack of 52 playing cards, is either an ace or a spade.

3.13 Determine the probability that a card, drawn at random from a pack of 52 playing cards, is neither an ace nor a spade.

3.14 We define the **difference between two events** by the equation

$$A - B = A \cap \overline{B}.$$

Intuitively, this consists of outcomes that are favourable to the event A but not favourable to the event B.

(a) Show that

$$P(A - B) = P(A) - P(A \cap B).$$

Draw a Venn diagram to illustrate your proof.

(b) Simplify $A - B$ in the case that A and B are mutually exclusive.

(c) A letter is chosen at random from the 26 letters of the alphabet. Let A denote the event that the selected letter is from the word *letter*; let B denote the event that the selected letter is from the word *alphabet*. Determine $P(A - B)$.

3.15 Given events A and B in a sample space $\mathcal{S}$ with $P(A) = 0.3$, $P(B) = 0.5$ and $P(A \cup B) = 0.6$, determine:

(a) $P(A \cap B)$;

(b) $P(\overline{A} \cup \overline{B})$;

(c) $P(A \cap \overline{B})$;

(d) $P(A \cup \overline{B})$.

Draw separate Venn diagrams to illustrate each of these results.

3.16 A shot is fired randomly at the target shown.

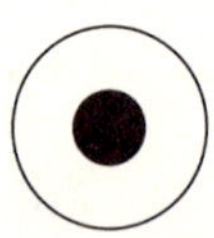

Assuming that the target is hit, determine the probability that the central bulls-eye is hit if the bulls-eye has radius r and the target has radius:

(a) $2r$;

(b) R, where $R > r$.

3.17 Consider two events A and B in a sample space $\mathcal{S}$. Suppose that A and B each contain two outcomes, $\mathcal{S}$ contains three outcomes, $P(A) = 0.5$ and $P(A \cap B) = 0.25$.

Discuss which of the following are true and which are false. Give reasons for your answers.

- Both $A \cap B$ and $A \cap \overline{B}$ are equally likely.
- Both $A \cap B$ and $A \cap \overline{B}$ are simple events.
- $P(A \cup B) = 1$.
- The events $A \cap B$ and $A \cap \overline{B}$ are mutually exclusive.

3.18 The following data from Example 2.8 give the observed time (x number of weeks) since each of the 45 numbers appeared in Tattslotto. (*Source: The Herald*, Melbourne, 22 August 1990.) A '0' indicates numbers just drawn and eight numbers are drawn each week. f_i is the number of times that x_i was observed.

x_i	0	1	2	3	4	5
f_i	8	6	6	6	4	5

x_i	6	7	9	18	21
y_i	3	2	3	1	1

(a) Is there any evidence in these data that some Tattslotto numbers are drawn two weeks in a row?

(b) Use these data to estimate the probability that it can be at least three weeks since your favourite number occurred.

(c) Can you use the data to estimate the probability that it can be at least three weeks between drawings of your favourite number?

(d) If we consider the data in Example 2.7 on the frequency of occurrence of Tattslotto numbers, is there evidence in these data that all points are equally likely?

(e) The high data value '21' occurring was the Tattslotto number 6. That is, it has been (a long) 21 weeks since the number 6 occurred. Will this give 6 a greater chance of being drawn in the next draw? Discuss.

3.19 For any two events A and B, define the event $A\triangle B$ by the equation

$$A\triangle B = (A - B) \cup (B - A).$$

(a) Explain in words the meaning of $A\triangle B$.

(b) Establish that

$$A\triangle B = (A \cup B) - (A \cap B).$$

(c) Determine an expression for $P(A\triangle B)$ in terms of the probabilities $P(A)$, $P(B)$ and $P(A \cap B)$.

(d) Show that $\triangle$ satisfies the 'triangle inequality': for any events A, B and C,

$$P(A\triangle C) \leq P(A\triangle B) + P(B\triangle C).$$

(e) A letter is chosen at random from the 26 letters of the alphabet. Let A denote the event that the selected letter is from the word *letter*; let B denote the event that the selected letter is from the word *alphabet*. Determine $P(A\triangle B)$.

3.20 If A, B and C are three events in a sample space $\mathcal{S}$, then

$$P(A \cup B \cup C)$$
$$= P(A) + P(B) + P(C) - P(A \cap B)$$
$$-P(A \cap C) - P(B \cap C) + P(A \cap B \cap C).$$

(a) Illustrate this result by using a Venn diagram.

(b) Prove the result by using the Addition Law

$$P(E \cup F) = P(E) + P(F) - P(E \cap F).$$

(c) During a TV ratings season, 100 people indicated which of the Channels 7, 9 or 10 they watched. The results gave:

Channel	**Number of viewers**
10	20
9	40
7	35
10 and 9	10
10 and 7	10
9 and 7	20
10 and 9 and 7	5

If one of these viewers is selected at random, determine the probability that they watched at least one of the three channels during the ratings season.

3.21 Let A, B and C be three events with $P(A) = a$, $P(B) = b$ and $P(A \cap B) = c$.

(a) Determine $P(A \cup B)$ in terms of a, b, and c.

(b) Show that c must satisfy the constraints

$$max\{0, a+b-1\} \leq c \leq min\{a, b\}.$$

3.3 LEARNING TO COUNT

In operating the classical definition of probability, as has been done often in this chapter, two operating instructions are apparent: that we be able to convince ourselves that the proposed simple outcomes in $\mathcal{S}$ are 'equally likely', and that we can count them. The first of these may be resolved with subjective assessment or from experience with relative frequencies. The second we now consider. In relatively complicated sample spaces, perhaps arising from several stages of experimentation, how do we count the outcomes? In $\mathcal{S}$? Those favourable to an event A? For then,

$$P(A) = \frac{n_A}{n_{\mathcal{S}}}.$$

The multiplication principle

Consider our probabilistic experiment faceted into several 'stages' of experimentation. Many complicated experiments may be multi-staged in this way. For example, in industrial quality control, Stage 1 may represent the random selection of a production line (from the four available, running side by side on the factory floor); Stage 2 may represent the sampling from the production line in Stage 1 of an item which may be regarded as defective (D) or non-defective (N). For *each* of the 5 production line outcomes there are 2 quality outcomes, making a total of $(5)(2) = 10$ outcomes for the whole experiment with

$$\mathcal{S} = \{(1, D), (1, N), (2, D), (2, N), (3, D), (3, N), (4, D), (4, N), (5, D), (5, N)\}$$

showing these 10 outcomes as ordered pairs.

> **Multiplication principle:**
> In a multi-stage experiment with r stages, if there are n_1 outcomes at Stage 1, n_2 outcomes at Stage 2, n_3 outcomes at Stage 3, $\ldots$, n_r outcomes at Stage r, then the whole process has $n_1 n_2 n_3 \ldots n_r$ outcomes.

The multiplication principle has wide application—to obtain full usefulness of the principle we must be liberal in our interpretation of 'experiment', 'outcome' and 'stage'. The applications which follow are basic to counting. If we enumerate each stage we obtain a sequence of positions that may be filled one at a time as outcomes are decided.

Position 1	Position 2	Position 3	$\ldots$	Position r

The multiplication principle says that for each of the n_1 ways of filling Position 1, there are n_2 ways of filling Position 2 and for each of these ways there are n_3 ways of filling Position 3, ..., n_r ways of filling Position r. The whole process is completed in $n_1 n_2 n_3 \ldots n_r$ ways.

Example 3.8

Motor vehicle registration in Victoria: watching plates

Standard Registration Plates for motor vehicles in Victoria are required by VICROADS to begin with three letters of the alphabet and end with three digits. For a price, registrants may secure *Customised Registration Plates* (that begin with 4–6 letters followed by digits) or *Personalised Registration Plates* (that begin with two letters, representing initials, followed by 3 or 4 digits).

X H F 7 3 6

Worked example:

How many different number plates are possible?

For *Standard Registration Plates* the multiplication principle immediately gives, by filling the positions one at a time, $n = (26)(26)(26)(10)(10)(10) =$ 17 576 000. That is enough number plates to keep Victoria well stocked with numbers. But there are more.

For *Customised Plates*, again by filling the positions one at a time:

(26)(26)(26)(26)(10)(10) = 45 697 600 begin with four letters;
(26)(26)(26)(26)(26)(10) = 118 813 760 begin with five letters;
(26)(26)(26)(26)(26)(26) = 308 915 776 begin with six letters.

These values sum to $n =$ 473 427 136 which is the total number of *Customised Registration Plates.*

The two kinds of *Personalised Registration Plates* may be summarised as:

(26)(26)(10)(10)(10) = 676 000 end with three digits;
(26)(26)(10)(10)10)(10) = 6 760 000 end with four digits;

giving a total number of $n =$ 7 436 000 in this final category. Adding these 3 values of n gives a total of nearly 500 million different possibilities.

Permutations or arrangements

The multiplication principle can be used to state a general result for arrangements of objects in a row where the order of the arrangement matters; thus, for example, a, f, g and g, a, f correspond to different *arrangements* of the letters a, f and g, but represent the same *selection* of three letters from the alphabet. 'Arrangements' will be considered first, and 'selections' in the next section.

The following theorem is proved immediately by assigning, from the multiplication principle,

$$\begin{aligned} & n_i = n, & i = 1, 2, 3, \ldots, r \text{ for (a)}, \\ \text{and } & n_i = n - i + 1, & i = 1, 2, 3, \ldots, r \text{ for (b)}. \end{aligned}$$

Theorem 3.5 *(a)* *With* *possible repetition, the number of arrangements of r objects from n distinct objects available is n^r.*
(b) *Without* *possible repetition, and for $r \leq n$, the number of arrangements of r objects from n distinct objects available is $n(n-1)(n-2)\ldots(n-r+1)$.*

■

A special term is used for arrangements without repetition.

Definition 3.9 $n^{(r)} = n(n-1)(n-2)\ldots(n-r+1)$ *is called the* **permutation of n distinct objects taken r at a time** *and is defined for $r \leq n$. The special permutation $n^{(n)}$ when $r = n$ is called* **n factorial** *and is denoted by $n! = n(n-1)(n-2)\ldots 3.2.1$ with the convention that* $0! = 1$.

□

Factorials grow recursively very quickly ...

$0! = 1$
$1! = 1$
$2! = (2)(1) = 2$
$3! = (3)(2)(1) = (3)2! = 6$
$4! = (4)(3)(2)(1) = (4)3! = 24$
$5! = (5)(4)(3)(2)(1) = (5)4! = 120$
$\vdots$
$n! = n[(n-1)!]$
$n! \sim \sqrt{2\pi} n^{n+\frac{1}{2}} e^{-n}.$

This last formula, known as **Stirling's formula**, may be used for an approximation to $n!$ for n large. The symbol '$\sim$' conveys this in the sense that

$$a_n \sim b_n \text{ if } \lim_{n\to\infty} \frac{a_n}{b_n} = 1.$$

For example,

$$100! \sim \sqrt{2\pi}\, 100^{100.5} e^{-100} = f, \text{ say.}$$

Then $\log_{10} f = \frac{1}{2}\log_{10}(2\pi) + 201 - 100\log_{10} e = 157.96964$. Hence, in scientific notation,

$$100! \text{ is approximately } 9.32 \times 10^{157}.$$

This is a very large number indeed.

By using the structure of factorials we can write

$$\begin{aligned} n^{(r)} &= n(n-1)(n-2)\ldots(n-r+1) \\ &= n(n-1)(n-2)\ldots(n-r+1)\frac{(n-r)(n-r-1)\ldots 3.2.1}{(n-r)(n-r-1)\ldots 3.2.1} \\ &= \frac{n(n-1)(n-2)\ldots(n-r+1)(n-r)(n-r-1)\ldots 3.2.1}{(n-r)(n-r-1)\ldots 3.2.1} \\ &= \frac{n!}{(n-r)!} \end{aligned}$$

so that permutations are expressed conveniently in factorial notation as

$$n^{(r)} = \frac{n!}{(n-r)!}. \tag{3.3}$$

(${}^nP_r$ is used by some authors as an alternative notation for $n^{(r)}$.) For example, the coded symbols $\oplus$ $\ominus$ $\otimes$ $\oslash$ $\odot$ may be arranged in a line in 5^5 ways with repetitions, $5! = 120$ ways without repetitions, and $5^{(3)} = \frac{5!}{2!} = 60$ ways when only three of the five symbols are used without repetition (five symbols permuted three at a time).

Combinations or selections

Suppose now that from our n distinct objects available, we are concerned with which r we choose without repetition and pay no attention to the order in which we could have selected them or the order in which they could be arranged in a row.

Let *selection* of the r objects be accomplished in c ways. We wish to determine c. By the multiplication principle, once the particular r objects have been selected, they may be arranged among themselves in $r!$ ways. Then, again by the multiplication principle and Definition 3.9, $n^{(r)} = c \times r!$. Equation 3.3 then gives:

Theorem 3.6 *The number of ways of selecting r distinct objects from n distinct objects is*

$$\frac{n!}{(n-r)!r!}$$

for $n \geq r$.

■

Definition 3.10 *We set*

$$\binom{n}{r} = \frac{n!}{(n-r)!r!}$$

so that $\binom{n}{r}$ *is the* **combination of n objects taken r at a time**.

□

$\binom{n}{r}$ is conveniently expressed verbally as 'n choose r'. (${}^nC_r$ is used by some authors as an alternative notation for $\binom{n}{r}$.) As an illustrative example, we may select two of the coded symbols $\oplus$ $\ominus$ $\otimes$ $\oslash$ $\odot$ in 'five choose two ways': that is, in $\binom{5}{2} = \frac{5!}{3!\,2!} = 10$ ways and these choices may be listed as:

$$\begin{array}{ccccc} \oplus\ominus & \oplus\otimes & \oplus\oslash & \oplus\odot & \ominus\otimes \\ \oslash\odot & \ominus\oslash & \ominus\odot & \otimes\oslash & \otimes\odot \end{array}$$

Any other listing corresponds to a rearrangement of one of these selections. There is another way of regarding $\binom{n}{r}$: as the number of *arrangements* of n objects in a row when r of the objects are of one kind and the remaining $(n-r)$ objects are of a second kind. The proof of this is a special case of Theorem 3.7 below. This introduces $\binom{n}{r}$ as a **binomial coefficient** giving the number of arrangements of two types of objects (such as 'success' and 'failure' in a simple experiment having these as its outcomes).

The arithmetic of binomial coefficients springs from **Pascal's Triangle**:

$$\begin{array}{c} \binom{0}{0} \\ \binom{1}{0}\ \binom{1}{1} \\ \binom{2}{0}\ \binom{2}{1}\ \binom{2}{2} \\ \binom{3}{0}\ \binom{3}{1}\ \binom{3}{2}\ \binom{3}{3} \\ \binom{4}{0}\ \binom{4}{1}\ \binom{4}{2}\ \binom{4}{3}\ \binom{4}{4} \\ \vdots \end{array} \qquad \begin{array}{c} 1 \\ 1\ \ 1 \\ 1\ \ 2\ \ 1 \\ 1\ \ 3\ \ 3\ \ 1 \\ 1\ \ 4\ \ 6\ \ 4\ \ 1 \\ 1\ \ 5\ \ 10\ \ 10\ \ 5\ \ 1 \\ \vdots \end{array}$$

In these triangular structures, two adjacent terms on a given line add to the term immediately below and between them on the following line: $\binom{3}{1} + \binom{3}{2} = \binom{4}{2}$ ... and in general line $n+1$ is constructed from line n by

$$\binom{n+1}{r} = \binom{n}{r-1} + \binom{n}{r}.$$

Each line in Pascal's Triangle gives the coefficients of the powers of x in the **binomial expansion**

$$(1+x)^n = \sum_{r=0}^{n} \binom{n}{r} x^r.$$

The **multinomial coefficients**

$$\frac{n!}{r_1!\, r_2!\, r_3!\, \ldots r_k!}$$

are a generalisation of the binomial coefficients from two categories to k categories.

Theorem 3.7 *The number of arrangements of n objects of different* kinds *consisting of r_1 identical objects of a 1st kind, r_2 identical objects of a 2nd kind, r_3 identical objects of a 3rd kind, ..., r_k identical objects of a kth kind, is*

$$\frac{n!}{r_1!\, r_2!\, r_3!\, \ldots r_k!}.$$

Proof: By the multiplication principle, c, the number of required arrangements, satisfies

$$n! = c \times r_1! \times r_2! \times \ldots \times r_k!$$

since all identical objects may be permuted among themselves without changing the arrangement.

■

For example, there are $\frac{14!}{4!\,3!\,3!\,1!\,1!\,1!\,1!}$ (about 10^8) distinct arrangements of letters from *intostatistics*. Anagram specialists may try to detect which may be found in a dictionary!

Counting for probability

We now give an example of probability calculations that use the counting techniques of this section.

Example 3.9

Counting koalas

Suppose that there is a population of n koalas in a particular nature reserve—n is unknown, but needs to be monitored (estimated) to ensure that the species does not die out in the area. Obviously a complete census is not possible. One frequently used method for estimating the size of animal populations is called 'capture/recapture'.

Suppose that m koalas are caught and tagged and then released back into the population. (This sorts the koala population into two types—'tagged' and 'not tagged'.) After thorough integration, a second sample of size k is captured. The probability that it contains exactly r *tagged* koalas is

$$p_r(n) = \frac{\binom{m}{r}\binom{n-m}{k-r}}{\binom{n}{k}},$$

since in the second catch we regard all $\binom{n}{k}$ selections of koalas (points in the sample space) as equally likely, and we simply count those containing r tagged koalas. This count is: the r tagged may be chosen from the m tagged 'out there' in the population in $\binom{m}{r}$ ways. This done, then the $k-r$ non-tagged may be chosen from the $n-m$ non-tagged 'out there' in the population in $\binom{n-m}{k-r}$ ways; now operate the multiplication principle to get the numerator for $p_r(n)$.

How do we estimate n from this? One way is to choose n to make $p_r(n)$ as large as possible, for then the r tagged koalas that we actually observed would be the most likely outcome. (This is called the method of *maximum likelihood.*) Since, by cancelling factorials,

$$\frac{p_r(n)}{p_r(n-1)} = \frac{(n-m)(n-k)}{(n-m-k+r)n},$$

so, by equating numerator and denominator of this expression in the search for the maximum value of n, we find that $nr = mk$. Therefore $p_r(n)$ is largest when the integer n is given by

$$n = \left[\frac{mk}{r}\right],$$

where $[x]$ denotes the integer part of x. Thus, if we observed $r = 10$ tagged koalas after securing samples of $m = k = 80$, the koala population would be estimated to contain approximately $\frac{(80)^2}{10} = 640$ individuals.

EXERCISES

3.22 A PIN number (personal identification number) for accessing a bank account with a plastic card through an ATM (automatic teller machine) consists of 4 digits chosen with replacement from: 0,1,2,3,4,5,6,7,8,9. If someone steals my card, what chance has he or she got of accessing my bank account by guessing my PIN number in one attempt?

3.23 The instructions on an exam paper read: 'Answer three of the following six questions'. If the questions may be answered in any order, how many different orderings are there? How many different exams are there? [*Note*: The orderings Q1,Q2,Q5 and Q2,Q5,Q1 are *different* but correspond to the *same* exam.]

3.24 How many different sums of money can be formed with at most five coins of different denominations?

3.25 In how many different sequences can five experiments be performed if two specified experiments must be performed one after the other in the same order?

3.26 How many different arrangements are there of the symbols $\spadesuit, \diamondsuit, \clubsuit, \heartsuit$?

3.27 How many different arrangements are there of the symbols

$$\spadesuit, \spadesuit, \spadesuit, \spadesuit, \diamondsuit, \diamondsuit, \clubsuit, \heartsuit, \heartsuit?$$

3.28 How many connecting cables are needed in order that any two of n offices in a building may communicate directly? Try this out first with $n = 2, 3, 4, 5, \ldots$ and then finally for a general n.

3.29 Fifteen people randomly take their seats at a round table. Show that the odds are six to one against two particular people sitting together.

3.30 I have five keys in my pocket. Only one of the keys will open my door. If I try them in the lock in random order one at a time without replacement, show that with probability $\frac{1}{5}$ I have to try the four wrong keys before finally trying the correct one. [*Hint*:

The arrangement of the keys is important here.]

Generalise the problem: If I have n keys in my pocket, what is the probability that I will need to try *exactly* $(k-1)$ wrong keys before finding the correct one? (Try with $k = 1, 2, 3, \ldots, n$.) What is your interpretation of the general result?

3.31 (a) Explain why a poker hand (of five cards) contains 'three tens' with probability

$$\frac{\binom{4}{3}\binom{12}{2}\binom{4}{1}\binom{4}{1}}{\binom{52}{5}}.$$

What assumptions are made in reaching this conclusion? Determine the probability that a poker hand contains 'three of a kind'. By 'kind' we mean one of the thirteen card types—
A K Q J 10 9 8 7 6 5 4 3 2.

(b) In the final rounds of a game of poker, you and Quickdraw are the only ones left in the game. You have 'three tens'. You very strongly suspect that Quickdraw has 'three of a kind'. Based on this information, how do you rate your chances of beating him? Give your reasoning. The hierarchy of ordering of cards within a suit is:
A K Q J 10 9 8 7 6 5 4 3 2,
so that three aces beat three kings which beat three queens ...

3.32 Explain why a poker hand contains 'two pairs' with probability

$$\frac{\binom{13}{2}\binom{11}{1}\binom{4}{2}\binom{4}{2}\binom{4}{1}}{\binom{52}{5}}.$$

3.33 Determine the probability of getting a 'full house' in a game of poker. A full house consists of three cards of one denomination and two cards of another denomination. [*Hint*: Use the multiplication principle and first choose the denominations and then the suits.]

3.34 Suppose that there are k people in a room.

Let E_k denote the event:

'At least two of the k people have the same birthday; that is, have their birthdays on the same day and month of a 365 day year.'

(a) Show that

$$P(E_k) = 1 - \frac{(365)(364)\ldots(365-k+1)}{365^k}.$$

Hence complete the following table:

k	P(E$_k$)
5	
10	0.117
20	0.411
30	0.706
40	0.891
50	0.970

(b) Use the table to determine the probability that 50 people in the room all have different birthdays.

(c) What is the smallest value of k (given in the table) such that the probability is 0.5 or more that at least two people in the room have the same birthday?

3.35 Show that

$$\binom{n}{r} = \binom{n}{n-r}.$$

What is the interpretation of this result in terms of Pascal's Triangle?

3.36 (a) Determine the number of 6 digit strings of integers between 000 000 and 999 999 inclusive which have all the digits different. For each $i = 2, 3, 4, 5, 6$ determine how many strings have *exactly* i digits the same.

(b) A lottery has a million tickets numbered from 000 000 to 999 999. The lottery winner is selected by drawing a single ticket.

i. Describe the sample space $\mathcal{S}$ for this experiment.

ii. For each $i = 1, 2, 3, 4, 5, 6$, let A_i denote the event that the winning ticket has exactly i digits the same. Describe in words, the event A_1.

iii. For each $i = 1, 2, 3, 4, 5, 6$, determine $P(A_i)$.

iv. Because it is clear that $P(A_3) > P(A_6)$, does this mean that ticket 333 987 has a better chance of winning than ticket 333 333? Explain briefly.

v. Discuss which of the following are true and which are false:

$A_1 \cap A_2 \cap A_3 \cap A_4 \cap A_5 \cap A_6 = \emptyset$.
$A_1 \cup A_2 \cup A_3 \cup A_4 \cup A_5 \cup A_6 = \mathcal{S}$.
$\sum_{i=1}^{6} P(A_i) = 1$.

3.37 Suppose that $\mathcal{S} = \{e_1, e_2, e_3, \ldots, e_n\}$ is a sample space.

(a) Show that $\mathcal{S}$ contains $\binom{n}{1}$ elementary events.

(b) Show that $\mathcal{S}$ contains $\binom{n}{2}$ compound events each with two elements.

(c) How many events each with r outcomes are there in $\mathcal{S}$?

(d) Show that there are a total of

$$\sum_{r=0}^{n} \binom{n}{r}$$

events in $\mathcal{S}$.

(e) Use the binomial expansion

$$(1 + x)^n = \sum_{r=0}^{n} \binom{n}{r} x^r$$

to show that the sample space $\mathcal{S} = \{e_1, e_2, e_3, \ldots, e_n\}$, contains 2^n events.

3.38 A wholesaler delivers batches of 100 test tubes to the CSIRO. As a quality control measure, 5 test tubes are selected at random from the batch and if these are all acceptable (not broken, cracked or chipped) the batch is accepted, otherwise the batch is rejected. If m denotes the number of unacceptable test tubes in a given batch, show that the probability, p, that the batch is accepted is given by

$$p = \frac{\binom{m}{0}\binom{100-m}{5}}{\binom{100}{5}}.$$

By taking logarithms, show that, if m is small in relation to 100,

$$\log_e p \approx -m \left(\sum_{i=0}^{4} \frac{1}{100 - i} \right).$$

$$\left[\textit{Hint:} \ \lim_{x \to 0} \frac{\log_e(1 + x)}{x} = 1. \right]$$

3.4 TREE DIAGRAMS AND BAYES' RULE

In Theorem 3.4 we presented the Addition Law for determining the probability of unions of events. The parallel of this is the Multiplication Rule which utilises conditional probabilities for determining the probability of intersections of events.

Conditional probabilities and the Multiplication Rule

Suppose that events A and B lie in a sample space $\mathcal{S}$.

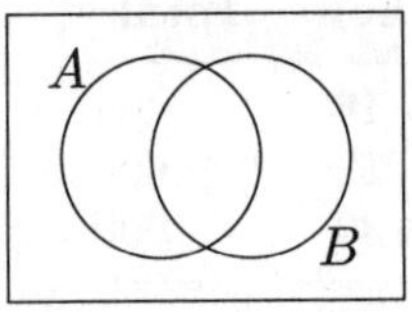

We will consider events $A \cap B$ and B to be non-empty. Notice in particular that $A \cap B \subset B$. If, as usual, n_U is the number of points in the sample space corresponding to an arbitrary event U, then the fraction

$$\frac{n_{A \cap B}}{n_B}$$

corresponds to the fraction of the points in $\mathcal{S}$ favourable to the outcome of B that are also favourable to the outcome of $A \cap B$. That is, the number of points in $A \cap B$ are relativised with respect to the number of points in B. Let us carefully compare this fraction with that given in Definition 3.3, the classical definition of probability. A revision of the classical definition of probability has occurred; the sample space $\mathcal{S}$ has been replaced by the event B. The event B has become the totality of all possible outcomes and the event $A \cap B$, being contained in B, is one collection of such outcomes. B is assuming the role of a **reduced sample space**. Therefore, we are in a position to know that B has occurred and we question the likelihood of A occurring. Knowing that B has occurred is 'inside information'; it allows us to update our assessment $P(A)$, the probability of A occurring, to $P(A|B)$, the 'conditional probability of A occurring *given that* B has occurred'. In the following example we can see just how easy conditional probabilities are to evaluate.

Example 3.10

Shooting stars

Each night a measure of the *intensity* or number of shooting stars observed in a given time period was recorded as 'low', 'medium' or 'high'. The intensity is known to show seasonal fluctuation and an assessment of the influence of solar activity (through 'high' and 'low' sunspot seasons) is of interest to astrophysicists.

Suppose that the following data, classified *intensity* by *solar activity*, was collected over 100 clear nights of observation.

		Intensity		
	High	**Medium**	**Low**	**Total**
Solar activity low	13	9	18	40
Solar activity high	26	22	12	60
Total	39	31	30	100

Let us label the events: A—'High shooting star intensity'; B—'High solar activity'.

On a randomly chosen night of observation, the table indicates that we should estimate: $P(A) = \frac{39}{100}, P(B) = \frac{60}{100}$. However, suppose that we know that B has occurred: that is, we are observing the sky on a night of high sunspot activity. How does this change our assessment of the likelihood of seeing a shooting star? In fact, by knowing that B has occurred, our attention shifts to the reduced sample space (with 60 points) defined by the 'Solar activity high' line of the table. This line is our new sample space, so that, with reference to this line, $P(A|B) = \frac{26}{60}$ since 26 points were favourable to both A and B occurring.

How does this compare with an uninfluenced prediction of shooting star intensity? From above, $P(A) = 0.39$ and now $P(A|B) = 0.43$ (2 dp). Therefore, our confidence in observing a high intensity of shooting stars has improved slightly by knowing that observation is occurring in a period of high solar activity.

A reduced sample space may sit in any column of the table as well: $P(B|A) = \frac{26}{39}$.

The cancellation occurring in the product of fractions

$$\frac{n_{A\cap B}}{n_S} = \frac{n_B}{n_S} \cdot \frac{n_{A\cap B}}{n_B}$$

is the motivation of the definition of conditional probability in the Multiplication Rule:

Definition 3.11 (Multiplication Rule) *For events A and B in a sample space $\mathcal{S}$, such that $P(B) > 0$, the* **conditional probability of A occurring given that B has already occurred**, *denoted by $P(A|B)$, is defined implicitly by the equation*

$$P(A \cap B) = P(B)P(A|B). \tag{3.4}$$

□

Equation 3.4 is used to write $P(A|B)$ *implicitly* in terms of $P(B)$ and $P(A \cap B)$ because often our intuition serves us better in assessing $P(A|B)$ than it does $P(A \cap B)$. Equation 3.4 provides a method for evaluating $P(A \cap B)$ by using the often easier $P(A|B)$. This is the main point of the Multiplication Rule.

Notice that, in this notation, '$A|B$' is not an event in the sample space $\mathcal{S}$. Rather, for an event B with $P(B) > 0$, the probability function P^*, defined by $P^*(A) = P(A|B)$, may be shown to satisfy the three axioms of Definition 3.7 in the reduced sample space defined by B. From this point of view, **all probabilities are conditional**: the effect is seen by putting $B = \mathcal{S}$ in Definition 3.11 with $P(\mathcal{S}) = 1$ and $P(A|\mathcal{S}) = P(A)$.

We should never blindly assume that the occurrence or non-occurrence of A remains completely unaffected by the fact that B has occurred. All successful subjective probabilists are quite prepared to re-assess their probability assessments when extra information comes to light. Like an insider with view to 'insider trading' in the Stock Market we should be prepared to reconsider our evaluation of $P(A)$ (probability of a dramatic stock rise for company A) in the event that we know B (record profits for company A to be announced) has occurred.

The Multiplication Rule is generalised in an obvious way to more than two events: For events $A_1, A_2, A_3, A_4, \ldots, A_n$ in a sample space $\mathcal{S}$, if $P(A_1 \cap A_2 \cap \ldots \cap A_{n-1}) > 0$,

$$\begin{aligned} P(A_1 \cap A_2 \cap A_3 \cap A_4 \cap \ldots \cap A_n) = \\ P(A_1)P(A_2|A_1)P(A_3|A_1 \cap A_2)P(A_4|A_1 \cap A_2 \cap A_3) \\ \ldots P(A_n|A_1 \cap A_2 \cap \ldots \cap A_{n-1}). \end{aligned} \tag{3.5}$$

Example 3.11

Obvious conditional probabilities: sampling without replacement from a small group

Suppose that a room contains 50 people, 10 of whom are 'smokers' and the remainder are 'non-smokers'. Suppose that we select three people at random to interview about their smoking habits. Naturally the sample is without replacement, because we don't want to interview the same person twice!

Worked example:

Consider the following probability questions:

1. **Determine the probability that the sample consists of a 'smoker', 'non-smoker', 'smoker' drawn in that order.**
2. **How many points are there in the sample space that lists the possible 'smoker/non-smoker' outcomes for the three people selected?**
3. **Determine the probability that the sample consists of a 'non-smoker' and two 'smokers'.**
4. **Is there any difference between:**
   - (a) **selecting sample members one after the other, each time making a random choice from those available, and**
   - (b) **placing the 50 names in a hat and randomly selecting three from the hat at the same time?**

To answer these inquiries it is best to begin by labelling the events of interest:

For $i = 1, 2, 3$, let S_i denote the event that the ith person sampled is a 'smoker'; so that $\overline{S_i}$ denotes the event that the ith person sampled is a 'non-smoker'.

1. First select a person at random and then, depending on the outcome for this first person selected (i.e. by conditioning on the outcome of this first person selected) choose a second person from those remaining ... select conditionally ...
   In terms of the specified events, we require

$$\begin{aligned} P(S_1 \cap \overline{S_2} \cap S_3) &= P(S_1)P(\overline{S_2}|S_1)P(S_3|S_1 \cap \overline{S_2}) \\ &= \left(\frac{10}{50}\right)\left(\frac{40}{49}\right)\left(\frac{9}{48}\right) = 0.031 \ \ (3 \text{ dp}). \end{aligned}$$

2. The sample space is ordered; by the Multiplication Rule it contains $(50)(49)(48) = 117\ 600$ points.

3. In this question, only the content of the sample is important:

$$\begin{aligned} P(\text{two smokers and one non-smoker}) &= \frac{\binom{10}{2}\binom{40}{1}}{\binom{50}{3}} \\ &= 3\left(\frac{10}{50}\right)\left(\frac{40}{49}\right)\left(\frac{9}{48}\right) \\ &= 0.092 \text{ (3 dp)}. \end{aligned}$$

4. The two methods of selection are the same. Notice that

$$\begin{aligned} P(\text{two smokers and one non-smoker}) &= 3P(S_1 \cap \overline{S_2} \cap S_3) \\ &= P(S_1 \cap \overline{S_2} \cap S_3) \\ &\quad +P(\overline{S_1} \cap S_2 \cap S_3) \\ &\quad +P(S_1 \cap S_2 \cap \overline{S_3}) \end{aligned}$$

under the assumption that all points in the sample space are equally likely.

To see how the Multiplication Rule is implemented in a useful way, we return to 'The partition principle' of Example 3.3. Suppose that A and B are events in a sample space S. A may be viewed as the disjoint union of two events $A \cap B$ and $A \cap \overline{B}$, according to whether or not B occurs.

The partition principle: $A = (A \cap B) \cup (A \cap \overline{B})$

Since the two sections of A are mutually exclusive, by Axiom 3

$$\begin{aligned} P(A) &= P(A \cap B) + P(A \cap \overline{B}) \\ &= P(A|B)P(B) + P(A|\overline{B})P(\overline{B}). \end{aligned} \tag{3.6}$$

Equation 3.6 is most useful. To determine the probability of an event A, simply condition according to whether or not a suitably chosen event B occurs. In a sequence of events, B may be the outcome of a prior stage; it is likely to be easier to count points in the present sample space for A if the immediate past is known (or rather is conditionally known).

The appropriate generalisation of Equation 3.6 involves generalising the partition principle:

Definition 3.12 *By a* **partition** *of the sample space $\mathcal{S}$ is meant a countable collection of mutually disjoint events with union $\mathcal{S}$.*

□

For example, in a sample space $\mathcal{S}$, B and $\overline{B}$ form a partition of $\mathcal{S}$ since $B \cap \overline{B} = \emptyset$ and $B \cup \overline{B} = \mathcal{S}$. Then Equation 3.6,

$$P(A) = P(A|B)P(B) + P(A|\overline{B})P(\overline{B}),$$

is a special case ($n = 2, B_1 = B, B_2 = \overline{B}$) of the following general partition theorem:

Theorem 3.8 (Partition Theorem) *If $B_1, B_2, B_3, \ldots, B_n$ is a sequence of mutually exclusive events in $\mathcal{S}$ that partition $\mathcal{S}$, then*

$$P(A) = \sum_{i=1}^{n} P(A|B_i)P(B_i)$$

for any event A in $\mathcal{S}$.

Proof: This result follows from Axiom 3 since $(A \cap B_i)$ for $i = 1, 2, 3, \ldots, n$ is a sequence of mutually disjoint events in $\mathcal{S}$ which 'cover' A. That is, $A = \cup_{i=1}^{n}(A \cap B_i)$.

■

Example 3.12

Recapturing koalas: Example 3.9 continued

At the recapture stage, sampling occurs without replacement from m tagged koalas and $n - m$ untagged koalas. Let us examine what happens for the *second* koala sampled (in the recapture stage). Let T_i denote the event that the ith koala is tagged, $i = 1, 2, 3, \ldots$ Then, the probability that the second koala caught is already tagged is

$$P(T_2) = P(T_2|T_1)P(T_1) + P(T_2|\overline{T_1})P(\overline{T_1}) = \left(\frac{m-1}{n-1}\right)\left(\frac{m}{n}\right) + \left(\frac{m}{n-1}\right)\left(\frac{n-m}{n}\right) = \frac{m}{n}.$$

This is, of course, the same as $P(T_1) = \frac{m}{n}$. The Partition Theorem may be used to show that $P(T_3) = \frac{m}{n}$. Therefore, this suggests that before recapture commences, there is the same chance of recapturing a tagged koala at any stage in the recapture procedure.

Tree diagrams

The conditional probabilities required for the Partition Theorem are often placed in a **tree diagram** to give a visual impression of how the probabilities are operating. In such a diagram, the branches of the tree spread laterally; the probability associated with each branch is a conditional probability—that is, conditional on being an off-shoot from the previous branch.

An $\boldsymbol{n_1 \times n_2}$ **tree diagram** shows in two sequential stages the possible outcomes for a partition $B_1, B_2, B_3, \ldots, B_{n_1}$ of S at Stage 1, and an n_2 element partition of the conditionally reduced sample space at Stage 2. The two stages comprise an 'experiment' which, by the multiplication principle of Section 3.3, has $n_1 n_2$ possible outcomes. This is the number of possible pathways in the tree diagram.

The 2×2 tree diagram illustrated shows the two sequential stages. At the first stage, B either occurs or it does not; at the second stage, depending on which member of the partition B, $\overline{B}$, occurs, the event A either occurs or it does not.

A 2×2 *tree diagram*

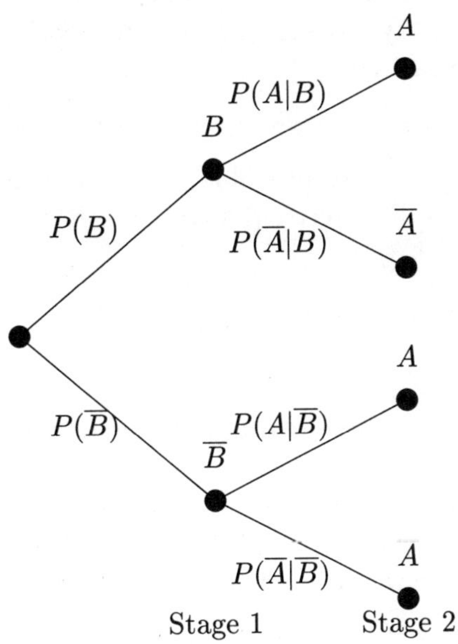

Since a partition of the (possibly reduced) sample space occurs at each stage, the probabilities arising from any branch point add to 1:

$$\begin{aligned} P(B) + P(\overline{B}) &= 1 \\ P(A|B) + P(\overline{A}|B) &= 1 \\ P(A|\overline{B}) + P(\overline{A}|\overline{B}) &= 1 \end{aligned}$$

The Multiplication Rule operates by the product of probabilities associated with a branch and its predecessor branches. That is:

$$\begin{aligned} P(B \cap A) &= P(B)P(A|B) \\ P(B \cap \overline{A}) &= P(B)P(\overline{A}|B) \\ P(\overline{B} \cap A) &= P(\overline{B})P(A|\overline{B}) \\ P(\overline{B} \cap \overline{A}) &= P(\overline{B})P(\overline{A}|\overline{B}) \end{aligned}$$

Adding together the probabilities associated with branches leading to the same eventual outcomes at Stage 2 automatically yields the Partition Theorem (Theorem 3.8) applied to the partition of $\mathcal{S}$ shown in the tree diagram at Stage 1.

$$\begin{aligned} P(A) &= P(A|B)P(B) + P(A|\overline{B})P(\overline{B}) \\ P(\overline{A}) &= P(\overline{A}|B)P(B) + P(\overline{A}|\overline{B})P(\overline{B}) \end{aligned}$$

Bayes' Theorem

The probabilities shown in an $n_1 \times n_2$ tree diagram are called **prior probabilities**. They represent known probability assessments. In terms of the two stages of an overall 'experiment', the results at Stage 2 are conditional on the outcome at Stage 1. If the order of the stages were reversed, the conditional probability assessments that we would then make are termed **posterior probabilities**.

Prior probability	$P(A\|B)$
Posterior probability	$P(B\|A)$

Example 3.13

Priors and posteriors—random breath testing: Example 3.4 continued

The following data is from Example 3.4 showing the *gender* by *breath test* result for fifty motorists. Motorists with blood alcohol levels exceeding 0.05 mg L^{-1} were classified as 'driving under the influence of alcohol'. We denote this event by '$\geq$'.

	Blood alcohol level ≥ 0.05	< 0.05
Men	6	26
Women	2	16

Reading from the table, for a randomly chosen individual with M denoting 'male' and W denoting 'female', we insert probabilities

$$P(\geq) = \tfrac{8}{50},\ P(<) = \tfrac{42}{50},$$
$$P(M|\geq) = \tfrac{6}{8},\ P(W|\geq) = \tfrac{2}{8},\ P(M|<) = \tfrac{26}{42},\ P(W|<) = \tfrac{16}{42},$$

on the tree diagram according to the likely status of the person selected. Stage 1 represents the breath test result; Stage 2 represents gender.

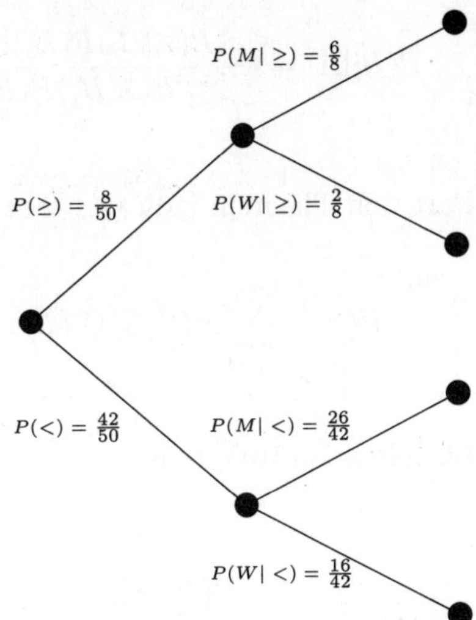

The information displayed in the tree diagram would probably support the intuitive probability assessments of an officer manning the breath testing station.

Of the people charged with driving while under the influence of alcohol, the proportions of men and women being loaded into the police van provide intuitive estimates of the **prior probabilities** $P(M|>), P(W|>)$. The **posterior probabilities** $P(>|M)$, $P(>|W)$, ..., are more difficult to resolve for the on-the-spot police officer: Given that a male/female driver is stopped, what is the probability that he/she is over the 0.05 limit?

From the survey files, the posterior probabilities may be read directly:

$$P(>|M) = \tfrac{6}{32} = 0.1875 \quad \text{(which is updated from } P(>) = \tfrac{8}{50} = 0.16);$$
$$P(>|W) = \tfrac{2}{18} = 0.1111 \quad \text{(which is updated, reduced, from } P(>) = \tfrac{8}{50} = 0.16).$$

Example 3.13 provided a complete dossier on $\geq, <, M$ and W. When such complete information is lacking, the posterior probabilities are simply determined from the prior probabilities according to the probability laws that we have so far developed as shown in **Bayes' Rule** below. Such posterior probabilities may be placed in the tree diagram that illustrates the stages in the 'experiment' reversed.

Theorem 3.9 (Bayes' Rule) *If $B_1, B_2, B_3, \ldots, B_{n_1}$ is a partition of the sample space S then for any event A and each $i = 1, 2, 3, \ldots, n_1$,*

$$P(B_i|A) = \frac{P(A|B_i)P(B_i)}{\sum_{j=1}^{n_1} P(A|B_j)P(B_j)}.$$

Proof: Since the Partition Theorem indicates that

$$P(A) = \sum_{j=1}^{n_1} P(A|B_j)P(B_j),$$

it follows by the Multiplication Rule that

$$\begin{aligned} P(B_i|A) &= \frac{P(B_i|A)P(A)}{P(A)} \\ &= \frac{P(B_i \cap A)}{P(A)} \\ &= \frac{P(A|B_i)P(B_i)}{\sum_{j=1}^{n_1} P(A|B_j)P(B_j)} \end{aligned}$$

and this completes the trivial proof of Bayes' Rule.

■

Example 3.14

The problem of false positives

Suppose that a proportion p of a population is known to carry the HIV virus (that leads to Acquired Immune Deficiency Syndrome). Test results over a period of time have shown that, of people carrying the virus, the test correctly returns a positive test result 99% of the time. Of people who are not carrying the virus, the test correctly returns a negative test result 98% of the time.

This prior probability information may be placed in the tree diagram below where Stage 1 indicates whether the virus is present, V, or absent, $\overline{V}$. The test result at Stage 2 is either positive $\oplus$, or negative $\overline{\oplus}$.

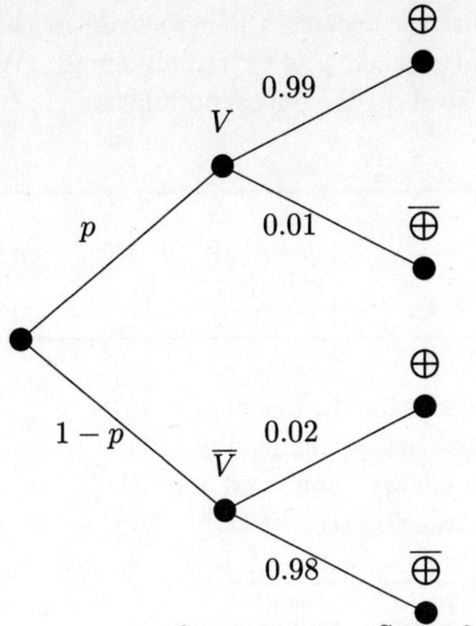

The question arises: If an individual from this population gives a positive test result when tested, what is the probability that the individual is carrying the virus?

This is crucial information. By Bayes' Rule we require

$$
\begin{aligned}
P(V|\oplus) &= \frac{P(\oplus|V)P(V)}{P(\oplus|V)P(V) + P(\oplus|\overline{V})P(\overline{V})} \\
&= \frac{0.99p}{0.99p + 0.02(1-p)} \\
&= \frac{0.99p}{0.02 + 0.97p}.
\end{aligned}
$$

Some values of p need to be suggested before this can be evaluated. Remember that p gives an indicator of the rarity of the disease in the population under consideration. In a 'high-risk' population, where p may be 0.1 for example, this gives a value of 0.846 (3 dp) for $P(V|\oplus)$. However the list

p	**P(V|⊕)**
0.1	0.846
0.01	0.333
0.001	0.047
0.0001	0.005

shows how $P(V|\oplus)$ changes as p changes. If the disease is very scarce, then we are faced with the surprising result that posterior probability $P(V|\oplus)$ is still small; the reason is that most of the test results are 'false positives' because the error rate in the testing procedure is much larger than p when p is extremely small. What if the testing procedure were released on the entire population ...?

EXERCISES

3.39 It is compulsory for the driver of a car to wear a seat belt while driving. The results of a survey show that not all drivers are wearing seat belts.

Age	Driver wearing seat belt	Driver not wearing seat belt
Under 40	375	52
40 and over	425	148

Use the data to estimate the probability that:

(a) a driver will use a seat belt;

(b) a driver who is under 40 will use a seat belt;

(c) a driver who uses a seat belt is not under 40.

3.40 Use the solar activity/shooting star activity data of Example 3.10 to estimate the probability that:

(a) a low intensity of shooting stars is observed when there is low sunspot activity;

(b) sunspot activity is low when a low intensity of shooting stars is observed;

(c) on a randomly chosen night there is a high intensity of shooting stars or there is a high sunspot activity.

3.41 Consider a batch of 100 items among which there are 5 defectives. Items are sampled from the batch one at a time, at random and without replacement. Determine the probability that the first five items sampled are in the order D, N, N, D, N where D denotes a defective item and N denotes a non-defective item.

3.42 The Saulwick Age Poll (*The Age*, Melbourne, 7 December 1987) reported the following opinions from a national telephone poll of 1000 people.

Do you approve of abortion if the mother's health is at risk?	
Yes	88%
No	7%
Don't know	5%

If three people are randomly selected (without replacement) from this survey determine the probability that:

(a) all three answered 'Yes';

(b) the opinion of the three selected was not unanimous.

3.43 A shift worker has an alarm clock to assist him in getting to work on time.

The alarm clock will ring at the preset time with probability 0.7. If it rings, the probability that he wakes is 0.8

and if it does not ring the probability that he wakes anyway is 0.3. Determine the probability that he awakes at the preset time.

3.44 In a statistics course, assume that a student who regularly answers assignments has a probability of 0.8 of passing the final exam, while a student who does not has a probability of 0.3 of passing the final exam. If 40% of students in the course regularly answer their assignments, show that only half the class is likely to pass.

3.45 One half of the fleet of cars in a rental car company are Toyotas. One-third of cars have been driven for more than 100 000 km. (Colloquially—have more than 100 000 km *on the clock.*) In total, two-thirds of the cars are either Toyotas or have more than 100 000 km on the clock. For a randomly selected car from the company's files let T denote the event 'selected car is a Toyota'; C denote the event 'selected car has more than 100 000 km on the clock'.

(a) Determine each of the probabilities

$$P(T|C), P(T|\overline{C}), P(\overline{T}|C), P(\overline{T}|\overline{C}),$$
$$P(C|T), P(C|\overline{T}), P(\overline{C}|T), P(\overline{C}|\overline{T}).$$

(b) $P(T|C)$ is the probability that the selected car is a Toyota when it is known to have more than 100 000 km on the clock. In a similar way, express the other probabilities from (a) in words.

3.46 For the breath test data of Example 3.13, repeated below, construct a 2×2 tree diagram which shows 'gender' at Stage 1 and 'breath test result' at Stage 2.

	Blood alcohol level	
	≥ 0.05	< 0.05
Men	6	26
Women	2	16

Determine the probabilities associated with each branch through the tree.

3.47 (a) For events L, B_1 and B_2 in $\mathcal{S}$ with $P(B_1) + P(B_2) = 1$,

i. Show that

$$P(L) = P(L|B_1)P(B_1) + P(L|B_2)P(B_2).$$

ii. Express each of $P(B_1|L)$ and $P(B_2|L)$ in terms of $P(L|B_1)$, $P(L|B_2)$, $P(B_1)$ and $P(B_2)$.

iii. Is it true that

$$P(B_1|L) + P(B_2|L) = 1?$$

Explain briefly.

(b) You have two boxes in front of you; Box 1 contains 5 live matches and 45 dead ones and Box 2 contains 49 live matches and a single dead one. You select a box by tossing a fair six-sided die with faces numbered $1, 2, 3, 4, 5, 6$. If the toss results in a '3' you choose Box 1; otherwise you choose Box 2. You then select a match at random from the chosen box.

i. Determine the probability that the match you get is live.

ii. What is the probability that you selected Box 1, given that the match you selected was live?

(c) Construct a tree diagram for this experiment. Label all branches in the tree with appropriate probabilities.

3.48 In the experience of a certain building society, customers who have sufficient funds in their cheque accounts postdate a cheque (by mistake) once for every thousand cheques written, while customers who write cheques against insufficient funds always postdate them; such customers constitute 1% of all cheque account holders.

(a) Identify the prior and posterior probabilities in this problem.

(b) If a cashier receives a postdated cheque from a customer, what is the probability that the customer has insufficient funds to cover the cheque?

(c) Display the prior probabilities in a tree diagram.

3.49

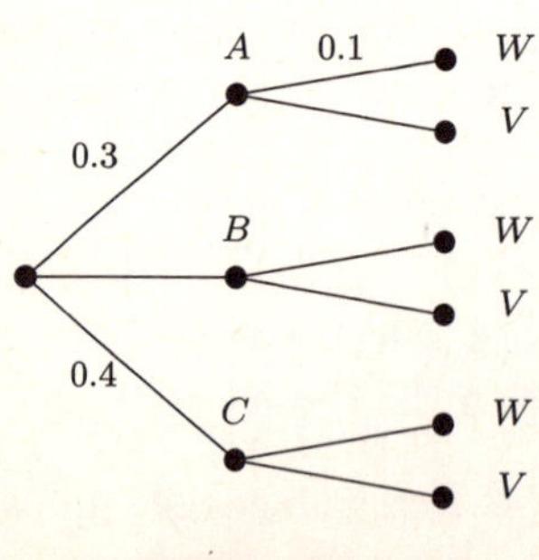

For the 3×2 tree diagram illustrated,

(a) Write down the values of

$$P(B), P(W|A), P(W \cup V|A), P(A \cap V)$$

(b) If we now use the extra information

$$P(W) = 0.38, P(W|B) = P(W|C):$$

i. compute the probability associated with each branch in the tree;

ii. construct the 2×3 tree diagram with the two stages reversed and compute the probabilities associated with each branch in the tree.

3.50 In a sample space $\mathcal{S}$, let A be an event with $P(A) > 0$.

(a) Show that the conditional probability function P^*, defined for any event $B \subset \mathcal{S}$ by

$$P^*(B) = P(B|A)$$

satisfies the three axioms of probability in the reduced sample space A.

(b) If $C \subset A$, show that

$$P^*(C) = \frac{P(C)}{P(A)}.$$

3.5 INDEPENDENCE

The assumption that events are 'independent' has led to many incorrect probability assessments. Our intuitive understanding of the term independent is often not sufficient justification for using an independence assumption. The essence of the probabilistic use of the term is that events are independent if they do not *influence* one another in the sense that the outcome of one of the events does not alter the likelihood of the other event occurring. This means that in a sample space $\mathcal{S}$, two non-null events A and B will be independent when $P(A|B) = P(A)$.

By interchanging the symbols for A and B we obtain that for independent (non-influencing) events A and B, $P(B|A) = P(A)$ and $P(A|B) = P(B)$. This allows the Multiplication Rule (Definition 3.11) the symmetric representation $P(A \cap B) = P(A)P(B)$ and we take this as the definition of independence since there is no added condition that $P(B) > 0$ as required from the conditional probability definition of $P(A|B)$.

Definition 3.13 *In a sample space $\mathcal{S}$, two events A and B are termed* **independent events** *if*

$$P(A \cap B) = P(A)P(B). \tag{3.7}$$

Otherwise events A and B are termed **dependent events.**

□

Example 3.15

Independence from a fair die

If we toss a fair die with faces labelled 1, 2, 3, 4, 5 and 6, then clearly $\mathcal{S} = \{1, 2, 3, 4, 5, 6\}$. Suppose that we define the following events:

$$A = \{1, 2, 3\}$$
$$C = \{1, 2, 4, 5\}$$

How is your intuition? Would you be prepared to assume that events A and C are independent? To check, use the classical approach to regard the six faces of the 'fair' die as equally likely. Then,

$$P(A \cap C) = P(\{1, 2\}) = \tfrac{1}{3} = (\tfrac{1}{2})(\tfrac{2}{3}) = P(A)P(C).$$

Hence A and C are independent. Was your intuition correct?

Notice that if A and B are independent, then $\overline{A}$ and $\overline{B}$ are also independent: in fact Definition 3.13 is satisfied for these complementary events since

$$\begin{aligned}
P(\overline{A}\cap\overline{B}) &= P(\overline{A\cup B}) && \text{(De Morgan)}\\
&= 1-P(A\cup B) && \text{(complement)}\\
&= 1-[P(A)+P(B)-P(A\cap B)] && \text{(Addition Law)}\\
&= 1-P(A)-P(B)+P(A)P(B) && \text{(independence)}\\
&= [1-P(A)][1-P(B)] && \text{(factorisation)}\\
&= P(\overline{A})P(\overline{B})
\end{aligned}$$

Independence as a consequence of the Multiplication Rule

We suggest that this (3.7) symmetric form of the multiplication rule, where the probabilities of individual events are multiplied, should be used with caution. For non-null events A and B, where $P(A\cap B)$ is required, it is surely better to proceed via careful determination of the product $P(A)P(B|A)$; if this product happens to equal (give the same numerical result as) $P(A)P(B)$, then, we can say in hindsight that independence is a justifiable assumption for the events A and B.

Where possible, try to visualise independence as a *consequence* of the multiplication rule rather than an *assumption* to make the multiplication rule easier to calculate. However, independence often creates a needed simplification of reality for entire subject areas in probability. (See, for example, problems in *reliability* discussed in the next section.)

Mutual independence

The definition of independence for two events generalises to the consideration of more than two events.

Definition 3.14 *The events $A_1, A_2, A_3, \ldots, A_n$ are termed* **pairwise independent** *if each and every pair of these events satisfies Definition 3.13. The events $A_1, A_2, A_3, \ldots, A_n$ are termed* **mutually independent** *if all subcollections $A_{i_1}, A_{i_2}, A_{i_3}, \ldots, A_{i_k}$, of these events satisfy*

$$P(A_{i_1}\cap A_{i_2}\cap A_{i_3}\cap\ldots\cap A_{i_k}) = P(A_{i_1})P(A_{i_2})P(A_{i_3})\ldots P(A_{i_k}).$$

□

The following example shows the necessity for this definition. Indeed, three events A, B and C, considered in pairs, may satisfy Equation 3.7, and yet not satisfy the 'obvious independence criterion' $P(A\cap B\cap C) = P(A)P(B)P(C)$.

Example 3.16

Pairwise independence does not imply mutual independence

Suppose that a particular building has three floors and that the night-watchman has four keys on his key-ring: key 1 opens the doors on floor 1 only, key 2 opens the doors on floor 2 only, key 3 opens the doors on floor 3 only, and key 4 is the master-key which opens all doors in the building. If one of the four keys is randomly selected, let A_i denote the event that the selected key opens the doors on floor i for $i = 1, 2, 3$.

We show that any pair of the events A_1, A_2, A_3 is independent, but these three events are not mutually independent. Indeed,

$$P(A_1) = P(A_2) = P(A_3) = \frac{2}{4}$$

since two of the four equally likely keys will work for each floor. However, only the master-key will work on two different floors, so that for $i \neq j$,

$$P(A_i \cap A_j) = \frac{1}{4}.$$

Therefore, for $i \neq j$, $P(A_i \cap A_j) = P(A_i)P(A_j)$ and so the events A_1, A_2, A_3 are pairwise independent. Finally, $P(A_1 \cap A_2 \cap A_3) = \frac{1}{4} \neq \frac{1}{8} = P(A_1)P(A_2)P(A_3)$, so that the three events are not mutually independent.

Sampling with/without replacement

Let us return to the capture/recapture of koalas that began in Example 3.9 with a discussion of counting techniques and continued in Example 3.12 illustrating an application of the Partition Theorem. Recall that m koalas are caught, tagged, then released back into the population. At the recapture stage, sampling occurs without replacement from the m tagged koalas 'at large' in the total population of size n.

Let T_i denote the event that the ith recaptured koala has a tag (i = 1,2). Then T_1 and T_2 are not independent since $P(T_1) = \frac{m}{n}$ and, by Example 3.12, $P(T_2) = \frac{m}{n}$ as well. But

$$P(T_1 \cap T_2) = P(T_1)P(T_2|T_1) = \left(\frac{m}{n}\right)\left(\frac{m-1}{n-1}\right) \neq \left(\frac{m}{n}\right)\left(\frac{m}{n}\right) = P(T_1)P(T_2).$$

Therefore T_1 and T_2 are not independent. The lack of independence is caused by the fact that the 'tagged to untagged' ratio m:n changes when the first koala sampled is not returned to the population before the second koala is sampled. If we replenish the population at each drawing, then

$$P(T_2|T_1) = \frac{m}{n} = P(T_2)$$

and T_1 and T_2 are independent. Therefore, sampling with replacement causes the results of successive drawings to be independent; sampling without replacement causes the results of successive drawings to be dependent.

Reliability

Suppose that the components of an electronic system either operate or fail to operate. We will consider the events

$$\begin{aligned} A_1 &= \{\text{ Component 1 is operational }\} \\ A_2 &= \{\text{ Component 2 is operational }\} \\ &\vdots \\ A_n &= \{\text{ Component } n \text{ is operational }\} \end{aligned}$$

to be mutually independent.

When the components of an electronic system are placed in 'series', then the **series system** will fail if and only if any one of its components fails.

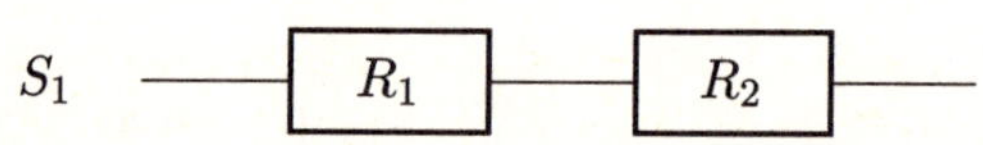

This is illustrated in the above diagram where two components are connected in series. Following Trivedi (1982) we define the **reliability** of Component i by $R_i = P(A_i)$. Then the reliability of system S_1 is given by

$$\begin{aligned} R_{S_1} &= P(S_1 \text{ is operational}) \\ &= P(\text{Component 1 is operational and Component 2 is operational}) \\ &= P(A_1 \cap A_2) \\ &= P(A_1)P(A_2) \qquad \text{(independence)} \\ &= R_1 R_2. \end{aligned}$$

This is called the **product law of reliabilities**. When Components 1 and 2 of a system are placed in 'parallel', then the **parallel system** will fail if and only if either Component 1 or Component 2 fails (or both components fail). This two-component system, S_2, may be represented schematically by:

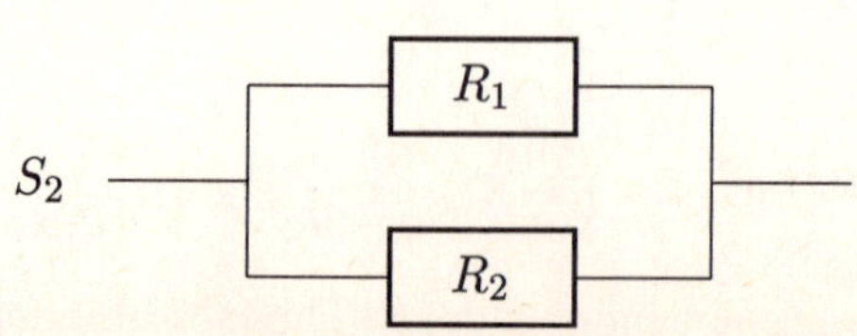

The reliability of S_2 is determined by

$$\begin{aligned}
R_{S_2} &= P(S_2 \text{ is operational}) \\
&= P(\text{Component 1 is operational or Component 2 is operational}) \\
&= P(A_1 \cup A_2) \\
&= P(A_1) + P(A_2) - P(A_1 \cap A_2) \\
&= P(A_1) + P(A_2) - P(A_1)P(A_2) \\
&= 1 - [1 - P(A_1) - P(A_2) + P(A_1)P(A_2)] \\
&= 1 - [1 - P(A_1)][1 - P(A_2)] \\
&= 1 - [1 - R_1][1 - R_2]
\end{aligned}$$

so indicating the **product law of unreliabilities**:

$$1 - R_{S_2} = [1 - R_1][1 - R_2].$$

The reliability of more complex systems follows similarly, from a reliability (or unreliability) product of 'series' and 'parallel' connections.

Example 3.17
Combined series and parallel systems

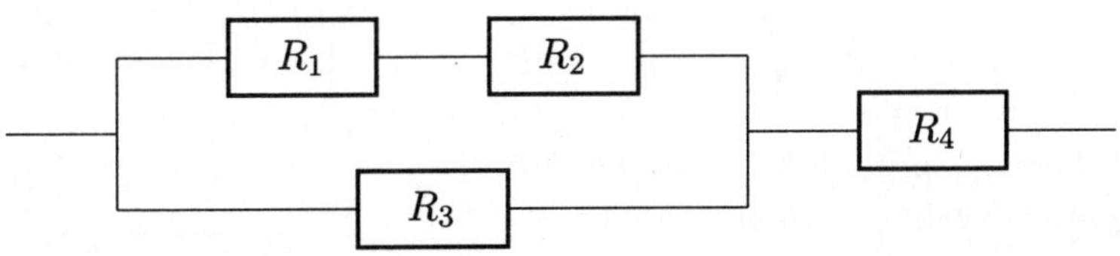

Worked example:
Determine the reliability of the system illustrated if four components in the system have reliabilities $R_1 = 0.9$, $R_2 = R_3 = R_4 = 0.95$.
Let S_1 (or System 1) consist of Components 1 and 2; S_2 of Components 1,2 and 3. Then

$$\begin{aligned}
R_S &= R_{S_2} R_4 \\
&= [1 - (1 - R_3)(1 - R_{S_1})]R_4 \\
&= [1 - (1 - R_3)(1 - R_1 R_2)]R_4 \\
&= [1 - (1 - 0.95)(1 - [0.9][0.95])]\, 0.95 \\
&= 0.943 \text{ (3 dp)}.
\end{aligned}$$

Note that this reliability is larger than if the four components had been placed in series for which

$$R_S = (0.9)(0.95)^3 = 0.772 \text{ (2 dp)}.$$

Placing components in parallel increases reliability compared with the same components being placed in series.

EXERCISES

3.51 Let A and B be events in a sample space $\mathcal{S}$ such that $P(A) = \frac{1}{3}$ and $P(A \cup B) = \frac{1}{2}$. Suppose that A and B are independent.

(a) Determine $P(B)$.

(b) Use the fact that A and $\overline{B}$ are independent to determine $P(A \cap \overline{B})$.

3.52 The two fire engines in a small town operate independently. (That is, the availability of Engine 1 is independent of the availability of Engine 2.) The probability that a specific engine is available when needed is 0.99. Determine the probability that neither engine is available when needed.

3.53 A player throws darts at a target. On each throw, independently of other throws, he hits the bulls-eye with probability 0.05.

What is the smallest number of throws needed to ensure that the player has at least a 50:50 chance of hitting the bulls-eye at least once?

3.54 In sampling k times with replacement from the population $\{1, 2, 3, \ldots, n\}$ determine the probability that 'n' is included at least once.

3.55 Can events be both mutually exclusive *and* independent?

3.56 Each of the three components in the circuit diagram below operates independently of the other components. Each component has the same reliability 0.99 (being the constant probability of successful operation). Determine the reliability of the system.

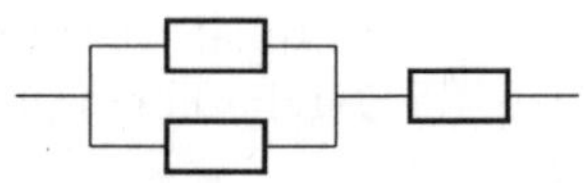

3.57 The diagram shows communication channels 1,2,3,4 between two command centres A and B. Each channel has reliability 0.9 and fails independently of other channels. The centres are able to communicate if at least one path of open channels exists.

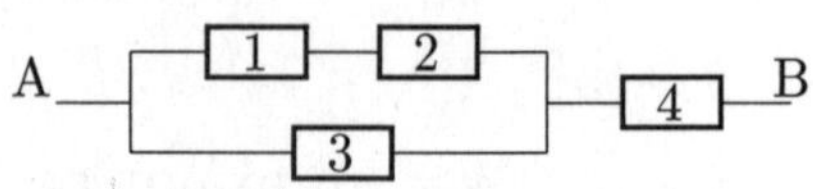

Determine the probability that communication is possible between the command centres. Discuss the same question with regard to the communication network:

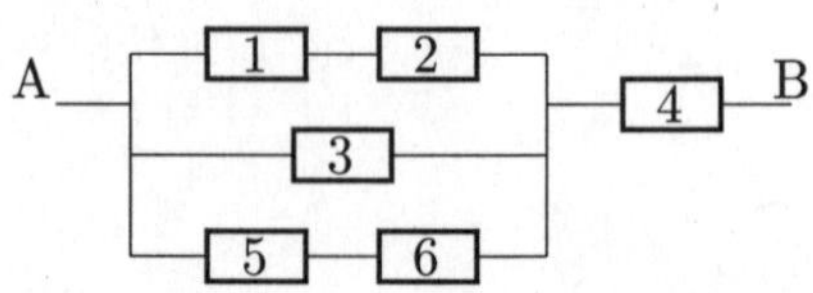

3.58 When two identical fair coins are tossed, Example 3.1 indicates that the sample space

$$\mathcal{S} = \{hh,\ ht,\ th,\ tt\}$$

contains equi-probable points. (Here, as usual, h denotes 'heads' and t denotes 'tails'.)

This serves as a model for two tosses of the *same* fair coin. For $i = 1, 2$, let H_i denote the event 'heads on toss number i'.

(a) Show that H_1 and H_2 are independent.

(b) If this same coin is tossed repeatedly, determine

$$P(H_i|H_{i-1}) \text{ for } i \geq 2.$$

(This example illustrates why a coin is described as *memoryless*.)

3.59 (a) Show that if the three events A_1, A_2, and A_3 are mutually independent then

$$P(A_1 \cup A_2 \cup A_3) = 1 - \prod_{i=1}^{3}[1 - P(A_i)].$$

(b) Three oil wells are drilled in a certain region. The events $A_i = \{$ oil struck at drill site i $\}$ for $i = 1, 2, 3$, are assumed to be mutually independent, each with a 60% chance of occurring. Determine the probability that oil is struck.

3.60 Three people A, B and C take turns (in the same order) to toss a fair coin starting with A in Round 1. The first to toss a 'head' wins the game. It is of interest to determine their respective chances of winning.

(a) Show that the event {A wins} may be written as the union of the outcomes

A_1
$A_2 \cap \overline{C_1} \cap \overline{B_1} \cap \overline{A_1}$
$A_3 \cap \overline{C_2} \cap \overline{B_2} \cap \overline{A_2} \cap \overline{C_1} \cap \overline{B_1} \cap \overline{A_1}$
$\ldots$

where A_i, for example, is notation for A winning at Round i (similarly for B_i and C_i).

(b) Use independence between each toss to show that $P(\text{A wins}) = \frac{4}{7}$.

(c) Use similar methods to determine the probabilities for each of B, C winning.

3.61 In a sample space $\mathcal{S}$, show that if A and B are independent, then A and $\overline{B}$ are independent.

CHAPTER FOUR

Random Variables

In this chapter you will find

4.1 RANDOM VARIABLES AND CUMULATIVE DISTRIBUTION FUNCTIONS

From the discussion in Section 1.1 concerning data comes the statement: 'In obtaining data we are observing the value of a **characteristic** or **variable**.' This is a first indication that an observed datum is one realisation from many (perhaps even uncountably many) potential possibilities. The datum may come from an experiment where, for example, a length is measured, or a value recorded from an apparatus, or a randomly chosen individual from a population has his/her opinion taken.

If our experimental measurements were recorded at another time or another place we would expect different results. Experimental results are not fixed steadfast; we do not know them beforehand—they are subject to variation. To clarify this variation we must step back from our data and question these other potential possibilities that might have been. We seek a clearer understanding of the **random variables** that generate data.

Random variables

Since the sample space $\mathcal{S}$ is a list of possible outcomes of the experiment under consideration, $\mathcal{S}$ is crucial to the understanding of the concept of 'random variable'. In the following definition we use the sample space to link the *abstract* notion of 'variable' to the *concrete* notion of 'data'.

Definition 4.1 *A* **random variable** X *is a rule which associates a real number with each point in the sample space* $\mathcal{S}$ *of an experiment.*

□

Paraphrasing this definition mathematically, X is a random variable if, for the sample space $\mathcal{S}$ of an experiment, $X : \mathcal{S} \to (-\infty, \infty)$. Since X is a function, according to the definition, for $\omega \in \mathcal{S}$, $X(\omega)$ is a real number. That is, $X(\omega) = x$ if x is an observed outcome of X. This is consistent with the standard practice of using upper case italic letters for random variables (for example, X) and lower case italic letters for the values that they may assume (for example, x).

For example, suppose that there are h dwellings (houses, units, apartments, etc.) in a suburb; we list them as $\omega_1, \omega_2, \omega_3, \ldots, \omega_h$. A dwelling is randomly selected by a market research team monitoring television ratings. This random selection constitutes the experiment, giving the sample space $\mathcal{S} = \{\omega_1, \omega_2, \omega_3, \ldots, \omega_h\}$. A monitor is connected to the TV set in the chosen residence to record automatically the time, X, that the television is tuned to Channel 10, say, on a particular day. This is an example of a random variable, and we would write $X(\omega_i) = 6.25$ if the family members at residence ω_i spent six hours and fifteen minutes watching Channel 10 on the day in question. Writing $X(\omega_i) = 6.25$ allows us to see X behaving explicitly as a function acting on the sample point ω_i giving an observed value 6.25 for the random variable. 6.25 constitutes a datum. A data set is built up from many records from many families over many days. Because the outcome of the random variable may be any value in a given range, we call X a **continuous random variable**. The market researchers will very likely wish to know other information about the selected dwelling. For example, we may define Y to be the number of people living at the dwelling. As before, we may write $Y(\omega_i) = 4$ if four people live at dwelling ω_i. This random variable is intuitively different from the random variable X; in fact Y may assume only the specified values 1, 2, 3, ... Such a random variable is termed a **discrete random variable** since it may assume at most countably many outcomes. Discrete random variables produce discrete data.

Example 4.1
Random variables from coin tosses

Experiment: Toss a single coin.
Sample space: $\mathcal{S} = \{h, t\}$
Random variable: X, the number of heads obtained, may be written as

$$X(h) = 1$$
$$X(t) = 0.$$

X is clearly seen to be acting on the sample space. If we toss two coins—
Experiment: Toss two identical coins.
Sample space: $\mathcal{S} = \{hh, ht, th, tt\}$.
Random variable: Y, the number of heads obtained, may be written as

$$Y(hh) = 2$$
$$Y(ht) = 1$$
$$Y(th) = 1$$
$$Y(tt) = 0.$$

Experiment: Toss a single coin repeatedly until the first 'h' is observed.
Sample space: $\mathcal{S} = \{h, th, tth, ttth, tttth, \ldots\}$
Random variable: Z, the number of tosses up till and including the first head, may be written as

$$Z(h) = 1$$
$$Z(th) = 2$$
$$Z(tth) = 3$$
$$Z(ttth) = 4$$
$$\vdots$$

The probability mass function for discrete random variables

Often the notation $X(\omega) = x$ is shortened to

$\{X = x\}$ by which we mean the **event** $\{\omega : \omega \in \mathcal{S} \text{ and } X(\omega) = x\}$

which, written in this form, is clearly a subset of $\mathcal{S}$. When such events are specified, we may define probabilities on them in the manner described in Chapter 3. We write $P(X = x)$ to represent the probability of the event $\{\omega : \omega \in \mathcal{S} \text{ and } X(\omega) = x\}$. In essence, this is the probability that the random variable X assumes the value x. For example, from the sample space $\mathcal{S} = \{hh, ht, th, tt\}$ of Example 4.1, where the random variable Y represents the number of heads in tossing two identical coins:

$$\begin{aligned} P(Y=0) &= P(\{\omega : \omega \in \mathcal{S} \text{ and } Y(\omega) = 0\}) \\ &= P(\{tt\}) \\ &= \tfrac{1}{4} \end{aligned}$$

$$\begin{aligned} P(Y=1) &= P(\{\omega : \omega \in \mathcal{S} \text{ and } Y(\omega) = 1\}) \\ &= P(\{ht, th\}) \\ &= \tfrac{1}{2} \end{aligned}$$

$$\begin{aligned} P(Y=2) &= P(\{hh\}) \\ &= \tfrac{1}{4}. \end{aligned}$$

Together, these probabilities show the distribution of probability across the sample space relevant to the random variable X. This provides motivation for the following definition.

Definition 4.2 *If X is a discrete random variable, then the function, p_X, defined on the outcomes of X by*

$$p_X(x) = P(X = x),$$

is called the **probability mass function** *(pmf) of the discrete random variable X. If X has outcomes $x_1, x_2, x_3, \ldots$, then for $p_i = P(X = x_i)$, a list such as*

x	x_1	x_2	x_3	$\cdots$
$p_X(x)$	p_1	p_2	p_3	$\cdots$

of outcomes of X together with the probabilities corresponding to those outcomes, is called the **probability distribution** *of the discrete random variable X.*

□

The probability distribution for the random variable Y, the total number of heads observed when two identical coins are tossed, is

y	0	1	2
$p_Y(y)$	$\frac{1}{4}$	$\frac{1}{2}$	$\frac{1}{4}$

It is very simple to draw a graph of this function by simply plotting the points $(y, p_Y(y))$ for $y = 0, 1, 2$. The graph shows quite dramatically where the probabilities are *massed* and hence the name *probability mass function.* Notice that the domain of p_X is the set of possible outcomes for the discrete random variable X.

Graph of probability mass function for the number of heads when two identical coins are tossed

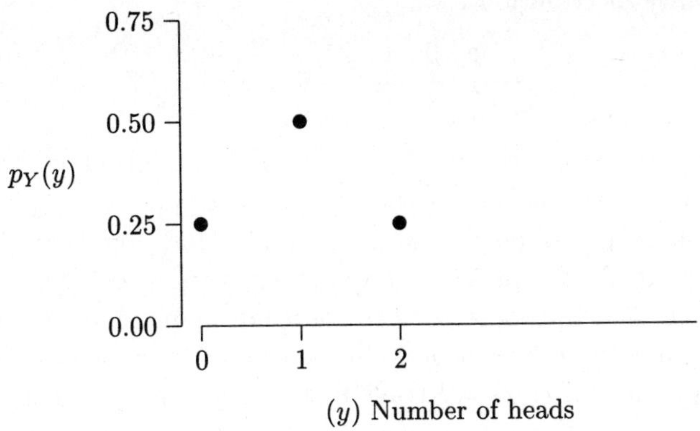

Probability histogram for the number of heads when two identical coins are tossed

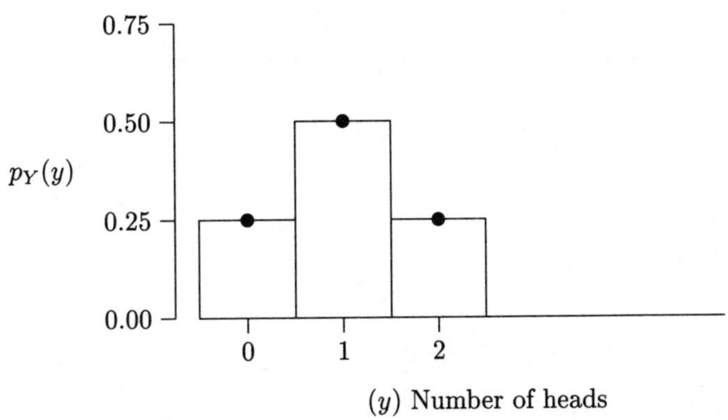

Many authors avoid the functional representation of the graph of p_X and simply offer a histogram equivalent where the histogram's rectangles are centred at the outcomes of X along the bottom axis and have heights given by the corresponding probabilities. Such a histogram is called a **probability histogram**. The main use for such a histogram is to give an immediate impression of the shape of the probability distribution.

Both the graph of the probability mass function and the corresponding probability histogram provide insight into the behaviour of the discrete random variable. All probability mass functions have countable domains and range [0, 1] (since they represent probabilities). Further, since $P(\mathcal{S}) = 1$, it follows that for a discrete random variable X with outcomes $x_1, x_2, x_3, \ldots$,

$$\sum_{\text{all } i} p_X(x_i) = \sum_{\text{all } x} p_X(x) = 1.$$

For the probability distribution

y	0	1	2
$p_Y(y)$	$\frac{1}{4}$	$\frac{1}{2}$	$\frac{1}{4}$

of the random variable Y, the total number of heads observed when two identical coins are tossed, it is clear that $\sum_{\text{all } y} p_Y(y) = \frac{1}{4} + \frac{1}{2} + \frac{1}{4} = 1$.

The same ideas apply when the random variable has countably infinitely many sample points. If Z represents the number of tosses required until the first 'head' is obtained in repeated tosses of a fair coin, then the first head may occur: on the first toss with probability $P(Z = 1) = P(\text{heads}) = \frac{1}{2}$; on the second toss with $P(Z = 2) = P(\text{tails first toss and heads second toss}) = (\frac{1}{2})(\frac{1}{2}) = \frac{1}{4}$; on the third toss with $P(Z = 3) = (\frac{1}{2})^3 = \frac{1}{8}$;... If we collect these probabilities together in the same table, then Z has probability distribution

z	1	2	3	4	...
$p_Z(z)$	$\frac{1}{2}$	$\frac{1}{4}$	$\frac{1}{8}$	$\frac{1}{16}$	...

for which the probabilities add in the geometric series

$$\sum_{\text{all } z} p_Z(z) = \tfrac{1}{2} + \tfrac{1}{4} + \tfrac{1}{8} + \tfrac{1}{16} + \ldots = \tfrac{1}{2} + (\tfrac{1}{2})^2 + (\tfrac{1}{2})^3 + (\tfrac{1}{2})^4 + \ldots = \frac{\frac{1}{2}}{1 - \frac{1}{2}} = 1.$$

In fact, these properties completely characterise discrete distributions. Given any function p defined on the countable set $\{x_1, x_2, x_3, \ldots\}$ satisfying:

1. for each $i = 1, 2, 3, \ldots$, $0 \leq p(x_i) \leq 1$; and
2. $\sum_{\text{all } i} p(x_i) = 1$;

then $p = p_X$ for some random variable X. Therefore pmf's may be used to *define* types of random variables. The geometric random variable is one such example.

Definition 4.3 *The discrete random variable X is said to have a* **geometric distribution,** *and we write $X \sim Geom(p)$, if X has probability mass function*

$$p_X(x) = pq^{x-1}, \quad x = 1, 2, 3, 4, \ldots$$

for some constants p and q satisfying $0 < p < 1$ and $q = 1 - p$.

□

We have seen that the discrete random variable Z representing 'the number of tosses required until the first head is obtained in repeated tosses of a fair coin' has probability distribution

z	1	2	3	4	...
$p_Z(z)$	$\frac{1}{2}$	$\frac{1}{4}$	$\frac{1}{8}$	$\frac{1}{16}$	...

and from the second line of this probability distribution, the pmf for Z may be expressed in powers of $\frac{1}{2}$. Therefore, an equivalent expression in terms of Definition 4.3, uses $p = q = \frac{1}{2}$, and

$$p_Z(z) = P(Z = z) = (\tfrac{1}{2})^z, \; z = 1, 2, 3, 4, \ldots.$$

Hence $Z \sim Geom(\frac{1}{2})$.

Suppose now that the coin was not 'fair', but that we believe the probability assessment $P(\text{heads}) = \frac{2}{3}$ at any particular toss. Then, the random variable W, representing 'the number of tosses required until the first head is obtained in repeated tosses of this biased coin', has probability distribution specified by the pmf

$$p_W(w) = P(W = w) = (\frac{2}{3})(\frac{1}{3})^{w-1}, \; w = 1, 2, 3, 4, \ldots.$$

Hence, for this case, $W \sim Geom(\frac{2}{3})$.

Clearly, there are infinitely many $Geom(p)$ distributions; every new choice of p creates a new pmf and therefore a new probability distribution. We call p a **parameter** of the geometric probability distribution. In realistic situations, the parameter p is usually unknown. One of the major tasks that statisticians face is the *estimation* of such parameters by using available sample data. Such ideas will be examined in detail in later chapters.

Example 4.2
Defect-free runs

On a production line a defect-free 'run' of items occurs between adjacent instances of defective items. Suppose that we represent a defective item by the symbol d and a non-defective item by the symbol n. Assume that, as events, defectives and non-defectives occur independently at any point in the production.

Suppose that we have just observed a defective item, so that the defect-free run which now occurs has sample points given by the table:

Sample point	**Defect-free run length**
d	0
nd	1
nnd	2
nnnd	3
nnnnd	4
nnnnnd	5
nnnnnnd	6
...	...

In this example we are examining the waiting time between occurrences of defectives on the production line. Let X denote 'number of sampled items up to and including the next defective'. If we set $P(\text{defective}) = p$ and $P(\text{non-defective}) = 1 - p = q$, then we have through the assumed independence:

$$\begin{aligned} P(X=1) &= p \\ P(X=2) &= qp \\ P(X=3) &= q^2 p \\ P(X=4) &= q^3 p \\ \vdots &= \vdots \\ P(X=x) &= q^{x-1}p. \end{aligned}$$

Hence $X \sim Geom(p)$. From this we may determine the probability that more than k non-defective items occur before a defective re-occurs is given by

$$\begin{aligned} P(X>k) &= P(X=k+1)+P(X=k+2)+P(X=k+3)+\dots \\ &= q^k p + q^{k+1}p + q^{k+2}p + \dots \\ &= q^k p(1+q+q^2+q^3+\dots) \\ &= q^k p(\frac{1}{1-q}) \\ &= q^k. \end{aligned}$$

Here we have used the standard result for the sum to infinity of a geometric series: for $|r| < 1$,

$$1 + r + r^2 + r^3 + r^4 + \ldots = \frac{1}{1-r}.$$

The cumulative distribution function

Following the idea of cumulative frequencies introduced in Chapter 2, we give the definition of the cumulative distribution function; this definition applies to random variables that are either discrete or continuous.

Definition 4.4 *The* **cumulative distribution function** *(cdf) of the random variable X, denoted by F_X, is given by*

$$F_X(x) = P(X \leq x)$$

for all $-\infty < x < \infty$.

□

Cumulative distribution functions provide an assessment of the fraction of the distribution lying below a certain value. For example, if the random variable W represents the length of life (days) of a heart transplant patient after surgery, then $P(W \leq 7)$ represents the probability that the patient dies in the first week after surgery.

The cdf of a random variable $X \sim Geom(p)$, by Example 4.2, is very easily determined. Since for any $k \geq 1$, $P(X > k) = q^k$, we find via the complementary event that for x a positive integer,

$$F_X(x) = P(X \leq x) = 1 - P(X > x) = 1 - q^x.$$

Hence,

$$F_X(x) = \begin{cases} 0 & \text{if } x \leq 0; \\ 1 - q^{[x]} & \text{if } x > 0, \end{cases}$$

defines F_X for any real number x with integer part $[x]$. (Recall that by the **integer part** of a real number x, we mean the greatest integer less than or equal to x.)

In Example 4.3 the cumulative distribution function is calculated for a continuous random variable.

Example 4.3

Shear stress on a bar at a random point

Suppose that a force impacts on a horizontal iron bar of length 1 metre at a randomly chosen point a distance X metres from the hinge at the left-hand end.

Worked example:

Calculate and sketch the cumulative distribution function of X.

Shear force on an iron bar X metres from the hinge

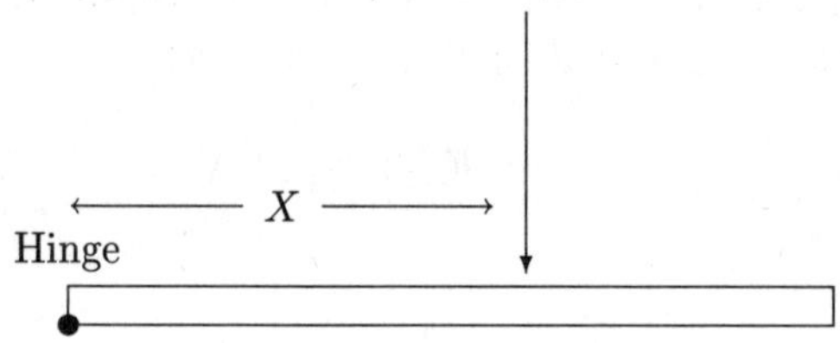

In this example, if the origin is set at the hinge end of the bar, then the continuous sample space associated with X is $\mathcal{S} = [0, 1]$. The event $\{X \leq x\}$ corresponds to the subinterval $[0, x)$ of $\mathcal{S}$. According to the discussion of probability in continuous sample spaces in Chapter 3, the probability of any subinterval of $\mathcal{S}$ is directly proportional to the length of the subinterval. Therefore, if $0 \leq x \leq 1$, $P(X \leq x) = \frac{x-0}{1-0} = x$, and $P(X \leq x) = 1$ if $x > 1$. Combining these separate results gives

$$F_X(x) = \begin{cases} 0 & \text{if } x < 0; \\ x & \text{if } 0 \leq x \leq 1; \\ 1 & \text{if } x > 1. \end{cases}$$

Cumulative distribution function

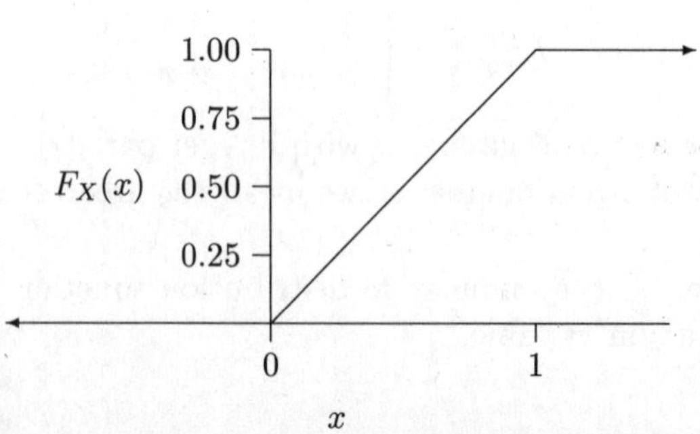

Cumulative distribution functions enjoy certain properties because of the manner in which they are defined in terms of probabilities. Notice that for a given random variable X, the events $\{X \leq a\}$ and $\{a < X \leq b\}$ have disjoint union $\{X \leq b\}$, so guaranteeing that

$$P(X \leq a) + P(a < X \leq b) = P(X \leq b).$$

This proves the following basic theorem for cumulative distribution functions:

Theorem 4.1 *Suppose that X is a random variable with cumulative distribution function F_X. Then for fixed values of a and b with $a < b$,*

$$P(a < X \leq b) = F_X(b) - F_X(a).$$

■

A most important property (on which the theory of continuous random variables relies heavily) is that cdf's are **non-decreasing functions**: for this it needs to be established, for an arbitrary random variable X and fixed values a and b, that if $a < b$ then $F_X(a) \leq F_X(b)$. This is immediately clear, since when $a < b$, by Theorem 4.1,

$$0 \leq P(a < X \leq b) = F_X(b) - F_X(a)$$

giving $F_X(a) \leq F_X(b)$. This establishes the following theorem.

Theorem 4.2 *Given a random variable X with cumulative distribution function F_X, then $F_X(x)$ is an increasing function of x on $(-\infty, \infty)$.*

■

The use of the term 'increasing' in this theorem does not refer to 'strictly increasing' but rather to 'non-decreasing'; indeed the cdf's of discrete random variables have piece-wise linear 'flat' sections. Note that if we use the notation

$$F_X(-\infty) = \lim_{r \to -\infty} F_X(r) \quad \text{and} \quad F_X(\infty) = \lim_{r \to \infty} F_X(r),$$

it follows that $F_X(-\infty) = 0$ and as x increases, $F_X(x)$ increases on $(-\infty, \infty)$ until $F_X(\infty) = 1$.

Example 4.4

Blowing up balloons

Suppose that we blow up a spherical balloon in such a way that we stop when a volume, V in cubic metres, is achieved, where V is randomly chosen from the interval $[0, \frac{4\pi}{3}]$. A volume of $\frac{4\pi}{3}$ cubic metres corresponds to a balloon radius of 1 metre since $V = \frac{4\pi}{3}X^3$, where X is the balloon's radius.

Let us determine the cdf of X. Following the ideas presented in Example 4.3, for $0 < x < 1$,

$$\begin{aligned} F_X(x) &= P(X \le x) \\ &= P(V \le \frac{4\pi}{3}x^3) \\ &= \frac{\frac{4\pi}{3}x^3 - 0}{\frac{4\pi}{3} - 0} \\ &= x^3. \end{aligned}$$

We may therefore write the cdf of X as

$$F_X(x) = \begin{cases} 0 & \text{if } x < 0; \\ x^3 & \text{if } 0 \le x \le 1; \\ 1 & \text{if } x > 1. \end{cases}$$

Notice that F_X possesses all the appropriate properties of a cdf: since the cubic function is an increasing function, it follows that F_X is an increasing function defined on the whole real line $(-\infty, \infty)$ with $F_X(-\infty) = 0$, and $F_X(\infty) = 1$.

Worked example:
Determine the probability that the radius of the balloon lies between 0.4 and 0.6 metres.
By Theorem 4.1,

$$P(0.4 < X \le 0.6) = F_X(0.6) - F_X(0.4) = (0.6)^3 - (0.4)^3 = 0.152.$$

Jump discontinuities in the graphs of discrete cdf's

The remaining discussion in this section concerns how to sketch the cdf for a discrete random variable X. Care needs to be taken to ensure that the graph is drawn with domain the entire interval $(-\infty, \infty)$. No real numbers should be omitted from the domain, since cdf's are defined for *all* real numbers. The task is surprisingly more difficult than for the case of continuous cdf's already considered in Example 4.3. For simple illustration, we return to the discrete random variable Y, the total number of heads observed when two identical coins are tossed. Y has probability distribution:

y	0	1	2
$p_Y(y)$	$\frac{1}{4}$	$\frac{1}{2}$	$\frac{1}{4}$

To construct $F_Y(y)$, we first notice that there is no probability massed below 0. This indicates that for all $y < 0$ we have $P(Y \le y) = 0$. The first probability mass encountered as y increases is at $y = 0$, so that $P(Y \leq -0.0000001) = 0$, but $P(Y \leq 0) = 0.25$ and $P(Y \leq 0.0000001) = 0.25$. Mathematically, we say that the graph of F_Y has a **jump discontinuity** at the point $y = 0$ and is **right-continuous**. We indicate this in the graph of F_Y by using a circle for excluded points; the circle is shaded to indicate included points. Immediately above $y = 0$, the graph then remains horizontal until the next probability mass is encountered (at $y = 1$) where there is another jump discontinuity. Finally, $F_Y(y)$ reaches the maximum value of 1 at *any* point beyond the largest possible outcome for Y.

Graph of cumulative distribution function for the number of heads when two identical coins are tossed

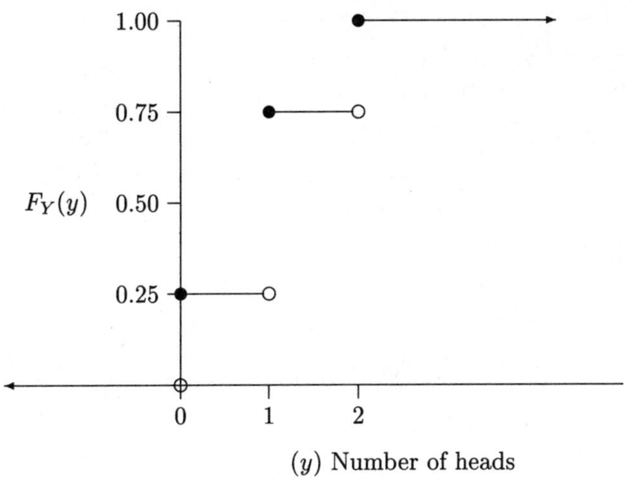

EXERCISES

4.1 A car pooling study shows that the number of passengers, X, in a car (excluding the driver) is likely to assume the values 0, 1, 2, 3, and 4 with probabilities given by the table:

x	0	1	2	3	4
$P(X = x)$	0.7	0.1	0.1	0.05	0.05

(a) Determine $P(X > 2)$.

(b) Determine $P(X \geq 2)$.

(c) What is the probability that a car will have no passengers?

(d) Determine the smallest value of k so that $P(X < k) > 0.85$.

(e) Evaluate $P(X \leq w)$ for

$$w = -12,\ 0,\ 0.5,\ 2.4,\ 103.$$

4.2 A Wellington car sales firm knows from past experience that they may sell up to 4 cars in a day. Consider the random variable, X, denoting the number of car sales in a day. The firm establishes the following probability mass function for X:

$$P(X = x) = \frac{c - x}{20}, \quad x = 0, 1, 2, 3, 4,$$

where c is a constant.

(a) Why should X be regarded as a discrete random variable rather than a continuous random variable?

(b) Find c.

(c) Write out the probability distribution of X in tabular form.

(d) Determine $P(1 < X \leq 3)$.

(e) Determine the probability that at least one car is sold in the day.

(f) Explain why the probability mass function

$$P(X = x) = \frac{k + x}{20},$$

$$x = 0, 1, 2, 3, 4,$$

where k is a constant, would be unsuitable to describe the likelihood of x car sales, $x = 0, 1, 2, 3$ and 4, during a period of economic recession when it is extremely difficult to sell cars.

4.3 A builder orders a shipment of brick pavers to construct a pathway. The shipment arrives in 5 lots with each lot containing a nominal count of 100 bricks. The random variable X, the number of broken bricks per lot, is estimated by the suppliers to have the following probability distribution:

x	0	1	2	3	4	≥ 5
$p_X(x)$	0.7	0.1	0.05	0.05	0.03	α

(a) Determine the value of α.

(b) Determine the probability that a given lot contains 3 or more broken brick pavers.

(c) Determine the probability that a given lot contains broken brick pavers.

(d) Assuming that the number of broken brick pavers in a given lot is independent of the number of such defectives in any other lot, determine the probability that the entire shipment of 5 lots is free of broken pavers.

(e) Use the given probability distribution to construct a probability distribution for the random variable Y, representing the number of brick pavers in a lot that are not broken.

4.4 (a) Suppose that X is the outcome when a six-sided die is tossed. Explain why X has probability mass function

x	1	2	3	4	5	6
$p_X(x)$	$\frac{1}{6}$	$\frac{1}{6}$	$\frac{1}{6}$	$\frac{1}{6}$	$\frac{1}{6}$	$\frac{1}{6}$

if the die is regarded as 'fair'.

(b) Now suppose that the die in (a) is loaded so that the outcome '6' is twice as likely as any other outcome. Give the probability mass function for X in this case.

4.5 A marine salvage firm is conducting two independent salvage operations: one operation on a shipwreck in Bass Strait and one on a shipwreck in Cook Strait. The firm rates its chances of success for the salvage operation at the Bass Strait wreck at 50%, but ascribes only a 20% chance for successful salvage at the Cook Strait wreck because of hazardous currents and weather conditions.

Construct a probability mass function for the number X of successful salvage operations. Determine the probability that at least one of the operations will be successful.

4.6 The random variable X giving the number of passengers (excluding the driver) per car in rush hour traffic has pmf

x	0	1	2	3	4
$p_X(x)$	0.7	0.1	0.1	0.05	0.05

(a) Sketch the cumulative distribution function of X.

(b) Define a new random variable Z by

$$Z = \begin{cases} 0 & \text{if} \quad X > 2; \\ 1 & \text{if} \quad X \leq 2. \end{cases}$$

Sketch a graph of the probability mass function for Z.

4.7 A Sydney commuter on the Liverpool train line believes that when the morning train is more than 1 minute late it will be Y minutes late where

$$P(Y > y) = \frac{1}{y^2} \quad \text{for } y > 1.$$

(a) Determine the cumulative distribution function $F_Y(y)$. Sketch a graph of this function.

(b) Use the cdf from (a) to determine the probability that the train will be between three and four minutes late.

(c) Use the definition of conditional probability to determine

$$P(Y > 3|Y > 2).$$

Explain in words what this probability means.

4.8 A point, X, is randomly chosen from the interval $[a, b]$.

(a) Determine the cumulative distribution function of X.

(b) Sketch a graph of the cumulative distribution function.

(c) Find a fixed value r in $[a, b]$ so that $P(X < r) = 0.5$.

4.9 After a defective has been observed on a production line, suppose that the number of items, X, sampled up to and including the next defective follows a geometric distribution with $X \sim Geom(p)$, where p is the probability of a defective item being sampled. (See Example 4.2.)

Use the definition of conditional probability to show that for fixed positive integers a and b,

$$P(X > a + b|X > a) = P(X > b).$$

Explain what this equation means in terms of the occurrence of defectives on the production line.

4.2 PROBABILITY DENSITY FUNCTIONS

The probability mass function (pmf) for a discrete random variable X is given by p_X, where $p_X(x) = P(X = x)$. For continuous random variables with outcomes occurring within a specified interval, probabilistic statements refer to intervals rather than points. Intuitively, there are uncountably many points in an interval, so that even if a minute probability were assigned to each point, then we would have no control over the probability content of the entire interval. In fact, we shall see that all fixed points have zero probability. Even so, an interval, no matter how short may have non-zero probability.

How then do we assess the likelihood that the outcome of a continuous random variable may be observed in a (possibly minute) interval? The answer is in detecting how the probability content, or probability density, of an interval changes as the interval changes.

Histograms with diminishing interval width

If a grandfather clock is left to wind down, it will stop at some point in the 24 hour cycle (for which we will use time-keeping units *hours* : *minutes* : *seconds*) between one midnight and the next. We may express the sample space for this experiment as the interval [00:00:00, 24:00:00] of length 24 hours. The probability that it stops at a *pre-specified* instant within that 24 hour period is, according to reasonable intuition, zero; we have seen already in the discussion of continuous sample spaces that the probability that it stops within a specific time interval is the relative length of the interval. In fact:

$$\begin{aligned}
&\text{P(clock stops between 12:00:00 and 13:00:00)} = \tfrac{1}{24}\\
&\text{P(clock stops between 12:00:00 and 12:01:00)} = \tfrac{1}{(24)(60)} = \tfrac{1}{1440}\\
&\text{P(clock stops between 12:00:00 and 12:00:01)} = \tfrac{1}{(24)(60)(60)} = \tfrac{1}{86400}\\
&\text{P(clock stops at 12:00:00)} = 0.
\end{aligned}$$

Clearly, as the time interval gets smaller and smaller the probability of that interval containing the stopping time reduces. Further, all 1 hour intervals have the same probability $\frac{1}{24}$ of containing the stopping time; there is no prior information to suggest that a given 1 hour interval should be accorded a higher probability assessment than any other 1 hour interval. There is nothing special about 1 hour: there is no prior information to suggest that a given interval of length t should be accorded a higher probability assessment than any other interval of length t. This means that the probability density of the random variable X = 'clock stopping time' is constant.

Relative frequency histogram
for the stopping times of a clock

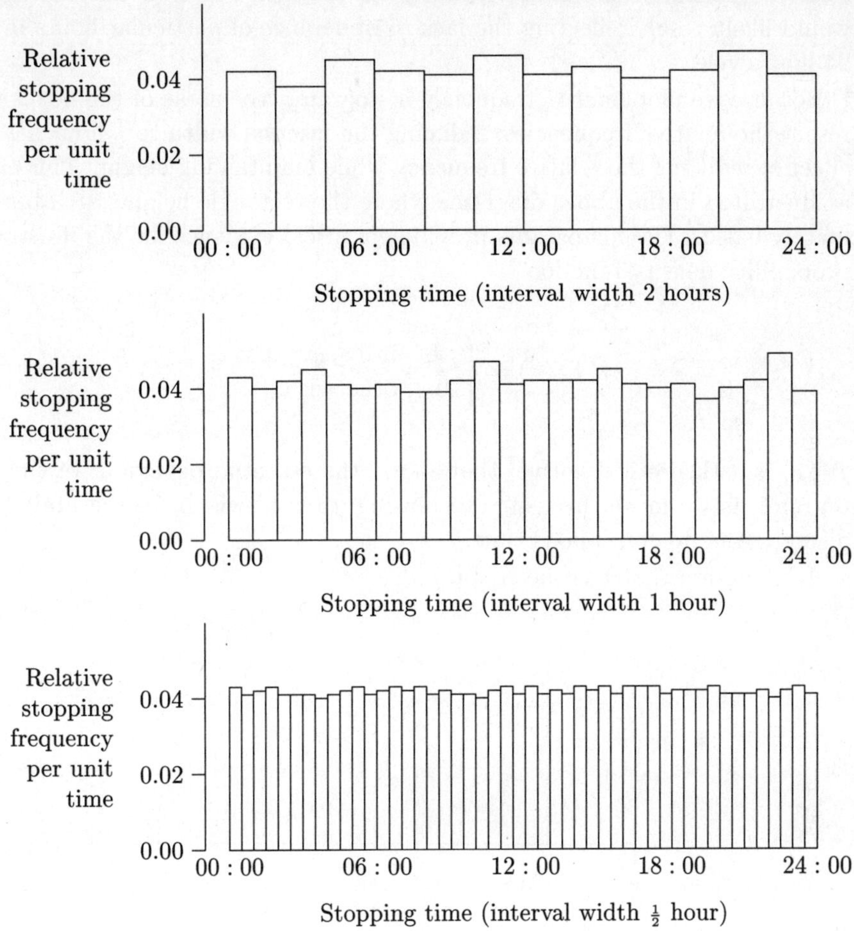

Graph of the probability density function
for the stopping times of a clock

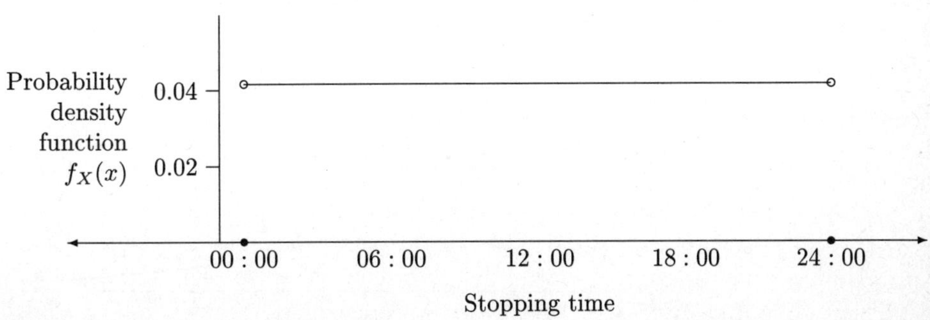

The constant probability density may be noticed empirically: if many observations of the clock's stopping time were taken, and the results plotted in a relative frequency histogram with intervals of width 1 hour, a 'flat' histogram would likely result reflecting the lack of preference of particular hours in the 24 hour cycle.

Because we plot *relative* frequency histograms, the areas of the boxes represent the relative frequencies; reducing the interval width to $\frac{1}{2}$ hour has the effect of reducing the relative frequency while maintaining height. This effect is illustrated in the above diagrams where the rectangle heights are approximately 0.042. The histograms provide empirical evidence for the illustrated probability density function

$$f_X(x) = \begin{cases} \frac{1}{24} & \text{if } 0 \leq x \leq 24; \\ 0 & \text{otherwise.} \end{cases}$$

$f_X(x)$ is a theoretical 'curve' that shows the essential behaviour of the histograms as we move, by using decreasing interval width, in the limit from discrete time to continuous time.

A histogram that has the shape,

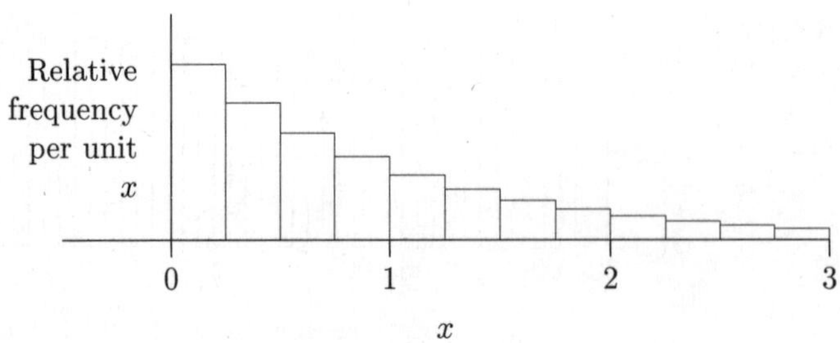

(which may occur for observations on the random variable X denoting the time till the first arrival of a telephone call through a switchboard once all lines have been opened) may be suggesting a probability density function with the shape

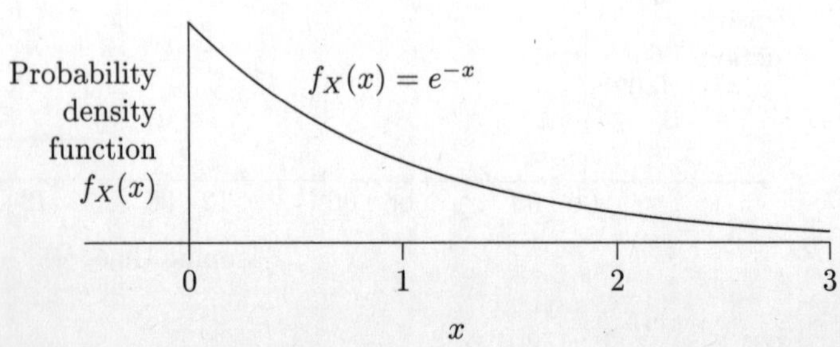

Probabilities as areas under the probability density function

The following definition both provides a working definition for probability density functions and demonstrates their role in probability calculations.

Definition 4.5 *For a continuous random variable X, a* **probability density function (pdf) of** X *is a function $f_X(x)$ of x such that for any $a \le b$,*

$$P(a \le X \le b) = \int_a^b f_X(x)dx.$$

The set of values x for which $f_X(x) > 0$ is called the **support of** X*. A specification of $f_X(x)$ together with the support of X is called the* **probability distribution of** X*.*

□

This definition shows that the probability that a continuous random variable lies in a specified interval is given by the interval's area under the probability density function. Some pdf's contain areas that are simple to evaluate (for example, the 'uniform' pdf), while some require (numerical) integration techniques to evaluate probabilities in the manner of Definition 4.5. We set up a dictionary of pdf's beginning with the simplest to define.

Definition 4.6 *For $\alpha < \beta$ we say that a continuous random variable X is* **uniformly distributed** *on the interval $[\alpha, \beta]$, and write $X \sim U[\alpha, \beta]$, if X has pdf*

$$f_X(x) = \begin{cases} \frac{1}{\beta-\alpha} & \text{if } \alpha \le x \le \beta; \\ 0 & \text{otherwise.} \end{cases}$$

□

Often α has value 0, in which case β gives the length of the support on which the pdf stands. So far we have seen two examples of uniformly distributed random variables. In Example 4.3, a force impacts on an iron bar at a random point X metres from the hinge; here $X \sim U[0, 1]$. At the start of this section, X, the clock stopping time, satisfies $X \sim U[0, 24]$. These uniform distributions have different supports. The shape of the pdf in each case is determined by the values of the **parameters** α and β; a new uniform distribution is generated each time the values of these parameters are changed.

Example 4.5
Random angles
Worked example:
If $X \sim U[-\frac{\pi}{2}, \frac{\pi}{2}]$, sketch the pdf of X and evaluate $P(|X| < 1)$.
According to Definition 4.6,

$$\begin{aligned} f_X(x) &= \begin{cases} \frac{1}{\frac{\pi}{2}-(-\frac{\pi}{2})} & \text{if } -\frac{\pi}{2} \le x \le \frac{\pi}{2}; \\ 0 & \text{otherwise} \end{cases} \\ &= \begin{cases} \frac{1}{\pi} & \text{if } -\frac{\pi}{2} \le x \le \frac{\pi}{2}; \\ 0 & \text{otherwise.} \end{cases} \end{aligned}$$

Therefore,

$$P(|X| < 1) = P(-1 < X < 1) = \int_{-1}^{1} \frac{1}{\pi} dx = \frac{x}{\pi}\Big]_{-1}^{1} = \frac{2}{\pi}.$$

Graph of the probability density function
for $X \sim U[-\frac{\pi}{2}, \frac{\pi}{2}]$

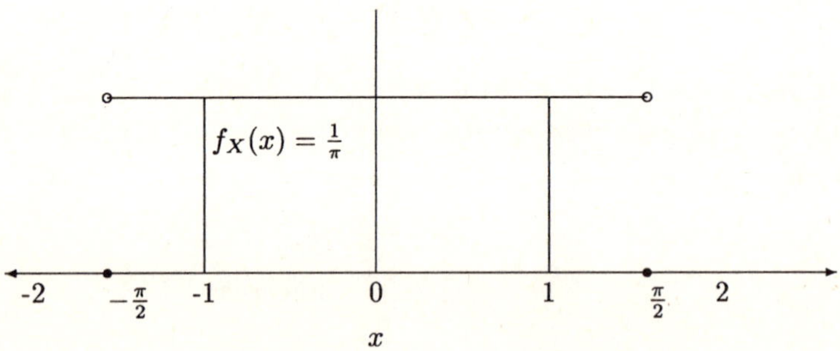

Notice that the area under this density representing $P(|X| < 1)$ is simply the area of the rectangle which has width 2 and height $\frac{1}{\pi}$.

Definition 4.7 *Let $\beta > 0$. A continuous random variable X is said to have an* **exponential distribution**, *and we write $X \sim exp(\beta)$, if X has pdf*

$$f_X(x) = \begin{cases} \frac{1}{\beta} e^{-\frac{x}{\beta}} & \text{if } x \ge 0; \\ 0 & \text{otherwise.} \end{cases}$$

□

Definition 4.7 indicates that exponential distributions have a single parameter β. We will use the exponential distribution as an example of evaluating probabilities according to Definition 4.5.

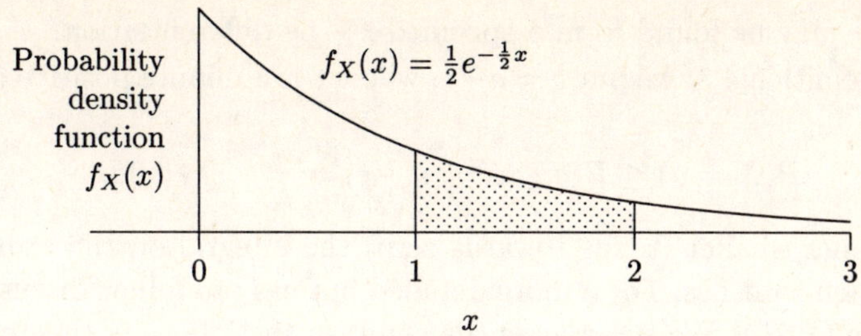

Suppose that $X \sim exp(2)$. Determine $P(1 \leq X \leq 2)$ and $P(X \geq 1)$.

$$\begin{aligned} P(1 \leq X \leq 2) &= \int_1^2 \tfrac{1}{2}e^{-\frac{1}{2}x}dx = -e^{-\frac{1}{2}x}\Big]_1^2 \\ &= -e^{-1} - (-e^{-\frac{1}{2}}) = e^{-\frac{1}{2}} - e^{-1} = 0.2387 \text{ (4 dp)}. \end{aligned}$$

Further,

$$P(X \geq 1) = \int_1^\infty \tfrac{1}{2}e^{-\frac{1}{2}x}dx = -e^{-\frac{1}{2}x}\Big]_1^\infty = 0-(-e^{-\frac{1}{2}}) = e^{-\frac{1}{2}} = 0.6065 \text{ (4 dp)}.$$

The relationship between pdf's and cdf's

For a continuous random variable X, since $P(-\infty \leq X \leq x) = P(X \leq x)$, we may link f_X and F_X through

$$F_X(x) \quad = \quad \int_{-\infty}^{x} f_X(t)dt. \tag{4.1}$$

In fact, (4.1) indicates how F_X may be calculated from f_X. If, for example, $X \sim U[0, 24]$, then

$$f_X(x) = \begin{cases} \frac{1}{24} & \text{if } 0 \leq x \leq 24; \\ 0 & \text{otherwise.} \end{cases}$$

Therefore, if $x \leq 0$, then $F_X(x) = 0$. If $0 \leq x \leq 24$, then

$$F_X(x) = \int_{-\infty}^{x} f_X(t)dt = \int_0^x \frac{1}{24}dt = \frac{t}{24}\Big]_0^x = \frac{x}{24}.$$

Finally, if $x \geq 24$, then $F_X(x) = 1$. Collectively,

$$F_X(x) = \begin{cases} 0 & \text{if } x \leq 0; \\ \frac{x}{24} & \text{if } 0 \leq x \leq 24; \\ 1 & \text{if } x \geq 1. \end{cases}$$

For an arbitrary continuous random variable X, if f_X is continuous at x, the Fundamental Theorem of Calculus implies that

$$F'_X(x) = f_X(x), \tag{4.2}$$

so that f_X may be found from a specified F_X by differentiation.

If, in Definition 4.5, we put $b = a+\epsilon$, where ϵ is a minute positive quantity, then

$$P(X = a) \leq P(a \leq X \leq a + \epsilon) = \int_a^{a+\epsilon} f_X(x)dx.$$

As ϵ becomes smaller (tends towards zero) the integral on the right-side of this equation vanishes. For a more detailed but easy to follow discussion, see Woodroofe (1975). We summarise the result in the following theorem:

Theorem 4.3 *If X is a continuous random variable and a is a constant, $P(X = a) = 0$.*

This theorem indicates that an infinitesimally thin vertical slice of zero thickness taken from anywhere under a pdf has zero area (integral). It also means that probability statements about intervals do not depend on the endpoints of the intervals. Thus for a continuous random variable X,

$$P(a < X < b) = P(a \leq X < b) = P(a < X \leq b) = P(a \leq X \leq b).$$

Our final result of this section concerns results which characterise the nature of probability density functions.

Theorem 4.4 *The pdf f_X of a continuous random variable X satisfies:*
(a) $f_X(x) \geq 0$;
(b) $\int_{-\infty}^{\infty} f_X(x)dx = 1$.

Proof: For (a), $f_X(x) = F'_X(x) \geq 0$ by (4.2) and (b) follows since

$$\int_{-\infty}^{\infty} f_X(x)dx = F_X(\infty) - F_X(-\infty) = 1 - 0 = 1.$$

■

The result is stronger than Theorem 4.4 indicates. Any function that is positive on $(-\infty, \infty)$ and integrates to 1 over $(-\infty, \infty)$—in the manner of (a) and (b) of Theorem 4.4—is the probability density function of some random variable. This idea is used to detect which functions are probability density functions and which functions are not.

Percentiles of continuous distributions

In Chapter 1 we discussed the determination of the sample percentiles x_p from a given data set $x_1, x_2, x_3, \ldots, x_n$. (See Theorem 1.2.) The notion of percentile is easily extended to continuous random variables and their probability distributions.

Definition 4.8 *Let $0 < p < 1$. The* **100pth percentile of X** *is the value ξ_p such that $F_X(\xi_p) = p$.*

□

Definition 4.8 indicates that $P(X \leq \xi_p) = p$ and $P(X > \xi_p) = 1 - p$. When $p = \frac{1}{2}$, the value $\xi_{0.5}$ divides the area under the pdf $f_X(x)$ exactly in half. We call $\xi_{0.5}$ the **median of** X, $\xi_{0.25}$ the **lower quartile** and $\xi_{0.75}$ the **upper quartile of** X.

If $X \sim U[0,1]$, then we have seen that

$$F_X(x) = \begin{cases} 0 & \text{if } x \leq 0; \\ \frac{x}{24} & \text{if } 0 \leq x \leq 24; \\ 1 & \text{if } x \geq 1. \end{cases}$$

Hence to find the 30th percentile $\xi_{0.3}$ of X, the equation $F_X(\xi_{0.3}) = 0.3$ is solved for $\xi_{0.3}$. This gives $\frac{\xi_{0.3}}{24} = 0.3$, for which $\xi_{0.3} = (24)(0.3) = 7.2$.

EXERCISES

4.10 The duration Z (hours) of my weekly run is a continuous random variable with pdf

$$f_Z(z) = \begin{cases} kz & \text{if } 0 \leq z \leq \frac{1}{k}; \\ 0 & \text{otherwise.} \end{cases}$$

(a) Find k.

(b) Determine $P(Z > \frac{1}{2})$.

(c) Determine the probability that my run will last longer than fifteen minutes.

(d) Is my run more likely to last more than one hour or less than one hour?

(e) What is the probability that my run will last *exactly* one hour?

(f) Determine the interquartile range

$$\xi_{0.75} - \xi_{0.25}$$

of Z.

(g) Sketch the cumulative distribution function of Z.

4.11 After the start of observation on a given summer evening, the time T, in minutes until the first shooting star is observed, follows an exponential distribution for which

$$P(T > t) = e^{-\frac{t}{10}},$$

where $t > 0$.

(a) Determine the probability that it takes between five and ten minutes for the first shooting star to be observed.

(b) Determine the probability density function of T.

4.12 (a) The life, X (in hours), of an electronic component is exponentially distributed with $X \sim exp(100)$. Determine the probability that a component will fail before 75 hours?

(b) A system is made up of two components of the type outlined in (a) under independent operation. The whole system is operational if and only if both components are operating. What is the probability that the system will fail before 75 hours?

4.13 The lifespan X, in hours, of a graphics card in a PC (personal computer) is thought to have a pdf inversely proportional to the cube of x:

$$f_X(x) = \begin{cases} \frac{c}{x^3} & \text{if } 4000 \le x \le 5000; \\ 0 & \text{otherwise,} \end{cases}$$

where c is a constant.

(a) Determine the value of c that ensures f_X will be a pdf.

(b) Determine the probability that the graphics card will last longer than 4500 hours after installation.

(c) What guarantee period should the manufacturers offer if they wish to replace (under guarantee) at most 1% of graphics cards sold?

4.14 (a) Verify that the function f defined by the equation

$$f(x) = \frac{1}{\pi}\frac{1}{1+x^2}, \quad -\infty < x < \infty,$$

is a probability density function for some random variable X. Sketch the graph of f.
[*Note:* $\frac{d}{dx}(tan^{-1}x) = \frac{1}{1+x^2}$.]

(b) If a random variable X has pdf as in (a), determine:

i. the median of X;

ii. $P(-1 < X < 1)$;

iii. $P(|X| > 1)$.

4.15 Assume that the No 67 tram operates along Glenhuntly Road, Melbourne, every 20 minutes.

(a) What is the probability that a person arriving at a particular stop in Glenhuntly Road will have to wait for at least 15 minutes for a tram?

(b) Special signals govern the movement of trams through Melbourne's intersections. The signal

T

set amid the ordinary traffic signals shows *red* for stop and *white* for proceed. If, at a particular intersection, the signal is white for 40 seconds and then red for 85 seconds, determine the probability that a tram arriving at the intersection will not have to wait.

4.16 A stick of length 2 metres is broken at a random point. Determine the probability that the break occurs at least 2 centimetres from either end.

The problem may be generalised: Suppose that the stick is of length l metres. Determine the probability that the break occurs at least m metres (where $m < \frac{1}{2}l$) from either end.

Give the probability distribution for the point X at which the l metre stick is broken. What is the probability that the stick is broken *exactly* in half?

4.17 Suppose that $f_X(x)$ is the pdf of a continuous random variable X.

(a) If f_X satisfies the condition

$$f_X(c+u) = f_X(c-u)$$

for all positive u and a fixed constant c, show that

$$\int_c^\infty f_X(x)dx = \int_{-\infty}^c f_X(x)dx.$$

(b) If $f_X(x)$ is symmetric about a constant c, show that c is the median of X.

4.3 EXPECTED VALUES

Random variables have been so named because we do not know in advance when a particular outcome will occur. Of course, if we knew this then we would be masters of blackjack, poker and dice. However, in considering the definition of a particular discrete random variable, it is likely that we are able to list all the potential outcomes of the random variable and, by making probability assessments about those outcomes, so construct a discrete probability distribution for the random variable.

The question of this section is: If we know of all potential outcomes of a random variable, how can we decide on the typical outcome that we *expect* to get, on average?

Expected values for discrete random variables

When a die is tossed, the outcomes of the random variable X denoting 'number of dots facing uppermost' are recorded in the sample space $\mathcal{S} = \{1, 2, 3, 4, 5, 6\}$. If we regard the die as 'fair', then equal probabilities are assigned to each of these outcomes resulting in the pmf

x	1	2	3	4	5	6
$p_X(x)$	$\frac{1}{6}$	$\frac{1}{6}$	$\frac{1}{6}$	$\frac{1}{6}$	$\frac{1}{6}$	$\frac{1}{6}$

Imagine repeatedly throwing the die 100 times and that (using the notation of Section 2.4 so that f_i is the observed frequency of outcome i) the following data are generated:

x	1	2	3	4	5	6
f_i	15	18	17	22	17	11

For this *particular* set of realisations, the sample mean is

$$\begin{aligned}
\overline{x} &= \frac{\sum_{i=1}^{6} f_i}{100} = \frac{(1)(15) + (2)(18) + (3)(17) + (4)(22) + (5)(17) + (6)(11)}{100} \\
&= (1)(\frac{15}{100}) + (2)(\frac{18}{100}) + (3)(\frac{17}{100}) + (4)(\frac{22}{100}) + (5)(\frac{17}{100}) + (6)(\frac{11}{100}) \\
&= 3.41.
\end{aligned}$$

Another set of 100 tosses would, very likely, produce a different sample mean. In the long run we would expect to get (with as many as ten million tosses of the die) a value close to

$$(1)(\frac{1}{6}) + (2)(\frac{1}{6}) + (3)(\frac{1}{6}) + (4)(\frac{1}{6}) + (5)(\frac{1}{6}) + (6)(\frac{1}{6}) = 3.5,$$

where the relative frequencies $\frac{15}{100}$, $\frac{18}{100}$, $\frac{17}{100}$, $\frac{22}{100}$, $\frac{17}{100}$, $\frac{11}{100}$ have been replaced by the probabilities $\frac{1}{6}$, $\frac{1}{6}$, $\frac{1}{6}$, $\frac{1}{6}$, $\frac{1}{6}$, $\frac{1}{6}$. The value of 3.5 is called the 'expected value' of the random variable X. It is not a value that we are likely to observe when we toss the die once; it is a value that we would expect, on average, in repeated tossings of the die.

Definition 4.9 *If $p_X(x)$ is the probability mass function of a discrete random variable X, then the* **expected value of** X*, denoted by $E(X)$, is given by*

$$E(X) = \sum_{\text{all } x} x p_X(x).$$

□

This definition applies to discrete random variables that have a finite number of outcomes, and to discrete random variables with uncountably many outcomes. The expected value of X is also called the **mean of** X which is conveniently denoted by μ_X; therefore the following notations are equivalent:

$$E(X) = \mu_X = \sum_{\text{all } i} x_i P(X = x_i).$$

Some examples will illustrate the use of these definitions. The probability distribution for X, 'the number facing uppermost when a fair die is tossed', is

x	1	2	3	4	5	6
$p_X(x)$	$\frac{1}{6}$	$\frac{1}{6}$	$\frac{1}{6}$	$\frac{1}{6}$	$\frac{1}{6}$	$\frac{1}{6}$

From the definition,

$$E(X) = (1)(\frac{1}{6}) + (2)(\frac{1}{6}) + (3)(\frac{1}{6}) + (4)(\frac{1}{6}) + (5)(\frac{1}{6}) + (6)(\frac{1}{6}) = 3.5.$$

In a similar way, the random variable Y, giving 'the maximum of the numbers facing uppermost when *two* dice are tossed', has probability distribution (which you should verify by counting points favourable to the events $\{Y = 1\}, \{Y = 2\}, \{Y = 3\}, \ldots$, from among the 36 equally likely points in the sample space):

y	1	2	3	4	5	6
$p_Y(y)$	$\frac{1}{36}$	$\frac{3}{36}$	$\frac{5}{36}$	$\frac{7}{36}$	$\frac{9}{36}$	$\frac{11}{36}$

Again from the definition,

$$E(Y) = (1)(\frac{1}{36})+(2)(\frac{3}{36})+(3)(\frac{5}{36})+(4)(\frac{7}{36})+(5)(\frac{9}{36})+(6)(\frac{11}{36}) = 4.47 \text{ (2 dp)}.$$

This follows our intuition that the expected value for the maximum of two dice should exceed the expected value for a single die.

Example 4.6

Expected profit from a spin of the roulette wheel

Consider playing roulette at the Burswood Island Casino. The roulette wheel is perfectly balanced and spins in the horizontal plane. A marble, released into the spinning wheel comes to rest in any one of 37 positions equally spaced about the rim of the wheel. The positions are labelled with numbers 1 to 36; the 37th position is labelled 0 and the 'house' takes all bets when the ball lands in this position. Because of the perfect construction of the wheel, all positions may be regarded as equally likely. Suppose that we bet with a single \$5 chip on a single number. If we win, our \$5 chip is returned (or at least the croupier leaves it on the betting table so that the chip is never endangered until a bet is lost), and the payout for our win is 35:1.

Let the random variable X denote our net profit, in dollars, on a single spin of the wheel. Since 36 of the 37 spaces cause us to lose our \$5 chip (or make a profit of $\$(-5)$) the probability distribution for X is given by

x	175	−5
$p_X(x)$	$\frac{1}{37}$	$\frac{36}{37}$

When playing the game, it is important that we know our expected profit and the interpretation of this for both casino and player. By definition,

$$E(X) = (175)(\frac{1}{37}) + (-5)(\frac{36}{37}) = \$(-\frac{5}{37}).$$

Therefore, in the long run averaged over many millions of spins, we expect to lose approximately 13 cents per spin. The winner in all this is the casino: it can expect to win on average 13 cents per spin for each of several wheels spinning every few minutes in an endless cycle, 24 hours per day at least 360 days a year. This is a lot of spins—more than enough for the casino to achieve the predicted expected profit. However, in our small slice of only a few spins of the wheel, we are very unlikely to achieve our expected loss; we could win or lose, big or small. The fact that we have a negative expected profit is an ominous warning for those who may care to keep on playing forever.

Expected values of functions of random variables

If X is a random variable and $h()$ a function, then $Y = h(X)$ is also a random variable.

To evaluate the expected value of a function $h(X)$ of the discrete random variable X, we sum values of the function weighted by probabilities from the probability distribution for X. It is not necessary to determine the distribution of $Y = h(X)$ first and then evaluate $E(Y)$, since the definition of expected value does not refer to the underlying sample space. That is,

$$E[h(X)] = \sum_{all\ x} h(x) p_X(x).$$

A simple choice for $h(X)$ could be $h(X) = X^2$. In this case, $E(X^2) = \sum_{all\ x} x^2 p_X(x)$.

We will evaluate this quantity for the random variable Y giving 'the maximum of the numbers facing uppermost when *two* dice are tossed'. Recall that Y has probability distribution

y	1	2	3	4	5	6
$p_Y(y)$	$\frac{1}{36}$	$\frac{3}{36}$	$\frac{5}{36}$	$\frac{7}{36}$	$\frac{9}{36}$	$\frac{11}{36}$

and therefore, according to the definition,

$$\begin{aligned} E(Y^2) &= (1)^2(\frac{1}{36}) + (2)^2(\frac{3}{36}) + (3)^2(\frac{5}{36}) + (4)^2(\frac{7}{36}) + (5)^2(\frac{9}{36}) + (6)^2(\frac{11}{36}) \\ &= 21.97 (2\ \text{dp}). \end{aligned}$$

Since $E(Y) = 4.47$, it is noticeable that $E(Y^2)$ and $[E(Y)]^2$ are unequal. In general, $E(Y^2) > [E(Y)]^2$ for random variables that are not 'constants'. The positive difference between the magnitudes of these important quantities provides a measure of the variability of the random variable about its mean.

Great expectations

As a further illustration, consider the expected value of a discrete random variable with countably infinitely many outcomes. If, when we toss a fair coin, X is the waiting time (number of tosses) until the first head is observed, then $X \sim Geom(\frac{1}{2})$. X has pmf $p_X(x) = (\frac{1}{2})^x,\ \ x = 1, 2, 3 \ldots,$ for which

$$E(X) = (1)(\tfrac{1}{2}) + (2)(\tfrac{1}{2})^2 + (3)(\tfrac{1}{2})^3 + (4)(\tfrac{1}{2})^4 + \ldots$$

This series may be summed by placing $r = \frac{1}{2}$ in the result of differentiating each side of the sum to infinity of the geometric series (a move valid for values of r in the range $|r| < 1$)

$$1 + r + r^2 + r^3 + r^4 + \ldots = \frac{1}{1-r}.$$

That is,

$$1 + 2r + 3r^2 + 4r^3 + 5r^4 + \ldots = \frac{d}{dr}\frac{1}{1-r} = \frac{1}{(1-r)^2},$$

which yields, after multiplication by r each side,

$$r + 2r^2 + 3r^3 + 4r^4 + \ldots = \frac{r}{(1-r)^2}.$$

Therefore, with $r = \frac{1}{2}$,

$$E(X) = (1)(\tfrac{1}{2}) + (2)(\tfrac{1}{2})^2 + (3)(\tfrac{1}{2})^3 + (4)(\tfrac{1}{2})^4 + \ldots = \frac{\frac{1}{2}}{(1-\frac{1}{2})^2} = 2.$$

Thus, with a fair coin, we can expect to wait 2 tosses until a head turns up.

If we start betting on this fair coin, we soon meet the great wall of the St Petersburg paradox. It unfolds as follows:

The St Petersburg game. I will win \$$2^x$ from the Reserve Bank if it takes x tosses for a head to appear when a fair coin is tossed successively.

(We use the Reserve Bank here as a government-guaranteed unlimited supply of money. As you will see, this much money is needed no matter the size of the tax levied on us to play the game.)

What is the expected profit from this venture? The profit function is $h(X) = 2^X$, with expected value

$$E[h(X)] = (2)(\tfrac{1}{2}) + (2^2)(\tfrac{1}{2})^2 + (2^3)(\tfrac{1}{2})^3 + (2^4)(\tfrac{1}{2})^4 + \ldots = 1+1+1+1+\ldots = \infty.$$

Therefore, no matter what finite amount we pay to play the game against the Reserve Bank, we will win. These are indeed great expectations.

Notice that if the Reserve Bank restricted its payout to a maximum claim of 2^{100} dollars or, equivalently, about 10^{30} dollars (a ten followed by twenty-nine zeroes) then, following the description in Arnold (1990) for example, the profit function is

$$h(X) = \begin{cases} 2^X & \text{if } \; X < 100; \\ 2^{100} & \text{if } \; X \geq 100. \end{cases}$$

The expected value of this function is

$$\begin{aligned} E[h(X)] &= (2)(\tfrac{1}{2}) + (2^2)(\tfrac{1}{2})^2 + (2^3)(\tfrac{1}{2})^3 + \ldots + (2^{99})(\tfrac{1}{2})^{99} \\ &\quad + (2^{100})\left[(\tfrac{1}{2})^{100} + (\tfrac{1}{2})^{101} + (\tfrac{1}{2})^{102} + \ldots\right] \\ &= 99 + 2^{100}\left[\frac{(\frac{1}{2})^{100}}{(1-\frac{1}{2})}\right] \\ &= 101. \end{aligned}$$

If you pay more than \$101 to play this game, you are paying too much!

Expected values for continuous random variables

The results of the previous section easily carry over to continuous random variables; we replace the summation sign by an integral sign and the probability mass function by a probability density function.

Definition 4.10 *The* **expected value of a continuous random variable** X*, with probability density function* f_X*, is given by*

$$E(X) = \int_{-\infty}^{\infty} x f_X(x)dx,$$

provided this integral exists.

□

For example, if X is a random variable with probability density function

$$f_X(x) = \begin{cases} 3x^2 & \text{if } 0 < x < 1; \\ 0 & \text{otherwise,} \end{cases}$$

then, by definition,

$$E(X) = \int_{-\infty}^{\infty} x f_X(x)dx = \int_0^1 x[3x^2]dx = \left.\frac{3x^4}{4}\right]_0^1 = \frac{3}{4}.$$

As a further example, we now evaluate the mean (expected value) of an exponential random variable. Suppose that a brand of streetlamp is manufactured to the specification that the lifetime, T, of the lamp will exceed t years with probability $e^{-\frac{t}{\beta}}$, where β is a positive constant. That is,

$$P(T > t) = e^{-\frac{t}{\beta}}.$$

By considering the complementary event $\{T \leq t\}$, we have

$$F_T(t) = 1 - P(T > t) = 1 - e^{-\frac{t}{\beta}}.$$

Differentiating this expression with respect to t gives (for $t > 0$)

$$f_T(t) = \frac{d}{dt}F_T(t) = \frac{1}{\beta}e^{-\frac{t}{\beta}}.$$

This is precisely the exponential density; $T \sim exp(\beta)$. The definition for finding the expected value is then applied to this density with the range of integration over the positive real numbers (since streetlamps have positive lifetimes).

$$E(T) = \int_{-\infty}^{\infty} t f_T(t)dt$$

$$\begin{aligned}
&= \int_0^\infty \frac{t}{\beta} e^{-\frac{t}{\beta}} dt \\
&= \beta \int_0^\infty x e^{-x} dx \qquad \text{(substituting } x = \frac{t}{\beta}\text{)} \\
&= \beta \int_0^\infty e^{-x} dx \\
&= \beta.
\end{aligned}$$

The integration $\int_0^\infty x e^{-x} dx = \int_0^\infty e^{-x} dx$ is a standard result that may be established by the integration process *integration by parts*. Summarising:

> If $T \sim exp(\beta)$, then $E(T) = \beta$.

There is no requirement that continuous random variables possess finite means or even that the means be defined. For example, if X has probability density function

$$f_X(x) = \frac{1}{\pi} \frac{1}{1+x^2}, \qquad -\infty < x < \infty,$$

then X is symmetric about 0 since $f_X(x) = f_X(-x)$. This indicates that the median of X is 0. We might expect the same 0 result for $E(X)$. By definition, the mean X is

$$E(X) = \int_{-\infty}^{\infty} x \frac{1}{\pi} \frac{1}{1+x^2} dx,$$

which does not exist. This particular probability density function may be described as being 'heavy-tailed'. That is, very large values of X may be observed, even though such events occur with small probability. This particular distribution is known as the **Cauchy distribution**.

Definition 4.11 *A continuous random variable X with probability density function*

$$f_X(x) = \frac{1}{\pi} \frac{\delta}{1+\delta^2 x^2}, \qquad -\infty < x < \infty,$$

is said to have a **Cauchy distribution** *for $\delta > 0$ and we write $X \sim Cauchy(\delta)$.*

□

As in the discrete case, the expected value of a function $h(X)$ of the continuous random variable X may be evaluated without first determining the probability distribution of $h(X)$:

$$E[h(X)] = \int_{-\infty}^{\infty} h(x) f_X(x) dx.$$

Expected values for special functions

Suppose that we consider a random variable X and simple functions $h(X)$ of X. We are particularly interested in finding the expected value of $h(X)$ in the following cases:

- $h(X) = c$
- $h(X) = cX$
- $h(X) = c_1h_1(X) + c_2h_2(X)$

in which the c, c_1 and c_2 are constants and $h(X)$, $h_1(X)$ and $h_2(X)$ are functions of X. The third such case is so important that we express it in theorem form.

Theorem 4.5 *If c_1, c_2 are constants and $h_1(X)$, $h_2(X)$ are functions of a random variable X (which may be either discrete or continuous) then, provided the expected values $E[h_1(X)]$ and $E[h_2(X)]$ exist,*

$$E[c_1h_1(X) + c_2h_2(X)] = c_1E[h_1(X)] + c_2E[h_2(X)].$$

Proof: (We will give the proof for the continuous case. The proof for the discrete case is similar.) The proof is structured from the idea that the integral of a sum of two functions is the sum of the two integrals.

$$\begin{aligned}
E[c_1h_1(X) + c_2h_2(X)] &= \int_{-\infty}^{\infty} [c_1h_1(x) + c_2h_2(x)]f_X(x)dx \\
&= \int_{-\infty}^{\infty} c_1h_1(x)f_X(x)dx + \int_{-\infty}^{\infty} c_2h_2(x)f_X(x)dx \\
&= c_1\left[\int_{-\infty}^{\infty} h_1(x)f_X(x)dx\right] + c_2\left[\int_{-\infty}^{\infty} h_2(x)f_X(x)dx\right] \\
&= c_1E[h_1(X)] + c_2E[h_2(X)].
\end{aligned}$$

■

This theorem shows how to obtain the expected value of a linear combination of random variables without actually having to determine the distribution of the linear combination first.

If, in Theorem 4.5 we put $c_1 = c$, $c_2 = 0$ and $h_1(X) = 1$, we obtain:

For c a constant, $\quad E(c) = c.$

If, in Theorem 4.5 we put $c_1 = c$, $c_2 = 0$ and $h_1(X) = X$, we obtain:

For c a constant, $\quad E(cX) = cE(X).$

EXERCISES

4.18 A fair coin is tossed 10 times. Suppose that at any stage in the tossing procedure the events H denoting 'heads' and T denoting 'tails' are independent.

(a) Determine the probability of obtaining 5 heads.

(b) Determine the probability of obtaining 5 heads in the *particular sequence* H T H T H T H T H T.

(c) Suppose that after 9 tosses we obtain the particular sequence H H H H H H H H H. Is the next toss more likely to result in a tail than a head? Explain briefly.

(d) A casino offers the following game. A player bets \$10. A fair coin is then tossed 10 times. If all tosses result in heads, the casino pays the player \$10 000 (plus his money back); otherwise the player loses the \$10. Suppose that the outcomes of a random variable X give the casino's profit, in dollars, when a game is played.

Construct a table for the probability mass function $p_X(x)$ and hence determine the casino's profit (in dollars) for each game played.

4.19 The random variable X represents the number of visitors that a hospital patient has during a visiting period. A maximum number of 4 visitors is permitted. X is found to have the probability distribution

x	0	1	2	3	4
$p_X(x)$	c	0.2	0.4	0.2	0.1

(a) Write down the value of c.

(b) Show that $E(X) = 2$ and $E(X^2) = 5.2$.

(c) Write down the probability distribution of the random variable $Y = (X-2)^2$ and hence, or otherwise, find

$$E[(X-2)^2].$$

(d) Sketch the cumulative distribution function of X.

4.20 Consider random number tables in which (we assume that) the digits $0, 1, 2, \dots, 9$ occur with equal probabilities.

Let X denote the random variable 'randomly chosen digit'.

(a) State the probability mass function for X.

(b) Determine the mean of X.

(c) Find $P(X^2 < 12)$.

4.21 A random variable X has probability mass function

$$p_X(x) = \begin{cases} \frac{x+c}{6} & \text{if } x = 0, 1, 2; \\ 0 & \text{otherwise.} \end{cases}$$

Evaluate the constant c and hence determine μ_X.

4.22 (a) Let X be the amount, in dollars, collected from a single free ticket in a lottery in which there are 1000 tickets one of which carries a \$100 prize, five carry \$20 prizes and ten carry \$5 prizes. Determine the probability distribution for X and hence $E(X)$.

(b) The Wattletree Childcare Co-operative, as part of their fundraising drive for new playground equipment, wished to make \$1750 profit on running the lottery in (a). How much should the Co-op charge per ticket? [Assume that there are no administrative expenses in running the lottery.]

4.23 In the game of 'Two-Up' at the Burswood Island Casino in Perth, two identical unbiased coins are spun (or tossed).

Let

- H denote the event that a spin results in two heads uppermost;
- T denote the event that a spin results in two tails uppermost;
- O denote the event that a spin results in a head and a tail uppermost. (Such an event is termed 'Odds'.)

Suppose that we bet with a single \$5 chip on the event H which pays $1:1$ (and we retain our bet if we win).

The coins are tossed repeatedly until our bet

(i) wins if H is spun;

(ii) loses if T is spun **or** 5 consecutive O are spun;

(iii) neither wins nor loses if O is spun, in which case our bet remains unaltered until a result is obtained in accordance with rules (i) and (ii).

Such a sequence of tosses is termed a 'play'. Let X denote our profit per 'play'.

(a) Explain, with reference to the above rules (i)–(iii), why the sequences

$$H,\ OH,\ OOH,\ OOOH,$$

all lead to our winning a play.

(b) By listing *all possible* sequences which lead to our winning a 'play', show that the probability that we win a 'play' is $\frac{31}{64}$.

(c) Construct a probability distribution for X.

(d) Show that $E(X) < 0$. Explain carefully what this result means for both player and casino.

4.24 A particular brand of wood-grained panel is manufactured in such a way that each panel is exactly twice as long as it is wide. The observed length of panels manufactured in this way is a random variable X which varies from panel to panel in such a way that $E(X) = \mu$.

(a) Determine, in terms of μ, the mean perimeter of a panel.

(b) If the cutting equipment is adjusted so that $E(X)$ is now given by $E(X) = \mu + 0.001$, what percentage change does this cause in the mean perimeter?

4.25 If I randomly think of a number (is that possible?) between 0 and 1, what is the mean absolute deviation of that number from $\frac{1}{2}$? That is: If $X \sim U[0,1]$, determine

$$E(|X - \tfrac{1}{2}|).$$

4.26 Square bathroom floor tiles are manufactured so that the true mean length equals the true mean width of 20 cm. The error in length (or width) is a random variable U (in cm), where $U \sim U[-1, 1]$.

(a) Explain why the expression $(20+U)^2$ represents the observed area of a tile.

(b) Hence determine the mean area of a tile.

(c) Explain why it is incorrect to determine the mean area of a tile according to

$(\text{mean length})^2 = 20^2 = 400 \text{ cm}^2$.

4.27 Let X be a continuous random variable with probability density function $f(x)$. If $\xi_{0.5}$ is the unique median of the distribution of X and b is a real constant, show that

$$E(|X-b|) = E(|X-\xi_{0.5}|) + 2\int_{\xi_{0.5}}^{b} (b-x)f(x)dx,$$

provided that the expectations exist. For what value of b is $E(|X-b|)$ a minimum?

4.4 MEAN, VARIANCE AND MOMENTS

The variance of a random variable

In the previous section we defined the mean μ_X by

$$E(X) = \sum_{\text{all } x} x p_X(x) \quad (X \text{ discrete})$$

and

$$E(X) = \int_{-\infty}^{\infty} x f_X(x)dx \quad (X \text{ continuous}).$$

These are special cases (with $h(X) = X$) of the formulas

$$E[h(X)] = \sum_{\text{all } x} h(x) p_X(x) \tag{4.3}$$

$$E[h(X)] = \int_{-\infty}^{\infty} h(x) f_X(x)dx, \tag{4.4}$$

for determining the expected value of the function $h(X)$ of the random variable X.

Our knowledge of data analysis indicates that any determination of the mean or centre of a distribution must be accompanied by a determination of the variability of the distribution about that centre. Similarly, when considering random variables, the mean and variance must be considered together.

Definition 4.12 *The* **variance of the random variable** X, *denoted by* σ_X^2, *or* $Var(X)$, *is given by*

$$\sigma_X^2 = E(X^2) - [E(X)]^2.$$

The **standard deviation of the random variable** X, *denoted by* σ_X *is the positive square root of* σ_X^2.

□

Simple discrete examples that we have often referred to are the random variables:

- X denoting 'the number facing uppermost when a fair die is tossed';
- Y denoting 'the maximum of the numbers facing uppermost when two dice are tossed'.

From Section 4.3, X and Y have probability distributions

x	1	2	3	4	5	6
$p_X(x)$	$\frac{1}{6}$	$\frac{1}{6}$	$\frac{1}{6}$	$\frac{1}{6}$	$\frac{1}{6}$	$\frac{1}{6}$

y	1	2	3	4	5	6
$p_Y(y)$	$\frac{1}{36}$	$\frac{3}{36}$	$\frac{5}{36}$	$\frac{7}{36}$	$\frac{9}{36}$	$\frac{11}{36}$

from which we find (to 2 dp) $E(X) = 3.50$, $E(X^2) = 15.17$ and so

$$Var(X) = \sigma_X^2 = E(X^2) - [E(X)]^2 = 15.17 - (3.5)^2 = 2.92.$$

Similarly, $E(Y) = 4.47$, $E(Y^2) = 21.97$, giving

$$Var(Y) = \sigma_Y^2 = E(Y^2) - [E(Y)]^2 = 21.97 - (4.47)^2 = 1.99.$$

Thus, X has a larger variance than Y; values 5 and 6 occur as outcomes for Y with high probability and lie relatively close to $E(Y)$. These same values occur for X with lower probability and they lie further from $E(X)$. This increased distance between outcome and mean increases the variance. The following theorem shows the variance to be a positive mean-squared deviation.

Theorem 4.6 *The variance* σ_X^2 *of a random variable* X *is given by* $\sigma_X^2 = E[(X - \mu_X)^2]$.

Proof: In this proof it is important to remember that $E(X)$ is a constant and can 'pass through' an expected value sign:

$$\begin{aligned} E[(X-\mu_X)^2] &= E(X^2 - 2\mu_X X + \mu_X^2) \\ &= E(X^2) - E(2\mu_X X) + E(\mu_X^2) \quad \text{(by Theorem 4.5)} \\ &= E(X^2) - 2\mu_X E(X) + \mu_X^2 \\ &= E(X^2) - 2\mu_X^2 + \mu_X^2 \\ &= E(X^2) - \mu_X^2 \\ &= \sigma_X^2. \end{aligned}$$

■

Example 4.7

Strategies at roulette: Example 4.6 continued

In a casino, roulette followers often sit beside the wheel, recording the result of every spin, perhaps waiting for a pattern to reveal itself, or, perhaps just waiting for a game plan to become apparent.

Should one bet on reds? on singles? on pairs? on blocks of four?

Consider this: All such strategies lead to the same *expected* loss of $\$(\frac{5}{37})$ per spin of the wheel. However, the variability about that expected loss is different for different strategies. Let X_i denote the net profit, in dollars, for a single spin of the wheel when we bet with a single \$5 chip on i squares. We will generally consider values of i as

i	**Location**
1	single number
2	pair numbers
4	block of four numbers
18	red (or black) numbers

When we bet on i squares the payout by the casino is $(\frac{36}{i} - 1) : 1$. (This is based on the spurious assessment that every square has equal probability of $\frac{1}{36}$ of winning.) Therefore X_i has probability distribution

x	$5(\frac{36}{i} - 1)$	-5
$p_{X_i}(x)$	$\frac{i}{37}$	$\frac{37-i}{37}$

Hence

$$E(X_i) = [5(\frac{36}{i} - 1)](\frac{i}{37}) + (-5)(\frac{37-i}{37}) = -\frac{5}{37}$$

and this, very noticeably, does not depend on i. That is, different strategies (with different values of i) give the same negative expected net profit per spin of the wheel. If we investigate the variance:

$$\begin{aligned} Var(X_i) &= [5(\frac{36}{i} - 1)]^2(\frac{i}{37}) + (-5)^2(\frac{37-i}{37}) - (-\frac{5}{37})^2 \\ &= (\frac{180}{37})^2(\frac{37}{i} - 1). \end{aligned}$$

(The last line requires care with the algebraic manipulation!) Note that as i increases, then $Var(X_i)$ decreases. That is, the more squares you bet on, the less the variability about the expected loss $\$(-\frac{5}{37})$.

Example 4.7 illustrates distributions with the same mean and different variances. In general, the mean and variance together measure the width or spread of a distribution; Chebychev's Inequality specifies a lower bound for the proportion of any probability distribution that lies in the interval $(\mu_X - k\sigma_X, \mu_X + k\sigma_X)$.

Theorem 4.7 (Chebychev's Inequality) *Any random variable X with mean μ_X and finite variance σ_X^2 satisfies*

$$P(|X - \mu_X| < k\sigma_X) \geq 1 - \frac{1}{k^2}$$

for any positive constant k.

Note that Chebychev's Inequality may be equivalently written in the form $P(|X - \mu_X| \geq k\sigma_X) \leq \frac{1}{k^2}$.
Proof:

$$\begin{aligned} \sigma_X^2 &= \int_{-\infty}^{\infty} (x - \mu_X)^2 f_X(x)dx \\ &\geq \int_{|x-\mu_X| \geq k\sigma_X} (x - \mu_X)^2 f_X(x)dx \\ &\geq \int_{|x-\mu_X| \geq k\sigma_X} k^2\sigma_X^2 f_X(x)dx \\ &= k^2\sigma_X^2 P(|X - \mu_X| \geq k\sigma_X). \end{aligned}$$

■

Example 4.8
Victoria Market crowds: how far have they come?
The Victoria Market lies in the heart of Melbourne's Central Business District. We suppose that the commuting distance (from home to market) for Saturday morning Victoria Market shoppers is represented by the continuous random variable X. Since there are few residents of the central business district who would commute to the market, we consider only commuting distances which exceed 1 km. Assume that X has probability density function inversely proportional to the fourth power of commuting distance—the further away you live from the market, the more unlikely you are to be a Saturday morning shopper.

$$f_X(x) = \begin{cases} \frac{k}{x^4} & \text{if } x > 1; \\ 0 & \text{otherwise.} \end{cases}$$

We wish to assess how frequently people are commuting great distances to come shopping at the market on Saturday morning. That is, what proportion of commuting distances lie further than two standard deviations of the mean commuting distance?

Worked example:
Determine also exact value of $P(|X - \mu_X| \geq c\sigma_X)$, for $c = 2$ and compare this with the upper bound value determined from Chebychev's Theorem.

We must first determine the values of k, μ_X and σ_X.
The value of the constant k is found by integrating the density:

$$1 = \int_1^\infty \frac{k}{x^4} dx = -\frac{k}{3x^3}\Big]_1^\infty = \frac{k}{3},$$

and therefore $k = 3$. The mean and variance of X are now found:

$$\mu_X = \int_1^\infty x\frac{3}{x^4} dx = \int_1^\infty \frac{3}{x^3} dx = -\frac{3}{2}\frac{1}{x^2}\Big]_1^\infty = 1.5 \text{ km},$$

$$\sigma_X^2 = E(X^2) - [E(X)^2] = \int_1^\infty x^2 \frac{3}{x^4} dx - (1.5)^2 = 3 - 2.25 = 0.75 \text{ (km)}^2.$$

Armed with the measures μ_X of centre and σ_X^2 of spread, the size of the centre of the distribution may be assessed by the probability calculation

$$\begin{aligned} P(|X - \mu_X| < 2\sigma_X) &= P(\mu_X - 2\sigma_X < X < \mu_X + 2\sigma_X) \\ &= \int_{\mu_X - 2\sigma_X}^{\mu_X + 2\sigma_X} f_X(x) dx \\ &= \int_1^{1.5+2\sqrt{0.75}} \frac{3}{x^4} dx. \end{aligned}$$

[Note that in the above calculation, since $f_X(x) = 0$ for values of x less than 1, the range of integration reduced from the interval $(1.5 - 2\sqrt{0.75},\ 1.5 + 2\sqrt{0.75})$ to the interval $(1,\ 1.5 + 2\sqrt{0.75})$.]

Then, since

$$\int_1^{1.5+2\sqrt{0.75}} \frac{3}{x^4} dx = -\frac{1}{x^3}\Big]_1^{3.232} = 0.9704,$$

approximately 97% of commuters to the market are from within two standard deviations of the mean commuting distance. This assessment indicates that few people are commuting from far outlying areas to the Victoria Market. The Chebychev Inequality gives

$$P(|X - \mu_X| < c\sigma_X) \geq 1 - \frac{1}{c^2}$$

so that with $c = 2$, the right-side of this inequality is 0.75. This conservatively blunt effort at obtaining the proportion of the distribution that is within two standard deviations of the mean is because the Chebychev bound holds firm for *all* distributional shapes.

A relative measure of spread is provided by the **coefficient of variation** of the random variable X. Following the lead of Definition 2.7 for the coefficient of variation of a data set, we use

$$\vartheta_X = \frac{\sigma_X}{\mu_X}$$

to denote the coefficient of variation of X.

Generating the moments of a random variable

In our early discussions of data analysis we found that the sample moments, defined as averages of powers of the data (see Definition 2.9), gave indications of centre, spread, skewness and kurtosis in the data set. For random variables, such questions of shape refer to the probability density function (continuous random variables) and to the probability mass function (discrete random variables).

Definition 4.13 *The* rth **moment** *of* X *(about 0),* μ'_r, *is given by*

$$\mu'_r = E(X^r).$$

The **moment generating function**, $m_X(t)$, *of* X *is given by*

$$m_X(t) = E(e^{tX}) \quad \text{for } t \in D,$$

where D *is the set of values of* t *such that* $m_X(t)$ *is finite.*

□

$m_X(t)$ exists if its domain D contains an interval of points around 0. Under these circumstances, it is easy to see heuristically how $m_X(t)$ quite explicitly generates the moments of X from the exponential expansion

$$e^{tX} = 1 + tX + \frac{(tX)^2}{2!} + \frac{(tX)^3}{3!} + \frac{(tX)^4}{4!} + \ldots,$$

so that on taking the expected value of each side and using an extended version of Theorem 4.5 to the sum of a countable number of functions of X,

$$\begin{aligned} m_X(t) &= E[1 + tX + \frac{(tX)^2}{2!} + \frac{(tX)^3}{3!} + \frac{(tX)^4}{4!} + \ldots] \\ &= 1 + E(X)t + E(X^2)\frac{t^2}{2!} + E(X^3)\frac{t^3}{3!} + E(X^4)\frac{t^4}{4!} + \ldots \\ &= 1 + \mu'_1 t + \mu'_2\frac{t^2}{2!} + \mu'_3\frac{t^3}{3!} + \mu'_4\frac{t^4}{4!} + \ldots \end{aligned}$$

The probability density function

$$f_X(x) = \begin{cases} \frac{3}{x^4} & \text{if } x > 1; \\ 0 & \text{otherwise,} \end{cases}$$

used to describe how the relative frequency of market commuters changes with commuting distance x (Example 4.8) has moments about 0 of

$$\mu_1' = \int_1^\infty \frac{3}{x^3} dx = 1.5$$

$$\mu_2' = \int_1^\infty \frac{3}{x^2} dx = 3$$

$$\mu_3' = \int_1^\infty \frac{3}{x} dx = \infty,$$

and $\mu_r = \infty$ for all $r \geq 3$. In this case, the moment generating function fails to exist. Only the first two moments have finite values.

We will leave this badly behaved distribution and now define a probability distribution that is extremely well behaved (in the sense that all the moments exist); a probability distribution that forms the basis for many areas of statistical modelling and decision making, a probability distribution which will be a constant source of reference.

Definition 4.14 *When a random variable Z has probability density function*

$$f_Z(z) = \frac{1}{\sqrt{2\pi}} e^{-\frac{1}{2}z^2}, \quad -\infty < z < \infty,$$

then we say that Z has a **standard normal distribution** *and we write $Z \sim N(0,1)$.*

□

It is possible to use double integration methods (Chapter 7 exercises) to show that this pdf integrates to 1. We will cleverly adopt the final result for our current use. That is, we will assume that $\int_{-\infty}^{\infty} f_Z(z) dz = 1$, so that

$$\int_{-\infty}^{\infty} \frac{1}{\sqrt{2\pi}} e^{-\frac{1}{2}z^2} dz = 1.$$

Example 4.9

Normal moments

Worked example:

Determine the moments of $Z \sim N(0,1)$.

First, the moment generating function is evaluated and then the moments may be essentially 'read' from the series expansion of the function.

$$\begin{aligned} m_Z(t) &= E(e^{tZ}) \\ &= \int_{-\infty}^{\infty} e^{tz} \frac{1}{\sqrt{2\pi}} e^{-\frac{1}{2}z^2} dz \end{aligned}$$

$$\begin{aligned}
&= \frac{1}{\sqrt{2\pi}} \int_{-\infty}^{\infty} e^{tz} e^{-\frac{1}{2}z^2} dz \\
&= \frac{1}{\sqrt{2\pi}} e^{\frac{1}{2}t^2} \int_{-\infty}^{\infty} e^{-\frac{1}{2}(z-t)^2} dz \\
&= e^{\frac{1}{2}t^2} \int_{-\infty}^{\infty} \frac{1}{\sqrt{2\pi}} e^{-\frac{1}{2}w^2} dw \quad \text{(substituting } w = z - t\text{)} \\
&= e^{\frac{1}{2}t^2}.
\end{aligned}$$

Now recall that for any real r

$$e^r = 1 + r + \frac{r^2}{2!} + \frac{r^3}{3!} + \frac{r^4}{4!} + \ldots,$$

and with $r = \frac{1}{2}t^2$,

$$\begin{aligned}
m_Z(t) &= 1 + \tfrac{1}{2}t^2 + \frac{(\frac{1}{2}t^2)^2}{2!} + \frac{(\frac{1}{2}t^2)^3}{3!} + \frac{(\frac{1}{2}t^2)^4}{4!} + \ldots \\
&= 1 + (0)(t) + (1)\frac{t^2}{2!} + (0)\frac{t^3}{3!} + (3)\frac{t^4}{4!} + (0)\frac{t^5}{5!} + \ldots
\end{aligned}$$

Comparing this with the moment generating function expansion

$$m_Z(t) = 1 + E(Z)t + E(Z^2)\frac{t^2}{2!} + E(Z^3)\frac{t^3}{3!} + E(Z^4)\frac{t^4}{4!} + \ldots$$

it is clear that

$$\begin{aligned}
E(Z) &= 0 \\
E(Z^2) &= 1 \\
E(Z^3) &= 0 \\
E(Z^4) &= 3 \\
E(Z^5) &= 0 \\
&\vdots
\end{aligned}$$

The probability density function of the standard normal distribution
$f_Z(z) = \frac{1}{\sqrt{2\pi}} e^{-\frac{1}{2}z^2}$

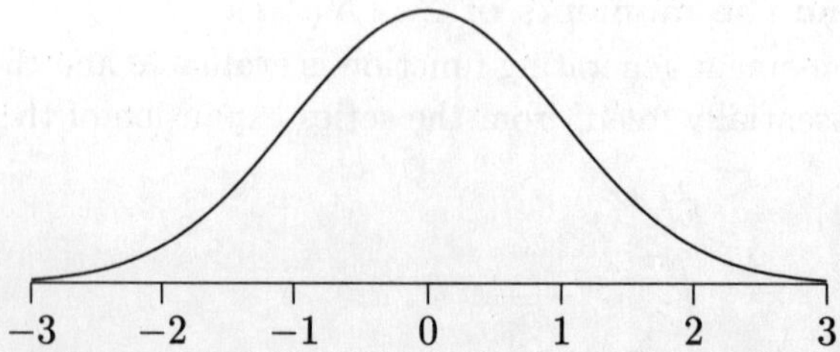

The mean and variance of a standard normal random variable Z can be trivially written down from Example 4.9; $E(Z) = 0$ and $Var(Z) = E(Z^2) - [E(Z)]^2 = 1 - 0 = 1$.

> If $Z \sim N(0,1)$, then $E(Z) = 0$ and $Var(Z) = 1$.

Example 4.9 illustrates the following important principle concerning the probability distribution of a random variable for which a moment generating function is defined:
The distribution of a random variable is completely identified by its moment generating function.
This indicates that it takes *all* the moments of a random variable X, as a minimum requirement, to completely specify the distribution of X. Consequently, the specification of only the first few moments $\mu'_1, \mu'_2, \mu'_3, \mu'_4, \mu'_5, \ldots$, of the random variable X does not determine the entire moment generating function $m_X(t)$. Mood, Graybill and Boes (1974) illustrate two different probability density functions with the same first four moments! Even so, the first few moments usually provide partial insight into the distributional shape of X; such shape refers to the shape of the graphs of the pdf (continuous case) or pmf (discrete case).

Definition 4.15 *The rth* **central moment**, μ_r, *of a random variable X about its mean value is given by,*

$$\mu_r = E[(X - \mu_X)^r]$$

provided the expectation exists.

□

Clearly, Definition 4.15 shows that the first two central moments of a random variable X have simple forms:

$$\mu_1 = E(X - \mu_X) = 0$$
$$\mu_2 = E[(X - \mu_X)^2] = \sigma_X^2.$$

The higher order central moments, in a form that does not depend on the scale of measurement used, are defined by

$$\zeta_r = \frac{\mu_r}{(\sqrt{\mu_2})^r}.$$

The values of ζ_r for $r = 2$ and $r = 3$ are useful measures of distributional shape.

Definition 4.16 *The* **coefficient of skewness**, ζ_3, *is given by*

$$\zeta_3 = \frac{\mu_3}{(\sqrt{\mu_2})^3}.$$

The **coefficient of kurtosis**, ζ_4, *is given by*

$$\zeta_4 = \frac{\mu_4}{(\sqrt{\mu_2})^4}.$$

□

ζ_3 is an indicator of skewness and when the third moment exists and the pdf (or pmf) is symmetric about its mean, then $\zeta_3 = 0$. Generally, when a pdf shows a long tail towards the right, then $\zeta_3 > 0$ and the distribution is said to be positively skewed. Similarly, when a pdf displays a long tail towards the left, then $\zeta_3 < 0$ and the distribution is said to be negatively skewed. These ideas run parallel to the discussion of the *sample* coefficient of skewness following Definition 2.10.

ζ_4 is a measure of the degree of central flatness of the pdf. This flatness near the centre (or, at the other end of the scale, peakedness near the centre) is measured on a scale in relation to the central shape of the standard normal density. By Example 4.9, if $Z \sim N(0,1)$ then $E(Z^4) = 3$ and since Z has mean 0 and variance 1, it follows that $\zeta_4 = 3$. Densities for which $\zeta_4 > 3$ are termed **leptokurtic** (more peaked than normal), and densities for which $\zeta_4 < 3$ are termed **platykurtic** (flatter than normal).

Cumulant generating function

A function defined in terms of the moment generating function that is often easier to determine is the *cumulant generating function.*

Definition 4.17 *If X is a random variable with moment generating function $m_X(t)$, then the* **cumulant generating function**, *$\kappa_X(t)$ of X is given by*

$$\kappa_X(t) = \log_e m_X(t).$$

□

As the name suggests, $\kappa_X(t)$ generates the **cumulants** of X. These are easily determined by differentiation.

First, we will look at a differentiation process applied to the moment generating function: recall that

$$m_X(t) = 1 + \mu_1' t + \mu_2' \frac{t^2}{2!} + \mu_3' \frac{t^3}{3!} + \mu_4' \frac{t^4}{4!} + \ldots \qquad (4.5)$$

By differentiating each side of this equation and substituting $t = 0$ into the result, we obtain

$$m_X^{(1)}(0) = \mu_1',$$

since all other terms (apart from the first) in the differentiated series have a factor of t and become zero after the substitution of $t = 0$. Similarly, differentiating $m_X^{(1)}(t)$ again term by term and using the value $t = 0$, we obtain

$$m_X^{(2)}(0) = \mu_2',$$

and, in general, by differentiating in this way often enough,

$$m_X^{(r)}(0) = \mu_r'.$$

Differentiating successively each side of the equation

$$\kappa_X(t) = \log_e m_X(t)$$

and using $t = 0$ provides even more interesting (and very useful) results:

$$\begin{aligned} \kappa_X^{(1)}(0) &= E(X) \\ \kappa_X^{(2)}(0) &= Var(X) \\ \kappa_X^{(3)}(0) &= \mu_3 \\ \kappa_X^{(4)}(0) &= \mu_4 - 3\sigma^4. \end{aligned}$$

In particular, these methods establish the following theorem.

Theorem 4.8 *The coefficient of kurtosis is determined from the cumulant generating function by*

$$\zeta_4 = \frac{\kappa_X^{(4)}(0)}{\sigma^4} + 3.$$

■

EXERCISES

4.28 Consider random number tables in which (we assume that) the digits $0, 1, 2, \ldots, 9$ occur with equal probabilities.

Let X denote the random variable 'randomly chosen digit'.

(a) Give the pmf for X.

(b) Determine the mean and variance of X.

4.29 If X is a continuous random variable, determine the value of b which minimises the expression $E[(X-b)^2]$ as a function of b.

4.30 (a) Show that if constants a, b and c satisfy

$$\int_a^b c(x-a)(b-x)dx = 1,$$

then $c = \frac{6}{(b-a)^3}$. Show that with this value of c, the function $f_X(x) =$

$$\begin{cases} c(x-a)(b-x) & \text{if } a < x < b; \\ 0 & \text{otherwise,} \end{cases}$$

is the probability density function of some random variable X.

(b) The manufacturing specification for a cask of Cabernet Sauvignon is 2 litres. The observed volume, X, at manufacture is a random variable with outcome in the interval from 1.9 litres to 2.1 litres. Suppose that X is assumed to have a quadratic pdf of the form

$f_X(x) =$

$$\begin{cases} c(x-1.9)(2.1-x) & \text{if } 1.9 < x < 2.1; \\ 0 & \text{otherwise,} \end{cases}$$

where c is a constant.

i. Write down the value of $E(X)$. (No integration should be necessary.)

ii. Determine $P(X < 2)$.

iii. Determine $Var(X)$.

4.31 The stopping time X of a clock when left to wind down within a 24 hour time period may be assumed to have probability density function

$$f_X(x) = \begin{cases} \frac{1}{24} & \text{if } 0 < x < 24; \\ 0 & \text{otherwise.} \end{cases}$$

(a) Determine the mean and variance of X.

(b) Determine the *coefficient of variation* of X which is defined by the ratio

$$\frac{\sigma_X}{\mu_X}.$$

(c) Sketch the pdf $f_X(x)$ and shade the area representing the probability

$$P(|X-\mu_X| < 1.2\sigma_X).$$

Show how to evaluate this probability as the area of a rectangle under $f_X(x)$.

4.32 A discrete random variable X has probability mass function

x	$-k$	0	k
$p_X(x)$	$\frac{1}{2k^2}$	$1-\frac{1}{k^2}$	$\frac{1}{2k^2}$

expressed in terms of the constant k which is greater than 1.

(a) Determine $E(X)$ and $Var(X)$.

(b) Evaluate the probability

$$P(|X-\mu_X| < k\sigma_X)$$

and compare the result with that obtained from the Chebychev Inequality.

4.33 If $T \sim exp(1)$, determine an expression for $m_T(t)$.

Show how to use $m_T(t)$ to determine the mean and the variance of T.

By determining the power series expansion of $m_T(t)$ show that

$$\int_0^\infty x^r e^{-x} dx = r!$$

4.34 At the World Swimming Championships held in Perth in the summer of 1991, considerable attention was paid to the maintenance of accurate time-keeping records. Suppose that the starting gun fires at time $t = 0$ for a particular race. However, it is fair to assume that not all swimmers in the race leave the blocks at $t = 0$ to begin their race. Some, in anticipation of the starting gun, may leave earlier (and risk disqualification), and some, in delayed reaction, leave the blocks later than $t = 0$.

Assume that the actual time, Y, in seconds, that a swimmer leaves the blocks is uniformly distributed in a two-second interval about $t = 0$. That is, assume that $Y \sim U[-1, 1]$.

(a) Evaluate the moments

$$\mu'_r = E(Y^r)$$

in terms of r.

(b) Interpret the magnitude of the first four moments in terms of the shape of the pdf $f_Y(y)$ of Y. You should use the terms *centre*, *spread*, *skewness* and *kurtosis*.

(c) Determine the moment generating function of Y by evaluating the expectation

$$E(e^{tY}).$$

Use the fact that

$$e^t = 1 + \frac{t}{1!} + \frac{t^2}{2!} + \frac{t^3}{3!} + \ldots$$

to show how to express the moment generating function as a power series in t from which the moments μ'_r may be 'read off' as coefficients of $\frac{t^r}{r!}$.

(d) Consider the following statement:

'More swimmers are likely to leave the blocks to begin their races after $t = 0$ than before $t = 0$.'

Does this statement support the suggestion that $Y \sim U[-1, 1]$?

4.35 The probability distribution for the random variable X giving 'the number of dots facing uppermost when a fair die is thrown' may be written as

x	1	2	3	4	5	6
$p_X(x)$	$\frac{1}{6}$	$\frac{1}{6}$	$\frac{1}{6}$	$\frac{1}{6}$	$\frac{1}{6}$	$\frac{1}{6}$

This is a simple example of a discrete uniform distribution. In general, a random variable X has a **discrete uniform distribution** on the integers $1, 2, 3, \ldots, n$ if X has probability mass function

$$p_X(x) = \begin{cases} \frac{1}{n} & \text{if } x = 1, 2, 3 \ldots, n; \\ 0 & \text{otherwise.} \end{cases}$$

In this case we write $X \sim U(n)$.

(a) If $X \sim U(n)$, determine $E(X)$. Here you may require the summation

$$1 + 2 + 3 + \ldots + n = \tfrac{1}{2}n(n+1).$$

(b) If $X \sim U(n)$, determine $Var(X)$. Here you may require the summation

$$1^2 + 2^2 + 3^2 + \ldots + n^2 = \frac{1}{6}n(n+1)(2n+1).$$

(c) Show that your results for mean and variance from (a) and (b) give the results for the mean and variance of the outcome of a fair die when $n = 6$.

(d) If $X \sim U(n)$, predict the value of the coefficient of skewness ζ_3.

(e) A lottery has tickets labelled consecutively from 0 000 001 to 1 000 000.

A single ticket is randomly selected as the lottery winner.

i. How many tickets are there?

ii. What is the expected number of the winning ticket? Does the ticket that you hold have a greater chance of winning if it is numerically close to this expected number? Explain.

iii. If X denotes number of the winning ticket, give the value of n so that $X \sim U(n)$.

iv. Write down $Var(X)$.

4.36 Suppose that a code-word may be regarded as a string of 0 and 1 symbols of various lengths. The following is a list of 12 code-words:

0 0 0 0 0 0 0 0 0 0
1 1 1 1
1 1 0 0 0 0 0 1
1 1 1 1 1 1 1
0 1 0 1 0 1 0 1 0
0 0 0 0 0 1
1 1 1 0 0 0 0 0
0 0 1 0 0 0 0 0 0 0
1 1
1
0 0 0 1 0 1 0 1 1 1 1
0 0 0 0 0

When a code-word is selected at random from this list, we record the observed value of the random variable X denoting the number of zeros in the selected code-word.

(a) Construct a graph of the probability mass function of X.

(b) Is X symmetrically distributed about μ_X? Use the pmf graph from (a) to justify your answer.

(c) Show that $\zeta_3 = 0$.

4.37 Compare the probability density functions

$$f_Z(z) = \frac{1}{\sqrt{2\pi}} e^{-\frac{1}{2}z^2}, \qquad -\infty < z < \infty,$$

and

$$f_X(x) = \begin{cases} 6(\frac{1}{2}+x)(\frac{1}{2}-x) & \text{if } -\frac{1}{2} < x < \frac{1}{2}; \\ 0 & \text{otherwise,} \end{cases}$$

with respect to *centre*, *spread*, *skewness* and *kurtosis*. For this comparison, use measures based on moments.

4.38 Given that the cumulant generating function $\kappa_X(t)$ is defined by

$$\kappa_X(t) = \log_e m_X(t),$$

show that:

$$\begin{aligned} \kappa_X^{(1)}(0) &= \mu \\ \kappa_X^{(2)}(0) &= \sigma^2 \\ \kappa_X^{(3)}(0) &= \mu_3 \\ \kappa_X^{(4)}(0) &= \mu_4 - 3\sigma^4, \end{aligned}$$

where $\mu = E(X)$ and $\sigma^2 = Var(X)$.

4.5 TRANSFORMATION OF RANDOM VARIABLES

Power transformation

We recall from Definition 1.2 that for an index κ and constants a and b, a **power transformation of a random variable** X creates a new random variable Y where

$$Y = \begin{cases} a + bX^{\kappa}, & \text{for } \kappa \neq 0; \\ a + b\log_e X, & \text{for } \kappa = 0. \end{cases}$$

In Chapter 1 such power transformations were used on data sets for the purpose of reducing long tails (by removing skewness) and improving symmetry. The rewards for this lie in our later study of statistical inference where much of the methodology relies on assumptions of symmetry in the data. The above power family of transformations results from the application of a power transformation of the form

$$W = f_{\kappa}(X) = \begin{cases} X^{\kappa}, & \text{for } \kappa \neq 0; \\ \log_e X, & \text{for } \kappa = 0. \end{cases}$$

followed by the linear transformation $Y = a + bW$. Such a composition of transformations changes X to W in a non-linear way and then W to Y in a linear way.

In this section we give some answers to the following questions. When a transformation is applied to a random variable X:

what is the effect on the probability distribution of X?
what is the effect on $E(X)$?
what is the effect on $Var(X)$?

The distribution function method

Suppose that X is a continuous random variable and define a new random variable Y by $Y = h(X)$. Typically, h may be a linear function, a power transformation, or even a general non-linear function of X (such as $tanX$, $e^X \ldots$). Using the result of (4.2), that is, $F'_Y(y) = f_Y(y)$, to specify the relationship between the cdf $F_Y(y)$ of Y and the pdf $f_Y(y)$, the so-called **distribution function method** for determining the density of Y involves:

- finding the range of h;
- determining the cdf $F_Y(y)$;
- differentiating the cdf to obtain the pdf $f_Y(y)$.

For example, if $X \sim exp(1)$, what is the distribution of βX for some constant $\beta > 0$? In this case, $Y = h(X) = \beta X$ corresponds to a scale change so that the range of h is still the positive real numbers. The pdf of X is $f_X(x) = e^{-x}$ for $x > 0$. The cdf of X is

$$F_X(x) = \int_0^x e^{-t}dt = 1 - e^{-x}, \quad \text{for } x \geq 0.$$

We use this information in the evaluation of the cdf of Y.

$$F_Y(y) = P(Y \leq y) = P(\beta X \leq y) = P(X \leq \frac{y}{\beta}) = F_X(\frac{y}{\beta}).$$

The final stage in the process is to differentiate each side of the equation $F_Y(y) = F_X(\frac{y}{\beta})$ with respect to y to obtain the pdf of Y:

$$f_Y(y) = \frac{d}{dy}F_Y(y) = F'_X(\frac{y}{\beta})\frac{1}{\beta} = f_X(\frac{y}{\beta})\frac{1}{\beta} = \frac{1}{\beta}e^{-\frac{y}{\beta}} \text{ for } y > 0.$$

Certainly we can recognise from the equation of the pdf of Y that Y has an exponential pdf. In fact $Y \sim exp(\beta)$. Therefore this scale change has had the effect of altering only the shape of the pdf of X and not the defining nature of X itself. That is, X has not changed from an exponential to a uniform, normal, Cauchy ... random variable, through simple multiplication by a constant. This is often true of the effect of linear transformations on continuous random variables.

If $X \sim exp(1)$, then $\beta X \sim exp(\beta)$.

Power transformations may inflict more fundamental change on X. Let us examine the effect of taking logarithms of uniformly distributed random variables. Suppose that $X \sim U[0,1]$ and let $Y = h(X) = -\log_e X$. Then since $0 < X < 1$, by the definition of the logarithm of fractions, $-\log_e X > 0$. Then, for $y > 0$,

$$\begin{aligned} F_Y(y) &= P(Y \leq y) \\ &= P(-\log_e X \leq y) \\ &= P(X \geq e^{-y}) \\ &= 1 - P(X < e^{-y}) \\ &= 1 - e^{-y}, \end{aligned}$$

implying, again by differentiating each side of $F_Y(y) = 1 - e^{-y}$ with respect to y, that $f_Y(y) = e^{-y}$ for $y > 0$. In summary

If $X \sim U[0,1]$, then $-\log_e X \sim exp(1)$.

Linear transformations: the effect on mean and variance

Suppose that the distribution of the random variable X is known. For constants a and b, the random variable $Y = a + bX$ has a distribution that may be determined by the 'distribution function method' of the previous section. However we may only require the mean, variance and possibly some higher order moments of Y and not a complete specification of the entire distribution of Y.

Recall from Theorem 4.5 that for functions h_1 and h_2 of X,

$$E[c_1 h_1(X) + c_2 h_2(X)] = c_1 E[h_1(X)] + c_2 E[h_2(X)],$$

provided expected values exist. With $c_1 = a, h_1(X) = 1, c_2 = b$ and $h_2(X) = X$, it follows that

$$E(a + bX) = a + bE(X).$$

Therefore, when $Y = a + bX$, $\mu_Y = a + b\mu_X$. This information assists in the variance calculation:

$$\begin{aligned} Var(Y) &= E(Y - \mu_Y)^2 \\ &= E[(a + bX) - (a + b\mu_X)]^2 \\ &= E(bX - b\mu_X)^2 \\ &= b^2 E(X - \mu_X)^2 \\ &= b^2 Var(X). \end{aligned}$$

The standard deviation is found by taking positive square roots of each side of this equation. That is $\sigma_X = |b|\sigma_X$. Finally,

$$E(Y - \mu_Y)^r = b^r E(X - \mu_X)^r$$

for any positive integer r. This shows that the rth central moments

$$\zeta_r = \frac{\mu_r}{(\sqrt{\mu_2})^r}$$

are scale independent (since the powers of b cancel from the numerator and the denominator). We have therefore established the following theorem:

Theorem 4.9 *If random variables X and Y are related by $Y = a + bX$, then*

$$\begin{aligned} \mu_Y &= a + b\mu_X \\ \sigma_Y^2 &= b^2\sigma_X^2 \\ \sigma_Y &= |b|\sigma_X \\ (\zeta_r)_Y &= (\zeta_r)_X. \end{aligned}$$

■

Example 4.10

Trucking down the Hume Highway from Sydney to Melbourne

A large freight-moving firm uses the Hume Highway as one of its many long-haul truck routes. The experience of the cost of many thousands of interstate journeys has indicated that the operating cost of a Mack Super Liner 500 V8 truck (excluding the wages of the driver) is $15.5+0.3X$ cents per kilometre when the truck travels at X kilometres per hour where X has pdf

$$f_X(x) = \begin{cases} \frac{1}{4000}x & \text{if } 80 < x < 120; \\ 0 & \text{otherwise.} \end{cases}$$

Worked example:

Determine the expected operating cost for the Hume Highway trip from Sydney to Melbourne of 1000 km.

The random variable X has a simple density to integrate to determine the mean driving speed:

$$E(X) = \int_{80}^{120} x\frac{x}{4000}dx = \left.\frac{x^3}{12000}\right]_{80}^{120} = \frac{304}{3} \text{ km hr}^{-1}.$$

The cost function Y, in dollars, for the Sydney/Melbourne trip of 1000 km is

$$Y = \frac{1}{100}(15.5 + 0.3X)1000 = 155 + 3X,$$

so that $E(Y) = E(155 + 3X) = 155 + 3E(X) = 155 + (3)(\frac{304}{3}) = 459$. This establishes that the expected cost for the journey is \$459.

Further non-linear transformations

The distribution function method may be used to successfully extract the distribution of $Y = h(X)$ from the distribution of X for a wide variety of non-linear functions h.

Example 4.11

Firing shots at a random angle: how far do they land from the target?

A gun fires bullets at an angle X to the horizontal, where $-\frac{\pi}{2} < X < \frac{\pi}{2}$ is a randomly chosen angle. That is $X \sim U[-\frac{\pi}{2}, \frac{\pi}{2}]$. If the target is the point 0 labelled on the vertical axis 1 unit away, then many bullets will land well clear of the target. The distance to the target is represented, through elementary trigonometry, by the transformed random variable $Z = tanX$.

Random shots at target 0

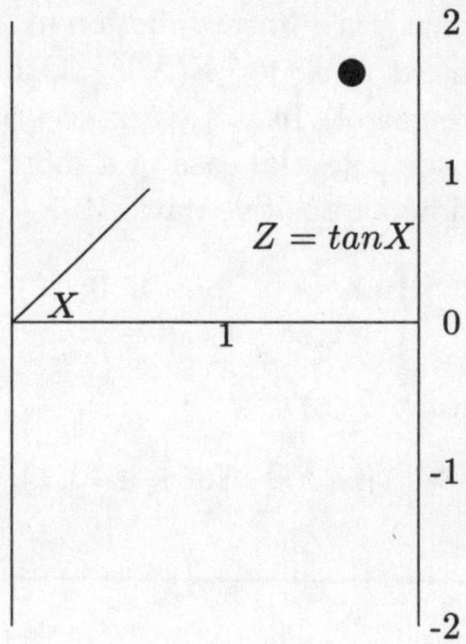

$$\begin{aligned} F_Z(z) &= P(Z \le z) \\ &= P(tanX \le z) \\ &= P(X \le tan^{-1}z) \text{ for } X \sim U[-\frac{\pi}{2}, \frac{\pi}{2}] \\ &= \frac{tan^{-1}z - (-\frac{\pi}{2})}{\frac{\pi}{2} - (-\frac{\pi}{2})} \\ &= \frac{tan^{-1}z}{\pi} + \tfrac{1}{2}. \end{aligned}$$

Since $F'_Z(z) = f_Z(z)$, we may differentiate the distribution function to obtain for $-\infty < z < \infty$,

$$f_Z(z) = \frac{d}{dz}\left[\frac{tan^{-1}z}{\pi} + \tfrac{1}{2}\right]$$

$$= \frac{1}{\pi}\frac{d}{dz}(tan^{-1}z)$$
$$= \frac{1}{\pi}\frac{1}{1+z^2} \quad \text{for} \quad -\infty < z < \infty.$$

This is the recognisable pdf of the Cauchy distribution from Definition 4.11. That is, $Z \sim Cauchy(1)$. For this distribution there is no finite mean, so that the higher moments do not exist either. This is hardly accurate shooting!
The median *does* exist however. By solving $F_Z(\xi_{0.5}) = \frac{1}{2}$, we find that $tan^{-1}\xi_{0.5} = 0$, giving $\xi_{0.5} = 0$ as the median of Z.

Theorem 4.9 indicated that if $h(X) = a + bX$ is a linear function of X, then

$$E[h(X)] = h[E(X)]. \tag{4.6}$$

In general, (4.6) holds only when h is a *linear* function of X. We have already seen this occurring in the context $E(X^2) \neq E(X)^2$. (Otherwise all variances $\sigma_X^2 = E(X^2) - E(X)^2$ would be zero!) In fact, variances are positive (or ≥ 0) so that $E(X^2) \geq E(X)^2$. This is a special case of a more general result that applies to power transformations of positive data: if

$$f_\kappa(X) = \begin{cases} X^\kappa & \text{for } \kappa \neq 0; \\ \log_e X & \text{for } \kappa = 0, \end{cases}$$

then we state without proof that

$$E[f_\kappa(X)] \geq f_\kappa[E(X)] \quad \text{for } \kappa \notin (0,1). \tag{4.7}$$

Example 4.12

Expected value of a non-linear function: Example 4.10 continued

When the driver's wages of \$30 per hour are included, then the time for the Hume Highway journey of 1000 km at X km per hour is $\frac{1000}{X}$ hours so that the total operating cost is

$$Y_{tot} = Y + 30(\frac{1000}{X}) = Y + \frac{30000}{X} \quad \text{dollars,}$$

where Y is the operating cost of the truck defined in Example 4.10. This total cost has expected value

$$E(Y_{tot}) = E(Y + \frac{30000}{X})$$
$$= E(Y) + E(\frac{30000}{X})$$

$$\begin{aligned} &= 459 + 30000E(\frac{1}{X}) \\ &= 459 + 30000\int_{80}^{120} \frac{1}{x}\frac{x}{4000}dx \\ &= 459 + \frac{30000}{4000}(120 - 80) \\ &= 759. \end{aligned}$$

It is important to remember in this calculation that the power transformation $f_{-1}(x) = x^{-1}$ is non-linear so that we expect

$$E\left(\frac{1}{X}\right) \neq \frac{1}{E(X)}.$$

In this example, $E(\frac{1}{X}) = 0.01$ while $\frac{1}{E(X)} = \frac{3}{304} < 0.01$ as is indicated by (4.7).

Simulating random variables

In this section we demonstrate how certain transformations may be used to simulate continuous data; that is, to artificially generate observations from a probability distribution of a specified type.

Theorem 4.10 *For any continuous random variable X with cdf F_X,*

$$F_X(X) \sim U[0,1].$$

Proof: The random variable $F_X(X)$ is simply a function $h(X)$ of X where $h = F_X$. The distribution function technique allows us to determine the distribution of $Y = F_X(X)$:

$$F_Y(y) = P(Y \leq y) = P[F_X(X) \leq y] = P[X \leq F_X^{-1}(y)]$$

since continuous distribution functions are strictly increasing. Therefore

$$F_Y(y) = F_X[F_X^{-1}(y)] = y \quad \text{for} \;\; 0 < y < 1$$

and we can recognise $F_Y(y) = y$ as the cdf of a $U[0,1]$ random variable.

■

Theorem 4.11 *(Monte Carlo simulation of X) Suppose that $U \sim U[0,1]$. Then $F_X^{-1}(U)$ has cdf F_X for any continuous random variable X.*

Proof: Set $Y = F_X^{-1}(U)$. Then

$$F_Y(y) = P(Y \leq y) = P[F_X^{-1}(U) \leq y] = P[U \leq F_X(y)] = F_U[F_X(y)] = F_X(y),$$

since $U \sim U[0,1]$ has cdf $F_U(u) = u$ for all $0 < u < 1$. Therefore Y has cdf F_X.

■

Theorem 4.11 can be applied in the following way: To construct a data set from a known specified (continuous) probability distribution (exponential, normal, ...), select randomly the data set $u_1, u_2, u_3, \ldots, u_n$ from the interval $[0,1]$, then transform each member of this uniform data set according to the transformation

$$y_i = F_X^{-1}(u_i), \quad i = 1, 2, 3 \ldots, n.$$

This artificially created data set (or **simulated data set**) $y_1, y_2, y_3, \ldots, y_n$ represents observations on the random variable X. The only difficulty in this procedure is that F_X must be inverted and this requires careful consideration if X is discrete or if there is no algebraic formula for F_X^{-1} (as in the case of the normal distribution).

Example 4.13

Simulating exponential data

In this example we take 36 random numbers from $[0,1]$ and transform them into observations on a random variable $X \sim exp(1)$. By Theorem 4.11, since $F_X(x) = 1 - e^{-x}$ for $x > 0$, we have that, if $u = F_X(x)$, then $x = F_X^{-1}(u)$. That is, $u = 1 - e^{-F_X^{-1}(u)}$. Solving this equation for $F_X^{-1}(u)$ we find that $F_X^{-1}(u) = -\log_e(1-u)$.

Random numbers from $[0,1]$
Scale: stems—tenths; leaves—hundredths
2|2 represents 0.22

	U
0	1 3 4 5
1	3 6 7
2	2 2 8
3	0 1 2 6 9
4	3 6 8
5	2 2 9 9
6	1 6 9
7	0 6 7 8
8	0 5 6 9
9	1 4 6

Random numbers from $exp(1)$
Scale: stems—units; leaves—tenths
1 * |1 represents 1.1

	$-\log_e(1-U)$
0*	0 0 0 0 1 1 1
t	2 2 3 3 3 3
f	4 4 5
s	6 6 7 7
.	8 8 9
1*	0 1
t	2
f	4 4 5
s	6
.	8 8
2*	
t	2
f	4
s	
.	8
3*	
t	2

EXERCISES

4.39 When a photocopier distributer can sell X machines the advertising and transportation costs are $C = 150X + 220$ dollars, where X has pmf

x	0	1	2	3	4
$p_X(x)$	0.1	0.1	0.4	0.3	0.1

Determine the mean and variance of the cost function C. Evaluate $E(C^2)$.

4.40 If $X \sim U[a, b]$, determine the values of c and d in terms of a and b so that $Y = c + dX \sim U[0, 1]$. How do the mean and variance of X and Y compare?

4.41 If $X \sim N(0, 1)$, use the distribution function method to find the pdf of $Y = X^2$.

4.42 If a particle of fixed mass m has variable velocity $V \sim N(0, 1)$, determine the probability density function of the kinetic energy $\frac{1}{2}mV^2$ of the particle.

4.43 The Leadbeater possum is an endangered species of possum restricted to areas of bushland in the Dandenong hills near Melbourne. In this habitat the species must compete with other species of possum for food.

Assume that the proportion P of the food supply available to an endangered species is a random variable P, where $P \sim U[0, 1]$. If the observed value of P is large, then any food shortages will have an adverse effect on the population of the species while other competing species that consume the same food will flourish.

(a) Determine the distribution of $1 - P$.

(b) Do the distributions of P and $1 - P$ have the same mean and variance?

(c) The distribution of

$$Y = max\{P, 1 - P\},$$

the maximum of the random variables P and $1 - P$, is uniformly distributed. Determine the support of the distribution.

(d) The ratio

$$\frac{P}{1 - P}$$

is called an **odds ratio**. Determine the probability that the odds ratio exceeds $\frac{1}{2}$.

(e) Is the odds ratio a power transformation?

(f) Determine the cdf of the odds ratio.

(g) Determine the pdf of the odds ratio.

(h) The quantity

$$L = \log_e \left(\frac{P}{1 - P} \right)$$

is called the **log odds ratio**. Determine the pdf of L.

4.44 There is enough fencing material to make a rectangular security enclosure for equipment storage with a perimeter of 2 km. If the length of one of the sides is a random variable X km where $X \sim U[0, 1]$, determine the pdf of the area of the enclosure.

4.45 Suppose that the income, Y (in dollars), of a randomly selected individual working in a large company has cdf

$$F_Y(y) = \begin{cases} 0, & y < 30\,000; \\ 1 - (\frac{30\,000}{y})^3, & y \geq 30\,000. \end{cases}$$

(a) Determine the probability that an individual's income is less than \$40 000.

(b) Determine the median income.

4.46 In Exercise 4.45:

(a) Determine the pdf of superannuation contributions which are set at 7% of income.

(b) If the company plans to give all employees a 10% wage rise determine the cdf of the income distribution that this wage rise would generate.

Show how the company could simulate the income distribution after the wage rise by first selecting 100 random numbers from the interval $[0, 1]$ and then transforming them to post wage rise incomes.

CHAPTER FIVE

Joint Distributions: Utilising Independence

In this chapter you will find

5.1 JOINT DISTRIBUTIONS

Often the process of obtaining data leads us to measure and record data for more than one variable. Frequently the main point to our inquiries is the collection of data for several variables for each experimental unit: when a laboratory mouse (experimental unit) is injected with one of several dose levels (Variable 1), how many days will it live (Variable 2)? If the researcher wishes to question whether the dose level is affecting survival time, the possible nature and strength of the relationship between Variable 1 and Variable 2 is examined. Experimental units may be expensive to maintain, so that it is worth recording as much information as possible.

In Example 3.4, the unfortunate motorist is the experimental unit when the police set up a road block for taking breath tests. For each of the motorists stopped the interviewing officer would (usually) have access to: name, gender, seatbelt wearing, passenger numbers, time of day, time of week, blood alcohol reading ... For one batch of 50 motorists, two such variables are cross-tabulated here:

	Blood alcohol level	
	≥ 0.05	< 0.05
Men	6	26
Women	2	16

By dividing by 50, the total sample size, the frequencies of occurrence of the four different categories in the above table are turned to relative frequencies of occurrence. Such relative frequencies sum to 1 across all cells in the table. If a different set of motorists had been stopped in the road block, then a different set of relative frequencies would probably have resulted.

If the exact conditions of the road block (same time, same place, same weather, ...) were repeated over many occasions we approach the concept of limiting relative frequencies in each of the 4 categories. As we have seen in Chapter 3, one of the many approaches to probability is through limiting relative frequencies. As far as the police manning the road block are concerned, their experience in seeing lots of relative frequencies leads them to probability assessments; a constable manning the road block may respond: *'yes, after midnight on a Thursday the chances of picking up drivers over the 0.05 limit is only 1 in 20'*. Thus, probabilities may be assigned to the 4 categories of the above table to give a **joint probability distribution.**

		X = **Blood alcohol level**	
		≥ 0.05	< 0.05
Y = Gender	Men	0.15	0.50
	Women	0.05	0.30

We will first be considering jointly distributed discrete random variables that have specific numerical outcomes. A single random variable X is called a **univariate** random variable. Such random variables were studied in Chapter 4. A pair (X, Y) of random variables taken together constitute a **bivariate random variable**. We will study bivariate random variables in this chapter.

More than two random variables considered together constitute a **multivariate random variable.**

Joint mass function

Definition 5.1 *A* **discrete joint probability distribution** *is a tabular listing of the values (x, y) that the bivariate random variable (X, Y) may assume, together with the* **joint probability mass function** $p_{X,Y}(x, y)$ *defined by*

$$p_{X,Y}(x, y) = P(X = x, Y = y).$$

□

The comma in the expression $P(X = x, Y = y)$ is suitably interpreted as 'and'. With appropriate category labels '0' for 'men' and '≥ 0.05' and '1' for 'women' and '< 0.05', we obtain the following probability distribution for the blood alcohol level investigation:

$p_{X,Y}(x, y)$ x	y	0	1
0		0.15	0.50
1		0.05	0.30

The bivariate random variable (X, Y) is defined on a sample space $\mathcal{S}_X \times \mathcal{S}_Y$ which is the **cartesian product** of the sample spaces $\mathcal{S}_X$ for X and $\mathcal{S}_Y$ for Y. That is,

$$\mathcal{S}_X \times \mathcal{S}_Y = \{(\omega_1, \omega_2) : \omega_1 \in \mathcal{S}_X \textit{ and } \omega_2 \in \mathcal{S}_Y\}.$$

Outcomes of the bivariate random variable (X, Y) are therefore ordered pairs $(X(\omega_1), Y(\omega_2)) = (x, y)$ where $\omega_1 \in \mathcal{S}_X$ and $\omega_2 \in \mathcal{S}_Y$. This (x, y) pair is simply a point in the plane (making no reference to the underlying sample spaces). If we place dots at heights $p_{X,Y}(x, y)$ above such points (x, y), then the resulting diagram is a graph of the discrete joint probability mass function $p_{X,Y}$. This structure of the sample space ensures the following properties for $p_{X,Y}(x, y)$:

1. for all pairs (x, y), $p_{X,Y}(x, y) \geq 0$;
2. $\sum_{all\ (x,y)} p_{X,Y}(x, y) = 1$.

These ideas generalise the concept of probability mass function used for a single discrete random variable: univariate random variables possess positive mass functions which sum to 1. Probabilities for bivariate discrete random variables are simply calculated in the product sample space according to: If $A \subset \mathcal{S}_X \times \mathcal{S}_Y$, then

$$P[(X,Y) \in A] = \sum_{(x,y)\in A} p_{X,Y}(x,y).$$

We often wish to examine the distribution of more than two discrete random variables at once. Definition 5.1 may be generalised to cater for n discrete random variables where $n > 2$. Suppose that $X_1, X_2, X_3, \ldots, X_n$ is a collection of n random variables. The **joint probability mass function**, $p_{X_1, X_2, X_3, \ldots, X_n}(x_1, x_2, x_3, \ldots, x_n)$, of $X_1, X_2, X_3, \ldots, X_n$ is defined by

$$p_{X_1, X_2, X_3, \ldots, X_n}(x_1, x_2, x_3, \ldots, x_n) = P(X_1 = x_1, X_2 = x_2, X_3 = x_3, \ldots, X_n = x_n).$$

This reduces to Definition 5.1 when $n = 2$.

Example 5.1
Winning off jacks and queens
An individual player in a game of poker is dealt 5 cards from a pack of 52 cards. If the hand consists of 'pairs' or 'three of a kind' of cards such as *jacks* and *queens*, considerable bluff may be required to avoid being out-ranked by an opponent's *aces* and *kings*. Let us define the discrete random variables X = the number of *jacks* in the hand; Y = the number of *queens* in the hand. Both the random variables X and Y may assume the values $0, 1, 2, 3, 4$ under the condition that there is at most a combined total of 5 *jacks* and *queens* in the hand: $x + y \leq 5$. Then

$$p_{X,Y}(x,y) = P(X = x, Y = y) = \frac{\binom{4}{x}\binom{4}{y}\binom{44}{5-x-y}}{\binom{52}{5}},$$

and on substitution of pairs of x and y values in succession we obtain

$p_{X,Y}(x,y)$ y / x	0	1	2	3	4
0	0.4179	0.2089	0.0306	0.0015	*
1	0.2089	0.0815	0.0087	0.0003	*
2	0.0306	0.0087	*	*	0
3	0.0015	0.0003	*	0	0
4	*	*	0	0	0

where results are expressed to 4 dp and * represents probabilities smaller than 10^{-4}. For example,

$$p_{X,Y}(0,0) = P(X=0, Y=0) = \frac{\binom{4}{0}\binom{4}{0}\binom{44}{5}}{\binom{52}{5}} = \frac{\binom{44}{5}}{\binom{52}{5}} = 0.4179.$$

This table certainly shows how the probabilities are massed in two dimensions: since $p_{X,Y}(0,0) = 0.4179$, and no other probabilities in the table exceed this value, the point $(0,0)$ is the most likely outcome. That is, when considering *jacks* and *queens* in a poker hand, the most likely outcome is none of each!

Other events defined in the cartesian product space have their probabilities readily counted point by point. The probability that the poker hand contains the same number of *jacks* and *queens* is

$$\begin{aligned} P(X=Y) &= p_{X,Y}(0,0) + p_{X,Y}(1,1) + p_{X,Y}(2,2) + p_{X,Y}(3,3) \\ &\quad + p_{X,Y}(4,4) \\ &= 0.4179 + 0.0815 + * + 0 + 0 \\ &= 0.50 \text{ (2 dp)} \end{aligned}$$

Therefore, $P(X \neq Y) = 0.50$ (2 dp), and, by symmetry, $P(X > Y) = 0.25$ (2 dp).

Marginal probability mass function

Given a discrete joint probability distribution, the marginal probabilities are found by adding across the (rows/columns) of the joint table.

Definition 5.2 *The* **marginal probability mass functions** $p_X(x)$ *for* X *alone and* $p_Y(y)$ *for* Y *alone are given by:*

$$p_X(x) = \sum_{\text{all } y} p_{X,Y}(x,y); \qquad p_Y(y) = \sum_{\text{all } x} p_{X,Y}(x,y).$$

□

The marginal mass functions, as their name would suggest, are written about the edges of the joint table. The marginal distribution is found by summing over the unwanted variable.

Example 5.2
Cans of tomatoes: price and quantity
Suppose that a shipment of cans of tomatoes arriving at a supermarket warehouse contains either 50 or 100 cartons, and that these cartons are sold to retail outlets at either \$6 per carton or at the discounted price of \$5 per carton. Random variables X, denoting 'number of cartons in shipment', and Y, denoting 'price per carton', are assumed to have joint pmf

$p_{X,Y}(x,y)$ y x	5	6
50	0.40	0.20
100	0.15	0.25

Worked example:

1. **Determine the probability that a shipment contains 100 cartons.**
2. **Construct the marginal probability mass functions for X and Y alone.**

1. A shipment of 100 cartons may be either full price at \$6 per carton, or at the discounted price of \$5 per carton. Therefore the probability that the shipment contains 100 cartons is

$$P(X = 100) = P_{X,Y}(100,5) + p_{X,Y}(100,6) = 0.15 + 0.25 = 0.40.$$

2. The marginal mass functions are constructed by adding along the rows and columns of the joint probability table:

$p_{X,Y}(x,y)$ y x		5	6	$p_X(x)$
50		0.40	0.20	0.60
100		0.15	0.25	0.40
	$p_Y(y)$	0.55	0.45	1

Then, reading the last row and last column, we have:

y	5	6
$p_Y(y)$	0.55	0.45

x	50	100
$p_X(x)$	0.60	0.40

Joint density functions

Suppose that X and Y are continuous random variables and comprise the bivariate random variable (X, Y).

Definition 5.3 *The* **joint probability density function** *for the bivariate random variable (X, Y) is the function $f_{X,Y}$ satisfying*

$$P[(X,Y) \in A] = \int\int_A f_{X,Y}(x,y)dxdy$$

for any two-dimensional subset A of the plane. The set of values (x, y) for which $f_{X,Y}(x, y) > 0$ is called the **support** *of the density.*

□

Given a point (x, y) in the plane, the pdf rises to a height $f_{X,Y}(x, y)$ above that point. Therefore the equation $z = f_{X,Y}(x, y)$ creates a surface in three dimensions as x and y vary in the horizontal plane. Definition 5.3 indicates that probabilities for bivariate random variables are volumes under that surface: $P[(X, Y) \in A]$ is the volume (represented by the integral) under that portion of the surface $z = f_{X,Y}(x, y)$ that stands on the set A in the horizontal plane. In particular,

$$P(a \leq X \leq b,\ c \leq Y \leq d) = \int_c^d \int_a^b f_{X,Y}(x,y)dxdy.$$

This is analogous to the case for a univariate random variable for which $P(a \leq X \leq b) = \int_a^b f_X(x)dx$. Pursuing the analogy further, since univariate densities are positive and integrate to 1 across their support, we find the same results govern joint densities:

1. $f_{X,Y}(x, y) \geq 0$ for all (x, y);
2. $\int_{-\infty}^{\infty} \int_{-\infty}^{\infty} f_{X,Y}(x, y)dxdy = 1$.

As in the discrete case we often wish to examine the distribution of more than two continuous random variables at once. Definition 5.3 may be generalised to cater for n continuous random variables where $n > 2$. Suppose that $X_1, X_2, X_3, \ldots, X_n$ is a collection of n continuous random variables. The **joint probability density function,** $f_{X_1, X_2, X_3, \ldots, X_n}(x_1, x_2, x_3, \ldots, x_n)$, of $X_1, X_2, X_3, \ldots, X_n$ satisfies

$$\begin{aligned} &P[(X_1, X_2, X_3, \ldots, X_n) \in A] \\ &= \int\int\int \cdots \int_A f_{X_1, X_2, X_3, \ldots, X_n}(x_1, x_2, x_3, \ldots, x_n)dx_1dx_2dx_3 \ldots dx_n \end{aligned}$$

for every n-dimensional subset A of n-dimensional space. This reduces to Definition 5.3 when $n = 2$.

Example 5.3

Uniform support

In a laboratory test for the presence of bacteria, the sample to be tested is thoroughly mixed in a medium and placed on a flat circular dish. After several days of storage at a controlled temperature, a small black dot appearing in the medium is the first sign of a positive test result.

If the coordinates of this first dot are (X, Y), then (X, Y) is uniformly distributed over the (unit) circle of the dish since the bacteria are well mixed in the medium. The disc $x^2 + y^2 \leq 1$ of area $\pi(1)^2 = \pi$ is the support of the joint density function. Hence

$$f_{X,Y}(x,y) = \begin{cases} \frac{1}{\pi} & \text{if } x^2 + y^2 \leq 1; \\ 0 & \text{otherwise.} \end{cases}$$

Hence, the probability that the first dot appears within the square with coordinates

$$(\frac{1}{\sqrt{2}}, \frac{1}{\sqrt{2}}), (\frac{1}{\sqrt{2}}, -\frac{1}{\sqrt{2}}), (-\frac{1}{\sqrt{2}}, \frac{1}{\sqrt{2}}) \text{ and } (-\frac{1}{\sqrt{2}}, -\frac{1}{\sqrt{2}})$$

is

$$P(\text{spot first appears in square}) = \frac{\text{area square}}{\text{area circle}} = \frac{(\frac{2}{\sqrt{2}})^2}{\pi} = \frac{2}{\pi}.$$

The same probability may be determined by integrating the joint density over the square:

$$\begin{aligned} P(-\frac{1}{\sqrt{2}} \leq X \leq \frac{1}{\sqrt{2}}, -\frac{1}{\sqrt{2}} \leq Y \leq \frac{1}{\sqrt{2}}) &= \int_{-\frac{1}{\sqrt{2}}}^{\frac{1}{\sqrt{2}}} \int_{-\frac{1}{\sqrt{2}}}^{\frac{1}{\sqrt{2}}} f_{X,Y}(x,y)dydx \\ &= \int_{-\frac{1}{\sqrt{2}}}^{\frac{1}{\sqrt{2}}} \int_{-\frac{1}{\sqrt{2}}}^{\frac{1}{\sqrt{2}}} \frac{1}{\pi} dydx \\ &= \int_{-\frac{1}{\sqrt{2}}}^{\frac{1}{\sqrt{2}}} \left. \frac{y}{\pi} \right|_{y=-\frac{1}{\sqrt{2}}}^{y=\frac{1}{\sqrt{2}}} dx \\ &= \int_{-\frac{1}{\sqrt{2}}}^{\frac{1}{\sqrt{2}}} \frac{\sqrt{2}}{\pi} dx \\ &= \left. \frac{\sqrt{2}x}{\pi} \right|_{x=-\frac{1}{\sqrt{2}}}^{x=\frac{1}{\sqrt{2}}} \end{aligned}$$

$$= \frac{\sqrt{2}}{\pi}\left(\frac{1}{\sqrt{2}} - (-\frac{1}{\sqrt{2}})\right)$$
$$= \frac{2}{\pi}.$$

Because this double integral is over a rectangular region in the plane, it is evaluated by determining each integral in sequence, first integrating with respect to y and then with respect to x. The *order* of integration, however, is not important.

As is the case for discrete bivariate random variables, the marginal densities are obtained by integrating out (rather than summing) the unwanted variable.

Definition 5.4 *Given a joint density $f_{X,Y}(x,y)$ of the continuous bivariate random variable (X,Y), the* **marginal probability density functions** *$f_X(x)$ and $f_Y(y)$ of X and Y respectively are*

$$f_X(x) = \int_{-\infty}^{\infty} f_{X,Y}(x,y)dy \quad \textit{and} \quad f_Y(y) = \int_{-\infty}^{\infty} f_{X,Y}(x,y)dx.$$

□

By definition, the marginal density for X is obtained by integrating out the unwanted variable Y from the joint density $f_{X,Y}(x,y)$. This coincides with the pdf for X when X is considered as a random variable in its own right.

EXERCISES

5.1 The value of the random variable X gives the number of absentees from the morning shift and the value of the random variable Y gives the number of absentees from the evening shift on the same day at a small factory that produces aerosol spray cans.

The following joint probability mass function of (X,Y) was ascertained from a long series of past attendance records.

$p_{X,Y}(x,y)$ y x	0	1	2	3
0	0.10	0.05	0.05	0.05
1	0.20	0.10	0.10	0.05
2	0.10	0.15	0.05	0

(a) Determine $p_{X,Y}(2,1)$.

(b) Find the probability that there is at least one absentee from the evening shift.

(c) Determine the probability that there is the same number of absentees in each shift.

(d) Find the probability that there is at least one absentee from either shift.

5.2 In Exercise 5.1, determine the marginal distributions for X and Y alone.

5.3 $p_{X,Y}(x,y) = \frac{1}{4}$ for

$$(x,y) = (0,2), (-2,0), (1,-1), (-1,1),$$

and is 0 elsewhere.

(a) Write out the joint probability distribution in tabular form.

(b) Determine the marginal distributions of X and Y alone.

(c) The outcomes for X and the outcomes for Y each have probability 0.25. Does this mean that X and Y have identical probability distributions?

5.4 Evaluate $P(X+Y=0)$ for the joint probability mass function of Exercise 5.3.

5.5 Does the table below define a discrete joint probability distribution for a pair of random variables (X,Y)? Explain.

$p_{X,Y}(x,y)$ y / x	-5	1	8
-3	0.01	0.05	0.05
1	0.05	0.10	0.10
2	0.01	0.15	0.05
6	0.05	0.05	0.01

5.6 Suppose that the bivariate random variable (X,Y) may assume the values (1,0), (2,0), (1,3), (2,3) with respective probabilities 0.3, 0.2, 0.1 and 0.4.

Evaluate $P(Y=3)$ and $P(X=1)$.

5.7 Use the data of Example 5.1 as a guide to determine the probability that a poker hand has the same number of *kings* as *queens*.

5.8 On a given weekday a rental car firm ascertains that it is likely to receive X requests for Volvos when there are Y available in stock where X and Y are jointly described by the bivariate pmf:

$p_{X,Y}(x,y)$ y / x	0	1	2
0	0.05	0.05	0.10
1	0.15	0.10	0.10
2	0.10	0.10	0.15
3	0.05	0.05	0.00

(a) Determine the probability distribution for X.

(b) Determine the probability distribution for Y.

(c) What is the expected number of requests for Volvos on a given day?

(d) Determine the probability that the number of requests for Volvos exceeds the available stock.

5.9 Two fair dice are tossed. Let X denote the *maximum* of the two numbers showing and Y denote the *minimum* of the two numbers showing.

(a) Construct the joint probability mass function of (X, Y) by copying and completing the missing entries marked * in the following table with entries $p_{X,Y}(x, y)$:

x \ y	1	2	3	4	5	6
1	$\frac{1}{36}$	0	0	0	0	0
2	$\frac{2}{36}$	$\frac{1}{36}$	0	0	0	0
3	*	*	*	*	*	*
4	*	*	*	*	*	*
5	*	*	*	*	*	*
6	$\frac{2}{36}$	$\frac{2}{36}$	$\frac{2}{36}$	$\frac{2}{36}$	$\frac{2}{36}$	$\frac{1}{36}$

(b) Determine $P(X > Y)$.

(c) Determine the marginal pmf for X alone.

(d) Find the expected maximum of the two numbers showing.

(e) Find the expected minimum of the two numbers showing.

(f) A student of probability thought carefully about the relationship between X and Y and then wrote:

$$P(X = k) = P(Y = 7 - k),$$
$$k = 1, 2, \ldots, 6.$$

Was the student correct? Discuss.

5.10 Given the discrete bivariate random variable (X, Y), with pmf $p_{X,Y}(x, y)$ having support $(x_1, y_1), (x_2, y_2), \ldots, (x_n, y_n)$, show that the function, $q(x)$, defined by

$$q(x) = \sum_{j=1}^{n} p_{X,Y}(x, y_j),$$

is a probability mass function in the sense that:

- $q(x_i) \geq 0$ for each $i = 1, 2, \ldots, n$;
- $\sum_{i=1}^{n} q(x_i) = 1$.

Give the marginal distribution of X in tabular form and in terms of the function q.

5.11 In Example 5.3, the point (X, Y) is uniformly distributed on the unit disc with centre $(0, 0)$. Use the cumulative distribution function technique to determine the probability density function of the distance

$$R = \sqrt{X^2 + Y^2}$$

of the point (X, Y) from $(0, 0)$.

Hence evaluate $E\left(\sqrt{X^2 + Y^2}\right)$.

5.12 In Example 5.3, the point (X, Y) is uniformly distributed on the unit disc with centre $(0, 0)$.

(a) Explain why the marginal distribution of X may be represented by the integral

$$f_X(x) = \int_{-\sqrt{1-x^2}}^{\sqrt{1-x^2}} \frac{1}{\pi} dy,$$

for $|x| \leq 1$.

(b) Hence evaluate:

i. $f_X(x)$;

ii. $f_Y(y)$;

iii. $E(X)$.

(c) What is the geometric interpretation of $E(Y)$?

5.13 If the support of the joint pdf

$$f_{X,Y}(x, y) = e^{-(x+y)}$$

in the first quadrant of the x–y plane where the x and y values are both positive, determine the marginal density functions.

5.14 (a) The area under the standard normal probability density curve is 1. Use this idea to determine the value of the integral

$$\int_{-\infty}^{\infty} e^{-\frac{1}{2}t^2}\,dt.$$

(b) If the support of the joint pdf

$$f_{X,Y}(x,y) = ke^{-\frac{1}{2}(x^2+y^2)}$$

is the entire $x-y$ plane, determine the value of the constant k.

5.15 Determine $P(X \geq Y)$ for the joint pdf of Exercise 5.13.

5.16 This problem shows that the marginal distributions are *not* sufficient to characterise the joint distribution of two random variables.

Let X and Y have pdfs $f_X(x)$ and $f_Y(y)$ respectively and cdfs $F_X(x)$ and $F_Y(y)$ respectively. Show that for any constant α with $|\alpha| < 1$, the function $f_{X,Y}(x,y)$ defined by

$$f_{X,Y}(x,y) = f_X(x)f_Y(y)\{1+\alpha[2F_X(x)-1][2F_Y(y)-1]\},$$

is a joint pdf with marginal density functions $f_X(x)$ and $f_Y(y)$. (Therefore there are uncountably infinitely many joint densities that have $f_X(x)$ and $f_Y(y)$ as marginals.)

5.2 FACTORISING THROUGH INDEPENDENCE

We have seen the term 'independence' applied to *events* in a sample space: Definition 3.13 characterises events A and B as independent if

$$P(A \cap B) = P(A) \cap P(B).$$

By suitable choice of the events A and B, we can observe the same type multiplication occurring for probabilities of events defined in terms of random variables.

Independence for discrete random variables

When random variables X and Y are taken as jointly distributed, the events $\{X \in A\}$ and $\{Y \in B\}$ may be regarded as subsets of the product sample space. Then, the events $\{X \in A\}$ and $\{Y \in B\}$ are independent precisely when

$$P(\{X \in A\} \cap \{Y \in B\}) = P(\{X \in A\})P(\{Y \in B\}). \tag{5.1}$$

If, specifically, in Equation 5.1 we take $A = \{x\}$, $B = \{y\}$, then

$$\{X \in A\} = \{X = x\} \text{ and } \{Y \in B\} = \{Y = y\}.$$

This shifts the primary focus from independent events to independent random variables. The fact that the joint probability mass function may be written $p_{X,Y}(x, y) = P(\{X = x\} \cap \{Y = y\})$, leads us naturally to the definition:

Definition 5.5 *Discrete random variables X and Y are termed* **independent** *if for all* (x, y),

$$p_{X,Y}(x, y) = p_X(x)p_Y(y).$$

Random variables that are not independent are said to be **dependent**.

□

When discrete random variables are independent, their joint probability mass function factorises as the product of the marginal mass functions. When two discrete mass functions $p_X(x)$ and $p_Y(y)$ represent the marginal mass functions of a discrete bivariate random variable (X, Y), then *every* entry in the joint probability table must factorise for X and Y to be independent. For example, if the discrete random variable Y with distribution

y	1	2	3
$p_Y(y)$	0.3	0.5	0.2

and the discrete random variable X with distribution

x	0	2	4
$p_X(x)$	0.4	0.3	0.3

are written in the joint probability table

$p_{X,Y}(x,y)$	y	1	2	3	
x					$p_X(x)$
0					0.4
2					0.3
4					0.3
	$p_Y(y)$	0.3	0.5	0.2	1

as the marginals of the joint probability mass function of (X, Y), then if X and Y are independent, the probabilities $p_{X,Y}(x, y)$ *must* be entered in the joint table as products:

$p_{X,Y}(x,y)$	y	1	2	3	
x					$p_X(x)$
0		0.12	0.20	0.08	0.4
2		0.09	0.15	0.06	0.3
4		0.09	0.15	0.06	0.3
	$p_Y(y)$	0.3	0.5	0.2	1

In Example 5.2 the random variables X, denoting 'number of cartons in shipment', and Y, denoting 'price per carton', are dependent since, by the enumeration of the joint pmf in Example 5.2,

$$p_{X,Y}(50, 5) = 0.40 \neq (0.60)(0.55) = p_X(50)p_Y(5).$$

Since the factorisation fails for at least one cell in the joint table, the random variables are dependent.

Independence for continuous random variables

Let us recall Equation 5.1, the independence criterion for random variables X and Y:

$$P(\{X \in A\} \cap \{Y \in B\}) = P(\{X \in A\})P(\{Y \in B\}).$$

Suppose now that X and Y are continuous random variables and if we make the identifications $A = (-\infty, x]$, $B = (-\infty, y]$, for $-\infty < x < \infty$, $-\infty < y < \infty$, then

$$\{X \in A\} = \{X \leq x\} \text{ and } \{Y \in B\} = \{Y \leq y\}.$$

Since $P(\{X \leq x\} \cap \{Y \leq y\}) = P(X \leq x, Y \leq y)$, by differentiating each side of the equation

$$P(X \leq x, Y \leq y) = P(X \leq x)P(Y \leq y)$$

with respect to both x and y, one after the other, it follows that continuous random variables X and Y are independent if their joint density factorises. Factorising joint densities may also be seen to imply Equation 5.1, so we have obtained a useful operating definition for independent continuous random variables.

Definition 5.6 *Continuous random variables X and Y are termed* **independent** *if, for all (x, y),*

$$f_{X,Y}(x, y) = f_X(x)f_Y(y).$$

When X and Y are not independent, they are termed **dependent**.

□

Example 5.4

Waiting for landing clearance

At Auckland International Airport, if two jumbo jets arrive within 5 minutes of one another, the second to arrive must wait for clearance to land while the first taxis from the runway. This delays the arrival time of the second jet.

Suppose that the arrival times X_1 and X_2 of two aircraft are independent and uniformly distributed over the next 20 minutes. That is: $X_i \sim U[0,20]$, $i = 1,2$. Because of the independence of the arrival times, the joint distribution (X_1, X_2) is uniform over the square $[0,20] \times [0,20] = \{0 \le x_1 \le 20, 0 \le x_2 \le 20\}$.

The support of (X_1, X_2)

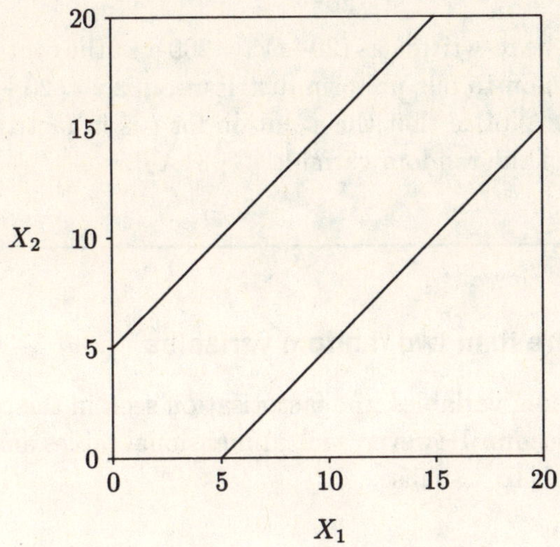

$$f_{X_1,X_2}(x_1,x_2) = f_{X_1}(x_1)f_{X_2}(x_2) = \begin{cases} \frac{1}{20^2} & 0 \le x_1 \le 20, 0 \le x_2 \le 20; \\ 0 & \text{otherwise.} \end{cases}$$

We wish to develop the probability that one of the aircraft will be delayed on landing. In terms of the random variables X_1 and X_2 this means

$$\begin{aligned} P(|X_1 - X_2| < 5) &= P(-5 < X_1 - X_2 < 5) \\ &= P(X_1 > X_2 - 5 \text{ and } X_1 < X_2 + 5). \end{aligned}$$

In the (x_1, x_2) coordinate system, the probability that we require is that volume standing on the region

$$\begin{aligned} x_1 &> x_2 - 5 \\ x_1 &< x_2 + 5 \\ 0 \le\ &x_1 \le 20 \\ 0 \le\ &x_2 \le 20 \end{aligned}$$

under the density $f_{X_1,X_2}(x_1,x_2)$. The area of the required region is the area of the entire square less the area of the two triangles in the corners of the square. This is $20^2 - (2)(\frac{1}{2})(15^2)$. Therefore

$$P(|X_1 - X_2| < 5) = \frac{1}{20^2}[20^2 - (2)(\tfrac{1}{2})(15^2)] = \frac{175}{400}.$$

This probability is quite substantial—almost $\frac{1}{2}$. It leads us to inquire further: What clearance time t is required for aircraft just landed so that there is a 50:50 chance that an incoming aircraft will not be given immediate landing clearance? We require this time a solution (for t) to the equation

$$\tfrac{1}{2} = P(|X_1 - X_2| < t) = \frac{1}{20^2}[20^2 - (2)(\tfrac{1}{2})(20 - t)^2)].$$

This quadratic may be re-written as $(20-t)^2 = 200$ and the root of interest in the support common to our uniform distribution is $t = 20 - \sqrt{200}$, or 5.86 minutes (2 dp). Notice that this solution for t is in fact the *median* of the distribution of the random variable $|X_1 - X_2|$.

Independence for more than two random variables

For more than two random variables, the factorisation seen in the case of two random variables also occurs. However, multidimensional tables and multiple integrals become difficult to visualise.

Definition 5.7 *Discrete random variables $X_1, X_2, X_3, \ldots, X_n$ are independent if*

$$p_{X_1, X_2, X_3, \ldots, X_n}(x_1, x_2, x_3, \ldots, x_n) = p_{X_1}(x_1)p_{X_2}(x_2)p_{X_3}(x_3)\ldots p_{X_n}(x_n)$$

for all values $(x_1, x_2, x_3, \ldots, x_n)$ of $(X_1, X_2, X_3, \ldots, X_n)$. Similarly, continuous random variables $X_1, X_2, X_3, \ldots, X_n$ are independent if

$$f_{X_1, X_2, X_3, \ldots, X_n}(x_1, x_2, x_3, \ldots, x_n) = f_{X_1}(x_1)f_{X_2}(x_2)f_{X_3}(x_3)\ldots f_{X_n}(x_n)$$

for all values $(x_1, x_2, x_3, \ldots, x_n)$ of $(X_1, X_2, X_3, \ldots, X_n)$.

□

There are many equivalent ways of phrasing the condition of independence expressed in Definition 5.7. The following independence criterion uses cumulative probabilities that we have encountered in the case $n = 2$. The random variables $X_1, X_2, X_3, \ldots, X_n$ are independent if

$$\begin{aligned} &P(X_1 \le x_1, X_2 \le x_2, \ldots, X_n \le x_n) = \\ &\quad P(X_1 \le x_1)P(X_2 \le x_2)\ldots P(X_n \le x_n) \end{aligned} \tag{5.2}$$

for all $(x_1, x_2, x_3, \ldots, x_n)$. This independence result is used in Example 5.5. Even more generally, the random variables $X_1, X_2, X_3, \ldots, X_n$ are independent if for n sets $A_1, A_2, A_3, \ldots, A_n$,

$$\begin{aligned} &P(X_1 \in A_1, X_2 \in A_2, \ldots, X_n \in A_n) = \\ &\quad P(X_1 \in A_1)P(X_2 \in A_2)\ldots P(X_n \in A_n). \end{aligned} \tag{5.3}$$

We will not concern ourselves here with establishing the equivalence of these 'versions' of independence. Instead, we will put these definitions into operation.

Example 5.5
Series and parallel component failure

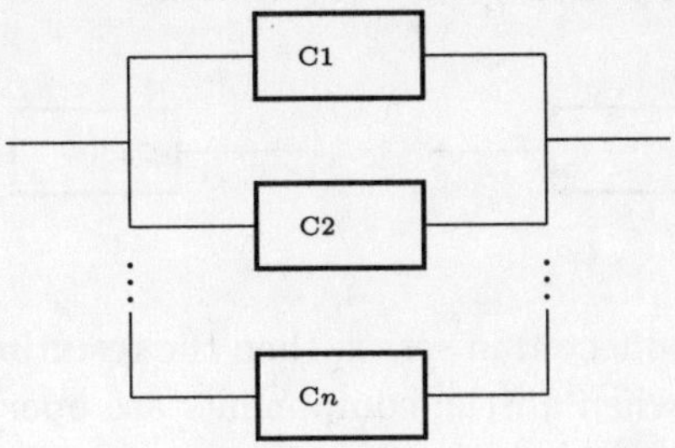

Suppose that the independent random variables $X_1, X_2, X_3, \ldots, X_n$ give the lifetimes (or times to failure) of n components which are labelled in the diagram as C1 C2, ..., Cn. This means that:
X_1 is the time to failure of Component 1;
X_2 is the time to failure of Component 2;
X_3 is the time to failure of Component 3;
...
X_n is the time to failure of Component n.
These times are assumed to be independent of one another. That is, knowing how long one component will last does not help us determine how long another component will last. Suppose that the components are of the same type, and that their lifetimes follow exponential distributions with common mean β. Thus, for each i, the random variables X_i each have $exp(\beta)$ distributions.
If the components are connected in parallel, then the resulting system will operate precisely when at least one of the components is operating. That is, the system fails to operate when the longest running component fails. Therefore the random variable, $Y_{parallel}$, describing the lifetime of the entire parallel system, is the maximum of all the possible lifetimes $X_1, X_2, X_3, \ldots, X_n$. To determine the distribution of

$$Y = Y_{parallel} = X_{(n)} = max\{X_1, X_2, X_3, \ldots, X_n\},$$

we may use the cumulative distribution function method of Section 4.5.

$$\begin{aligned} F_Y(y) &= P(Y \leq y) \\ &= P(X_{(n)} \leq y) \end{aligned}$$

$$
\begin{aligned}
&= P(max\{X_1, X_2, X_3, \ldots, X_n\} \leq y) \\
&= P(X_1 \leq y \text{ and } X_2 \leq y \text{ and } \ldots, \text{ and } X_n \leq y) \\
&= P(X_1 \leq y, X_2 \leq y, \ldots, X_n \leq y) \\
&= P(X_1 \leq y)P(X_2 \leq y)\ldots P(X_n \leq y) \quad \text{(by independence)} \\
&= (1 - e^{-\frac{y}{\beta}})(1 - e^{-\frac{y}{\beta}})\ldots(1 - e^{-\frac{y}{\beta}}) \\
&= (1 - e^{-\frac{y}{\beta}})^n.
\end{aligned}
$$

The pdf for $Y_{parallel}$ may be found by differentiation.

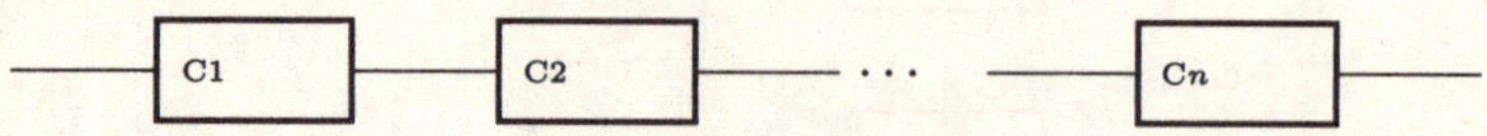

If the components are all connected in series, then the resulting series system will operate precisely when all the components are operating. That is, the system fails to operate when the first component fails. Therefore the random variable, Y_{series}, describing the lifetime of the entire series system is the minimum of all the possible lifetimes $X_1, X_2, X_3, \ldots, X_n$. To determine the distribution of

$$Y = Y_{series} = X_{(1)} = min\{X_1, X_2, X_3, \ldots, X_n\},$$

we may again use the cumulative distribution function method.

$$
\begin{aligned}
F_Y(y) &= P(Y \leq y) \\
&= 1 - P(Y > y) \\
&= 1 - P(X_{(1)} > y) \\
&= 1 - P(min\{X_1, X_2, X_3, \ldots, X_n\} > y) \\
&= 1 - P(X_1 > y \text{ and } X_2 > y \text{ and } \ldots, \text{ and } X_n > y) \\
&= 1 - P(X_1 > y, X_2 > y, \ldots, X_n > y) \\
&= 1 - P(X_1 > y)P(X_2 > y)\ldots P(X_n > y) \quad \text{(by independence)} \\
&= 1 - (e^{-\frac{y}{\beta}})(e^{-\frac{y}{\beta}})\ldots(e^{-\frac{y}{\beta}}) \\
&= 1 - e^{-n\frac{y}{\beta}}.
\end{aligned}
$$

Hence $Y_{series} \sim exp(\frac{\beta}{n})$. Since, by Section 4.5, $E(Y_{series}) = \frac{\beta}{n}$, it follows that the mean lifetime has been *reduced* by a factor of $\frac{1}{n}$ by having the components connected in series.

---

i.i.d. random variables

The lifetimes of the components in Example 5.5 are examples of what we call call *i.i.d.* random variables (independent and identically distributed), each with the same distribution as a random variable $X \sim exp(\beta)$. In general:

Definition 5.8 *Jointly distributed random variables $X_1, X_2, X_3, \ldots, X_n$ are termed* **independent and identically distributed,** *or* ***i.i.d.*** *random variables, if $X_1, X_2, X_3, \ldots, X_n$ are independent and for each $i = 1, 2, 3, \ldots, n$, the random variable X_i has the same probability distribution as a specified random variable X.*

□

Therefore, if $X_1, X_2, X_3, \ldots, X_n$ are *i.i.d.* random variables, then the joint density is

$$f_{X_1, X_2, X_3, \ldots, X_n}(x_1, x_2, x_3, \ldots, x_n) = \prod_{i=1}^{n} f_{X_i}(x_i) = \prod_{i=1}^{n} f(x_i),$$

where f is the pdf common to each of $X_1, X_2, X_3, \ldots, X_n$ and $\prod$ denotes 'product'.

In Example 5.5, $X_1, X_2, X_3, \ldots, X_n$ are *i.i.d.* $exp(\beta)$, for which

$$f_{X_i}(x_i) = \frac{1}{\beta} e^{-\frac{x_i}{\beta}},$$

so that

$$f_{X_1, X_2, X_3, \ldots, X_n}(x_1, x_2, x_3, \ldots, x_n) = \prod_{i=1}^{n} f_{X_i}(x_i) = \prod_{i=1}^{n} \frac{1}{\beta} e^{-\frac{x_i}{\beta}} = \frac{1}{\beta^n} e^{-\frac{\sum_{i=1}^{n} x_i}{\beta}}.$$

EXERCISES

5.17 It is thought that the number of packets of breakfast cereal sold in a day depends on which shelf (or rather the height of the shelf off the floor) the packets are located. The joint probability distribution tabulates probabilities for X, the number of packets sold (to the nearest 10) in a day, against Y, the height of the shelf off the floor in $\frac{1}{2}$-metre units.

$p_{X,Y}(x,y)$ $\quad y$ x	1	2	3
0	0	0.05	0
10	0.1	0.20	0.05
20	0.05	0.35	0.05
30	0	0.15	0

Are X and Y independent? Is there evidence for your answer in the way that the probabilities are arranged in the table?

For example, is there evidence in the table that there is a greater chance of selling more packets from a high shelf?

5.18 The number, X, of overnight calls to a fire station between 6 p.m. and 6 a.m. has been found over time to follow the pmf

x	0	1	2	3	4
$p_X(x)$	0.4	0.3	0.1	0.1	0.1

Assume that the number of calls on one night is independent of the number of calls received on any other night.

(a) Construct the joint probability distribution of (X, Y), where X and Y denote the number of night calls to the station on two consecutive nights.

(b) Determine the probability that there are more calls on a Monday night than a Tuesday night.

(c) Determine the probability that there are the same number of calls two nights in a row.

5.19 If

$$p_{X,Y}(x,y) = \begin{cases} \frac{1}{4} & (x,y) = (0,2), (-2,0), (1,-1), (-1,1); \\ 0 & \text{elsewhere}, \end{cases}$$

determine whether X and Y are independent.

5.20 Suppose that the bivariate random variable (X, Y) has joint pdf $f_{X,Y}(x, y)$. If

$$f_{X,Y}(x,y) = h_1(x)h_2(y),$$

for known functions h_1 and h_2, determine the marginal distributions of X and Y. Are X and Y independent? Explain briefly.

5.21 (a) If X and Y are *i.i.d.* $U[0, 1]$, give the joint density $f_{X,Y}(x, y)$ of X and Y.

(b) If X and Y are *i.i.d.* $U[0, 1]$, determine

$$P(|X - Y| \leq \frac{1}{4}).$$

(c) Two people agree to meet for coffee at a bar in High Street at any time between 3 p.m. and 4 p.m. They agree that the first to arrive at the coffee bar should wait no longer than 15 minutes for the other. Determine the probability that they will meet.

5.22 Suppose that two electronic components have lifetimes X_1 and X_2 that are *i.i.d.* $exp(\beta)$ for an unknown constant $\beta > 0$. Determine the value of $P(X_1 > X_2)$.

5.23 Suppose that two electronic components have lifetimes X_1 and X_2 that are independent random variables with

$X_1 \sim exp(\beta_1)$ and $X_2 \sim exp(\beta_2)$.

Give the joint density of (X_1, X_2).

5.24 In Exercise 5.23, determine the value of $P(X_1 > X_2)$.

5.25 The diagram shows an electronic unit which consists of two components, labelled C1 and C2, arranged in series:

The component lifetimes are independent and exponentially distributed. On average, C2 lasts twice as long as C1 which has a mean time to failure of 1000 hours.

(a) Determine the probability density function for the time to failure of the unit.

(b) What is the mean time to failure of the unit?

(c) Determine the probability that the unit will fail after 1000 hours of operation.

5.26 Discrete random variables X_1 and X_2 are *i.i.d.*, each having the following probability distribution:

x	-1	1
$p_X(x)$	$\frac{1}{2}$	$\frac{1}{2}$

(a) Construct the joint pmf of X_1 and X_2.

(b) Determine the pmf of X_1X_2.

(c) Construct the joint pmf of X_1 and X_1X_2.

(d) Are X_1 and X_1X_2 independent? Discuss.

5.27 The diagram shows an electronic unit which consists of two components, labelled C1 and C2, arranged in parallel:

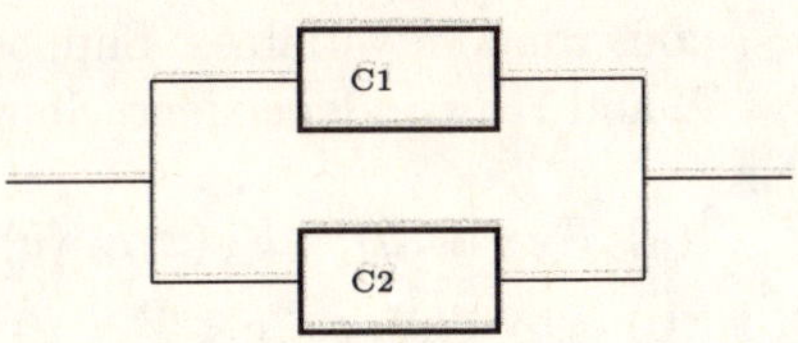

Suppose that X_1 denotes the lifetime of C1 and X_2 denotes the lifetime of C2. Assume that X_1 and X_2 are independent.

(a) If X_1, X_2 are *i.i.d.* $exp(\beta)$ random variables, determine the cumulative distribution function of the lifetime of the unit.

(b) If $X_i \sim exp(\beta_i)$, $i = 1, 2$, determine the cumulative distribution function of the lifetime of the unit.

(c) If X_1, X_2 are *i.i.d.* $exp(\beta)$ random variables, and two such units are connected in series, determine the cumulative distribution function of the lifetime of the resulting system.

5.28 The diagram shows an electronic unit which consists of two components, labelled C1 and C2, arranged in series:

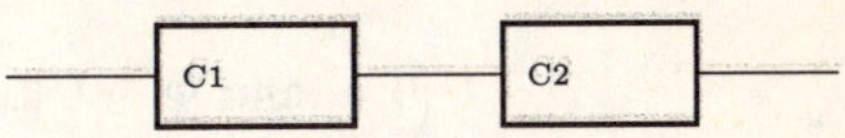

Suppose that X_1 denotes the lifetime of C1 and X_2 denotes the lifetime of C2. Assume that X_1 and X_2 are independent.

(a) If X_1, X_2 are *i.i.d.* $exp(\beta)$ random variables, determine the cumulative distribution function of the lifetime of the unit.

(b) If $X_i \sim exp(\beta_i)$, $i = 1, 2$, determine the cumulative distribution function of the lifetime of the unit.

(c) If X_1, X_2 are *i.i.d.* $exp(\beta)$ random variables, and two such units are connected in parallel, determine the cumulative distribution function of the lifetime of the resulting system.

5.29 Four numbers are chosen at random from the unit interval $[0, 1]$. Determine the expected value of the largest.

[*Hint*: Let X_i denote the ith number chosen. We require the pdf of $X_{(4)} = max\{X_1, X_2, X_3, X_4\}$.]

Can you generalise your solution to cover the case of the expected size of the largest of n numbers selected at random from the unit interval?

5.30 (a) Show that the area under the curve $y = cos\phi$ between $\phi = 0$ and $\phi = \frac{\pi}{2}$ is 1.

(b) Suppose that a and b are constants with $a > b$. Suppose further that random variables Y and Φ are independent with

$$Y \sim U[0, a] \text{ and } \Phi \sim U[0, \frac{\pi}{2}].$$

i. Give the joint pdf $f_{Y,\Phi}(y, \phi)$ of Y and Φ.

ii. Draw a diagram of the support of $f_{Y,\Phi}(y, \phi)$.
Shade the region of this support corresponding to the event $\{Y < bcos\Phi\}$.

iii. Show that

$$P(Y < bcos\Phi) = \frac{2b}{\pi a}.$$

(c) *Buffon's Needle Problem.*

A needle of length $2b$ is thrown on a board covered in parallel lines a distance $2a$ apart.

Show that, under suitable assumptions, the probability that the needle intersects one of the lines is

$$\frac{2b}{\pi a}.$$

(d) Suppose that you repeat Buffon's Needle Experiment 1000 times and record the number of times that the needle is observed to land on a line. How could you use this information to get an estimate of π?

5.31 Suppose that the height H, in metres, reached by a river in flood exceeds a fixed height h metres with probability

$$P(H > h) = e^{-\frac{h}{10}}.$$

The height reached in one flood is assumed to be independent of the height reached in any other flood.

Determine the probability that a 20 metre stop bank will contain the worst of 20 successive floods.

5.32 For jointly distributed random variables X and Y, the function $F_{X,Y}(x, y)$ defined for all $-\infty < x < \infty$ and $-\infty < y < \infty$ by the probability

$$F_{X,Y}(x, y) = P(X \leq x, Y \leq y),$$

is called the **joint cumulative distribution function** of X and Y.

This definition applies equally well to jointly distributed discrete or continuous random variables. Suppose that X and Y are independent. Show that:

(a) $F_{X,Y}(x, y) = F_X(x)F_Y(y)$;

(b) $P(a < X < b, c < Y < d)$
$= F_X(b)F_Y(d) - F_X(b)F_Y(c)$
$-F_X(a)F_Y(d) + F_X(a)F_Y(c)$,
for constants $a < b$ and $c < d$;

(c) $F_{X,Y}(x, \infty) = F_X(x)$;

(d) $F_{X,Y}(\infty, y) = F_Y(y)$;

(e) $F_{X,Y}(\infty, \infty) = 1$.

5.3 INDEPENDENT PRODUCTS AND INDEPENDENT SUMS

One of the main aims of the current chapter and the two which follow is to develop an understanding of the distributions of sums of independent random variables from knowledge of the component random variables making up that sum. For example, in a sum of quantities which vary, the mean and variance of the components in the sum each contribute explicitly to the overall mean and variance of the entire sum.

We may seek only the mean and the variance of a sum of independent random variables. If this is so then we are lucky, in a sense, because these may be extracted without determining the definitive distribution of the entire sum. To determine the distribution of a sum is possibly quite a daunting task, as the next section shows.

Determining the distribution of a sum of independent random variables

We begin the concept of a sum of random variables with a very simple discrete example.

Example 5.6

When two dice are rolled, how high is the sum?

Let X denote the outcome for the face showing uppermost when the first die is rolled, and Y denote the outcome for the second die, rolled independently of the first. The joint probability table for (X, Y) is given below, where $*$ denotes $\frac{1}{36}$, the probability assigned to the 36 equally likely for the two dice.

To determine the probability distribution for $X+Y$, the sum of X and Y, the required probabilities may be determined by summing the joint probabilities along the diagonals of the table. Illustrated is

$$
\begin{aligned}
p_{X+Y}(8) &= P(\{X+Y=8\}) \\
&= p_{X,Y}(6,2)+p_{X,Y}(5,3)+p_{X,Y}(4,4) \\
&\quad +p_{X,Y}(3,5)+p_{X,Y}(2,6) \\
&= \frac{1}{36}+\frac{1}{36}+\frac{1}{36}+\frac{1}{36}+\frac{1}{36} \\
&= \frac{5}{36}.
\end{aligned}
$$

Adding random variables along the diagonal

x \ y	1	2	3	4	5	6
1	*	*	*	*	*	*
2	*	*	*	*	*	*
3	*	*	*	*	*	*
4	*	*	*	*	*	*
5	*	*	*	*	*	*
6	*	*	*	*	*	*

$\{X+Y=8\}$

Notice that $X+Y$ may assume values between 2 and 12 inclusive, and the probabilities for each of these points is similarly determined by adding across different but parallel diagonals in the joint table. The discrete probability distribution for $Z = X+Y$ is then

z	2	3	4	5	6	7	8	9	10	11	12
$p_Z(z)$	$\frac{1}{36}$	$\frac{2}{36}$	$\frac{3}{36}$	$\frac{4}{36}$	$\frac{5}{36}$	$\frac{6}{36}$	$\frac{5}{36}$	$\frac{4}{36}$	$\frac{3}{36}$	$\frac{2}{36}$	$\frac{1}{36}$

Graph of probability mass function
for the total showing when two identical dice are tossed

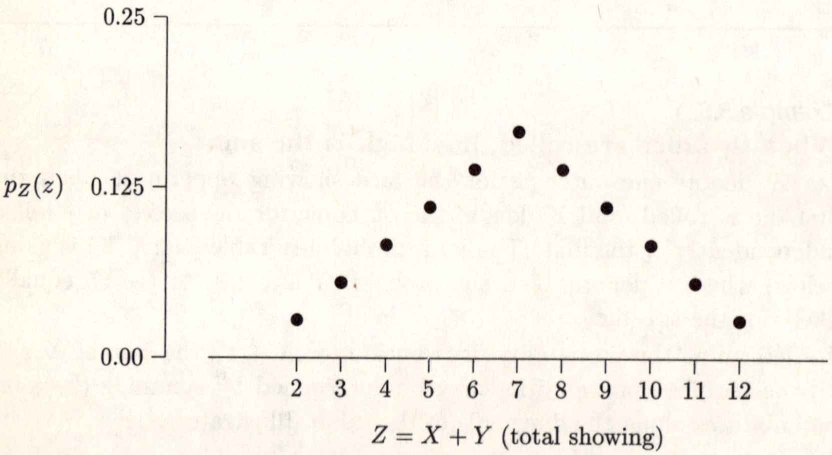

In the next example we repeat this procedure of searching along the diagonals of a square to determine the distribution of a sum. This, the case for adding independent continuous uniform random variables, follows the blueprint of the discrete case.

Example 5.7

When two random numbers are added, how big is the sum?

Suppose that I randomly select a number, X, from the interval $[0,1]$. Without knowledge of my number, you randomly select a number, Y, from the same interval. The aim of this example is to construct the pdf of the continuous random variable $Z = X + Y$, denoting the sum of the numbers selected.

By construction X and Y are *i.i.d.* $U[0,1]$ and through this independence, $f_{X,Y}(x,y) = f_X(x)f_Y(y)$ with support the unit square $[0,1] \times [0,1]$.

The cumulative distribution function method (of Section 4.5) is applied as follows: There are essentially two cases to consider—whether our sum is larger than 1, or whether it is smaller than 1. These two cases carve out different-shaped areas from the support of $f_{X,Y}(x,y)$.

Suppose initially that $z \leq 1$. Then

$$\begin{aligned} F_Z(z) &= P(Z \leq z) \\ &= P(X+Y \leq z) \\ &= \int\int_A f_{X,Y}(x,y)dxdy \\ &= \text{Area of the triangular Region } A \\ &= \tfrac{1}{2}z^2. \end{aligned}$$

The subset A of the support of (X,Y) where $x+y \leq z$ and $z \leq 1$

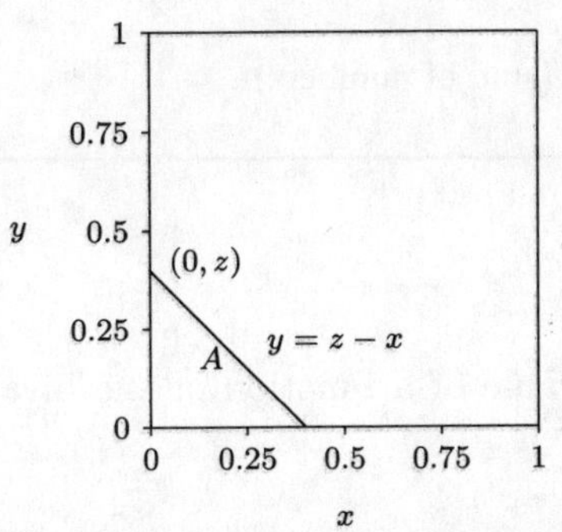

The subset B of the support of (X,Y) where $x+y > z$ and $z > 1$

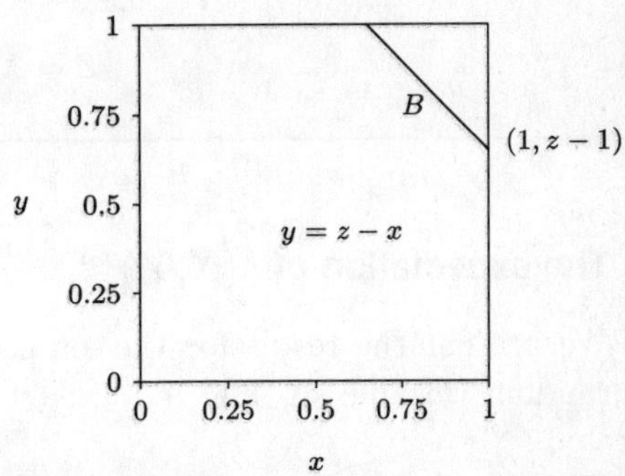

Now suppose that $1 < z \leq 2$. Then

$$\begin{aligned} F_Z(z) &= P(X+Y \leq z) \\ &= 1 - P(X+Y > z) \end{aligned}$$

$$
\begin{aligned}
&= 1 - \int\int_B f_{X,Y}(x,y)dxdy \\
&= 1 - \text{Area of the triangular Region } B \\
&= 1 - \tfrac{1}{2}[1-(z-1)]^2 \\
&= 1 - \tfrac{1}{2}(2-z)^2.
\end{aligned}
$$

Clearly z cannot exceed 2. Combining both these regions of support for Z, gives the structure

$$F_Z(z) = \begin{cases} \frac{1}{2}z^2 & \text{if } 0 \le z \le 1; \\ 1 - \frac{1}{2}(2-z)^2 & \text{if } 1 < z \le 2. \end{cases}$$

The pdf, obtained from $F_Z(z)$ by differentiation, is the triangular density

$$f_Z(z) = \begin{cases} z & \text{if } 0 \le z \le 1; \\ 2-z & \text{if } 1 < z \le 2. \end{cases}$$

Graph of probability density function
for the sum of two random numbers from $[0,1]$

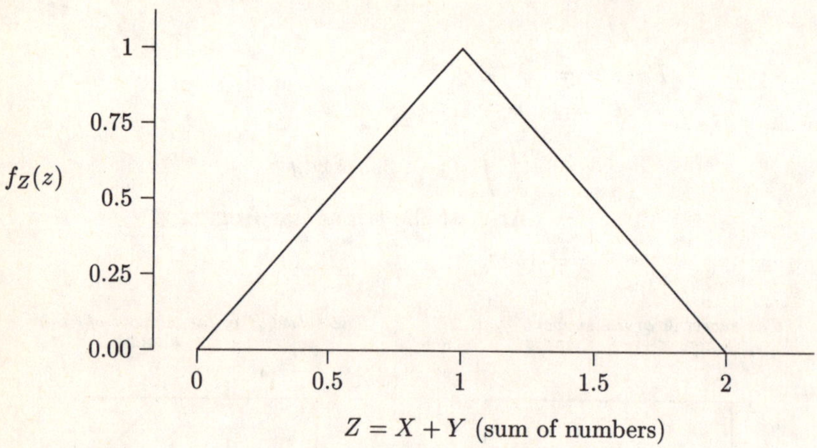

The expectation of $h(X,Y)$

We present the result for the expected value of a function of the bivariate random variable (X, Y).

$$E[h(X,Y)] = \int_{-\infty}^{\infty}\int_{-\infty}^{\infty} h(x,y)f_{X,Y}(x,y)dxdy \quad \text{(continuous case).}$$

$$E[h(X,Y)] = \sum_{\text{all } x}\sum_{\text{all } y} h(x,y)p_{X,Y}(x,y) \quad \text{(discrete case).} \tag{5.4}$$

These are obvious generalisations of formulas $E[h(X)] = \int_{-\infty}^{\infty} h(x)f_X(x)dx$ and $E[h(X)] = \sum_{all\ x} h(x)p_X(x)$ corresponding to the respective continuous and discrete cases when averaging over only one variable. We will be using these univariate formulas in this section. In fact it would be useful now to revise the expected value concepts of Section 4.3. Note that we do not need to determine the probability distribution of $h(X, Y)$ in order to determine $E[h(X, Y)]$.

Let us first consider the expected value of products of independent random variables. This corresponds to the choice $h(x, y) = xy$ of the function h. That is, we will begin by determining $E(XY)$ when X and Y are independent. The result is simply stated in Theorem 5.1 below. The importance of this theorem is not so much to characterise the occasions when random variables X and Y are independent, but rather to emphasise the consequences of the assumption that X and Y are independent.

Theorem 5.1 *If X and Y are independent random variables, then $E(XY) = E(X)\ E(Y)$.*

Proof: The proof is given here for the continuous case. The adjustments required for the discrete case are obvious and left to the reader. The main thrust of the proof is that, under independence,

$$f_{X,Y}(x, y) = f_X(x)f_Y(y),$$

for all points (x, y). This factorisation allows the integrals to factorise:

$$\begin{aligned} E(XY) &= \int_{-\infty}^{\infty}\int_{-\infty}^{\infty} xyf_{X,Y}(x, y)dxdy \\ &= \int_{-\infty}^{\infty}\int_{-\infty}^{\infty} xyf_X(x)f_Y(y)dxdy \\ &= \int_{-\infty}^{\infty} xf_X(x)\left(\int_{-\infty}^{\infty} yf_Y(y)dy\right)dx \\ &= \left(\int_{-\infty}^{\infty} xf_X(x)dx\right)\left(\int_{-\infty}^{\infty} yf_Y(y)dy\right) \\ &= E(X)E(Y). \end{aligned}$$

■

Theorem 5.1 is simply generalised from the expected value of the product of X and Y to the expected value of the product $g(X)h(Y)$, where g and h are functions. The separation of the x terms and y terms into different factors is what allows the integral to factorise.

If X and Y are independent, $E[g(X)h(Y)] = E[g(X)]E[h(Y)]$.

This important result has immediate implications for moment generating functions. Given independent random variables X and Y, suppose that we choose $g(x) = e^{tx}$ and $h(y) = e^{ty}$ for a fixed value of t. Then, $g(x)h(y) = e^{tx}e^{ty} = e^{t(x+y)}$. By the definition of moment generating functions, provided $m_X(t)$ and $m_Y(t)$ both exist, we may write

$$m_{X+Y}(t) = E[e^{t(X+Y)}] = E[e^{tX}e^{tY}] = E[e^{tX}]E[e^{tY}] = m_X(t)m_Y(t).$$

This establishes a further major consequence of independence. The mgf of a sum of independent random variables is the *product* of the mgf's for each random variable. Having given the proof, we more formally state the result for future reference.

Theorem 5.2 *For independent random variables X and Y with respective moment generating functions $m_X(t)$ and $m_Y(t)$, the moment generating function of the sum, $X+Y$, is given by*

$$m_{X+Y}(t) = m_X(t)m_Y(t).$$

■

The mean and variance of a sum of independent random variables

Example 5.7 revealed how difficult it may be to determine the distribution of a sum of independent random variables. We now develop strategies for determining the mean and variance of the sum, without first establishing the distribution of the sum.

Theorem 5.3 *Suppose that a_1 and a_2 are finite constants. Then for two random variables X_1 and X_2, not necessarily independent,*

$$E(a_1X_1 + a_2X_2) = a_1E(X_1) + a_2E(X_2).$$

Further, **when X_1 and X_2 are independent,**

$$Var(a_1X_1 + a_2X_2) = a_1^2Var(X_1) + a_2^2Var(X_2).$$

Proof: We give the proof that $E(a_1X_1+a_2X_2) = a_1E(X_1)+a_2E(X_2)$ for the discrete case only. From this, the reader, as usual, may fill in the details of the continuous case. Using the function $h(x_1, x_2) = a_1x_1 + a_2x_2$, we obtain, by definition,

$$\begin{aligned} E(a_1X_1 + a_2X_2) &= \sum_{x_1}\sum_{x_2}(a_1x_1 + a_2x_2)p_{X_1,X_2}(x_1, x_2) \\ &= \sum_{x_1}\sum_{x_2}[a_1x_1p_{X_1,X_2}(x_1, x_2) + a_2x_2p_{X_1,X_2}(x_1, x_2)] \end{aligned}$$

$$\begin{aligned}
&= \sum_{x_1} a_1x_1 \sum_{x_2} p_{X_1,X_2}(x_1,x_2) + \sum_{x_2} a_2x_2 \sum_{x_1} p_{X_1,X_2}(x_1,x_2) \\
&= \sum_{x_1} a_1x_1p_{X_1}(x_1) + \sum_{x_2} a_2x_2p_{X_2}(x_2) \\
&= a_1 \sum_{x_1} x_1p_{X_1}(x_1) + a_2 \sum_{x_2} x_2p_{X_2}(x_2) \\
&= a_1E(X_1) + a_2E(X_2).
\end{aligned}$$

The derivation of the variance formula

$$Var(a_1X_1 + a_2X_2) = a_1^2Var(X_1) + a_2^2Var(X_2),$$

is given in a form that applies equally well to either discrete or continuous random variables. The concept of independence is used in the proof to factorise the expected value of a product as the product of the expected values. That is, since X_1 and X_2 are independent,

$$E[(X_1 - \mu_{X_1})(X_2 - \mu_{X_2})] = E[(X_1 - \mu_{X_1})]E[(X_2 - \mu_{X_2})] = (0)(0) = 0,$$

so that the 'cross product' terms disappear in the variance expansion:

$$\begin{aligned}
Var(a_1X_1 + a_2X_2) &= E[(a_1X_1 + a_2X_2) - E(a_1X_1 + a_2X_2)]^2 \\
&= E[(a_1X_1 + a_2X_2) - (a_1\mu_{X_1} + a_2\mu_{X_2})]^2 \\
&= E[a_1(X_1 - \mu_{X_1}) + a_2(X_2 - \mu_{X_2})]^2 \\
&= E[a_1^2(X_1 - \mu_{X_1})^2 + 2a_1a_2(X_1 - \mu_{X_1})(X_2 - \mu_{X_2}) \\
&\quad + a_2^2(X_2 - \mu_{X_2})^2] \\
&= a_1^2E[(X_1 - \mu_{X_1})^2] + 2a_1a_2E[(X_1 - \mu_{X_1})(X_2 - \mu_{X_2})] \\
&\quad + a_2^2E[(X_2 - \mu_{X_2})^2] \\
&= a_1^2Var(X_1) + 2a_1a_2E[(X_1 - \mu_{X_1})]E[(X_2 - \mu_{X_2})] \\
&\quad + a_2^2Var(X_2) \\
&= a_1^2Var(X_1) + a_2^2Var(X_2).
\end{aligned}$$

■

Example 5.8

Mean and variance of a difference

An industrial engineer takes two independent observations X_1 and X_2 on the temperature of effluent water that has been used to cool a boiler at the plant. These observations are *i.i.d.* since they are simple repeated measurements of the same quantity. Suppose that $E(X_i) = \mu$ and $Var(X_i) = \sigma^2$, $i = 1, 2$.

The engineer wishes to determine the mean and variance of the difference $X_1 - X_2$ between the two temperature observations. Notice that $X_1 - X_2 =$

$(1)X_1 + (-1)X_2$, so that Theorem 5.3 may be applied to the *difference* $X_1 - X_2$ with $a_1 = 1$ and $a_2 = -1$. Hence, Theorem 5.3 gives

$$\begin{aligned} E(X_1 - X_2) &= E[(1)X_1 + (-1)X_2] \\ &= (1)E(X_1) + (-1)E(X_2) \\ &= \mu - \mu \\ &= 0. \end{aligned}$$

Also, from Theorem 5.3 we obtain

$$\begin{aligned} Var(X_1 - X_2) &= Var[(1)X_1 + (-1)X_2] \\ &= (1)^2 Var(X_1) + (-1)^2 Var(X_2) \\ &= Var(X_1) + Var(X_2) \\ &= 2\sigma^2. \end{aligned}$$

The importance of this result is that when independent measurements are subtracted, their variances are added.

These results may be generalised to give formulae that we will refer to many times in the remainder of this text. If $X_1, X_2, X_3, \ldots, X_n$ are random variables with finite expectations and $a_1, a_2, a_3, \ldots, a_n$ are constants, then

SUM 1:

$$E(a_1X_1 + a_2X_2 + \ldots + a_nX_n) = a_1E(X_1) + a_2E(X_2) + \ldots + a_nE(X_n). \tag{5.5}$$

When $X_1, X_2, X_3, \ldots, X_n$ are independent, then

SUM 2:

$$Var(a_1X_1 + a_2X_2 + \ldots + a_nX_n) = a_1^2Var(X_1) + a_2^2Var(X_2) + \ldots + a_n^2Var(X_n). \tag{5.6}$$

Immediately, let us put **SUM 1** and **SUM 2** into operation.

Example 5.9

Fencing

Suppose that 10 timber railings are needed to make a fence. Each railing is supposed to be 3 metres long but, due to variations in the lengths of the original flitches (or trimmed logs) from which the railings are cut, it is reasonable to assume that the observed lengths $X_1, X_2, X_3, \ldots, X_{10}$ of the 10 railings are *i.i.d.* $U[2.9, 3.1]$ random variables.

Worked example:
Determine the mean and variance of the total length of the fence formed by placing these 10 railings end to end.
Here we can apply the formulas for **SUM 1** and **SUM 2** with $n = 10$ and the constants $a_1, a_2, \ldots, a_{10}$ all equal to 1. Note that by symmetry, if $X \sim U[a, b]$, then $E(X) = \frac{a+b}{2}$ and therefore, our particular uniform distribution for the length of a railing has mean length $\frac{2.9+3.1}{2} = 3.0$ metres. This means that $E(X_1) = 3.0$ and $E(X_2) = 3.0$ and $\ldots E(X_{10}) = 3.0$. (Certainly if the timber lengths are identically distributed, then they must all have the same mean.)

$$\begin{aligned} E(X_1 + X_2 + \ldots + X_{10}) &= E(X_1) + E(X_2) + \ldots + E(X_{10}) \\ &= 3.0 + 3.0 + 3.0 + \ldots + 3.0 \quad \text{(10 terms)} \\ &= 30.0. \end{aligned}$$

This means that, on average, our fence will have a total length of 30 metres (which is hardly surprising).
According to **SUM 2**, the key to determining the variance in the total length is in adding the variances of the individual railings that make up the fence. For this we will need to determine the variance of an individual railing. If $X \sim U[a, b]$, then $Var(X) = E(X^2) - [E(X)]^2 = \int_a^b x^2 (\frac{1}{b-a}) dx - (\frac{a+b}{2})^2 = \frac{(b-a)^2}{12}$. Therefore, our particular uniform distribution for the length of a railing has variance $\frac{(3.1-2.9)^2}{12} = 0.0033 \text{ m}^2$. That is, for each $i = 1, 2, 3, \ldots, 10$, $Var(X_i) = 0.0033$.

$$\begin{aligned} Var(X_1 + X_2 + \ldots + X_{10}) &= Var(X_1) + Var(X_2) + \ldots + Var(X_{10}) \\ &= 0.0033 + \ldots + 0.0033 \quad \text{(10 terms)} \\ &= 0.033. \end{aligned}$$

It is noticeable that, by adding measurements together, the errors in each measurement add together to give a much larger error for the total. In the next example, a simple way of counteracting this effect is examined.

Example 5.10

Combining length measurements for greater precision

Two pieces (1 and 2) of timber railing have respective true lengths L_1 and L_2. We wish to measure these true lengths as accurately as possible. The rules of our ability to measure length accurately are as follows: Every time we make a measurement of length, the variance of the amount by which we are in error is σ^2, a constant; further, we may make only two measurements.

There are probably several methods by which to proceed. Two methods are:

Method 1: Measure each piece once and record measured lengths X_1 of piece 1 and X_2 of piece 2.
Method 2: Join the two pieces end to end and measure their total length X_{tot}.
Place the two pieces side by side and measure the difference in their lengths X_{diff}.

Measuring length of pieces individually
Method 1

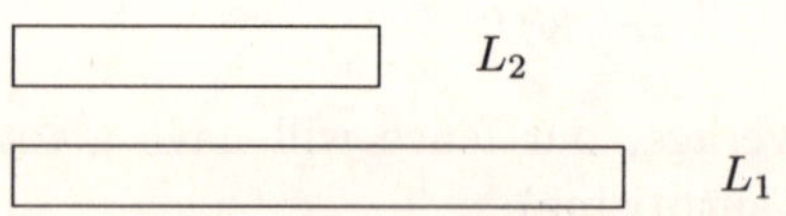

Measuring length difference and length sum
Method 2

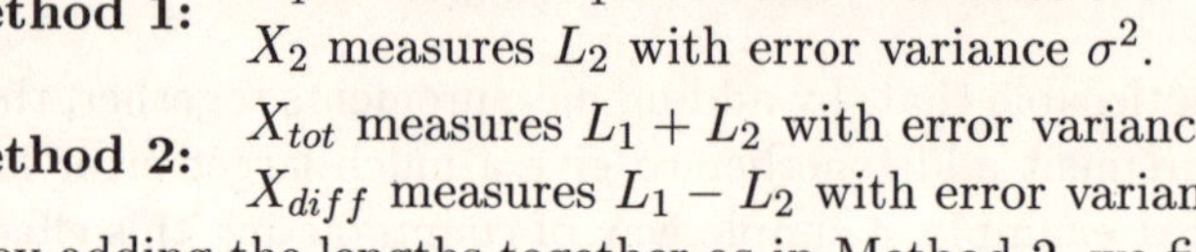

Under Method 1: X_1 measures L_1 with error variance σ^2.
X_2 measures L_2 with error variance σ^2.

Under Method 2: X_{tot} measures $L_1 + L_2$ with error variance σ^2.
X_{diff} measures $L_1 - L_2$ with error variance σ^2.

Therefore, by adding the lengths together as in Method 2, we find that

$$\frac{X_{tot} + X_{diff}}{2} \text{ measures } \frac{(L_1 + L_2) + (L_1 - L_2)}{2} = L_1 \text{ with error variance}$$

$$Var(\frac{X_{tot} + X_{diff}}{2}) = (\tfrac{1}{2})^2 Var(X_{tot}) + (\tfrac{1}{2})^2 Var(X_{diff}) = \frac{1}{4}\sigma^2 + \frac{1}{4}\sigma^2 = \tfrac{1}{2}\sigma^2.$$

That is, $\frac{X_{tot}+X_{diff}}{2}$ measures L_1 with error variance $\frac{1}{2}\sigma^2$, which is a 50% reduction on what could be achieved in Method 1 for measuring L_1.
Because of the arbitrary way the pieces are labelled as L_1 and L_2, the same results apply to the measurement of L_2:

$$\frac{X_{tot} - X_{diff}}{2} \text{ measures } \frac{(L_1 + L_2) - (L_1 - L_2)}{2} = L_2 \text{ with error variance}$$

$$Var(\frac{X_{tot} - X_{diff}}{2}) = (\tfrac{1}{2})^2 Var(X_{tot}) + (-\tfrac{1}{2})^2 Var(X_{diff}) = \frac{1}{4}\sigma^2 + \frac{1}{4}\sigma^2 = \tfrac{1}{2}\sigma^2.$$

That is $\frac{X_{tot}-X_{diff}}{2}$ measures L_2 with error variance $\frac{1}{2}\sigma^2$, which is again a 50% reduction on what could be achieved by Method 1 for measuring L_2.

■

The power of moment generating functions

Theorem 5.2 indicates that for independent random variables X and Y with respective moment generating functions $m_X(t)$ and $m_Y(t)$, the moment generating function of the sum, $X + Y$, is given by $m_{X+Y}(t) = m_X(t)m_Y(t)$. In general:

Theorem 5.4 *If $X_1, X_2, X_3, \ldots, X_n$ are independent random variables where, for each i, X_i has moment generating function $m_{X_i}(t)$, then the moment generating function of the sum $X_1 + X_2 + \ldots + X_n$ is given by*

$$m_{X_1+X_2+\ldots+X_n}(t) = \prod_{i=1}^{n} m_{X_i}(t).$$

■

An obvious special case of this theorem is: The moment generating function for a sum of *i.i.d.* random variables is a *power* of the moment generating function common to all components in the sum. This may be stated as follows:

Theorem 5.5 *If $X_1, X_2, X_3, \ldots, X_n$ are i.i.d. random variables each with common moment generating function $m(t)$, then the moment generating function of the sum $X_1 + X_2 + \ldots + X_n$ is given by*

$$m_{X_1+X_2+\ldots+X_n}(t) = m(t)^n.$$

■

We will apply this result to assist in the construction of probability distributions in the next chapter.

EXERCISES

5.33 Suppose that in the manufacture of moulded plastic toys, the preparation time, X, and the drying time in the mould, Y, had the following probability mass functions expressed in time units of multiples of six minutes (there being 10 such time units in an hour).

x	1	2	3	4	5
$p_X(x)$	0.4	0.3	0	0.2	0.1

y	1	2	3	4	5
$p_Y(y)$	0	0	0.2	0.3	0.5

The total time $T = X + Y$ for both phases of manufacture was of interest to the management.

(a) Assuming that X and Y are independent, determine the probability mass function $p_T(t)$ for T.

(b) Determine $E(T)$ by the formula

$$E(T) = \sum_{all\ t} tp_T(t).$$

Demonstrate that this gives the same result as that value determined by

$$E(T) = E(X) + E(Y).$$

5.34 In Exercise 5.33, determine the variance of T. Would you expect that

$$Var(T) \geq max\{Var(X),\ Var(Y)\}?$$

5.35 In Exercise 5.33, determine the probability that the drying time in the mould takes longer than the preparation time of the material for the mould.

5.36 A production process has two heating/cooling stages which are conveniently labelled as Stage 1 and Stage 2. At each stage shrinkage defects occur independently of any other stage in the heating/cooling process. The number, Y, of defects at any stage is thought to follow a discrete probability distribution for which

y	0	1	2	3	4
$p_Y(y)$	0.4	0.2	0.2	0.1	0.1

(a) Determine the expected number of defects at Stage 2.

(b) Determine the expected number of defects for Stage 1 and Stage 2 combined.

(c) Determine the expected difference in the number of defects at Stage 1 and Stage 2.

(d) Determine the variance of the difference between the number of defects at Stage 1 and the number of defects at Stage 2.

(e) Construct the joint pmf for (Y_1, Y_2), where Y_i gives the number of defects at Stage i, $i = 1, 2$.

(f) Determine the pmf for $Y_1 + Y_2$.

(g) Determine the probability that there are more defects in Stage 1 than there are in Stage 2. Is this the same as the probability that there are more defects in Stage 2 than there are in Stage 1? Explain briefly.

5.37 Two fair dice are tossed so that the outcomes X_1 and X_2 are independent random variables. Determine the mean and the variance of the:

(a) sum $X_1 + X_2$;
(b) difference $X_1 - X_2$;
(c) average $\frac{X_1+X_2}{2}$.

5.38 Discrete random variables X and Y have the following joint pmf:

x \ y	0	1
0	0.2	0.2
1	0.3	0.3

Determine the value of:

(a) $E(X)$;
(b) $Var(Y)$;
(c) $E(X+1)$;
(d) $E(XY)$;
(e) $E[Y(X+1)]$;
(f) $Var(XY)$;
(g) $Var[Y(X+1)]$.

5.39 When submitting a tender for a construction job, the manager of a firm needs to examine carefully both the amount and the variability in the amount of building products needed to complete the job.

Suppose that the volume of sand (in cubic metres) required is a random variable Y with $\mu_Y = 10$ and $\sigma_Y = 0.5$. The amount of cement mix needed, in hundreds of kg, is also a random variable, X, with $\mu_X = 4$ and $\sigma_X = 0.2$.

The sand costs \$25 per cubic metre and the cement mix costs \$3 per kilogram.

Adding \$500 for other costs, the manager computes the total cost to be

$$C = 500 + 300X + 25Y.$$

(a) Explain how C was determined.

(b) Determine the mean and variance of the total cost C assuming that X and Y are independent. Is the independence assumption reasonable? Explain.

5.40 Liquid pesticide is made up in batches in which the concentration has a mean of 8 parts per million and a standard deviation of 1.5 parts per million.

If two batches are mixed equally by volume, the resulting concentration, C, in parts per million, is the average of independent concentrations C_1 and C_2 in the constituent batches. This means that

$$C = \frac{1}{2}C_1 + \frac{1}{2}C_2.$$

(a) Show that $E(C) = 8$.

(b) Decide whether the standard deviation of C is $\frac{3}{2}$ or $\frac{3}{2\sqrt{2}}$. Justify your choice.

(c) 'The concentration is reduced on average when batches are mixed equally by volume.' Do you agree or disagree with this statement? Justify your choice.

(d) 'The concentration is less variable when batches are mixed equally by volume.' Do you agree or disagree with this statement? Justify your choice.

5.41 If the two batches in Exercise 5.40 were mixed in proportions p and $1-p$, then the resulting concentration, C, is given by

$$C = pC_1 + (1-p)C_2.$$

Determine the proportion p that should be chosen so as to minimise $Var(C)$.

5.42 Consider the following information concerning the mean and standard deviation of operating cost random variables for an 'average sized' family car; X_1 denotes fuel cost, X_2 the cost of tyre maintenance and X_3 the cost of service and repairs (all measured in cents/km):

Operating costs X_i	**Mean**	**Standard deviation**
Fuel	6.6	0.2
Tyres	0.7	0.1
Service/Repairs	3.6	1.0

Assuming that the three cost types X_1, X_2, X_3 are independent, determine the mean and variance in total operating cost per week over 15 000 km per annum. [These figures do not include 'standing costs' such as interest, depreciation, registration and insurance.]

5.43 Assume that an ambulance will attend calls within a distance of 5 km of the station and that an ambulance travels at a constant average speed of v km hr^{-1}, in light traffic.

(a) Determine the mean and variance of time taken for an ambulance to reach an accident that occurs X_1 km from the ambulance station, where $X_1 \sim U[0,5]$.

(b) The time, T, that the ambulance is in attendance at the scene of the accident is independent of the distance that the ambulance had to travel from the station. (Does this seem reasonable?) Suppose that T is exponentially distributed with a mean of 10 minutes.

Determine the mean and variance of the total time that the ambulance is on call (that is, away from the station) when $v = 30$.

5.44 Perfectly square tiles are manufactured with a length of side, in cm, that is uniformly distributed on the interval [19.5, 20.5]. The tiles are selected from a very large shipment so that the length of side of different tiles is independent from one tile to another.

(a) Show that the areas of the tiles are independent.

(b) Determine the mean and variance of the area of an individual tile.

(c) 100 tiles are used to tile a bathroom floor. Determine the mean and variance of the area of the floor actually covered by tiles.

5.45 Consider three different chemical balances. The recorded weight of an object weighed on any of the balances has a probability distribution with mean the true weight of the object.

However, the variances in recorded weight differ for each of the three machines; such variances for Balances 1, 2 and 3 are in the ratio 1:2:3.

Assuming that the weighing errors are independent, in order to determine the true weight of an object, would it be more accurate to weigh the object on Balance 1, or to weigh the object on Balance 2 and on Balance 3 and use the average of the two readings?

5.46 Fuses are manufactured to have a mean lifetime of 100 hours. The lifetimes X_1 and X_2 of two such fuses have *i.i.d.* exponential distributions.

(a) Determine the joint pdf of X_1 and X_2.

(b) The fuses are connected into a back-up system, so that the second fuse is switched into operation only after the first fuse fails. This means that the lifetime of the entire system may be represented by the random variable $Z = X_1 + X_2$.

i. Determine the mean and variance of Z.

ii. By integrating the joint pdf, determine the cumulative distribution function $F_Z(z)$ of Z.

iii. Differentiate this cdf with respect to z to obtain the probability density function $f_Z(z)$ of Z.

CHAPTER SIX

Distributions of Counts

In this chapter you will find

6.1 BERNOULLI TRIALS AND BINOMIAL STRUCTURE

We have seen that random variables associate a real number with each point in the sample space (of potential outcomes of an experiment). This means that in the data-gathering process we are observing realised outcomes of random variables. We now start to question seriously the nature of random variables in real situations which interest us. What kind of random variables are they likely to be? What are their typical values? How do they vary on average?

Probability models for populations

We use the term **probability model** to describe how we anticipate a population will behave. The probability model is often specified by ascribing a random variable (or random variables) to the population of interest. The random variable is selected according to the nature of its probability distribution. We have met this concept previously: the stopping time of a clock is uniformly distributed over a specified time interval; the waiting time till a six appears when we successively toss a fair die follows a geometric distribution; the lifetime of a lightbulb is exponentially distributed. Such random variables were introduced in Chapter 4.

The probability distribution of a random variable is likely to be described in terms of **parameters**; these are characteristics in the population that are of interest to us. We have interest in the *mean* lifetime of a lightbulb, or the probability that a six appears when we toss a die (is it really a fair die?). Generally the type of data that we can collect from a given population will assist our building of a probability model somewhat like a police identikit of the perpetrator of a crime. The model must be in agreement with observed data, for it is likely that we may wish to use the model for prediction of how the population is likely to behave under similar circumstances or on other occasions. This is real detective work. To make the system work we need a close link between probability models and data.

This chapter is called 'distributions of counts'. We will discuss several random variables that model well data that measure how often a given event occurs.

Bernoulli random variable

We now consider the simplest and most fundamentally useful of all the discrete random variables; one that may assume only two different values, that we associate with the experimental outcomes *success/failure, on/off, present/absent.* This is numerically *indicated* by the numerical values '1' representing 'success', 'on', 'present' ... and '0' representing 'failure', 'off', 'absent'.... Such **0–1 random variables** are called **Bernoulli random variables** or **indicator random variables**. A simple experiment that results in a 'success/failure' outcome is called a **Bernoulli trial**.

We take a simple example from the property market. When a vendor attempts to sell a house at auction (where the highest bidder is deemed the purchaser), then the house will be sold if the highest bid received exceeds the reserve price. The likelihood of this depends on how realistic the vendor has been at setting the reserve price, the state of the housing market, etc. From many points of view it is reasonable to label the event { house sold } as a 'success'. Then the discrete random variable defined by:

$$X = \begin{cases} 1 & \text{if house sold;} \\ 0 & \text{if house not sold,} \end{cases}$$

is a Bernoulli random variable and the conduct of the auction is a Bernoulli trial.

We will use Bernoulli random variables as building blocks for counting. Thus a large real estate firm conducting house auctions right across Sydney on a given weekend can measure its overall sales success by adding together the outcomes for each Bernoulli trial; every house sold counts as 1 while every house not sold counts as 0 and so does not add to the weekend tally.

Definition 6.1 *A discrete random variable X is said to have a* **Bernoulli distribution** *with parameter p, and we write $X \sim Bin(1,p)$ if X has probability distribution*

x	0	1
$p_X(x)$	$1-p$	p

□

The interpretation here is that p is the single parameter of a Bernoulli distribution; change p and you change the distribution. In fact, $p = P(X = 1) = P(\text{success})$.

If X denotes the number of heads observed when a 'fair' coin is tossed, then $X \sim Bin(1, \frac{1}{2})$ with probability distribution

x	0	1
$p_X(x)$	$\frac{1}{2}$	$\frac{1}{2}$

If Y denotes the number of heads observed when a 'biased' coin is tossed (for which we assess $P(\text{heads}) = 0.6$), then $Y \sim Bin(1, 0.6)$ with probability distribution

y	0	1
$p_Y(y)$	0.4	0.6

The '1' appearing in the notation $Bin(1,p)$ reminds us of the fact that we are observing the outcome of *one* Bernoulli trial. The '*Bin*' appearing in the notation is a reference to the binomial distribution of the next section; Bernoulli distributions are a special case of binomial distributions.

Define the **indicator function** of an event A as

$$I_A(x) = \begin{cases} 1 & \text{if } x \in A; \\ 0 & \text{otherwise.} \end{cases}$$

A 'occurs' if the experimental outcome x is in A. If we define a random variable X by $X = I_A$, then $X \sim Bin(1,p)$ for $p = P(X \in A)$.

Since Bernoulli random variables occur very frequently, it is important to be familiar with their mean and variance. Suppose that $X \sim Bin(1,p)$ so that X has probability distribution

x	0	1
$p_X(x)$	$1-p$	p

According to the definition of expected value,

$$E(X) = \sum_{all\ x} x p_X(x) = (0)(1-p) + (1)(p) = p$$

and

$$E(X^2) = \sum_{all\ x} x^2 p_X(x) = (0^2)(1-p) + (1^2)(p) = p.$$

Then, $Var(X) = E(X^2) - [E(X)]^2 = p - p^2 = p(1-p)$. By a similar simple calculation we may obtain the moment generating function

$$m_X(t) = E(e^{tX}) = \sum_{all\ x} e^{tx} p_X(x) = (e^0)(1-p) + (e^t)(p) = 1 - p + pe^t.$$

The cumulant generating function, given by $\kappa_X(t) = \log_e m_X(t)$, is less useful than the moment generating function for this particular distribution. These results are summarised in the following theorem.

Theorem 6.1 *If* $X \sim Bin(1,p)$, *then*

$$\begin{aligned} E(X) &= p, \\ Var(X) &= p(1-p), \\ m_X(t) &= 1 - p + pe^t. \end{aligned}$$

■

Bernoulli random variables provide the basic units from which to build binomial random variables and we investigate this in the next section.

Binomial distributions

Instead of considering a single Bernoulli random variable, we consider a sequence of Bernoulli random variables. We will require this sequence of Bernoulli trials to have the following three properties:

Properties of a sequence of Bernoulli trials

1. The outcomes of each trial are denoted by 'success' and 'failure'.
2. $p = P(\text{success})$ is the same at every trial.
3. The trials are independent.

These ideas are put more succinctly in the following definition which uses the concept of *i.i.d.* random variables (independent and identically distributed random variables) introduced in the last chapter.

Definition 6.2 *By a* **sequence of Bernoulli trials** *we mean i.i.d. random variables* $X_1, X_2, X_3, \ldots$ *where for each* $i = 1, 2, 3, \ldots,$ $X_i \sim Bin(1, p)$.

□

A binomial distribution is then simply defined as the sum of a fixed number of Bernoulli trials from such a sequence.

Definition 6.3 *For* $0 < p < 1$ *and a fixed positive integer n, the discrete random variable X has a* **binomial distribution** *with parameters n and p, and we write* $X \sim Bin(n, p)$, *if*

$$X = X_1 + X_2 + X_3 + \ldots + X_n,$$

where $X_1, X_2, X_3, \ldots,$ *is a sequence of Bernoulli trials with probability of success p.*

□

The role of X as a counting random variable is now quite obvious: when the outcomes (either 0 for 'failure' or 1 for 'success') of each of $X_1, X_2, X_3, \ldots, X_n$ are known, then the observed value of X is simply a sum of zeroes and ones. The ones add to the total and the zeroes do not. Thus X is the total number of successes in n Bernoulli trials. This is exactly what the defining equation $X = X_1 + X_2 + X_3 + \ldots + X_n$ expresses.

Example 6.1
Identifying sequences of Bernoulli trials

In this example we discuss sequences of events, some of which fit exactly the Bernoulli trials probability model, some of which may reasonably be assumed to satisfy the Bernoulli trials model, and some of which do not satisfy the assumptions of the Bernoulli trials model.

1. Suppose that we successively toss a fair coin with $P(\text{heads}) = p$ for 100 tosses. Since the coin is memoryless, the event { heads on toss 31 } is independent of the event { tails on toss 57 }. On each toss, the coin has the same chance of upturning 'heads' (unless chewing gum becomes lodged on the side of the coin!). Each toss constitutes a trial with only two possible outcomes; successive tosses constitute a sequence of Bernoulli trials. If we (arbitrarily) classify 'heads' as a 'success', then for each $i = 1, 2, 3, \ldots, 100$, the random variable X_i denoting the number of heads at toss number i has a $Bin(1, p)$ distribution and

$$X = X_1 + X_2 + X_3 + \ldots + X_{100} \sim Bin(100, p).$$

2. The masculinity ratio of Australia's population is defined to be the number of males per 100 females. The following data from *Year Book Australia* (1985) show that this ratio stands at approximately 105 at birth.

Year	Live births males	Live births females	Masculinity ratio
1978	114 964	109 217	105.3
1979	114 613	108 516	105.6
1980	115 948	109 579	105.8
1981	121 170	114 672	105.7
1982	123 254	116 649	105.7
1983	124 558	118 012	105.5

   The question is: Is it reasonable to model the next 100 live births as a sequence of Bernoulli trials with $P(\text{success}) = P(\text{male}) = 0.5$? With some assumptions, essentially, yes. Indeed 100 births constitute 100 trials and at each trial there are two possible outcomes—a boy or a girl. If the masculinity ratios in the data from 1985 remain current, then there is reasonably a 50:50 chance for a live birth being a boy or a girl. (The data indicate that boys are marginally more likely.) The constancy of the masculinity ratio over the years indicates that $P(\text{success})$ is relatively constant.

3. Consider the front 100 seats on an Air New Zealand Boeing 747-400B on a particular flight on the trans-Tasman route. Each of these seats is either empty or occupied (denoting 'failure' and 'success'). There are a fixed number (100) of seats under consideration.
   Does the occupancy of each seat constitute a sequence of Bernoulli trials? No! The trials are not independent—whether or not one seat is occupied depends very much on whether its nearest neighbours are occupied. Family members and friends tend to sit together and strangers tend to sit as far away from one another as possible. This also means that the probability that a seat is occupied does not remain constant from seat to seat either.

Mean and variance of binomial distributions

The *i.i.d.* structure of binomial random variables allows us very simply to determine the mean and variance of $X \sim Bin(n,p)$ by using the formulas **SUM 1** (Equation 5.5) and **SUM 2** (Equation 5.6).

Suppose that X_i are *i.i.d.* $Bin(1,p)$, $i = 1,2,3,\ldots,n$. These Bernoulli random variables each have mean $E(X_i) = p$ and $Var(X_i) = p(1-p)$. This is part of our understanding of 'identically distributed' random variables: the means and variances are all the same (and so do not depend on i). Let

$$X = X_1 + X_2 + X_3 + \ldots + X_n,$$

so that by definition, $X \sim Bin(n, p)$. Then, by **SUM 1**,

$$\begin{aligned} E(X) &= E[X_1 + X_2 + X_3 + \ldots + X_n] \\ &= E(X_1) + E(X_2) + E(X_3) + \ldots + E(X_n) \\ &= p + p + p + \ldots + p \qquad (n \text{ terms}) \\ &= np. \end{aligned}$$

Similarly, **SUM 2**, and the fact that the Bernoulli trials are independent, gives for the variance,

$$\begin{aligned} Var(X) &= Var[X_1 + X_2 + X_3 + \ldots + X_n] \\ &= Var(X_1) + Var(X_2) + Var(X_3) + \ldots + Var(X_n) \\ &= p(1-p) + p(1-p) + p(1-p) + \ldots + p(1-p) \qquad (n \text{ terms}) \\ &= np(1-p). \end{aligned}$$

This proves:

Theorem 6.2 *If $X \sim Bin(n, p)$, then X has mean and variance given by $E(X) = np$ and $Var(X) = np(1-p)$.*

■

Example 6.2

Spot welds and Bernoulli trials

On an automated production line in a car plant, a spot weld placed by a robot in a specified position on a car chassis is found to be malaligned only 2% of the time.

Worked example:

Over a time period for which 10 000 welds are performed, determine the expected number of welds that are correctly aligned. Find also the standard deviation in the number of welds that are correctly aligned.

Let X denote the number of successful welds by the robot in a sequence of 10 000 Bernoulli trials for which at each trial the robot either successfully or unsuccessfully positions the weld on the given chassis. Assuming that the robot functions independently for each weld, then

$$P(\text{successful weld}) = 1 - 0.02 = 0.98 \quad \text{so that} \quad X \sim Bin(10\,000, 0.98).$$

Therefore,

$$E(X) = np = (10\ 000)(0.98) = 9\ 800$$
$$Var(X) = np(1-p) = (10\ 000)(0.98)(0.02) = 196$$
$$\sigma_X = \sqrt{196} = 14.$$

Binomial probability mass function

We know a great deal about binomial distributions simply from their additive structure. However it is time to establish a formula for the probability mass function of $X \sim Bin(n,p)$.

In n Bernoulli trials suppose that there are x successes. In keeping with the additive binomial notation, we will denote the successes by '1' and the failures by '0'. Suppose that (somewhat unrealistically) the x successes all precede the $n-x$ failures. The pattern of resulting data may be represented as

$$\underbrace{111111111\ldots1}_{x\text{ successes}}\underbrace{00000000\ldots0}_{n-x\text{ failures}}$$

Because of the independence of trials, this single data string has probability

$$\underbrace{ppppppppp\ldots p}_{x\text{ successes}}\underbrace{(1-p)(1-p)(1-p)(1-p)(1-p)\ldots(1-p)}_{n-x\text{ failures}} = p^x(1-p)^{n-x}.$$

Now consider that the 0's and 1's in the given data string may be arranged among themselves in $\binom{n}{x}$ ways. (This is the number of selections of the positions for the x 1's from the n available positions.) When we observe, then, x successes in n Bernoulli trials, we observe exactly *one* of those $\binom{n}{x}$ strings. Therefore,

$$\begin{aligned} P(X=x) &= \binom{n}{x}(\text{probability of a given string}) \\ &= \binom{n}{x}p^x(1-p)^{n-x}. \end{aligned}$$

In summary:

If $X \sim Bin(n,p)$, then X has probability mass function $p_X(x) = \binom{n}{x}p^x(1-p)^{n-x}$, for $0 < p < 1$ and $x = 0, 1, 2, \ldots, n$.	(6.1)

Example 6.3

Scud missiles and their Patriot interceptors

During the Gulf War in the summer of 1991, Scud missiles were fired from mobile launchers in Iraq towards targets in Israel and Saudi Arabia. The Gulf Alliance protected themselves against attack by using Patriot (anti)missiles. We will make some hypothetical assumptions regarding targeting and interception success rates and then assess the probability of a direct hit on an important military target.

Suppose that a Scud missile fired at a target is intercepted by a Patriot missile 99% of the time. However, given that a Scud evades the Patriot security net, we assume that it is only 10% likely to reach its target. (In the Gulf War the Scud missiles did not prove to be accurate in striking *specific* targets.) Suppose that 4 Scud missiles are fired independently by Iraq. If we denote by 'success' a direct hit of the required target, then

$$\begin{aligned} P(\text{success}) &= P(\text{evades interceptor})P(\text{hits target} \mid \text{evades interceptor}) \\ &= (1-0.99)(0.1) \\ &= 0.001. \end{aligned}$$

Therefore, X, the number of 'successes' from four Scud launchings, satisfies $X \sim Bin(4, 0.001)$. This discrete random variable has pmf

$$p_X(x) = \binom{4}{x}(0.001)^x(0.999)^{4-x}, \quad x = 0, 1, 2, 3, 4.$$

The probability that all four missiles reach their required target is

$$P(X=4) = p_X(4) = \binom{4}{4}(0.001)^4(0.999)^{4-4} = 10^{-12}.$$

These are not strong odds! The probability that at least one will hit the target, is determined as is usual, by

$$\begin{aligned} P(\text{at least one Scud hits target}) &= 1 - P(\text{no Scuds hit the target}) \\ &= 1 - p_X(0) \\ &= 1 - \binom{4}{0}(0.001)^0(0.999)^4 \\ &= 1 - (0.999)^4 \\ &= 0.00399 \ \ (5 \text{ dp}). \end{aligned}$$

Worked example:

How many Scud missiles should be fired so that there is at least a 95% chance that at least one escapes interception by a Patriot (anti)missile?

Suppose that n Scuds are fired.

$$\begin{aligned} 0.95 &\le P(\text{at least one Scud escapes interception}) \\ &= 1 - P(\text{no Scuds escape interception}) \\ &= 1 - p_Y(0) \ \ \text{where } Y \sim Bin(n, 0.01) \\ &= 1 - (0.99)^n. \end{aligned}$$

Hence we must determine the smallest n for which $0.95 \le 1 - (0.99)^n$ or $(0.99)^n \le 0.05$. By inspection, by taking logarithms, or even simply 'trying out' different values of n gives $n = 299$.

Using binomial tables

The values of the cumulative distribution function

$$F_X(x) = P(X \le x) \quad \text{for} \quad X \sim Bin(n,p)$$

are given in Appendix 1 for convenient values of n and p: $n = 5, 10, 15, 20, 25$; $p = 0.01, 0.05, 0.1, \ldots, 0.99$. We will use upper-case notation $B(x;n,p)$ for these cumulative probabilities:

Notation: If $X \sim Bin(n,p)$, $\quad B(x;n,p) = F_X(x) = P(X \le x)$.

For example, from reading the tables we find $B(4;10,0.3) = 0.850$ and $B(3;10,0.3) = 0.650$, so that if $X \sim Bin(10,0.3)$, then

$$P(X=4) = p_X(4) = B(4;10,0.3) - B(3;10,0.3) = 0.850 - 0.650 = 0.200.$$

This means that individual probability mass calculations are simple subtractions of adjacent lines in the table.

Moment generating function for the binomial distribution

The result of Theorem 5.4 may be applied directly to determine the moment generating function for the binomial. Suppose that $X_1, X_2, X_3, \ldots, X_n$ are *i.i.d.* $Bin(1,p)$ random variables. Since these random variables all have the same distribution, they all have the same moment generating function which we will write (by Theorem 6.1) as $m(t) = 1 - p + pe^t$. If we then set

$$X = X_1 + X_2 + X_3 + \ldots + X_n,$$

then $X \sim Bin(n,p)$ and has moment generating function

$$\begin{aligned} m_X(t) &= m_{X_1+X_2+X_3+\ldots+X_n}(t) \\ &= m(t)^n \qquad \text{(Theorem 5.4)} \\ &= (1 - p + pe^t)^n. \end{aligned}$$

Thus the mgf of a $Bin(n,p)$ random variable is the nth power of the mgf of a $Bin(1,p)$ random variable. Notice here that the cumulant generating function

$\kappa_X(t)$ of a $Bin(n, p)$ random variable is n times the cumulant generating function of a $Bin(1, p)$ random variable. That is:

$$\begin{aligned}\kappa_X(t) &= \log_e m_X(t) \\ &= \log_e(1 - p + pe^t)^n \\ &= n \log_e(1 - p + pe^t).\end{aligned}$$

This makes the central moments of the binomial distribution easy to obtain from the central moments of the Bernoulli distribution. The moments may be used to determine the coefficient of skewness which, in the case of $X \sim Bin(n, p)$, is

$$\zeta_3 = \frac{1 - 2p}{\sqrt{np(1 - p)}}.$$

$\zeta_3 = 0$ when $2p - 1 = 0$ or $p = \frac{1}{2}$. The diagrams in Example 6.4 visibly demonstrate how the skewness changes with p.

Example 6.4

Graphs of probability mass function $p_Y(y)$ for $Y \sim Bin(10, p)$

These diagrams indicate exactly what the binomial probability mass function looks like for different values of the parameter p. When $p = \frac{1}{2}$, $p_X(x)$ is symmetric. The mass function becomes negatively skewed (having a long tail towards the left) as p increases towards 1. Similarly, the mass function is positively skewed (long tail towards the right) for values of p below $\frac{1}{2}$.

Probability p	Coefficient of skewness ζ_3
0.1	0.843
0.5	0.000
0.7	−0.276
0.9	−0.843

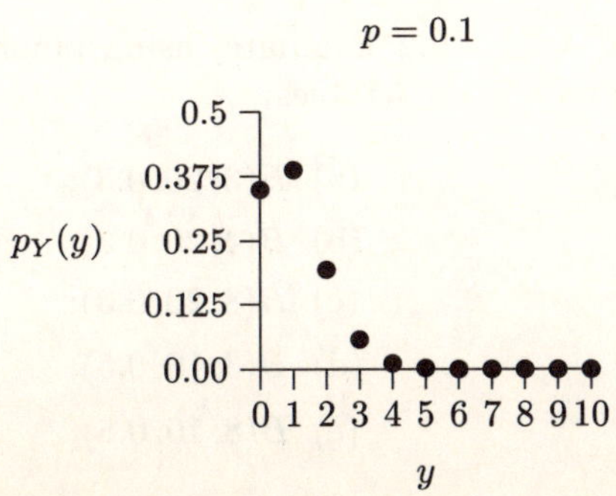

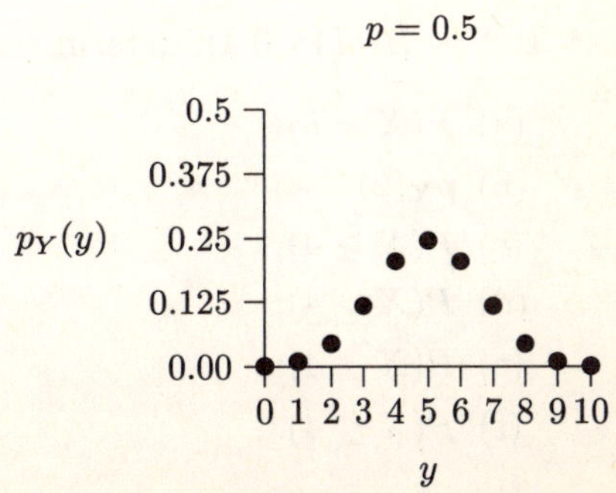

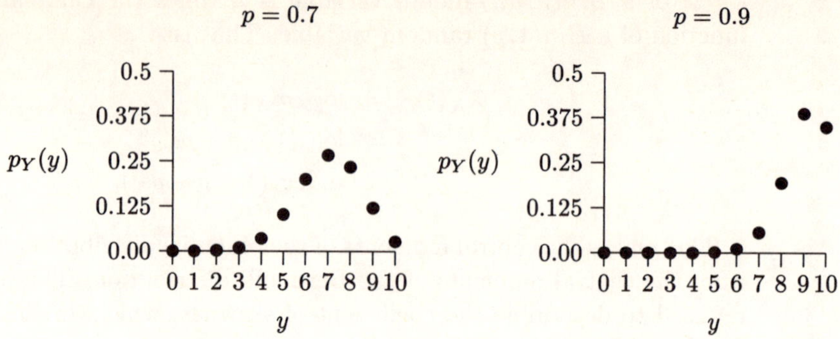

PROPERTIES OF THE BINOMIAL DISTRIBUTION
$X \sim Bin(n, p)$ (6.2)

Probability mass function $p_X(x) = \binom{n}{x} p^x (1-p)^{n-x}$
$x = 0, 1, 2, 3, \ldots, n; \quad 0 < p < 1.$

Mean	$E(X) = \mu = np$
Variance	$Var(X) = \sigma^2 = np(1-p)$
Standard deviation	$SD(X) = \sigma = \sqrt{np(1-p)}$
Coefficient of skewness	$\zeta_3 = \frac{1-2p}{\sqrt{np(1-p)}}$
Coefficient of kurtosis	$\zeta_4 = 3 + \frac{1-6p(1-p)}{np(1-p)}$
Moment generating function	$m_X(t) = (1 - p + pe^t)^n$
Cumulant generating function	$\kappa_X(t) = n \log_e(1 - p + pe^t)$

EXERCISES

6.1 If $X \sim Bin(15, 0.4)$, determine:

(a) $P(X = 3)$;

(b) $p_X(2)$;

(c) $P(X \geq 4)$;

(d) $P(X > 4)$;

(e) $P(X < 4)$;

(f) $P(X \leq 4)$.

6.2 Evaluate, using binomial probability tables:

(a) $B(3; 20, 0.3)$;

(b) $B(4; 20, 0.3)$;

(c) $B(8; 20, 0.3)$;

(d) $B(3; 10, 0.5)$;

(e) $B(8; 10, 0.8)$.

6.3 Evaluate $\sum_{r=0}^{9} \binom{15}{r}(0.4)^r(0.6)^{15-r}$.

6.4 Evaluate $\sum_{r=7}^{20} \binom{20}{r}(0.25)^r(0.75)^{20-r}$.

6.5 Suppose that 10% of jarrah railway sleepers must be discarded because the presence of a knot weakens the load-bearing capacity of the sleeper. Determine the probability that there is

(a) at least one discard;

(b) at most one discard;

(c) one discard;

in a shipment of 10 sleepers.

6.6 What conditions are assumed to hold in Exercise 6.5 so that the binomial distribution applies to the total number of discards in a shipment of 10 sleepers?

6.7 Suppose that there is a 70% chance that a speed camera will record an identifiable picture of a motorist who is exceeding the legal speed limit. Suppose that while speeding illegally down the freeway, I pass four 'speed traps' where cameras are installed and operate independently of one another.

Determine the probability that I am caught speeding.

(An identifiable photographic record from a speed camera is sufficient grounds for prosecution and constitutes being 'caught' speeding.)

6.8 Example 6.1 shows that boys and girls are born in approximately equal numbers.

Consider the random variable X denoting the number of girls in a family of size 6 children.

(a) Under reasonable assumptions, which you should outline, $X \sim Bin(n, p)$. State the appropriate values for the parameters n and p in this model.

(b) Determine $P(X = 3)$.

(c) Determine the probability that there are more girls than boys in a family of size 6 children.

(d) If 1000 families of 6 children are surveyed, determine the expected number of families containing all boys.

(e) Consider the random variable Y denoting the ratio of the number of boys to the number of girls in families of size 6 children with at least one girl.

Is Y a discrete random variable or a continuous random variable? Discuss.

6.9 In an experiment to test the effectiveness of a new drug, 10 patients are carefully monitored on a regime of the new drug. The manufacturers claim that only 20% of patients will experience side-effects while on the drug. In this clinical trial, X is the number of patients who experience side-effects.

(a) What aspects of this clinical trial make it reasonable to assume that

$$X \sim Bin(10, 0.2)?$$

Do the independence assumptions seem reasonable?

(b) Write down an expression for the probability mass function for X.

(c) Determine $P(X = 3)$.

(d) Use binomial probability tables to find the following probabilities:

i. $P(X \leq 1)$;

ii. $P(1 < X \leq 3)$;

iii. the probability that two or more patients show side-effects from the new drug.

6.10 In Exercise 6.9, if the manufacturer's claim that only 20% of patients experience side-effects is indeed true, we would expect on average 2 of the 10 patients to experience side-effects. What happens if many more than two patients experience side-effects? Assuming the manufacturer's figure of $p = 0.2$, what is the probability that five or more patients experience side-effects? Suppose that 5 patients *do* experience side-effects. In this case, is it reasonable to start to doubt the manufacturer's claim?

6.11 It is estimated that in criminal trials, the jury will reach the correct decision (guilty or not guilty) 90% of the time. Consider a group of 10 cases that are brought to trial before a Supreme Court jury.

(a) What is the expected number of cases where the jury will reach the correct decision?

(b) Determine the probability that an incorrect decision is reached in 3 or more cases.

(c) What is the probability that an incorrect decision is reached in 3 or more cases if the jury system reaches the correct decision 95% of the time?

6.12 The total number of defectives Y in a sample of n items taken from a large lot of manufactured items of which a proportion p are defective is closely approximated by a binomial distribution when the lot-size is large.

A quality control engineer may base judgment of lot quality (whether or not the lot is acceptable) on the following 'sampling inspection plan':

accept lot if $Y \leq a$;
reject lot if $Y > a$.

If we assume that $Y \sim Bin(5, 0.05)$, determine the probability of accepting a lot for each of $a = 0, 1, 2$.

6.13 In Exercise 6.12, suppose that a is fixed at the value 1, n is fixed at 5, but p may vary: that is, we are unsure of the proportion defective in the lot. Determine the probability of accepting the lot for $p = 0.1, 0.2, 0.3, 0.4, 0.5, 0.6, 0.7$.

Construct a graph of

$$h(p) = P(\text{accept lot for given } p)$$

against p.

[Such a graph is called an **operating characteristic curve** for the sampling plan.]

6.14 Consider the following sampling inspection plan used by a manufacturer of door fittings: hinges are dispatched in carton lots of 200 items. Prior to dispatch, a quality control engineer applies the following inspection plan.

Randomly select 20 items from the carton. Dispatch carton only if the sample contains at most one defective item.

If a carton is rejected, then all components in the carton are checked and the information is then used back on the production line.

(a) Explain why a Bernoulli trials model is appropriate in this sampling scheme if the sample is selected *with* replacement. What specific conditions in the Bernoulli trials model fail when the sample is selected *without* replacement?

(b) When the sample is selected with replacement, find the probability of rejecting a carton containing:

i. 20 defective components;

ii. 10 defective components.

(c) The production process used to manufacture the door hinges is said to have a 5% defective rate. If this is indeed the case, decide whether (with reasons) the current sampling inspection plan currently used is suitable.

6.15 60% of City of Caulfield ratepayers are in favour of the redevelopment of Elsternwick railway station. If X is the number of ratepayers who favour the proposal in a sample of 50 randomly chosen ratepayers, find the mean and variance of X.

6.16 In Exercise 6.9, determine the probability that two or more patients show side-effects from the new drug, given that at least one patient exhibits side-effects from the new drug.

6.17 Consider Example 6.3 where Scud missiles are intercepted in flight by Patriot (anti)missiles. What is the smallest number of missiles that need be fired to ensure that at least one successfully hits its target with probability at least 0.99?

6.18 A telephone company anticipates that there will be, on average, five lines in use at any given time between two towns during the peak hours. What is the smallest number of lines that should be installed so that there is less than one chance in twenty of a subscriber finding all lines in use when attempting to make a call?

6.19 This problem demonstrates that the function $p_X(x)$ defined for $0 < p < 1$ and $x = 0, 1, 2, \ldots, n$ by

$$p_X(x) = \binom{n}{x} p^x (1-p)^{n-x},$$

satisfies the conditions of being a probability mass function.

(a) Explain why $p_X(x) \geq 0$ for all $x = 0, 1, 2, \ldots, n$.

(b) Write out the binomial expansion of $(p + r)^n$.

(c) By putting $r = 1 - p$ show that

$$1 = \sum_{x=0}^{n} \binom{n}{x} p^x (1-p)^{n-x}. \quad (6.3)$$

6.20 Show that by putting $p = \frac{1}{2}$ in Equation 6.3 it follows that

$$2^n = \sum_{x=0}^{n} \binom{n}{x}.$$

What is the interpretation of this result in terms of Pascal's Triangle? (See Section 3.3.) The first few lines of Pascal's Triangle are

$$\begin{array}{c} 1 \\ 1 \quad 1 \\ 1 \quad 2 \quad 1 \\ 1 \quad 3 \quad 3 \quad 1 \\ 1 \quad 4 \quad 6 \quad 4 \quad 1 \\ \cdots \end{array}$$

6.21 (a) Show that if $0 \le x \le n$, then

$$\binom{n}{x} = \binom{n}{n-x}.$$

(b) Give an interpretation of this result in terms of Pascal's Triangle. (See Section 3.3.)

(c) If $X \sim Bin(n, \frac{1}{2})$, show that

$$P(X = x) = P(X = n - x).$$

6.22 A Bernoulli random variable X has probability distribution given by the table:

x	0	1
$p_X(x)$	$1-p$	p

Show that for any value of r, the moments $\mu_r^{'} = E(X^r)$ about 0 are constant and determine the value of that constant. [*Hint*: Determine the required expected values directly from the table of the probability distribution.]

6.23 Show that a $Bin(1, \frac{1}{2})$ random variable has smaller variance than a $Bin(1, p)$ random variable when $p \neq \frac{1}{2}$. Can this result be extended to $Bin(n, p)$ random variables for $n > 1$?

6.24 A blood sample is taken from each of 100 people to test for the presence of a disease. Suppose that the probability that the test is positive (disease present) is 0.01 for all people, and that the presence of the disease in any person does not affect the likelihood of anyone else in the group having the disease. Consider two methods of testing.

Method 1: Each person is tested separately.

Method 2: The blood sample of 4 people are pooled and analysed together. If this test is negative, this one test suffices for these 4 people. If the test is positive, each of the 4 people is called back and tested separately, and, in all, 5 tests are required for these 4 people.

Let the discrete random variable N denote the number of tests required for all 100 people.

(a) What is the value of $E(N)$ when **Method 1** is used?

(b) Show that, when **Method 2** is used,

$$N = 25 + 4X$$

where X is a binomial random variable with parameters $n = 25$ and

$$p = 1 - (0.99)^4.$$

(c) Use the result for the mean of a binomial distribution to show that with **Method 2**, $E(N)$ is approximately 29.

(d) Briefly discuss the relative merits of the two methods.

6.25 A multiple choice test with 5 choices per question (and one correct answer per question) is marked by giving 1 mark for a correct answer and subtracting $\frac{1}{2}$ a mark for a wrong answer. A question not attempted scores zero.

A student doing the test finds that in each question she can eliminate some choices as being wrong, but then must guess between those choices remaining.

Let n_i be the number of questions in which this student can eliminate $5 - i$

of the choices; that is, must guess between i possibilities, $1 \leq i \leq 5$. For a given value of i, let R_i be the number of questions (from the n_i available) that she answers correctly, assuming that she attempts all n_i questions.

(a) Explain carefully why R_i has a binomial distribution and give the parameters, the mean and the variance of this distribution.

(b) Show that the mark that she will obtain from the n_i questions is

$$\tfrac{1}{2}(3R_i - n_i)$$

and give the mean and variance of this mark.

(c) Show that the student neither expects to gain or lose marks by attempting questions where she is guessing from 3 alternatives.

(d) In a 50 question test a student finds that there are 18 questions to which she knows the answer for sure, 10 where she must guess from 2 choices and 10 where she must guess from 3 choices.

i. Show that the student must expect to lose marks by attempting the remaining 12 questions which involve guessing from 4 or 5 alternatives.

ii. If the pass mark is set at 25, determine the probability of failing if she does not attempt the questions where she is guessing from 3 or more alternatives.

6.26 A Bernoulli random variable X has probability distribution given by the table:

x	0	1
$p_X(x)$	$1-p$	p

(a) Show that the coefficient of skewness for X (Definition 4.16) is given by

$$\zeta_3 = \frac{1-2p}{\sqrt{p(1-p)}}.$$

(b) For what value of p does the Bernoulli distribution have a symmetric pmf ?

(c) Show that the coefficient of kurtosis is given by

$$\zeta_4 = \frac{1}{\sigma_X^2} - 3.$$

(d) Show that all Bernoulli distributions have coefficient of kurtosis at least 1.

(e) For what value of p does the Bernoulli distribution have minimum coefficient of kurtosis?

6.27 Let A be the event that the first trial is a success in a sequence of n Bernoulli trials which has X successes. Use the definition of the probability mass function to show that

$$P(A|X = k) = \frac{k}{n},$$

$k = 1, 2, 3, \ldots, n$.

[*Hint*: The left-side is simply a conditional probability as defined in Chapter 3.]

6.28 If $X \sim Bin(n,p)$, show that

$$n - X \sim Bin(n, 1-p).$$

What interpretation (in terms of 'successes' and/or 'failures') may be placed on the random variable $n - X$?

Give an expression for the variance of the random variable $n - X$.

Evaluate $E[(n - X)^2]$.

6.29 A fair coin is repeatedly tossed giving a sequence of Bernoulli trials. Determine the probability that the first head appears on the nth trial, given that at least one head occurs during the first $(m + n)$ trials.

6.30 Suppose that $X \sim Bin(n, p)$.

(a) Show that for $k \geq 1$,

$$\frac{P(X = k)}{P(X = k - 1)} \geq 1$$

if and only if

$$k \leq (n + 1)p.$$

(b) Hence show that the value of x which maximises $P(X = x)$ is

$$x = [(n + 1)p],$$

where $[t]$ denotes the largest integer less than or equal to t. Under what circumstances is this maximum achieved for more than one value of x?

6.2 RARE EVENTS AND THE POISSON MODEL

What do data from the outcome of Bernoulli trials look like? The first of the data sets which follow resulted from the tossing of a coin 250 times; the outcome 'heads' is denoted by 1 and the outcome 'tails' is denoted by 0. We assumed the coin to be fair, in which case the Bernoulli trials model for Y, the number of heads in 250 coin tosses, is the $Bin(250, \frac{1}{2})$ distribution. Since this data set produced $y = 126$ heads, there seems little reason to doubt the 'fairness' assumption. Notice that there are several so-called **runs** 00000... of zeroes and 11111... of ones although overall the data set shows a random scatter of 0 and 1.

250 observations on $X \sim Bin(1, \frac{1}{2})$

10101001001101000000110100001110111110010000101100
01010110001000110010010110110010100000110101010110
11000000100111110110010101111000110111000111101101
100000101000000011010100000110100110010100010110111
11100010011101111111111101110100100111101101110110

$$\textit{250 observations on } \boldsymbol{X \sim Bin(1, \frac{1}{36})}$$

```
00000100000000000000000000000000010000000000000000
00000000000000000000000000000000000000000000000000
00000100000100000000010000000000000000000000010000
00000000000000000010000000000000000010000000000000
00000000000000000000000000000000000000000000010000
```

The second data set shows the result of decreasing the probability of success for each Bernoulli trial. This data set was generated by 250 simulated throws of a pair of 'fair' dice. We wished to record as a success, '1', the event {pair of sixes}, which for fair dice has a probability of $\frac{1}{36}$ of occurrence. All other outcomes for the pair of dice are recorded as '0'. The Bernoulli trials model is again appropriate for Y, the number of successes in 250 throws: $Y \sim Bin(250, \frac{1}{36})$. The *expected* number of successes in 250 Bernoulli trials with $p = \frac{1}{36}$ is $E(Y) = np = (250)(\frac{1}{36}) = 6.9$ (2 dp). This was (not unusually) exceeded by an observed value $y = 9$.

This data set illustrates the behaviour of a sequence of n Bernoulli trials for which n is large and p is very small—still giving a moderate value for the mean np. The number of successes is an example of a *limiting binomial.*

Limiting binomials

The tables for $Bin(n, p)$ random variables may be applied immediately (for wide-ranging probabilities p) with values of $n \leq 25$. For large values of n, modern calculators and computers may easily determine binomial probabilities. Further, in the circumstances of Theorem 6.3, the Poisson distribution may be used to approximate binomial probabilities.

Suppose that the number of trials, n, increases at the same rate that the probability of success, p_n, decreases, so that the product np_n is constant. Then the outcomes from the Bernoulli trials become scattered, less clustered, and assume the character of outcomes from the Poisson distribution. For the algebraically adept, the limiting process is as follows:

Theorem 6.3 *If* $p_n = \frac{\lambda}{n}$ *where* λ *is a constant, then*

$$\lim_{n\to\infty} \binom{n}{y} p_n^y (1 - p_n)^{n-y} = \frac{\lambda^y e^{-\lambda}}{y!}.$$

Proof: On substitution of the success probability p_n directly into the binomial formula, it follows that

$$\begin{aligned}
&\lim_{n\to\infty} \binom{n}{y} p_n^y (1 - p_n)^{n-y} \\
&= \lim_{n\to\infty} \frac{n(n-1)\ldots(n-y+1)}{y!}\left(\frac{\lambda}{n}\right)^y \left(1 - \frac{\lambda}{n}\right)^{n-y} \\
&= \lim_{n\to\infty} \frac{\lambda^y}{y!} \frac{n(n-1)\ldots(n-y+1)}{n^y}\left(1 - \frac{\lambda}{n}\right)^n \left(1 - \frac{\lambda}{n}\right)^{-y}
\end{aligned}$$

$$= \frac{\lambda^y}{y!} \lim_{n\to\infty} (\frac{n}{n})(1-\frac{1}{n})(1-\frac{2}{n})\dots(1-\frac{y-1}{n})(1-\frac{\lambda}{n})^n(1-\frac{\lambda}{n})^{-y}$$

$$= \frac{\lambda^y}{y!} \lim_{n\to\infty} \underbrace{(\frac{n}{n})}_{=1}\underbrace{(1-\frac{1}{n})}_{\to 1}\underbrace{(1-\frac{2}{n})}_{\to 1}\dots\underbrace{(1-\frac{y-1}{n})}_{\to 1}\underbrace{(1-\frac{\lambda}{n})^n}_{\to e^{-\lambda}}\underbrace{(1-\frac{\lambda}{n})^{-y}}_{\to 1}$$

$$= \frac{\lambda^y e^{-\lambda}}{y!}.$$

In the last line of this derivation we have used the limiting form of the exponential function: $lim_{n\to\infty}(1+\frac{x}{n})^n = e^x$.

■

Theorem 6.3 shows that limiting binomial probabilities are Poisson probabilities.

Definition 6.4 *Suppose that λ is a positive constant. A discrete random variable X is said to have a* **Poisson distribution** *with parameter λ, and we write $X \sim Poi(\lambda)$, if X has probability mass function given by*

$$p_X(x) = \frac{\lambda^x e^{-\lambda}}{x!}, \qquad x = 0, 1, 2, 3, \dots$$

□

$p_X(x)$ is satisfies the two requirements of a probability mass function: $p_X(x) \geq 0$ and $\sum_{all\ x} p_X(x) = 1$. Indeed, since λ is a positive constant, the quantities $\frac{\lambda^x e^{-\lambda}}{x!}$ are positive for all x and the first condition is trivially satisfied. In the discussion of moment generating functions in Chapter 4, we used the exponential series expansion: for any real r,

$$e^r = 1 + r + \frac{r^2}{2!} + \frac{r^3}{3!} + \dots = \sum_{k=0}^{\infty} \frac{r^k}{k!}. \tag{6.4}$$

Hence,

$$\sum_{all\ x} p_X(x) = \sum_{x=0}^{\infty} \frac{\lambda^x e^{-\lambda}}{x!} = e^{-\lambda} \sum_{x=0}^{\infty} \frac{\lambda^x}{x!} = e^{-\lambda} e^{\lambda} = 1,$$

and the second condition is satisfied.

Example 6.5

Bad reaction to smallpox vaccine

To obtain a working visa, some countries require that the applicant be vaccinated against specific diseases such as smallpox. In rare cases (with probability estimated at only 0.002) an individual may suffer a bad reaction to smallpox vaccine. If 2000 individuals are vaccinated against smallpox, determine the probability that 5 of the individuals will show side-effects from the vaccine.

Under assumptions of independence of individuals and a constant likelihood of bad reaction from person to person, X, the number of individuals suffering side-effects, has a binomial distribution: $X \sim Bin(2000, 0.002)$. We require $P(X = 5) = \binom{2000}{5}(0.002)^5(1 - 0.002)^{2000-5} = 0.1564$. Notice that the mean of this binomial distribution is $np = (2000)(0.002) = 4$. If $Y \sim Poi(4)$, then

$$P(Y = 5) = \frac{4^5 e^{-4}}{5!} = 0.1563.$$

Thus, $P(Y = 5)$ is a reasonable approximation to the desired probability $P(X = 5)$.

The Poisson approximation to the binomial may be usefully applied if n is large, p very small and np moderate:

> **Poisson approximation to the binomial**
> If $n \geq 100$, $p \leq 0.01$ and $\lambda = np \leq 20$, then
> $P(X = k) \approx P(Y = k)$, $k = 1, 2, \ldots, n$,
> for $X \sim Bin(n, p)$ and $Y \sim Poi(\lambda)$. (6.5)

Example 6.6

Shape of the Poisson probability mass function $p_Y(y)$ for $Y \sim Poi(\lambda)$

The following diagrams indicate the shape of the Poisson probability mass function for different values of the parameter λ. Clearly the pmf moves towards symmetry as the value of λ increases.

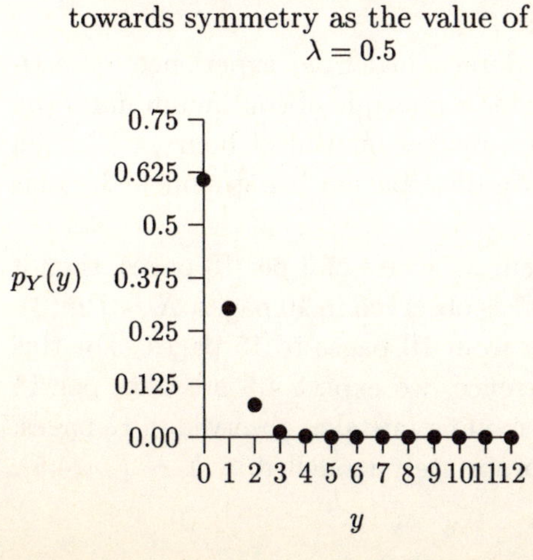

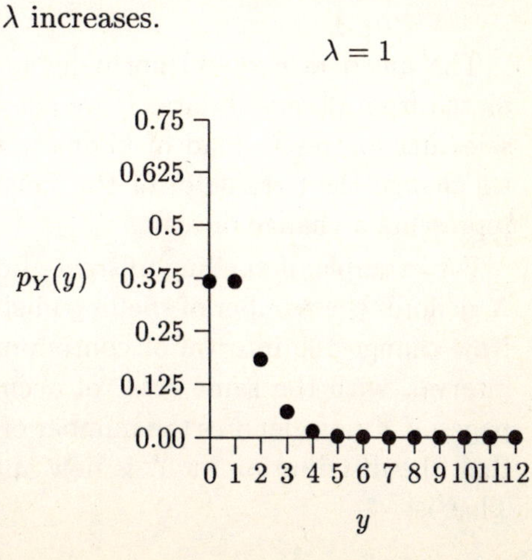

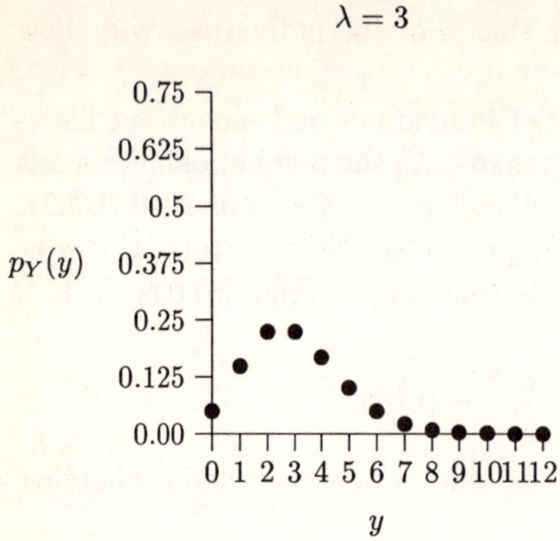

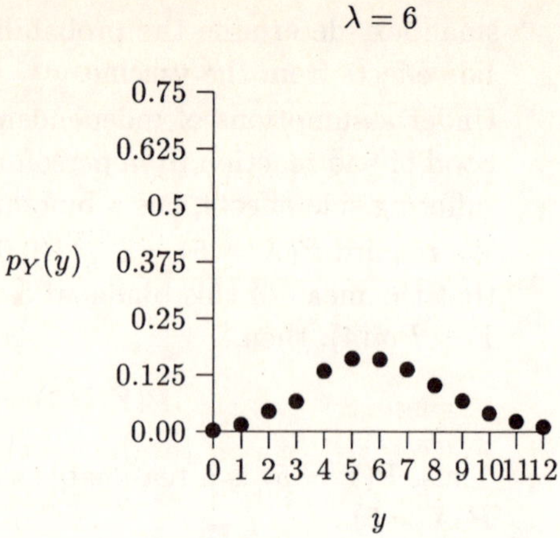

The Poisson distribution is most frequently applied to the frequency of occurrence of an event in a continuum. The continuum may be time, length, area, volume .... Of particular interest is the *mean rate* of the occurrence of the event per unit of the continuum. X has a Poisson distribution if X counts the number of times that the event occurs in a unit of the continuum. For example:

Event	Continuum	Rate per unit	Poisson model
telephone calls	time	10 per hour	$X \sim Poi(10)$
car accidents	time	4 per week	$X \sim Poi(4)$
flaws in rolled canvas	area	1 per m^2	$X \sim Poi(1)$
defects in nylon wire	length	1 per 10 000 m	$X \sim Poi(1)$
starfish on coral reef	area	2 per 100 m^2	$X \sim Poi(2)$
spelling mistakes in book	page	3 per 10 pages	$X \sim Poi(3)$

The mean rate would normally be determined from experience, or estimated from observed data. If we consider a multiple of continuum units (by selecting metres instead of kilometres, minutes instead of hours, ...) then we change the parameter of the Poisson distribution being modelled. This represents a change of scale.

For example, if spelling mistakes occur at a rate of 3 per 10 pages, then if X denotes the number of spelling mistakes observed in 10 pages, $X \sim Poi(3)$. Now change the interval of continuum from 10 pages to 15 pages. For this interval, with the same *rate* of occurrence, we expect 4.5 mistakes per 15 pages; if Y now denotes the number of spelling mistakes observed in 15 pages, then the distribution for Y is now appropriately modelled by $Y \sim Poi(4.5)$. That is:

Event	Continuum	Rate	Poisson model
textbook mistakes	1000 pages	3 per 10 pages	$X \sim Poi(3)$
textbook mistakes	1000 pages	4.5 per 15 pages	$Y \sim Poi(4.5)$

Example 6.7

Fitting a Poisson model to the data of Rutherford and Geiger

Larsen and Marx (1981) report the fit of a Poisson probability model to Rutherford and Geiger's data from their early experiments on alpha particle emissions. This pioneering work dates back to 1910.

The experimenters recorded the presence of alpha particles released from a radioactive source over time. In the continuum 'time', the basic unit selected was an '8 minute time interval' in which to (repeatedly) observe the event 'alpha particle recorded' in the decay process. The left-side data below gave a sample mean of 3.87 emissions per 8 minutes based on $n = 2608$ data points (observations of 8 minute intervals). For example, this data shows that emissions of 3 or 4 particles per 8 minutes was most common, there being 532 time intervals for which there were counts of 4.

Alpha particles recorded per 8 minutes (x_i)	Frequency f_i
0	57
1	203
2	383
3	525
4	532
5	408
6	273
7	139
8	45
9	27
10	10
≥ 11	6

Alpha particles recorded per 8 minutes (x_i)	Expected frequency $X \sim Poi(3.87)$
0	54
1	211
2	407
3	526
4	508
5	394
6	254
7	140
8	68
9	29
10	11
≥ 11	6

The expected frequencies for each observed value $0, 1, 2, 3, 4, \ldots$ of X are reported in the right-side table. These are the emission frequencies that we would expect if the $Poi(3.87)$ model is appropriate. Thus we hope for a suitable fit of the model to the data by equating the model mean λ (see Equation 6.6) to the sample mean 3.87. That is, of all the Poisson models, the $Poi(\lambda)$ model should provide the best fit: if $X \sim Poi(3.87)$, then $np_X(x_i)$ gives the fitted values

$$2608P(X=0)=54, \quad 2608P(X=1)=211, \quad 2608P(X=3)=407\ldots$$

as we move down the column. The observed and expected frequencies are close, especially when any discrepancies are considered relative to the magnitudes of the frequencies involved.

The Poisson process

We now give an intuitive discussion of how Bernoulli trials are generated in the context of a continuum. If we represent t units of our continuum diagrammatically by the interval

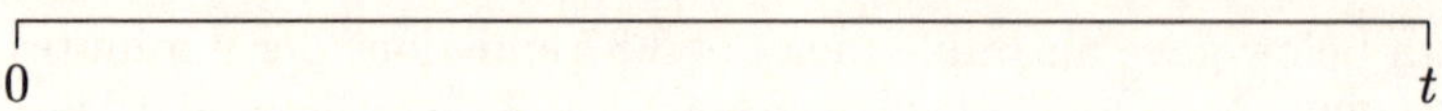

then arriving phonecalls, customers, flaws, ..., are scattered, and not too clustered, throughout the interval. We represent the on-going occurrence of such events in the interval by black dots:

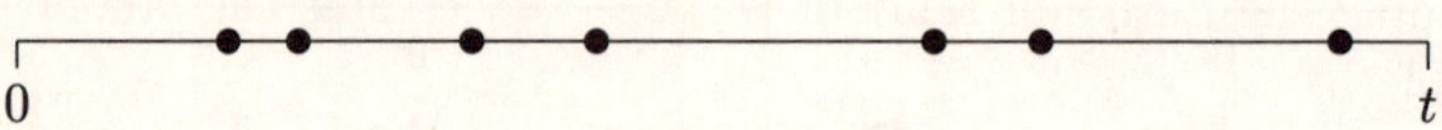

Now splitting the interval into 10 subintervals we find that:

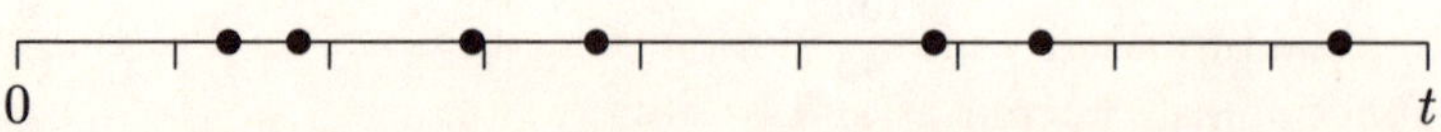

Notice that the second subinterval so created contains two occurrences of the event in question. If we carve the interval into even more subintervals, then the continuum has become 'discretised' by adjacent non-overlapping subintervals:

Suppose that we have created n subintervals, where n is large. There must be sufficient subintervals so that each event occurs in its own subinterval (without doubling up with another event in the same subinterval). The random variables

$$W_i = \begin{cases} 1 & \text{if } A \text{ occurs in the } i\text{th subinterval;} \\ 0 & \text{otherwise,} \end{cases}$$

which count a 1 every time A occurs, form a sequence of Bernoulli trials: $W_i \sim Bin(1, p_n)$, where p_n is proportional to the subinterval length $\frac{t}{n}$:

$$p_n = P(A \text{ occurs in a specified subinterval}) = \frac{\omega t}{n}.$$

Here ω, the constant of proportionality, is the mean rate of occurrence of A per unit of the continuum, or ωt is the mean rate of occurrence of A per t units of the continuum. The requirements for such a sequence of Bernoulli trials are:

- $P($ A occurs in a specified subinterval$) = p_n$ is constant from subinterval to subinterval;
- $P($ A occurs more than once in a specified subinterval$) = 0$;
- the occurrence of A in one subinterval is independent of the occurrence of A in any other subinterval.

Adding the Bernoulli random variables W_i creates the binomial random variable

$$\sum_{i=1}^{n} W_i \sim Bin(n, \frac{\omega t}{n}).$$

Now, as n gets larger and p_n gets smaller in such a way that the binomial mean $n(p_n) = n(\frac{\omega t}{n}) = \omega t$ is constant, then by Theorem 6.3 it follows that the binomial probability mass function tends towards the Poisson probability mass function as n increases and we may write

$$X_t \sim Poi(\lambda) \quad \text{for} \quad \lambda = \omega t,$$

where X_t denotes the number of occurrences of the event A in t units of the continuum. The random variable X_t is called a **Poisson process**. We use 'length' to denote our measure of an interval of the continuum—for a small 'interval' of time, for a small 'area' of coral reef, for a small 'volume' of sea-water. In general there is considerable flexibility in choosing the subintervals in our above arguments. They need not be of identical 'length' $\frac{t}{n}$:

The random variable X_t denoting the number of occurrences of an event A in t units of a continuum is a **Poisson process with rate** $\omega > 0$ if:

1. The probability of A occurring exactly once in a small subinterval of length ϵ is $\omega\epsilon + o(\epsilon)$ [where $o(\epsilon)$ is of small order relative to ϵ so that $\lim_{\epsilon \to 0} \frac{o(\epsilon)}{\epsilon} = 0$].
2. The probability of A occurring more than once in a subinterval of length ϵ is $o(\epsilon)$.
3. The occurrence of A in a subinterval of length ϵ is independent of the occurrence of A in any other non-overlapping subinterval of length ϵ.

When these conditions are satisfied, $X_t \sim Poi(\lambda)$ for $\lambda = \omega t$.

Using Poisson tables

The values of the cumulative distribution function $F_X(x) = P(X \leq x)$ for $X \sim Poi(\lambda)$ are given in Appendix 2 for values of λ between 0.1 and 20. Therefore, if $x \sim Poi(15)$, then the tables give $P(X \leq 14) = 0.466$ and $P(X \leq 15) = 0.568$. If $X \sim Poi(15)$, then

$$P(X = 15) = P(X \leq 15) - P(X \leq 14) = 0.568 - 0.466 = 0.102.$$

As was the case for binomial tables, individual probability mass calculations are simple subtractions of adjacent lines in the table.

Example 6.8

Early morning newspaper customers at Wynyard

Early morning customers arrive at a newspaper stand at Wynyard Station at an average rate of 2 per minute after the morning rush starts at 7 a.m. (defined as time $t = 0$).

Worked example:

Determine the probability that no customers arrive to purchase a newspaper during an early morning period of:

1. **1 minute;**
2. **2 minutes;**
3. **30 seconds.**

Determine also the probability that more than five customers arrive during a 4 minute period after 7 a.m.

We assume that X_t, the number of newspaper customers in a period of t minutes is a Poisson process with rate $\omega = 2$. That is, $X_t \sim Poi(2t)$.

1. When the time interval is 1 minute, $t = 1$ and $X_1 \sim Poi(2)$. Hence,

$$P(X_1 = 0) = \frac{2^0 e^{-2}}{0!} = \frac{1}{e^2} = 0.1353 \quad \text{(3 dp)}$$

2. When the time interval is 2 minutes, the Poisson parameter changes and, in terms of this new $Poi(4)$ distribution,

$$P(X_2 = 0) = \frac{4^0 e^{-4}}{0!} = \frac{1}{e^4} = 0.0183 \ \ (3 \text{ dp})$$

3. $P(X_{0.5} = 0) = \frac{1^0 e^{-1}}{0!} = \frac{1}{e} = 0.3679$ (3 dp).

Finally, by using the Poisson tables from Appendix 2, we find

$$\begin{aligned} P(X_4 > 5) &= 1 - P(X_4 \leq 5) \quad \text{where } X_4 \sim Poi(8) \\ &= 1 - 0.191 \\ &= 0.809. \end{aligned}$$

Poisson moments

Our motivational approach to the Poisson process counts observations of a rare event unfolding in the appropriate continuum. How many occurrences of this event do we *expect*?

Inside information comes from studying limiting binomials. That is, if $Y \sim Bin(n,p)$, then in the limit as n increases and p decreases, so that $\lambda = np$ remains constant and $np^2 \to 0$:

$$\begin{aligned} &E(Y) = np = \lambda; \\ &Var(Y) = np(1-p) = np - np^2 \to \lambda. \end{aligned}$$

If $X \sim Poi(\lambda)$, then $E(X) = \mu_X = \lambda$ and $Var(X) = \sigma_X^2 = \lambda$. (6.6)

This means that the rate ω, in a Poisson process X_t, is the mean number of events per unit interval of the continuum. Indeed, since $E(X_t) = \omega t$, for a unit interval for which $t = 1$, $E(X_1) = (\omega)(1) = \omega$. To obtain (6.6) directly from the probability mass function $p_X(x)$ is not trivial. Of similar difficulty is the derivation of the moment generating function, which naturally unleashes all the moments.

Theorem 6.4 *The moment generating function and the cumulant generating function of the Poisson random variable $X \sim Poi(\lambda)$ are given respectively by*

$$m_X(t) = e^{\lambda(e^t - 1)} \quad \text{and} \quad \kappa_X(t) = \lambda(e^t - 1).$$

Proof: By definition, the mgf is $m_X(t) = E(e^{tX})$. This expected value is easily evaluated by using the exponential series (6.4):

$$\begin{aligned} m_X(t) &= E(e^{tX}) \\ &= \sum_{x=0}^{\infty} e^{tx} \frac{\lambda^x e^{-\lambda}}{x!} \\ &= e^{-\lambda} \sum_{x=0}^{\infty} \frac{(e^t)^x \lambda^x}{x!} \\ &= e^{-\lambda} \sum_{x=0}^{\infty} \frac{(\lambda e^t)^x}{x!} \\ &= e^{-\lambda} e^{\lambda e^t} \\ &= e^{\lambda(e^t-1)}. \end{aligned}$$

Finally, by definition, $\kappa_X(t) = \log_e m_X(t) = \lambda(e^t - 1)$.

■

The cumulant generating function is in a simple form for the evaluation of moments: from the cumulant generating function we obtain

$$\kappa'_X(0) = E(X) = \lambda \quad \text{and} \quad \kappa''_X(0) = Var(X) = \lambda.$$

Independent Poisson sums are again Poisson

Armed with the moment generating function, we can now observe that independent Poisson processes add to produce a new Poisson process; it seems intuitively reasonable that the rates of different Poisson processes should add to produce the rate for the overall process. This is indeed the case.

Suppose that $X_1, X_2, X_3, \ldots, X_n$ are independent Poisson random variables with $X_i \sim Poi(\lambda_i)$. These random variables have moment generating function

$$m_{X_i}(t) = e^{\lambda_i(e^t-1)}.$$

By the independence results of Chapter 5 (specifically Theorem 5.4), the moment generating function of the sum $X_1 + X_2 + X_3 + \ldots + X_n$ is given by the product of the moment generating functions:

$$\begin{aligned} m_{X_1+X_2+X_3+\ldots+X_n}(t) &= m_{X_1}(t) m_{X_2}(t) m_{X_3}(t) \ldots m_{X_n}(t) \\ &= e^{\lambda_1(e^t-1)} e^{\lambda_2(e^t-1)} e^{\lambda_3(e^t-1)} \ldots e^{\lambda_n(e^t-1)} \\ &= e^{\lambda_1(e^t-1)+\lambda_2(e^t-1)+\lambda_3(e^t-1)+\ldots+\lambda_n(e^t-1)} \\ &= e^{(\lambda_1+\lambda_2+\lambda_3+\ldots+\lambda_n)(e^t-1)} \\ &= e^{\lambda(e^t-1)} \end{aligned}$$

This is the moment generating function of a Poisson random variable with mean

$$\lambda = \lambda_1 + \lambda_2 + \lambda_3 + \ldots + \lambda_n.$$

This result may be summarised as :

Theorem 6.5 *If Poisson random variables $X_1, X_2, X_3, \ldots, X_n$ are independent with $X_i \sim Poi(\lambda_i)$, then,*

$$X_1 + X_2 + X_3 + \ldots + X_n \sim Poi(\lambda_1 + \lambda_2 + \lambda_3 + \ldots + \lambda_n).$$

Example 6.9

Combining rates of newspaper sales: Wynyard news-stand revisited

Example 6.8 examined the number of newspapers sold at Wynyard Station early in the morning. Now note that the news-stand sells many papers and customers usually purchase their *usual* or favourite paper.

Suppose then that the morning newspapers *The Australian*, *The Telegraph Mirror* and *The Sydney Morning Herald* are purchased according to independent (does one customer's purchase really influence another?) Poisson processes with respective average rates per minute (for this time of the morning) of 1, 4 and 5.

Therefore we assume that the number of sales X_1, X_2 and X_3, of *The Australian*, *The Telegraph Mirror* and *The Sydney Morning Herald* respectively occur independently in a 1 minute period with

$$\begin{aligned} \textit{The Australian} \quad & X_1 \sim Poi(1), \\ \textit{The Telegraph Mirror} \quad & X_2 \sim Poi(4), \\ \text{and } \textit{The Sydney Morning Herald} \quad & X_3 \sim Poi(5). \end{aligned}$$

Theorem 6.5 indicates that the total, X, of *all* newspaper sales in a 1 minute early morning period follows a $Poi(1+4+5)$ or $Poi(10)$ distribution. Therefore the probability that, in total, at least 15 newspapers are sold is

$$P(X \geq 15) = 1 - P(X \leq 14) = 0.083.$$

The expected number of newspapers sold in such a 1 minute period is $1+4+5 = 10$ papers. Similarly the standard deviation in the number of newspapers sold is $\sqrt{1+4+5} = \sqrt{10}$.

Finally, the probability that in a 1 minute period no copies of *The Australian* are sold, given that a total of 15 papers are sold :

$$\begin{aligned} P(X_1 = 0 | X_1 + X_2 + X_3 = 15) &= \frac{P(X_1 = 0, X_1 + X_2 + X_3 = 15)}{P(X_1 + X_2 + X_3 = 15)} \\ &= \frac{P(X_1 = 0, X_2 + X_3 = 15)}{P(X_1 + X_2 + X_3 = 15)} \\ &= \frac{P(X_1 = 0)P(X_2 + X_3 = 15)}{P(X_1 + X_2 + X_3 = 15)} \\ &= \frac{P(X_1 = 0)P(Y = 15)}{P(X = 15)} \end{aligned}$$

$$\begin{aligned} &\text{where } Y \sim Poi(9), \text{ and } X \sim Poi(10) \\ &= \frac{\left(\frac{1^0 e^{-1}}{0!}\right)\left(\frac{9^{15} e^{-9}}{15!}\right)}{\left(\frac{10^{15} e^{-10}}{15!}\right)} \\ &= (0.9)^{15} \\ &= 0.2059 \quad (4 \text{ dp}). \end{aligned}$$

PROPERTIES OF THE POISSON DISTRIBUTION
$X \sim Poi(\lambda)$ (6.7)

Probability mass function $p_X(x) = \frac{\lambda^x e^{-\lambda}}{x!}$
$x = 0, 1, 2, 3, \ldots, \quad \lambda > 0.$

Mean	$E(X) = \mu = \lambda$
Variance	$Var(X) = \sigma^2 = \lambda$
Standard deviation	$SD(X) = \sigma = \sqrt{\lambda}$
Coefficient of skewness	$\zeta_3 = \frac{1}{\sqrt{\lambda}}$
Coefficient of kurtosis	$\zeta_4 = 3 + \frac{1}{\lambda}$
Moment generating function	$m_X(t) = e^{\lambda(e^t - 1)}$
Cumulant generating function	$\kappa_X(t) = \lambda(e^t - 1)$

EXERCISES

6.31 X follows a Poisson probability distribution with $\lambda = 3$. Determine: $P(X \leq 4)$; $P(X \leq 3)$; $P(X < 4)$; $P(X = 4)$.

6.32 If $X \sim Poi(10)$, determine:

(a) $P(5 \leq X \leq 10)$;
(b) $P(5 < X \leq 10)$;
(c) $P(5 \leq X < 10)$;
(d) $P(5 < X < 10)$.

6.33 Suppose that in late summer, the Fremantle Surf Lifesaving Club makes an average of one surf rescue per day. Use Poisson probability distributions to determine the probability that:

(a) 3 surf rescues are made in a 3-day period;
(b) 1 rescue is made on each of 3 consecutive days.

6.34 Workers in a large factory suffer accidents at a rate of two per week. Determine the probability that there will be more than 3 accidents in a two-week period.

6.35 The number of machine failures per day in a certain plant has a Poisson distribution with mean rate 3 per day. Present maintenance facilities can repair 3 machines per day. Failures in excess of 3 are repaired by a contractor.

(a) On a given day, what is the probability of having machines repaired by the contractor?

(b) Let the random variable Y denote the number of machines repaired by the contractor. Determine the probability distribution for Y.

6.36 The following ranked data give the number of mainframe computer breakdowns recorded each month in a large tertiary institution over a 2-year period. It is assumed that only one breakdown can occur per day!

3	5	5	6	6	7	7	7
7	8	8	8	8	8	9	9
9	9	10	10	10	11	12	14

(a) Find the proportion of the period for which there were:

i. exactly 7 breakdowns per month;

ii. more than 12 breakdowns per month;

iii. at least 4 breakdowns per month.

(b) Suppose that the random variable X representing the number of breakdowns per month is assumed to have a binomial distribution with $n = 20$ (approximately the number of weekdays per month) and $p = 0.4$ (the probability of a breakdown occurring during a day).

Determine the probability of each event in (a) occurring.

(c) Now assume that the random variable representing the number of breakdowns per month is assumed to have a Poisson distribution with $\lambda = 8$. Determine the probability of each event in (a) occurring.

(d) Both the binomial model of (b) and the Poisson model of (c) are possible models for computer breakdowns.

Give the mean and variance for each model.

(e) Suggest which might be the more appropriate model for this situation. Justify your answer.

6.37 The Poisson probability model has been applied to the number of surface defects on polished steel plates examined in a metallurgical laboratory as a routine inspection measure.

(a) Write down a formula for the probability that a polished steel plate has r defects, explaining the meaning of the symbols that you use.

(b) Simplify the formula in the case $r = 0$.

(c) In a large homogeneous collection of polished steel plates, 10% have one or more defects. Determine the percentage having exactly 2 defects.

(d) Why may the Poisson probability model be applied to data of this kind? Are there any model assumptions that are unlikely to be met?

6.38 A Geiger counter detects particles emitted from a radioactive source at an average rate of 2 particles per minute; the number of particles X_t detected in any t minute interval may be assumed to have a Poisson distribution for which

$$P(X_t = k) = \frac{e^{-\mu}\mu^k}{k!}, \quad k = 1, 2, \ldots,$$

where $\mu = 2t$.

(a) If $t = 3$, what is the value of μ ?

(b) Determine $P(X_3 = 0)$.

(c) X_t may be interpreted as a Poisson process. What is the rate of this process?

(d) Use Poisson tables to determine the probability that:

i. no particles are emitted during a 1 minute interval;

ii. exactly 4 particles are emitted in a 3 minute interval;

iii. at most 5 particles are emitted during a 30 second time interval;

iv. at least 3 particles are emitted during a one and a half minute time interval.

6.39 For traffic turning left at a particular intersection, the number of accidents occurring per week, X, is assumed to follow a discrete probability distribution for which

$$P(X \geq k) = \sum_{x=k}^{\infty} \frac{e^{-1}}{x!}, \quad k = 0, 1, 2, \ldots$$

(a) Determine $P(X \geq 1)$.

(b) Determine $P(X > 1)$.

(c) State the value of $E(X)$ and $Var(X)$.

At the same intersection, for traffic turning right, Y, the number of accidents occurring per week is assumed to follow a discrete probability distribution for which

$$P(Y \geq k) = \sum_{y=k}^{\infty} \frac{2^y e^{-2}}{y!}, \quad k = 0, 1, 2, \ldots$$

(d) Assuming that X and Y are independent, determine the probability that in a given week there is at least one accident involving *turning* traffic.

(e) Is the independence assumption reasonable? Explain very briefly.

(f) The following data of 10 weeks of traffic accidents is for **either** left-turning traffic **or** right-turning traffic at the intersection. Are these most likely to be observations on the random variable X or the random variable Y? Explain very briefly.

0, 1, 2, 3, 2, 4, 3, 0, 4, 2.

6.40 An aeroplane wing is assembled with a large number of rivets. The number, X, of defective rivets is assumed to follow a Poisson distribution with mean rate of defective rivets λ per aeroplane wing.

(a) Write down the probability mass function of X.

(b) Independent observations were taken on 12 aeroplane wings and the number of defective rivets were:

4	3	5	4	8	4
12	5	2	5	4	4

Determine $\overline{x}$ for this data set.

(c) If the Poisson distribution is going to fit these data well, then the fitted model should have the mean λ estimated by the value for $\overline{x}$. (At least then the mean of the data and the mean of X agree.)

Use this fitted version of the probability model to estimate the probability of the dangerous event occurring: more than 10 rivets are defective on an aeroplane wing.

6.41 If $X \sim Poi(3)$, determine $E(X^2)$.

6.42 The number of requests for glass pipettes received by a laboratory supplier averages 2 per week, varying each week according to a Poisson probability model.

At the beginning of a particular week, the supplier has 5 items in stock.

(a) Determine:

i. the probability that there is a demand for precisely 3 pipettes in that week;

ii. the probability that there are more requests than can be supplied from stock;

iii. the probability distribution of the number of pipettes in stock at the end of the week.

(b) If and when the stock of pipettes is reduced to two items during the week, replacement stock is ordered. Determine the probability that such an order is required.

6.43 If $X \sim Poi(9)$, determine an integer k such that $P(X \geq k)$ is as large as possible without exceeding 0.05. For this value of k, determine the value of the probability α defined by $\alpha = P(X \geq k)$.

6.44 2% of all manufactured computer chips are defective, the defectives occurring at random during production. If the chips are packed in boxes of 100, what is the probability, p_r, that a given box will contain r defectives? Construct a table to compare the binomial and the Poisson evaluations of p_r for values $r = 0, 1, 2, 3, \ldots, 10$.

Does the Poisson approximation seem reasonable for these low values of r?

6.45 A 500 m roll of heavy duty canvas will be cut into 5 m lengths to make tarpaulins. Flaws in the fabric occur on average at a rate of five flaws per 100 m. Determine the expected number of flawless pieces.

Brass eyelets will be riveted, one per corner, into the flawless pieces at a cost of \$1.25 per eyelet. Determine the expected cost of this operation.

6.46 Sketch the probability mass function of $X \sim Poi(7)$. Does the graph appear to be symmetric?

6.47 A company with newly installed conveyor belts on its production line makes a profit of k hundred dollars per hour, where $k > 2$. However, the equipment breaks down unexpectedly, creating additional repair costs and reducing overall profit.

Suppose that, X_t, the number of breakdowns during any period of length t hours, is a Poisson process with rate 1. The X_t breakdowns cost $X_t^2 + X_t$ hundreds of dollars in lack of production and repair bills.

The problem confronting the plant management is to decide how long the equipment should be maintained in operation before a preventive service overhaul. This involves choosing an optimal value for the time frame t.

(a) Write down an expression in terms of X_t and k for the realised profit during any period of t hours.

(b) Determine the value of t that maximises the expected profit.

(c) What value of t causes the company to expect to break even?

6.48 A specified event E does not occur in the time interval $(t, t + \Delta t)$ with probability $1 - [\omega\Delta t + o(\Delta t)]$, where ω is a constant and $o(\Delta t)$ is of small order compared with Δt so that $\lim_{\Delta t \to 0} \frac{o(\Delta t)}{\Delta t} = 0$.

The occurrence of E is independent of any previous occurrences of E.

Show that if $p_o(t)$ is the probability that E does not occur in time $(0, t)$, then

(a) $p_o(t + \Delta t) = p_o(t)\{1 - [\omega\Delta t + o(\Delta t)]\}$;

(b) $\frac{d}{dt} p_o(t) = -\omega p_o(t)$;

(c) $p_o(t) = e^{-\omega t}$.
Show that this result is in agreement with $P(X_t = 0)$ for $X_t \sim Poi(\omega t)$.

6.49 Independent runs of a carpet-making process produce a standard length of carpet in which the random variable, X, the number of flaws, has a Poisson distribution with rate λ. Suppose that 2 independent production runs are made.

(a) If the cost of removal of flaws is \$1 per flaw for the first production run, and \$2 per flaw for the second run, determine a random variable C which gives the total cost for the removal of flaws from both runs.

(b) Show that the variance of C is $\frac{5}{3}$ times the expected value of C.

6.50 (a) Show that the *coefficient of skewness* of $X \sim Poi(\lambda)$ is

$$\zeta_3 = \frac{1}{\sqrt{\lambda}}.$$

[*Hint*: Use the cumulant generating function.]

(b) For what values of λ is the *coefficient of skewness* negative? Use this result to discuss the skewness of the graphs of the pmf of X for various values of λ in Example 6.6.

(c) Explain why you would expect a $Poi(\lambda)$ distribution to be almost symmetric for large values of λ.

6.51 Let X denote the number of taxis arriving at a certain taxi stand in Darling Harbour, and Y denote the number of (parties of) people at the taxi stand wishing to catch a cab.

Suppose that X and Y are independent Poisson distributions, each with parameter λ.

Determine the cumulant generating function for the excess of users over taxis; that is, for $Z = Y - X$, which may be interpreted as assuming either positive or negative values.

6.52 Show that the *coefficient of kurtosis* of $X \sim Poi(\lambda)$ is

$$\zeta_4 = 3 + \frac{1}{\lambda}.$$

Use this result to discuss the kurtosis of the graphs of the pmf of X for various values of λ in Example 6.6.

6.3 WAITING FOR THE COUNT

The geometric distribution revisited

The geometric distribution was used extensively in Chapter 4 to introduce discrete random variables. In Example 4.2 the distribution was used to model 'defect-free' runs. Our interest now in the distribution is in the context of a discrete waiting-time distribution. This is in contrast to the continuous exponential waiting-time distribution of Chapter 4.

We begin by revising the formula for the geometric probability mass function and examine the mean, variance and moment generating function. Recall that $X \sim Geom(p)$ if X has probability mass function $p_X(x) = pq^{x-1}$, $x = 1, 2, 3, \ldots$; $0 < p < 1$, $q = 1 - p$. From this the moment generating function is straightforward to determine:

Theorem 6.6 *If X has a geometric distribution with parameter p, $X \sim Geom(p)$, then X has moment generating function*

$$m_X(t) = \frac{pe^t}{1 - qe^t}, \quad \textit{with} \quad q = 1 - p.$$

Proof: The geometric distribution is quite naturally related to the **geometric series:**

$$\frac{1}{1-w} = 1 + w + w^2 + w^3 + w^4 + \ldots \quad \text{for} \quad |w| < 1,$$

which can be used to show that geometric probabilities sum to 1. However, in this proof, we use the geometric series to find a simple form for the mgf:

$$\begin{aligned} m_X(t) &= E(e^{tX}) \\ &= \sum_{x=1}^{\infty} e^{tx} pq^{x-1} \\ &= \sum_{x=1}^{\infty} (qe^t)^{x-1} pe^t \\ &= pe^t \underbrace{\sum_{x=1}^{\infty} w^{x-1}}_{=\frac{1}{1-w}} \quad \text{for} \quad w = qe^t \\ &= \frac{pe^t}{1-qe^t}. \end{aligned}$$

■

The cumulant generating function is $\kappa(t) = \log_e m_X(t) = \log_e(pe^t) - \log_e(1-qe^t)$. Placing $t = 0$ in this expression and its derivatives gives:

$$\kappa(0) = 0; \quad \kappa'(0) = \tfrac{1}{p}; \quad \kappa''(0) = \tfrac{q}{p^2}.$$

Since $E(X) = \kappa'(0)$ and $Var(X) = \kappa''(0)$, we may summarise:

If $X \sim Geom(p)$, then $E(X) = \frac{1}{p}$ and $Var(X) = \frac{1-p}{p^2}$.

Careful manipulation of the geometric series is required to obtain these mean and variance results directly.

The binomial distribution results from counting the number of successes in n Bernoulli trials, where n is fixed in advance. For the geometric distribution, a sequence of Bernoulli trials is observed until the first success is recorded. Since we do not know at what trial this will occur, n is no longer fixed. The mean of a sum of independent geometric distributions is cleverly applied to solve the following problem.

Example 6.10

Adding geometric waiting times

Worked example:

Determine the expected number of tosses required of a fair die until each face has appeared at least once.

Define independent geometric random variables as follows:

Let: X_1 be the number of tosses until the first face appears;
X_2 be the number of tosses after the first face has appeared until a second face appears which is different from the first;
X_3 be the number of tosses after the second face has appeared until a 3rd face appears which is different from the first two;
X_4 be the number of tosses after the 3rd face has appeared until a 4th face appears which is different from the first three;
...
X_6 be the number of tosses after the 5th face has appeared until the 6th face appears which is different from the first five faces.

These definitions ensure that the number of tosses, X, required until each face has appeared at least once is given by $X = X_1 + X_2 + X_3 + X_4 + X_5 + X_6$, where $X_1 = 1$ with probability 1;

$X_2 \sim Geom(\frac{5}{6})$; $X_3 \sim Geom(\frac{4}{6})$; $X_4 \sim Geom(\frac{3}{6})$; $X_5 \sim Geom(\frac{2}{6})$;
$X_6 \sim Geom(\frac{1}{6})$.

Finally, by **SUM 1**,

$$\begin{aligned} E(X) &= E(X_1) + E(X_2) + E(X_3) + E(X_4) + E(X_5) + E(X_6) \\ &= 1 + \frac{1}{\frac{5}{6}} + \frac{1}{\frac{4}{6}} + \frac{1}{\frac{3}{6}} + \frac{1}{\frac{2}{6}} + \frac{1}{\frac{1}{6}} \\ &= 14.7. \end{aligned}$$

Therefore we expect to wait at least 15 tosses of a fair die until each face has appeared once. This answers the question as posed. However, little interpretation may be placed on a mean without simultaneous consideration of the variance. By **SUM 2**,

$$\begin{aligned} Var(X) &= Var(X_1) + Var(X_2) + Var(X_3) + Var(X_4) + Var(X_5) \\ &\quad + Var(X_6) \\ &= 0 + \frac{\frac{1}{6}}{(\frac{5}{6})^2} + \frac{\frac{2}{6}}{(\frac{4}{6})^2} + \frac{\frac{3}{6}}{(\frac{3}{6})^2} + \frac{\frac{4}{6}}{(\frac{2}{6})^2} + \frac{\frac{5}{6}}{(\frac{1}{6})^2} \\ &= 38.99. \end{aligned}$$

It follows that the standard deviation of X is 6.24 (2 dp). Trying this out *once* on a (possibly) fair die gave the data

$$\underbrace{1}_{X_1=1}\ \underbrace{5}_{X_2=1}\ \underbrace{1\ 2}_{X_3=2}\ \underbrace{5\ 2\ 2\ 3}_{X_4=4}\ \underbrace{3\ 3\ 6}_{X_5=3}\ \underbrace{1\ 1\ 5\ 2\ 1\ 6\ 3\ 6\ 4}_{X_6=9}$$

with an observed value 20 for X. Thus, given that $x = 20$ corresponds to an observation lying approximately one standard deviation above the mean, it is hardly a surprising result and we are inclined to believe that the die is 'fair'.

Negative binomials

A sum of independent (but differently distributed) geometric random variables in Example 6.10 represented the waiting time for all faces of a die to appear when it is tossed repeatedly. We continue this theme of sums of independent random variables: Negative binomial random variables are the sums of independent geometric random variables each having the *same* distribution.

Definition 6.5 *Given a positive integer $r \geq 1$ and $0 < p < 1$, the discrete random variable X has a* **negative binomial distribution** *with parameters r and p, and we write $X \sim Neg\ Bin(r,p)$ if*

$$X = X_1 + X_2 + X_3 + \ldots + X_r,$$

where $X_1, X_2, X_3, \ldots, X_r$ are i.i.d $Geom(p)$ random variables.

□

There is a Bernoulli trials interpretation that may be applied in Definition 6.5. Consider a sequence of Bernoulli trials with probability of success p. Then we may interpret X_1 as the number of trials until the 1st success occurs. Resetting our counters to zero, then X_2 is the number of trials until the 2nd success occurs. Therefore $X_1 + X_2$ is the total number of trials from the beginning of the sequence until the second success occurs. Adding an independent wait X_3 for the third success gives a total number of trials $X_1 + X_2 + X_3 \ldots$ and the process continues until

$$X_1 + X_2 + X_3 + \ldots + X_r$$

records the total number of trials until the rth success. By **SUM 1** and **SUM 2** the mean and variance of a negative binomial distribution are easily determined.

Theorem 6.7 *For $q = 1 - p$, the mean and variance of $X \sim NegBin(r,p)$ are given by*

$$E(X) = \frac{r}{p} \quad \text{and} \quad Var(X) = \frac{rq}{p^2}.$$

Proof: Let $X_i \sim Geom(p)$, $i = 1, 2, 3, \ldots, r$, and $X = X_1 + X_2 + X_3 + \ldots + X_r$. Then $X \sim NegBin(r,p)$ and applying **SUM 1** directly:

$$\begin{aligned} E(X) &= E(X_1 + X_2 + X_3 + \ldots + X_r) \\ &= E(X_1) + E(X_2) + E(X_3) + \ldots + E(X_r) \\ &= \frac{1}{p} + \frac{1}{p} + \frac{1}{p} + \ldots + \frac{1}{p} \\ &= \frac{r}{p} \end{aligned}$$

Similarly, **SUM 2** and independence of the geometric random variables X_i gives

$$\begin{aligned} Var(X) &= Var(X_1 + X_2 + X_3 + \ldots + X_r) \\ &= Var(X_1) + Var(X_2) + Var(X_3) + \ldots + Var(X_r) \\ &= \frac{q}{p^2} + \frac{q}{p^2} + \frac{q}{p^2} + \ldots + \frac{q}{p^2} \\ &= \frac{rq}{p^2} \end{aligned}$$

■

This proof has been accomplished without knowledge of the probability density function for X. (In fact this was how we began the discussion of binomial random variables in Section 6.1.)

To construct $p_X(x)$ we proceed as follows: Suppose that the event $\{X = x\}$ occurs, so that the rth success occurs on trial number x. Necessarily, this means that the prior $r - 1$ successes occur, along with the $x - r$ failures, at any of the trials numbered from 1 to $x - 1$. The probability of this happening is

$$P(W = r - 1), \quad \text{where} \quad W \sim Bin(x - 1, p).$$

Hence

$$\begin{aligned} P(X = x) &= P(r - 1 \text{ successes in } x - 1 \text{ trials, success on } x\text{th trial}) \\ &= P(r - 1 \text{ successes in } x - 1 \text{ trials}) P(\text{success on } x\text{th trial}) \\ &= P(W = r - 1)p \quad \text{where} \quad W \sim Bin(x - 1, p) \\ &= \left[\binom{x-1}{r-1} p^{r-1}(1 - p)^{x-r}\right] p \\ &= \binom{x-1}{r-1} p^r (1 - p)^{x-r}. \end{aligned}$$

In summary:

> If $X \sim Neg\,Bin(r, p)$, then X has probability mass function $p_X(x) = \binom{x-1}{r-1} p^r (1 - p)^{x-r}$ for $x = r, r + 1, r + 2, \ldots, \infty$.

Example 6.11

Testing booster rockets

A new booster rocket is being tested for reliability in sending satellites into orbit. The probability of a successful launch is estimated at 0.8. Because of the enormous value of the payload satellite, the testing regime requires that booster rockets are launched repeatedly until 3 successful launches (with no satellite payload) have been achieved.

Worked example:

Determine the probability that:

1. exactly 6 attempts;

2. fewer than 6 attempts

will be necessary to satisfy the testing requirements.

1. If we let X denote the number of launch attempts until the third success, the negative binomial parameters for X are $r = 3$ and $p = 0.8$. Thus, $X \sim NegBin(3, 0.8)$ and

$$p_X(6) = \binom{6-1}{3-1}(0.8)^3(1-0.8)^{6-3} = 0.041 \quad (3 \text{ dp}).$$

2. The minimum number of attempts for 3 successful launches is 3. Hence, $P(\text{fewer than 6 attempts}) = p_X(3) + p_X(4) + p_X(5)$, or $\binom{3-1}{3-1}(0.8)^3(1-0.8)^{3-3} + \binom{4-1}{3-1}(0.8)^3(1-0.8)^{4-3} + \binom{5-1}{3-1}(0.8)^3(1-0.8)^{5-3}$ which simplifies to 0.942 (3 dp).

EXERCISES

6.53 In Example 6.8, early morning customers arrive at a newspaper stand at Wynyard Station at an average rate of 2 per minute after the stand opens at 7 a.m. (defined as time $t = 0$).

Let X_t denote the number of customers arriving in the time interval $[0, t]$; that is, in the first t minutes after counting starts.

Let Y denote the time, in minutes, that elapses before the first customer arrives in the morning after the newspaper stand opens.

(a) State whether each of the random variables X_t and Y is discrete or continuous.

(b) Give the expected number of customers in a 3 minute interval.

(c) Determine $P(X_t = 0)$.

(d) Show that, for $t > 0$,

$$P(Y > t) = e^{-2t}.$$

(e) Hence determine an expression for $F_Y(y)$, the cumulative distribution function of Y.

6.54 The carnage on Australian roads is well documented in publications such as *Year Book Australia* (1985). In 1983 there were 966 traffic deaths on NSW roads.

(a) Estimate the probability that in NSW during this period:

i. there are no accidents on a given day;

ii. there are two accident-free days in a given week;

iii. more than 2 days will elapse between successive fatal accidents.

(b) Determine the expected number of accident-free days in a month.

(c) Determine the expected number of days to elapse (starting with New Years Day) before there have been 10 accident-free days.

(d) Suppose that, in a particular year, by 26 January, there had been 57 deaths on NSW roads. Based on this information alone, determine the expected number of deaths for the calendar year, and decide whether it is likely that the 1983 figure would be exceeded.

6.55 In a weekly lottery there is one chance in 20 of winning a prize. Suppose that I purchase a ticket each week.

(a) Write down the probability mass function of the number of weekly tickets that I need to purchase up to and including my first winning ticket.

(b) Write down the probability mass function of the number of weekly tickets that I need to purchase up to and including my *second* winning ticket.

6.56 Suppose that in a board game involving two dice, a pair of sixes must be thrown before a player may start the game. At each turn, the dice are thrown simultaneously and this denotes a 'toss'.

(a) If the dice are fair, determine the mean and variance in the number of tosses required before a player may start.

(b) Suppose that one die is fair, but the other die is loaded so that a '6' is twice as likely to be thrown as any other face. In this case, determine the expected number of tosses of pair of dice until a 'double 6'.

6.57 If $Y \sim Geom(p)$, give an expression for the probability mass function of $X = Y - 1$.

[X corresponds to a waiting time where the first possible outcome is 0.]

6.58 Consider a small post office with a single staff member operating a single postal counter. Suppose that the probability p_k, that there are k customers in the post office, is given by

$$p_k = p_0 p^k, \quad k = 0, 1, 2, 3, \ldots,$$

where p is a constant having a value between 0 and 1.

(a) Show that $p_0 = 1 - p$.

(b) Determine the probability that a newly arriving customer will have to wait to be served.

(c) Determine the expected number of customers in the post office.

6.59 Suppose that p is the probability of a head occurring when a biased coin is tossed. Let the random variable T represent the total number of tosses up to and including the first head to occur *after* the first tail occurs.

(a) What values may T have?

(b) Express T as the sum of two clearly defined random variables.

(c) Show that the mean of T is

$$\frac{1}{p(1-p)}.$$

(d) Determine the variance of T.

6.60 The following chart, published in *The Herald*, Melbourne, 22 August 1990, shows the number of weeks since each number was drawn in Tattslotto.

Each week, 8 balls are randomly selected from the Tattslotto barrel. The chart exhibits the 'times since last drawn' for all 45 numbers that were in the game as of 22 August 1990. Note that data entries recorded as 0 correspond to Tattslotto numbers that have been drawn in the current week, so that it is '0 weeks' since these numbers last appeared.

1	2	3	4	5	6	7	8
0	3	5	5	0	21	5	3
9	10	11	12	13	14	15	16
4	1	18	6	3	7	2	1
17	18	19	20	21	22	23	24
7	1	2	1	1	3	5	9
25	26	27	28	29	30	31	32
3	6	5	2	0	9	1	0
33	34	35	36	37	38	39	40
0	4	0	9	0	2	0	6
41	42	43	44	45			
2	2	3	4	4			

A frequency distribution for this data is as follows:

x_i	0	1	2	3	4	5
f_i	8	6	6	6	4	5

x_i	6	7	9	18	21
f_i	3	2	3	1	1

(a) If the mechanism is truly random, give the probability that a particular number is among the 8 selected in a given week.

(b) Determine the probability mass function for the number of weeks, W_1, that I have to wait (commencing with week 0) up to and including the week that my favourite number is drawn.

(c) Determine the probability mass function for the number of weeks, W_2, since my favourite number was drawn.

[Interpret this week as week 0 and last week as 1 week ago.]

(d) Are W_1 and W_2 identically distributed?

(e) Are W_1 and W_2 independent?

(f) Give the expected number of weeks that I have to wait until my favourite number is drawn. How does this theoretical expected value $E(W_1)$ compare with the observed sample mean $\overline{x}$ calculated from the above frequency distribution?

(g) Give the variance in the number of weeks that I have to wait before my favourite number is drawn. How does this theoretical expected value $Var(W_1)$ compare with the observed sample variance s^2 calculated from the above frequency distribution?

(h) Do the data seem to agree with the probability model specified in (c)?

[Only an intuitive judgment is required at this stage; formal testing procedures for 'fitting' a data set to a probability model are discussed later.]

6.61 At one stage in my childhood, packets of Weet-bix breakfast cereal contained one plastic animal per packet.

This was always hidden in the bottom of the packet so that one had to invent ways of seeking the plastic animal out quickly, because it was simply too long to wait till all the contents were eaten.

This animal series had 5 different members: a rhinoceros, a lion, an elephant, an antelope and a giraffe.

Assuming that I am allowed to buy one packet of Weet-bix per week, what is the expected number of weeks that I have to wait before I have the entire animal series?

Carefully outline your assumptions.

6.62 A used car salesman boasts that he makes a sale to about 30% of his Saturday morning customers. Determine the probability that on a particular Saturday morning, his third sale occurs with the 10th customer.

6.63 Determine the probability that a couple must have at least 4 children in order to have a second son. [Assume that boys and girls are equally likely.]

6.64 Suppose that 90% of people are right-handed. A psychologist requires 5 right-handed people to participate in an experiment on eye-hand coordination.

(a) Determine the probability that the psychologist must interview 7 applicants before finding the requisite number of right-handed people.

(b) If the psychologist interviews X people before finding the requisite number of right-handed people, sketch the probability mass function of X.

6.65 If $X \sim NegBin(r,p)$, determine the moment generating function of X.

6.66 A bag contains w white marbles and b black marbles, indistinguishable except for colour. Marbles are drawn at random one at a time and **replaced** in the bag after each drawing, the process continuing until both colours have appeared.

Let X denote the number of drawings required for the process to terminate.

(a) Show that the probability mass function of X is given by $P(X=x) = p^{x-1}q + q^{x-1}p$, $x = 2,3,4,\ldots$; where $p = \frac{w}{w+b}$ and $q = 1-p$.

(b) Determine the moment generating function of X.

(c) Use the moment generating function to show that with the same definitions of p and q as in (a),

$$E(X) = \frac{1-pq}{pq}.$$

Hence show that $E(X) \geq 3$ no matter what the ratio of white marbles to black marbles in the bag.

Does this mean that the process can never stop *before* three selections have been made? Explain briefly.

6.67 In a sequence of Bernoulli trials with probability of success p, Y represents the number of failures until (and including) the rth success.

(a) What is the probability distribution of $Y + r$?

(b) Determine the moment generating function of Y.

6.68 Show that if $X \sim Geom(p)$, then the cumulant generating function of X has

$$\kappa'''(0) = \frac{q+q^2}{p^3}.$$

Hence show that when p is small, the geometric distribution is highly positively skewed.

6.69 Suppose that $X_1, X_2, X_3, \ldots, X_n$ are independent random variables with

$$X_i \sim Neg\ Bin(r_i, p).$$

Use moment generating functions to determine the distribution of $\sum_{i=1}^{n} X_i$.

CHAPTER SEVEN

Modelling with Probability Density Functions

In this chapter you will find

7.1 THE FAMILY OF NORMAL DISTRIBUTIONS

Example 6.4 showed graphs of the probability mass function for binomial random variables. Each graph shows a mass function drawn with a different value of $p = P(success)$. The graphs change shape as the parameter p changes. For example, as p increases from $\frac{1}{2}$ towards 1, the mean increases, the variance increases, and the distribution becomes left-skewed. These effects are reflected in the behaviour of the first moments of the distribution. Essentially, $E(X) = \mu$ relates to 'central location', $E(X-\mu)^2$ to 'spread about the centre', $E(X-\mu)^3$ to 'skewness from the centre' and $E(X-\mu)^4$ to 'peakedness at the centre'.

The moments of a continuous random variable, X, help us understand the shape of the probability density function of X. The normal probability distribution was introduced in Chapter 4 to illustrate the calculation of moments and to record its classical distributional shape. It is against this distribution that the shapes of others are measured. In this sense the normal probability distribution sets a benchmark for centre, spread, skewness and kurtosis.

The standard normal distribution

We begin with a recap of what is known from Chapter 4 by carefully defining the standard normal probability distribution.

Definition 7.1 *If a continuous random variable Z has probability density function*

$$f_Z(z) = \frac{1}{\sqrt{2\pi}} e^{-\frac{1}{2}z^2} \qquad -\infty < z < \infty,$$

then we say that Z is a **standard normal random variable** *with a* **standard normal distribution** *and we write $Z \sim N(0,1)$.*

□

The probability density function of the standard normal distribution
$f_Z(z) = \frac{1}{\sqrt{2\pi}} e^{-\frac{1}{2}z^2}$

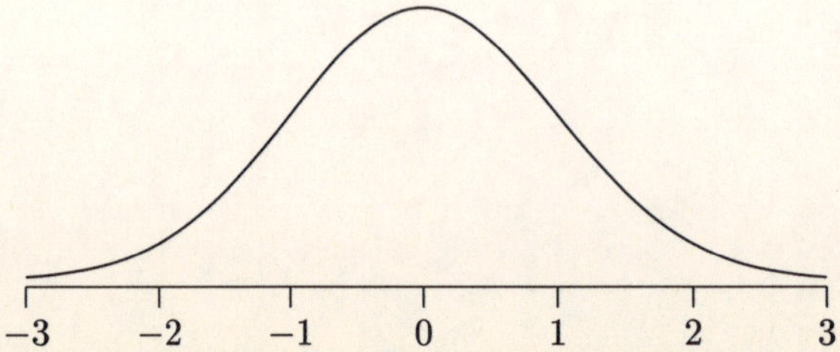

In Example 4.9, it was shown that Z has moment generating function $m_Z(t) = e^{\frac{1}{2}t^2}$. The cumulant generating function, $\kappa_Z(t)$ is the logarithm of the moment generating function, so that

$$\kappa_Z(t) = \tfrac{1}{2}t^2.$$

The moments of Z may be derived directly from the moment generating function (Example 4.9), or, by differentiating the cumulant generating function:

$$\kappa'_Z(0) = 0; \quad \kappa''_Z(0) = 1; \quad \kappa_Z^{(r)}(0) = 0 \ \text{ for } \ r \geq 3.$$

Theorem 7.1 *The standard normal random variable $Z \sim N(0,1)$ has mean $E(Z) = 0$, variance $Var(Z) = 1$, coefficient of skewness $\zeta_3 = 0$ and coefficient of kurtosis $\zeta_4 = 3$.*

Proof: The mean and variance may be found by direct integration, but more easily follow from the cumulant generating function: $E(Z) = \kappa'_Z(0) = 0$ and $Var(Z) = \kappa''_Z(0) = 1$. The moments of skewness and kurtosis follow directly from the cumulants as well:

$$\begin{aligned} \zeta_3 &= \frac{\mu_3}{(\sqrt{\mu_2})^3} = \frac{\kappa'''_Z(0)}{(\sqrt{\mu_2})^3} = 0; \\ \zeta_4 &= \frac{\mu_4}{(\sqrt{\mu_2})^4} = \frac{\kappa''''_Z(0) + 3\mu_2^2}{(\sqrt{\mu_2})^4} = \frac{0 + 3(1)^2}{(1)^4} = 3. \end{aligned}$$

■

Definition 7.2 *The* **standard normal cumulative distribution function**, *denoted by Φ, is defined on $-\infty < z < \infty$ by $\Phi(z) = P(Z \leq z)$ when $Z \sim N(0,1)$.*

□

The properties of cdf's of arbitrary random variables that were discussed at the outset of Chapter 4 may now be usefully applied. In particular, $\Phi(-\infty) = 0$, and since it has a positive derivative, Φ is increasing on $(-\infty, \infty)$ until $\Phi(\infty) = 1$. Notice that $\Phi(z)$ represents the area under the standard normal probability density function to the left of the point z.

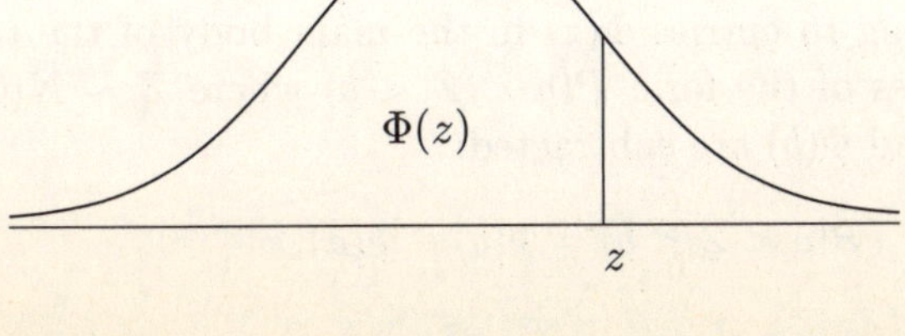

Since Φ is continuous, the inverse function Φ^{-1} is uniquely determined by

$$\Phi^{-1}(y) = z \quad \text{if} \quad y = P(Z \leq z).$$

The following definition gives a special notation which is reserved for the percentiles of a standard normal distribution.

Definition 7.3 *For $0 < \alpha < 1$, the* **upper alpha point**, *z_α, of a standard normal distribution is the $100(1-\alpha)$th percentile of the distribution and is determined by*

$$z_\alpha = \Phi^{-1}(1-\alpha).$$

□

This definition shows that z_α is determined implicitly by the equation $\Phi(z_\alpha) = 1 - \alpha$. It follows that for a standard normal curve, z_α is that value on the horizontal axis beyond which probability α is trapped under the curve. Consequently, the area under the curve to the left of z_α is $1-\alpha$. This simple idea has been phrased in terms of the inverse cdf in preparation for the construction of normal probability plots.

The upper alpha point, z_α, of the standard normal distribution

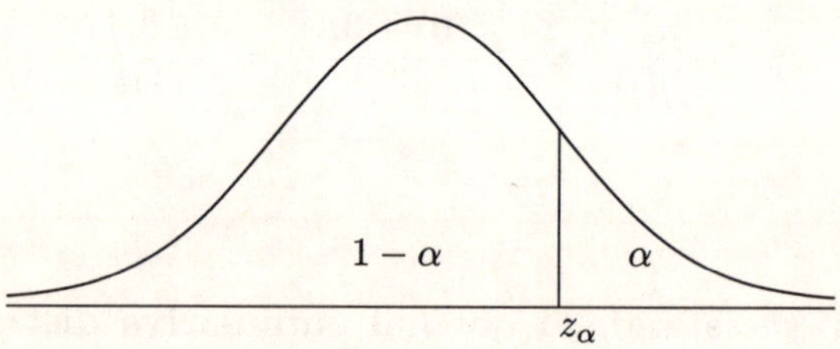

Using normal probability tables

By definition,

$$\Phi(z) = \int_{-\infty}^{z} \frac{1}{\sqrt{2\pi}} e^{-\frac{1}{2}t^2} dt.$$

This integral has no closed form algebraic solution for an arbitrary choice of z. Instead, the integral must be evaluated using numerical integration techniques. Fortunately, the numerical integrations have been done for us with results presented in tabular form in Appendix 3.

In these standard normal tables, z values are specified on the left-side of the table corresponding to entries $\Phi(z)$ in the main body of the table. To determine probabilities of the form $P(a < Z < b)$ where $Z \sim N(0,1)$, the tabled entries $\Phi(a)$ and $\Phi(b)$ are subtracted:

$$P(a < Z < b) = \Phi(b) - \Phi(a).$$

Recall that for a continuous random variable such as Z, the probabilities of fixed points are 0; the probabilities $P(a < Z < b)$, $P(a < Z \leq b)$, $P(a \leq Z < b)$ and $P(a \leq Z \leq b)$ are equal. Reading from the table we find:

$$\Phi(-2.1) = 0.0179;\ \Phi(-1.96) = 0.0250;\ \Phi(0) = 0.5;\ \Phi(2.1) = 0.9821.$$

Therefore

$$P(-2.1 < Z \leq -1.96) = \Phi(-1.96) - \Phi(-2.1) = 0.0250 - 0.0179 = 0.0071.$$

Similarly,

$$P(Z \geq 2.1) = 1 - P(Z < 2.1) = 1 - \Phi(2.1) = 1 - 0.9821 = 0.0179.$$

Notice that $P(Z \geq 2.1) = P(Z \leq -2.1)$ is to be expected since the standard normal pdf is symmetric about 0. This symmetry also ensures that the following results hold: if c is a constant

$$P(-c < Z < c) = \Phi(c) - \Phi(-c) = \Phi(c) - [1 - \Phi(c)] = 2\Phi(c) - 1, \tag{7.1}$$

and

$$P(|Z| > c) = 1 - P(-c < Z < c) = 1 - [2\Phi(c) - 1] = 2[1 - \Phi(c)].$$

Note that (7.1) allows the entries of the following table to be simply determined.

c	$P(-c < Z < c)$	**Conclusion**
2.58	0.99	99% of a $N(0,1)$ distribution lies in the interval $(-2.58, 2.58)$
1.96	0.95	95% of a $N(0,1)$ distribution lies in the interval $(-1.96, 1.96)$
1.65	0.90	90% of a $N(0,1)$ distribution lies in the interval $(-1.65, 1.65)$
0.675	0.50	50% of a $N(0,1)$ distribution lies in the interval $(-0.675, 0.675)$

These tabled probability values indicate a great deal about the standard normal distribution: the first line of the table shows that the range of $Z \sim N(0,1)$ may be approximated by a slightly larger value than $2.58 - (-2.58) = 5.16$; in fact the range of Z is often approximated by 6. Similarly, the interquartile range is $z_{0.25} - z_{0.75} = 0.675 - (-0.675) = 1.35$. Interquartile ranges of other distributions may be compared to the standard normal distribution benchmark by dividing by 1.35. This creates a measure of spread, the **pseudo-standard deviation** which was defined in Chapter 2 as

$$p_s = \frac{q_s}{1.35},$$

where q_s is the interquartile range of the data under consideration.

This shows us how to use the tables to evaluate areas (as probabilities) under the standard normal curve. Some probability questions are asked in reverse: a probability is specified and the upper alpha point corresponding to that probability is sought. That is: given a specified probability p, determine $\Phi^{-1}(p)$. This, according to Definition 7.3 is z_{1-p}. For such problems, simply locate the probability p in the main body of the probability table and read across to the left-side of the table for the specification of the appropriate z value (namely z_{1-p}).

For example, directly from the tables,

$$\Phi^{-1}(0.025) = -1.96; \ \Phi^{-1}(0.5) = 0; \text{ and } \Phi^{-1}(0.9955) = 2.61.$$

Such results provide solution to questions of the form: 'For what value of x is $P(Z \leq x) = 0.025$?' Answer: $x = \Phi^{-1}(0.025) = -1.96$.

Why is there only one table of normal probabilities?

Earlier we examined the $N(0,1)$ distribution extensively. This considerable emphasis catalogues the standard normal. But the standard normal distribution is only one of uncountably many normal distributions comprising a vast family of distributions.

Definition 7.4 *When a random variable X has probability density function*

$$f_X(x) = \frac{1}{\sigma\sqrt{2\pi}} e^{-\frac{1}{2}(\frac{x-\mu}{\sigma})^2}, \quad -\infty < x < \infty,$$

for $-\infty < \mu < \infty$ and $\sigma > 0$, then we say that X has a normal distribution and we write $X \sim N(\mu, \sigma^2)$.

□

The members of this family are generated in the following theorem. Each new value of the parameters μ and σ creates a new normal distribution by a linear transformation of the standard normal.

Theorem 7.2 *If $Z \sim N(0,1)$ and $X = \mu + \sigma Z$, then $X \sim N(\mu, \sigma^2)$.*

Proof: Since

$$\Phi(z) = \int_{-\infty}^{z} \frac{1}{\sqrt{2\pi}} e^{-\frac{1}{2}t^2}\, dt,$$

it follows, by differentiating under the integral sign, that

$$\Phi'(z) = \frac{1}{\sqrt{2\pi}} e^{-\frac{1}{2}z^2}.$$

Now applying the cumulative distribution function technique to finding the probability density function of X, write:

$$F_X(x) = P(X \leq x) = P(\mu + \sigma Z \leq x) = P(Z \leq \frac{x-\mu}{\sigma}) = \Phi(\frac{x-\mu}{\sigma}). \quad (7.2)$$

Differentiating $F_X(x)$ with respect to x gives

$$f_X(x) = F'_X(x) = \Phi'(\frac{x-\mu}{\sigma})(\frac{1}{\sigma}) = \frac{1}{\sqrt{2\pi}}e^{-\frac{1}{2}(\frac{x-\mu}{\sigma})^2}(\frac{1}{\sigma}).$$

■

The results concerning the mean and variance of a linear transformation (Theorem 4.9) establish that:

$$E(X) = \mu + \sigma E(Z);$$
$$Var(X) = \sigma^2 Var(Z).$$

But since Z has a standard normal distribution, $E(Z) = 0$ and $Var(Z) = 1$. Therefore:

> If $X \sim N(\mu, \sigma^2)$, then $E(X) = \mu$ and $Var(X) = \sigma^2$.

The linear transformation approach to the normal distribution family avoids integrations for determining the higher moments of $X \sim N(\mu, \sigma^2)$. Again by Theorem 4.9, $X \sim N(\mu, \sigma^2)$ and $Z \sim N(0, 1)$ have the same coefficient of skewness $\zeta_3 = 0$ and the same coefficient of kurtosis $\zeta_4 = 3$.

The following diagram shows two members of the family of normal pdf's corresponding to random variables with different means and different variances.

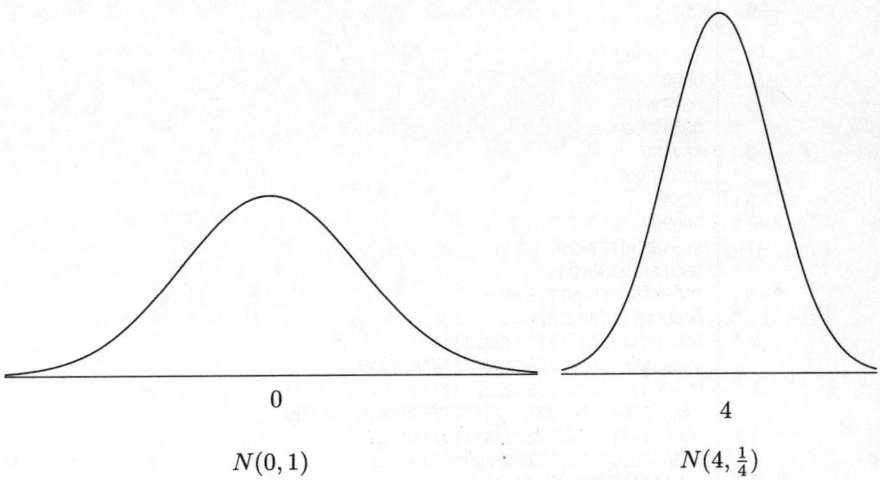

Clearly, if $X \sim N(\mu, \sigma^2)$, increasing the mean μ of the distribution increases the centre of location of the pdf on the horizontal measurement axis; increasing the variance σ^2 of the distribution makes the pdf appear more spread out with less central height.

Example 7.1

Looking for normally distributed data

When certain data sets are plotted in a S/L plot the result is a symmetric bell shape. Such symmetric data often arises from the measurement of certain types of errors. The S/L plot below displays an 'error' data set, where 400 observations are recorded on the discrepancy $X - 25$, where X is the observed weight and 25 kg is the 'label weight' of containers of PVC powder.

The so-called 'bell' shape is distorted somewhat by the group of negative data values (at the top of the S/L plot). These values were observed towards the end of the recorded production run and indicate that the filling apparatus possibly needed adjusting.

Weights 25 kg PVC powder corrected for label weight

Scale: stems—10^{-1} kg; leaves—10^{-2} kg

2.|5 denotes 250 g

```
−10 * | 3
 −9 . |
    * |
 −8 . |
    * | 0
 −7 . |
    * | 00
 −6 . |
    * | 3
 −5 . | 56
    * | 0000
 −4 . | 55779
    * | 000123334
 −3 . | 557889
    * | 02233334
 −2 . | 58889
    * | 000234
 −1 . | 55555556777899
    * | 000111123333344
 −0 . | 555555555566677788899
    * | 0000001111222333344
  0 * | 000000111111111122222233444
    . | 555555555666666667777777788888999
  1 * | 000000000000111122222222223333333344444444
    . | 5555555555566667777777777777888888888999999
  2 * | 0000111111112222222233334444444
    . | 5555556667777777788889999999
  3 * | 0001111122222222444
    . | 5566677888889999
  4 * | 000001111122233344
    . | 5566666777888889
  5 * | 0000123
    . | 55
  6 * | 112233
    . | 69
  7 * | 1
```

The following theorem shows how a given member, X, of the family of normal random variables may be re-scaled under linear translation $Z = \frac{X-\mu}{\sigma}$ into a standard normal random variable. From this it follows that only tables of the standard normal distribution are required for all normal probability calculations.

Theorem 7.3 *Suppose that* $X \sim N(\mu, \sigma^2)$. *Then if*

$$Z = \frac{X - \mu}{\sigma},$$

it follows that $Z \sim N(0, 1)$.

Proof: Again, using the cumulative distribution function technique,

$$F_Z(z) = P(Z \leq z) = P(\frac{X - \mu}{\sigma} \leq z) = P(X \leq \mu + \sigma z) = \Phi(z)$$

where the last equality is obtained by placing $x = \mu + \sigma z$ in (7.2). Therefore Z has the cumulative distribution function of a standard normal random variable; that is, $Z \sim N(0, 1)$.

■

If $X \sim N(\mu, \sigma^2)$, then the random variable $Z = \frac{X-\mu}{\sigma}$ is called a **z–score**. Normal probabilities are evaluated by a process of constructing z-scores.

Example 7.2

Developing photographs in the darkroom

Suppose that the development time, X, for a particular weight of Kodak photographic printing paper after exposure to a 5 second light source in the darkroom is normally distributed with a mean of 25 seconds and a standard deviation of 1.3 seconds.

Worked example:

1. **What are the values of a and b so that $X \sim N(a, b)$?**
2. **Determine $P(X \geq 25)$.**
3. **Determine $P(X > 25)$.**
4. **Determine the probability that a particular print will take more than 26.5 seconds to develop.**
5. **What is the probability that the development time is less than 23 seconds?**
6. **Compute the probability that the development time differs from the expected time by more than 1.5 seconds.**

1. $X \sim N(25, 1.3^2)$, so that $a = 25$ and $b = 1.69$.
2. $P(X \geq 25) = P(\frac{X-25}{1.3} \geq 0) = P(Z \geq 0) = \frac{1}{2}$; here $Z \sim N(0, 1)$.

3. Since X is a continuous random variable, $P(X = 25) = 0$. Hence
$$P(X \geq 25) = P(X > 25) = \tfrac{1}{2}.$$
4. The required probability is symbolically expressed as $P(X > 26.5)$. Then,
$$\begin{aligned} P(X > 26.5) &= P\left(\frac{X-25}{1.3} > \frac{26.5-25}{1.3}\right) = P(Z > 1.15) \\ &= 1 - P(Z < 1.15) = 0.1251. \end{aligned}$$
5. $P(X < 23) = P(\frac{X-25}{1.3} < \frac{23-25}{1.3}) = P(Z < -1.54) = 0.0618.$
6. The expected development time is $E(X) = 25$. The question requires us to find
$$\begin{aligned} P(|X-25| > 1.5) &= P\left(\left|\frac{X-25}{1.3}\right| > \frac{1.5}{1.3}\right) = P(|Z| > 1.15) \\ &= 2P(Z > 1.15) = 0.2502. \end{aligned}$$

Example 7.3
Guillotining bolts to length

A guillotine device that cuts long lengths of steel rods into bolts is adjusted so that 1% of manufactured bolts have lengths outside the specifications 2.560 ± 0.025 cm and such bolts are just as likely to be too long as too short. Experience in the industry has shown that X, the 'length of bolt', is a normally distributed random variable. Determine the values of the mean and standard deviation of X.

The sketch shows the pdf of X, which, from the information given, sits symmetrically about the 'tolerance' interval with endpoints 2.560 ± 0.025 trapping probability 0.005 in each tail of the distribution. In this diagram, x_L represents the lower endpoint $2.560 - 0.025 = 2.535$ and x_U represents the upper endpoint $2.560 + 0.025 = 2.585$.

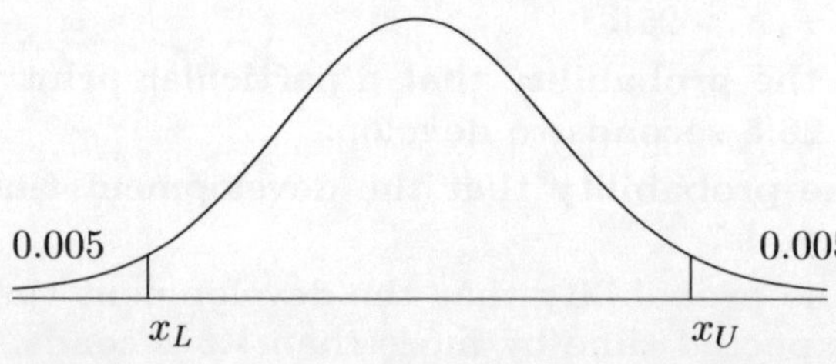

Notice that the symmetry of the problem automatically ensures that $\mu = 2.560$. Writing x_U as a z-score, an equation is formed in the standardising process and this may be solved for σ:

$$\frac{x_U - \mu}{\sigma} = z_{0.005}.$$

That is,

$$\frac{2.585 - 2.560}{\sigma} = 2.575,$$

giving $\sigma = 0.0097$.

The moment generating function applied to adding normal distributions

Theorem 7.4 *The normal random variable $X \sim N(\mu, \sigma^2)$ has moment generating function given by*

$$m_X(t) = e^{\mu t + \frac{1}{2}\sigma^2 t^2}.$$

Proof: Example 4.9 established that $Z \sim N(0,1)$ has moment generating function $m_Z(t) = e^{\frac{1}{2}t^2}$. Hence $X = \mu + \sigma Z$ has mgf

$$m_X(t) = E(e^{tX}) = E(e^{\mu t + \sigma t Z}) = e^{\mu t} E(e^{(\sigma t)Z}) = e^{\mu t} e^{\frac{1}{2}(\sigma t)^2} = e^{\mu t + \frac{1}{2}\sigma^2 t^2}.$$

■

Normally distributed random variables are characterised through having this moment generating function. In fact, from the moment generating function, the mean and variance may be identified. For example, the mgf e^{2t+8t^2} is of the form $e^{\mu t + \frac{1}{2}\sigma^2 t^2}$ (with $\mu = 2$ and $\sigma = 4$) and therefore is the mgf of a $N(2, 4^2)$ random variable. This characterisation idea is used in the proof that the sum of independent normally distributed random variables is again normally distributed.

Theorem 7.5 *If $X_1, X_2, X_3, \ldots, X_n$ are independent normally distributed random variables with $X_i \sim N(\mu_i, \sigma_i^2)$, for each $i = 1, 2, 3, \ldots, n$, then*

$$X_1 + X_2 + X_3 + \ldots + X_n \sim N(\mu_1 + \mu_2 + \ldots + \mu_n, \sigma_1^2 + \sigma_2^2 + \ldots + \sigma_n^2).$$

Proof: The properties of moment generating functions for independent random variables ensure that

$$\begin{aligned} m_{X_1+X_2+\ldots+X_n}(t) &= m_{X_1}(t) m_{X_2}(t) \ldots m_{X_n}(t) \\ &= e^{\mu_1 t + \frac{1}{2}\sigma_1^2 t^2} e^{\mu_2 t + \frac{1}{2}\sigma_2^2 t^2} \ldots e^{\mu_n t + \frac{1}{2}\sigma_n^2 t^2} \\ &= e^{(\mu_1 + \mu_2 + \ldots + \mu_n)t + \frac{1}{2}(\sigma_1^2 + \sigma_2^2 + \ldots \sigma_n^2)t^2}. \end{aligned}$$

This is of the form $e^{\mu t+\frac{1}{2}\sigma^2 t^2}$ for $\mu = \mu_1 + \mu_2 + \ldots + \mu_n$ and $\sigma^2 = \sigma_1^2 + \sigma_2^2 + \ldots + \sigma_n^2$.

■

PROPERTIES OF THE NORMAL DISTRIBUTION $X \sim N(\mu, \sigma^2)$

(7.3)

Probability density function $f_X(x) = \frac{1}{\sigma\sqrt{2\pi}} e^{-\frac{1}{2}(\frac{x-\mu}{\sigma})^2}$
$-\infty < x < \infty$;
$-\infty < \mu < \infty,\ 0 < \sigma < \infty$.

Mean	$E(X) = \mu$
Variance	$Var(X) = \sigma^2$
Standard deviation	$SD(X) = \sigma$
Coefficient of skewness	$\zeta_3 = 0$
Coefficient of kurtosis	$\zeta_4 = 3$
Moment generating function	$m_X(t) = e^{\mu t+\frac{1}{2}\sigma^2 t^2}$
Cumulant generating function	$\kappa_X(t) = \mu t + \frac{1}{2}\sigma^2 t^2$

Example 7.4

How long does it take to develop five photographs in the darkroom?

Here we continue Example 7.2 on the development time of photographs. Suppose that we have five photographs to develop independently; that is, X_1, X_2, X_3, X_4 and X_5 are *i.i.d.* $N(25, 1.3^2)$ with X_i representing the development time (seconds) of photograph i. ($i = 1, 2, 3, 4, 5$.) Let us assume that between each photograph we need 4 minutes observation time and washing time. This allows us to check under a safe light whether a print is sufficiently 'burned in'. This 4 minutes we take as a constant requirement for each photograph.

What is the distribution of time that we must remain in the darkroom?

The total darkroom time, T, in seconds, for the printing of the five photographs is given by

$$T = X_1 + X_2 + X_3 + X_4 + X_5 + 1200.$$

The additive constant of 1200 simply shifts the location on the time axis of the pdf for the sum of the 5 development times. (This may be checked formally by the moment generating function technique.) Thus, using the result of Theorem 7.5 with $n = 5$ independent random variables each with the same mean and variance, we find that

$$T \sim N(5(25) + 1200, 5(1.3)^2) \quad \text{or} \quad T \sim N(1325, 8.45).$$

EXERCISES

7.1 If Z is a standard normal random variable, determine:

(a) $P(Z < 1)$;
(b) $P(Z \leq 1)$;
(c) $P(Z = 1)$.

7.2 If Z is a standard normal random variable, determine:

(a) $P(Z \geq -0.5)$;
(b) $P(-0.5 < Z \leq 0.5)$;
(c) $P(|Z - 1| \geq 0.5)$;
(d) $P(|Z - 1| \leq 0.5)$.

7.3 X is normally distributed with mean 2 and variance 4. Find:

(a) $P(X \geq 3)$;
(b) $P(-1 < X < 1)$;
(c) $P(-2 < X)$;
(d) $P(X > 1 \text{ and } X > 2)$;
(e) $P(X > 1 \text{ or } X > 2)$.

7.4 On separate sketches of a $N(2, 4)$ probability density function, shade regions representing the probabilities in Exercise 7.3.

7.5 If $Z \sim N(0, 1)$ and has cdf Φ, determine from normal probability tables the value of the constant ω which makes each statement correct:

(a) $\Phi(\omega) = 0.2296$;
(b) $P(0 \leq Z \leq \omega) = 0.1700$;
(c) $P(\omega \leq Z) = 0.0132$;
(d) $P(-\omega \leq Z < \omega) = 0.762$.

7.6 Determine the value of z_α for

$$\alpha = 0.01,\ 0.10,\ 0.80,\ 0.90,\ 0.99.$$

7.7 Solve the equation

$$P(|Z| < z_\alpha) = 0.1$$

for α if Z represents a standard normal random variable.

7.8 Draw a diagram that explains why

$$z_{1-\alpha} = -z_\alpha.$$

7.9 Determine the 60th percentile of a standard normal distribution.

7.10 Suppose that $X \sim N(0, \frac{1}{9})$ and $Y \sim N(\frac{3}{2}, \frac{1}{4})$.

(a) On the same diagram sketch the probability density functions for each of X and Y.

(b) Explain how to use the diagram to determine whether or not

$$P(Y > 0) > P(X > 0).$$

7.11 The diagram shows the pdf for a normally distributed random variable with mean μ and standard deviation σ. Determine the magnitude of the shaded area.

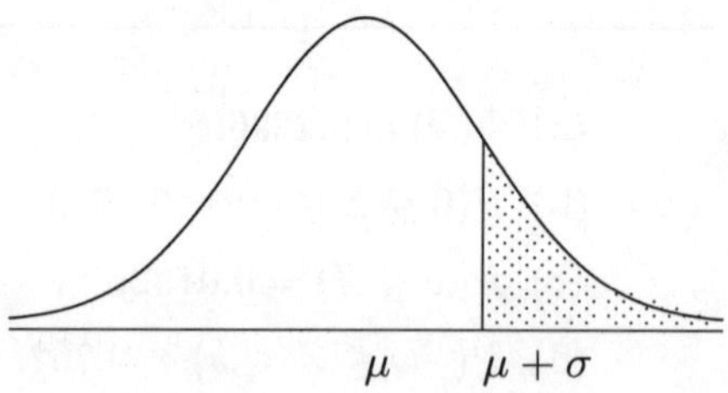

7.12 Bolt diameters, X, are normally distributed with a mean of 1 cm and a standard deviation 0.01 cm.

(a) Above what diameter do the thickest 10% of bolts lie?

(b) Give the values of the constants a and b such that

$$Z = a + bX \sim N(0, 1).$$

7.13 When certain building materials containing fibreglass are disturbed during the process of building renovation or restoration, particles are released into the air. If the renovator wears a breathing filter, the number of particles inhaled diminishes.

Suppose that the particle diameters are approximately normally distributed about a mean of 1.2 micrometres (μm) with a standard deviation of 0.6 μm.

Determine the percentage of particles that would pass through a filter having a porosity of 0.4 μm which allows particles smaller than 0.4 μm to pass through.

7.14 Police records show that the average speed of cars on the Kwinana Freeway is $\mu = 98$ km hr^{-1} with a standard deviation of 10 km hr^{-1}. Suppose that the current speed limit is 100 km hr^{-1}.

A government report proposes that the speed limit be increased to a value L km hr^{-1}, where $L > 100$.

(a) Determine the value of L so that at most 1% of drivers may be caught speeding.

(b) The opening of a new lane on the freeway has no effect on the average speed, but results in a 50% reduction in the variance in speed on the freeway.
Determine the value of L in this case, so that at most 1% of drivers may be caught speeding.

7.15 An automatic coffee-vending machine is set for dispensing hot coffee into standard 250 mL cups. The actual volume (mL) dispensed by the machine is a random variable which is normally distributed with mean 200 mL and standard deviation 20 mL. This means that sometimes the machine will dispense more coffee than the standard cup will hold, causing the cup to overflow. The actual volume dispensed is independent from one cup to another.

For a period in which 1000 cups of coffee are dispensed by the machine, let X denote the number of cups that overflow.

(b) Explain briefly why the binomial model is suitable for the random variable X. Using this model, show that to one decimal place,

$$E(X) = Var(X) = 6.2.$$

State, with reasons, another probability model for X.

(c) The company owning the machine makes a loss of 15 cents on a dispensed cup of coffee that overflows and a profit of 25 cents on a dispensed cup of coffee that does not overflow. Show that Y, the total profit in dollars for the period, is related to X by

$$Y = -0.40X + 250.$$

Hence determine $E(Y)$ and $Var(Y)$.

7.16 A company sells naturally sprouted bean shoots to a fruit and vegetable wholesaler in packages labelled 100 g each. The price is 67 cents per package. Data from an automatic check weigher show that the package weights are normally distributed with a standard deviation of 5 g.

In the weighing process, packages weighing less than 100 g are separated out and sold at a discount of 63 cents each. The production cost is 25 cents per 100 g of product.

(a) If the target mean is set at 106 g (that is, at 6% overfill) determine:

i. the percentage of packets sold at discount;

ii. the expected profit for a batch of 100 randomly chosen packets.

(b) The company accountant claims that the expected profit can be increased by reducing the overfill from 6% to 3%. Repeat the calculations in (a) for this proposed target mean. Hence confirm or refute the accountant's claim.

7.17 Suppose that bolts are produced with diameter B, such that

$$B \sim N(1, 0.0001).$$

Nuts to fit on these bolts are produced to an internal diameter A, where

$$A \sim N(1.06, 0.0003).$$

(a) Write down the probability distribution for the discrepancy $A - B$.

(b) A randomly selected nut will fit a randomly selected bolt if $A - B > 0$. Determine the probability of such a fit.

7.18 On a production line, plastic components used in the assembly of PC's are guillotined to a desired length of 40 mm. The production line is constantly monitored to ensure that a quality product is manufactured to a specified standard.

At regular intervals, components are sampled from a hopper that collects the mixed output from two different guillotines and the length of each of the components is measured.

The results are as follows:

Length (mm)					
34.1	41.7	44.9	39.0	34.5	34.6
44.7	43.8	40.2	42.2	38.5	43.0
39.8	36.1	42.7	44.4	35.0	40.0
33.2	36.2	39.7	35.2	37.7	41.9
35.1	42.3	40.1	36.9	82.5	67.0
40.1	36.0	40.3	39.2	38.5	35.6
40.6	35.7	40.0	35.8	40.8	34.2
36.3	41.5	44.9	35.5	38.5	37.6
41.5	41.6	36.7	43.6	41.1	37.6
34.6	39.2	38.5	42.4	40.0	41.6
35.5	40.0	38.9	36.0	40.5	35.7
39.0	43.7	33.0	40.0	39.0	42.7
41.6	35.2	39.4	36.1	41.3	40.0
36.5	40.2	37.8	42.9	41.4	41.8
33.8	37.8	35.0	34.0	38.5	45.7
43.2	43.0	40.6	43.9	39.0	39.8
42.1	45.1	40.7	39.8	41.9	41.2
41.5	42.2	34.8	42.8	42.3	39.5
43.4	40.4				

(a) Construct a stem-and-leaf plot for these data. [Let the leaf digits be 10^{-1} mm.]

(b) Construct a box-and-whisker plot for these data. Include an outlier analysis.

(c) What information do the plots give about this particular production process?

(d) Suppose that $X_1 \sim N(35, 2^2)$ and $X_2 \sim N(40, 2^2)$ have respective probability density functions $f_1(x)$ and $f_2(x)$.

Show that if α is a constant with

$$0 < \alpha < 1,$$

then the function defined by

$$f(x) = \alpha f_1(x) + (1 - \alpha) f_2(x)$$

is a probability density function.

(e) Suppose that $\alpha = \frac{1}{3}$ and that X is the random variable with probability density function f as in (d). Determine $P(X > 40)$.

(f) Is f a useful probability density function to model the length of components in the hopper on this production line? Explain.

7.19 The sector flight time between Melbourne and Auckland is normally distributed with a mean of 180 minutes and a standard deviation of 10 minutes. The sector flight time between Auckland and Wellington is normally distributed with a mean of 50 minutes and a standard deviation of 5 minutes.

Suppose that I leave Melbourne on a flight bound for Auckland at 12 noon (New Zealand time). I intend to have a brief stopover in Auckland before catching a connecting same-day flight to Wellington at 4 p.m. (New Zealand time).

Assume that all flights leave on time.

(a) Determine the expected length of the Auckland stopover.

(b) What time should I tell my friends in Wellington to meet my flight so that there is only a 5% chance that they are kept waiting for me to arrive?

(c) If the Wellington flight boards and is closed 10 minutes prior to departure at 4 p.m., is there any danger that I may miss this connecting flight?

(d) Give the mean, variance and distribution of the length (total time) of my journey.

7.20 Let X_1, X_2 be two independent normally distributed random variables such that

$$X_i \sim N(\mu_i, \sigma_i^2) \quad \text{for} \quad i = 1, 2.$$

State the mean, variance and distribution of the random variable $X_1 - X_2$.

7.21 (a) Apply the result of Exercise 7.20 to the following. When an Australian citizen applies for a passport, certain distinguishing characteristics ('age', 'height', 'eye colour' ...) are recorded. Files on the 'height' variable show distributions that are gender based. Suppose that the heights (cm) of female passport holders are normally distributed with mean 165 and variance 41, while those of male passport holders are normally distributed with mean 178 and variance 59. One male and one female are selected at random from the files. Determine the probability that the male is shorter than the female.

(b) A flight from London arriving at Kingsford Smith Airport contains a large number of Australian passport holders. Among those travelling on Australian passports on this particular flight, female passengers outnumber male passengers in the ratio 2:1.

From those travelling on Australian passports, one male and one female are selected for a tourist promotion venture. Use the result of (a) to determine the probability that the male is shorter than the female. What assumptions have you made?

7.22 In the assembly of knitting machines, four fibreglass sections are fitted side by side into a gap as shown below. For *satisfactory fit*, the sections must fit into the gap and the clearance, W, must not exceed 0.19 cm (where overlap is taken as negative clearance). Any machine is termed *defective* if the fit is not satisfactory.

Component assembly

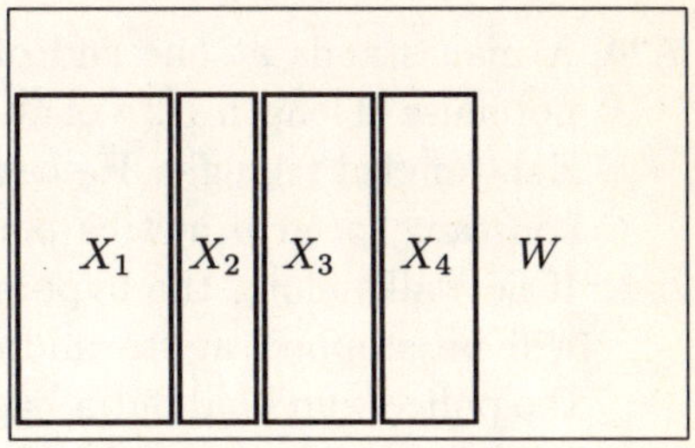

Y (width of gap)

The widths (in cm) of the sections are independently normally distributed random variables X_i where, for each $i = 1, 2, 3, 4$,

$$X_i \sim N(4.000, 0.000225).$$

Y, the width (in cm) of the gap, is normally distributed independently of the section widths, where $Y \sim N(16.066, 0.0007)$.

(a) Express W in terms of Y, X_1, X_2, X_3 and X_4.

(b) Find, on first assembly, the distribution of the clearance W.

(c) Find the proportion of first assemblies that are defective. [*Hint*: What are the limitations on W?]

7.23 After first assembly in Exercise 7.22, any machines with negative clearance are automatically discarded. The remaining machines are gathered in lots of 1000 ready for wholesale distribution.

Let N denote the number of defectives in a given lot. Give the appropriate binomial distribution for N and show that

$$P(N < 2) \approx \frac{2}{e}.$$

7.24 A man stands at one end of the hypotenuse of length $\sqrt{2}\,a$ of an isosceles right-angled triangle. He can walk at constant speed p metres per second. If he walks along the hypotenuse, he will be stopped at its midpoint (by the police who hold out a bag for him to blow into) for a variable time T seconds, before proceeding to the other end of the hypotenuse.

On average, the total time T_2 that he takes to walk the hypotenuse is the same as the time t_1 that he would take to walk the other two sides. Further, the probability that T_2 is within

$$\frac{0.12a}{p}$$

of t_1 is 0.80.

Assuming that T is normally distributed, determine its mean and variance in terms of a and p.

7.25 If $X \sim N(-1, 8)$, determine the mean, median, variance, standard deviation, upper quartile and probability density function of $W = \frac{1-X}{2}$.

7.26 (a) Draw a freehand sketch of the function $y = \Phi(x)$, where Φ is the standard normal cdf.

[*Hint*: There is no need to use graph paper, but use normal probability tables to plot some points so that the *shape* of your curve is reasonably accurate.]

(b) On your sketch, mark the 95th percentile of the distribution as a z_α value.

(c) Give the maximum value of the derivative of $\Phi(x)$, and the value of x for which this maximum occurs.

(d) Sketch the function

$$y = \Phi(\frac{x-1}{2})$$

on the same set of axes as in (a).

7.27 Give the mean and variance of the random variable which has cdf

$$\Phi(\frac{x-1}{2}).$$

7.28 **Unit area under a standard normal curve.** Let I denote the area under the standard normal probability density function

$$fz(z) = \frac{1}{\sqrt{2\pi}}e^{-z^2},$$

for $Z \sim N(0, 1)$. In this problem we show that $I = 1$.

(a) Show that

$$I^2 = \frac{1}{2\pi}\int_{-\infty}^{\infty}\int_{-\infty}^{\infty} e^{-\frac{1}{2}(x^2+y^2)}dydx.$$

(b) In (a), show that the substitution

$$x = r\cos\theta, \ y = r\sin\theta$$

gives

$$I^2 = \frac{1}{2\pi}\int_0^{\infty}\int_0^{2\pi} e^{-\frac{1}{2}r^2} rd\theta dr.$$

7.29 Use the fact that the area under a certain pdf is 1 to evaluate the integral

$$\int_0^\infty e^{-x^2}\,dx.$$

7.30 Draw a freehand sketch of the function

$$y = \Phi^{-1}(x).$$

[*Hint*: There is no need to use graph paper, but use normal probability tables to plot some points so that the *shape* of your curve is reasonably accurate.]

7.2 LIFETIME DISTRIBUTIONS: EXPONENTIAL

Exponential waiting times

The Poisson distribution has been used to model the number of events, E, occurring in an interval of a continuum, when E occurs at a certain average rate.

The question now is: If observation commences at a specified point, how much of the continuum do we have to observe before E occurs? If the continuum is 'time' (as is often the case), then our question reduces to: How long do we have to wait after $t = 0$ before E occurs? This 'waiting time' may be found by a simple argument based on the equality of two events.

If radioactive emissions are detected by a Geiger counter according to a Poisson process with rate ω, then X_t, the number of emissions in t minutes, satisfies $X_t \sim Poi(\omega t)$. Let T denote the elapsed time, after counting starts ($t = 0$), until the first emission is detected. Then, if it takes *longer* than t seconds for the first particle to be detected, the number of particles occurring in the interval $[0, t]$ is clearly 0! That is, the events

$$\{T > t\} \quad \text{and} \quad \{X_t = 0\}$$

are equivalent. Therefore,

$$P(T > t) = P(X_t = 0) = \frac{(\omega t)^0 e^{-\omega t}}{0!} = e^{-\omega t}.$$

This shows that the cumulative distribution function for T is

$$F_T(t) = P(T \leq t) = 1 - P(T > t) = 1 - e^{-\omega t}.$$

Differentiating $F_T(t)$ with respect to t, the probability density function $f_T(t)$ is given by

$$f_T(t) = \frac{d}{dt}F_T(t) = \frac{d}{dt}[1 - e^{-\omega t}] = \omega e^{-\omega t}.$$

Differentiating $F_T(t)$ with respect to t, the probability density function $f_T(t)$ is given by

$$f_T(t) = \frac{d}{dt}F_T(t) = \frac{d}{dt}[1 - e^{-\omega t}] = \omega e^{-\omega t}.$$

We recognise this (Chapter 4) as an exponential distribution with mean $\beta = \frac{1}{\omega}$. This is how long we *expect* to have to wait until the first particle is detected.

Exponential lifetimes

We have seen exponential distributions used to describe the lifetimes of components in Section 5.2. Examples in Section 5.2 then discussed the lifetime (reliability) of components in both 'series' and 'parallel'.

In this section, however, we are discussing the *modelling* of such probability density functions as the exponential. Therefore we ask: Is there intuitive justification for the exponential model to describe the distribution of lifetimes of electronic components in a system?

Essentially, yes. The environment that a component operates in subjects the component to stress—on rare occasions the event $A = \{$ component experiences peak stress $\}$ will occur. Rare events occurring at an average rate over time are likely to follow a Poisson process with rate ω for which the number of occurrences, X_t, of A in the interval $[0, t)$ have a Poisson distribution with mean ωt. If the first occurrence of A is fatal, then according to the previous section, the time T to failure is a waiting time with exponential probability density function

$$f_T(t) = \frac{1}{\beta}e^{-\frac{t}{\beta}},$$

where the parameter β is defined by $\beta = \frac{1}{\omega}$. Here a **re-parametrisation** from ω to $\beta = \frac{1}{\omega}$ is intuitively appropriate: if incoming phone calls arrive at a switchboard at an average rate of 10 per hour ($\omega = 10$), then the expected time to the first call is $\frac{1}{10}$ hour (or 6 minutes). Thus there seems to be a natural inverse relationship between the Poisson and the exponential means.

Useful properties of the exponential distribution

The exponential distribution has been used extensively in previous chapters; we summarise here the shape characteristics of this distribution—that is, the mean, variance, skewness, ...

PROPERTIES OF THE EXPONENTIAL DISTRIBUTION $X \sim exp(\beta)$

(7.4)

Probability density function $f_X(x) = \frac{1}{\beta}e^{-\frac{x}{\beta}}$
$0 < x < \infty; \quad 0 < \beta < \infty.$

Mean	$E(X) = \beta$
Variance	$Var(X) = \beta^2$
Standard deviation	$SD(X) = \beta$
Coefficient of skewness	$\zeta_3 = 2$
Coefficient of kurtosis	$\zeta_4 = 9$
Moment generating function	$m_X(t) = \frac{1}{1-\beta t} \quad t < \frac{1}{\beta}$
Cumulant generating function	$\kappa_X(t) = -\log_e(1 - \beta t)$

In this list of properties, notice in particular that the scale parameter β does not appear in the expressions for the coefficients of skewness and kurtosis: these measures of shape are scale invariant. Recall from Section 4.5 that if

$$X \sim exp(1), \quad \text{then} \quad Y = \beta X \sim exp(\beta).$$

Example 7.5

Replacing starter motors

The time, T, between replacements of the starter motor of a car is exponentially distributed with a mean of 4 years.

Worked example:

1. **Write down the probability density function for T.**
2. **Determine $E(T^2)$.**
3. **Determine the probability that having just replaced the starter motor in my car it will be between 1 and 2 years before I must replace it again.**

1. By the summary at (7.4), we write the exponential density with mean 4 as

$$f_T(t) = \frac{1}{4}e^{-\frac{t}{4}}, \quad t > 0.$$

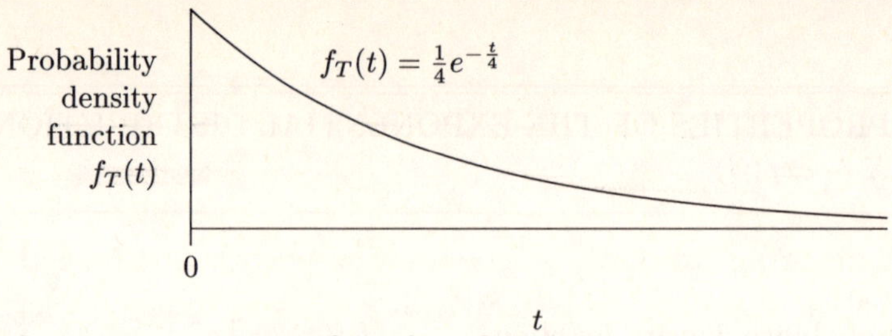

2. $E(T^2) = var(T) + [E(T)]^2 = 4^2 + 4^2 = 32.$
3. $P(1 < T < 2) = \int_1^2 f_T(t)dt = \int_1^2 \frac{1}{4}e^{-\frac{t}{4}}dt = -e^{-\frac{t}{4}}\Big|_1^2 = e^{-\frac{1}{4}} - e^{-\frac{1}{2}} =$ 0.172 (3 dp).

Exponential distributions are used extensively in reliability studies to model lifetimes. An important feature of such a model is what we term the 'memory-less property'. That is, the future distribution of the lifetime of a component depends on the past only through the present.

Theorem 7.6 *If $X \sim exp(\beta)$, then for $s > 0$ and $t > 0$,*

$$P(X > s + t|X > s) = P(X > t).$$

Proof: The probability $P(X > s + t|X > s)$ may be interpreted as a conditional probability $P(A|B)$ as discussed in Chapter 3, with the identification of the events A and B as $A = \{X > s + t\}$ and $B = \{X > s\}$.
Then, since $t > 0$, in order that a lifetime be simultaneously longer than both s and $s + t$, it must exceed $s + t$. Therefore, the event $A \cap B$ may be written as $\{X > s + t\}$.
Finally, since $P(X > r) = e^{-\frac{r}{\beta}}$ for any positive value of r,

$$\begin{aligned} P(X > s + t|X > s) &= \frac{P(\{X > s + t\} \cap \{X > s\})}{P(\{X > s\})} = \frac{P(\{X > s + t\})}{P(\{X > s\})} \\ &= \frac{e^{-\frac{s+t}{\beta}}}{e^{-\frac{s}{\beta}}} = e^{-\frac{t}{\beta}} = P(X > t). \end{aligned}$$

■

Theorem 7.6 states that for a component with an exponentially distributed lifetime, the probability that a 3-month old a component lasts 3 more weeks in operation is the same as the probability that a 6-month old component lasts 3 more weeks in operation. This indicates that the component's lifetime does not pass through a period of 'old age', where there is an increased risk of mortality. In this sense, an exponential lifetime is different from a human lifetime, where the survival distribution shrinks with age.

Example 7.6

Comparing human life expectancy with that of a lightbulb

The left-side data table showing the life expectancy at age t for men and women and various values of t was reported in *The Age*, Melbourne, 4 March 1991, from information supplied by the Australian Bureau of Statistics.

Life expectancy			**Life expectancy**	
Age now t	**Male (years)**	**Female (years)**	**Age now** t	**Lightbulb**
0	73.3	79.5	0	73.3
10	74.2	80.3	10	83.3
20	74.7	80.5	20	93.3
30	75.2	80.7	30	103.3
40	75.9	81.0	40	113.3
50	76.7	81.6	50	123.3
60	78.3	82.8	60	133.3
70	81.5	84.8	70	143.3
80	86.5	88.4	80	153.3
90	93.5	94.1	90	163.3

These data show that the survival distributions for men and women are slightly different. When we are born, the expected value of the length of life variable, X, is 73.3 years for men and 79.5 years for women. Given that we live to age t (the left-side column of the table), the future life expectancy beyond t reduces. That is, a man's life expectancy at age 10 is only $74.2 - 10 = 64.2$ more years. This is less than the average lifetime 73.3 years from age 0. It keeps declining: the life expectancy at age 20 is $74.7 - 20 = 54.7$ years.

On the other hand the life expectancies in the right-side table are for a theoretical lightbulb with survival time random variable $X \sim exp(73.3)$. When this lightbulb is turned on, it has an expected lifetime of 73.3 units of time (and such units will obviously not be years). However, since the lifetime distribution is exponential, it is, in particular, memoryless. That is, the life expectancy of the lightbulb at age 10 units is $83.3 - 10 = 73.3$ units more—still! It does not wear out (unlike human beings, who do). The survival distribution remains the same at any time period.

The future life expectancy at age t is given in the table below for men and lightbulbs. It is obtained by subtracting t from the life expectancy at age t.

Future life expectancy at age t		
Age now t	Male (years)	Lightbulb (time units)
0	73.3	73.3
10	64.2	73.3
20	54.7	73.3
30	45.2	73.3
40	35.9	73.3
50	26.7	73.3
60	18.3	73.3
70	11.5	73.3
80	6.5	73.3
90	3.5	73.3

These results demonstrate that for humans the expected future survival from age t reduces as t increases, but typically for lightbulbs and certain types of electronic components this does not happen.

A mixture distribution involving the exponential

Often an electronic component is likely to fail when it is switched into operation. There is a small, but measurable, probability that a lightbulb will fail when it is switched on; a VCR manufacturer must guard against a recorder failing at its first use, when it is first plugged in. Such failures cost a manufacturer, since they occur under guaranteed replacement.

In each of these situations the lifetime described may indeed follow an exponential distribution, but after an initial probability of failure at the discrete time point $t = 0$. The lifetime random variable, X, is a **mixture distribution** having both discrete and continuous parts. If the probability of initial failure is p_0, and the component's lifetime beyond $t = 0$ is exponentially distributed with mean β, then the mixture probability density function for X is

$$f_X(x) = \begin{cases} p_0 & \text{if} \quad x = 0; \\ (1 - p_0)\frac{1}{\beta}e^{-\frac{x}{\beta}} & \text{if} \quad x > 0. \end{cases}$$

This function is sketched in the diagram below.

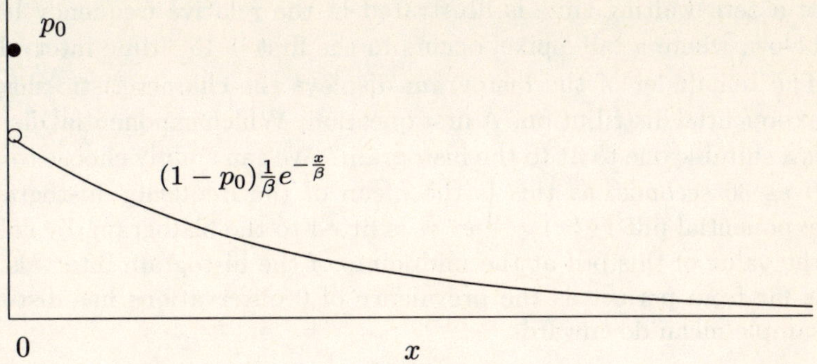

Notice that the area under the continuous portion of f_X is $1-p_0$, and this, added to the probability mass $P(X=0)=p_0$ situated at $t=0$, gives total probability 1. Combining this result with the obvious fact that $f_X(x) \geq 0$ for all x shows f_X to be a valid probability model for describing the component's lifetime. The graph is drawn for a large value of $p_0 > \frac{1}{1+\beta}$, for it is in this circumstance that the point mass (dot) lies above the height of the curve near $x=0$.

Example 7.7
Waiting for a lift

The following frequency distribution records 471 observations on the 'waiting time', X (in seconds), for a lift to arrive when called to the ground floor (street level) for the Royal Melbourne Institute of Technology's Science and Engineering building on Swanston Street. Four lifts are in simultaneous operation in a building with 13 floors.

Interval	Width	Frequency
0–15^-	15	206*
15–30^-	15	84
30–45^-	15	49
45–60^-	15	38
60–75^-	15	29
75–90^-	15	15
90–120^-	30	28
120–150^-	30	9
150–180^-	30	8
180–270^-	90	5
Total		471

The asterisk (*) denotes that this frequency includes 57 waiting times of 0;

that is, a lift had already arrived once counting began. This high chance of a zero waiting time is illustrated in the relative frequency histogram below, where a tall 'spike' occurs in the first $0\text{–}15^-$ time interval.

The remainder of the histogram displays the characteristic shape of an exponential distribution. A first question: Which exponential distribution is a suitable one to fit to the histogram? We can simply choose to estimate β as 36 seconds, as this is the mean of the frequency histogram. The exponential pdf $f_X(x) = \frac{1}{36}e^{-\frac{x}{36}}$ is fitted to the histogram, by calculating the value of this pdf at the midpoints of the histogram intervals. The fit is far from perfect as the prevalence of 0 observations has distorted the sample mean downwards.

(Relative) frequency histogram of the waiting time for a lift

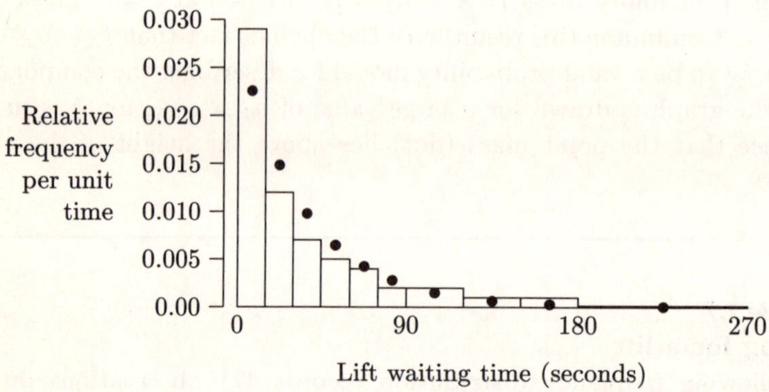

The fraction of the data set at 0 is $\frac{57}{471} = 0.121$. If these points are separated out and the mean recalculated, the new sample mean value determined from the frequency histogram increases to 40 seconds. An improved fit may be seen in the histogram below.

(Relative) frequency histogram of the waiting time for a lift

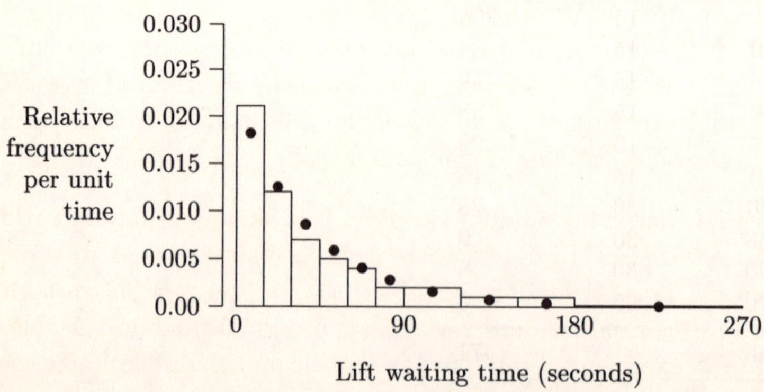

The estimate of 0 waiting time is taken as 0.121 in the mixture model

$$f_X(x) = \begin{cases} p_0 & \text{if} \quad x = 0; \\ (1-p_0)\frac{1}{\beta}e^{-\frac{x}{\beta}} & \text{if} \quad x > 0. \end{cases}$$

The fitted model is

$$\begin{aligned} f_X(x) &= \begin{cases} 0.121 & \text{if} \quad x = 0; \\ (1-0.121)\frac{1}{40}e^{-\frac{x}{40}} & \text{if} \quad x > 0. \end{cases} \\ &= \begin{cases} 0.121 & \text{if} \quad x = 0; \\ 0.022e^{-\frac{x}{40}} & \text{if} \quad x > 0. \end{cases} \end{aligned}$$

Using this probability model we find that

1. the probability that it takes less than 5 seconds for the lift to arrive when called is given by

$$\begin{aligned} P(X < 5) &= 0.121 + 0.022 \int_0^5 e^{-\frac{x}{40}}\,dx \\ &= 0.121 + (0.022)(40)[1 - e^{-\frac{5}{40}}] = 0.224 \ \text{(3 dp)}. \end{aligned}$$

2. The probability that the lift waiting time exceeds 10 seconds is given by

$$P(X > 10) = 0.022 \int_{10}^{\infty} e^{-\frac{x}{40}}\,dx = (0.022)(40)e^{-\frac{10}{40}} = 0.685 \quad \text{(3 dp)}.$$

The sum of i.i.d. exponentials is Erlang

In this section we examine the probability distribution for the random variable which arises when independent and identically distributed exponential random variables are added. The catalogue of sums of random variables (discussed in previous sections) reads as follows:

The sum of *i.i.d.* Bernoulli random variables is a binomial random variable.
The sum of *i.i.d.* binomial random variables is a binomial random variable.
The sum of *i.i.d.* Poisson random variables is a Poisson random variable.
The sum of *i.i.d.* normal random variables is a normal random variable.
The sum of *i.i.d.* exponential random variables is an Erlang random variable.

The new probability distribution here, in the last line, is the Erlang distribution.

Definition 7.5 *If a random variable X has probability density function*

$$f_X(x) = \frac{x^{n-1}e^{-\frac{x}{\beta}}}{\beta^n(n-1)!}, \quad x \geq 0,$$

then X has an **Erlang distribution** *with parameters n and β. In this case we write $X \sim Erlang(n, \beta)$.*

□

This generalisation of the exponential distribution is constructed in the following way.

When a rare event A occurs according to a Poisson process with rate ω, we have seen at the outset of this chapter that if the random variable X_1 represents the waiting time from the start of observation until A is first observed, then $X_1 \sim exp(\beta)$, where $\beta = \frac{1}{\omega}$. We have seen too in Theorem 7.6 the 'memoryless' nature of the exponential distribution. The life expectancy beyond age t is constant for all t. Therefore, having observed the first occurrence of A, all clocks are set back to 0 and we note that the continuous waiting time, X_2, until A re-occurs follows an $exp(\beta)$ distribution as before. This is summarised by $X_2 \sim exp(\beta)$. Through the 'memorylessness', this exponential distribution is independent of the first. As random variables, X_2 is independent of X_1. Again, by putting the clocks back to 0, we find that a new and independent exponential random variable describes the time between the 2nd and 3rd occurrence of A, 3rd and 4th occurrence of A, ..., $(n-1)$th and nth occurrence of A.

In this framework, X_i represents the time between the $(i-1)$th and ith occurrences of A and the random variables X_i are *i.i.d.* $exp(\beta)$.

Waiting time for the nth occurrence of A
is the sum of n exponential random variables

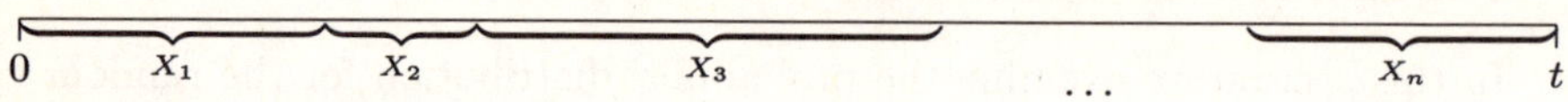

Now let us define a new random variable

$$X = X_1 + X_2 + X_3 + \ldots + X_n$$

which represents the time to the nth occurrence of A in the Poisson process. The Poisson probability distribution is the key to determining the probability distribution for X.

Let $x > 0$ be fixed on the time axis. Then the number of occurrences, Y, of A in the interval $[0, x)$ has a Poisson distribution with rate ωx. That is: $Y \sim Poi(\omega x)$. Notice that the random variables X and Y are inextricably related: if the nth occurrence of A occurs at a time prior to x, then the number of occurrences of A in the time interval $[0, x)$ is at least n. This means that the events

$$\{X \leq x\} \quad \text{and} \quad \{Y \geq n\}$$

are equivalent. Using the cumulative distribution function technique for generating the probability density function for X we obtain

$$\begin{aligned} F_X(x) &= P(X \leq x) \\ &= P(Y \geq n) \\ &= 1 - P(Y \leq n-1) \\ &= 1 - \sum_{k=0}^{n-1} \frac{(\omega x)^k e^{-\omega x}}{k!}. \end{aligned}$$

By differentiating each side of this expression with respect to x it follows that

$$f_X(x) = F_X'(x) = \frac{\omega^n x^{n-1} e^{-\omega x}}{(n-1)!}.$$

This is the Erlang pdf of Definition 7.5 with $\beta = \frac{1}{\omega}$. This establishes the following theorem:

Theorem 7.7 *If independent and identically distributed random variables $X_1, X_2, X_3, \ldots, X_n$ each have an exp(β) distribution, then*

$$X = X_1 + X_2 + X_3 + \ldots + X_n \sim \mathit{Erlang}(n, \beta).$$

and X has cdf

$$F_X(x) = 1 - \sum_{k=0}^{n-1} \frac{(\omega x)^k e^{-\omega x}}{k!}, \qquad \omega = \frac{1}{\beta}.$$

■

From the formula for $F_X(x)$ in this theorem, it follows that Erlang probabilities may be evaluated using tables of the Poisson distribution.

Example 7.8

Customer arrivals: are you being served?

The suit department of a large menswear store must answer customer inquiries at an average rate of 1 inquiry per 5 minutes during lunch hour. Determine the probability density function of the time, T, in minutes, until the 3rd inquiry.

If we assume that the lunchtime inquiries follow a Poisson process with rate 0.2 (inquiries per minute), then $T \sim \mathit{Erlang}(3, \frac{1}{0.2})$, which has pdf

$$f_T(t) = \frac{t^2 e^{-\frac{t}{5}}}{5^3 2!} = 0.004t^2 e^{-0.2t}.$$

Areas under the pdf may be evaluated using Poisson probability tables. For example, if $Y \sim Poi(3)$,

$$P(T < 15) = 1 - P(Y \leq 2) = 1 - 0.423 = 0.577.$$

PROPERTIES OF THE ERLANG DISTRIBUTION $X \sim Erlang(n, \beta)$

(7.5)

Probability density function $f_X(x) = \frac{x^{n-1}e^{-\frac{x}{\beta}}}{\beta^n(n-1)!}$
$n = 1, 2, 3, \ldots;$
$0 < x < \infty; \quad 0 < \beta < \infty.$

Mean	$E(X) = n\beta$
Variance	$Var(X) = n\beta^2$
Standard deviation	$SD(X) = \sqrt{n}\beta$
Coefficient of skewness	$\zeta_3 = \frac{2}{\sqrt{n}}$
Coefficient of kurtosis	$\zeta_4 = 3 + \frac{6}{n}$
Moment generating function	$m_X(t) = \left(\frac{1}{1-\beta t}\right)^n \quad t < \frac{1}{\beta}$
Cumulant generating function	$\kappa_X(t) = -n \log_e(1 - \beta t)$

EXERCISES

7.31 After the start of observation on a given summer evening, the time T in minutes until the first shooting star is observed follows an exponential distribution for which

$$P(T > t) = e^{-\frac{t}{10}},$$

where $t > 0$. Determine the probability that it takes between five and ten minutes for the first shooting star to be observed.

7.32 In Exercise 7.31, how long do we expect to have to wait after the start of observation for the first shooting star to be observed?

7.33 Early in the morning, calls arrive at a telephone switchboard at a rate of

10 per hour. If we measure time from the point when the switchboard first opens, determine:

(a) the probability distribution of the time to the first call;

(b) the probability that it takes longer than 20 minutes for the first call to arrive;

(c) the median time to the first call.

7.34 The lifetime X, in hours, of a cog used in the roller bearing of escalators has pdf

$$f_X(x) = \frac{1}{1000} e^{-\frac{x}{1000}}, \quad x > 0.$$

(a) Determine $\mu = E(X)$ and $\sigma^2 = Var(X)$.

(b) Determine tha probability that a new cog will need replacing before 1000 hours of operation.

(c) Find, in terms of α, the time beyond which $\alpha\%$ of cogs will fail.

(d) An escalator has 4 identical cogs; two in the roller at the base of the escalator and two in the bearing at the top of the escalator. The escalator breaks down precisely when either both 'top cogs' or both 'bottom cogs' fail. Assuming that the cogs operate independently, determine the probability that the escalator will break down within 1000 hours after all cogs have been replaced.

7.35 The RACV (Royal Automobile Club of Victoria) receives calls from motorists requiring, on average:

towing assistance—5 per hour
petrol—3 per hour
battery recharge—12 per hour,

for weekday periods after 7 a.m.

(a) Determine the probability distribution of the time (in minutes) to the first request for assistance with a battery recharge. Give the mean and variance of this distribution.

(b) Determine the probability distribution of the time (in minutes) to the third request for petrol. Give the mean and variance of this distribution.

(c) Determine the probability distribution of the total number of towing, petrol and battery requests in a given half hour period after 7 a.m.

State clearly the assumptions that you have used to answer these questions.

7.36 As a member of the RACV, I am charged a fee (over and above the basic membership fee) if I request their assistance 5 or more times in a calendar year.

Suppose that requests from individual members form a Poisson process with rate $\omega = 4$ requests per year for a given member.

Determine the probability distribution of the time T, measured from 1 January, until my 5th request for RACV assistance. Hence determine the probability that I am charged the extra fee.

7.37 Lightbulbs often fuse at the instant they are turned on. When a lightswitch is turned on, the light will go on with probability 0.99; thereafter it lasts T hours until failure, where

$$P(T > t) = 0.99e^{-0.01t} \quad \text{for} \quad t > 0.$$

(a) Determine $P(T > 200)$.

(b) Determine the probability that the lightbulb fails in less than 100 hours when it is switched on.

(c) Give a mixture distribution for the time to failure of the lightbulb.

7.38 If $X \sim exp(\beta)$, show that the moments of X about 0 are given by $\mu'_r = \beta^r r!$.

[*Hint*: Consider individual terms in the moment generating function of X.]

7.39 Determine the mean time to failure for the lightbulbs of Exercise 7.37.

7.40 Accidents at a certain coal mine have been occurring at a long-run average rate of 2 per month when a new safety measure is introduced. Determine the 99.9th percentile of the probability distribution of the time, X, to the next accident under the hypothesis that the safety measure has no effect.

Suppose that the first accident does not occur for $3\frac{1}{2}$ months. How likely is it that the time to the first accident exceeds the $3\frac{1}{2}$ month figure assuming that the safety measure has had no effect on the accident rate?

7.41 Three components with respective mean time to failure 300, 400 and 500 hours are connected in series under independent operation. If the times to failure of the components may be modelled by independent exponential distributions, determine the probability distribution, mean and variance of the time to failure of the series system.

Suggest a generalisation of this result to the case of m components connected in series and X_i, the time to failure of the ith component has an exponential distribution with mean β_i, $i = 1, 2, \ldots m$.

7.42 The length of time, X, to complete a certain key task in house construction is an exponentially distributed random variable with a mean of 10 hours. The cost, C, of completing this task is related to the square of the time to completion by the formula $C = 100 + 40X + 4X^2$.

Determine:

(a) the expected value of C;

(b) the variance of C;

(c) $P(C < c)$ in terms of c, and by differentiating this expression with respect to c, determine the probability density function of C.

[*Hint*: Express C as a perfect square.]

7.3 GAMMA AND WEIBULL DISTRIBUTIONS

The gamma distribution

If $T \sim Erlang(\alpha, 1)$, then the pdf of T may be written in the form

$$f_T(t) = (\text{constant})t^{\alpha-1}e^{-t}, \quad t > 0,$$

where $(\text{constant}) = \left(\int_0^\infty t^{\alpha-1}e^{-t}dt\right)^{-1}$ is that value which ensures that the pdf integrates to 1. If the distribution of T is generalised in such a way that α is no longer required to be an integer, then (constant) may be determined through the gamma function evaluated at α.

Definition 7.6 *The* **gamma function,** $\Gamma(\alpha)$ *is defined for all* $\alpha > 0$ *by the integral*

$$\Gamma(\alpha) = \int_0^\infty t^{\alpha-1}e^{-t}dt.$$

□

The gamma function enjoys many interesting properties and special features which we outline in the following theorem. The establishment of 1 and 2 are discussed in the problems at the end of this section; 3 follows through integration by parts; and 4 is the result of repeated application of 3 using $x = n$.

Theorem 7.8 *Properties of the gamma function.*

1. $\Gamma(1) = 1$.
2. $\Gamma(\frac{1}{2}) = \sqrt{\pi}$.
3. $\Gamma(x+1) = x\Gamma(x)$, *for* x *any positive real number.*
4. $\Gamma(n+1) = n!$, *for* $n = 1, 2, 3, \ldots$, *a positive integer.*

■

By Definition 7.6, if α is a positive real number, the function f_X defined for all $x > 0$ by

$$f_X(x) = \frac{1}{\Gamma(\alpha)}x^{\alpha-1}e^{-x}, \tag{7.6}$$

integrates to 1, is positive, and so is the pdf corresponding to some random variable X. The random variable which has pdf f_X is the **gamma random variable**.

Definition 7.7 *X has a* **gamma distribution** *with parameters $\alpha > 0$ and $\beta > 0$ if X has pdf*

$$f_X(x) = \begin{cases} \frac{1}{\beta^\alpha \Gamma(\alpha)} x^{\alpha-1} e^{-\frac{x}{\beta}} & \text{if } x > 0; \\ 0 & \text{otherwise.} \end{cases}$$

In this case we write $X \sim \gamma(\alpha, \beta)$.

□

Clearly, according to this definition, equation (7.6) gives the probability density function for $T \sim \gamma(\alpha, 1)$. In fact *all* gamma distributions differ from the standard $\gamma(\alpha, 1)$ distribution by a **change of scale.** This is expressed in:

Theorem 7.9 *If $X \sim \gamma(\alpha, 1)$, and a random variable Y is defined by the linear equation $Y = \beta X$ for $\beta > 0$, then $Y \sim \gamma(\alpha, \beta)$.*

Proof: Suppose that X has pdf given by (7.6). Then for $Y = \beta X$, by using the cumulative distribution function method we obtain

$$F_Y(y) = P(Y \leq y) = P(\beta X \leq y) = P(X \leq \frac{y}{\beta}) = F_X(\frac{y}{\beta}).$$

Differentiating each side of this equation with respect to y gives $f_Y(y) = f_X(\frac{y}{\beta})\frac{1}{\beta}$, and this is the pdf specified in Definition 7.7.

■

Gamma moments

If $X \sim \gamma(\alpha, 1)$, then $\mu_r' = E(X^r)$ is given by

$$\begin{aligned} E(X^r) &= \int_0^\infty x^r \frac{1}{\Gamma(\alpha)} x^{\alpha-1} e^{-x} dx \\ &= \int_0^\infty \frac{1}{\Gamma(\alpha)} x^{r+\alpha-1} e^{-x} dx \\ &= \frac{\Gamma(r+\alpha)}{\Gamma(\alpha)} \underbrace{\int_0^\infty \frac{1}{\Gamma(r+\alpha)} x^{r+\alpha-1} e^{-x} dx}_{=1} \\ &= \frac{\Gamma(r+\alpha)}{\Gamma(\alpha)}. \end{aligned}$$

The integral in a brace is equal to 1 since it represents the area under the pdf of a $\gamma(r+\alpha, 1)$ probability distribution. Now, if $Y = \beta X$ represents a scale change, then the moments become scaled accordingly.

$$E(Y^r) = \frac{\beta^r \Gamma(r+\alpha)}{\Gamma(\alpha)}$$

gives the moments of a $\gamma(\alpha, \beta)$ distribution. The mean and variance may be determined from the first and second moments:

$$E(Y) = \frac{\beta\Gamma(1+\alpha)}{\Gamma(\alpha)} = \frac{\beta\alpha\Gamma(\alpha)}{\Gamma(\alpha)} = \alpha\beta.$$

and similarly,

$$E(Y^2) = \frac{\beta^2\Gamma(2+\alpha)}{\Gamma(\alpha)} = \frac{\beta^2(1+\alpha)\Gamma(1+\alpha)}{\Gamma(\alpha)} = \frac{\beta^2(1+\alpha)\alpha\Gamma(\alpha)}{\Gamma(\alpha)} = \alpha\beta^2(1+\alpha).$$

Therefore,

$$Var(Y) = E(Y^2) - E(Y)^2 = \alpha\beta^2(1+\alpha) - \alpha^2\beta^2 = \alpha\beta^2.$$

These details are collected in the following table of properties of the gamma distribution.

PROPERTIES OF THE GAMMA DISTRIBUTION (7.7)
$X \sim \gamma(\alpha, \beta)$

Probability density function $\quad f_X(x) = \frac{1}{\beta^\alpha\Gamma(\alpha)} x^{\alpha-1} e^{-\frac{x}{\beta}}$
$0 < x < \infty$;
$0 < \alpha < \infty; \quad 0 < \beta < \infty.$

Mean	$E(X) = \alpha\beta$
Variance	$Var(X) = \alpha\beta^2$
Standard deviation	$SD(X) = \sqrt{\alpha}\beta$
Coefficient of skewness	$\zeta_3 = \frac{2}{\sqrt{\alpha}}$
Coefficient of kurtosis	$\zeta_4 = 3 + \frac{6}{\alpha}$
Moment generating function	$m_X(t) = (\frac{1}{1-\beta t})^\alpha, \quad t < \frac{1}{\beta}$
Cumulant generating function	$\kappa_X(t) = -\alpha \log_e(1 - \beta t)$

The gamma as a family of distributions

A particular member of the gamma family is called the **chi-squared distribution**. This probability distribution is related to the normal distribution and occupies a prominent role in later chapters on statistical inference.

Definition 7.8 *In terms of a parametric value ν, called the* **degrees of freedom,** *which may assume only positive integer values, a random variable X is said to have a* **chi-squared distribution with ν degrees of freedom** *if $X \sim \gamma(\frac{1}{2}\nu, 2)$. In this case we write $X \sim \chi^2_\nu$.*

□

We have already encountered several special cases of the gamma distribution $\gamma(\alpha,\beta)$; these are the exponential, the Erlang, and the chi-squared.

> A $\gamma(1,\beta)$ distribution is the same as an $exp(\beta)$ distribution.
> A $\gamma(n,\beta)$ distribution is the same as an $Erlang(n,\beta)$ distribution.
> A $\gamma(\frac{1}{2}\nu,2)$ distribution is the same as a χ^2_ν distribution.

We now examine a method for determining such probabilities as $P(a < X < b)$, when $X \sim \gamma(\alpha,\beta)$, representing areas under the general gamma pdf. There is a difficulty in obtaining a closed form expression for

$$F_X(x) = \int_0^x \frac{1}{\beta^\alpha \Gamma(\alpha)} t^{\alpha-1} e^{-\frac{t}{\beta}}\, dt, \tag{7.8}$$

the cumulative distribution function of $X \sim \gamma(\alpha,\beta)$ for arbitrary α and β. Equation 7.8 defines the **incomplete gamma function** which has been extensively tabulated for various values of α and β; such tables present the results of numerical integrations. (See, for example, Pearson (1922).) However, if $\alpha = n$, an integer, and $\beta = 1$, then by Theorem 7.7, X has an $Erlang(n,1)$ distribution with cdf

$$F_X(x) = 1 - \sum_{k=0}^{n-1} \frac{x^k e^{-x}}{k!}.$$

Hence

> If $X \sim \gamma(n,1)$, then $P(a < X < b)$ may be determined from tables of the $Poi(a)$ and $Poi(b)$ distributions.

Using Poisson tables to evaluate gamma-type probabilities is illustrated in Example 7.9.

Example 7.9

Poisson tables for the gamma distribution

Suppose that the time X_1 between arrivals of cars at a freeway entry point, On-ramp 1, is modelled as a gamma distribution with parameters $\alpha = 4$ and $\beta = 2$; that is $X_1 \sim \gamma(4,2)$. At a second entry point, On-ramp 2, to the same freeway, in the same vicinity but independently feeding in from a different suburb, the time X_2 between arrivals of cars is modelled as $X_2 \sim \gamma(6,2)$.

Worked example:

1. **Determine $P(6 < X_1 < 8)$.**
2. **Determine the probability distribution of $X_1 + X_2$, the total number of cars that feed into the freeway in the vicinity of the on-ramps.**

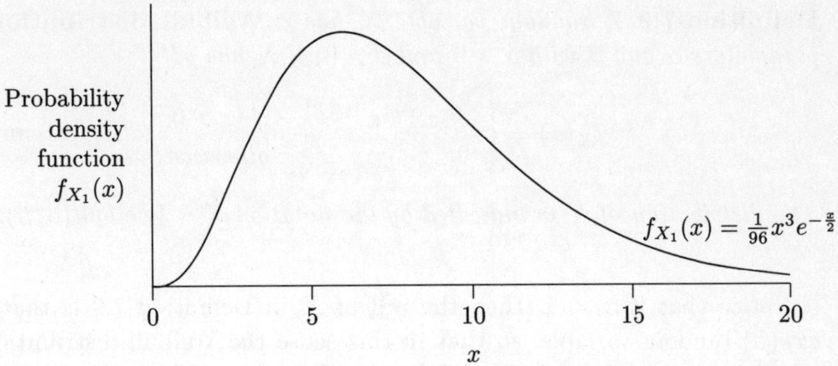

1. Since $X_1 \sim \gamma(4, 2)$, it follows by Theorem 7.7 that $Y = \frac{1}{2}X_1 \sim \gamma(4, 1)$. Therefore,

$$P(6 < X_1 < 8) = P(3 < Y < 4) = F_Y(4) - F_Y(3)$$

Evaluating F_Y by cumulative $Poi(4)$ and $Poi(3)$ tables we obtain

$$\begin{aligned} F_Y(4) - F_Y(3) &= (1 - \sum_{k=0}^{3} \frac{4^k e^{-4}}{k!}) - (1 - \sum_{k=0}^{3} \frac{3^k e^{-3}}{k!}) \\ &= (1 - 0.433) - (1 - 0.647) = 0.214. \end{aligned}$$

2. We use the moment generating function technique to determine the probability distribution of $X_1 + X_2$. The properties listed in (7.7) give

$$m_X(t) = \left(\frac{1}{1 - \beta t}\right)^{\alpha},$$

for $X \sim \gamma(\alpha, \beta)$. Hence,

$$m_{X_1+X_2}(t) = m_{X_1}(t)m_{X_2}(t) = \left(\frac{1}{1-2t}\right)^4 \left(\frac{1}{1-2t}\right)^6 = \left(\frac{1}{1-2t}\right)^{10}.$$

Therefore $X_1 + X_2 \sim \gamma(10, 2)$.

The Weibull family

In this section we study the Weibull distribution. As in the case of the gamma distribution, a wide variety of shapes are possible for the pdf by a judicious choice of parameters. This accounts for the frequent use of the Weibull distribution to model failure of materials or electronic equipment under test conditions (Weibull, 1951).

Definition 7.9 *A random variable X has a* **Weibull distribution** *with parameters α and β with $\alpha > 0$ and $\beta > 0$, if X has pdf*

$$f_X(x) = \begin{cases} \frac{\alpha}{\beta^\alpha} x^{\alpha-1} e^{-(\frac{x}{\beta})^\alpha}, & \text{if } x > 0; \\ 0, & \text{otherwise.} \end{cases}$$

The distribution of X is indicated by the notation $X \sim Weibull(\alpha, \beta)$.

□

Notice that if $\alpha = 1$, then the pdf of X in Definition 7.9 is that of an $exp(\beta)$ random variable, so that in this sense the Weibull distribution is a generalisation of the exponential distribution. β is a scale parameter so that the standard Weibull distribution is that of $X \sim Weibull(\alpha, 1)$. As usual, such scaling is recorded in the following theorem. The proof is similar to that of Theorem 7.9.

Theorem 7.10 *If $X \sim Weibull(\alpha, 1)$ and $Y = \beta X$, then $Y \sim Weibull(\alpha, \beta)$.*

■

Using the distribution of $X \sim Weibull(\alpha, 1)$, the moments about 0 are given by

$$\begin{aligned} E(X^r) &= \int_0^\infty x^r \alpha x^{\alpha-1} e^{-x^\alpha} dx \\ &= \int_0^\infty u^{\frac{r}{\alpha}} e^{-u} du \quad (\text{substitution } u = x^\alpha) \\ &= \Gamma(1 + \frac{r}{\alpha}). \end{aligned}$$

Now if $Y = \beta X \sim Weibull(\alpha, \beta)$, then the rth moment about 0 is given by

$$E(Y^r) = \beta^r E(X^r) = \beta^r \Gamma(1 + \frac{r}{\alpha}).$$

The first and second moments about 0 are used to write down the mean $E(Y)$ and variance $E(Y^2) - E(Y)^2$ of Y:

> If $Y \sim Weibull(\alpha, \beta)$,
> then $E(Y) = \beta\Gamma(1 + \frac{1}{\alpha})$ and $Var(Y) = \beta^2[\Gamma(1 + \frac{2}{\alpha}) - \Gamma(1 + \frac{1}{\alpha})^2]$.

The Weibull distribution has an advantage over the gamma distribution in the sense that there exists a closed form for the cumulative distribution function. This ensures that Weibull probabilities are very easy to calculate.

PROPERTIES OF THE WEIBULL DISTRIBUTION
$X \sim Weibull(\alpha, \beta)$ (7.9)

Probability density function $f_X(x) = \frac{\alpha}{\beta^\alpha} x^{\alpha-1} e^{-(\frac{x}{\beta})^\alpha}$
$0 < x < \infty; \quad 0 < \alpha < \infty; \quad 0 < \beta < \infty.$

Mean	$E(X) = \beta\Gamma(1 + \frac{1}{\alpha})$
Variance	$Var(X) = \beta^2[\Gamma(1 + \frac{2}{\alpha}) - \Gamma(1 + \frac{1}{\alpha})^2]$
Standard deviation	$SD(X) = \beta[\Gamma(1 + \frac{2}{\alpha}) - \Gamma(1 + \frac{1}{\alpha})^2]^{\frac{1}{2}}$
Coefficient of skewness	*no simple expression*
Coefficient of kurtosis	*no simple expression*
Central moments	$E(X^r) = \beta^r\Gamma(1 + \frac{r}{\alpha})$
Cumulant generating function	*no simple expression*

When considering probability models for the distributions of lifetimes, survival times, times to breakdown, times to death, times between arrivals, etc., the random variable, X, representing such time is a positive-valued random variable called a **survival random variable**. In the scenario where X represents the lifetime of a money-back guaranteed item, it is crucial to be able to assess the likelihood that the item will last longer than a guarantee time t. In this context, of particular importance is the **survival function** or **reliability function**.

Definition 7.10 *The* **survival function** *or* **reliability function** $S_X(x)$ *of a positive-valued random variable* X *is defined by*

$$S_X(x) = P(X > x).$$

□

Definition 7.10 implies that the survival function may be written $S_X(x) = 1 - F_X(x)$, where $F_X(x)$ is the cdf of X.

Theorem 7.11 *The survival function for* $Y \sim Weibull(\alpha, \beta)$ *is given by*

$$S_Y(y) = e^{-(\frac{y}{\beta})^\alpha}.$$

Proof: Suppose that $X \sim Weibull(\alpha, 1)$. Then

$$\begin{aligned} S_X(x) &= P(X > x) \\ &= \int_x^\infty \alpha t^{\alpha-1} e^{-t^\alpha} dt \\ &= \int_{x^\alpha}^\infty e^{-u} du \quad \text{(substitute } u = t^\alpha) \\ &= -e^{-u}\Big|_{x^\alpha}^\infty \\ &= e^{-x^\alpha}. \end{aligned}$$

The result then follows by the scale-change $Y = \beta X$ since $P(Y > y) = P(X > \frac{y}{\beta}) = S_X(\frac{y}{\beta})$.

■

The reliability function examines the chance that breakdowns occur beyond a given point in time. To monitor the lifetime of a component across the support of the lifetime distribution, the **hazard function** is used.

Definition 7.11 *A positive-valued random variable X has* **hazard function**, $h_X(x)$ *defined by*

$$h_X(x) = \frac{f_X(x)}{S_X(x)}.$$

□

The hazard function $h_X(x)$ describes the imminent risk of death or failure per unit time conditional on survival to time x:

$$h_X(x) = \lim_{\Delta x \to 0^+} \frac{P(x \leq X \leq x + \Delta x | X \geq x)}{\Delta x}$$

and this expression is equivalent to that of Definition 7.11. According to Definition 7.11, if $X \sim Weibull(\alpha, \beta)$, then

$$h_X(x) = \frac{f_X(x)}{S_X(x)} = \frac{\frac{\alpha}{\beta^\alpha} x^{\alpha-1} e^{-(\frac{x}{\beta})^\alpha}}{e^{-(\frac{x}{\beta})^\alpha}} = \frac{\alpha}{\beta^\alpha} x^{\alpha-1}.$$

This hazard is constant if $\alpha = 1$; in this case the Weibull distribution is really an exponential distribution. This agrees with the idea of constant life expectancy for the exponential model for the lifetime of a lightbulb in Example 7.6. More generally, the hazard function for a Weibull random variable is a power of x. For $\alpha > 1$, $h_X(x)$ is increasing and X is said to have **increasing failure rate**.

Example 7.10
Weibull batteries
Suppose that the length of life, Y, in years, for a car battery has a probability density function

$$f_Y(y) = \begin{cases} \frac{2}{3}ye^{-\frac{y^2}{3}}, & \text{if } y > 0; \\ 0, & \text{otherwise.} \end{cases}$$

Worked example:
Identify this member of the Weibull family, determine the hazard function, reliability function, the mean and standard deviation, and the probability that a battery will last at least 4 years given that it is now 2 years old.
A direct exponent by exponent comparison of the stipulated pdf with the general Weibull pdf

$$f_X(x) = \begin{cases} \frac{\alpha}{\beta^\alpha}x^{\alpha-1}e^{-(\frac{x}{\beta})^\alpha}, & \text{if } x > 0; \\ 0, & \text{otherwise.} \end{cases}$$

for $X \sim Weibull(\alpha, \beta)$ shows that $Y \sim Weibull(2, \sqrt{3})$. That is, $\alpha = 2$ and $\beta = \sqrt{3}$.

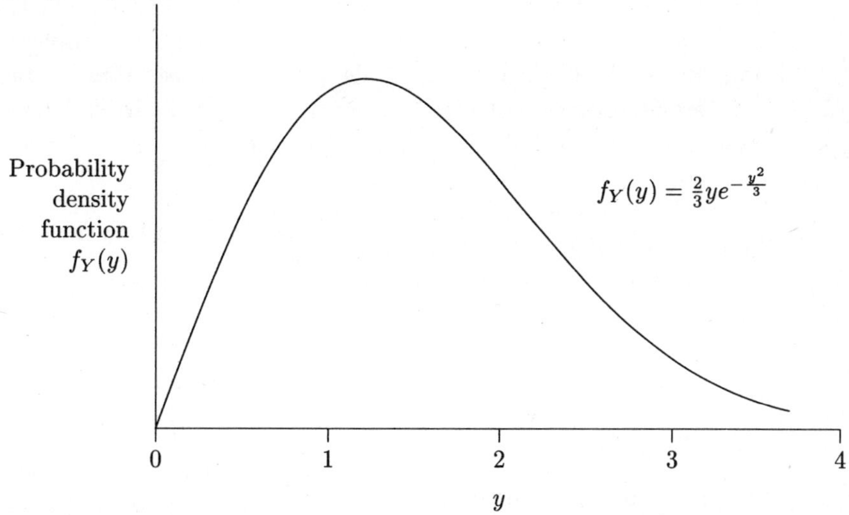

Therefore,

$$\mu = E(Y) = \sqrt{3}\Gamma(1 + \tfrac{1}{2}) = \sqrt{3}[\tfrac{1}{2}\Gamma(\tfrac{1}{2})] = 1.53 \quad (2 \text{ dp})$$

and

$$\sigma = SD(Y) = \sqrt{3}[\Gamma(2) - \Gamma(1 + \tfrac{1}{2})^2]^{\frac{1}{2}} = \sqrt{3}[1 - \frac{\pi}{4}]^{\frac{1}{2}} = 0.80 \quad (2 \text{ dp})$$

and the reliability function is simply (Theorem 7.11) $S_Y(y) = e^{-\frac{y^2}{3}}, y > 0$. The required probability is then $P(Y \geq 4|Y \geq 2)$ which may be evaluated as the conditional probability of the event $\{Y \geq 4\}$ given the event $\{Y \geq 2\}$. Using the expression just derived for the reliability function,

$$P(Y \geq 4|Y \geq 2) = \frac{P(Y \geq 4)}{P(Y \geq 2)} = \frac{e^{-\frac{4^2}{3}}}{e^{-\frac{2^2}{3}}} = e^{-4} = 0.018 \quad (3 \text{ dp}).$$

The size of this probability is not unexpected, given the size of the mean and variance of Y. Finally note that the hazard function is linear:

$$h_Y(y) = \frac{2}{3}y, \quad \text{for} \quad y > 0,$$

Under this Weibull model, the batteries show a linearly increasing risk of failure.

EXERCISES

7.43 In a psychologist's test to assess an individual's reaction time, a subject monitors a TV screen on which the same image is flashed at intervals of 1 second; occasionally the image flashed up is presented with a slight distortion. Suppose that the time, X, until the subject records seeing the distorted image (by pressing an electronic buzzer) is found to follow a gamma distribution with parameters $\alpha = 1.5$ and $\beta = 4$.

(a) Determine the mean reaction time $E(X)$.

(b) Determine the coefficient of variation

$$\vartheta_X = \frac{\sigma_X}{\mu_X}.$$

(c) After obtaining the data for different subjects, it was found that by taking the square root of reaction time, the distributions of reaction times became more comparable. The question is: How is the mean affected? That is, determine $E(\sqrt{X})$.

7.44 Use the results of Theorem 7.8 to determine numerical values for the gamma function $\Gamma(\alpha)$ for

$$\alpha = 0.5, 1, 1.5, 2, 2.5, 3, 3.5, 4.$$

By plotting points $(\alpha, \Gamma(\alpha))$, sketch the gamma function over the interval (0, 4).

7.45 Suppose that the rainfall rate R, in mm h^{-1}, has probability density function of the form

$$f_R(r) = cr^{-b}e^{-ur},$$

where c is a constant, u is a parameter dependent on climate and geographical features and b (with $0 < b < 1$)

is a parameter dependent on the rain rate exceeded 0.01 per cent of the time.

(This density is a simplified version of a density proposed by Moupfouma et al. in *The International Journal of Satellite Communications*, 8, 151-161, concerning rainfall rate in certain tropical climates.)

If, based on experimental evidence, b is taken as 0.8 and u as 0.02 for typically rolling terrain, determine the mean and variance of R.

7.46 State a random variable X for which

$$P(X > 0) = \Gamma(1).$$

Hence obtain the numerical value of $\Gamma(1)$.

7.47 Jobs arriving on a mainframe computer system have been found to have a CPU time, X, following a gamma distribution with mean of 20 seconds and a variance of 200 seconds^2.

(a) Determine the values of α and β so that $X \sim \gamma(\alpha, \beta)$.

(b) Determine $E(\frac{1}{X})$.

(c) Suppose that jobs are queued as they arrive for processing with jobs having CPU time of less than 10 seconds being given Priority A and those with CPU time of between 10 and 20 seconds being given a Priority B rating (and wait longer for processing). Jobs with CPU time longer than 20 seconds are given no priority.

Determine the probability that a randomly chosen job arriving for processing receives:

i. Priority A;

ii. Priority B;

iii. no priority.

7.48 Suppose that $Y \sim N(0, \frac{1}{2})$.

(a) Write down the pdf of Y.

(b) Hence state the value of

$$\int_0^\infty e^{-u^2}\,du.$$

[*Hint*: The area under the pdf of Y is 1.]

(c) Express $\Gamma(\frac{1}{2})$ in integral form, and by making the substitution $u = \sqrt{x}$ into this integral, show that $\Gamma(\frac{1}{2}) = \sqrt{\pi}$.

7.49 Example 7.9 is a special case of the following more general result:

If $X_1, X_2, X_3, \ldots, X_n$ are independent random variables with $X_i \sim \gamma(\alpha_i, \beta)$, then

$$\sum_{i=1}^{n} X_i \sim \gamma(\sum_{i=1}^{n} \alpha_i, \beta).$$

Prove this result by using Example 7.9 as a guide.

7.50 Assume that for a certain grade of steel beam, the yield strength Y, in kg mm^{-2}, follows a Weibull distribution with parameters $\alpha = 3$ and $\beta = 8$ for values above a threshold level of $t_0 = 30$ kg mm^{-2}. This means that the Weibull distribution applies to values $W = Y - 30$, so we assume that $W = Y - 30 \sim Weibull(3, 8)$.

(a) Determine $P(8 < W < 10)$.

(b) Determine the probability that the yield strength is between 38 and 46 kg mm^{-2}.

(c) Determine the mean and variance in yield strength.

(d) Suppose that the threshold value t_0 for a particular grade of steel is unknown. How would you propose to estimate the threshold from 100 observations of yield strength? For example, would the average, maximum or minimum of the observations provide an appropriate estimate of the threshold?

(e) The threshold value t_0 is often built into the Weibull distribution as an extra parameter to create the 3-parameter Weibull distribution. From

$$W = Y - t_0 \sim Weibull(\alpha, \beta),$$

determine the cdf of Y and by differentiating this expression, determine the pdf of Y.

7.51 Consider the problem of designing a life test in which 20 electric motors are run to destruction, giving independent observations $Y_1, Y_2, Y_3, \ldots, Y_{20}$ (measured in hours) from a $Weibull(\frac{1}{2}, 10)$ distribution. (This distribution may be suggested by previous test results for motors of a similar design.) Because of time and cost considerations, the test cannot run indefinitely.

(a) What test fixed duration, t, should be scheduled in advance, to ensure that 95% of the motors will have burned out by the test termination date? [*Hint*: Determine the upper 5% point of the Weibull distribution.]

(b) Determine the probability that the first motor fails before t. That is, find $P(Y_1 < t)$.

(c) Determine $P(Y_i < t)$ for each $i = 1, 2, 3, \ldots, 20$.

(d) Determine the probability that all the 20 test motors fail before t.

(e) An observation of the failure time of a test motor is termed **censored** if the motor does not fail before the test termination date t. Determine the probability that 2 of the 20 motors are censored by the fixed test termination date.

7.52 Suppose that Y denotes the 'time to failure' of a component in manufacturing equipment. A replacement policy that is designed to avoid excess cost in replacing failed components, ensures that components are replaced regularly (hopefully before they fail).

The **expected service life**, $ESL(t)$, of the component when it is automatically replaced after t hours of usage is defined as the expected value of the mixture random variable

$$Z = min\{Y, t\}.$$

(a) Explain heuristically why Z represents the 'service life' of a component.

(b) Z may be regarded as a function of the random variable Y with a single point discontinuity at t. Let f denote the pdf of Y. Use the methods for determining the expected value of a function of a random variable, namely,

$$E[g(Y)] = \int_0^\infty g(y) f(y) dy,$$

to show that

$$ESL(t) = \int_0^t y f(y) dy + t P(Y > t).$$

[This is the formula implied by Brick et al. (1989) in a paper entitled: *Using Statistical Thinking to Solve Maintenance Problems.*]

(c) Determine $ESL(2)$, for a 2 year service life policy for the batteries of Example 7.10, where

$$Y \sim Weibull(2, \sqrt{3}).$$

7.53 The random variable Y, denoting the survival time of an individual, has hazard function $h(y)$. Show that the survival function is given by

$$S(y) = e^{-\int_0^y h(t)dt}.$$

7.54 The survival random variable Y has hazard function $h(y)$. Suppose that constants a and b satisfy $a < b$.

(a) If A and B represent the respective events $\{Y > a\}$ and $\{Y > b\}$, use the definition of conditional probability to show that

$$P(B|A) = \frac{S(b)}{S(a)},$$

where S denotes the survival function.

(b) Show that the probability that an a year-old individual reaches age b is given by

$$e^{-\int_a^b h(t)dt}.$$

7.55 Suppose that the death rate for a person that smokes is, at each age, twice that of a non-smoker.

(a) If $h_s(y)$ denotes the hazard function for (the survival time of) a smoker at age y and $h_n(y)$ that of a non-smoker at age y, write down an equation relating $h_s(y)$ and $h_n(y)$.

(b) Use the result of Exercise 7.54 to show that:

For two individuals of the same age, one of whom is a smoker and the other is a non-smoker, the probability that the smoker survives to any given age is the square of the corresponding probability for a non-smoker.

7.56 If $Y \sim Weibull(\alpha, 1)$, then the random variable T defined by $T = \log_e Y$ is said to have an **extreme value distribution.**

The extreme value distribution is important since lifetimes are often measured on a logarithmic scale.

Determine the pdf of T. Simplify the pdf for the case $\alpha = 1$.

7.4 CHECKING DISTRIBUTIONAL ASSUMPTIONS WITH PROBABILITY PLOTS

Probability plots provide a simple effective means for determining whether a sample of continuous data is from a population having a specified probability distribution. The probability models most frequently checked are the normal, the exponential and the Weibull, and these continuous distributions have been the focus of earlier sections in this chapter.

For example, to assess whether the inter-arrival times of cars at a service station follow an exponential distribution, we first obtain a data set by direct observation of the service station, then decide through a stem-and-leaf plot whether an exponential distribution is worthy of contemplation as a possible model. (Recall that the stem-and-leaf plot provides insight into the shape of the data set.) This decision process is sharpened through the simple construction of a normal probability plot.

We will find that many statistical tests require an assumption of normally distributed data; now such a premise may be graphically checked.

Sample percentiles

In order to understand what is measured along each axis in a probability plot the notions of *sample percentile* and *rank* must be retrieved and digested from Chapter 1.

An observed data set is written $x_1, x_2, x_3, \ldots, x_n$ in the order in which the observations are recorded. The notation indicates that the data set has n data points. The notation for the **ranked data set**, which represents $x_1, x_2, x_3, \ldots, x_n$ arranged in rank order from smallest to largest, is

$$x_{(1)}, x_{(2)}, x_{(3)}, \ldots x_{(n-1)}, x_{(n)},$$

for which

$$x_{(1)} \leq x_{(2)} \leq x_{(3)} \leq \ldots \leq x_{(n-1)} \leq x_{(n)}.$$

We now ask the question: What **percentile** does each ranked data point in the sample correspond to? According to the development of sample percentiles in Chapter 1, we will use the convention that:

$$\boxed{x_{(i)} \text{ is the } 100(\tfrac{i-\frac{1}{2}}{n})\text{th sample percentile.}} \qquad (7.10)$$

This notation is illustrated in Example 7.11, where the sample percentiles are determined for a data set of size 20.

Example 7.11

Determining percentiles for an ordered sample

The following list of 20 'prices realised' resulted from auction realisations of Melbourne properties on the weekend of 17 August 1991. These prices appear after 18 months of sliding property values in Melbourne; at this time, commentators were trying to detect whether the market had 'bottomed-out'.

520 000	90 000	120 000	119 000	(U) 87 500
(U) 88 000	147 000	(U) 137 000	158 500	337 000
(U) 60 000	170 000	165 000	211 500	120 000
130 000	133 000	232 000	(U) 137 000	200 000

(*Source*: *The Sunday Age*, 18 August 1991.)

The sample percentiles are found from the ranked data—the results are suitably reported in the form of a table. (U) signifies a price realised for a 'flat', 'apartment' or 'unit'.

Data x_i	**Ranked data** $x_{(i)}$		**Sample percentiles** $100(\frac{i-\frac{1}{2}}{n})\%$
520 000	(U) 60 000	$x_{(1)}$	2.5
90 000	(U) 87 500	$x_{(2)}$	7.5
120 000	(U) 88 000	$x_{(3)}$	12.5
119 000	90 000	$x_{(4)}$	17.5
(U) 87 500	119 000	$x_{(5)}$	22.5
(U) 88 000	120 000	$x_{(6)}$	27.5
147 000	120 000	$x_{(7)}$	32.5
(U) 137 000	130 000	$x_{(8)}$	37.5
158 500	133 000	$x_{(9)}$	42.5
337 000	(U) 137 000	$x_{(10)}$	47.5
(U) 60 000	(U) 137 000	$x_{(11)}$	52.5
170 000	147 000	$x_{(12)}$	57.5
165 000	158 500	$x_{(13)}$	62.5
211 500	165 000	$x_{(14)}$	67.5
120 000	170 000	$x_{(15)}$	72.5
130 000	200 000	$x_{(16)}$	77.5
133 000	211 500	$x_{(17)}$	82.5
232 000	232 000	$x_{(18)}$	87.5
(U) 137 000	337 000	$x_{(19)}$	92.5
200 000	520 000	$x_{(20)}$	97.5

This determines the sample percentiles at each ranked data point.

Notice that by interpolating between the 47.5th percentile and the 52.5th percentile (both of which have the value 137 000) we obtain a 50th percentile of 137 000. Thus the sample median house price for this sample is 137 000. This is lower than median values for post-1989 Melbourne property auction realisations.

What points are plotted in a probability plot?

A probability plot provides evidence as to whether data has come from a population having a stated probability model. This probability model is specified through the cumulative distribution function F.

If the probability model fits the data well, then for $p = \frac{i-\frac{1}{2}}{n}$, the 100$p$th percentile of the theoretical probability distribution defined by F should be in close agreement with the 100pth sample percentile $x_{(i)}$. That is, the proportion of the data that is less than the ith largest data point should be close to p. In Equation 7.10 is the key to finding the 100pth sample percentile for $p = \frac{i-\frac{1}{2}}{n}$. The 100$p$th population percentile, ξ_p, is defined implicitly as the solution to the equation $F(\xi_p) = p$. However, since F is the cumulative distribution function of a continuous random variable, it is in particular increasing; this is an appropriate criterion for invertibility. Therefore, taking F^{-1} of each side of the equation $F(\xi_p) = p$ gives

$$\xi_p = F^{-1}(p) = F^{-1}\left(\frac{i-\frac{1}{2}}{n}\right),$$

which describes the theoretical percentiles that must be calculated for each i. This leads to the following definition.

Definition 7.12 *A* **probability plot** *is a plot of the points*

$$\left(F^{-1}\left(\frac{i-\frac{1}{2}}{n}\right),\ x_{(i)}\right), \quad i = 1, 2, 3 \ldots, n,$$

in the usual rectangular coordinate system.

□

In a probability plot, Definition 7.12 indicates that the **ranked data** is plotted on the vertical axis and **percentiles from a ... distribution** on the horizontal axis. The boldface terms provide appropriate labels for the axes.

If the points plotted according to Definition 7.12 fall close to a 45 degree line through the origin (0,0) of the plot, then the probability model fits the data well. Often it is difficult to obtain perfect linear agreement at all points along the horizontal axis. Points in the plot lying above (respectively below) the 45 degree line represent data points that are larger (respectively smaller) than the probability model predicts they should be. If this occurs in the tails of the ranked data set, it may lead to an 'S-shaped plot'; that is, the bulk of the data set follows the proposed distributional form but the tails are shorter or longer than they should be.

We will use flexibility in the interpretation of Definition 7.12. For example, a probability plot of Weibull data is most suitably presented on the logarithm scale (for each axis). The interpretation of a straight line fit is then relative to the scale involved.

Normal probability plots

In this section we concentrate on checking whether data may be assumed to be normally distributed.

To determine whether data may follow a *standard normal distribution*, the cumulative distribution function, F, is taken as

$$F(x) = \Phi(x) \quad \text{where} \quad \Phi(x) = \int_{-\infty}^{x} \frac{1}{\sqrt{2\pi}} e^{-\frac{1}{2}t^2}\, dt.$$

For plotting purposes, we have $F^{-1}(x) = \Phi^{-1}(x)$. Thus, for example, tables of the standard normal distribution show for a sample of size $n = 5$,

i	$\frac{i-\frac{1}{2}}{n}$	$\Phi^{-1}(\frac{i-\frac{1}{2}}{n})$
1	0.1	−1.28
2	0.3	−0.52
3	0.5	0.00
4	0.7	0.52
5	0.9	1.28

To check whether data has come from a $N(\mu, \sigma^2)$ distribution, the cumulative distribution function of $X \sim N(\mu, \sigma^2)$ is required: (7.2) shows that this cdf is given by

$$F(x) = \Phi(\frac{x-\mu}{\sigma}).$$

Placing $x = x_{(i)}$ in each side of this equation we obtain

$$F(x_{(i)}) = \Phi(\frac{x_{(i)}-\mu}{\sigma}).$$

Therefore if F is the cdf for a probability model that fits the data well, then $F(x_{(i)}) = \frac{i-\frac{1}{2}}{n}$. This identification on the left-side of our equation gives

$$\frac{i-\frac{1}{2}}{n} = \Phi(\frac{x_{(i)}-\mu}{\sigma}).$$

On taking Φ^{-1} of each side of this equation,

$$\Phi^{-1}(\frac{i-\frac{1}{2}}{n}) = \frac{x_{(i)}-\mu}{\sigma},$$

or

$$x_{(i)} = \mu + \sigma\Phi^{-1}(\frac{i-\frac{1}{2}}{n}). \qquad (7.11)$$

This means that, for normally distributed data, the plot of points

$$\left(\Phi^{-1}(\frac{i-\frac{1}{2}}{n}),\ x_{(i)}\right), \quad i = 1, 2, 3, \ldots, n,$$

would follow a straight line with intercept μ and slope σ. In fact, such parameters may then be *estimated* from the slope and intercept of the observed plotted line when it is agreed that the data is from a normal population, but the true values μ and σ^2 of the mean and variance of the population are unknown. This estimation process is best seen as a guide to the magnitudes of mean and variance rather than an accurate estimation procedure.

When normal probability plotting is done by hand, the process may be shortened by the use of a type of graph paper called **normal probability paper**. In this case, Φ^{-1} does not need to be calculated, but is automatically adjusted for in the non-linear scale on one axis of the paper. The scale on the axis where the ranked data is plotted is not adjusted.

Example 7.12

A probability plot for bolt lengths: Example 7.3 continued

A guillotine device that cuts long lengths of steel rods into bolts is adjusted so that 1% of manufactured bolts have lengths outside the specifications 2.560 ± 0.025 cm. Experience in the industry has shown that 'length of bolt' is a normally distributed random variable.

Twenty bolts produced by the machine were selected at random and measured with the following results:

2.543	2.553	2.554	2.556	2.545
2.562	2.573	2.576	2.550	2.570
2.564	2.551	2.559	2.578	2.557
2.572	2.560	2.566	2.567	2.561

Worked example:

1. **Construct a normal probability plot to demonstrate that this data is consistent with having come from a normal population.**
2. **Use the probability plot to assess whether the apparent normal distribution is compatible with the manufacturing specifications.**

The points to be included in the probability plot are determined in the following table. The last two columns in the table are the respective $x-$ and $y-$ coordinates.

$\frac{i-\frac{1}{2}}{n}$	**Percentile from normal distribution** $\Phi^{-1}(\frac{i-\frac{1}{2}}{n})$	**Ranked data** $x_{(i)}$
0.025	−1.96	2.543
0.075	−1.44	2.545
0.125	−1.15	2.550
0.175	−0.94	2.551
0.225	−0.75	2.553
0.275	−0.60	2.554
0.325	−0.45	2.556
0.375	−0.32	2.557
0.425	−0.19	2.559
0.475	−0.06	2.560
0.525	0.06	2.561
0.575	0.19	2.562
0.625	0.32	2.564
0.675	0.45	2.566
0.725	0.60	2.567
0.775	0.75	2.570
0.825	0.94	2.572
0.875	1.15	2.573
0.925	1.44	2.576
0.975	1.96	2.578

Normal probability plot of bolt lengths

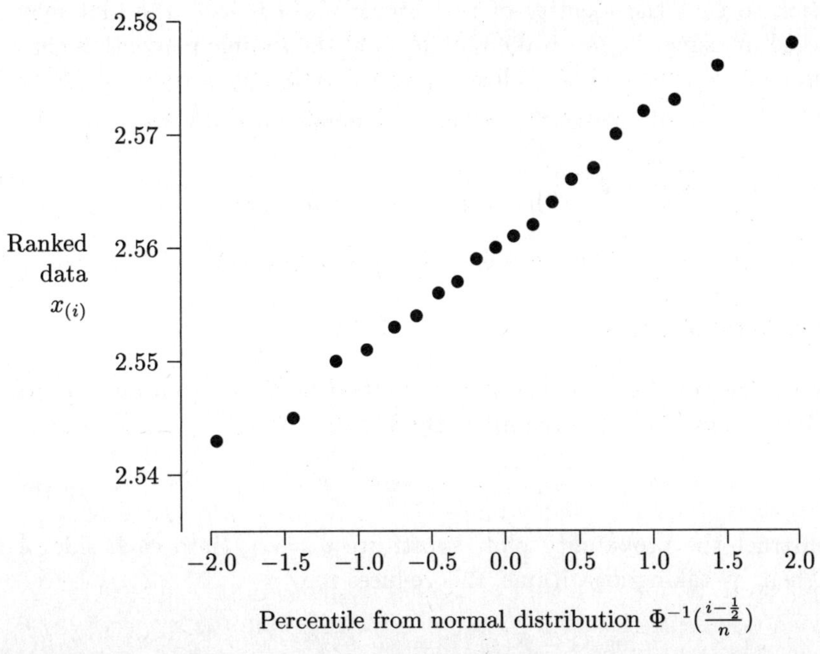

If the random variable X denotes 'bolt length', then the straight line in the normal probability plot indicates that it is reasonable to assume that $X \sim N(\mu, \sigma^2)$. The values of μ and σ which allow the specifications to be achieved are determined as follows: If 1% of bolt lengths are to lie outside the specification 2.560 ± 0.025 cm, then these limits of observation lie 2.58 standard deviations from the mean $\mu = 2.560$ which lies, by symmetry, at the centre of the specification interval. Hence, the z-score for the upper specification limit is

$$\frac{2.585 - 2.560}{\sigma} = 2.58,$$

giving a solution $\sigma = 0.0097$.
Are these required population values for μ and σ observable in the probability plot?
An estimate of the slope of the line indicated in the plot is

$$\frac{2.576 - 2.545}{1.44 - (-1.44)} = 0.010.$$

This provides confirmation that the data have come:

- from a normal population;
- with mean 2.56 cm;
- with standard deviation 0.0097 cm.

Notice that when data have been grouped into intervals in a frequency distribution so that the identity of the original data is lost, the plot may still proceed by using the interval endpoints and the sample percentiles that these endpoints correspond to. Thus if interval endpoint e_i corresponds to the $100(\frac{j-\frac{1}{2}}{n})$th sample percentile (where j depends on the value of i), the points

$$\left(F^{-1}(\frac{j-\frac{1}{2}}{n}),\ e_{(i)}\right), \quad i = 1, 2, 3 \ldots, k$$

are plotted; here k denotes the number of interval endpoints.

Weibull probability plots

To check whether data have come from a $Weibull(\alpha, \beta)$ distribution, the cdf $F(y)$ of $Y \sim Weibull(\alpha, \beta)$ is required. By Theorem 7.11,

$$F(y) = 1 - e^{-(\frac{y}{\beta})^{\alpha}}. \tag{7.12}$$

To construct the probability plot, substitute $y = x_{(i)}$ into each side of (7.12). Then, by taking logarithms, this reduces to

$$\log_e(1 - F(x_{(i)})) = -(\frac{x_{(i)}}{\beta})^{\alpha}.$$

Again, taking logarithms to create a new variable *log-life* yields

$$\log_{10}[-\log_e(1 - F(x_{(i)})] = \alpha[\log_{10} x_{(i)} - \log_{10}\beta].$$

(Note that base 10 logarithms are traditionally used to study lifetimes, however natural logarithms are equivalent, differing only by a scale-change.) When log-life is made the subject of the last formula, the straight line format

$$\log_{10} x_{(i)} = \log_{10}\beta + \frac{1}{\alpha}\log_{10}[-\log_e(1 - F(x_{(i)})] \qquad (7.13)$$

is obtained. If the data fits the Weibull model well, then $F(x_{(i)}) = \frac{i-\frac{1}{2}}{n}$. If, in (7.13), we make this substitution, then when the model fits the data well,

$$\log_{10} x_{(i)} = \log_{10}\beta + \frac{1}{\alpha}\log_{10}[-\log_e(1 - \frac{i-\frac{1}{2}}{n})].$$

This means that for Weibull data, the plot of points

$$\left(\log_{10}[-\log_e(1 - \frac{i-\frac{1}{2}}{n})],\ \log_{10} x_{(i)}\right), \qquad i = 1, 2, 3, \ldots, n,$$

would follow a straight line with slope $\frac{1}{\alpha}$ and intercept $\log_{10}\beta$. Notice that this particular probability plot differs from the normal probability plot in that linearity occurs on the log scale. In agreement with the normal probability plot, however, the slope and intercept of a well-fitting line may provide estimates for $\frac{1}{\alpha}$ and $\log_{10}\beta$ (and hence for α and β.) Notice that if the observed slope of a fitted line is close to 1, then the estimate of α is close to 1, and it is then reasonable to assume that the data has come from an *exponential population.*

Example 7.13

Weibull age-replacements

In an article entitled *Using Statistical Thinking to Solve Maintenance Problems*, Brick et al. (1989) use a Weibull probability plot to detect whether their lifetime data on the 'time to replacement of sinker rolls in a sheet metal galvanising process' may be appropriately modelled by the Weibull distribution.

This was part of a cost-cutting study. The \$2000 cost of roll replacement before failure was less than half the cost of on-line failure.

$n = 17$ rollers were observed to have the following lifetimes (measured as a number of 8-hour shifts) ranked from smallest to largest in column 3 of the table. In the final column the log-lifetimes are given—these are plotted on the vertical axis in the probability plot against the appropriate Weibull percentiles determined in Column 1 and Column 2.

$\frac{i-\frac{1}{2}}{n}$	$\log_{10}[-\log_e(1-\frac{i-\frac{1}{2}}{n})]$	**Lifetime** $x_{(i)}$	**Log-lifetime** $\log_{10} x_{(i)}$
0.03	−1.53	10	1.00
0.09	−1.03	12	1.08
0.15	−0.80	15	1.18
0.21	−0.64	17	1.23
0.26	−0.51	18	1.26
0.32	−0.41	18	1.26
0.38	−0.32	20	1.30
0.44	−0.24	20	1.30
0.50	−0.16	21	1.32
0.56	−0.09	21	1.32
0.62	−0.02	23	1.36
0.68	0.05	25	1.40
0.74	0.12	27	1.43
0.79	0.20	29	1.46
0.85	0.28	29	1.46
0.91	0.39	30	1.48
0.97	0.55	35	1.54

Weibull probability plot for age-replacement

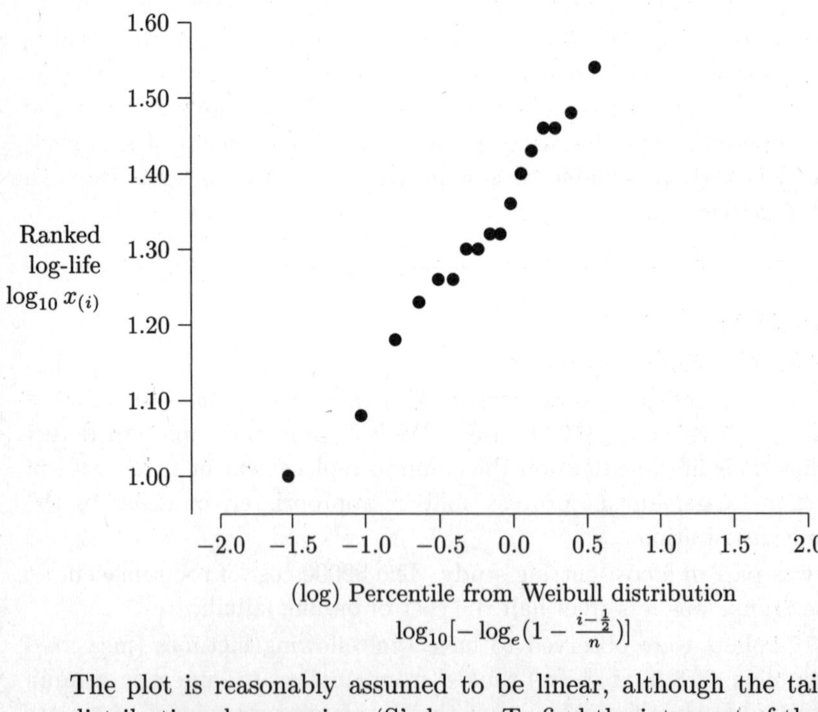

The plot is reasonably assumed to be linear, although the tails of the distribution show a minor 'S' shape. To find the intercept of the straight line, set the value plotted on the horizontal axis to zero. Solving

$$\log_{10}[-\log_e(1-\frac{i-\frac{1}{2}}{n})] = 0,$$

it follows that

$$\frac{i-\frac{1}{2}}{17} = 1 - \frac{1}{e} = 0.63 \text{ (2 dp)}.$$

The 63rd sample percentile is approximately 23 and this provides a reasonable estimate of β, the scale parameter. For estimating α, we take the inverse of the slope of the estimated straight line: this gives an α value of approximately 3.33. Since Weibull models with $\alpha > 1$ have hazard functions that increase with time, and because of the costs involved, the authors of the article indicate that it is inappropriate to wait for the sinker rolls to fail. A policy of scheduled replacement should be used.

EXERCISES

7.57 The data give the weight (grams) of portions of 'bacon eye' in 20 prepackaged airline meals.

18.9	13.5	27.5	22.9	26.2
48.0	17.1	16.3	15.2	16.4
16.0	13.3	17.4	16.2	12.8
16.3	17.0	15.4	15.9	15.9

Is it reasonable to assume that these data are normally distributed? Construct a B/W plot to examine both symmetry and the possible presence of outliers in the data set.

7.58 A company produces wood-grained panels to be used for the lining of interior walls. The production line is constantly monitored to ensure that a quality product is manufactured to a specified standard. The manufacturing process is such that the panels are always twice as long as they are wide. However, the width of produced panels is subject to variability. The following data give the widths observed (in metres) for 50 panels randomly selected from the production line in a particular week.

0.980	0.997	1.030	1.016	0.954	1.007
1.013	0.960	1.050	0.972	1.022	1.020
1.010	0.944	1.004	0.990	0.996	0.976
1.008	0.964	1.028	0.966	1.042	1.024
1.034	0.974	1.018	0.995	0.995	0.992
1.000	1.000	0.998	1.014	0.991	0.983
1.032	0.988	0.995	0.970	1.060	1.006
0.984	0.982	1.038	1.012	1.026	1.002
1.011	1.001				

(a) Construct a stem-and-leaf plot for these data.

(b) Construct a normal probability plot for these data.

(c) Let the random variable X, in metres, denote the width of the timber panels produced. Manufacturing specifications require that $X \sim N[1, (0.025)^2]$.

List all the evidence contained in the normal probability plot as to whether or not the manufacturing specifications are achieved.

(d) Regardless of your answer to (c), assume for the remainder of this question that $X \sim N[1, (0.025)^2]$. Assume also that panels are *exactly* twice as long as they are wide.

i. Determine $E(X-1)^2$.

ii. Write down the expected length of a randomly chosen panel.

iii. Calculate the expected area (or mean area) of a randomly chosen panel.

iv. Determine the probability that a randomly chosen panel has area exceeding 1.96 square metres.

v. If the equipment is adjusted so that the mean width is increased by 20%, and the variance in width is reduced by 20% determine the percentage increase in mean area.

7.59 There is not complete agreement between authors on the 'optimal' representation of sample percentiles. One family of representations may be described as follows:

$x_{(i)}$ is the $100(\frac{i-b}{n+1-2b})$th sample percentile. Popular choices of b are $b = 0, \frac{1}{2}, \frac{3}{8}$.

For large data sets there appears to be little difference between the choices for b.

Which value of b corresponds to the definition of sample percentiles given in (7.10) and used throughout this book?

7.60 If the data

2.3 4.2 4.7 4.9 6.1 8.3

represent 6 observations taken from a normal distribution, determine the sample percentiles

$$\Phi^{-1}(\frac{i-\frac{1}{2}}{n})$$

for each ranked data point.

7.61 The following S/L plot records the waiting time (in minutes) at a bus stop:

Bus stop waiting time
Stems—mins; leaves—10^{-1} mins.
3|2 represents 3.2 minutes

Stem	Leaves
1	1
2	3 7
3	2 5 8
4	3 7 7 9
5	6 6 8 9 9
6	0 0 0 0 3 4
7	0 1 1 3 7
8	3 6 7 7
9	4 5 5
10	6 6
11	7

(a) Determine the median of the data. Without precise determination, where would you anticipate that the sample mean would lie?

(b) Show that the function f defined by

$$f(x) = \begin{cases} \frac{1}{36}x & \text{if} \quad 0 < x < 6; \\ \frac{1}{3} - \frac{1}{36}x & \text{if} \quad 6 \leq x < 12. \end{cases}$$

satisfies the requirements for being a probability density function. f is the pdf for a random variable X that has a **triangular distribution**.

(c) Sketch a graph of this pdf and comment on the suitability of f as a probability model to describe the bus stop waiting times.

(d) Determine the lower quartile of the probability distribution described by the pdf f. Compare this with the lower quartile of the data set.

(e) Construct a normal probability plot for the data. Is it possible to detect in the plot any evidence of non-normality?

7.62 Consider the pdf f described in the previous exercise.

(a) Determine a formula for the cdf $F(x)$ for all real values of x.
[You will need to determine F over each of the domains $0 < x < 6$ and $6 \le x < 12$.]

(b) Hence determine $F^{-1}(x)$.

(c) Construct a 'triangular' probability plot

$$(F^{-1}(\frac{i-\frac{1}{2}}{n}),\ x_{(i)})$$

using the 36 data points in the bus waiting time data set of the previous exercise. How should this probability plot be interpreted?

7.63 The cdf of a $Weibull(\alpha, \beta)$ distribution is given by

$$F(y) = 1 - e^{-(\frac{y}{\beta})^{\alpha}}.$$

(a) Substitute $y = F^{-1}(x)$ into each side of this equation and hence obtain a formula for $F^{-1}(x)$ in terms of x.

(b) Show that $\log_{10} F^{-1}(x)$ is a linear function of $\log_{10}[-\log_e(1-x)]$.

(c) Discuss these results in the context of a Weibull probability plot.

7.64 The following data (Chatfield, 1983) resulted from a life test of refrigerator motors (hours to burnout):

104.3	158.7	193.7	201.3	206.2
227.8	249.1	307.8	311.5	329.6
358.5	364.3	370.4	380.5	394.6
426.2	434.1	552.6	594.0	691.5

(a) Construct a Weibull probability plot for these data.

(b) Estimate the parameters α and β in the fitted probability model.

(c) Use the fitted model to estimate the probability that a refrigerator motor will fail in less than 300 hours.

(d) If there is a threshold lifetime t_0 beyond which all motors will last, how could t_0 be estimated from the data? Subtract this estimated value of t_0 from all the data and reconstruct a Weibull probability plot for all non-zero data points.

(e) Estimate the values of t_0, α_0 and β_0 in the model

$$P(T > t) = e^{-(\frac{t-t_0}{\beta_0})^{\alpha_0}},$$

specifying the probability that the lifetime, T, of a refrigerator motor exceeds the time t.

7.5 VARIANCE STABILISERS AND HANGING ROOTOGRAMS

The normal probability plot gives immediate indication as to whether the fitted probability model furnishes a good description of the data. The 'goodness-of-fit', or lack of it, is immediately apparent through the linearity of the probability plot. In this section we present a histogram method for assessing fit of probability model to data. At the heart of these methods is the concept of **residual**.

Residuals

Residuals quantify the discrepancy between observed data values and values that are fitted by using a probability model. These fitted values are really estimates of 'how big the data should be' if the suggested probability model is assumed to be correct. Residuals provide a simple numerical measure of discrepancy between data and model.

Definition 7.13 *Suppose that particular realised values of the random variables $Y_1, Y_2, Y_3, \ldots, Y_n$ constitute the data set $y_1, y_2, y_3, \ldots, y_n$ in corresponding order and $\widehat{y}_1, \widehat{y}_2, \widehat{y}_3, \ldots, \widehat{y}_n$ the values, in corresponding order, hypothesised by a particular probability model. Then, the ith* **residual**, *e_i, is defined by*

$$e_i = y_i - \widehat{y}_i, \quad i = 1, 2, 3, \ldots, n.$$

□

Clearly, some residuals will be positive and some residuals will be negative, depending on whether the estimated data point is larger or smaller than the observed data point. Further, as a general principle, small residuals indicate a good fit, and large residuals point to a bad fit.

For example, if a mean is fitted to the data set $x_1, x_2, x_3, \ldots, x_n$, using a probability model which attaches equal importance (or weight) to each point in the data set, then

$$\widehat{x}_i = \overline{x} = \frac{\sum_{i=1}^{n} x_i}{n}.$$

In this case $e_i = x_i - \widehat{x}_i = x_i - \overline{x}$ is the ith residual. In our discussion of data analysis in Chapter 2, such residuals were termed **mean deviations** and they summed to 0. That is, $\sum_{i=1}^{n}(x_i - \overline{x}) = 0$. This is generally the case with residuals $e_i = y_i - \widehat{y}_i$, the positive residuals cancel with the negative residuals when all are added: $\sum_{i=1}^{n} e_i = 0$ indicates that these residuals have mean 0.

Variance stabilising transformations

Unfortunately, the variance of residuals is often not so well behaved. This is a difficulty, because it is desirable to have the fitted model equally accurate at all points where the model is fitted, otherwise the model could not be used for predicting how future data sets may behave.

One way to stabilise the variance of the e_i is to divide by the standard deviation. Recall from Chapter 2 that the standardised data

$$z_i = \frac{x_i - \overline{x}}{s_x} \quad \text{has} \quad \overline{z} = 0 \quad \text{and} \quad s_z = 1.$$

Thus, by standardising the data $x_1, x_2, x_3, \ldots, x_n$, the new standard deviation is 1 irrespective of the original standard deviation.

More generally, we may ask: How can a random variable be transformed so that it achieves constant variance? For example, if $X \sim Poi(\lambda)$, then $Var(X) = \lambda$ is not constant but depends on the value of the parameter λ. Or, if $X \sim Bin(n,p)$, then $Var(X) = np(1-p)$ is not constant but depends on the value of the parameter p.

Suppose that f is a differentiable function of a random variable X and that $Y = f(X)$. If, near μ_X we approximate the function f by a linear function, then it follows by Taylor's Expansion (see, for example, Thomas and Finney (1984)) of the function f near $\mu = E(X)$ that

$$f(X) \approx f(\mu) + (X - \mu)f'(\mu). \tag{7.14}$$

Taking the variance of each side of this equation, and recalling that for constants a and b, $Var(a + bX) = b^2 Var(X)$,

$$\begin{aligned} Var(Y) &= Var[f(X)] \\ &\approx Var[f(\mu) + (X - \mu_X)f'(\mu)] \\ &= [f'(\mu)]^2 Var(X). \end{aligned}$$

This process for determining the approximate variance of a function of a random variable is called the **delta method** or **δ–method**. The δ-method eliminates the need of first determining the probability distribution for $Y = f(X)$ and then calculating the variance of that distribution. The δ-method will offer an advantage if only an approximate version of the variance of Y is required or if the pdf of Y is difficult to determine. If the expected value is taken on each side of Equation 7.14, then $E[f(X)] \approx f(\mu) + f'(\mu)E(X - \mu)$, and since $E(X - \mu) = 0$, it follows that, at least approximately, $E[f(X)] \approx f(\mu)$.

We now apply the δ-method to the problem of variance stabilisation. Given the random variable X, we wish to determine a function f so that $Var[f(X)]$ is constant. We will assume that $Var(X) = \mu$ to cover the case $X \sim Poi(\lambda)$

or $X \sim Bin(n,p)$ for n large and p small. This is the situation which we shall be applying in the next section. Then, approximately, $Var(Y)$ is constant when $[f'(\mu)]^2 Var(X)$ is constant:

$$\begin{aligned} f'(\mu)^2 Var(X) &= constant \\ \Rightarrow \quad f'(\mu)^2 \mu &= constant \\ \Rightarrow \quad f'(\mu) &= \frac{constant}{\sqrt{\mu}} \\ \Rightarrow \quad f(\mu) &= (constant)\sqrt{\mu}. \end{aligned}$$

This is the solution: To stabilise the variance of a random variable with variance a multiple of the mean, use the square root transformation (which lies on the ladder of power transformations with $p = \frac{1}{2}$). Therefore, if $X \sim Poi(\lambda)$, then $[\frac{d}{d\lambda}(\sqrt{\lambda})]^2 Var(X) = \frac{1}{4}$. In summary:

$$\begin{aligned} E(\sqrt{X}) &\approx \sqrt{\lambda} \\ Var(\sqrt{X}) &\approx \frac{1}{4}. \end{aligned}$$

Hanging rootograms

Suppose that we use n observations to construct a (relative) frequency histogram standing on the intervals $[e_0, e_1), [e_1, e_2), [e_2, e_3), \ldots, [e_{k-1}, e_k)$, where adjustments $e_0 = -\infty$ and $e_k = \infty$ may be made for presenting data on the whole real line. Histograms are surprisingly variable objects: histograms constructed from different samples taken from the same population may show fluctuation in the heights of the boxes. In fact, the random variable f_i denoting the frequency of observation in the ith interval $[e_{i-1}, e_i)$ satisfies

$$f_i \sim Bin(n, p_i),$$

with a variance $np_i(1-p_i)$ depending on p_i denoting the probability that an observation will land in this interval. For a fixed sample size, intervals near the centre of the histogram with larger values of p_i (near $\frac{1}{2}$) will experience more variable frequencies than intervals located in the tails of the histogram. This is because the function $p(1-p)$ of p has maximum value at $p = \frac{1}{2}$.

Suppose that a pdf is fitted to the relative frequency histogram. If the corresponding fitted cdf is denoted by $\widehat{F}$, then the area under the pdf standing on the interval $[e_{i-1}, e_i)$ is given by

$$\widehat{p}_i = \widehat{F}(e_i) - \widehat{F}(e_{i-1}).$$

Thus $\widehat{f_i} = n\widehat{p_i}$ gives the estimated expected frequency in $[e_{i-1}, e_i)$.

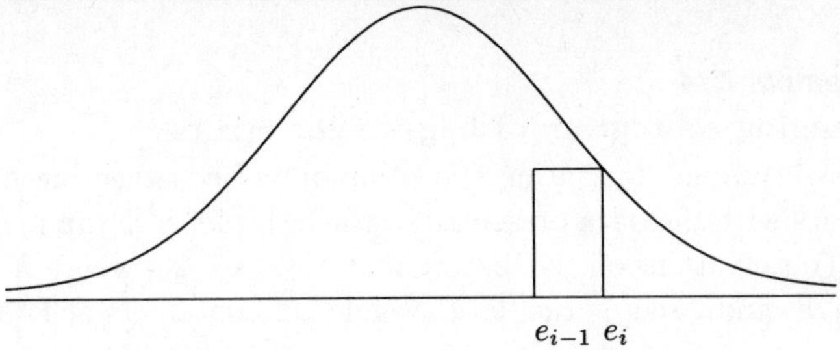

Since relative frequencies in a relative frequency histogram are represented by the areas of the boxes, the residual

$$f_i - \widehat{f_i},$$

represents the shaded difference in area between the density curve on $[e_{i-1}, e_i)$ and the histogram box on $[e_{i-1}, e_i)$ scaled by a factor of n. This then raises the question: If a residual is large, is this due to a bad fit or to chance? Since most of the variability in residuals is apparent in histogram box heights being different from sample to sample, we may write

$$Var(f_i - \widehat{f_i}) \approx Var(f_i) = np_i(1 - p_i) \approx np_i$$

when p_i is small. A square root transformation is then in order to help clarify the difference between 'bad fit' and 'random fluctuation'.

Definition 7.14 *A* **hanging rootogram** *is a histogram showing, for each $i = 1, 2, 3, \ldots k$, the height*

$$\sqrt{f_i} - \sqrt{\widehat{f_i}}$$

of the rectangle standing on $[e_{i-1}, e_i)$. The plotted values are **rootogram residuals**. *Here f_i is the frequency of occurrence and $\widehat{f_i}$ is the estimated expected frequency of occurrence for a fitted probability model.*

□

Obviously, in rootograms, frequency is no longer proportional to area. The standard deviation of the quantities $\sqrt{f_i}$ is approximately $\frac{1}{2}$. This invites the following rule:

A residual is *large* if the corresponding rootogram residual is larger than 1.5 units away from 0 (1.5 units represents 3 standard deviations).

Velleman and Hoaglin (1981) show how to create so-called **double root residuals** to improve on the square root transformation when expected interval frequencies are very small.

Example 7.14

Hanging rootogram of bag-weight errors

This example illustrating the technique of constructing rootograms continues with the data presented in the S/L plot in Example 7.1 comprising 400 observations on the discrepancy $Y = X - 25$, where X is the observed weight and 25 kg is the 'label weight' of containers of PVC powder.

The sample mean and sample standard deviation of the 400 data points are $\overline{y} = 0.0948$ and $s_y = 0.2671$ and therefore the best-fitting normal pdf is

$$f_Y(y) = \frac{1}{\sqrt{2\pi}0.2671} e^{-\frac{1}{2}(\frac{y-0.0948}{0.2671})^2}.$$

This density has cdf

$$F_Y(y) = \Phi(\frac{y - 0.0948}{0.2671}),$$

where Φ is the standard normal cdf. To determine the area under the fitted pdf standing on the interval between e_{i-1} and e_i, we use

$$\widehat{p}_i = \Phi(\frac{e_i - 0.0948}{0.2671}) - \Phi(\frac{e_{i-1} - 0.0948}{0.2671}), \quad i = 1, 2, 3, \ldots, k.$$

For the hanging rootograms illustrated we have used $k = 15$ intervals with endpoints:

$$e_0 = -\infty, e_1 = -0.6, e_2 = -0.5, \ldots, e_{13} = 0.6, e_{14} = 0.7, e_{15} = \infty.$$

The observed frequencies f_i for each of these intervals are readily counted from the S/L plot (as each interval occupies two lines of the S/L plot). In the hanging rootograms which follow, these observed frequencies are compared with the estimated expected interval frequencies $\widehat{f}_i = n\widehat{p}_i$ through the rootogram residuals $\sqrt{f_i} - \sqrt{\widehat{f}_i}$. In Plot A, the original $n = 400$ data points are used in the fit, and in Plot B, the 4 worst outliers are removed and the fit obtained using 396 points. There is a slightly improved fit after outlier removal as the rootogram residuals in Plot B lie within 1.5 units of 0. The lighter-than-normal bag weights in the lower tail cause an initial group of positive rootogram residuals. This is serially followed by a negative group of rootogram residuals, showing that there are fewer data in this area of the distribution than the normal curve would predict. Overall, the fit is reasonable, as no rootogram residuals are extraordinarily large—the deficiencies evident in the rootograms may be explained through filling machine malfunctions.

Plot A: Hanging rootogram
for bag weight errors of 25 kg bags PVC powder

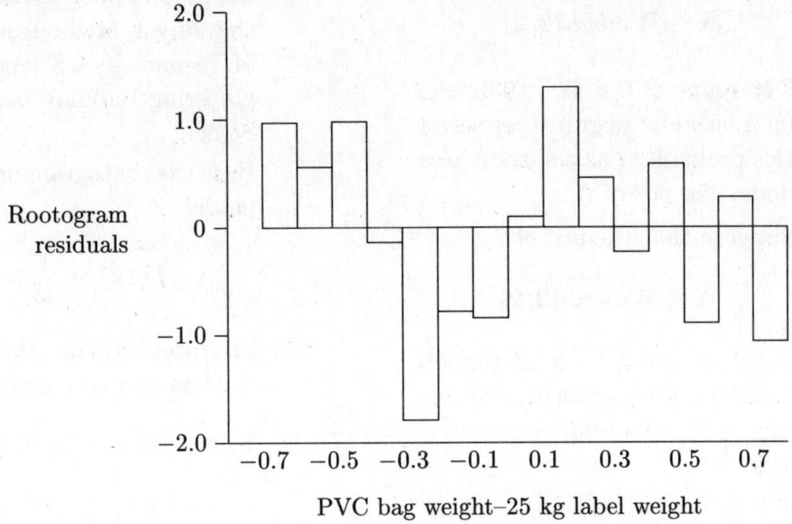

Plot B: Hanging rootogram
for bag weight errors of 25 kg bags PVC powder
after the removal of outliers

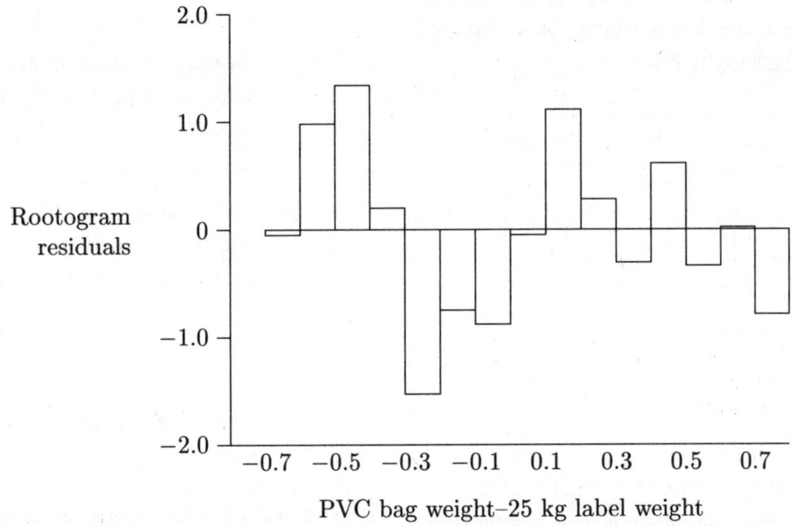

EXERCISES

7.65 (a) Write down the cdf of

$$X \sim Weibull(2, 2).$$

(b) Determine $P(0 < X < 0.5)$ and in a sketch diagram represent this probability as a shaded area under the pdf of X.

(c) Suppose that the pdf of

$$X \sim Weibull(2, 2)$$

is to be fitted to a histogram based on 100 observations occurring in the 8 equal-width intervals

$[0, 0.5), [0.5, 1), [1.0, 1.5), \ldots [3.5, 4)$.

Determine the estimated expected counts occurring in each interval.

7.66 (a) Construct a hanging rootogram for the lift waiting time data of Example 7.7.

Interval	Width	Frequency
0–15⁻	15	206*
15–30⁻	15	84
30–45⁻	15	49
45–60⁻	15	38
60–75⁻	15	29
75–90⁻	15	15
90–120⁻	30	28
120–150⁻	30	9
150–180⁻	30	8
180–270⁻	90	5
Total		471

* This frequency includes 57 waiting times of 0; that is, a lift had already arrived once counting began.

The frequency distribution records 471 observations on the 'waiting time', X (in seconds), for a lift to arrive when called to the ground floor (street level) for the Royal Melbourne Institute of Technology's Science and Engineering building on Swanston Street.

Base the rootogram on the fitted model

$$f_X(x) = \frac{1}{36} e^{-\frac{x}{36}}.$$

(b) Interpret/explain the shape of the hanging rootogram.

7.67 (a) Show that the function

$$f(p) = sin^{-1}\sqrt{p}$$

satisfies the equation

$$f'(p)^2 Var(X) = constant$$

when $X \sim Bin(n, p)$.

(b) Suggest a transformation that may be applied to a binomial random variable to create approximately constant variance.

(c) Suggest a transformation that could be used to stabilise the variability in proportions

$$\widehat{p} = \frac{X}{n},$$

where $X \sim Bin(n, p)$, for fixed n.

7.68 The following example illustrates the variable nature of frequency histograms.

Two samples of 100 random numbers were selected from the unit interval. The results in a truncated 2 decimal place form are recorded in a back-to-back S/L plot as follows:

Comparing samples from $[0, 1]$
Stems—10^{-1}; leaves—10^{-2}
2 | 2 represents 0.22

Sample A		Sample B
776544211	0	0114699
986666432	1	111234667
9988877777654332	2	2333335799
6554322	3	234456777789
955554333221	4	00135588
975432	5	1233455666
988766533	6	0011123334445799
98776655430	7	355557
777775300	8	222344456999
998877643220	9	1223555799

These S/L plots show the diagrammatic shapes of frequency histograms. It is difficult to imagine that both data sets originated from the same population! Both data sets show considerable variability.

(a) Construct a relative frequency histogram for each of Sample A and Sample B.

(b) Specify the probability model for the population by stating the pdf for the random variable X denoting a randomly chosen number from the interval $[0, 1]$.

(c) Construct a hanging rootogram for each data set.

(d) Are any of the rootogram residuals large? Is the frequency of 16 observations in the interval $[0.2, 0.3)$ unusually large, or simply due to chance fluctuation?

CHAPTER EIGHT

Joint Distributions: When Independence Fails

In this chapter you will find

8.1 CONDITIONAL DISTRIBUTIONS

Conditional probability mass function

Given a discrete joint probability distribution, the marginal probabilities are found by adding across the (rows/columns) of the joint table. This removes the presence of the unwanted variable in the table. We have seen in Chapter 3 that, by restricting possible outcomes to a (row/column) of the joint table, conditional probabilities may be defined in terms of such a restriction relative to the marginal probability of the (row/column).

Example 8.1

Preferred Prime Minister: conditional along party lines

Suppose that the following joint probability table was constructed from information provided by a large survey of voters across the country who were asked to indicate their electoral support (Labor/Liberal) and then asked: Who would make the best Prime Minister—the current Prime Minister or the current Leader of the Opposition?

y x	**Prime Minister**	**Leader Opposition**	**Total**
Labor voter	0.42	0.18	0.6
Liberal voter	0.12	0.28	0.4
Total	0.54	0.46	1

The poll is fairly evenly divided along the last 'Total' row: 54% favour the current Prime Minister as the preferred leader. This is an appropriate estimate of the probability that a randomly chosen person would favour the Prime Minister as the preferred leader. But, among Labor voters? Suppose we know that the randomly chosen person is a Labor voter. Do we change our assessment of their preference for leader? Restricting attention to the 'Labor voter' line of the table, we estimate that 42% of the 60% of Labor voters favour the current Prime Minister. Therefore, in terms of conditional probabilities,

$$P(\text{Prime Minister favoured leader}|\text{Labor voter}) = \frac{0.42}{0.60} = 0.7.$$

Interpretation: While approximately half the general population supports the Prime Minister, a much larger 70% of Labor voters do.

Note that with appropriate '0' and '1' labels the joint probability table may be written as:

$p_{X,Y}(x,y)$ x	y	0	1	$p_X(x)$
0		0.42	0.18	0.6
1		0.12	0.28	0.4
	$p_Y(y)$	0.54	0.46	1

The conditional probability mass functions are simply defined by using the ideas of *conditional probability* as outlined in Example 8.1.

Definition 8.1 *The* **conditional mass function of** Y **given** $X = x$, *denoted by* $p_{Y|x}$, *is the function of* y *defined for a fixed value of* x *by the conditional probability*

$$p_{Y|x}(y) = P(Y = y|X = x).$$

Similarly $p_{X|y}(x) = P(X = x|Y = y)$ *is the conditional mass function of* X *given* $Y = y$.

□

Notice that the conditional mass functions may be written in the form of a joint mass function divided by a marginal mass function:

$$p_{Y|x}(y) = P(Y = y|X = x) = \frac{P(Y = y, X = x)}{P(X = x)} = \frac{p_{X,Y}(x, y)}{p_X(x)}.$$

A similar result holds for $p_{X|y}(x)$. From this representation it follows that if X and Y are independent random variables (in which case the joint mass function factorises), then

$$p_{Y|x}(y) = p_Y(y) \quad \text{and} \quad p_{X|y}(x) = p_X(x).$$

In Example 8.1, this definition is applied as follows:

$$p_{Y|0}(0) = P(Y = 0|X = 0) = \frac{p_{X,Y}(0, 0)}{p_X(0)} = \frac{0.42}{0.60} = 0.7;$$

$$p_{Y|0}(1) = P(Y = 1|X = 0) = \frac{p_{X,Y}(0, 1)}{p_X(0)} = \frac{0.18}{0.60} = 0.3.$$

This indicates that, conditionally, $Y|0$ has probabilities which may be summarised in the table

y	0	1
$p_{Y|0}(y)$	0.7	0.3

This shows that $p_{Y|0}(y)$ satisfies the positivity and summing to 1 requirements of a probability mass function. The steps in the above calculation for $p_{Y|0}(y)$ provide the pattern that may be followed for determining $p_{Y|1}(y)$.

Definition 8.2 *When* X *and* Y *have a joint discrete distribution, the* **conditional expected value**, $E(Y|x)$, *of* Y *given that* $X = x$ *is the function of* x *defined by*

$$E(Y|x) = E(Y|X = x) = \sum_{\text{all } y} y p_{Y|x}(y).$$

Similarly,

$$E(X|y) = E(X|Y = y) = \sum_{\text{all } x} x p_{X|y}(x).$$

□

Example 8.2

Is price related to quantity ordered?

Example 5.2 continued. Suppose that a shipment of cans of tomatoes contains either 50 or 100 cartons, and that these cartons are sold to retail outlets at either \$6 per carton or at the discounted price of \$5 per carton. Random variables X, denoting 'number of cartons in shipment', and Y, denoting 'wholesale price per carton', are assumed to have joint pmf

$p_{X,Y}(x,y)$ $\quad y$ x	5	6
50	0.40	0.20
100	0.15	0.25

Worked example:

Determine the mean wholesale price per carton for a bulk delivery of 100 cartons.

The question asks us to determine $E(Y|X = 100)$. This requires the joint (given) marginal and conditional distributions. We begin by rewriting the joint table to display the marginal mass functions:

$p_{X,Y}(x,y)$ x	y	5	6	 $p_X(x)$
50		0.40	0.20	0.60
100		0.15	0.25	0.40
	$p_Y(y)$	0.55	0.45	1.00

Then,

$$\begin{aligned}
E(Y|X = 100) &= \sum_{all\ y} y p_{Y|100}(y) \\
&= (5)\left(p_{Y|100}(5)\right) + (6)\left(p_{Y|100}(6)\right) \\
&= (5)\left(\frac{p_{X,Y}(100,5)}{p_X(100)}\right) + (6)\left(\frac{p_{X,Y}(100,6)}{p_X(100)}\right) \\
&= (5)\left(\frac{0.15}{0.40}\right) + (6)\left(\frac{0.25}{0.40}\right) \\
&= 5.625.
\end{aligned}$$

Hence the mean wholesale price per carton is \$5.63 (2 dp).

Conditional density functions

Consider now the case where the random variables (X, Y) are jointly distributed continuous random variables. By definition, the marginal density for X is obtained by integrating out the unwanted variable Y from the joint density $f_{X,Y}(x, y)$. This coincides with the pdf for X when X is considered as a random variable in its own right.

Definition 8.3 *The* **conditional probability density function,** $\boldsymbol{f_{Y|X}(y|x)}$, **of** Y **given** X, *is defined by*

$$f_{Y|X}(y|x) = \frac{f_{X,Y}(x,y)}{f_X(x)}$$

if $f_X(x)$ is positive, and is 0 if $f_X(x) = 0$. Similarly, the conditional probability density function, $f_{X|Y}(x|y)$, of X given Y, is defined by

$$f_{X|Y}(x|y) = \frac{f_{X,Y}(x,y)}{f_Y(y)}$$

if $f_Y(y)$ is positive, and is 0 if $f_Y(y) = 0$. These conditional densities have **conditional expected values***:*

$$E(Y|x) = \int_{-\infty}^{\infty} y f_{Y|X}(y|x)dy; \quad E(X|y) = \int_{-\infty}^{\infty} x f_{X|Y}(x|y)dx.$$

□

As in the discrete case, the notation $E(Y|X = x)$ is also commonly used to denote $E(Y|x)$.

Example 8.3

An exponential distribution with a conditionally fixed mean

Suppose that T, the time between phone calls through a switchboard, is exponentially distributed once a randomly varying mean has been determined. That is, conditionally on a random variable Ω, the time between calls through the switchboard is exponential:

$$f_{T|\Omega}(t|\omega) = \omega e^{-\omega t}, \quad \text{for } \omega > 0.$$

We want to determine the *unconditional* time between adjacent calls through the switchboard. To determine this we must imagine how the mean rate of calls is likely to vary. For example, late at night, large periods of inactivity on the switchboard would be more likely than small periods. A pdf that reflects this is

$$f_\Omega(\omega) = e^{-\omega}, \quad \text{for} \quad \omega > 0.$$

From this we may determine the joint density of T and Ω: for $\omega > 0$,

$$f_{T,\Omega}(t,\omega) = f_{T|\Omega}(t|\omega) f_\Omega(\omega) = \omega e^{-\omega t} e^{-\omega} = \omega e^{-(t+1)\omega}.$$

Then, integrating out Ω, we obtain the marginal distribution of T:

$$\begin{aligned} f_T(t) &= \int_0^\infty f_{T,\Omega}(t,\omega) d\omega \\ &= \int_0^\infty \omega e^{-(t+1)\omega} d\omega \\ &= \frac{1}{(t+1)^2} \int_0^\infty x e^{-x} dx \quad (\text{substitute } x = (t+1)\omega) \\ &= \frac{1}{(t+1)^2}, \end{aligned}$$

where the last integral is 1 since the expected value of an *exp*(1) random variable is 1.

Definition 8.4 *When random variables X and Y have a joint distribution, the function of x defined by*

$$\mu(x) = E(Y|X = x),$$

having domain the support of X, is called the **regression function of Y on X**.

□

In some modelling situations for bivariate data, the regression function is assumed to be of a specified functional form. A graph of $\mu(x)$ for various values of x displays how the mean values of Y change with x. One commonly modelled function is the linear function, in which case constants α and β are sought so that $\mu(x) = \alpha + \beta x$. Such a modelling process is termed **linear regression**, and is discussed fully in Chapter 14.

When the regression function $\mu(x)$ is evaluated at the random variable X, the function $\mu(X)$ of the random variable X is obtained. This new random variable $\mu(X)$ has a particularly simple mean: for the jointly continuous case this conditional mean is determined as

$$\begin{aligned} E[\mu(X)] &= E[E(Y|X)] \\ &= \int \{E(Y|x)\} f_X(x)dx \\ &= \int \left\{\int y f_{Y|X}(y|x)dy\right\} f_X(x)dx \\ &= \int\int y f_{Y|X}(y|x) f_X(x)dy\ dx \\ &= \int\int y f_{X,Y}(x,y)dy\ dx \\ &= \int y \left\{\int f_{X,Y}(x,y)dx\right\} dy \\ &= \int y f_Y(y)dy \\ &= E(Y). \end{aligned}$$

The jointly discrete case is established in a similar way with summation signs replacing integrals. This establishes the theorem:

Theorem 8.1 *Jointly distributed random variables X and Y satisfy the equation $E(Y) = E[E(Y|X)]$.*

■

Example 8.4

Expected time for a journey by conditioning on the path taken

In driving from Silverdale to Auckland, there are two routes:

- via East Coast Bays (30 km);
- directly via the freeway (40 km).

Suppose that on the East Coast Bays route, because it passes through a built-up area, my average speed is 40 km per hour while travelling; however I always stop for half an hour at my favourite roadhouse (for a pie and a coffee). On the freeway route, there are no stops, and I usually achieve an average speed of 100 km per hour. Experience has shown that (because I like pies?) I am twice as likely to travel to Auckland via East Coast Bays than via the freeway.

Worked example:

Determine the expected time for my trip from Silverdale to Auckland.

Let the Bernoulli random variable X have labels '0' for a journey via East Coast Bays and '1' for a journey via the freeway. Then, since the East Coast Bays' route is twice as likely, X has pmf

x	0	1
$p_X(x)$	$\frac{2}{3}$	$\frac{1}{3}$

Let Y denote the journey time (hours). We wish to determine $E(Y)$. By conditioning on the route,

$$\begin{aligned} E(Y) &= E[E(Y|X)] \\ &= E(Y|X=0)P(X=0) + E(Y|X=1)P(X=1) \\ &= (\frac{30}{40} + \frac{1}{2})(\frac{2}{3}) + (\frac{40}{100})(\frac{1}{3}) \\ &= \frac{29}{30}. \end{aligned}$$

This indicates that the expected time to complete the journey is 58 minutes. A key to the mathematical terms: $E(Y|X=0)$ is the expected time for the journey via East Coast Bays; this is evaluated as $\frac{30}{40} + \frac{1}{2}$ to allow for the half-hour roadhouse stop. Similarly, $E(Y|X=0)$ is the expected time via the freeway.

EXERCISES

8.1 The joint probability mass function given in the table below was applied to the random variables X (number of cartons in shipment) and Y (wholesale price per carton) in Example 8.2.

$p_{X,Y}(x,y)$ y / x	5	6
50	0.40	0.20
100	0.15	0.25

Determine the conditional distribution of Y for a given shipment of 50 cartons.

8.2 Suppose that the bivariate random variable (X, Y) may assume the values

$(1,0)$, $(2,0)$, $(1,3)$, $(2,3)$

with respective probabilities $0.3, 0.2, 0.1$ and 0.4. Evaluate $E(Y|X=1)$.

8.3 On a given weekday a rental car firm ascertains that it is likely to receive X requests for Volvos when there are Y available in stock where X and Y are jointly described by the following bivariate pmf.

$p_{X,Y}(x,y)$ y / x	0	1	2
0	0.05	0.05	0.10
1	0.15	0.10	0.10
2	0.10	0.10	0.15
3	0.05	0.05	0.00

(a) What is the expected number of requests for Volvos on a given day?

(b) Determine the probability that the number of requests for

Volvos exceeds the available stock.

(c) Determine the expected number of requests for Volvos on a given day when none are in stock.

(d) Determine the expected number of Volvos in stock on a given day when the firm receives 1 request for hiring a Volvo.

8.4 If

$$p_{X,Y}(x,y) = \frac{x+y}{36}$$

for $x = 1,2,3$; $y = 1,2,3$, and is 0 elsewhere:

(a) write out the joint probability distribution in tabular form;

(b) determine the marginal distributions of X and Y alone;

(c) evaluate $P(Y = 1|X = 1)$;

(d) show that $p_X(x) = \frac{2+x}{12}$, for $x = 1,2,3$;

(e) determine $p_{Y|x}(y)$.

8.5 A box contains 10 fuses of which 2 are defective. Two fuses are randomly selected from the box; the first fuse is not replaced before the second fuse is selected.

Let random variables X_1 and X_2 be defined as follows: for each $i = 1,2$, $X_i = 1$ if the ith selected fuse is defective; $X_i = 0$ otherwise.

Construct the joint probability mass function for (X_1, X_2). Are X_1 and X_2 independent?

8.6 Two fair dice are tossed. Let X denote the *difference* between the two numbers showing and Y denote the *sum* of the two numbers showing.

(a) Construct the joint probability mass function of (X, Y).

(b) Determine $P(Y = 5|X = 3)$.

(c) Determine $P(Y = 3|X = 5)$.

8.7 In an archery competition, the target is a circle of radius 1 metre. When an arrow is fired at the target, assume that it will hit at a random point (X, Y) relative to the bulls-eye at the centre of the target; the point (X, Y) is uniformly distributed on the unit disc with centre $(0,0)$.

(a) Determine each of the following:

i. $f_{Y|X}(y|x)$;

ii. $E(Y|x)$;

iii. $E(X|y)$.

(b) What is the geometric interpretation of $E(Y|x)$? What is the geometric interpretation of $E(X|y)$?

8.8 An individual player in a game of poker is dealt 5 cards from a pack of 52 cards. Let us define the discrete random variables:

X = the number of *jacks* in the hand;

Y = the number of *queens* in the hand.

Both the random variables X and Y may assume the values $0,1,2,3,4$ under the condition that there is at most a combined total of 5 *jacks* and *queens* in the hand: $x + y \leq 5$.

(a) Give a combinatorial argument as to why the the joint probability mass function $p_{X,Y}(x,y)$ is given by

$$p_{X,Y}(x,y) = \frac{\binom{4}{x}\binom{4}{y}\binom{44}{5-x-y}}{\binom{52}{5}}.$$

(b) Show how to re-express the formula in (a) as a joint probability table and determine the missing entries labelled *. (Results are expressed to 4 dp and 0.0000 represents probabilities smaller than 10^{-4}.)

x \ y	0	1	2	3	4
0	*	*	0.0306	0.0015	0.0000
1	*	*	0.0087	0.0003	0.0000
2	*	*	0.0000	0.0000	0
3	*	*	0.0000	0	0
4	*	*	0	0	0

(c) Determine the marginal probability mass functions of X and Y as combinatorial formulae. Show how to obtain these distributions by summing rows or columns in the joint table.

Do X and Y have the same probability distribution?

(d) Determine the conditional probability mass functions of $X|y$ and $Y|x$ as combinatorial formulae. Show how to obtain these distributions by restricting consideration to rows and/or columns in the joint table.

Do $X|y$ and $Y|x$ have the same probability distribution?

8.9 Given the discrete bivariate random variable (X, Y), show that the function, $q(x)$, defined by $q(x) = p_{X|y}(x)$ for y in the support of Y, is a probability mass function in the sense that:

- $q(x) \geq 0$;
- $\sum_{all\ x} q(x) = 1$.

8.10 Recall Example 8.3. Suppose that T, the time between phone calls through a switchboard, is exponentially distributed once a randomly varying mean has been determined. That is, conditionally on a random variable Ω, the time between calls through the switchboard is exponential with

$$f_{T|\Omega}(t|\omega) = \omega e^{-\omega t}, \quad \text{for } \omega > 0.$$

If the 'call rate' random variable, Ω, is modelled as a gamma distribution with pdf

$$f_{\Omega}(\omega) = \omega e^{-\omega}, \quad \text{for } \omega > 0,$$

determine:

(a) the parameters of the gamma distribution being used as a model for the 'call rate';

(b) $E(\Omega)$;

(c) $E(T|\Omega)$;

(d) $E[E(T|\Omega)]$;

(e) $E(T)$.

8.11 The word 'operation' is present in a list of 10^6 words with probability 0.5. Given that the word is in the list, it is equally likely to be in any position on the list. A computer scanner reads through the list from the first word until 'operation' is reached, or until all 10^6 words have been scanned. What is the expected number of words that must be searched through?

8.12 In 1983, X_1, the number of people killed on Northern Territory roads per 10 000 motor vehicles registered had an observed value of 8; X_2, the number killed on ACT roads per 10 000 motor vehicles registered had an observed value of 2.

Is there a simple explanation why the road death rates for the two territories are so different?

Suppose that these figures were used to model future accident rates: $X_1 \sim Poi(8)$ independently of $X_2 \sim Poi(2)$.

(a) Give the probability distribution of the random variable $Z = X_1 + X_2$, giving the road death rates for both territories combined.

(b) Given that $Z = 15$, determine the probability distribution for X_1.

(c) Determine

$$E(X_1|Z = n) \text{ and} E(X_2|Z = n)$$

for n a positive integer. What is the interpretation of each of these expected values?

8.13 The number of flaws, X, occurring in sheet metal follows a Poisson distribution with mean rate λ per square metre. Given a particular value x of X, the conditional distribution of Y, the number of flaws that may be repaired by burnishing, is binomial with parameters x and p.

Show that Y is marginally Poisson with mean λp.

8.14 The number of cars per day that travel down a country road connecting the Inland Highway to the Great Ocean Road follows a Poisson distribution with mean 20. The only point of interest on this connecting road is a roadhouse; the number of cars using the connecting road that stop at the roadhouse is binomially distributed with $p = 0.4$. Determine the probability distribution of the number of cars that stop at the roadhouse on a given day.

8.15 Suppose that the regression function of Y on X is given by $\mu(x) = 2x - 1$.

(a) As the observed values of the random variable X increase, does the mean value of Y increase, decrease or stay the same?

(b) Determine $E(Y)$.

8.16 In experiments where the data are **right-censored**, instead of observing outcomes of the random variable Y of interest to us, we are only able to observe outcomes of the mixture variable $min(Y, t)$, where t is a fixed constant. (t may, for example, be the termination date of an experiment monitoring the time to failure, Y, of a battery in a personal computer (PC); if the battery hasn't failed once the experiment has terminated, we observe only t but know that $Y > t$.)

Suppose that the random variable δ, called a **censor indicator**, defined by

$$\delta = \begin{cases} 1 & \text{if} \quad Y < t; \\ 0 & \text{if} \quad Y > t, \end{cases}$$

is also known. Censored data occur when $\delta = 0$.

Use conditional expectation to show that the variable Y^* defined by

$$Y^* = Y\delta + E(Y|\delta = 0)(1 - \delta),$$

satisfies

$$E(Y^*) = E(Y).$$

8.2 COVARIANCE

When two continuous random variables X and Y are independent, by definition their joint density (or mass function in the discrete case) factorises at *all* points in the support of the joint distribution. This extremely strong condition on the behaviour of the random variables may not hold in realistic situations. In this section we examine a weaker restriction on the behaviour of two random variables: that they be uncorrelated.

Covariance: is X acting with Y or against Y?

Recall that the expected value of a function $h(X, Y)$ of a bivariate random variable (X, Y) is given by:

$$E[h(X,Y)] = \int_{-\infty}^{\infty}\int_{-\infty}^{\infty} h(x,y) f_{X,Y}(x,y)\,dx\,dy \quad \text{(continuous case);}$$

$$E[h(X,Y)] = \sum_{\text{all } x}\sum_{\text{all } y} h(x,y) p_{X,Y}(x,y) \quad \text{(discrete case).}$$

Our purpose now is to choose a function h whose expectation will measure the strength of the relationship between X and Y. The answer is to choose h to be the product

$$h(X,Y) = (X - \mu_X)(Y - \mu_Y)$$

of the mean corrected variables. Then, on average, h will provide some measure of association between X and Y. 'Some measure' means: an assessment of when association between the variables is '+', '−', or '0'.

When X and Y have a positive association, then large observed values for X (relative to μ_X) tend to accompany large observed values for Y (relative to μ_Y), and small observed values for X tend to accompany small observed values for Y. Thus, X is acting *with* Y and observations of the product $(X - \mu_X)(Y - \mu_Y)$ tend to show:

(positive) times (positive) = positive, or,
(negative) times (negative) = positive.

The other two possible patterns,

(positive) times (negative) = negative, and,
(negative) times (positive) = negative,

produce a negative product for h, representing the negative association when large values of one variable accompany small values of the other. Such negative association indicates that X acts *against* Y. The expected value of this choice of $h(X,Y)$ has a particularly simple structure.

Theorem 8.2 *If random variables X and Y are jointly distributed with respective means μ_X and μ_Y, then*

$$E[(X-\mu_X)(Y-\mu_Y)] = E(XY) - \mu_X\mu_Y.$$

Proof: For the pair (X,Y) either jointly discrete or jointly continuous, it follows by the linearity of E as described by **SUM 1** in (5.5) that

$$\begin{aligned} E[(X-\mu_X)(Y-\mu_Y)] &= E[XY - X\mu_Y - \mu_X Y + \mu_X\mu_Y] \\ &= E(XY) - E(X\mu_Y) - E(\mu_X Y) + E(\mu_X\mu_Y) \\ &= E(XY) - \mu_Y E(X) - \mu_X E(Y) + \mu_X\mu_Y \\ &= E(XY) - \mu_Y\mu_X - \mu_X\mu_Y + \mu_X\mu_Y \\ &= E(XY) - \mu_X\mu_Y. \end{aligned}$$

■

This provides justification for the following definition of the covariance between X and Y.

Definition 8.5 *The* **covariance**, $Cov(X,Y)$, *between the random variables X and Y is given by*

$$Cov(X,Y) = E(XY) - E(X)E(Y).$$

□

The covariance is defined in a form that is easy to calculate. By Theorem 8.2, we could equivalently write

$$Cov(X,Y) = E[(X-\mu_X)(Y-\mu_Y)]. \tag{8.1}$$

Example 8.5

Visualising covariance

The covariance between discrete random variables X and Y is shown according to sign in the three diagrams below which chart the positions of five equally likely points (X,Y) of mass relative to axes positioned at the centroid (μ_X, μ_Y).

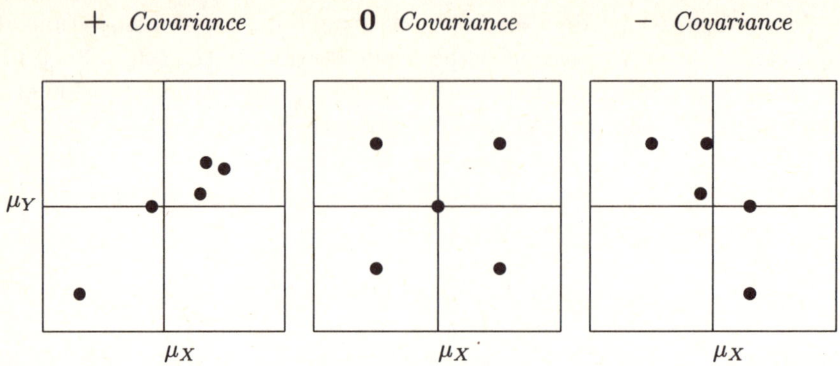

Example 8.6

Tracing the steps taken by a drunkard

Picture a lonely soul, struggling along Kings Cross after a long night of drinking. The size, Y, of the next step, is 1 if the drunkard staggers forward; 0 if there is no current move, and -1 if a step backwards is taken. Compare this with the size, X, of the *previous* move: record 1 if moved forward and 0 otherwise.

Suppose that the joint probability mass function $p_{X,Y}(x,y)$ is described by the table:

$p_{X,Y}(x,y)$ / x	y	-1	0	1	$p_X(x)$
0		$\frac{1}{4}$	0	$\frac{1}{4}$	$\frac{1}{2}$
1		$\frac{1}{8}$	$\frac{1}{4}$	$\frac{1}{8}$	$\frac{1}{2}$
	$p_Y(y)$	$\frac{3}{8}$	$\frac{1}{4}$	$\frac{3}{8}$	1

Worked example:

Determine the value of $Cov(X,Y)$.

It is convenient to begin by determining $E(X)$ and $E(Y)$:

$$E(X) = (0)\frac{1}{2} + (1)\frac{1}{2} = \frac{1}{2}; \quad E(Y) = (-1)\frac{3}{8} + (0)\frac{1}{4} + (1)\frac{3}{8} = 0.$$

Using these directly in the covariance formula we find:

$$\begin{aligned} Cov(X,Y) &= E(XY) - E(X)E(Y) \\ &= (0)(-1)\frac{1}{4} + (0)(0)0 + (0)(1)\frac{1}{4} + (1)(-1)\frac{1}{8} + (1)(0)\frac{1}{4} \\ &\quad +(1)(1)\frac{1}{8} - (\tfrac{1}{2})(0) \end{aligned}$$

$$= (0) - (\frac{1}{2})(0)$$
$$= 0$$

Therefore the drunkard's next move Y has no association with his previous move X. This has occurred because of the balance (symmetry) of the probabilities and achievable values for the random variables in the table. Indeed, if the symmetry of the table is altered by changing the largest value that Y may assume from 1 to 2, this results in the joint probability table for a new pair (U, V) of bivariate discrete random variables

$p_{U,V}(u,v)$ / u	v	-1	0	2	$p_U(u)$
0		$\frac{1}{4}$	0	$\frac{1}{4}$	$\frac{1}{2}$
1		$\frac{1}{8}$	$\frac{1}{4}$	$\frac{1}{8}$	$\frac{1}{2}$
	$p_V(v)$	$\frac{3}{8}$	$\frac{1}{4}$	$\frac{3}{8}$	1

It is easily verified that $Cov(U, V)$ is now given by

$$Cov(U, V) = E(UV) - E(U)E(V) = (\frac{1}{8}) - (\frac{1}{2})(\frac{3}{8}) = -\frac{1}{16}.$$

Notice that there is not a high probability of a step forward at the current stage in this random walk given that there was a step forward at the previous stage. It will be hard for the drunkard to make forward progress while the negative covariance is in action! An increase in the *size* of the next step forward could change that.

Uncorrelated random variables

When $Cov(X, Y) = 0$, the random variables X and Y are said to be **uncorrelated.** Notice that when X and Y are independent it follows that $E(XY) = E(X)E(Y)$ (this is Theorem 5.1), so that $Cov(X, Y) = 0$. It is important to remember that:

> If X and Y are independent then $Cov(X, Y) = 0$.

Being uncorrelated is a weaker requirement of random variables than their being independent. Example 8.6 exhibits a jointly distributed pair of random variables (X, Y) satisfying $Cov(X, Y) = 0$, but the joint probability table fails to factorise, for example, in the cell $(0, -1)$: $p_{X,Y}(0, -1) = \frac{1}{4} \neq (\frac{1}{2})(\frac{3}{8}) = p_X(0)p_Y(-1)$. Therefore:

If $Cov(X,Y) = 0$ then X and Y are not necessarily independent.

The next example demonstrates this result for a continuous pair of random variables.

Example 8.7
Zero covariance from normal moments
The central moments of the standard normal random variable were obtained in Example 4.9; in particular, the odd moments are zero.
Suppose that $Z \sim N(0,1)$. Then Z and Z^2 are dependent, by definition, since one variable is determined by the other. However, Z and Z^2 are uncorrelated:

$$Cov(Z, Z^2) = \underbrace{E(Z.Z^2)}_{E(Z^3)} - E(Z)E(Z^2) = (0) - (0)(1) = 0.$$

Properties of covariance

It is necessary to understand how the covariance between a pair of random variables will change as the random variables themselves change. For summary purposes we list the important results in the form of a theorem, but leave the pieces comprising the proof to be picked up in the exercises at the end of this section.

Theorem 8.3 *For given random variables X, Y, Z and constants a, b, c and d:*

1. $Cov(X,X) = Var(X)$;
2. $Cov(a,X) = 0$;
3. $Cov(X,Y) = Cov(Y,X)$;
4. $Cov(X+Y,Z) = Cov(X,Z) + Cov(Y,Z)$;
5. $Cov(a+bX, c+dY) = bdCov(X,Y)$.

■

Our main purpose for Theorem 8.3 is to provide building blocks for determining the variability of a general sum of random variables where the components in the sum are not necessarily independent. For example, if X_1

and X_2 are random variables and a_1 and a_2 are constants, a direct application of Theorem 8.3 gives

$$\begin{aligned} Var(a_1X_1 + a_2X_2) &= Cov(a_1X_1 + a_2X_2, a_1X_1 + a_2X_2) \\ &= a_1a_1Cov(X_1, X_1) + a_1a_2Cov(X_1, X_2) \\ &\quad +a_2a_1Cov(X_2, X_1) + a_2a_2Cov(X_2, X_2) \end{aligned}$$

by parts *4* and *5*. In order to detect a pattern emerging, it is convenient to write this variance as

$$Var(a_1X_1 + a_2X_2) = \begin{array}{rcr} a_1^2Var(X_1) & + & a_1a_2Cov(X_1, X_2) \\ + \quad a_2a_1Cov(X_2, X_1) & + & a_2^2Var(X_2), \end{array}$$

for then the result is simple to generalise inductively: the variance of the sum of 3 terms,

$$Var(a_1X_1 + a_2X_2 + a_3X_3),$$

is equal to the sum of $3^2 = 9$ terms

$$\begin{array}{rrrrr} & a_1^2Var(X_1) & + \quad a_1a_2Cov(X_1, X_2) & + & a_1a_3Cov(X_1, X_3) \\ + & a_2a_1Cov(X_2, X_1) & + \qquad\quad a_2^2Var(X_2) & + & a_2a_3Cov(X_2, X_3) \\ + & a_3a_1Cov(X_3, X_1) & + \quad a_3a_2Cov(X_3, X_2) & + & a_3^2Var(X_3). \end{array}$$

Following this pattern, even the monstrous $Var(a_1X_1 + a_2X_2 + \ldots + a_nX_n)$ may be written as the sum of n^2 terms:

$$\begin{array}{rrrrr} & a_1^2Var(X_1) & + \quad a_1a_2Cov(X_1, X_2) & +\ldots+ & a_1a_nCov(X_1, X_n) \\ + & a_2a_1Cov(X_2, X_1) & + \qquad\quad a_2^2Var(X_2) & +\ldots+ & a_2a_nCov(X_2, X_n) \\ + & a_3a_1Cov(X_3, X_1) & + \quad a_3a_2Cov(X_3, X_2) & +\ldots+ & a_3a_nCov(X_3, X_n) \\ \vdots & & \vdots & \ddots & \\ + & a_na_1Cov(X_n, X_1) & + \quad a_na_2Cov(X_n, X_2) & +\ldots+ & a_n^2Var(X_n). \end{array}$$

Notice that in this structure, the variances of the individual terms X_1, X_2, X_3 ... lie along the main diagonal of the square that the $n \times n = n^2$ terms neatly fit into. This square is essentially an array where the term in row i and column j is the term involving $Cov(X_i, X_j)$. This leaves all the covariance terms positioned away from the main diagonal. There is more: Since, by Theorem 8.3, $Cov(X_i, X_j) = Cov(X_j, X_i)$ for all rows i and columns j, the terms below the main diagonal are mirror images of the terms above the main diagonal.

Having seen this structure, we collapse it to a single line summary, to a simple record of the result referred to in this text as **SUM 3**; that is, the third important result relating to sums of random variables.

SUM 3:

$$Var(a_1X_1 + a_2X_2 + a_3X_3 + \ldots + a_nX_n)$$
$$= \sum_{i=1}^{n} a_i^2 Var(X_i) + 2\sum_{i<j} a_i a_j Cov(X_i, X_j). \quad (8.2)$$

SUM 3 is for determining the variance of a sum of random variables **when they are not independent**. This occurs, for example, when sampling from a finite population without replacement. We take up that discussion in the next chapter.

Example 8.8

Total income for a working couple

Suppose that the income of working couples is jointly distributed (in thousands of dollars) according to

$p_{X,Y}(x,y)$ / x	y	20	30	40	50	$p_X(x)$
20		0.00	0.04	0.01	0.00	0.05
30		0.00	0.36	0.09	0.00	0.45
40		0.20	0.05	0.10	0.00	0.35
50		0.10	0.00	0.00	0.05	0.15
	$p_Y(y)$	0.30	0.45	0.20	0.05	1.00

Here X denotes husband's income; Y denotes wife's income. If tax is levied at 20% of X and 30% of Y, we wish to determine the mean and variance of the total family income, $T = 0.8X + 0.7Y$, after taxes.
The means of X and Y are determined from the marginal distributions:

$$E(X) = (20)(0.05) + (30)(0.45) + (40)(0.35) + (50)(0.15) = 36.0;$$
$$E(Y) = (20)(0.30) + (30)(0.45) + (40)(0.20) + (50)(0.05) = 30.0.$$

From these, we may write directly by **SUM 1**:

$$E(T) = 0.8E(X) + 0.7E(Y) = (0.8)(36) + (0.7)(30) = 49.8.$$

This corresponds to a mean after-tax income of \$49 800. To determine the variance of T, we first need the variance of each of the random variables X and Y which are components in the total:

$$Var(X) = (20)^2(0.05) + (30)^2(0.45) + (40)^2(0.35) + (50)^2(0.15) - (36)^2$$
$$= 64.0;$$
$$Var(Y) = (20)^2(0.30) + (30)^2(0.45) + (40)^2(0.20) + (50)^2(0.05) - (30)^2$$
$$= 70.0.$$

These variances, along with the covariances

$$E(XY) = (20)(30)(0.04) + (20)(40)(0.01) + \ldots + (50)(50)(0.05) = 1069$$
$$Cov(X,Y) = E(XY) - E(X)E(Y) = 1069 - (30)(36) = -11,$$

may now be placed directly in **SUM 3**:

$$\begin{aligned} Var(T) &= Var(0.8X + 0.7Y) \\ &= (0.8)^2 Var(X) + (0.7)^2 Var(Y) + 2(0.7)(0.8)Cov(X,Y) \\ &= (0.8)^2(64) + (0.7)^2(70) + 2(0.7)(0.8)(-11) \\ &= 62.94 \end{aligned}$$

When there is independence between the random variables in the sum $a_1X_1 + a_2X_2 + a_3X_3 + \ldots + a_nX_n$, then in determining the variance of such a sum, all the off-diagonal terms in the array structure evaporate:

$$\begin{array}{cccccccc} \multicolumn{8}{l}{Var(a_1X_1 + a_2X_2 + a_3X_3 + \ldots + a_nX_n) =} \\ & a_1^2 Var(X_1) & + & 0 & + & 0 & +\ldots+ & 0 \\ + & 0 & + & a_2^2 Var(X_2) & + & 0 & +\ldots+ & 0 \\ + & 0 & + & 0 & + & a_3^2 Var(X_3) & +\ldots+ & 0 \\ \vdots & & \vdots & & \vdots & & \ddots & \\ + & 0 & + & 0 & + & 0 & +\ldots+ & a_n^2 Var(X_n). \end{array}$$

Therefore the case when $X_1, X_2, X_3, \ldots, X_n$ are independent reduces to **SUM 2** which we recognise as

$$Var(a_1X_1 + a_2X_2 + \ldots + a_nX_n) = a_1^2 Var(X_1) + a_2^2 Var(X_2) + \ldots + a_n^2 Var(X_n).$$

The array of covariance terms that we have studied in the above formulae has a special name.

Definition 8.6 *The* **variance-covariance matrix** *for random variables $X_1, X_2, X_3, \ldots, X_n$, is the array of terms*

$$\begin{pmatrix} Var(X_1) & Cov(X_1,X_2) & Cov(X_1,X_3) & \ldots & Cov(X_1,X_n) \\ Cov(X_2,X_1) & Var(X_2) & Cov(X_2,X_3) & \ldots & Cov(X_2,X_n) \\ Cov(X_3,X_1) & Cov(X_3,X_2) & Var(X_3) & \ldots & Cov(X_3,X_n) \\ \vdots & & \vdots & \ddots & \vdots \\ Cov(X_n,X_1) & Cov(X_n,X_2) & Cov(X_n,X_3) & \ldots & Var(X_n) \end{pmatrix}$$

where $Cov(X_i, X_j)$ is the term in row i and column j, $i = 1,2,3,\ldots,n$, $j = 1,2,3,\ldots,n$.

□

To understand this structure, it is useful to try and count how many terms occupy the different locations in the array:

n	terms down the main diagonal (the variance terms)
$\frac{1}{2}(n^2 - n)$	terms above the main diagonal (covariance terms)
$\frac{1}{2}(n^2 - n)$	terms below the main diagonal (covariance terms)
n^2	total number of terms

Note that this array of n^2 terms is symmetric about the main diagonal. Further,

> $Var(X_1 + X_2 + X_3 + \ldots + X_n)$ is the sum of all terms in the variance-covariance matrix.

Example 8.9

Using the variance-covariance matrix for working couples: Example 8.8 continued

A joint distribution of the income (X, Y) of husband and wife was suggested in Example 8.8. In this example, $Var(X)$ and $Var(Y)$ were determined, and also the covariance between X and Y. The variance-covariance matrix then becomes

$$\begin{pmatrix} Var(X) & Cov(X,Y) \\ Cov(X,Y) & Var(Y) \end{pmatrix} = \begin{pmatrix} 64 & -11 \\ -11 & 70 \end{pmatrix}$$

Then, $Var(X+Y)$, the variance of a couple's total income before taxes is the sum of the four terms in the variance-covariance matrix: $Var(X+Y) = 64 + (-11) + (-11) + 70 = 112$. Notice that this variance is almost twice as large as the variance in after-tax income given in Example 8.8.

The final covariance result of this section follows by repeated application of *4.* and *5.* of Theorem 8.3.

Theorem 8.4 *Given random variables $X_1, X_2, \ldots, X_n$ and $Y_1, Y_2, \ldots, Y_n$, and constants $a_1, a_2, a_3, \ldots, a_n$ and $b_1, b_2, b_3, \ldots, b_n$, the covariance between the linear combinations $\sum_{i=1}^n a_i X_i$ and $\sum_{j=1}^n b_j X_j$ may be expressed as*

$$Cov(\sum_{i=1}^n a_i X_i, \sum_{j=1}^n b_j X_j) = \sum_{i=1}^n \sum_{j=1}^n a_i b_j Cov(X_i, X_j).$$

■

We will see an example of such covariance calculations in the next section.

The great multinomial experiment

Suppose that two bins stand side-by-side at the front of the room. One bin is labelled BIN 1 'success', and the other bin is labelled BIN 2 'failure'. By standing at the back of the room, a sequence of Bernoulli trials is performed: n balls are thrown at the BINS, one at a time and independently, so that all balls land in a BIN. The result, U_i, of trial number i is:

$$U_i = \begin{cases} 1 & \text{if ball number } i \text{ lands in the success BIN 1;} \\ 0 & \text{if ball number } i \text{ lands in the failure BIN 2.} \end{cases}$$

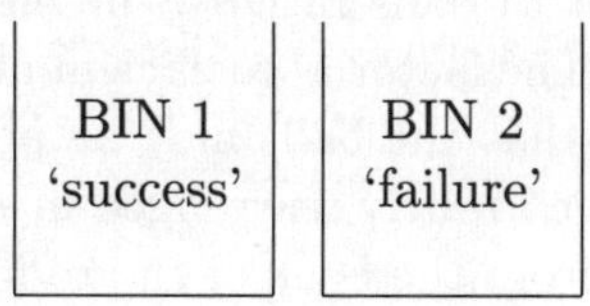

Then

$$X = U_1 + U_2 + U_3 + \ldots + U_n \sim Bin(n, p),$$

where $p = P(\text{ball lands in success BIN})$, gives the number of 'successes' in the n Bernoulli trials; that is, the number of balls that land in the 'success' BIN. Therefore the Binomial result is obtained essentially by throwing the balls one at a time.

This situation is obviously simply generalised: put k BINS at the front of the room and fire away. Record the number of balls, X_i, landing in BIN number i.

If we label $p_i = P(\text{ball lands in BIN } i)$, then, under the rules of no spillage, so that all the balls land in *some* BIN, it follows that

$$X_1 + X_2 + X_3 + \ldots + X_k = n,$$

the total number of balls to be allocated (thrown). For the same reason,

$$p_1 + p_2 + p_3 + \ldots + p_k = 1.$$

The same type of thinking that gave the binomial probability mass function is applied again:

$$P(X_1 = x_1, X_2 = x_2, \ldots, X_k = x_k) = \frac{n!}{x_1! x_2! \ldots x_k!} p_1^{x_1} p_2^{x_2} \ldots p_k^{x_k}.$$

represents the probability that x_1 balls land in BIN 1, x_2 balls land in BIN 2, $\ldots x_{k-1}$ balls land in BIN $(k-1)$ and $x_k = n - (x_1 + x_2 + \ldots + x_{k-1})$ balls land in BIN k.

Definition 8.7 *The collection $X_1, X_2, X_3, \ldots, X_k$ of discrete random variables has a* **multinomial distribution** *with parameters n and $p_1, p_2, \ldots, p_{k-1}$, and we write $(X_1, X_2, X_3, \ldots, X_k) \sim Multinomial(n, p_1, p_2, p_3, \ldots, p_{k-1})$, if the joint probability mass function of $X_1, X_2, X_3, \ldots, X_k$ may be written in the form*

$$p_{X_1,X_2,X_3,\ldots,X_k}(x_1, x_2, x_3, \ldots, x_k) = \frac{n!}{x_1!x_2!\ldots x_k!} p_1^{x_1} p_2^{x_2} \ldots p_k^{x_k},$$

where $\sum_{i=1}^{k} x_i = n$ and $\sum_{i=1}^{k} p_i = 1$.

□

Clearly the arrangement of balls in BINS in the front of the room has a multinomial distribution. The factorial coefficient at the front of the joint pmf gives the number of ways that the balls may be permuted in the sense that the x_i balls in BIN i may be linearly rearranged in $x_i!$ ways without changing the BIN contents as seen from the back of the room. $(i = 1, 2, 3, \ldots, k.)$

Example 8.10

Inspection plan: classifications of 'defective', 'non-defective' and 're-inspect'

Computer keyboards are randomly selected from the output of a production line to check on quality. This sampling inspection plan tests 10 keyboards. The acceptance (for non-defectives) and rejection (for defectives) rates have been monitored previously at the levels:

$$\begin{aligned} \text{P(non-defective)} &= 0.75 \\ \text{P(defective)} &= 0.15 \\ \text{P(sent for re-inspection)} &= 0.10 \end{aligned}$$

so that one in 10 keyboards are sent for re-inspection and further testing.

Worked example:

Determine the probability that of 10 keyboards selected, 6 are non-defective, 2 are defective and 2 are sent for re-inspection.

Let X_1, X_2 and X_3 denote the numbers in respective categories: non-defective/defective/re-inspect. Then $X_1 + X_2 + X_3 = 10$, and substitution into the point mass function for the multinomial distribution with three categories (BINS) gives

$$p_{X_1,X_2,X_3}(6, 2, 2) = \frac{10!}{6!2!2!}(0.75)^6(0.15)^2(0.10)^2 = 0.050 \quad (3\text{ dp}).$$

Multinomials are marginally binomial

Suppose that we concentrate on BIN i. As we throw the balls at the lineup of BINS, we will have successes and failures with this particular BIN. As before, define Bernoulli random variables

$$U_r = \begin{cases} 1 & \text{if ball number r lands in BIN } i; \\ 0 & \text{if ball number r lands in any BIN other than i,} \end{cases}$$

for each $r = 1, 2, \ldots, n$. Then

$$X_i = U_1 + U_2 + U_3 + \ldots + U_n \sim Bin(n, p_i),$$

is the marginal distribution of X_i. Therefore $E(X_i) = np_i$ and $Var(X_i) = np_i(1 - p_i)$.

The same reasoning allows us to determine the covariance between the contents of two different BINS, say BIN i and BIN j. To be able to tell when we have successes with getting balls in BIN j, define more Bernoulli random variables. For each $s = 1, 2, \ldots, n$,

$$V_s = \begin{cases} 1 & \text{if ball number s lands in BIN } j; \\ 0 & \text{if ball number s lands in any BIN other than } j. \end{cases}$$

Then $X_j = V_1 + V_2 + V_3 + \ldots + V_n \sim Bin(n, p_j)$, is the marginal distribution of X_j. The Bernoulli trials are independent, and importantly, $U_r V_r = 0$ for all values of r, since a ball cannot be in two BINS at once. Therefore,

$$Cov(U_r, V_r) = E(U_r V_r) - E(U_r)E(V_r) = 0 - p_i p_j = -p_i p_j.$$

These terms now combine to give

$$\begin{aligned} Cov(X_i, X_j) &= Cov(U_1 + U_2 + U_3 + \ldots + U_n,\ V_1 + V_2 + V_3 + \ldots + V_n) \\ &= \sum_{r=1}^{n} Cov(U_r, V_r) + \sum_{r \neq s} Cov(U_r, V_s) \quad \text{(Theorem 8.4)} \\ &= \sum_{r=1}^{n} (-p_i p_j) + \sum_{r \neq s} 0 \\ &= -np_i p_j. \end{aligned}$$

What does the minus sign mean for this covariance expression? If there are more balls thrown in one BIN, then there are less balls for the other BINS. In this sense X_i is acting against X_j when $i \neq j$.

Example 8.11

Example 8.10 continued: probability keyboard re-inspected

For the keyboard sampling inspection plan of Example 8.10, X_1, X_2 and X_3 denote the numbers of sampled keyboards in respective categories: non-defective/defective/re-inspect.

Worked example:

1. **Determine the probability that no keyboard is sent for re-inspection.**
2. **Determine the covariance between the number of keyboards that are re-inspected and the number of keyboards that are not.**

Since the joint distribution of X_1, X_2, X_3 is trinomial, it follows that marginally, X_3 is binomial; $X_3 \sim Bin(10, 0.1)$. We are required to find

$$P(X_3 = 0) = \binom{10}{0}(0.1)^0(0.9)^{10-0} = 0.3487 \quad (4 \text{ dp}).$$

The second of the questions posed requires the covariance between the total number accepted/rejected, $X_1 + X_2$ and the number X_3 that are re-inspected.

$$\begin{aligned} Cov(X_1 + X_2, X_3) &= Cov(X_1, X_3) + Cov(X_2, X_3) \\ &= (-10)(0.75)(0.10) + (-10)(0.15)(0.10) \\ &= -0.9. \end{aligned}$$

EXERCISES

8.17 If random variables X and Y have the following joint probability distribution, determine the value of $Cov(X, Y)$.

$p_{X,Y}(x,y)$ x	y	-2	0	1
0		$\frac{2}{9}$	0	$\frac{2}{9}$
1		$\frac{1}{9}$	$\frac{1}{3}$	$\frac{1}{9}$

8.18 Consider the joint probability distribution in Exercise 8.17. Is the value of $Cov(X, Y)$ automatically

(a) increased by 10 if the constant 10 is added to all the X values?

(b) increased by 10 if the constant 10 is added to all the Y values?

(c) increased by 10 if the constant 10 is added to all the X values and all the Y values?

(d) multiplied by 10 if all the X values are multiplied by 10?

(e) multiplied by 100 if all the X values are multiplied by 10 and all the Y values are multiplied by 10?

8.19 Suppose that a **signal** consists of a series of pulses of magnitude X; **noise** is a series of pulses of magnitude Y. Suppose that X and Y are jointly distributed according to the table

$p_{X,Y}(x,y)$ x	y -2	0	2
-1	$\frac{1}{3}$	0	0
0	$\frac{1}{6}$	$\frac{1}{6}$	0
1	$\frac{1}{6}$	0	$\frac{1}{6}$

From the theory of the superposition of pulses, if noise and signal are mixed with the pulses synchronised, the magnitude, M, of the resulting pulses is the sum of the noise and signal magnitudes. That is, $M = X+Y$.

Determine the mean and variance of M.

8.20 On a 500-seat aircraft each seat is booked for a particular flight with probability 0.9. Assume that the bookings occur independently. Determine the covariance between the number of seats that are occupied and the number of seats remaining empty.

Explain why one would expect this covariance to be negative.

8.21 Suppose that $X \sim Bin(n,p)$.

Determine $Cov(X,\ n - X)$.

8.22 If X and Y are independent and identically distributed random variables, show that $X + Y$ and $X - Y$ are uncorrelated.

8.23 Show that $Var(X - Y)$

$= Var(X) + Var(Y) - 2Cov(X,Y)$.

by first simplifying the left-side as

$$Cov(X - Y, X - Y).$$

Explain how the result is really a special case of **SUM 3**.

8.24 Assume that the weight, X, in kg, of a randomly chosen man is normally distributed with a mean of 73 kg and a standard deviation of 8 kg. Women, on the other hand, weigh on average 5 kg less than men and have 25% less variance in their weight than men.

(a) Determine the probability distribution for Y, the weight of a randomly chosen woman.

(b) If a man and a woman are randomly chosen, determine the mean and variance of the excess of the man's weight over the woman's weight.

(c) Suppose that the weights, (X,Y), of a randomly chosen married couple are jointly distributed with covariance $Cov(X,Y) = 6$. Use this information to determine the mean and variance of the excess of a husband's weight over his wife's weight for a randomly chosen married couple.

Write down the variance-covariance matrix of (X,Y).

8.25 If X is a random variable and c is a constant, show that $Cov(c,X) = 0$.

8.26 In Example 8.8, a married couple's pre-tax income has a joint distribution (X,Y), where X denotes husband's income and Y denotes wife's income. Use the joint probability mass function given in this example to determine the mean and variance of the total tax $T = 0.2X+0.3Y$ that the couple pays (to the government).

Would you expect T to be positively correlated with the couple's total pre-tax income? Explain.

8.27 Suppose that in the manufacture of moulded plastic toys, the preparation time X, and the drying time in the mould, Y, had the following probability distributions of 6 minute time units:

x	1	2	3	4	5
$p_X(x)$	0	0.4	0.2	0	0.4

y	1	2	3	4	5
$p_Y(y)$	0.4	0.4	0	0	0.2

Two questions were of interest to the management:

- the total time, T, of both phases of manufacture;
- the cost C of preparation time, at rates of \$200 per hour plus a fixed cost of \$300.

Suppose that X and Y are correlated with

$$Cov(X,Y) = -0.6.$$

(a) Explain the practical significance of the fact that $Cov(X,Y) = -0.6$.

(b) State the value of $Cov(Y,X)$.

(c) How is the covariance affected if the 'preparation time' is doubled and the 'drying time' is halved?

(d) Complete the following table:

Variable	Expected Value	Variance
X		
Y		
T		
C		

8.28 Suppose that the distance X that a seed is broadcast from the parent plant is assumed to have pdf

$$f_X(x) = \begin{cases} \frac{k}{x^4} & \text{if } x \geq 1; \\ 0 & \text{otherwise,} \end{cases}$$

where k is a constant.

(a) Determine the value of k.

(b) Suppose that a new random variable Y representing the possibility that a seed broadcast X metres from the parent plant fails to germinate. When the parent plant is in an isolated position, the seed has more chance of germinating when X is large. If

$$Y = \frac{1}{X},$$

determine the cumulative distribution function and the probability density function for Y.

Verify that $E(XY) \neq E(X)E(Y)$ for these definitions of X and Y. Is this a result that you could have predicted in advance?

Determine $Cov(X,Y)$.

8.29 The price movements of three stocks in the stock market, Stock 1—banking, Stock 2—property and Stock 3—manufacturing, do not occur independently. The gains (\$) from \$100 invested in each of the stocks on a 1-year holding are denoted by X_1, X_2 and X_3. Assume that the variance-covariance matrix for the three stocks is given by

$Cov(X_i, X_j)$		
250	200	150
200	400	−50
150	−50	500

(a) Based on the information in the covariance matrix, explain whether Stock 3 is likely to show strong gains when Stock 2 shows strong gains.

(b) Based on the information in the covariance matrix, explain whether Stock 3 is likely to show strong gains when Stock 1 shows strong gains.

(c) Determine the standard deviation in total gain from a \$100 investment in each of the three stocks.

(d) Determine the standard deviation in total gain from a \$200 investment in Stock 1 and \$100 investment in each of Stocks 2 and 3.

8.30 For given random variables X, Y, Z and constants a, b, c and d, show that:

(a) $Cov(X, X) = Var(X)$;

(b) $Cov(X, Y) = Cov(Y, X)$;

(c) $Cov(X + Y, Z) = Cov(X, Z) + Cov(Y, Z)$;

(d) $Cov(a + bX, c + dY) = bdCov(X, Y)$.

Show how to implement these results in Exercise 8.27 for the determination of $Cov(C, T)$, the covariance between 'cost' and 'total time'.

8.31 Let X_1 and X_2 be two random variables with finite means and positive variances:

$$E(X_i) = \mu_i;\ Var(X_i) = \sigma_i^2,\ i = 1, 2.$$

(a) Show that $X_2 - \frac{Cov(X_1,X_2)}{\sigma_1^2}X_1$ and X_1 are uncorrelated.

(b) Determine the constant c so that $X_1 - cX_2$ and X_2 are uncorrelated.

(c) Determine the mean and variance of

$$X_2 - \frac{Cov(X_1, X_2)}{\sigma_1^2}X_1.$$

8.32 A fair die is tossed 10 times.

(a) Explain how the outcomes may be modelled with a multinomial distribution. State the parameters of the distribution being modelled.

(b) Determine the probability that 5 ones and 5 sixes are obtained.

(c) In 10 tosses, if X_1 denotes the number of ones and X_6 denotes the number of sixes, determine $Cov(X_1, X_6)$.

8.33 How do the results of Exercise 8.32 change if the die is loaded so that a six is twice as likely to occur as any other face?

8.34 The joint pmf of a discrete bivariate random variable (X, Y) is given by

$p_{X,Y}(x, y)$	y	0	1	
x				$p_X(x)$
0		p_{00}	p_{01}	$p_{0.}$
1		p_{10}	p_{11}	$p_{1.}$
	$p_Y(y)$	$p_{.0}$	$p_{.1}$	1

This joint table uses the '.' notation to indicate that a variable has been summed over.

(a) Find an expression for $Cov(X,Y)$.

(b) Show that $Cov(X,Y) = 0$ if and only if $p_{11}p_{00} = p_{10}p_{01}$.

(c) Prove that

$$p_{1.}p_{.1} = p_{11} + (p_{10}p_{01} - p_{11}p_{00}).$$

Write down similar expressions for

$$p_{1.}p_{.0},\ p_{0.}p_{.1},\ p_{0.}p_{.0}.$$

(d) Hence show that X and Y are independent if $Cov(X,Y) = 0$.

8.3 CORRELATION AND THE BIVARIATE NORMAL

When a pair of random variables has non-zero covariance, it is important to know whether this covariance is positive or negative; the magnitude or size of the covariance between a pair of random variables conveys little. It is not possible to say whether a given numerical value for covariance is 'big' or whether it is 'small'. To introduce a sense of scale beyond '+', '0' and '−' for covariance, we must assess when two covariances are alike. This is done through correlation.

Definition 8.8 *The* **correlation coefficient**, $\rho_{X,Y}$, *between two random variables* X *and* Y *with variances satisfying* $0 < Var(X) < \infty$ *and* $0 < Var(Y) < \infty$, *is given by*

$$\rho_{X,Y} = \frac{Cov(X,Y)}{\sqrt{Var(X)}\sqrt{Var(Y)}}.$$

□

Example 8.12

Checking out at the supermarket

A supermarket has four check-out counters arranged in a line along the front of the store. Studies have shown that customers are twice as likely to use either of the two centre check-outs as either of the end check-outs.

Worked example:

Determine the correlation between the number of customers using a high-volume check-out counter and the number of customers opting for the low-volume counter at the end of the line.

If n customers finish their shopping and proceed to the check-out area, then with probability p_i a customer will choose check-out Counter i, where

$$p_1 = \frac{1}{6},\ p_2 = \frac{2}{6},\ p_3 = \frac{2}{6},\ p_4 = \frac{1}{6},$$

so that Counters 2 and 3 are high-volume counters and Counters 1 and 4 are low-volume counters. If we let X_i represent the number of customers choosing check-out Counter i, then

$$(X_1, X_2, X_3, X_4) \sim Multinomial(n, \frac{1}{6}, \frac{2}{6}, \frac{2}{6}).$$

By the results in Section 8.2,

$$Var(X_1) = n(\frac{1}{6})(\frac{5}{6}), Var(X_2) = n(\frac{2}{6})(\frac{4}{6}), \text{and} Cov(X_1, X_2) = -n(\frac{1}{6})(\frac{2}{6}).$$

Then, according to Definition 8.5,

$$\rho_{X_1,X_2} = \frac{Cov(X_1, X_2)}{\sqrt{Var(X_1)}\sqrt{Var(X_2)}} = \frac{-n(\frac{1}{6})(\frac{2}{6})}{\sqrt{n(\frac{1}{6})(\frac{5}{6})}\sqrt{n(\frac{2}{6})(\frac{4}{6})}} = -\frac{1}{\sqrt{10}}.$$

Notice that the correlation coefficient does not depend on the sample size n in this example. The negative sign, which occurs because the covariance is negative, means that with a fixed number of customers, if more choose check-out Counter 1, then relatively less will choose check-out Counter 2.

The effect on correlation of a change of scale

We focus in this section on the very important result that the square of a correlation coefficient is **scale invariant**. In particular, when variables are standardised, their correlation does not change.

Theorem 8.5 *If two random variables X and Y have a correlation $\rho_{X,Y}$, then the standardised random variables $\frac{X-\mu_X}{\sigma_X}$ and $\frac{Y-\mu_Y}{\sigma_Y}$ also have correlation $\rho_{X,Y}$.*

Proof: The standardised variables $Z_X = \frac{X-\mu_X}{\sigma_X}$ and $Z_Y = \frac{Y-\mu_Y}{\sigma_Y}$ satisfy $Var(Z_X) = 1$ and $Var(Z_Y) = 1$ by definition. Further,

$$Cov(Z_X, Z_Y) = Cov(\frac{X-\mu_X}{\sigma_X}, \frac{Y-\mu_Y}{\sigma_Y}) = \frac{1}{\sigma_X\sigma_Y} Cov(X, Y) = \rho_{X,Y},$$

by Theorem 8.3. Therefore

$$\rho_{Z_X,Z_Y} = \frac{Cov(Z_X, Z_Y)}{\sqrt{Var(Z_X)}\sqrt{Var(Z_Y)}} = \frac{\rho_{X,Y}}{(1)(1)} = \rho_{X,Y}.$$

■

Theorem 8.5 is a particular instance of scale invariance. The same type of result holds for random variables X and Y each subjected to a general linear transformation. However, there may be a reversal of sign in the transformation process.

Since $\sqrt{Var(a+bX)} = \sqrt{b^2Var(X)} = |b|\sqrt{Var(X)}$, and $Cov(a+bX, c+dY) = bdCov(X,Y)$, it follows that if random variables X and Y have correlation $\rho_{X,Y}$, then the random variables $a+bX$ and $c+dY$ have correlation

$$\rho_{a+bX,c+dY} = \frac{bd}{|bd|}\rho_{X,Y}.$$

Consequently, if the product bd is positive, then $\rho_{a+bX,c+dY} = \rho_{X,Y}$. In general, however, $\frac{bd}{|bd|} = \pm 1$, so that $\rho_{a+bX,c+dY} = \pm\rho_{X,Y}$ and the sign may be eliminated by squaring each side:

For any constants a, b, c, and d, $\rho^2_{a+bX,c+dY} = \rho^2_{X,Y}$.

How large are correlations?

In this section we show that correlations always lie between $+1$ and -1. This creates a scale by which we can assess the degree of association of a pair of variables. Variables X and Y which have $\rho_{X,Y}$: near $+1$ or -1 are said to be strongly correlated; near 0 are said to be weakly correlated.

Theorem 8.6 *Suppose that X and Y are random variables with finite positive variances σ^2_X and σ^2_Y respectively. Then:*

1. $-1 \le \rho_{X,Y} \le 1$;
2. $|\rho_{X,Y}| = 1$ *if and only if there are constants a and b such that* $Y = a + bX$.

Proof: The proof of this result depends on the idea that if W is a random variable, then

$$Var(W) \ge 0 \text{ and } Var(W) = 0 \text{ if and only if } W \text{ is constant}$$

(with probability 1) . Now let $W = Z_X \pm Z_Y$, where Z_X and Z_Y are the standardised random variables $Z_X = \frac{X-\mu_X}{\sigma_X}$ and $Z_Y = \frac{Y-\mu_Y}{\sigma_Y}$. Then, by **SUM 3**,

$$\begin{aligned} 0 &\le Var(W) \\ &= Var(Z_X \pm Z_Y) \\ &= Var(Z_X) + Var(Z_Y) \pm 2Cov(Z_X,\ Z_Y) \\ &= 1 + 1 \pm 2\rho_{X,Y}. \end{aligned}$$

By using the '+' sign we obtain $\rho_{X,Y} \geq -1$; the '$-$' sign shows that $\rho_{X,Y} \leq 1$. Taken together, this establishes that $|\rho_{X,Y}| \leq 1$. Further, $|\rho_{X,Y}| = 1$ when $Var(W) = 0$, which, in turn means that W is constant. For our choice of W, this condition ensures that

$$\frac{X - \mu_X}{\sigma_X} \pm \frac{Y - \mu_Y}{\sigma_Y} = (\text{constant}).$$

That is, for constants a and b, $Y = a + bX$.

■

Recall that when two random variables X and Y are independent, they are **uncorrelated**. This means that $Cov(X, Y) = 0$, and therefore by definition, $\rho_{X,Y} = 0$.

If random variables X and Y are independent, then $\rho_{X,Y} = 0$.

When $\rho_{X,Y} > 0$, the random variables X and Y are said to be **positively correlated**; similarly $\rho_{X,Y} < 0$ implies that X and Y are **negatively correlated**.

The bivariate normal distribution

The joint distribution of a pair of (possibly) correlated normal random variables is said to have a bivariate normal distribution. Such a distribution is simply constructed from independent standard normal distributions. There are two purposes to the presentation of the building of the bivariate normal from basic $N(0, 1)$ components: the building process shows clearly the role of correlation in the final joint distribution, and the final structure, once built, is a suitable model for the method of linear regression presented in Chapter 14.

Suppose that random variables U and V are **independent** $N(0, 1)$ random variables. Let ρ be a constant with $|\rho| < 1$ and, in terms of ρ, U and V, construct a new random variable

$$Z = \rho U + \sqrt{1 - \rho^2} V.$$

Then, since linear combinations of independent normally random variables are again normally distributed, with a mean given by **SUM 1** and a variance by **SUM 2**, such reasoning gives $Z \sim N(0, 1)$. In particular, our construction has ensured that

$$\begin{aligned} &Var(U) = 1, Var(Z) = 1, \\ &\text{and } \; Cov(U, Z) = Cov(U, \rho U + \sqrt{1 - \rho^2} V) = \rho Cov(U, U) = \rho. \end{aligned}$$

Placing these in a calculation for correlation gives

$$\rho_{U,Z} = \frac{Cov(U, Z)}{\sqrt{Var(U)}\sqrt{Var(Z)}} = \frac{\rho}{(1)(1)} = \rho.$$

That is, the random variables U and Z have correlation ρ. Utilising the distribution of this pair of random variables still further, it is clear that when U is conditionally fixed, Z is simply a linear transformation of the $N(0,1)$ random variable V. Therefore,

$$f_{Z|U}(z|u) = \frac{1}{\sqrt{2\pi}\sqrt{1-\rho^2}} e^{-\frac{1}{2}(\frac{z-\rho u}{\sqrt{1-\rho^2}})^2}$$

indicates that the conditional distribution of Z given $U = u$ is $N(\rho u, 1-\rho^2)$. The main goal is to determine the joint density of Z and U.

$$\begin{aligned} f_{Z,U}(z,u) &= f_{Z|U}(z|u) f_U(u) \\ &= \frac{1}{\sqrt{2\pi}\sqrt{1-\rho^2}} e^{-\frac{1}{2}(\frac{z-\rho u}{\sqrt{1-\rho^2}})^2} \frac{1}{\sqrt{2\pi}} e^{-\frac{1}{2}u^2} \\ &= \frac{1}{2\pi\sqrt{1-\rho^2}} e^{-\frac{1}{2}(\frac{z^2-2\rho uz+u^2}{1-\rho^2})} \end{aligned}$$

Definition 8.9 (Z,U) *has a* **standard bivariate normal distribution** *and we write* $(Z,U) \sim N(0,0,1,1,\rho)$ *if, for* $|\rho| < 1$, *the joint pdf of* (Z,U) *is*
$f_{Z,U}(z,u) = \frac{1}{2\pi\sqrt{1-\rho^2}} e^{-\frac{1}{2}(\frac{z^2-2\rho uz+u^2}{1-\rho^2})} \quad -\infty < z < \infty, \; -\infty < u < \infty.$ □

By the construction, it follows that if $(Z,U) \sim N(0,0,1,1,\rho)$, then

$$\rho_{Z,U} = \rho, \quad Z \sim N(0,1), \quad U \sim N(0,1);$$

that is, if (Z,U) have a standard bivariate normal distribution, then Z and U are correlated, and marginally standard normal. Definition 8.9 simplifies dramatically in the case $\rho = 0$. For then $f_{Z,U}(z,u)$ factorises as a product of standard normal densities; this is precisely the criterion for Z and U to be independent. This result is summarised for emphasis:

$$f_{Z,U}(z,u) = \frac{1}{2\pi} e^{-\frac{1}{2}(z^2+u^2)} = \frac{1}{\sqrt{2\pi}} e^{-\frac{1}{2}z^2} \frac{1}{\sqrt{2\pi}} e^{-\frac{1}{2}u^2}$$

for $-\infty < z < \infty, -\infty < u < \infty$.

Bivariate normal regression function of Y on X

We may scale the standard bivariate normal distribution to an arbitrary 5-parameter distribution as follows: replace Z and U by standardised versions of $X \sim N(\mu_X, \sigma_X^2)$ and $Y \sim N(\mu_Y, \sigma_Y^2)$.

$$(X, Y) \sim N(\mu_X, \mu_Y, \sigma_X^2, \sigma_Y^2, \rho_{X,Y}) \text{ if } (\frac{X - \mu_X}{\sigma_X}, \frac{Y - \mu_Y}{\sigma_Y}) \sim N(0, 0, 1, 1, \rho)$$

Put in this form, the 5 parameters in the bivariate normal distribution are:

$$\mu_X,\ \mu_Y,\ \sigma_X^2,\ \sigma_Y^2,\ \rho_{X,Y}$$

These enter the regression function through $E(Z|U = u) = \rho u$, and this may be rewritten in terms of $U = \frac{X-\mu_X}{\sigma_X}$ and $Z = \frac{Y-\mu_Y}{\sigma_Y}$ as

$$\mu(x) = E(\frac{Y - \mu_Y}{\sigma_Y}|X = x) = \rho(\frac{x - \mu_X}{\sigma_X})$$

By relegating the parameters μ_Y and σ_Y to the right-side of this equation, the most useful form of the regression function is obtained:

$$\mu(x) = E(Y|X = x) = \mu_Y + \sigma_Y \rho(\frac{x - \mu_X}{\sigma_X}).$$

This regression function is noticeably a straight line. Regrouping coefficients gives

$$\begin{aligned} \mu(x) &= \mu_Y + \sigma_Y \rho(\frac{x - \mu_X}{\sigma_X}) \\ &= [\mu_Y - \rho(\frac{\sigma_Y}{\sigma_X})\mu_X] + [\rho\frac{\sigma_Y}{\sigma_X}]x \\ &= \alpha + \beta x, \end{aligned}$$

where $\beta = \rho\frac{\sigma_X}{\sigma_Y}$ is the slope of the regression line and $\alpha = \mu_Y - \beta\mu_X$ is the intercept of the regression line. We will examine how these quantities are estimated from sample data in Chapter 14.

Correlation measuring linear association

When $\rho_{X,Y} = 0$, it does not mean that there is *no association* between the random variables X and Y. It means that there is no **linear** relationship between the variables.

Obviously, since $\beta = \rho\frac{\sigma_X}{\sigma_Y}$ is the slope of the regression line, then if $\rho = 0$ it follows immediately that $\beta = 0$; the straight line representing the regression function has zero slope and therefore there is no linear relationship between X and Y.

Example 8.13

Functional relationships and correlation

Suppose that $X \sim N(0,1)$ and define two random variables $Y_1 = a + bX$ and $Y_2 = X^2$, where a and b are constants. These equations show respective linear and quadratic relationships between random variables.

$$\rho_{X,Y_1} = 1 \text{ by Theorem 8.6;} \quad \rho_{X,Y_2} = 0 \text{ by Example 8.7.}$$

This result indicates that if variables show zero correlation, there is no linear relationship between them; however, the possibility of a different functional relationship existing between the variables should be investigated. This could be done by plotting the outcomes of one variable against the outcomes of the other.

EXERCISES

8.35 Determine the correlation coefficient between X denoting husband's income and Y denoting wife's income by using the bivariate distribution for (X,Y) in Example 8.8.

8.36 The following data (Lapin, 1990) apply to fatigue tests to be run on bronze strips. (X,Y) represents cycles to failure ($\times 10^5$) when alternate strips are bent at respective (*high*, *low*) levels of deflection.

$p_{X,Y}(x,y)$ y / x	20	30	40	50
4	0.01	0.03	0.05	0.02
5	0.03	0.10	0.08	0.04
6	0.02	0.08	0.12	0.11
7	0.02	0.04	0.07	0.18

(a) Explain why the random variable Y is likely to assume large values compared with the random variable X.

(b) How can you be sure that the random variables X and Y are not independent?

(c) Are large values of X more likely to accompany large values of Y than small values of Y?

(d) Would you expect X and Y to be positively correlated?

(e) Determine $\rho_{X,Y}$.

8.37 A box contains 10 fuses of which 2 are defective. Two fuses are randomly selected from the box; the first fuse is not replaced before the second fuse is selected.

Let random variables X_1 and X_2 be defined as follows: for each $i = 1, 2$, $X_i = 1$ if the ith selected fuse is defective; $X_i = 0$ otherwise.

Determine ρ_{X_1,X_2}.

8.38 A fair die is tossed. Construct the discrete bivariate mass function of (X, Y), where

X denotes the number facing uppermost;

$$Y = \begin{cases} X & \text{if } X \leq 3; \\ 6 & \text{if } X > 3. \end{cases}$$

Hence determine $\rho_{X,Y}$.

8.39 The pre-diet weight X of individuals enrolling at a health clinic has mean 85 kg and variance 10.0 (kg)^2. The clinic claims that the post-diet weight Y of these same individuals has mean 80 kg and variance 8.1 (kg)^2.

(a) Determine the mean and variance in the improvement in weight $T = X - Y$ if $\rho_{X,Y} = -0.3$.

(b) Are the weight improvement T and the pre-diet weight X uncorrelated?

(c) Calculate $\rho_{X,T}$.

8.40 Let X denote the minimum and Y denote the maximum of the scores showing when two fair dice are tossed. Determine $\rho_{X,Y}$.

8.41 Random variables X and Y are normally distributed with mean 1, variance 2 and correlation 0.75.

(a) Determine $Cov(X, Y)$.

(b) Determine $Cov(\frac{X-1}{\sqrt{2}}, \frac{Y-1}{\sqrt{2}})$.

(c) Determine the mean and variance of $3X + 4Y$.

8.42 Suppose that the outcomes (x, y) of the bivariate random variable (X, Y) lie on the straight line $y = \alpha + \beta x$. Show that

$$\rho_{X,Y} = \begin{cases} 1 & \text{if } \beta > 0; \\ -1 & \text{if } \beta < 0. \end{cases}$$

8.43 Suppose that random variables X and Y satisfy

$$\left(\frac{X-2}{3}, \frac{Y+1}{2}\right) \sim N(0, 0, 1, 1, 0.5).$$

Write down the conditional normal distribution for Y given $X = x$.

8.44 Suppose that the period of heat treatment X (in seconds) and the depth of hardening Y (in cm) of an enamel underlay follow a bivariate normal distribution where

$$\left(\frac{X-20}{3}, \frac{Y-0.2}{0.09}\right) \sim N(0, 0, 1, 1, 0.9).$$

(a) Determine the regression function $\mu(x)$.

(b) Determine the variance of Y given that $X = 30$.

[*Hint:* What is the variance of $\frac{Y-0.2}{0.09}$ given that $X = 30$?]

(c) Given that $X = 30$, determine the probability that the enamel underlay will be hardened to a depth of between 0.4 and 0.45 cm.

8.45 Show that if (X, Y) has a bivariate normal distribution given by

$$(X, Y) \sim N(\mu_X, \mu_Y, \sigma_X^2, \sigma_Y^2, \rho),$$

then $Var(Y|X = x) = \sigma_Y^2(1 - \rho^2)$.

CHAPTER NINE

Sampling

In this chapter you will find

9.1 SAMPLING DISTRIBUTION OF THE MEAN

The material of Chapters 1 and 2 is devoted to the analysis of data. Described were methods for the pictorial description (histograms, bar charts, S/L plots, B/W plots ...) and summary measure description (sample mean, sample median, sample variance, sample standard deviation, interquartile range, ...) of actual realised data.

These methods are immediately applicable to data sets after the hard work of obtaining those data sets is over. A statistical summary may be all that is required. If so, these methods, especially the graphical methods, are easily implemented and easy to interpret. However the statistician is often called upon to see behind the scenes, to see beyond. This calls for an interpretation of what the data are saying; it calls for an investigation of the population from which the data came; an assessment of what what is happening in the population; an estimation of the parameters that define our interest in the population, and, strategically, a conclusion or recommendation. Such is the process of **statistical inference**.

Random sample

Having obtained data, statistical inference begins with questions about the nature of the random variables that gave us the data. The data set is a slice of 'what might have been'. Therefore in defining the concept of a random sample, our definition must encompass not just the data set that may now be before us, but also the data sets that may be obtained at another time, on another day, during another similar investigation.

Definition 9.1 *By a* **random sample** *of size n is meant a collection X_1, $X_2, \dots, X_n$ of n independent and identically distributed (i.i.d.) random variables. A* **realised random sample** *is a set $x_1, x_2, x_3, \dots, x_n$ of outcomes for the random sample, where x_i is the observed outcome for the ith random variable X_i in the sample, $i = 1, 2, 3, \dots, n$.*

□

A realised random sample is therefore a data set. However, the variables providing the data in a random sample must be *i.i.d.* and therefore, when sampling from a finite population, the sampled items must be replaced in the population after selection for the sample. This is the characterising feature of **sampling with replacement**—the population remains identical throughout the sampling procedure. Therefore, if $X_1, X_2, X_3, \dots, X_n$ is a random sample of size n, taken with replacement from a population of size N, then the probability that any particular population member is included in the sample is $\frac{1}{N}$ throughout the sampling procedure. Since, by the definition of independence, the joint mass function of the random sample factorises,

$$\begin{aligned} P(X_1 = x_1, X_2 = x_2, \dots, X_n = x_n) &= P(X_1 = x_1) \dots P(X_n = x_n) \\ &= (\frac{1}{N})(\frac{1}{N}) \cdots (\frac{1}{N}) \\ &= \frac{1}{N^n}. \end{aligned}$$

This establishes the following theorem:

Theorem 9.1 *When selecting a random sample with replacement from a finite population of size N equally likely elements, all possible samples have the same probability $\frac{1}{N^n}$ of being selected as the sample.*

■

Sampling without replacement will be discussed in the next section. There are also relevant comments from a probability standpoint in Section 3.5.

If, however, the population is not finite, our realised sample very likely consists of measurements of a continuous random variable X. (Recall that a continuous random variable has outcomes occurring in a given range—there will be uncountably many such outcomes.) The random sample consists of *i.i.d.* random variables with the same probability distribution as X. X is called the **population random variable**. The probability distribution that we assign to X is the probability model for the population.

For example, if X denotes 'wind velocity', then an appropriate probability model for X may be $X \sim N(\mu, \sigma^2)$. If this probability model for wind velocities is indeed appropriate, then a random sample of size 10 is a collection $X_1, X_2, X_3, \ldots, X_{10}$ of 10 *i.i.d.* $N(\mu, \sigma^2)$ random variables. The realised random sample, taken under a fixed set of experimental conditions, may look like:

71.3, 71.6, 67.7, 66.1, 75.4, 77.0, 81.3, 69.3, 70.4, 70.5.

We have seen in Section 7.4 how to check such a distributional assumption with a probability plot.

The **parameters** μ and σ in the probability model represent aspects of the population that are of interest to us and yet they often remain unknown. (We will often use Greek letters to denote parameters.) In the wind velocity example, μ, the true average windspeed, is elusive; it changes with the day; it changes with the experimental conditions. It remains unknown. With the aid of a random sample of wind velocities, we hope to estimate it as precisely as possible. Perhaps of even more importance to the researcher is the unknown parameter σ—a quantified measure of the variability in wind speed. Again, with the aid of a random sample of wind velocities, we hope to estimate it as precisely as possible. The estimation is achieved by using a statistic.

Definition 9.2 *A* **statistic,** $\widehat{T}$, *is a function* $\widehat{T} = t(X_1, X_2, X_3, \ldots, X_n)$ *of the random sample alone.*

□

Since a statistic is a function of random variables, it itself is a random variable. The main feature of a statistic is that it can be completely evaluated once the sample is realised; it is completely determined from the data. If its expression is in terms of any other constants, then these must be *known* constants and not unknown parameters. There are some simple obvious examples: Given a random sample $X_1, X_2, X_3, \ldots, X_n$, define in terms of this sample the functions $\overline{X}$ and S^2 by

$$\overline{X} = \frac{\sum_{i=1}^{n} X_i}{n},$$
$$S^2 = \frac{\sum_{i=1}^{n}(X_i - \overline{X})^2}{n-1}.$$

Notice how each of these statistics is defined in terms of the (upper case) random sample $X_1, X_2, \ldots, X_n$: $\overline{X}$ is really a function $t_1(X_1, X_2, \ldots, X_n) = \frac{\sum_{i=1}^{n} X_i}{n}$ of $X_1, X_2, \ldots, X_n$ and S^2 is really a function $t_2(X_1, X_2, \ldots, X_n) = \frac{\sum_{i=1}^{n}(X_i-\overline{X})^2}{n-1}$ of $X_1, X_2, \ldots, X_n$. When the values of the random variables $X_1, X_2, \ldots, X_n$ are observed and data $x_1, x_2, \ldots, x_n$ are obtained (where x_i is the observed value of X_i), then these statistics have realised values corresponding to the sample mean and sample variance

$$\overline{x} = \frac{\sum_{i=1}^{n} x_i}{n},$$
$$s^2 = \frac{\sum_{i=1}^{n}(x_i - \overline{x})^2}{n-1},$$

as defined in Chapter 2. The purpose of the statistic $\overline{X}$ is to estimate μ. The purpose of the statistic S is to estimate σ.

A probability distribution for $\overline{X}$

Having established that statistics such as $\overline{X}$ and S^2 are random variables, then we must accept that they follow their own probability distributions. Such a probability distribution is called a **sampling distribution**. In this chapter the sampling distribution for $\overline{X}$ is examined in detail.

The discussion of sampling distributions begins with a simple example of sampling with replacement from a finite population.

Example 9.1

The distribution of the average number of defects

Suppose that a population consists of 20 manufactured items of which:

5	have no defects
5	have one defect
5	have two defects
5	have three defects
20	total items

Clearly, in this population, the item types are **equally likely.** A probability model for the population random variable X = 'number of defects observed for a randomly chosen item' is

x	0	1	2	3
$p_X(x)$	$\frac{1}{4}$	$\frac{1}{4}$	$\frac{1}{4}$	$\frac{1}{4}$

Notice that the probabilities for each outcome of X are the same. The mean $E(X) = \mu_X = \mu$ and variance $VarX = \sigma_X^2 = \sigma^2$ of X are easily determined through the definitions of the mean and variance of a discrete probability distribution:

$$\begin{aligned}\mu &= E(X) = \sum x p_X(x) = (0)(\tfrac{1}{4}) + (1)(\tfrac{1}{4}) + (2)(\tfrac{1}{4}) + (3)(\tfrac{1}{4}) = 1.5,\\ \sigma^2 &= E(X^2) - E(X)^2 = (0)^2(\tfrac{1}{4}) + (1)^2(\tfrac{1}{4}) + (2)^2(\tfrac{1}{4}) + (3)^2(\tfrac{1}{4}) - (1.5)^2\\ &= 1.25.\end{aligned}$$

When $n = 2$ items are selected from the population of 20 items, with replacement, the list of all possible equally likely samples X_1, X_2 is, in the coded form

(number of defects for first item , number of defects for second item) :

(0, 0) (0, 1) (0, 2) (0, 3)
(1, 0) (1, 1) (1, 2) (1, 3)
(2, 0) (2, 1) (2, 2) (2, 3)
(3, 0) (3, 1) (3, 2) (3, 3)

All these samples are listed in the table below together with their sample means and their probabilities of selection:

Sample (x_1, x_2)	**Sample mean**	**Probability of selection**
(0,0)	0.0	$\frac{1}{16}$
(0,1)	0.5	$\frac{1}{16}$
(0,2)	1.0	$\frac{1}{16}$
(0,3)	1.5	$\frac{1}{16}$
(1,0)	0.5	$\frac{1}{16}$
(1,1)	1.0	$\frac{1}{16}$
(1,2)	1.5	$\frac{1}{16}$
(1,3)	2.0	$\frac{1}{16}$
(2,0)	1.0	$\frac{1}{16}$
(2,1)	1.5	$\frac{1}{16}$
(2,2)	2.0	$\frac{1}{16}$
(2,3)	2.5	$\frac{1}{16}$
(3,0)	1.5	$\frac{1}{16}$
(3,1)	2.0	$\frac{1}{16}$
(3,2)	2.5	$\frac{1}{16}$
(3,3)	3.0	$\frac{1}{16}$

Notice that all possible samples have the same chance $\frac{1}{16}$ of selection as the realised random sample. This illustrates Theorem 9.1.

The sampling distribution of the mean may then be written as a probability distribution by collecting together all possible mean values from the above list and, for each value, combining together the associated probabilities:

Sampling distribution of the mean

$\overline{x}$	0	0.5	1.0	1.5	2.0	2.5	3.0
$p_{\overline{X}}(\overline{x})$	$\frac{1}{16}$	$\frac{2}{16}$	$\frac{3}{16}$	$\frac{4}{16}$	$\frac{3}{16}$	$\frac{2}{16}$	$\frac{1}{16}$

It must be agreed that sampling with replacement from a finite population is not commonly done. However, Example 9.1 and Example 9.2 which follows, both serve to illustrate the construction of the sampling distribution of the mean.

Example 9.2

Distribution of the average number of defectives: Example 9.1 continued

As in Example 9.1, consider 20 manufactured items of which:

5	have no defects
5	have one defect
5	have two defects
5	have three defects
20	total items

However, of interest now is the average number of **defectives** rather than 'defects' when a sample is taken, with replacement, from these 20 items. A 'defective' item is one which has one, or two, or three defects. We shall denote such an item by '1'. A 'non-defective' item is one with no defects; such an item is to be denoted by '0'. This changes the target population from which we sample: it now consists of the 20 items

0 0 0 0 0 0 0 0 0 0 0 0 0 0 0 1 1 1 1 1 .

A probability model for the population random variable Y = 'number of defectives observed for a randomly chosen item' is now an appropriate model for the population. The probability distribution for Y is

y	0	1
$p_Y(y)$	$\frac{1}{4}$	$\frac{3}{4}$

Hence $Y \sim Bin(1, \frac{3}{4})$. It is important to realise that the elements of the population are no longer equally likely. The mean and variance for

Y may be determined directly from the Bernoulli distribution: $E(Y) = \mu_Y = \mu = (1)(\frac{3}{4}) = 0.75$, and $Var(Y) = \sigma_Y^2 = \sigma^2 = (1)(\frac{3}{4})(1-\frac{3}{4}) = \frac{3}{16}$. When $n = 2$ items are selected from the population of 20 items, with replacement, the list of all possible samples Y_1, Y_2 is, in the coded form

(number of defectives first item , number of defectives second item) :
$(0,0)$ $(0,1)$ $(1,0)$ $(1,1)$

All these samples are listed in the table below together with their sample means and their probabilities of selection:

Sample (y_1, y_2)	Sample mean	Probability of selection
(0,0)	0.0	$\frac{1}{16}$
(0,1)	0.5	$\frac{3}{16}$
(1,0)	0.5	$\frac{3}{16}$
(1,1)	1.0	$\frac{9}{16}$

Since the elements in the population are no longer equally likely, we cannot expect all possible samples of size 2 to be equally likely. Indeed, we are most likely in our realised sample to obtain 2 defective items. The assumptions of Theorem 9.1 are not met. However, the pair Y_1, Y_2 is a random sample in the sense of Definition 9.1.
The sampling distribution of the mean may then be written as a probability distribution by collecting together all possible mean values from the above list, and, for each value, combining together the associated probabilities:

Sampling distribution of the mean

$\overline{y}$	0	0.5	1.0
$p_{\overline{Y}}(\overline{y})$	$\frac{1}{16}$	$\frac{6}{16}$	$\frac{9}{16}$

Example 9.3 which follows gives a simple determination of the sampling distribution of the mean (in the case $n = 2$) when the population random variable is continuous.

Example 9.3

What results when two random numbers are averaged?

Example 5.7 continued. Suppose that I randomly select a number, X_1, from the interval $[0, 1]$. Without knowledge of my number, you randomly select a number, X_2, from the same interval. Then, X_1, X_2 constitutes a random sample of size 2 from the interval $[0, 1]$. Here the population random variable is $X \sim U[0, 1]$, and the $U[0, 1]$ distribution is the probability model for the population.

In Example 5.7 the probability density function for $Z = X_1 + X_2$, denoting the sum of the numbers selected was determined as the triangular density

$$f_Z(z) = \begin{cases} z & \text{if } 0 \le z \le 1; \\ 2 - z & \text{if } 1 < z \le 2. \end{cases}$$

Now, the pdf of $\overline{X} = \frac{1}{2}Z$ is simply found, by the cumulative distribution function method, to be given by

$$f_{\overline{X}}(\overline{x}) = \begin{cases} 4\overline{x} & \text{if } 0 \le \overline{x} \le 0.5; \\ 4(1 - \overline{x}) & \text{if } 0.5 < \overline{x} \le 1. \end{cases}$$

Graphically:

Graph of probability density function for the average of two random numbers from $[0, 1]$

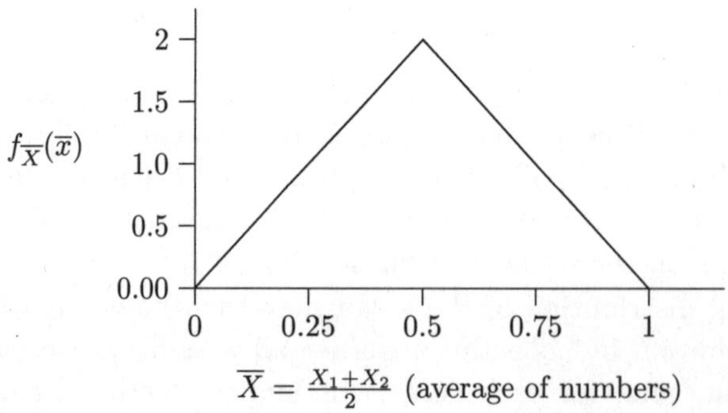

This is a graph of the pdf of the sampling distribution of the mean. This graph differs by a scale change from the graph of the pdf of the sum. This is to be expected since $\overline{X}$ is a linear change-of-scale of Z.

Properties of the sampling distribution of the mean

The previous section showed that $\overline{X}$ is a random variable with its own probability distribution, the sampling distribution of the mean. We shall see that this sampling distribution is often (at least approximately) normally distributed, especially for large values of n. (Examples 9.1, 9.2 and 9.3 all considered the case $n = 2$.)

Therefore we now concentrate on the mean and variance of $\overline{X}$ for a general sample size n. The sampling distribution of the mean has, itself, a mean.

For this we require the formulae **SUM 1** and **SUM 2** from Section 5.3. For reference, and because of their importance, we repeat these formulae for the current context. If $X_1, X_2, X_3, \ldots, X_n$ is a random sample from a population, and $a_1, a_2, a_3, \ldots, a_n$ is a sequence of constants:

SUM 1:

$$E(a_1X_1 + a_2X_2 + \ldots + a_nX_n) = a_1E(X_1) + a_2E(X_2) + \ldots + a_nE(X_n). \tag{9.1}$$

SUM 2:

$$Var(a_1X_1 + a_2X_2 + \ldots + a_nX_n) = a_1^2Var(X_1) + a_2^2Var(X_2) + \ldots + a_n^2Var(X_n). \tag{9.2}$$

The independence property of a random sample is required for **SUM 2** but not **SUM 1** which holds for *any* sequence $X_1, X_2, X_3, \ldots, X_n$ of n random variables.

If in **SUM 1** and **SUM 2** we choose constants

$$a_1 = a_2 = a_3 = \ldots = a_n = \frac{1}{n}$$

so that

$$\overline{X} = (\frac{1}{n})X_1 + (\frac{1}{n})X_2 + (\frac{1}{n})X_3 + \ldots + (\frac{1}{n})X_n,$$

then we obtain directly:

$$\begin{aligned}
\mu_{\overline{X}} &= E(\overline{X}) \\
&= (\frac{1}{n})E(X_1) + (\frac{1}{n})E(X_2) + (\frac{1}{n})E(X_3) + \ldots + (\frac{1}{n})E(X_n) \\
&= (\frac{1}{n})\mu + (\frac{1}{n})\mu + (\frac{1}{n})\mu + \ldots + (\frac{1}{n})\mu \\
&= (\frac{1}{n})(n\mu) \\
&= \mu
\end{aligned}$$

$$\begin{aligned}
\sigma^2_{\overline{X}} &= Var(\overline{X}) \\
&= (\frac{1}{n})^2Var(X_1) + (\frac{1}{n})^2Var(X_2) + (\frac{1}{n})^2Var(X_3) + \ldots + (\frac{1}{n})^2Var(X_n) \\
&= (\frac{1}{n})^2\sigma^2 + (\frac{1}{n})^2\sigma^2 + (\frac{1}{n})^2\sigma^2 + \ldots + (\frac{1}{n})^2\sigma^2 \\
&= (\frac{1}{n^2})(n\sigma^2) \\
&= \frac{\sigma^2}{n}.
\end{aligned}$$

This establishes the following theorem concerning the mean and variance of the sampling distribution of means:

Theorem 9.2 *If $\overline{X}$ denotes the sample mean of the random sample $X_1, X_2, X_3, \ldots, X_n$ taken from a population with mean μ and variance σ^2, then*

$$E(\overline{X}) = \mu \quad \text{and} \quad Var(\overline{X}) = \frac{\sigma^2}{n}.$$

■

This says that the mean of the sampling distribution of means is the population mean μ, and the variance of the sampling distribution of means is the population variance σ^2 divided by n, the sample size. This shows that averages are *less variable* than the original data being averaged. That is, when $n > 1$,

$$\frac{\sigma^2}{n} < \sigma^2 \quad \text{so that} \quad Var(\overline{X}) < Var(X_i),$$

for any $i = 1, 2, 3, \ldots, n$. This ensures that the support of the sampling distribution of the mean (the probability distribution for $\overline{X}$) is in a narrow interval about the population mean; when n is large, $Var(\overline{X})$ is likely to be quite small. We will return to this concept of small variability of the sample mean in the next chapter. The standard deviation of $\overline{X}$ has a special name, the standard error.

Definition 9.3 *The* **standard error**, *$S.E.(\widehat{T})$, of a statistic $\widehat{T}$ is the standard deviation of the sampling distribution of $\widehat{T}$.*

□

The standard error of $\overline{X}$ is the standard deviation of the sampling distribution of the mean. By Theorem 9.2,

$$\boxed{S.E.(\overline{X}) = \frac{\sigma}{\sqrt{n}}.} \qquad (9.3)$$

Example 9.4

The mean of a sampling distribution of means: Example 9.2 continued

Of 20 manufactured items under consideration,

5	have no defects
5	have one defect
5	have two defects
5	have three defects
20	total items

The population random variable Y = 'number of defectives observed for a randomly chosen item' has a Bernoulli distribution: $Y \sim Bin(1, \frac{3}{4})$. As outlined in Example 9.2, the population mean and variance are then $\mu = E(Y) = 0.75$ and $\sigma^2 = Var(Y) = \frac{3}{16}$.
Now examine the sampling distribution of the mean: this probability distribution for $\overline{Y}$, again from Example 9.2, is

$\overline{y}$	0	0.5	1.0
$p_{\overline{Y}}(\overline{y})$	$\frac{1}{16}$	$\frac{6}{16}$	$\frac{9}{16}$

Therefore,

$$E(\overline{Y}) = (0)(\frac{1}{16}) + (0.5)(\frac{6}{16}) + (1)(\frac{9}{16}) = \frac{12}{16} = 0.75 = \mu.$$

That is, the mean of the sampling distribution of means is demonstrated to be the population mean. For the case of the variance of the sampling distribution of means,

$$\begin{aligned} Var(\overline{Y}) &= E(\overline{Y}^2) - [E(\overline{Y})]^2 \\ &= (0)^2(\frac{1}{16}) + (0.5)^2(\frac{6}{16}) + (1)^2(\frac{9}{16}) - (0.75)^2 = \tfrac{1}{2}(\frac{3}{16}), \end{aligned}$$

which is the same as $\frac{\sigma^2}{n}$ when $n = 2$.

The role of the sampling distribution of $\overline{X}$

When a random sample is taken from a population and the sample mean is recorded, this recording is completely equivalent to selecting a single point (call it $\overline{x}$) from the sampling distribution of the mean. The sample mean that we record is just one of many that might have been. The role of the sampling distribution for $\overline{X}$ is to tell us how likely different observed values of $\overline{x}$ are. The study of the probability distribution for $\overline{X}$ is continued in the next section.

EXERCISES

9.1 A large population of recycled small glass beer bottles (usually referred to as stubbies) contains 10% that are cracked or chipped. These are removed for meltdown rather than being cleaned and washed.

Defective cracked or chipped bottles are assigned the score 0 and the rest are scored 1. Let the random variable X denote the assigned score.

(a) Determine the expected value and variance of X.

(b) Calculate the standard error of the mean score of a random sample of 2 bottles.

(c) Calculate the standard error of the mean score of a random sample of 100 bottles.

(d) As the sample size increases, the standard error of the mean score (decreases/increases/stays same). Select one.

(e) As the sample size increases, the expected value of the mean score (decreases/increases/stays same). Select one.

9.2 Suppose that four marbles (identical in size, weight, ...) labelled with numbers

2, 4, 6, 8

are placed in a hat. Let X denote the number appearing on a marble drawn at random from the hat.

(a) Explain why $p_X(2) = p_X(4) = p_X(6) = p_X(8) = \frac{1}{4}$.

(b) Show that $\mu_X = 5$ and $\sigma_X^2 = 5$.

(c) List the 16 equally likely samples of size 2 which can be drawn with replacement from this population.

(d) Determine the mean $\overline{x}$ for each sample in (c). Hence construct the probability distribution for $\overline{X}$, for samples of size 2.

(e) Draw a graph of the probability mass function for $\overline{X}$.

(f) Determine $P(\overline{X} > 3)$.

(g) Determine $E(\overline{X})$ directly as the expected value of the probability distribution for $\overline{X}$. Similarly, determine $Var(\overline{X})$ directly from the probability distribution for $\overline{X}$.

(h) Determine $E(\overline{X})$ and $Var(\overline{X})$ by using the results of Theorem 9.2.

What conclusions can you make about the sampling distribution of the mean?

9.3 When a random sample is taken from a population with known variance, the standard error of the mean is found to be 2.0. By what percentage must the sample size increase if the standard error of the mean is to be reduced to 1.2?

9.4 In a simulation trial, if 100 numbers are randomly chosen from the interval $[0, 1]$, how variable is their mean?

9.5 In a simulation trial, if 100 numbers are randomly chosen with replacement from the list $1, 2, 3, \ldots, 98, 99, 100$, how variable is their mean?

9.6 Suppose that when playing a game of blackjack at Jupiter's casino, I believe my chances to be summarised as follows: At any game, I will lose a chip with probability 0.4, win a chip with probability 0.3 and neither win nor lose (in which case my chip count remains the same) with probability 0.3.

If the change to my fortunes (change in chip count as the result of one game) is denoted by X, then X is a discrete random variable with probability mass function

x	-1	0	1
$p_X(x)$	0.4	0.3	0.3

(a) Determine $E(X)$ and $Var(X)$.

(b) Let X_1 and X_2 be independent random variables with the same distribution as X—possibly the results of two independent games played. Give the expectations and variance for each of X_1 and X_2.

(c) Give the expectation and variance of

$$\overline{X} = \frac{X_1 + X_2}{2},$$

where X_1 and X_2 are as in (b).

(d) Construct in a two-way table the joint probability distribution for X_1 and X_2.

(e) By adding along the diagonals of the distribution in (d), obtain the probability distribution of

$$\overline{X} = \frac{X_1 + X_2}{2},$$

and hence $E(\overline{X})$ and $Var(\overline{X})$ (and check that these results agree with those in (c)).

(f) Determine $P(\overline{X} > 0)$. How would this probability statement be usefully expressed in words?

(g) Determine $(-\frac{1}{2} \leq \overline{X} \leq \frac{1}{2})$.

9.7 The following is the graph (from Example 9.3) of the pdf for the average of two numbers randomly chosen from the interval $[0, 1]$.

Graph of probability density function for the average of two random numbers from $[0, 1]$

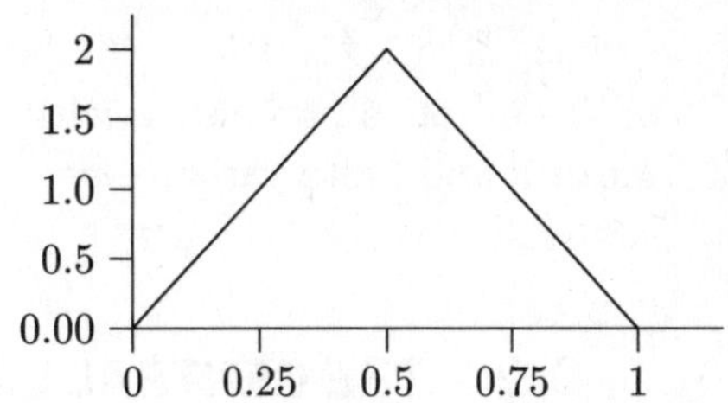

$\overline{X} = \frac{X_1+X_2}{2}$ (average of numbers)

(a) Use symmetry considerations to determine the value of $\mu_{\overline{X}}$, the mean of the sampling distribution of the mean.

(b) Show how to use the graph to determine $P(\overline{X} > 0.75)$.

(c) Copy the graph and shade an area representing the probability

$$P(|\overline{X} - \mu_{\overline{X}}| < 0.25).$$

(d) Hence determine the probability in (c).

9.8 Give a lower bound for the probability in Exercise 9.7(c) as determined by Chebychev's Inequality.

9.9 The outcome of a discrete random variable X is equally likely to be any of the n numbers $\alpha_1, \alpha_2, \alpha_3, \ldots, \alpha_n$. Show that $Var(X) = \frac{\sum_{i=1}^{n}(\alpha_i - \overline{\alpha})^2}{n}$, where $\overline{\alpha} = \frac{\sum_{i=1}^{n} \alpha_i}{n}$.
How may this result be applied to determining the variance of a finite population?

9.10 Suppose that

$$\overline{X_n} = \frac{X_1 + X_2 + X_3 + \ldots + X_n}{n}$$

is the mean of a random sample

$$X_1, X_2, X_3, \ldots, X_n$$

of size n from a population with finite mean μ and finite variance σ^2.

A new data point X_{n+1} is added to the data set creating a random sample

$$X_1, X_2, X_3, \ldots, X_{n+1}$$

of size $n+1$ having sample mean

$$\overline{X_{n+1}} = \frac{X_1 + X_2 + \ldots + X_n + X_{n+1}}{n+1}.$$

What change does this added data point have on the sample average?

Determine the mean and variance of the difference $\overline{X_n} - \overline{X_{n+1}}$ between the two sample means.

[*Hint*: Use **SUM 1** and **SUM 2**.]

9.2 THE CENTRAL LIMIT THEOREM

One of the most fundamental results in statistics, the so-called **Central Limit Theorem**, dictates that $\overline{X}$ has a (approximately) normal distribution when $\overline{X}$ represents the sample mean of a large sample of size n. This normal distribution occurs irrespective of the type of population being sampled, provided that the population variance is finite.

Building symmetric bell-shaped distributions

Theorem 9.3 *If $\overline{X}$ is the mean of the random sample $X_1, X_2, X_3, \ldots, X_n$ of size n from a $N(\mu, \sigma^2)$ population, then*

$$\overline{X} \sim N(\mu, \frac{\sigma^2}{n}).$$

Proof: Theorem 7.5 shows that sums $\overline{X} = (\frac{1}{n})X_1 + (\frac{1}{n})X_2 + (\frac{1}{n})X_3 + \ldots + (\frac{1}{n})X_n$ of independent normally distributed random variables are again normally distributed. The required mean and variance for this normal distribution are the mean and variance of the sampling distribution of means in Theorem 9.2.

■

Thus, when the data are normally distributed, $\overline{X}$ is normally distributed. It seems intuitively reasonable that, through the averaging process, $\overline{X}$ is likely to be closer to μ than a randomly chosen data point. This is shown in Theorem 9.3, for example, by a comparison of variances: $Var(X_i) = \sigma^2$ for any i, while $Var(\overline{X}) = \frac{\sigma^2}{n}$. The factor of $\frac{1}{n}$ in the formula for the variance of $\overline{X}$ ensures that the mean is very accurate (in the sense of having small variability) when n is large. If $n = 100$, the variability (variance) in the sample mean is 100 times smaller than the variability (variance) in an individual observation in the random sample.

Before we state and prove the Central Limit Theorem, we shall see it in operation.

Example 9.5

The Central Limit Theorem for random samples from [0,1]

The plotted points in the S/L plots are observations on the random variable $\overline{X}$, the sample mean for a sample of size n chosen randomly from the interval $[0, 1]$. Plots with $n = 1, 2, 5, 20$, were constructed for 100 plotted points with the assistance of the computer package MINITAB.

Averages of n uniform random numbers

Stems—10^{-1} units; leaves—10^{-2} units

3 | 2 represents 0.32 units

```
0 | 011247888
1 | 2344555
2 | 0133455666689
3 | 01255889
4 | 1233455566688
5 | 13334579                      n = 1
6 | 0114667
7 | 001113445557889999
8 | 002556788
9 | 00234689
```

```
0 | 56
1 | 0002345566899
2 | 145569
3 | 2234466789
4 | 011223346778
5 | 011122233333456778899         n = 2
6 | 11124677889
7 | 1222255677889
8 | 01244456789
9 | 0
```

```
0 |
1 |
2 | 0258899
3 | 023344555677889
4 | 0000113344455566667788888899
5 | 00000111222222345566777899    n = 5
6 | 000112233333555566677
7 | 02234
8 |
9 |
```

```
0 |
1 |
2 |
3 | 458899
4 | 0112222233455555566666666677777777778888888888999
5 | 0000011111111122223333444444555556666677789        n = 20
6 | 12
7 | 0
8 |
9 |
```

Immediately apparent is the shrinking variance with increasing sample size. For each of the four data sets, however, the sample mean has remained relatively constant—near the mean of the $U[0,1]$ population. Note that if $Y \sim U[0,1]$, then $E(Y) = \mu = 0.5$ and $Var(Y) = \sigma^2 = \frac{1}{12}$. These population values compare closely with the realised sample values in the following table:

n	μ	$\frac{\sigma}{\sqrt{n}}$	**Sample mean**	**Sample standard deviation**
1	0.5	0.2887	0.5108	0.2805
2	0.5	0.2041	0.5116	0.2317
5	0.5	0.1291	0.4985	0.1242
20	0.5	0.0645	0.4962	0.0593

By Theorem 9.3, the random variable $\overline{X}$ is normally distributed when sampling is from a normal population. When $\overline{X}$ is standardised by subtracting off the mean $E(\overline{X})$ and dividing by $S.E.(\overline{X})$, it follows that

$$\frac{\overline{X} - \mu}{\frac{\sigma}{\sqrt{n}}} \sim N(0,1).$$

When n is large, a $N(0,1)$ distribution may be constructed in this way with a high degree of accuracy. This is the substance of the **Central Limit Theorem**, which is fundamental to many statistical processes that are based on averaging.

Theorem 9.4 (The Central Limit Theorem) *If $X_1, X_2, X_3, \ldots, X_n$ is a random sample from* **ANY** *population with mean μ and finite variance σ^2, then the probability distribution of*

$$\frac{\overline{X} - \mu}{\frac{\sigma}{\sqrt{n}}}$$

tends to that of the standard normal probability distribution as $n \to \infty$.

Proof: To show that $\frac{\overline{X}-\mu}{\frac{\sigma}{\sqrt{n}}}$ has a limiting normal distribution, we show that in the limit, as $n \to \infty$, the moment generating function of this standardised mean collapses to the simple expression $e^{\frac{1}{2}t^2}$ which is the moment generating function of a standard normal random variable. Consider first the case $\mu = 0$ and $\sigma = 1$. Then, the standardised sample mean of interest is

$$\frac{\overline{X} - \mu}{\frac{\sigma}{\sqrt{n}}} = (\frac{X_1}{\sqrt{n}}) + (\frac{X_2}{\sqrt{n}}) + (\frac{X_3}{\sqrt{n}}) + \ldots + (\frac{X_n}{\sqrt{n}}).$$

If the variables $X_1, X_2, X_3, \ldots, X_n$ each have moment generating function $m(t)$ and cumulant generating function $\kappa(t) = \log_e m(t)$, then, in terms of this notation, the *i.i.d.* variables

$$\frac{X_1}{\sqrt{n}}, \frac{X_2}{\sqrt{n}}, \frac{X_3}{\sqrt{n}}, \ldots, \frac{X_n}{\sqrt{n}},$$

each have moment generating function $m(\frac{t}{\sqrt{n}})$. By independence, their sum has moment generating function $m(\frac{t}{\sqrt{n}})^n$ (Theorem 5.4) and cumulant generating function

$$\log_e m(\frac{t}{\sqrt{n}})^n = n\kappa(\frac{t}{\sqrt{n}}).$$

As n gets large, we show that this cumulant generating function tends to $\frac{1}{2}t^2$ which is the cumulant generating function of the standard normal distribution. The required limit follows by two applications of L'Hopital's rule (see, for example, Thomas and Finney, 1984) to detect the relative rates at which the numerator and denominator of

$$n\kappa(\frac{t}{\sqrt{n}}) = \frac{\kappa(\frac{t}{\sqrt{n}})}{\frac{1}{n}},$$

when regarded as functions of n, tend to 0. By using the definition of κ we may write $\kappa(0) = 0$, $\kappa'(0) = 0$ and $\kappa''(0) = 1$. From this, in the limit,

$$\begin{aligned}
\lim_{n\to\infty} n\kappa(\frac{t}{\sqrt{n}}) &= \lim_{n\to\infty} \frac{\kappa(\frac{t}{\sqrt{n}})}{\frac{1}{n}} \\
&= \lim_{n\to\infty} \frac{\kappa'(\frac{t}{\sqrt{n}})(-\frac{t}{2n^{\frac{3}{2}}})}{(-\frac{1}{n^2})} \\
&= \lim_{n\to\infty} \frac{\kappa'(\frac{t}{\sqrt{n}})(-\frac{t}{2})}{(-\frac{1}{n^{\frac{1}{2}}})} \\
&= \lim_{n\to\infty} \frac{\kappa''(\frac{t}{\sqrt{n}})(-\frac{t}{2})(-\frac{t}{2n^{\frac{3}{2}}})}{(\frac{1}{2n^{\frac{3}{2}}})} \\
&= \tfrac{1}{2}t^2.
\end{aligned}$$

The result then follows, since a distribution is completely identified by its moment generating function (or, equivalently, by its cumulant generating function).

In general, the above proof may be applied to the *i.i.d.* standardised variables $X_i = \frac{Y_i - \mu}{\sigma}$ if the original data $Y_1, Y_2, Y_3, \ldots, Y_n$ is from a population with mean μ and variance σ^2.

■

Example 9.6

More central limits: random samples from exp(2)

The S/L plots display plotted points $\overline{x}$, based on samples of size $n = 1, 2, 5, 20$ from an exponential distribution with mean 2. Recall that if $Y \sim exp(2)$, then $E(Y) = \mu = 2$ and $Var(Y) = \sigma^2 = 4$. The plots below were constructed with the assistance of the computer package MINITAB.

Averages of n exponential observations

Stems—units; leaves—10^{-1} units

3 | 2 represents 3.2 units

	$n = 1$
0	0011112222333344
0	55566677777889999
1	000111233334
1	55566667777788889
2	0144
2	555789
3	0001233
3	577
4	014
4	5669
5	
5	58
6	133
6	
7	
7	
8	133
8	
9	
9	
10	03
10	
11	1

	$n = 2$
0	22223344444
0	5555555666666677777889
1	0000001112222334444444
1	5567777777999999
2	000124
2	55678
3	0124
3	55567899
4	11
4	
5	2
5	778
6	01
6	
7	
7	8
8	
8	
9	
10	
10	
11	

	$n = 5$
0	5667788889999
1	000011222222223333444
1	55555666666667777777888899999999
2	01222233333444444
2	5566777889
3	0133
3	55668
4	
4	9
5	
5	
6	
6	
7	
7	
8	
8	
9	
10	
10	
11	

	$n = 20$
0	
1	0022333444
1	55555555566666666777777778888888899999999999
2	000000111111112222222222223334444444
2	55566677788
3	012
4	
4	
5	
5	
6	
6	
7	
7	
8	
8	
9	
10	
10	
11	

The highly positively skewed exponential distribution (with a coefficient of skewness of 1) rapidly symmetrises when sampled with large sample sizes. The four diagrams are drawn on the same scale for comparison purposes.

The interpretation of the CLT (Central Limit Theorem) most useful for our purposes follows from Theorem 9.4 by a continuity argument:

$$\text{If } n \text{ is large and } \Phi \text{ is the } N(0,1) \text{ cdf, then} P\left(\frac{\overline{X}-\mu}{\frac{\sigma}{\sqrt{n}}} < z\right) \approx \Phi(z). \tag{9.4}$$

When can the CLT be applied? When the sample size n exceeds 25; this estimate of sample size tends to be conservative (more cautious than necessary) and good results are often obtained for even smaller sample sizes.

Example 9.7

Overloading aircraft

Airline records show that on a certain long-haul route, luggage weight averages 17.3 kg per passenger, with a standard deviation of 3.1 kg per passenger. Passengers weigh on average 70 kg with a standard deviation of 8 kg.

Worked example:

1. **Determine the probability that the average luggage weight per passenger for those in a queue of 25 at the baggage check-in counter will be less than 16 kg.**
2. **Determine the probability that the total weight of 400 passengers booked on a B747 aircraft will exceed 28 480 kg.**
3. **Determine the probability that the 400 passengers and their luggage will exceed a maximum payload requirement of 35 450 kg.**
   Outline any assumptions made.

1. If the random variable X denotes 'luggage weight', then $\mu_X = 17.3$ and $\sigma_X = 3.1$. If $\overline{X}$ denotes the mean luggage weight for a random sample of size 25 passengers, then

$$\begin{aligned} P(\overline{X} < 16) &= P\left(\frac{\overline{X}-\mu}{\frac{\sigma}{\sqrt{n}}} < \frac{16-17.3}{\frac{3.1}{\sqrt{25}}}\right) \\ &\approx P(Z < -2.097) \quad \text{where} \quad Z \sim N(0,1) \\ &= 0.0179. \end{aligned}$$

2. Let W_i denote the weight of Passenger i, for $i = 1, 2, 3, \ldots, 400$. Then $\sum_{i=1}^{400} W_i$ is the total weight of all 400 passengers. We must determine $P(\sum_{i=1}^{400} W_i > 28480)$. By dividing the total weight of all 400 passengers by 400, the average passenger weight is obtained:

$$\begin{aligned} P(\sum_{i=1}^{400} W_i > 28\,480) &= P(\frac{\sum_{i=1}^{400} W_i}{400} > \frac{28\,480}{400}) \\ &= P(\overline{W} > 71.2) \\ &= P\left(\frac{\overline{W}-\mu}{\frac{\sigma}{\sqrt{n}}} > \frac{71.2-70}{\frac{8}{\sqrt{400}}}\right) \\ &\approx P(Z > 3) \quad \text{where} \quad Z \sim N(0,1) \\ &= 0.0013. \end{aligned}$$

3. Assumption: Passenger weights are independent from one passenger to another, luggage weights are independent, and passenger weights are independent of luggage weights.
   If we let $T_i = X_i + W_i$ denote the total combined weight for Passenger i, combining Passenger i's weight W_i and Passenger i's luggage weight X_i, then we need to determine $P(\sum_{i=1}^{400} T_i > 35\,450)$. By the summation methods of Chapter 2, it follows that $\overline{T} = \overline{X} + \overline{W}$. By **SUM 1**,

$$E(\overline{T}) = E(\overline{X}) + E(\overline{W}) = \mu_X + \mu_W = 17.3 + 70 = 87.3.$$

   SUM 2 is used for the variance calculation:

$$Var(\overline{T}) = Var(\overline{X}) + Var(\overline{W}) = \frac{(3.1)^2}{400} + \frac{(8)^2}{400} = 0.184025.$$

   The calculation then follows as for parts 1. and 2. through the Central Limit Theorem applied to a standardised version of $\overline{T}$.

$$\begin{aligned} P(\sum_{i=1}^{400} T_i > 35\,450) &= P(\overline{T} > 88.625) \\ &= P\left(\frac{\overline{T}-E(\overline{T})}{S.E.(\overline{T})} > \frac{88.625-87.3}{\sqrt{0.184025}}\right) \\ &\approx P(Z > 3.1) \quad \text{where} \quad Z \sim N(0,1) \\ &= 0.001. \end{aligned}$$

Equation 9.4 indicates that when $X_1, X_2, X_3, \ldots, X_n$ is a random sample of size n from a population with mean μ and variance σ^2, if n is large, then the standardised version of the sample mean $\overline{X} = \frac{\sum_{i=1}^{n} X_i}{n}$ has an approximate standard normal distribution. Specifically:

$$P\left(\frac{\overline{X}-\mu}{\frac{\sigma}{\sqrt{n}}} < z\right) \approx \Phi(z)$$

where Φ represents the cdf of the standard normal distribution.

> The approximating normal distribution for $\overline{X}$ is $N(\mu, \frac{\sigma^2}{n})$ when n is large ($n \geq 25$).

We wish to be able to use this result quite generally for averages (or sums) of independent random variables; probability calculations transfer to the normal distribution. This is illustrated in the following cases. The normal approximations are reasonably accurate even for sample sizes less than 25 if the population random variable is symmetrically distributed.

CLT and Poisson data with a large mean

$\boldsymbol{X_i \sim Poi(\lambda); \quad \mu = \lambda, \quad \sigma^2 = \lambda.}$

If $X_1, X_2, X_3, \ldots, X_n$ are n independent observations from a Poisson process with rate λ, then, for each i, $X_i \sim Poi(\lambda)$. The mean and variance of this distribution are: $E(X_i) = \lambda$ and $Var(X_i) = \lambda$. We apply the CLT with $\mu = \lambda$ and $\sigma^2 = \lambda$ and n large:

$$P\left(\frac{\overline{X}-\lambda}{\frac{\sqrt{\lambda}}{\sqrt{n}}} < z\right) \approx \Phi(z).$$

If we consider the case $\lambda = 1$ and set $Y = \sum_{i=1}^{n} X_i$, then by Theorem 6.5, $Y \sim Poi(n)$; that is, the sum of n independent Poisson random variables is again Poisson. The standardised score for $\overline{X}$ simplifies as

$$\frac{\overline{X}-\lambda}{\frac{\sqrt{\lambda}}{\sqrt{n}}} = \frac{\frac{Y}{n}-1}{\frac{\sqrt{1}}{\sqrt{n}}} = \frac{Y-n}{\sqrt{n}}.$$

Summarising:

> CLT normal approximation to the Poisson:
> $P\left(\frac{Y-n}{\sqrt{n}} < z\right) \approx \Phi(z)$ for n large and $Y \sim Poi(n)$.

The approximating normal distribution for Y is $N(n, n)$. The diagrams of the Poisson pmf in Example 6.6 show that the pmf becomes more symmetric as λ increases. The shape of the pmf even for $\lambda = 6$ is remarkably symmetric and bell-shaped.

Example 9.8

Assessing infestation rates with a normal approximation

Suppose that the incidence of caterpillar infestation of an area of pasture-land is estimated to be 90 per square metre. Such an infestation can strip not only all leaf growth within days, but also destroy the sustaining roots of the pasture.

Worked example:

Determine the probability that there are more than 1000 caterpillars in 10 square metres of pasture.

If we let Y denote the number of caterpillars per 10 square metres, then $Y \sim Poi(900)$. The CLT operates on the independent sum of 900 independent $Poi(1)$ random variables to give

$$\begin{aligned} P(Y > 1000) &= P\left(\frac{Y - 900}{\sqrt{900}} > \frac{1000 - 900}{\sqrt{900}}\right) \\ &\approx P(Z > 3.33) \quad \text{where} \quad Z \sim N(0,1) \\ &= 0.0004. \end{aligned}$$

Therefore, overall, $P(Y > 1000) \approx 0.0004$.

CLT and binomial data from a large number of trials

$\boldsymbol{X_i \sim Bin(1, p); \quad \mu = p, \quad \sigma^2 = p(1-p).}$

If $X_1, X_2, X_3, \ldots, X_n$ are n independent Bernoulli trials with probability of success p, then, for each i, $X_i \sim Bin(1, p)$. The mean and variance of this distribution are: $E(X_i) = p$ and $Var(X_i) = p(1-p)$. We apply the CLT with $\mu = p$ and $\sigma^2 = p(1-p)$ and n large:

$$P\left(\frac{\overline{X} - p}{\frac{\sqrt{p(1-p)}}{\sqrt{n}}} < z\right) \approx \Phi(z).$$

If we set $Y = \sum_{i=1}^{n} X_i$, then by definition of the binomial distribution as being the sum of a sequence of n Bernoulli trials, $Y \sim Bin(n, p)$. The approximating normal distribution for Y is $N(np, np(1-p))$.

CLT normal approximation to the binomial:

$$P\left(\frac{Y - np}{\sqrt{np(1-p)}} < z\right) \approx \Phi(z) \text{ for } n \text{ large and } Y \sim Bin(n,p).$$

Notice that the CLT applies directly to $\frac{Y}{n} = \frac{\sum_{i=1}^{n} X_i}{n}$, which is the *proportion of successes* in n Bernoulli trials. The interpretation here is that a proportion is really an average of 0's and 1's. Therefore, under this interpretation, the CLT may be used to determine probabilities concerning proportions.

The normal approximation to the binomial is most accurate when the binomial distribution being approximated is relatively symmetric; in practice this requires more than 'n large'. A useful rule of thumb is to use the approximation when $np \geq 5$ and $n(1-p) \geq 5$. The approximation may be improved by using the **continuity correction** to help smooth off the discrete boxes of the binomial histogram and approximate the area in the tail of a binomial histogram with the area in the tail of a smooth normal curve.

Suppose that for illustration purposes we select from the histogram of the $Bin(n, p)$ distribution the rectangle centred at the point $y = k$ representing k successes. The area of this rectangle is $P(Y = k)$ and the base of the rectangle is 1 unit long. The approximating normal curve is that of a random variable $W \sim N(np, np(1-p))$. The best approximation to $P(Y \geq k)$ is the area under the approximating normal curve to the right of $k - \frac{1}{2}$. The best approximation to $P(Y > k)$ is the area under the approximating normal curve to the right of $k + \frac{1}{2}$. The best approximation to the area of the rectangle is the area under the approximating normal curve between $k - \frac{1}{2}$ and $k + \frac{1}{2}$. This is illustrated in the following diagram.

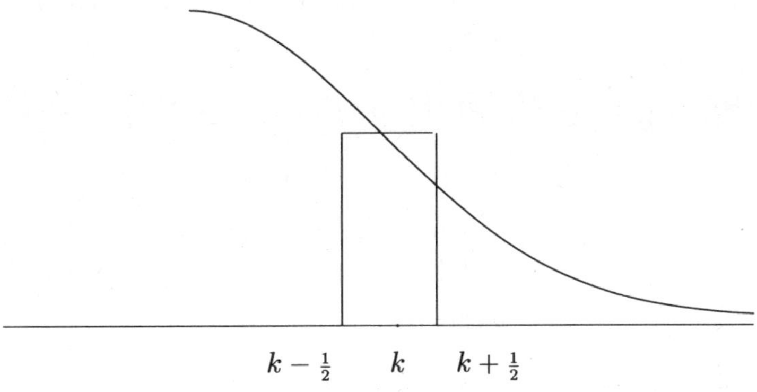

If n is large, $Y \sim Bin(n, p)$ and $W \sim N(np, np(1-p))$, then the *continuity correction* adjusts by $\frac{1}{2}$ the amount included in the approximating normal curve in the following situations:

$$\begin{aligned} P(Y \geq k) &\approx P(W > k - \tfrac{1}{2}); \\ P(Y > k) &\approx P(W > k + \tfrac{1}{2}); \\ P(Y = k) &\approx P(W > k - \tfrac{1}{2}) - P(W > k + \tfrac{1}{2}). \end{aligned}$$

The probabilities for W are evaluated by normal standardisation.

Example 9.9

Spinning for '2'

When I sit at the roulette table, I wait in vain for my lucky number '2'.

Worked example:

In 1000 spins of a roulette wheel, determine (approximately) the probability that '2' occurs 25 or more times.

Let X_i return 1 or 0 depending on whether a '2' does or does not appear on a single spin of the roulette wheel. Given that there are 37 numbers (1 to 36 plus 0 for the house) equally spaced about the wheel in such a way that all numbers are equally likely to occur on a given spin,

$$X_i \sim Bin(1, \frac{1}{37})$$

and $X_1, X_2, X_3, \ldots, X_{1000}$ constitute a sequence of Bernoulli trials. Therefore, $Y = \sum_{i=1}^{1000} X_i \sim Bin(1000, \frac{1}{37})$; the required probability follows by the CLT:

$$\begin{aligned} P(Y \geq 25) &= P\left(\frac{Y-(1000)(\frac{1}{37})}{\sqrt{1000(\frac{1}{37})(1-\frac{1}{37})}} \geq \frac{25-(1000)(\frac{1}{37})}{\sqrt{1000(\frac{1}{37})(1-\frac{1}{37})}}\right) \\ &\approx P(Z \geq -0.40) \quad \text{where} \quad Z \sim N(0,1) \\ &= 0.655. \end{aligned}$$

Using the *continuity correction* we repeat the calculation as

$$\begin{aligned} P(Y \geq 25) &= P\left(\frac{Y-\frac{1}{2}-(1000)(\frac{1}{37})}{\sqrt{1000(\frac{1}{37})(1-\frac{1}{37})}} \geq \frac{25-\frac{1}{2}-(1000)(\frac{1}{37})}{\sqrt{1000(\frac{1}{37})(1-\frac{1}{37})}}\right) \\ &\approx P(Z \geq -0.49) \quad \text{where} \quad Z \sim N(0,1) \\ &= 0.688. \end{aligned}$$

The exact probability is (to 3 dp)

$$P(Y \geq 25) = 1 - P(Y \leq 24) = 1 - \sum_{k=0}^{24} \binom{1000}{k} (\frac{1}{37})^k (\frac{36}{37})^{1000-k} = 0.681.$$

The error in the approximation is caused by the small probability of success $p = \frac{1}{37}$ creating a skewed distribution.

If a more symmetric example is considered, the approximation improves. The colour composition of the roulette wheel is: 18 squares are red, 18 squares are black and the zero square for the house is green. Suppose that we bet on the black squares, so that $p = P(\text{black}) = \frac{18}{37}$. This probability of success is much closer to $\frac{1}{2}$. Now, in 1000 spins of the roulette wheel, constituting the sequence of Bernoulli trials $X_1, X_2, \ldots, X_{1000}$ where $X_i \sim Bin(1, \frac{18}{37})$, the probability that 'black' occurs 500 times is approximated by

$$P\left(Z \leq \frac{500+\frac{1}{2}-(1000)(\frac{18}{37})}{\sqrt{1000(\frac{18}{37})(1-\frac{18}{37})}}\right) - P\left(Z \leq \frac{500-\frac{1}{2}-(1000)(\frac{18}{37})}{\sqrt{1000(\frac{18}{37})(1-\frac{18}{37})}}\right)$$

$$= \Phi(0.8866) - \Phi(0.8233) = 0.0167.$$

EXERCISES

9.11 Parts of inland New South Wales have a mean annual rainfall of 500 mm with a standard deviation of 60 mm. For a sample of 36 randomly selected years give the:

(a) variance of the mean annual rainfall in the area;

(b) standard error of the mean annual rainfall in the area.

9.12 In Exercise 9.11, determine the approximate probability that the mean annual rainfall for 36 randomly selected years will exceed 520 mm.

9.13 Police records show that the average speed of cars on the Kwinana Freeway is $\mu = 98$ km hr^{-1} with a standard deviation of 10 km hr^{-1}.

In regular random speed checking procedures, police record the speeds of 100 freeway drivers. Let $\overline{X}$ denote the average speed for one such sample.

(a) State the values of the mean and standard error of $\overline{X}$.

(b) Explain why $\overline{X}$ has an approximate normal distribution.

(c) Determine the probability that the average speed in the sample exceeds the speed limit of 100 km hr^{-1}.

9.14 Suppose that the mean tar content in a Marlboro cigarette is 12 mg with a standard deviation of 1 mg.

(a) Give the approximate probability distribution for the average tar content (per cigarette) in a pack of 30 cigarettes.

(b) Give the approximate probability distribution for the total tar content of a pack of 30 cigarettes.

(c) Give the approximate probability distribution for the total tar content of a carton containing 10 packs of cigarettes.

9.15 The time spent waiting for attention in the outpatient section of a hospital is exponentially distributed with a mean of 30 minutes.

(a) Determine the values of a and b so that if $\overline{X}$ represents the mean waiting time (measured in hours) for a random sample of

100 patients, $\overline{X} \sim N(a,b)$ approximately.

(b) Hence determine the probability that the total waiting time for the 100 patients exceeds 59 hours 48 minutes.

9.16 The Central Limit Theorem is not generally used unless n is large ($\geq$ 25). Even so, this example shows that probabilities associated with the sampling distribution of $\overline{X}$ may well be 'approximated' by normal probabilities even with n as small as 2.

The population random variable X is uniform and discrete with probability mass function

x	2	4	6	8
$p_X(x)$	$\frac{1}{4}$	$\frac{1}{4}$	$\frac{1}{4}$	$\frac{1}{4}$

(a) Show that by considering all possible samples of size 2 that may be taken from this population with replacement, the sampling distribution of $\overline{X}$ is

$\overline{x}$	2	3	4	5	6	7	8
$p_{\overline{X}}(\overline{x})$	$\frac{1}{16}$	$\frac{2}{16}$	$\frac{3}{16}$	$\frac{4}{16}$	$\frac{3}{16}$	$\frac{2}{16}$	$\frac{1}{16}$

(b) Show that the mean and variance of the sampling distribution in (a) are 5 and 2.5 respectively.

(c) Sketch the probability density function $f_Y(y)$, where $Y \sim N(5, 2.5)$. On the same diagram, sketch the probability mass function $p_{\overline{X}}(\overline{x})$. Note that the triangular shape of the mass function fits the smooth curve reasonably well. (To see this your diagrams will need to be accurately drawn.)

9.17 Each day a relief primary school teacher working for the Department of Education travels X km for emergency teaching at one of several nearby schools. The teacher has kept a log book of distance travelled and this indicates that an appropriate pmf for X is

x	0	5	10	15
$p_X(x)$	0.1	0.2	0.4	0.3

For an academic primary school year of four 10 week terms, determine the approximate probability that the relieving teacher travels more than 2000 km.

9.18 The weight, X, in kg, of a randomly selected man is assumed to follow a normal distribution with mean 73 kg and standard deviation of 8 kg.

(a) If we write $X \sim N(a,b)$, give numerical values for a and b.

(b) If men, on average, are 10 kg heavier than women, but women's weights show 25% less variance than men's weights, specify the probability distribution for the weight, Y, of a randomly chosen woman.

(c) 9 men and 7 women independently climb into a lift that has a sign:

MAXIMUM CAPACITY
16 PEOPLE OR 1149 kg

What is the probability that the lift will be overloaded?

9.19 Let X denote the lifetime of a colour television tube. Suppose that the logarithm, $\log_e X$, of the lifetime is normally distributed with a mean of 5 and a variance of 2. (Such a distribution is called a **lognormal distribution** since the logarithm function transforms the distribution to normality.)

(a) Show that the cumulative distribution function of X is given by

$$F_X(x) = \Phi\left(\frac{\log_e x - 5}{\sqrt{2}}\right),$$

where Φ, as usual, denotes the cdf of the standard normal distribution.

(b) Under a guarantee system, the TV manufacturers are willing to replace the picture tubes in a maximum of 1% of the TV sets sold.

What guarantee period can the manufacturers offer?

(c) Draw a sketch of the pdf of a $N(5,2)$ distribution. This curve is the pdf of the population random variable $Y =$ 'log(lifetime)'. In terms of our notation

$$Y = \log_e X \sim N(5,2).$$

On the same axes sketch the pdf for the random variable $\overline{Y}$, representing the sample mean of 16 observations of 'log(lifetime)'. Determine the probability that the average 'log(lifetime)' in the sample exceeds 6.

9.20 Electric conduit (plastic pipe to house electric wiring) is manufactured to a specified diameter of 2 cm with a standard deviation of 0.3 cm.

(a) If a sample of 100 measurements is taken on different pipes, determine the probability that the average diameter in the sample exceeds 2.05 cm.

(b) Repeat (a) to determine the probability that the average diameter in the sample exceeds 2.1 cm.

(c) If the observed sample mean was 2.11 cm, would you describe as 'likely' or 'unlikely' that the true mean diameter of all such pipes is 2 cm?

9.21 (Khazanie, 1979) A computer is programmed in such a way that the error involved in carrying out one computation has a distribution with mean 10^{-7} and standard deviation 10^{-8}. If 10 000 independent computations are made, determine the approximate probability that the total error will be less than 0.001001.

9.22 If $\overline{X}$ is the sample mean of a random sample of 50 measurements from a population with mean 0 and variance 4, determine the value of c so that $P(|\overline{X}| > c) = 0.05$.

9.23 Errors occur in the first edition of a 700 page textbook at an average rate of 1 per page. Approximate the probability that there are more than 750 errors in the entire book.

9.24 Voters arrive at a polling booth at an average rate of 25 per hour. Approximate the probability that in any one hour period, 35 or more voters will arrive at the polling booth.

9.25 On a weekday evening 2% of a large television audience is tuned to Channel 33. In a random sample of 100 viewers, determine the probability that 5% or more of the sample is watching Channel 33.

9.26 On the eve of the July 1987 Federal Election, the Morgan Gallup poll showed that 46% of people would vote for the ALP. (*Source:* *The Weekend Australian*, 11-12 July 1987.)

If this poll is taken as a true reflection of people's voting intentions, determine the probability that in a random sample of 1000 voters, *more than* half of them would vote for the ALP.

How does this probability assessment change if only 100 voters are asked their voting intentions? Explain briefly.

Would you expect that continuity correction would have much effect in this problem?

9.27 Use normal probability tables to approximate

$$\binom{1000}{340}\left(\frac{1}{3}\right)^{340}\left(\frac{2}{3}\right)^{660}.$$

9.28 A taxi-fleet owner finds that tyre punctures are reported at the rate of 3 per week.

(a) Determine the probability that less than 5 punctures are reported in a given week.

(b) Estimate the probability that more than 175 punctures are reported in a year.

(c) Estimate the probability that there are 10 or more puncture-free weeks over a 2-year period.

9.29 Lengths of timber are produced in a sawmill to a specified length L. However, due to variability in the milling process, the actual length, X, of a randomly selected piece of timber, has a normal distribution, where

$$X \sim N(L+\theta, \frac{L^2}{100})$$

for a parameter θ. A random sample

$$X_1, X_2, X_3, \ldots, X_n$$

of n pieces of timber is selected.

(a) Explain, in terms of the physical situation, the meaning of the parameter θ.

(b) Suppose now that θ is known:

$$\theta = \frac{L}{100}.$$

A railing of specified length $102L$ is to be constructed by placing 100 pieces of randomly selected timber end-to-end.

Determine the probability that the 100 pieces of timber will be too short.

9.30 If X has mgf $m(t)$, show that $\frac{X}{\sqrt{n}}$ has mgf

$$m(\frac{t}{\sqrt{n}}).$$

This result was used in the proof of the Central Limit Theorem.

9.31 100 random numbers

$$X_1, X_2, X_3, \ldots, X_{100}$$

are selected by a computer program from the interval (0,1) for simulation purposes. Use the CLT to determine

$$P(X_1X_2X_3\ldots X_{100} > e^{-100}).$$

[*Hint:* Take logarithms and use the result that $-\log_e X$ has an exponential distribution when X is uniformly distributed in the unit interval.]

9.3 SAMPLING WITHOUT REPLACEMENT

Sampling without replacement may be necessary (and realistic) for finite populations. In reality, often when we sample things, we cannot put them back unscarred into the population environment from which they came; new defective Kelvinator refrigerators, once sampled and ascertained to be defective, would not be placed back into a factory's output as if nothing were wrong with them. Sampled lightbulbs observed till burnout are examples of items destroyed in the sampling process, and not subject to put-back. An interviewer does not want to hear the opinions of the same person twice. To overcome this a plan or sampling scheme needs to be decided on in advance so that a sample is extracted from the population that is representative of that population.

Definition 9.4 *A proposal for the selection of a sample of size n from a population of size $N > n$ is called a* **sampling scheme**.

□

Sampling without replacement is natural, simple, and an important sampling scheme where independence between sample members fails. This is in contrast to our previous definition of a random sample (Definition 9.1) in which sample members are independent and identically distributed—*i.i.d.*

Suppose that α_i $(i = 1, 2, 3, \ldots, N)$ is the measurement recorded when the ith experimental unit (or element of the population) is sampled. The distinction between the units comprising the population and the datum to be recorded from each unit is often blurred. We therefore write a finite population as a data list $\{\alpha_1, \alpha_2, \alpha_3, \ldots, \alpha_N\}$.

Definition 9.5 *If the population of items to be sampled is represented as $\{\alpha_1, \alpha_2, \alpha_3, \ldots, \alpha_N\}$ where the α's are not necessarily distinct, then a* **without-replacement random sample** *of size $n < N$, is one in which each of the $\binom{N}{n}$ possible selections of n α's from the population has the same chance*

$$\frac{1}{\binom{N}{n}}$$

of selection as the sample.

□

If we denote the without-replacement random sample by the usual notation $X_1, X_2, X_3, \ldots, X_n$, then it follows as a consequence of the definition of the random sample that each point in the population has the same chance $\frac{1}{N}$ of being among those selected for the sample. (For if the contrary were true, and some population items had higher chance of sample inclusion than others, the equally likely nature of all possible samples, and hence the definition of a 'without-replacement random sample' would be violated.) That is, the marginal distribution of X_j, the jth member of the sample is given by:

$$X_j = \begin{cases} \alpha_1 & \text{with probability} \quad \frac{1}{N}; \\ \alpha_2 & \text{with probability} \quad \frac{1}{N}; \\ \ldots & \ldots \qquad\qquad\qquad \ldots \\ \alpha_N & \text{with probability} \quad \frac{1}{N}. \end{cases}$$

and this is true for each $j = 1, 2, 3, \ldots, n$. Therefore the sample members are identically distributed. Notice that each α_i having the same chance of inclusion in a sample does not necessarily imply that the sample is random in the sense of Definition 9.5.

Example 9.10

A sampling scheme for tyres with equal probability of selection

In a road test to detect 'tread wear' on *Good Year* tyres, 10 cars of the same year and make travelled 50 000 km on a new set of tyres. The front tyres of each car were maintained at 25 p.s.i. and the rear tyres at 30 p.s.i. It is conceivable that different tyre pressure would cause different tyre wear, all other factors unchanged.

Consider the following sampling scheme for selecting a sample of size 20 tyres from the 40 used in the tyre-wear trial.

Sampling scheme: *Select the 10 pairs of front tyres with probability $\frac{1}{2}$ and measure the wear, or, select the 10 pairs of rear tyres with probability $\frac{1}{2}$ and measure the tread wear.*

This sampling scheme has the property that each member of the population (in this case, of 40 tyres) has the same probability ($\frac{1}{2}$) of selection for the sample. Therefore a criterion such as 'each population element has the same chance of selection for the sample' is not sufficient to guarantee a sample that truly reflects the nature of the population.

In 'without-replacement random sampling' of the car tyres, samples containing mixtures of front and back tyres are taken into consideration. All $\binom{40}{20}$ possible collections of 20 tyres are given the same chance of being tread-measured.

Realising a without-replacement random sample

The logistics of this most obvious of sampling plans must not only be fully understood in theory from a probability standpoint, but also, vitally, be easily carried out. This is intended to be a *practical* plan. Let us represent the finite population of size N in the usual notation: $\{\alpha_1, \alpha_2, \alpha_3, \ldots, \alpha_N\}$. From this, in order to select a 'without-replacement random sample' of size n, select $n < N$ items according to the following procedure:

- Use **random number tables** or computer-generated random numbers to randomly select one point from $\{1, 2, 3, \ldots, N\}$. If the point j is selected, then α_j is included in the sample.
  This assigns to each of the N elements of $\{\alpha_1, \alpha_2, \alpha_3, \ldots, \alpha_N\}$ the same probability $\frac{1}{N}$ of selection and identifies the first member of the sample.
- Use **random number tables** or computer-generated random numbers to randomly select one item from the $N-1$ remaining in the population. Conditional on the first selection, this gives the remaining items each a probability $\frac{1}{N-1}$ of selection.
- Repeat the above steps, determining one member of the population at a time for inclusion in the sample, until $n < N$ have been selected.

Computer implementation speeds the process up, so that a sample may be simultaneously selected from a population.

Random number tables exist in extensive forms (see, for example, *A Million Random Digits with 10,000 Normal Deviates*, Rand Corporation, Santa Monica, California). Such tables have the property that: if a digit k is randomly chosen within the table then the next digit is l with probability $\frac{1}{10}$ for each $l = 0, 1, 2, 3, \ldots, 9$. Human minds cannot concoct such tables; to us a string 333 of three threes in a row is not random enough, even if the sequence is embedded in a sea of other digits.

Many computer packages contain **pseudo-random number generators** in the form of equations which, when supplied with a starting value (called a **seed**), produce sequences of digits iteratively. The randomness property alluded to above is very nearly satisfied, so that the output has the appearance of being random. Such output is suitable for most needs. One such sequence of digits output from the MINITAB package is:

17698 66724 04739 95076 28408 92272 16940 12263 40578 92491 84553 51002 34033 43089 32922
72284 61370 69225 75364 54228 56052 15312 35578 71251 53185 20524 78607 88029 14100 47347
06372 48153 65613 25931 19381 62515 18485 96540 19220 01635 41476 17054 14387 10713 67498
02566 04424 84503 08315 98380 39231 37289 20760 53067 62971 73890 97587 03909 61222 95108
33370 71433 10743 98390 88622 52842 88554 71290 29144 14185 98762 77803 05260 69345 11312
43034 85861 45352 25442 57597 68205 14006 79245 89168 69098 80357 99638 25699 50847 05691
44295 37302 44679 54778 12481 55936 11469 13041 62760 84917 47245 60169 92008 33658 53234
45067 14662 05646 21159 01056 07303 01901 33380 32762 05859 44881 32009 12952 31034 72542
95267 32372 10161 01171 19039 28644 67774 08470 46576 70168 46426 79910 22792 71778 58770

22091 71108 03280 88795 56654 72314 46323 61450 88566 10047 99440 47191 39302 90468 81260
70865 55890 30003 64166 12822 08545 57510 58195 86322 50411 90942 02856 68127 88864 74422
90270 01464 60778 57078 15878 08085 02834 83846 83109 05713 34746 84812 10709 54257 80793
88739 71921 43331 01510 38735 82148 99132 92457 83106 16479 88791 41882 68496 91580 57167
17927 59882 98966 35362 62030 47581 45985 62948 85329 70771 20364 53796 47721 48138 22908
66151 46398 45608 24493 65223 17329 61392 68945 99766 01034 61736 52899 66196 47687 18177
70835 29322 17094 12458 74555 95740 72242 54431 65336 36799 57254 19463 03546 30254 03918
67066 99929 56825 32968 66287 81824 89820 83274 91224 03212 21115 95678 28106 27560 09321
03088 01602 39414 72146 13331 45024 65142 86050 00329 26756 34537 66411 28047 42814 56287
41212 44615 62621 01444 05901 51820 35898 51517 76986 96031 80606 37696 77493 87290 39634
23362 58844 75815 12099 86720 11305 54298 20308 38747 76913 12470 40088 13179 87342 38545
97385 14234 58832 11040 47470 17814 18140 06973 79252 54079 80108 33836 26783 67610 18191
06509 59975 13516 29610 47883 51344 80577 13728 96880 89568 01331 85464 18069 56089 16059
10073 95995 66471 83116 58679 29135 07405 59238 99383 08910 89566 34875 29637 25652 04858
22953 13791 95834 59398 65661 06571 07310 69194 23953 79267 24782 62491 21444 13770 49300
94234 08382 97852 34559 80536 21298 62562 79312 47862 31586 85299 67098 62311 20279 14109
82867 48363 12047 87266 88897 34870 63677 58457 45398 61545 08323 12140 58812 59673 07190
74860 73643 40464 17535 51580 59168 98790 57586 56973 58064 91945 47609 96092 48781 98359
56515 58056 93206 51186 11516 97693 94963 71320 57091 50768 98288 39519 11626 70378 10118
36391 46003 86017 39881 53350 97291 70993 48967 50369 52232 85233 68788 61420 83659 20979
96167 29072 54193 98611 77540 15642 22382 20939 34079 74172 26403 92594 06990 97753 17455
59982 71608 00372 74343 87279 19221 81963 97774 51084 46607 92869 94778 02740 45420 21832
85592 25980 08706 43648 42553 77538 29075 99044 47600 88315 84844 19226 92304 34477 80547
50050 39389 62788 03330 38021 09714 68479 56354 23730 54856 16314 52672 14352 59933 44324
66312 66189 39319 84080 94013 91721 40616 89088 03281 14879 10033 51645 65226 77018 36101
26927 59901 54200 55736 53279 27341 12668 68661 87703 75098 67072 59955 17634 83576 02956
62870 87278 84106 94435 04319 47001 99136 58862 09866 13256 04413 39946 75212 50795 57817
33248 57090 51734 93311 01593 03985 27126 56033 36360 66849 63933 99122 42592 90799 04149
45056 56233 91649 90337 19347 49882 18739 32034 29179 56218 92237 32845 35290 42395 04345
47423 27310 29648 32488 11290 99400 43169 27980 13848 06052 17672 11257 25081 96201 97591
31021 56557 54625 07189 35180 25195 98899 77687 69649 51470 98687 97512 49427 62465 10874
06181 44130 35929 89056 78387 08159 59249 72650 30395 91161 32002 98407 19981 06186 81306
80942 95220 00255 00984 97672 73271 63246 60678 02573 31873 23073 09092

How do we use such tables? Suppose that the experimental units in the population are numbered consecutively from 1 to N. (Such a list is called a **sampling frame**.) Starting from an arbitrarily chosen position in random number tables (such as those above) choose strings of digits in an ordered way (perhaps from left to right in the table), disregarding any strings not corresponding to a number between 1 and N. Strings of length 2 are sufficient if $10 \leq N \leq 99$; of length 3 if $101 \leq N \leq 999$. Then, α_i is a sampled data point when i is selected in this way.

Example 9.11

A without-replacement random sample of tyres: Example 9.10 continued

Listed below are the 40 tyres belonging to the 10 cars used in the 50 000 km road test in Example 9.10. This list constitutes the sampling frame.

Experimental unit	Tyre number
Car 1, Tyre 1	1
Car 1, Tyre 2	2
Car 1, Tyre 3	3
Car 1, Tyre 4	4
Car 2, Tyre 1	5
Car 2, Tyre 2	6
...	...
Car 10, Tyre 3	39
Car 10, Tyre 4	40

Suppose that we wish to check the tread wear on nine of these cars. After all, if the tyres are hardly worn at all, it may be useful to run the trial for another 20 000 km (for example). That is, we wish to examine a small sample size to determine whether the experiment should be continued into a second phase.

Worked example:

Use random number tables to select 9 tyres from the 10 cars.

17698 66724 04739 95076 28408 92272 16940 12263 40578 92491 84553 51002 34033 43089
32922 72284 61370 69225 75364 54228 56052 15312 35578 71251 53185 20524 78607 88029
14100 47347 06372 48153 65613 25931 19381 62515 18485 96540 19220 01635 41476 17054
14387 10713 67498 02566 04424 84503 08315 98380 [39]231 37289 20760 53067 62971 73890
97587 03909 61222 95108 33370 71433 10743 98390 88622 52842 88554 71290 29144 14185
98762 77803 05260 69345 11312 43034 85861 45352 25442 57597 68205 14006 79245 89168
69098 80357 99638 25699 50847 05691 44295 37302 44679 54778 ...

A randomly chosen 2 digit string is shown in the random number table reproduced above. From this position, move left to right across the table producing the 2 digit strings, in order, disregarding any strings representing numbers greater than 40 or representing numbers already selected:

... [39] [23] [1 3] 7289
[20] 76 [0 5] [30] 67 6297 [1 7] [38] 90 97587 039 [09] 61222 95108 33370 71433 10743 98390 88622
52842 88554 71290 29144 14185 98762 77803 05260 69345 11312 43034 85861 45352 25442
57597 68205 14006 79245 89168 69098 80357 99638 25699 50847 05691 ...

This identifies the nine selected tyres as those with numbers

39 23 13 20 5 30 17 38 9.

The observed sample values are $x_1, x_2, x_3, \ldots, x_9$, with $x_1 = \alpha_{39}$, $x_2 = \alpha_{23}$, $x_3 = \alpha_{13}$, $x_4 = \alpha_{20}, \ldots, x_9 = \alpha_9$, where α_i is the tread wear on Tyre i.

Covariance in a without-replacement random sample

A without-replacement random sample $X_1, X_2, X_3, \ldots, X_n$ taken from the population $\alpha_1, \alpha_2, \alpha_3, \ldots, \alpha_N$ has the probability distribution

$$X_j = \begin{cases} \alpha_1 & \text{with probability } \frac{1}{N}; \\ \alpha_2 & \text{with probability } \frac{1}{N}; \\ \cdots & \cdots \qquad\qquad\qquad \cdots \\ \alpha_N & \text{with probability } \frac{1}{N}. \end{cases}$$

for each member of the sample. That is, each sample member is identically distributed. Each sample member has the same probability distribution as the population random variable. In particular, the mean of the population is given by

$$E(X_j) = \sum_{i=1}^{N} \alpha_i P(X_j = \alpha_i) = \sum_{i=1}^{N} \alpha_i \frac{1}{N} = \overline{\alpha} = \mu. \tag{9.5}$$

The variance of the population is given by

$$Var(X_j) = E(X_j^2) - E(X_j)^2 = \sum_{i=1}^{N} \alpha_i^2 \frac{1}{N} - \overline{\alpha}^2 = \frac{\sum_{i=1}^{N}(\alpha_i - \overline{\alpha})^2}{N} = \sigma^2,$$

but ...different sample members are not independent.

Theorem 9.5 *Suppose that $\alpha_1, \alpha_2, \alpha_3, \ldots, \alpha_N$ is a finite population with mean μ and variance σ^2. If $X_1, X_2, X_3, \ldots, X_n$ is a without-replacement random sample from this population, then*

$$Cov(X_i, X_j) = -\frac{\sigma^2}{N-1}, \qquad j \neq k.$$

Proof: The following proof avoids complicated conditional probability calculations involving the α_j. Define new random variables $Y_1, Y_2, Y_3, \ldots, Y_N$ which are a random permutation of the population values $\alpha_1, \alpha_2, \alpha_3, \ldots, \alpha_N$. Then, by the definition of the without-replacement random sample, the mean and variance of the Y values is the same as the mean and variance of the X values, that is, the population mean and variance: for each $i = 1, 2, 3, \ldots, n$ and $j = 1, 2, 3, \ldots, N$,

$$E(Y_j) = E(X_i) = \mu \quad \text{and} \quad Var(Y_j) = Var(X_i) = \sigma^2.$$

One way to think of the random variables $Y_1, Y_2, Y_3, \ldots, Y_N$ is as being a without-replacement random sample size N, of all elements of the population. Essentially, extra points are sequentially selected to raise the sample size from n to N. Therefore, the covariance between different Y sample members is the same as the covariance between different X sample members: if we label this common covariance as γ, then

$$\gamma = Cov(X_i,\ X_j) = Cov(Y_l,\ Y_m), \quad i \neq j, \ \ l \neq m.$$

Since, $Y_1 + Y_2 + Y_3 + \ldots + Y_N = \alpha_1 + \alpha_2 + \alpha_3 + \ldots + \alpha_N = N\mu$ is constant, it follows from **SUM 3** (Equation 8.2) that

$$\begin{aligned} 0 &= Var(N\mu) \\ &= Var(Y_1 + Y_2 + Y_3 + \ldots + Y_N) \\ &= \sum_{i=1}^{N} Var(Y_i) + 2\sum_{i<j} Cov(Y_i, Y_j) \\ &= \sum_{i=1}^{N} \sigma^2 + 2\sum_{i<j} \gamma \\ &= N\sigma^2 + (N^2 - N)\gamma. \end{aligned}$$

The result then follows by solving this equation for γ.

■

This result enables us to determine the behaviour of the sampling distribution of the mean when sampling is performed without replacement. In particular we are able to write down the mean and variance of this sampling distribution in a 'without-replacement' random sampling version of Theorem 9.2.

Theorem 9.6 *Suppose that $\overline{X}$ denotes the mean of a without-replacement random sample $X_1, X_2, X_3, \ldots, X_n$ taken from a finite population of N elements with mean μ and variance σ^2. Then*

$$E(\overline{X}) = \mu \quad \textit{and} \quad Var(\overline{X}) = \frac{\sigma^2}{n}\left(\frac{N-n}{N-1}\right).$$

Proof: By Equation 9.5, $E(X_i) = \mu$ for each i, so that $E(\overline{X}) = \mu$ follows exactly the same argument as the *with* replacement case given in the proof of Equation 9.2. Further,

$$Var(\overline{X}) = \frac{1}{n^2} Var(X_1 + X_2 + X_3 + \ldots + X_n),$$

and by **SUM 3** again $Var(X_1 + X_2 + X_3 + \ldots + X_n)$ is the sum of the n^2 terms in the variance-covariance matrix

$$\begin{pmatrix} \sigma^2 & -\frac{\sigma^2}{N-1} & -\frac{\sigma^2}{N-1} & \cdots & -\frac{\sigma^2}{N-1} \\ -\frac{\sigma^2}{N-1} & \sigma^2 & -\frac{\sigma^2}{N-1} & \cdots & -\frac{\sigma^2}{N-1} \\ -\frac{\sigma^2}{N-1} & -\frac{\sigma^2}{N-1} & \sigma^2 & \cdots & -\frac{\sigma^2}{N-1} \\ \vdots & \vdots & \vdots & \ddots & \vdots \\ -\frac{\sigma^2}{N-1} & -\frac{\sigma^2}{N-1} & -\frac{\sigma^2}{N-1} & \cdots & \sigma^2 \end{pmatrix}$$

Therefore,

$$\begin{aligned} Var(\overline{X}) &= \frac{1}{n^2}\left[n\sigma^2 + (n^2 - n)(-\frac{\sigma^2}{N-1})\right] \\ &= \frac{\sigma^2}{n}\left(1 - \frac{n-1}{N-1}\right) \\ &= \frac{\sigma^2}{n}\left(\frac{N-n}{N-1}\right). \end{aligned}$$

■

The standard error of the mean is the positive square root of $Var(\overline{X})$. If N is large so that $N - 1 \approx N$, then

$$Var(\overline{X}) = \frac{\sigma^2}{n}(1 - f),$$

where $f = \frac{n}{N}$ is called the **sampling fraction**. Clearly if n is very small relative to N, then the sampling fraction is small, and $Var(\overline{X}) \approx \frac{\sigma^2}{n}$. This is the the formula from Theorem 9.2 for the variance of the sample mean for sampling from a finite population with replacement or for sampling from an infinite population.

It makes sense to regard the quantity $(1 - f)$ as the **finite population correction factor**. The effect of this correction factor is often ignored if $f < 0.05$.

Example 9.12

SA FM radio station listeners

On a campus with 10 000 students, 60% listen to SA FM radio station at least once a week. Assuming that this proportion of student listeners hasn't changed, determine the standard error in the proportion of a sample of 1000 students who acknowledge listening to SA FM at least once a week.

For a random variable X_i record a 1 if Student i listens to SA FM at least once a week; 0, otherwise, $i = 1, 2, 3 \ldots, 1000$. Then, $X_i \sim Bin(1, 0.6)$

and $X_1, X_2 \ldots, X_{1000}$ constitute a without-replacement random sample of size $n = 1\,000$ from a population of size $N = 10\,000$. The proportion of the sample listening to SA FM is

$$\overline{X} = \frac{\sum_{i=1}^{1000} X_i}{1000}.$$

This sample mean has variance

$$Var(\overline{X}) = \frac{0.6(1-0.6)}{1000}\left(1 - \frac{1000}{10000}\right) = 0.000216$$

and therefore standard error 0.0147 (4 dp).
(Here we have used the approximation $N-1 \approx N$.) If the finite population correction factor is ignored, the standard error is 0.0155 (4 dp).

The hypergeometric distribution

Apply the results of the previous section to the following finite population:

$$\{\alpha_1, \alpha_2, \alpha_3, \ldots, \alpha_N\} \equiv \{\ \underbrace{1, 1, 1, \ldots, 1}_{S \text{ identical items}}\ ,\ \underbrace{0, 0, 0 \ldots 0}_{N-S \text{ identical items}}\ \}.$$

That is, the population consists of two distinct types of items: 1 (possibly denoting 'success') and 0 (possibly denoting 'failure'). Let $X_1, X_2, X_3, \ldots, X_n$ denote a without-replacement random sample of size $n < N$ from this population. Then, marginally, for each j, the X_j have the following probability distribution:

$$X_j = \begin{cases} 1 & \text{with probability} \quad p = \frac{S}{N}; \\ 0 & \text{with probability} \quad 1-p. \end{cases}$$

Thus the X_j are Bernoulli random variables and $E(X_j) = \mu = p$ and $Var(X_j) = \sigma^2 = p(1-p)$. The random variable $\overline{X}$ denotes the proportion of successes in the without-replacement random sample. According to Theorem 9.6, the mean and variance of this proportion is given by

$$E(\overline{X}) = \mu = p \quad \text{and} \quad Var(\overline{X}) = \frac{\sigma^2}{n}\left(\frac{N-n}{N-1}\right) = \frac{p(1-p)}{n}\left(\frac{N-n}{N-1}\right),$$

where $p = \frac{S}{N}$.

The probability distribution of the total number of successes is called the hypergeometric distribution.

Definition 9.6 *If* $X = X_1 + X_2 + X_3 + \ldots + X_n$ *denotes the total number of* $1'$*s in a without-replacement sample of size* n *from the dichotomous population*

$$\{\alpha_1, \alpha_2, \alpha_3, \ldots, \alpha_N\} \equiv \{\ \underbrace{1, 1, 1, \ldots, 1}_{S \text{ identical items}}\ ,\ \underbrace{0, 0, 0 \ldots 0}_{N-S \text{ identical items}}\ \}$$

containing S 1's and N − S 0's, then we say that X has a **hypergeometric distribution** *with parameters N, S and n and we write*

$$X \sim Hypergeom(N, S, n).$$

□

The hypergeometric distribution is the parallel of the binomial distribution to the case where the sequence of Bernoulli trials being monitored is **not** independent. The population values 0 and 1 are labels to indicate that the population consists of items of two types only. This is the same type of labelling for 'success' and 'failure' that was used in Definition 6.3.

The following theorem gives the probability mass function, the mean and the variance of a hypergeometric random variable.

Theorem 9.7 *If $X \sim Hypergeom(N, S, n)$, then X has probability mass function*

$$P(X = x) = \frac{\binom{S}{x}\binom{N-S}{n-x}}{\binom{N}{n}} \quad \text{for} \quad max\{0, n - N + S\} \leq x \leq min\{n, S\},$$

with mean $E(X) = np$ and $Var(X) = np(1-p)\left(\frac{N-n}{N-1}\right)$, where $p = \frac{S}{N}$.

Proof: The mass function is derived using the combinatorial arguments given in Example 3.9. The mean and variance follow from Theorem 9.6:

$$E(X) = E(n\overline{X}) = nE(\overline{X}); \quad Var(X) = Var(n\overline{X}) = n^2 Var(\overline{X}).$$

■

Notice that when the sampling fraction is small this mean and variance reduce to that of the binomial distribution. When the sampling fraction is less than 5%, hypergeometric probabilities may be approximated by binomial probabilities.

Example 9.13

Checking for exhaust pollution

As part of a campaign to reduce city smog levels, the Environmental Protection Agency (EPA) has the fleet of a small bus company inspected for exhaust pollution. The company has a fleet of 16 buses, 2 of which require engine overhauls before their pollutant emission levels would satisfy EPA guidelines.

Worked example:

If the inspector examines 8 buses chosen randomly from the fleet, determine the probability that the 2 substandard buses are 'caught' by the inspector.

Sampling in this example is, of course, without-replacement, since the inspector will not want to look at the same bus twice!

If we denote by X the number of defective buses 'nabbed' by the inspector, then, regarding the 'nabbing' as a 'success', $S = 2, N = 16$ and $n = 8$ so that the exact probability distribution for X is $X \sim Hypergeom(16, 2, 8)$. By substituting directly into the formula for the pmf for X, we find that

$$P(X = 2) = \frac{\binom{2}{2}\binom{16-2}{8-2}}{\binom{16}{8}} = 0.233 \quad (3 \text{ dp}).$$

For the binomial approximation to this probability we imagine that the inspector samples *with* replacement. On any of the $n = 8$ trials, the inspector selects a defective bus with probability $\frac{2}{16}$. With this independence assumption, for the random variable Y denoting the number of defective buses nabbed by the inspector, $Y \sim Bin(8, \frac{1}{8})$. Hence,

$$P(Y = 2) = \binom{8}{2}(\frac{1}{8})^2(1 - \frac{1}{8})^{8-2} = 0.196 \quad (3 \text{ dp}).$$

The probabilities are surprisingly close with a sampling fraction of 50%.

EXERCISES

9.32 **Drawing a without-replacement random sample sequentially**.

Suppose that we wish to draw a random sample of size n without replacement from a population of size $N > n$. When using random number tables for this purpose, at each stage an item from the population is selected in an 'equally likely' manner from those items not already in the sample.

When there are already i items in the sample, a specific population item is then included with probability

$$\frac{n-i}{N-i}, \quad i = 0, 1, 2, 3, \ldots, n-1.$$

Show that this sampling scheme is in fact without-replacement random sampling in the sense (of Definition 9.5) that every possible subset of size n in the population has the same chance

$$\frac{1}{\binom{N}{n}}$$

of selection as the sample.

9.33 Members of the electoral roll are regularly balloted for jury duty. In this way a pre-jury of 32 electors is assembled for a given trial. From this pre-jury 12 'good men and women' are randomly selected to serve as the jury for the trial. (The pre-jury is assembled to provide alternatives when defence and prosecution object to the inclusion of a given individual during the sequential selection of the jury 12 from the pre-jury 32.)

For the purpose of this question, assume that all allocations are done randomly.

(a) How many different juries of 12 may be chosen from the pre-jury 32?

(b) Show that each possible jury chosen from the 32 pre-jury members has (approximate) probability 4×10^{-9} of final jury status.

(c) Does (b) indicate that each of the 32 pre-jury members has the same chance 4×10^{-9} of being among the final 12 jury members?

(d) Suppose that I have been balloted into the pre-jury of 32. Explain why there are $\binom{31}{11}$ possible juries which have me as the 'twelfth man'. Show that the probability that I find myself in the final jury 12 is

$$\frac{\binom{31}{11}}{\binom{32}{12}}$$

and show that this simplifies to $\frac{12}{32}$.

9.34 In Example 9.13, a pollution inspector checks the exhaust emissions in a fleet of 16 buses. Construct a sampling frame for the fleet (that is, provide an identifying label for each bus).

Use random number tables to select the sample of size 8 for the inspector:

(a) with replacement;

(b) without replacement.

Describe carefully how you have used the tables in each case.

9.35 Two fencing staples are drawn without replacement from a box containing N staples, D of which are defective, $0 < D < N$. (A fencing staple is U-shaped, made from bent wire with pointed ends. The staple is defective when the sides of the U are different lengths.)

Let X denote the number of defectives in the first draw and Y denote the number of defectives in the second draw.

Show that

$$\rho_{X,Y} = -\frac{1}{N-1}.$$

Note that this correlation does not depend on how many defectives there are in the box.

9.36 A research committee is to consist of 5 people chosen randomly from 15 staff of the Department of Statistics and 15 senior students from the department.

Determine the probability that the committee contains:

(a) no students;

(b) no staff;

(c) exactly one student.

9.37 The support of the hypergeometric distribution occurs at integer values of x in the range

$$max\{0, n-N+S\} \le x \le min\{n, S\},$$

when $X \sim Hypergeom(N, S, n)$.

Explain how this formula works. It may be helpful to consider two cases for a population of size 20, half of which is sampled without replacement:

- Case 1: 90% of the population are successes;
- Case 2: 10% of the population are successes.

Give the support of the hypergeometric distribution in each case.

9.38 A company dispatches inexpensive digital watches in cartons containing 200 watches. Before dispatch, a carton is subjected to the following inspection plan:

- select 20 watches from the carton;
- check whether each watch battery is run down (most defective watches are due to a flat battery);
- dispatch carton if there is at most one flat battery; reject carton otherwise.

(a) Determine the probability of rejecting a carton where the batteries of 20 watches are flat. (Use sampling with replacement.)

(b) Determine the probability of rejecting a carton where the batteries of 10 watches are flat. (Use sampling with replacement.)

(c) Determine the probability of accepting a carton where the batteries of 5 watches are flat when the sample is taken without replacement.

(d) The manufacturers are aware that overall 5% of the batteries that they use in their watches are flat.

In this case, is the inspection plan suitable in terms of maintaining the quality of dispatched cartons?

9.39 A list of 10 000 computer-generated 'pseudo-random' digits contained 1100 zeros.

(a) If the digits are indeed 'random', how many zeros would you expect there to be?

(b) Give the probability distribution of Y, the number of zeros in 10 000 'random' digits.

(c) Use a normal approximation to determine $P(Y \geq 1100)$.

(d) Is the count of 1100 zeros (or more) consistent with the digits being 'random' ?

(e) Using output from your favourite statistics computer package, test the pseudo-random number generator for randomness in the manner described above. (This could be done manually by counting the number of zeros in the MINITAB output in this section.)

9.4 SAMPLING METHODS: STRATIFIED/PROPORTIONAL/SYSTEMATIC

Random sampling, either with or without replacement, is easily applied to a finite population. In the previous two sections, properties of the sampling distribution of the mean were derived for both sampling methods. In particular, the size of $S.E.(\overline{X})$ gives an impression of the accuracy of $\overline{X}$, the sample mean, in estimating the parameter μ, the population mean.

In this section we suggest that by splitting the population into disjoint sections, it may be possible to adopt a sampling scheme that enables the population mean to be estimated more precisely, in the sense that $S.E.(\overline{X})$ is smaller.

The disjoint sections (or **strata**) should be chosen so that the magnitude of the data being measured within a given stratum is relatively homogeneous, but is as variable as possible between strata. For example, in examining the average tax paid by companies in Tasmania, more precision may be gained by stratifying the population of all Tasmanian companies by size: that is, split the population into two disjoint pieces, large companies/small companies, where size may be measured by annual turnover. Sampling then within each strata would equitably ensure that the impact of smaller companies within the sample is not swamped by an overabundance of larger companies. We may represent the k strata for a population diagrammatically as:

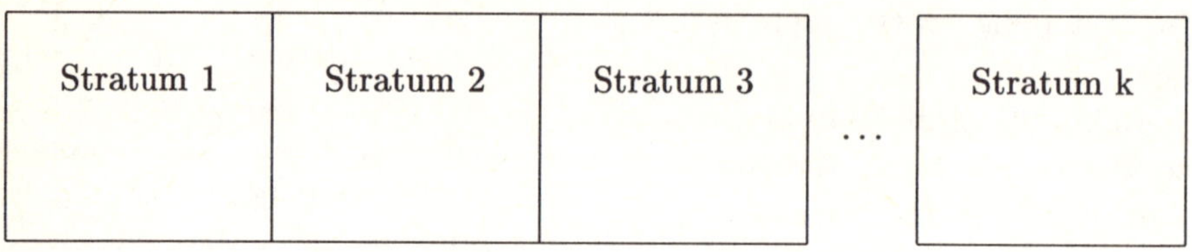

If there are N_i items in Stratum i, then $N_1 + N_2 + N_3 + \ldots + N_k = N$, the total population size.

Definition 9.7 *A* **stratified random sample** *from k strata is a (without-replacement) random sample from each stratum; n_i are sampled from Stratum i, so that $n_1 + n_2 + n_3 + \ldots + n_k = n$, the total sample size.*

□

Here is a summary of all the notation that we will use:

population size of Stratum i	N_i
sample size from Stratum i	n_i
population mean of Stratum i	μ_i
sample mean from Stratum i	$\overline{X_i}$
population variance of Stratum i	σ_i^2
sample variance from Stratum i	s_i^2

The ideas of this section are simply demonstrated with $k = 2$ strata. Recall from the start of this chapter on sampling, that sample means may be used to estimate population means, and sample variances may be used to estimate unknown population variances. With this in mind we will examine the mean of the stratified population and see how to estimate it. Since

$$\begin{aligned}\mu &= \frac{\text{population total}}{N} \\ &= \frac{\text{Stratum 1 total} + \text{Stratum 2 total}}{N} \\ &= \frac{N_1}{N}\left(\frac{\text{Stratum 1 total}}{N_1}\right) + \frac{N_2}{N}\left(\frac{\text{Stratum 2 total}}{N_2}\right) \\ &= \frac{N_1}{N}\mu_1 + \frac{N_2}{N}\mu_2 \\ &= \omega_1\mu_1 + \omega_2\mu_2,\end{aligned}$$

we may relate the true mean of the population and the true mean of the strata as follows:

> The population mean μ and the strata means μ_i are related by $\mu = \omega_1\mu_1+\omega_2\mu_2$, where $\omega_i = \frac{N_i}{N}$ is the **natural weight** of the ith stratum.

How do we estimate all these population and stratum means? Simply by using the corresponding sample means: the stratified sample mean, $\overline{X}_{strat}$ is defined by

$$\overline{X}_{strat} = \omega_1\overline{X_1} + \omega_2\overline{X_2}. \tag{9.6}$$

How do we estimate the population total? Simply by using the the stratified sample mean: the estimated population total, $\widehat{T}$, of the variable being sampled is found by

$$\widehat{T} = N(\overline{X}_{strat}). \tag{9.7}$$

The concepts of estimating means and totals will be examined more closely in the next chapter.

To obtain a measure of the standard error of the estimate $\overline{X}_{strat}$ of the population mean μ, we assume that the sample means from different strata are uncorrelated. Then, by **SUM 2** and the results of Theorem 9.6 for without replacement random sampling applied to each stratum,

$$\begin{aligned} Var(\overline{X}_{strat}) &= Var(\omega_1\overline{X_1} + \omega_2\overline{X_2}) \\ &= \omega_1^2 Var(\overline{X_1}) + \omega_2^2 Var(\overline{X_2}) \\ &= \omega_1^2 \frac{\sigma_1^2}{n_1}\left(\frac{N_1 - n_1}{N_1 - 1}\right) + \omega_2^2 \frac{\sigma_2^2}{n_2}\left(\frac{N_2 - n_2}{N_2 - 1}\right). \end{aligned}$$

In **proportional stratified sampling** the sample sizes $n_1, n_2, \ldots$ are chosen proportional to the stratum sizes (and therefore proportional to the natural weights). That is,

$$\frac{n_1}{n} = \frac{N_1}{N} = \omega_1 \quad \text{and} \quad \frac{n_2}{n} = \frac{N_2}{N} = \omega_2.$$

This will ensure that the fraction of a stratum sampled is the same as the fraction of the whole population sampled. Symbolically, for each i,

$$\frac{n_i}{N_i} = \frac{n}{N} = f,$$

the sampling fraction. Using these results in $Var(\overline{X}_{strat})$, it follows that

$$\begin{aligned} Var(\overline{X}_{strat}) &\approx (\omega_1^2 \frac{\sigma_1^2}{n_1} + \omega_2^2 \frac{\sigma_2^2}{n_2})(1 - f) \\ &= (\omega_1^2 \frac{\sigma_1^2}{\omega_1 n} + \omega_2^2 \frac{\sigma_2^2}{\omega_2 n})(1 - f) \\ &= \frac{\omega_1\sigma_1^2 + \omega_2\sigma_2^2}{n}(1 - f). \end{aligned}$$

Example 9.14

Selective milling

In the diagram below is a plantation of *Pinus radiata* where the trees are planted in rows to permit maximum growth per square km. Plantation 1 is older than Plantation 2, and trees with a girth of 2 metres are painted with a large X ready for selective milling. As the larger trees are removed from the plantation, the smaller trees have more light and ground space to encourage growth.

We will use stratified random sampling to determine the total number of trees that are suitable for milling; stratification by plantation keeps the younger and older plantations separate.

Plantation 1

ooXoooXoooooooXooooXoXXooXoooXoXoXXXooooXooXXoXooooXooXooooo
oXooXXXXooXooXXoooooXXoooXXoooooooooXooXoooXoooXoooooooooXoo
• oXooXooXoooooXXoXooXoXXoooXXoXoooooooooooooooXoooXoXooXoooXXoo
oXooXoXoooXoooXoooooXoooXoXoXoXoXoXooooXoooXoooooooXooooXoXX
oooooooooXooooXoooooXooooXoooXoXXXXXoooXoooXooooooooooXooXoo
• oXoXXXXoXooooXXooooXXXoooXXooXoooooXXooXoooooooooooooooooooXoX
• ooooooooooXXooooXXoooXooooXXXXoooooooooXoooXXooXooooXoXoXXoX
oXooXXXXoooooooooooooooooooooooooooooooooooooooXXoXoXoXX
ooooooXoooXooooooooXXXXooooXoXoooXXXoooXXooXXoXooooXooXooXoo
oXooXXXXXoXooXXooooXXXoooXXXoXXooooXXooXoooooooXooooooXooXoo
XXXXXXoXoooooXooXoooooooooooooooooooooooooooooXoooXXoXXXooXXoX
oXooXoXXooXoooXooooXXXooXXooXoXoXoXooooXoooXooooooooXoooXoXX
ooooooooooXoooXoooooXooooXoXoXoXXXXXoooooooXoooooooooooXooooo
• oXooXXXoXooooXXooooXXXoooXXXoooooooXXoXXoooXooooooXooooooooX
oXXXXXXXooXXoXXoXXoXoXooooXXXXooXooooooXoooXoooooooooooooooX
ooooXXXXoXXoXoXooooXXXXooXooooXoooooooooooooooooooooooooooXoo

Plantation 2

oooooooooooXooooXooooXooooXoXXoooooooooooooXXoooooooXoXooXoX
• ooooXXoXooooooooooooooooooooooooooooooooooooooooXooXoXooX
ooooooXooooooooooooXXoXooooXoooooooXoooooooXXoXoooooooXooXoo
ooooXXoXooXoooXooooXoooooXXoooXoooooooooooooooooooooooooXoo
ooooooooooooooooooooXooooooooXoooooXoooXoooooooooooooooooooooo
• ooooooooooooooooooooooooooooXoooooXoooooooooooooooooooooXoo
ooooooooooXooooooXoooXooooooXooooooooooXooooooooooooXoooXooo
ooooXoXoooooooooooooooooooooooooooooooooooooooooooooXoooXo

The variable to be measured is row count of X's. Using random number tables, the rows marked • were sampled with proportional allocation: Both plantations combined contain $N_1 + N_2 = 16 + 8 = 24 = N$ rows; financial resources allow only $n = 6$ rows to be counted. Sample sizes n_1 and n_2 are chosen so that

$$\frac{n_1}{16} = \frac{6}{24} \quad \text{and} \quad \frac{n_2}{8} = \frac{6}{24}.$$

Therefore $n_1 = 4$ rows and $n_2 = 2$ rows are selected from Plantations 1 and 2 respectively. This produced the following data sets for row counts:

Plantation 1: 18, 19, 18, 20;

Plantation 2: 7, 3.

The natural weights are $\omega_1 = \frac{16}{24} = \frac{2}{3}$ and $\omega_2 = \frac{8}{24} = \frac{2}{3}$. According to Equation 9.6,

$$\begin{aligned} \overline{X}_{strat} &= \omega_1 \overline{X_1} + \omega_2 \overline{X_2} \\ &= (\frac{2}{3})(\frac{18+19+18+20}{4}) + (\frac{1}{3})(\frac{7+3}{2}) \\ &= 14.16667. \end{aligned}$$

To estimate the total number of 'millable' trees, by Equation 9.7,

$$\textit{Estimated total} = \widehat{T} = N(\overline{X}_{strat}) = (24)(14.16667) \approx 340.$$

(In fact there are 343 trees in both plantations combined but such information on the total population count would normally not be available.) Further,

$$Var(\overline{X}_{strat}) \approx \frac{\omega_1\sigma_1^2 + \omega_2\sigma_2^2}{n}(1-f).$$

The best that we have available for estimating the strata variances are the sample variances $s_1^2 = 0.92$ and $s_2^2 = 8$ for each of Strata 1 and 2 respectively. Therefore, the estimated variance of the stratified mean is given by

$$\frac{(\frac{2}{3})s_1^2 + (\frac{1}{3})s_2^2}{6}(1 - \frac{6}{24}) = \frac{(\frac{2}{3})(0.92) + (\frac{1}{3})(8)}{6}(0.75) = 0.4096.$$

The estimated standard error in the population total is

$$\begin{aligned} S.E.(\text{estimated total}) &= N(\text{estimated } S.E.(\overline{X}_{strat})) \\ &= (24)(\sqrt{0.4096}) \approx 15.36. \end{aligned}$$

Gains produced by stratification

The standard error for the sample mean is smaller for a stratified random sample than it is for a simple random sample. The demonstration of this main advantage of stratified sampling depends on several applications of

$$\sum_{i=1}^{N}(\alpha_i - a)^2 = \sum_{i=1}^{N}(\alpha_i - \mu)^2 + N(\mu - a)^2. \tag{9.8}$$

This equation was discussed in Chapter 2 in the context of the sample variance.

Theorem 9.8 *If $\overline{X}$ denotes the sample mean of a (without-replacement) random sample of size n from the population $\{\alpha_1, \alpha_2, \alpha_3, \ldots, \alpha_N\}$ and $\overline{X}_{strat}$ denotes the stratified sample mean when proportional stratified sampling is used in $k = 2$ strata, then*

$$Var(\overline{X}) = Var(\overline{X}_{strat}) + \frac{w_1(\mu_1 - \mu)^2 + w_2(\mu_2 - \mu)^2}{n}(1-f),$$

where f is the sampling fraction, w_i are the natural weights and μ_i are the stratum means.

Proof: Since $Var(\overline{X}) = \frac{\sigma^2}{n}$ when sampling fractions are ignored, the proof depends on the re-structuring of σ^2 according to Equation 9.8.

$$\begin{aligned}
\sigma^2 &= \frac{\sum_{i=1}^{N}(\alpha_i-\mu)^2}{N} \\
&= \frac{\sum_{Stratum\ 1}(\alpha_i-\mu)^2+\sum_{Stratum\ 2}(\alpha_i-\mu)^2}{N} \\
&= \frac{[\sum_1(\alpha_i-\mu_1)^2+N_1(\mu_1-\mu)^2]+[\sum_2(\alpha-\mu_2)^2+N_2(\mu_2-\mu)^2]}{N} \\
&= \frac{N_1}{N}\left[\frac{\sum_{Stratum\ 1}(\alpha_i-\mu_1)^2}{N_1}+(\mu_1-\mu)^2\right] \\
&\quad +\frac{N_2}{N}\left[\frac{\sum_{Stratum\ 2}(\alpha_i-\mu_2)^2}{N_2}+(\mu_2-\mu)^2\right] \\
&= \omega_1\sigma_1^2+\omega_2\sigma_2^2+w_1(\mu_1-\mu)^2+w_2(\mu_2-\mu)^2.
\end{aligned}$$

The stated result follows by multiplying each side of the last equation by $\frac{1-f}{n}$.

■

The result of Theorem 9.8 shows that when the population mean μ differs from the stratum means μ_1 and μ_2, then $Var(\overline{X}) > Var(\overline{X}_{strat})$, so that the standard error of the mean is smaller for stratified random sampling; that is, $\overline{X}_{strat}$ is less variable (and more accurate) in estimating μ.

Systematic sampling

The experimental units in a finite population may be listed in a sampling frame. Such a labelled list of the population may only exist in theory, simply because the target population of interest may be difficult to locate. For example, a researcher studying a cattle disease which is most prevalent in Friesian heifers would have trouble assembling a sampling frame of all Friesian heifers in Australia. Such practical problems may seem overwhelming to the researcher. However, now we discuss a sampling scheme that is effective when a sampling frame *is* at call; a list of the entire population may exist in a filing cabinet right beside your desk!

Definition 9.8 *A* **systematic sample** *of size n from* $\{\alpha_1, \alpha_2, \ldots, \alpha_N\}$, *is a sample selected according to the following sampling scheme:*

- *Randomly select a single item,* α_l*, from the first k items in the population, where* $k = \left[\frac{N}{n}\right]$ *is the greatest integer less than or equal to* $\frac{N}{n}$.
- *Select, in order, items* $\alpha_l, \alpha_{l+k}, \alpha_{l+2k}, \alpha_{l+3k}, \ldots$ *until a sample of size n is obtained.*

□

Example 9.15
Sampling account holders systematically
In a publicity campaign, the Commonwealth Bank on campus offers \$500 deposits to 10 lucky account holders who opened accounts with the bank within a period of 3 months following the first day of classes. The accounts that qualified for the draw had account numbers serially numbered from

$$022\ 1223\ 2369 \quad \text{to} \quad 022\ 1223\ 9115.$$

Worked example:
Show how to select a systematic random sample of size 10 from the eligible account holders.
There are $N = 6747$ accounts between 022 1223 2369 and 022 1223 9115 inclusive. We begin by selecting an account randomly from the first $\left[\frac{N}{n}\right] = \left[\frac{6747}{10}\right] = 674$ accounts. Selecting the first account: start at a random point in random number tables and then sequentially examine 3 digit strings left to right in the table until a 3 digit number is located in the range 001 to 674 inclusive. Suppose that this random first selection step yields the 3 digit string 347. This corresponds to the lucky account number 022 1223 2369 + 347 = 022 1223 2716. Then every 674th account is selected, giving the following sample from within the required range of account numbers:

$$\begin{aligned}
022\ 1223\ 2369 + 347 &= 022\ 1223\ 2716\\
022\ 1223\ 2716 + (1)(674) &= 022\ 1223\ 3390\\
022\ 1223\ 2716 + (2)(674) &= 022\ 1223\ 4064\\
022\ 1223\ 2716 + (3)(674) &= 022\ 1223\ 4738\\
022\ 1223\ 2716 + (4)(674) &= 022\ 1223\ 5412\\
022\ 1223\ 2716 + (5)(674) &= 022\ 1223\ 6086\\
022\ 1223\ 2716 + (6)(674) &= 022\ 1223\ 6760\\
022\ 1223\ 2716 + (7)(674) &= 022\ 1223\ 7434\\
022\ 1223\ 2716 + (8)(674) &= 022\ 1223\ 8108\\
022\ 1223\ 2716 + (9)(674) &= 022\ 1223\ 8782
\end{aligned}$$

In Example 9.15, the account numbers provide a sampling frame. Even though the accounts are serially numbered, this does not apply to the owners of the accounts; the account holders are a random cross-section of students who had become clients of the bank during the stated period. It is the *account holders* that are the target population. Therefore, even though the account numbers are drawn systematically, a (without replacement) random sample of account holders results.

In using the mean $\overline{X}_{sys}$ of the systematic sample to estimate the population mean μ, where the population items are listed in the sampling frame in random order,

$$Var(\overline{X}_{sys}) = \frac{\sigma^2}{n}(1-f),$$

where f is the sampling fraction. Therefore the standard error of $\overline{X}_{sys}$ is the same as the standard error of a without replacement random sample. The accuracy of the two sampling methods is the same and the systematic sample may have an advantage in being cost effective and easily implemented.

However, the population may not be in a random order; there may be an inherent ordering to items in the population, or the magnitudes of the items in the population may change periodically. That is, there may be a trend effect or a seasonal effect (or both) present in the population. In this case, every kth member of the population (comprising the systematic sample) may be correlated. This will have an effect on our ability to estimate the population mean precisely.

Suppose that $X_1, X_2, X_3, \ldots, X_n$ represents a systematic sample of size n taken from a finite population which has mean μ and variance σ^2. In a simple case, assume that there is a constant correlation acting between different sample members:

$$\rho_{X_i,X_j} = \rho, \quad \text{for} \quad i \neq j.$$

Then, in a simple application of **SUM 3** that was used in the proof of Theorem 9.6, and ignoring the effect of the sampling fraction,

$$\begin{aligned}
Var(\overline{X}_{sys}) &= Var\left(\frac{X_1 + X_2 + X_3 + \ldots + X_n}{n}\right) \\
&= \frac{1}{n^2} Var(X_1 + X_2 + X_3 + \ldots + X_n) \\
&= \frac{1}{n^2}\left[\sum_{i=1}^{n} Var(X_i) + 2\sum_{i<j} Cov(X_i, X_j)\right] \\
&= \frac{1}{n^2}\left[\sum_{i=1}^{n} \sigma^2 + 2\sum_{i<j} \rho(\sigma)(\sigma)\right] \\
&= \frac{\sigma^2}{n^2}\left[n + (n^2 - n)\rho\right] \\
&= \frac{\sigma^2}{n}\left[1 + (n-1)\rho\right].
\end{aligned}$$

Therefore, only when the sampling fraction is particularly small and the correlation between sample items is negligible, do the formulas for $Var(\overline{X})$ and $Var(\overline{X}_{sys})$ agree. If $\rho < 0$ then, if the sampling fraction is ignored, systematic sampling may produce a sample mean with a smaller standard error than for without replacement random sampling.

EXERCISES

In the four problems that follow, our task is to compare the accuracy of different sampling methods for estimating the average monthly rental for business premises in Melbourne. We will sample from the 240 regions labelled in the 'Melways' map of Melbourne.

(You may, for example, take a map of your home town and cover it in a grid of 240 non-overlapping squares.)

9.40 Explain how you would select a systematic sample of size 50 regions from the 240 regions listed in the sampling frame. Write down the numbers of the regions selected for your sample. (One business could be randomly selected from each region chosen so that the data would consist of a list of 50 monthly rentals from across the city.)

9.41 (continuation) Suppose that a stratified random sample is selected using the same map of Melbourne stratified according to:

Stratum 1: Regions 1–24 (Central Business District);
Stratum 2: Regions 25–240 (Greater Metropolitan Melbourne).

Assume that the variance of the stratified sample mean is given by

$$w_1^2\frac{\sigma_1^2}{n_1}\left(1-\frac{n_1}{N_1}\right)+w_2^2\frac{\sigma_2^2}{n_2}\left(1-\frac{n_2}{N_2}\right),$$

where N_i, n_i, w_i and σ_i denote the population size, sample size, chosen weight, and sample standard deviation for Stratum i, for $i = 1, 2$.

Use this formula to show that the mean monthly rental may be estimated more precisely by using proportional stratified sampling than by systematic sampling provided

$$w_1\sigma_1^2 + w_2\sigma_2^2 < \sigma^2,$$

where σ^2 is the entire population variance.

9.42 (continuation) Obtain a proportional stratified random sample of total size 50 from the two strata. Write down the numbers of the regions selected for your sample. Explain how the regions were selected from the sampling frame.

9.43 (continuation) Estimate the variance of the stratified sample mean when a proportional stratified sample of 50 regions is selected using natural weights. Take the sample standard deviations for each stratum as $s_1 = 300$ and $s_2 = 200$ and use these as best estimates of the population stratum standard deviations σ_1 and σ_2.

9.44 Describe a basis of stratification likely to be useful in selecting 100 railway

9.44 Describe a basis of stratification likely to be useful in selecting 100 railway sleepers from the rail line between Adelaide and Perth, to estimate the average remaining service life of the sleepers.

9.45 Give a sampling frame for the pine plantation data of Example 9.14 for both plantations combined. Use the frame to generate a without-replacement random sample of $n = 6$ rows from the $N = 24$ rows aggregated from both plantations.

Use the sample to estimate the mean number of 'millable' trees per row. Estimate the standard error of this estimate, and compare your results with those presented in Example 9.14 applying to the stratified random sample.

9.46 An experimenter testing a new growth hormone wishes to estimate the weight of 90 rats (of which 60 are male) after one month on the hormone diet. The weights of the rats are apparently gender-related. The experimenter decides to weigh a proportionally allocated sample of 9 rats stratified by gender.

(a) Determine the sampling fraction, the natural weights, and the required male/female sample sizes that the experimenter will use.

(b) In what manner are the sample average weights for male/female rats combined to obtain an estimate of the mean diet-induced weight of all such rats?

(c) If the sample standard deviation of the male rats was 2.7 grams and the sample standard deviation of the female rats was 1.9 grams, give an estimate of the variability in the mean weight of rats on the hormone diet.

9.47 A hardware store wishes to estimate the average amount its customers spend on hardware products in a year. The store classifies three groups as a percentage of their total customers (estimated to be 15 000 different customers in a year):

Classification	**% Customers**
A: Tradesmen	16
B: Home handymen	54
C: Others	30

A sample of 75 customers is selected, stratified by the above classification.

(a) If proportional stratified sampling is to be used, how many customers from each group should be selected?

(b) Compare the variability in the sample mean by two methods:

1: proportional stratified sampling;
2: sample 25 from each stratum.

Assume that the Strata A, B, C have population variances in the ratio 2 : 1 : 1.

9.48 In an agricultural field trial, an area of land is divided into the grid as shown. Since soil quality varies across the field, in order to estimate average crop yield for the region, a sample of plots bounded by the grid is taken and the yield from each plot is recorded.

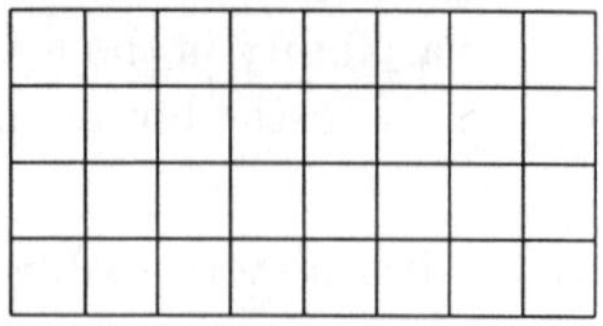

Show how to use tables of random numbers to extract a systematic sample of size $n = 8$.

Would you expect different plots to show correlated yields?

Would you expect adjacent plots to show correlated yields?

9.49 Equation 9.7 indicated that the population total may be estimated by $\widehat{T} = N(\overline{X}_{strat})$.

(a) How is the variance of $\widehat{T}$ related to the variance of $\overline{X}_{strat}$?

(b) If sampling is carried out with 2 strata, what is being estimated by $\widehat{T}_i$, where

$$\widehat{T}_i = N_i(\overline{X}_i), \quad i = 1, 2?$$

(c) Determine the observed values of $\widehat{T}_1$ and $\widehat{T}_2$ for the pine plantation data of Example 9.14.

9.50 **Randomised response** is a sampling procedure designed to avoid bias in sample surveys, when the question being asked is sensitive so that the respondent may wish to hide his/her true views.

It works as follows: Each respondent is given two questions which require a yes/no response (such as):

Question A: Have you ever used heroin?

Question B: Were you born in the first half of the year?

Question A is the sensitive question of interest. The interviewer does not know the answer to either question. However, overall it is known that p_0 is the proportion of all people who were born in the first half of the year. The respondent tosses a coin secretly and answers (truthfully) yes/no to A if 'heads'; to B if 'tails'.

From n respondents, X yes responses were recorded. An estimate of q, the proportion of heroin users in society is required.

(a) Show that

$$p = P(\text{yes}) = \tfrac{1}{2}q + \tfrac{1}{2}p_0.$$

(b) Explain why it is reasonable to assume that $X \sim Bin(n, p)$.

(c) If p is estimated by the fraction $\frac{X}{n}$, how should q be estimated?

CHAPTER TEN

Parameter Hunting with Points and Intervals

In this chapter you will find

10.1 POINT ESTIMATION

Parameter values in a population are those quantities such as mean, median, variance, standard deviation, population total, probability of 'success' ... that are of interest to us in our role as market researcher/experimenter/

consulting statistician/quality engineer. We model population random variables in terms of such parameters $\mu, \sigma, \sigma^2, \xi_p, \lambda, \ldots$ These characteristics of the population are usually unknown.

In Chapter 9, we investigated sampling strategies for obtaining data to be used in a process of *estimating* the values of these parameters. We obtained samples randomly, with replacement, without replacement, with stratification, systematically, but seldom ventured to actually estimate parameter values from data generated by such strategies.

It is time to address this practicality and concentrate on the actual estimation process. We **use** the sample data to hunt for (most likely) values for parameters. The result of this hunt may take two forms: we may quote a single number as our best 'educated guess' as to the value of the parameter in question—this is called a *point estimate*—or we may quote a range of values within which we believe the parameter to lie—this is called an *interval estimate* of the parameter.

We will begin with point estimation: how to do it and when is it working well?

Point estimators and point estimates

Refer back to the concept of a **statistic**. This is a function of the random sample alone (Definition 9.2). A statistic therefore may be evaluated once the sample has been realised. The value that we obtain for the statistic will depend on its sampling distribution. When we obtain the value of a statistic we are plucking a single number outcome from the sampling distribution of the statistic. The following definition indicates that we often use a 'hat' notation to represent estimators.

Definition 10.1 *A* **point estimator,** $\widehat{\theta}$, *of a parameter* θ, *is a statistic to be used for estimating* θ. *Once the sample has been observed, the resulting value of the point estimator is called a* **point estimate**.

□

Thus, a point estimator is a statistic with a purpose. Chapter 9 was essentially devoted to the sampling distribution of the important point estimator, the sample mean $\overline{X}$. The purpose to which we put $\overline{X}$ is to estimate μ. $\overline{X}$ and other (probably familiar) estimators are listed below. This list helps clarify notation; since point estimators are random variables, we use upper case letters for point estimators, lower case letters for point estimates. Parameter values are usually denoted by Greek letters, our exception being $p = P(\text{success})$ in n Bernoulli trials $X_i = Bin(n,p)$, $i = 1,2,3,\ldots,n$. Many texts denote $P(\text{success})$ by the Greek letter π, however intuition is better served by using p for probability parameters. More important than such labelling is the recognition that a population proportion, p, is estimated by the sample mean which is an average of zeros and ones; that is, $X_i \sim Bin(1,p)$

implies that $Y = \sum_{i=1}^{n} X_i \sim Bin(n,p)$. In n Bernoulli trials, the probability of success is estimated by the sample proportion of successes

$$\widehat{P} = \frac{Y}{n} = \frac{\sum_{i=1}^{n} X_i}{n} = \overline{X}.$$

θ	Parameter represents population ...	Point estimator $\widehat{\theta}$	Point estimate
μ	mean	$\overline{X} = \frac{\sum_{i=1}^{n} X_i}{n}$	$\overline{x} = \frac{\sum_{i=1}^{n} x_i}{n}$
σ^2	variance	$S^2 = \frac{\sum_{i=1}^{n} X_i^2 - n\overline{X}^2}{n-1}$	$s^2 = \frac{\sum_{i=1}^{n} x_i^2 - n\overline{x}^2}{n-1}$
σ	standard deviation	$S = \sqrt{\frac{\sum_{i=1}^{n} X_i^2 - n\overline{X}^2}{n-1}}$	$s = \sqrt{\frac{\sum_{i=1}^{n} x_i^2 - n\overline{x}^2}{n-1}}$
$\xi_{0.5}$	median	$\widetilde{X} = \begin{cases} X_{\left(\frac{n+1}{2}\right)} & n \text{ odd;} \\ \frac{X_{\left(\frac{n}{2}\right)} + X_{\left(\frac{n}{2}+1\right)}}{2} & n \text{ even} \end{cases}$	$\widetilde{x} = \begin{cases} x_{\left(\frac{n+1}{2}\right)} & n \text{ odd;} \\ \frac{x_{\left(\frac{n}{2}\right)} + x_{\left(\frac{n}{2}+1\right)}}{2} & n \text{ even} \end{cases}$
p	proportion	$\widehat{P} = \overline{X}$	$\widehat{p} = \overline{x}$

The standard error of a statistic (Definition 9.3) is the standard deviation of the sampling distribution of the statistic. This concept is very important for point estimators because it reflects their accuracy in estimating the unknown parameter.

The **standard error**, $S.E.(\widehat{\theta})$, of the estimator $\widehat{\theta}$ of θ is the standard deviation of the sampling distribution of $\widehat{\theta}$. That is,

$$S.E.(\widehat{\theta}) = \sqrt{Var(\widehat{\theta})}.$$

A standard error may, itself, require estimation with a point estimator before a known numerical value can be assigned to it.

The **estimated standard error**, $\widehat{S.E.}(\widehat{\theta})$, of the estimator $\widehat{\theta}$ of θ is a point estimator of $S.E.(\widehat{\theta})$ with realised value called the **standard error of estimate**.

Therefore, when estimating a population mean when the population variance is unknown:

parameter	μ
point estimator	$\widehat{\mu} = \overline{X}$
standard error of estimator	$S.E.(\widehat{\mu}) = \frac{\sigma}{\sqrt{n}}$
estimated standard error	$\widehat{S.E.}(\widehat{\mu}) = \frac{S}{\sqrt{n}}$
standard error of estimate	$\widehat{S.E.}(\widehat{\mu}) = \frac{s}{\sqrt{n}}$.

In this case the unknown population standard deviation σ is point estimated with the sample standard deviation S and, for simplicity, reference to the finite population correction factor is omitted.

Example 10.1

Driver fatigue

Many accidents on interstate highways are caused by drivers falling asleep at the wheel and then driving off the road or into oncoming traffic. In an experiment to investigate how often tired drivers fall asleep, the number of eye-closures per minute was monitored for sleep-deprived drivers while driving on a test circuit.

By an 'eye-closure' is meant a period longer than 500 milliseconds for which the eyes are closed.

Suppose that a certain subject gave the following frequencies of eye-closure under similar test conditions between 2 a.m. and 6 a.m.

4, 4, 5, 9, 6, 16, 4, 5, 9, 7.

Worked example:

Give a point estimator for, and a point estimate of, the mean eye-closures per minute for a fatigued driver. Estimate the standard error of your estimate.

Parameter of interest: μ denoting mean frequency of eye-closures per minute for fatigued drivers.

Point estimator: $\widehat{\mu} = \overline{X} = \frac{\sum_{i=1}^{10} X_i}{n}$, the sample mean.

Point estimate: $\overline{x} = \frac{4+4+5+9+6+16+4+5+9+7}{10} = 6.9$.

Standard error of the estimator: $S.E.(\overline{X}) = \frac{\sigma}{\sqrt{10}}$.

Standard error of estimate: $\widehat{S.E.(\overline{X})} = \frac{\widehat{\sigma}}{\sqrt{10}} = \frac{s}{\sqrt{10}} = \frac{3.725}{\sqrt{10}} = 1.18$.

Comment: As a simple data summary the statistic $\widetilde{X}$, the sample median, may well suit this type of data set (right-skewed). However properties of the sample median are difficult to obtain (in particular the standard error). The standard error of the mean involved the unknown 'nuisance parameter' σ; this parameter must also be point estimated with the observed sample standard deviation s, in order to obtain a numerical value for the standard error of the estimate.

When is a point estimator good? Using an 'unbiasedness' criterion

To determine when an estimator is 'good', whatever that means, is really quite difficult. It is really exactly like hunting with only one bullet in your gun and you must shoot at prey that always remain unseen! By this we mean that the parameter in question always remains unknown, and with a single numerical estimate, we must try to get close to its undisclosed value.

What criterion can be used to assess how successful the estimation process has been? There are several. The first discussed, and possibly the most famous criterion, is that of unbiasedness. Essentially, this criterion says that our estimator, in the long run, aims neither too high nor too low at the target parameter.

Definition 10.2 *The point estimator $\widehat{\theta}$ of a parameter θ is* **unbiased** *for θ, if for all values of the parameter θ,*

$$E(\widehat{\theta}) = \theta.$$

When $\widehat{\theta}$ is not an unbiased estimator, it is called a **biased estimator**.

□

An illustration of unbiasedness: Suppose that we are able to fire four shots at the circle in the centre of the target below:

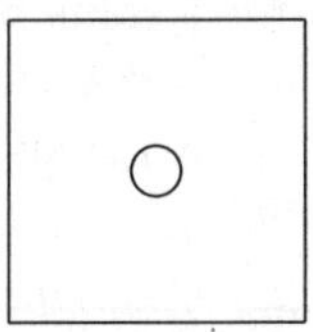

If we fire at the target in an unbiased way, then, over a period of time, there is an even spread of bullets across the vicinity of the target. Our four shots may look like:

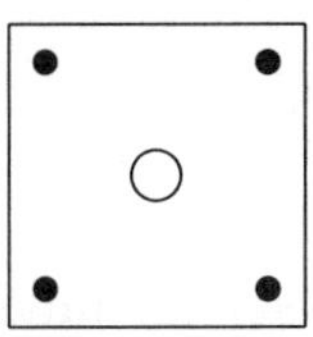

We have, without knowing it, seen many examples of unbiased estimators: the sample mean $\widehat{\mu} = \overline{X}$ is an unbiased estimator of the population mean μ. That

$$E(\widehat{\mu}) = \mu$$

is easy to see since $E(\overline{X}) = \mu$ as the mean of the sampling distribution of means is the population mean. (This is Theorem 9.2.) Since we have studied proportions as averages of zeros and ones, this also establishes that when $X_i \sim Bin(1, p)$ is a sequence of Bernoulli trials with probability of success p (so that each trial has mean p) and $\widehat{P} = \frac{\sum_{i=1}^{n} X_i}{n} = \overline{X}$ is the proportion of successes in the n Bernoulli trials, then $E(\widehat{P}) = p$. That is, the sample proportion is an unbiased estimator of the population proportion.

Example 10.2
Unbiased stratified sample mean
From (9.6) the mean $\overline{X}_{strat}$ of a stratified random sample is defined by

$$\overline{X}_{strat} = \omega_1 \overline{X_1} + \omega_2 \overline{X_2},$$

where $\overline{X}_1$ and $\overline{X}_2$ are the sample means of random samples selected from Strata 1 and 2 respectively. $\overline{X}_{strat}$ is an estimator of the population mean

$$\mu = \frac{N_1}{N}\mu_1 + \frac{N_2}{N}\mu_2 = \omega_1\mu_1 + \omega_2\mu_2.$$

Worked example:
Show that $\overline{X}_{strat}$ is an unbiased estimator of μ.
By **SUM 1**,
$E(\overline{X}_{strat}) = E(\omega_1\overline{X_1} + \omega_2\overline{X_2}) = \omega_1 E(\overline{X_1}) + \omega_2 E(\overline{X_2}) = \omega_1\mu_1 + \omega_2\mu_2 = \mu$.
Thus $\widehat{\mu} = \overline{X}_{strat}$ is an unbiased estimator of μ. That means that no bias in the estimation process is caused by stratifying the random sample.
By the calculation of Example 9.14, the observed value of the stratified sample mean number of trees per row in the *pinus radiata* plantations is 14.16667. Hence, a point estimate of μ is 14.2 (2dp) trees per row.

Theorem 10.1 *When $X_1, X_2, X_3, \ldots, X_n$ is a random sample from a population with mean μ and variance σ^2, the sample variance*

$$S^2 = \frac{\sum_{i=1}^{n} X_i^2 - n\overline{X}^2}{n-1}$$

is an **unbiased estimator** *of σ^2 and S is a* **biased estimator** *of σ.*

Proof: Since for any random variable Y, $\sigma_Y^2 = Var(Y) = E(Y^2) - E(Y)^2 = E(Y^2) - \mu_Y^2$, it follows that

$$E(Y^2) = \sigma_Y^2 + \mu_Y^2.$$

Using this to determine $E(X_i^2)$ and $E(\overline{X}^2)$, gives

$$E(S^2) = \frac{1}{n-1} E\left[\sum_{i=1}^{n} X_i^2 - n\overline{X}^2\right]$$

$$\begin{aligned}
&= \frac{1}{n-1}\left[\sum_{i=1}^{n} E(X_i^2) - nE(\overline{X}^2)\right] \\
&= \frac{1}{n-1}\left[\sum_{i=1}^{n}(\sigma^2+\mu^2) - n(\frac{\sigma^2}{n}+\mu^2)\right] \\
&= \frac{1}{n-1}\left[n(\sigma^2+\mu^2) - n(\frac{\sigma^2}{n}+\mu^2)\right] \\
&= \sigma^2.
\end{aligned}$$

Therefore the sample variance is an unbiased estimator of the population variance. This only occurs because the denominator of the sample variance contains $n-1$. A divisor of n would cause bias.

Finally, since for 'non-constant' random variables,

$$0 < Var(S) = E(S^2) - E(S)^2 = \sigma^2 - E(S)^2,$$

it follows that

$$E(S) < \sigma.$$

Therefore, since equality doesn't always hold, S is a biased estimator of σ. With reasonable sample sizes however, the bias is not sufficient to make the sample standard deviation an unreliable estimator of the population standard deviation.

■

Example 10.3

Estimating the variability of the proportion in favour of mining national parks

Saulwick (1991) reports on the question posed to 1000 voters nationally by telephone in 1989:

Would you say that mineral deposits that are in national parks should be mined or should not be mined? The response was as follows:

Mineral resources in national parks	%
Should be mined	30
Should not be mined	61
Don't know	9

(*Source*: Saulwick, I. (1991). *The Saulwick Poll Reprints, 1989.* David Syme and Co Limited: Melbourne.)

Worked example:

Determine a point estimate of the proportion of all voters that

are in favour of mining mineral deposits that are in national parks. Determine the variance of the estimator used. Show how to estimate this variance with an unbiased estimator.
The sample proportion of successes

$$\widehat{P} = \frac{Y}{1000} = \frac{\sum_{i=1}^{1000} X_i}{1000} = \overline{X},$$

where the X_i are *i.i.d.* $Bin(1,p)$ random variables

$$X_i = \begin{cases} 1 & \text{if Voter i is in favour of national park mining;} \\ 0 & \text{if Voter i is not in favour of national park mining,} \end{cases}$$

is an unbiased estimator of p, the true proportion of all voters in favour of mining mineral deposits in national parks. Notice that these Bernoulli trials X_i have common mean $\mu = p$ and common variance $\sigma^2 = p(1-p)$.
Point estimate:

$$\widehat{p} = \frac{y}{1000} = \frac{(\text{observed sample number in favour})}{1000} = \frac{300}{1000} = 0.3.$$

(The figure quoted of 300 in favour out of 1000 interviewed is after rounding to the nearest 10 voters since evidently the data reported in the table has been rounded to the nearest percentage.)
Variance of estimator:

$$Var(\widehat{P}) = \frac{\sigma^2}{1000} = \frac{p(1-p)}{1000}.$$

How is the variance estimated unbiasedly? As a first attempt, the quadratic $p(1-p)$ may be point estimated by the estimator $\widehat{P}(1-\widehat{P})$. Now check for unbiasedness:

$$\begin{aligned} E[\widehat{P}(1-\widehat{P})] &= E[\widehat{P} - \widehat{P}^2] \\ &= E[\widehat{P}] - E[\widehat{P}^2] \quad (\text{by } \textbf{SUM 1}) \\ &= E[\widehat{P}] - [Var(\widehat{P}) + E(\widehat{P})^2] \\ &= p - [\frac{p(1-p)}{1000} + p^2] \\ &= \frac{999}{1000} p(1-p). \end{aligned}$$

Dividing each side of this expression by 999 gives the surprising result:

$$E[\frac{\widehat{P}(1-\widehat{P})}{999}] = \frac{p(1-p)}{1000},$$

so that $\frac{\widehat{P}(1-\widehat{P})}{999}$ is an unbiased estimator of $\frac{p(1-p)}{1000}$. For this example, the estimated variance is

$$\frac{\widehat{p}(1-\widehat{p})}{999} = \frac{0.3(1-0.3)}{999} = 0.00021.$$

A choice between divisors $n = 1000$ and $n - 1 = 999$ makes little difference for large sample sizes.

Having seen some estimators that are unbiased (and some that are not), there is a nagging question: is unbiasedness the perfect criterion anyway? Somehow the question of long-term fairness has been addressed, but not *accuracy*. After all, both of the targets

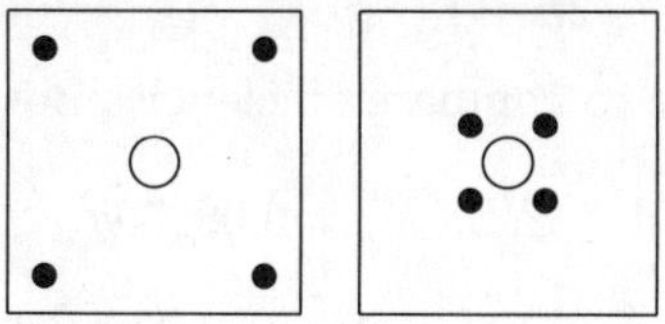

have been struck by black bullets in an unbiased way, but the central circle has been 'estimated' more accurately in one case than the other.

When is a point estimator good? Using an efficiency criterion

Since the standard error of a point estimator is the standard deviation of the estimator (when it is regarded as a random variable), it is a measure of how spread out the observed values of the estimator are likely to be. When choosing between two different point estimators the one with the smallest standard error is likely to be more accurate in the sense of being less variable about its expected value.

Definition 10.3 *If* $\widehat{\theta}_1$ *and* $\widehat{\theta}_2$ *are both point estimators of a parameter* θ, *then* $\boldsymbol{\widehat{\theta}_1}$ **is more efficient than** $\boldsymbol{\widehat{\theta}_2}$ *if* $S.E.(\widehat{\theta}_1) < S.E.(\widehat{\theta}_2)$.

□

A frequently used criterion of goodness of an estimator combines both unbiasedness and efficiency: from the collection of all unbiased estimators of the required parameter, the one with the smallest variance (being more efficient than the rest) is chosen as the 'best'. Such an estimator is termed **MVUE**, denoting **m**inimum **v**ariance **u**nbiased **e**stimator.

The sample mean and sample median may be investigated in terms of efficiency. Freund (1971) shows that if $X_1, X_2, X_3, \ldots, X_{2n+1}$ is a random sample of large size $(2n+1)$ from a symmetric distribution with pdf $f_X(x)$, then the sample median $\widetilde{X}$ has standard error

$$S.E.(\widetilde{X}) \approx \frac{1}{f_X(\mu)\sqrt{8n}}, \tag{10.1}$$

where X is the population random variable with mean μ and variance σ^2. If we sample from a normal population, $X \sim N(\mu, \sigma^2)$, and the value of

$f_X(\mu)$ is found by substituting $x = \mu$ into the $N(\mu, \sigma^2)$ pdf. The result: $f_X(\mu) = \frac{1}{\sigma\sqrt{2\pi}}$, and inserting this in Freund's (10.1) gives:

Standard error of the sample median: $S.E.(\widetilde{X}) \approx \frac{\sigma}{\sqrt{2n}}\sqrt{\frac{\pi}{2}}.$

Now recall that the equivalent result for the sample mean $\overline{X}$ for a sample of size $2n+1$ is:

$$S.E.(\overline{X}) = \frac{\sigma}{\sqrt{2n+1}}.$$

This puts us in a position to be able to compare efficiencies: for $n = 50$,

$$S.E.(\widetilde{X}) \approx 0.12533\sigma \quad \text{and} \quad S.E.(\overline{X}) = 0.09950\sigma.$$

Clearly, the sample mean is more efficient than the sample median in the sense that the standard error is smaller for all possible values of σ. This is an advantage. The same result may also be interpreted as follows: To achieve $S.E.(\overline{X}) = S.E.(\widetilde{X})$, that is, to estimate the centre of a symmetric distribution with the same accuracy by using a sample mean and a sample median, the sample size used to determine the sample median must be $\frac{\pi}{2} \approx 1.57$ times as large as the sample size used to determine the sample mean.

When is a point estimator good? Using a mean squared error criterion

Accuracy seems to be important. But, is it more important than unbiasedness? That is, as the shooter, in which of the following two ways would you prefer to hit the target?

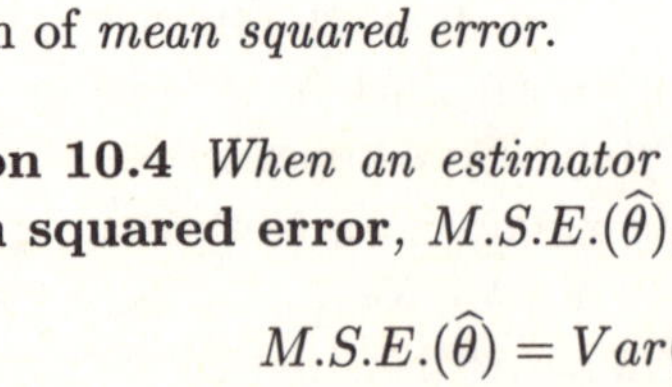

One way to combine the effects of both accuracy and unbiasedness is by using a criterion of *mean squared error.*

Definition 10.4 *When an estimator $\widehat{\theta}$ is used to estimate a parameter θ, the* **mean squared error**, *$M.S.E.(\widehat{\theta})$, of $\widehat{\theta}$ is determined by*

$$M.S.E.(\widehat{\theta}) = Var(\widehat{\theta}) + [E(\widehat{\theta}) - \theta]^2.$$

The term $E(\widehat{\theta}) - \theta$ is called the **bias** *of $\widehat{\theta}$.*

□

Recall that we may write the sample variance in the convenient form $S^2 = \frac{1}{n-1}\sum_{i=1}^{n}(X_i - \overline{X})^2$. Theorem 10.1 showed that $E(S^2) = \sigma^2$, so that S^2 is an unbiased estimator of σ^2. However, it may be shown that when sampling from a normal population, if we were forced to choose between the three point estimators

$$\frac{1}{n-1}\sum_{i=1}^{n}(X_i - \overline{X})^2, \quad \frac{1}{n}\sum_{i=1}^{n}(X_i - \overline{X})^2, \quad \frac{1}{n+1}\sum_{i=1}^{n}(X_i - \overline{X})^2,$$

of the population variance σ^2, then the estimator with smallest mean squared error (after a lengthy algebraic proof) is

$$\widehat{\sigma^2} = \frac{1}{n+1}\sum_{i=1}^{n}(X_i - \overline{X})^2.$$

So, in this measure, the usual sample variance S^2, with denominator $n-1$, is *not* the best performer! $\widehat{\sigma^2}$ is slightly biased but more accurate than the other options listed. Of course the difference is negligible when n is large.

EXERCISES

10.1 A sudden decrease in blood circulation to the fingertips is a symptom of a medical condition termed Raynaud's Syndrome. The reduction in blood supply causes a decrease in temperature. Ten patients afflicted with symptoms of Raynaud's Syndrome immersed their right forefingers in water maintained at 19°C.

The heat output (in cal/cm^2/min) of the forefinger was measured with a calorimeter giving the following data:

0.82 0.71 0.73 0.37 0.74
0.55 0.62 0.83 0.39 0.32.

Estimate the mean rate of heat loss from the forefinger for Raynaud's Syndrome patients. Determine the standard error of your estimate.

10.2 The lengths of 1032 pressed bricks from an old-style brickworks showed a mean length of 229.6 mm with a standard deviation of 1.8 mm. (*Brick Development Research Institute, Design Note 7A, 1980.*) A normal probability curve fitted well to a histogram of the data, indicating that the 'length of old-style pressed bricks' is a normally distributed random variable.

Estimate the median length of pressed bricks. Give the (estimated) standard error of your estimate.

10.3 A large department store has 5000 account holders and a random sample of 25 such customers showed that on average they spent \$78 per month with a sample standard deviation of \$20.

Give a point estimate of the total amount spent at the store each month by all account holders.

Estimate the standard error of the estimate.

10.4 The output file from a computer package contained the following summary of the analysis of a sample of n data points.

MEAN	STDEV	SEMEAN
9.645	4.112	0.582

(In this notation MEAN refers to the sample mean, STDEV refers to the sample standard deviation, and SEMEAN refers to the (estimated) standard error of estimate.)

(a) Give a point estimate of the population mean.

(b) On what sample size were the calculations based?

(c) If the sample size were doubled, which of the above three figures would you expect to have changed in the new output? Explain.

10.5 Fifty electrical components sampled randomly from a production line have the following resistances (in ohms):

2.91	2.98	2.95	3.10	3.12
2.95	3.10	3.00	3.12	2.93
3.17	2.92	2.95	3.09	2.93
3.07	3.05	2.90	2.98	2.90
2.98	3.09	3.05	2.91	3.04
3.03	2.88	2.85	3.18	3.00
2.94	2.90	3.12	3.03	2.86
2.96	3.05	2.95	2.92	2.93
2.98	2.91	3.04	2.92	2.94
2.95	3.00	3.04	2.99	2.95

(a) Construct a stem-and-leaf plot for these data.

(b) Estimate the mean resistance of all components of this type.

(c) Decide which of the following is true: The sample variance of this data set is 0.0815. The sample variance of this data set is $(0.0815)^2 = 0.0066$.

(d) Estimate the standard error of the estimate in (b).

(e) A very useful initial estimate of the standard deviation of a symmetrically distributed data set is (approximately) $\frac{1}{4}$ of the sample range.

Give a point estimate of σ, the standard deviation of resistance for components of this type.

10.6 Saulwick (1991) reports that of 1000 voters telephoned nationally in a survey (29 May 1989) 40% indicated that atmospheric pollution is the most serious environmental issue facing Australians.

Determine a point estimate, and estimated standard error of estimate, for the proportion of all Australians who do not think that atmospheric pollution is the most serious environmental issue.

Is the estimated standard error of estimate the same when estimating the proportion of all Australians who *do* think that atmospheric pollution is the most serious environmental issue?

Suppose that the pollsters had solicited the opinion of only the first 500 people telephoned (so that the sample size is halved) but still found that 40% felt that atmospheric pollution is the most serious environmental issue facing Australians. Is the estimated standard error of estimate doubled?

10.7 In the manufacture of seat belts, red 'press-here' buttons for the belt assembly are rejected by an inspector if the colour differs from a set standard.

Suppose that the number of buttons rejected per 100 inspected may be approximated by a Poisson distribution with parameter λ.

Based on n independent observations

$$X_1, X_2, X_3, \ldots, X_n$$

of the process, the estimator

$$\widehat{\lambda} = \frac{1}{n}\sum_{i=1}^{n} X_i$$

is proposed as an estimator of λ.

(a) Show that $\widehat{\lambda}$ is an unbiased estimator of λ.

(b) Determine $S.E.(\widehat{\lambda})$.

(c) When the production line is monitored on 15 independent occasions, the number per 100 buttons non-conforming due to colour were:

7 6 2 7 8
4 5 4 5 5
4 2 6 1 3

Discuss the estimation of λ by using these data.

10.8 In an attempt to check specifications for the thickness of manufactured hollow steel pipes, a random sample of 16 pipes produced a sample mean internal diameter $\overline{i} = 80$ mm with a standard deviation $s_i = 3$ mm.

An independent random sample of 16 of the same brand of pipes produced a sample mean external diameter $\overline{e} = 101$ mm with a standard deviation of $s_e = 5$ mm.

(a) Give an unbiased estimate of the mean wall thickness of the pipe.

(b) Show that the estimator used to produce the estimate in (a) is indeed unbiased.

(c) Estimate the standard error of estimate.

(d) If the *same* 16 pipes had been used for both internal and external measurements (as a time and cost saving measure), would the data set for the internal measurements be correlated with the data set for external measurements? If present, suggest a way that such correlation could be removed from this specification check.

10.9 Two chemists, working independently in different laboratories, produce point estimators $\widehat{\mu_1}$ and $\widehat{\mu_2}$ respectively, to estimate the average setting time, μ, of a newly developed epoxy resin. Although both estimators are unbiased, the first chemist, being aware of the benefit of good laboratory procedure, is able to control the size of the standard error of $\widehat{\mu_1}$ so that it is only one-quarter the size of the standard error of $\widehat{\mu_2}$.

In an attempt to produce a precise estimator of μ, the chemists combine their work. They decide on the following linear combinations of their individual estimators:

$$\begin{aligned}
\widehat{T_1} &= (\tfrac{1}{2})\widehat{\mu_1} + (\tfrac{1}{2})\widehat{\mu_2};\\
\widehat{T_2} &= (\tfrac{1}{3})\widehat{\mu_1} + (\tfrac{2}{3})\widehat{\mu_2};\\
\widehat{T_3} &= (\tfrac{1}{4})\widehat{\mu_1} + (\tfrac{3}{4})\widehat{\mu_2};\\
\widehat{T_4} &= (\tfrac{1}{5})\widehat{\mu_1} + (\tfrac{4}{5})\widehat{\mu_2}.
\end{aligned}$$

(a) Are these estimators unbiased for μ?

(b) Which estimator is more efficient than the rest?

10.10 Show that if $\widehat{\theta}$ is an estimator of θ, then the mean squared error may be written in the form $M.S.E.(\widehat{\theta}) = E(\widehat{\theta} - \theta)^2$.

10.11 More generally in Exercise 10.9, if for some $0 < p < 1$, the estimator $\widehat{T}$ of μ is defined by $\widehat{T} = p\widehat{\mu_1} + (1-p)\widehat{\mu_2}$, determine the value of p which minimises $M.S.E.(\widehat{T})$.

10.12 Let $a_1, a_2, a_3, \ldots, a_n$ be positive constants.

(a) Show that

$$\frac{\sum_{i=1}^n a_i^2}{(\sum_{i=1}^n a_i)^2} \geq \frac{1}{n},$$

with equality when $a_1 = a_2 = a_3 = \ldots = a_n$. [*Hint*: Begin with the formula for the sample variance of $a_1, a_2, \ldots, a_n$, and the fact that the sample variance must be positive.]

(b) A random sample $X_1, X_2, \ldots, X_n$ is taken from a population with unknown mean μ and unknown variance σ^2. The estimator

$$\widehat{T} = \frac{\sum_{i=1}^n a_i X_i}{\sum_{i=1}^n a_i}$$

is proposed as an estimator of μ.

i. Show that $\widehat{T}$ is an unbiased estimator of μ.

ii. Determine an expression for $Var(\widehat{T})$.

iii. Use (a) to establish that among all estimators of μ of the form $\frac{\sum_{i=1}^n a_i X_i}{\sum_{i=1}^n a_i}$, the estimator $\overline{X}$ has minimum variance.

10.13 We have seen that $\overline{X}$ is an unbiased estimator of the population mean μ. That is, $E(\overline{X}) = \mu$. Is $\overline{X}^2$ an unbiased estimator of μ^2?

[*Hint*: Consider $Var(\overline{X})$.]

10.14 Suppose that $X_1, X_2, X_3, \ldots, X_n$ is a random sample from a $N(\mu, \sigma^2)$ population. Let S denote the usual sample standard deviation.

(a) Show that S is a biased estimator of σ and that

$$E(S) = \frac{\sqrt{2}\Gamma(\frac{n}{2})}{\sqrt{n-1}\Gamma(\frac{n-1}{2})}\sigma.$$

(b) Determine the constant k so that kS is an unbiased estimator of σ.

(c) Determine the value of the bias $E(S) - \sigma$ for $n = 2, 4, 6, 10$.

10.15 Suppose that $X_1, X_2, X_3, \ldots, X_n$ is a random sample from a population with mean μ, variance σ^2 and fourth central moment $\mu_4 = E(X-\mu)^4$. Assume that

$$Var(S^2) = \frac{1}{n}\left[\mu_4 - \frac{n-3}{n-1}\sigma^4\right].$$

For the particular case where the data are normally distributed:

(a) show that $\mu_4 = 3\sigma^4$;

(b) show that $Var(S^2) = \frac{2\sigma^4}{n-1}$;

(c) determine the value of k so that kS^2 has smallest mean squared error.

10.16 A without-replacement random sample of size n is drawn from a finite population of size $N > n$. Show that

$$(\frac{1}{n} - \frac{1}{N})S^2,$$

where S^2 is the sample variance, is an unbiased estimator of the variance of the sample mean.

[*Hint*: Follow the proof of Theorem 10.1.]

10.2 SELECTING A POINT ESTIMATOR

Several criteria for assessing the 'goodness' of an estimator have been discussed. They are by no means a complete list of criteria and no one criteria seems to provide all the answers. The criterion of unbiasedness has remained well-used and popular (compared with smallest mean squared error criterion) simply because it is often easy to operate. In fact, we often make educated 'guesses' for an estimator and then adjust for unbiasedness. For example, if we guessed that σ^2 could be estimated by a multiple of $\sum_{i=1}^{n}(X_i - \overline{X})^2$, then

$$Guess = k\sum_{i=1}^{n}(X_i - \overline{X})^2,$$

where k is a constant. Then, by taking the expected value of each side, we find (Theorem 10.1)

$$E(Guess) = k(n-1)\sigma^2,$$

so that if 'Guess' is to be an unbiased point estimator, $k = \frac{1}{n-1}$. We now give two ways (better than guessing?) of inventing point estimators.

Method 1 for finding point estimators: equating moments

In Definition 4.13, the rth moment of the random variable X was given as

$$\mu_r' = E(X^r).$$

When X is a population random variable, then μ_r' is called the **rth population moment**.

To create suitable point estimators for the parameters of the distribution of X, we simply **equate** the population moments, μ_r', and the sample moments, $\overline{X^r}$, determined from a random sample $X_1, X_2, X_3, \ldots, X_n$ where each member of the sample has the same distribution as the population random variable X.

r	rth population moment	rth sample moment
	μ'_r	$\overline{X^r}$
1	$E(X)$	$\frac{1}{n}\sum_{i=1}^{n} X_i$
2	$E(X^2)$	$\frac{1}{n}\sum_{i=1}^{n} X_i^2$
3	$E(X^3)$	$\frac{1}{n}\sum_{i=1}^{n} X_i^3$
$\vdots$	$\vdots$	$\vdots$

Method of moments: To determine point estimators for k parameters, solve the system of k equations

$$E(X^r) = \frac{1}{n}\sum_{i=1}^{n} X_i^r \quad r = 1, 2, 3, \ldots, k,$$

for the 'unknown parameters'. The solution gives point estimators for the unknown parameters. Usually applications require estimation of only 1 or 2 parameters ($k = 1$ or $k = 2$).

Example 10.4

Fitting washers to a fixed 1 cm bolt

A washer is designed to fit a fixed 1 cm diameter bolt. Data from the production line are collected on the internal diameters of washers (the crucial dimension for a good fit of washer to the bolt shaft). In the production process, washers are drilled on a machine which has been set so that on average the washers fit fixed 1 cm test bolts perfectly, but the washers show considerable variability about this desired 1 cm average.

Worked example:

Determine a method of moments estimate for the clearance variability σ^2 based on a random sample

0, 0.1, 0.1, −0.1, −0.15, 0, 0, 0.1, 0.1, 0.12

of 10 observed clearances from a given day's output.

(Clearance = internal washer diameter − 1.)

Since the mean internal diameter of the washers is known to be 1 cm because of machine setting, the mean clearance is 0 and is therefore not an unknown parameter.

If $X_1, X_2, X_3, \ldots, X_{10}$ is a random sample of 10 clearances, then each X_i has mean 0 and variance σ^2, and this variance may be point estimated in the method of moments by equating second moments. That is, we must solve

$$E(X^2) = \frac{1}{10}\sum_{i=1}^{10} X_i^2.$$

However, since $Var(X) = E(X^2) - E(X)^2$, it follows that the population random variable X satisfies $E(X^2) = Var(X) + E(X)^2 = \sigma^2 + 0^2 = \sigma^2$. Therefore, the method of moments estimator $\widehat{\sigma^2}$, is the (trivial) solution

$$\widehat{\sigma^2} = \frac{1}{10}\sum_{i=1}^{10} X_i^2.$$

This is not the sample variance of the data 0, 0.1, 0.1, −0.1, −0.15, 0, 0, 0.1, 0.1, 0.12. The point estimate of σ^2 based on $\widehat{\sigma^2}$ is

$$\frac{0^2 + 0.1^2 + 0.1^2 + (-0.1)^2 + \ldots + 0.12^2}{10} = 0.00869.$$

Method 2 for finding point estimators: maximum likelihood

Estimators determined by this popular method tend to have small (or zero) bias, and small standard errors, especially for large sample sizes.
The joint density of a random sample $X_1, X_2, X_3, \ldots, X_n$ from a continuous population with parameter θ is (because of independence between sample members)

$$f_{X_1, X_2, X_3, \ldots, X_n}(x_1, x_2, x_3, \ldots, x_n) = \prod_{i=1}^{n} f_{X_i}(x_i) = \prod_{i=1}^{n} f(x_i; \theta),$$

where $f(x;\theta)$ is the probability density function of the population random variable and is the probability density function common to each sample member. This pdf is written in a form which shows the dependence of f on, first, the data x, and, second, on the unknown parameter θ.

The likelihood function reverses this first ...data, second ...parameter dependence. The likelihood is a function of the parameter first, and the data second.

Definition 10.5 *When $X_1, X_2, X_3, \ldots, X_n$ is a random sample from a population with parameter θ, the* **likelihood**, *$l(\theta)$, of the parameter θ, is any function of θ which a constant multiple of the joint probability density function (joint probability mass function in the discrete case) of $X_1, X_2, X_3, \ldots, X_n$.* □

Therefore,

$$l(\theta) = c \prod_{i=1}^{n} f(x_i; \theta) \quad \text{(continuous case)},$$

where f is the common density of each member of the random sample $X_1, X_2, X_3, \ldots, X_n$. (For the discrete case, f is replaced by the probability mass function p.) The constant multiplier c is usually chosen to eliminate (or cancel out) constant multiples in f, since such constant multiples do not assist in the estimation of θ.

Example 10.5
A Poisson likelihood
A likelihood based on 5 month-long observations x_1, x_2, x_3, x_4, x_5 of the number of earth tremors (exceeding 6 on the Richter scale) with epicentres in or near countries of the Pacific Basin, when such tremors are assumed to follow a Poisson process with rate λ, is found by the product of the Poisson mass functions

$$p(x_i; \lambda) = \frac{\lambda^{x_i} e^{-\lambda}}{x_i!} = (constant)\lambda^{x_i} e^{-\lambda} \quad (i = 1, 2, \ldots, 5).$$

Conveniently omitting the constant multiplier, we may express the likelihood in the multiplicative form

$$l(\lambda) = \left(\lambda^{x_1} e^{-\lambda}\right)\left(\lambda^{x_2} e^{-\lambda}\right)\left(\lambda^{x_3} e^{-\lambda}\right)\left(\lambda^{x_4} e^{-\lambda}\right)\left(\lambda^{x_5} e^{-\lambda}\right) = \lambda^{\sum_{i=1}^{5} x_i} e^{-5\lambda}.$$

Note that calculus shows that the maximum value of l as a function of λ occurs at $\lambda = \frac{\sum_{i=1}^{5} x_i}{5}$.

In the process of attempting to estimate a parameter, we must be conscious of estimates that are valid for a given parameter and estimates that are not. For example, it is not reasonable to estimate a probability with a point estimate of 2.6!

Definition 10.6 *The* **parameter space**, Θ, *of a parameter* θ *is the set of possible values that* θ *may assume.*

□

For example, p, the parameter in a $Bin(n,p)$ distribution, has parameter space $\Theta = [0, 1]$. Also, β, the parameter in an $exp(\beta)$ distribution must be positive, so that $\Theta = [0, \infty)$. It is advisable to check that point estimators of a parameter give rise to values in the parameter space.

Definition 10.7 *A* **maximum likelihood estimator** $\widehat{\theta}$ *of parameter* θ *is any value of* θ *in the parameter space for which* $l(\theta)$ *is maximum.*

□

The shortened expression MLE is often used to denote 'maximum likelihood estimator'. The maximisation of l as a function of θ may often be accomplished with the usual calculus maximising techniques of solving the first derivative of l set to zero and checking that the second derivative is negative at the solution.

Equivalently, it is often easier to consider the **log-likelihood** function $L(\theta) = \log_e l(\theta)$ for this maximising process since the maxima of L and l occur at the same values of θ. Indeed, this is the case in Example 10.5: for this example,

$$L(\lambda) = \log_e l(\lambda) = \log_e \left(\lambda^{\sum_{i=1}^5 x_i} e^{-5\lambda}\right) = \sum_{i=1}^{5} x_i \log_e(\lambda) - 5\lambda.$$

In this case, the equation $L'(\lambda) = 0$ is

$$\frac{\sum_{i=1}^5 x_i}{\lambda} - 5 = 0$$

and this equation has solution

$$\widehat{\lambda} = \frac{\sum_{i=1}^5 x_i}{5}$$

and since $L''(\widehat{\lambda}) < 0$,

$$\widehat{\lambda} = \overline{X}$$

is the maximum likelihood estimator of λ. This is a sensible answer for this problem; it agrees with our intuition that the mean of the sample should be used as a point estimator for the mean λ of the Poisson process.

One of the most important features of maximum likelihood estimation is the **invariance principle**. This principle states that functions of maximum likelihood estimators are again maximum likelihood estimators.

Invariance principle:
If $\widehat{\theta}$ is the MLE of θ, then $g(\widehat{\theta})$ is the MLE of $g(\theta)$ for any function g.

A proof of the invariance principle may be found in Mood, Graybill and Boes (1974). The principle is easily put into practice: Continuing Example 10.5, observe that for the population random variable $X \sim Poi(\lambda)$, $\mu_2' = E(X^2)$, the second population moment about 0 is given by $\mu_2' = \lambda + \lambda^2$ (because $E(X^2) = Var(X) + E(X)^2$). Since $\widehat{\lambda} = \overline{X}$ is the MLE of λ, according to the invariance principle, with $g(\lambda) = \lambda + \lambda^2$, the MLE of μ_2' is $\widehat{\mu_2'} = \widehat{\lambda} + \widehat{\lambda^2} = \overline{X} + \overline{X}^2$.

EXERCISES

10.17 If $X_1, X_2, X_3, \ldots, X_n$ is a random sample from a $N(\mu, 1)$ population, show that the likelihood may be written as a constant multiple of

$$e^{\mu \sum_{i=1}^{n} x_i - \frac{1}{2} n\mu^2}.$$

What value of μ maximises this function?

Is it the same value of μ that may be obtained by the method of moments?

10.18 Silicone sealants are available from hardware shops in a plastic dispensing cartridge and are applied by extrusion through a nozzle at one end of the cartridge.

Tests have shown that for a particular 'leak-proof' formulation the variable 'extrudability', X, in grams per minute has a known standard deviation of $\sigma = 20$.

Determine the maximum likelihood estimator of the mean extrudability rate μ, based on 10 observations $X_1, X_2, \ldots, X_{10}$, on the random variable $X \sim N(\mu, 20^2)$.

Evaluate the maximum likelihood estimate for μ, based on the observed 10 data points:

216	220	210	163	192
183	222	229	196	236

10.19 Consider fitting washers to a fixed 1 cm bolt as discussed in Example 10.4. Use the observed clearance values

0.00	0.10	0.10	−0.10	−0.15
0.00	0.00	0.10	0.10	0.12

to determine the method of moments estimate of the standard deviation of clearances.

10.20 Determine the maximum likelihood estimate of the standard deviation of clearance for the data in Exercise 10.19. Assume that the clearances are from a $N(0, \sigma^2)$ population.

10.21 Let X be a single observation from an exponential population with mean β. Suppose that we define an estimator $\widehat{T_c}$ of β by the equation $\widehat{T_c} = cX$, where c is a constant.

(a) Determine the value of c if $\widehat{T_c}$ is to be an unbiased estimator of β.

(b) Determine the value of c so that $\widehat{T_c}$ has smallest mean squared error among all estimators of β that are a multiple of X.

10.22 Suppose that the 10 clearance values in Example 10.4 are denoted by $X_1, X_2, \ldots, X_{10}$. Show that the method of moments estimator of the variance of the internal diameters is given by

$$\tfrac{1}{10} \sum_{i=1}^{10} (X_i + 1)^2 - 1.$$

Estimate the variance of the internal diameters by using the data of Example 10.4.

10.23 Suppose that $X_1, X_2, X_3, \ldots, X_n$ is a random sample from a population with continuous cumulative distribution function F.

Suppose that c is an unknown constant. In terms of c, define for each $i = 1, 2, 3, \ldots, n$ the indicator random variables

$$\delta_i = \begin{cases} 1 & \text{if } X_i > c; \\ 0 & \text{if } X_i \le c. \end{cases}$$

Suppose that we may observe the outcomes for the δ_i, even though the value of c is unknown.

(a) Show that
$$\widehat{\theta} = \frac{\sum_{i=1}^{n} \delta_i}{n}$$
is an unbiased estimator of
$$\theta = 1 - F(c).$$

(b) Complete the following: $\widehat{\theta}$ measures the proportion of the sample that is ...

(c) Determine $Var(\widehat{\theta})$.

(d) Show that δ_i has probability mass function
$$p(\delta_i; \theta) = \theta^{\delta_i}(1-\theta)^{1-\delta_i}.$$

(e) Determine the likelihood of θ based on the data $\delta_1, \delta_2, \ldots, \delta_n$.

10.24 On my way to work each day I must take a tram and a train. The total, X, of the time that I spend waiting for the tram plus the time that I spend waiting for the train follows a triangular probability distribution with density function expressed in terms of the parameter α:

$$f_X(x) = \begin{cases} k(\alpha - x) & 0 \leq x \leq \alpha; \\ 0 & \text{otherwise.} \end{cases}$$

(a) Show that $k = \frac{2}{\alpha^2}$.

(b) On 10 randomly chosen days I record the 10 'total times' $X_1, X_2, \ldots, X_{10}$, which I assume to be a random sample from the above probability distribution. Obtain a method of moments estimator of α. Is there a physical interpretation which may be placed on the parameter α?

(c) Is the method of moments estimator unbiased for α?

(d) Obtain a method of moments estimate of α if the observed times are (in minutes):
$$x_1 = 3, \quad x_2 = 4, \quad x_3 = 1,$$
$$x_4 = 2, \quad x_5 = 1, \quad x_6 = 2,$$
$$x_7 = 5, \quad x_8 = 7, \quad x_9 = 4,$$
$$x_{10} = 1.$$
How would you estimate the standard error of your estimate?

10.25 Nine observations on the number of faults observed per 100 metres in canvas supplied to a sailmaker were:

0 2 1 0 0 1 3 0 0.

Use an appropriate probability model to describe the number of faults that may be observed in 100 metres of this canvas. What assumptions are necessary for the model?

The sailmaker wishes to order 300 metres of the same fabric from the same supplier. Determine the maximum likelihood estimate of the probability that the 300 metres will contain no faults.

10.26 Let $X_1, X_2, X_3, \ldots, X_n$ be a random sample from the discrete uniform distribution which has probability mass function

$$p(x; \theta) = \begin{cases} \frac{1}{\theta} & \text{for } x = 1, 2, 3, \ldots, \theta; \\ 0 & \text{otherwise.} \end{cases}$$

Let $T = X_{(n)}$ denote the largest member of the sample.

(a) Show that the likelihood for θ may be written as
$$l(\theta) = \begin{cases} \frac{1}{\theta^n} & \text{if } \theta = t, t+1, \ldots, \\ 0 & \text{otherwise,} \end{cases}$$
where t is the observed value of T.

(b) Show that the maximum likelihood estimator of θ is T.

(c) Show that for $t = 1, 2, 3, \ldots, \theta$,

$$P(T \leq t) = \frac{t^n}{\theta^n}.$$

(d) By considering the expression

$$P(T \leq t) - P(T \leq t-1)$$

determine the probability mass function of T.

(e) Show that

$$\frac{T^{n+1} - (T-1)^{n+1}}{T^n - (T-1)^n}$$

is an unbiased estimator of θ. (It may be shown that this unusual estimator is the MVUE of θ.)

(f) At Auckland International Airport there is a fleet of mini-buses which run a door-to-door taxi service for airline passengers between hotel (or home) and the airport.

The buses are sequentially numbered 1,2,3,.... A tourist who is a regular visitor to the city notes that the largest number that he has observed on all his 7 visits to the airport is 14.

How should the tourist estimate the number of mini-buses in service?

10.27 For the indicator variables $\delta_1, \delta_2, \ldots, \delta_n$ of Exercise 10.23:

(a) show that $\widehat{\theta}$ is the maximum likelihood estimator of θ;

(b) give the maximum likelihood estimator of c.

10.28 Suppose that $X_1, X_2, \ldots, X_n$ is a random sample from the exponential density

$$f(x;\lambda) = \begin{cases} \lambda e^{-\lambda x} & \lambda > 0, \quad x > 0; \\ 0 & \text{otherwise.} \end{cases}$$

Determine the maximum likelihood estimator of λ.

Determine the method of moments estimator of λ.

10.3 BUILDING CONFIDENCE INTERVALS FOR μ FROM A PIVOT

In point estimation, the observed data will provide a single outcome (the estimate) that is our best 'guess' at the unknown value of the parameter. It is our best shot. But how accurate will we be on this one occasion? We may only have one random sample, have determined one sample mean $\overline{x}$, and hope that this one value lies close to the population mean μ.

We know that as a random variable, $\overline{X}$ has all the right properties, is unbiased, has small mean squared error, is efficient. However, with our one observed value $\overline{x}$, what can we do? How can we measure its accuracy in terms of closeness to μ? It is either close or not close. Certainly the standard error provides a yardstick measure of closeness; if standard error is large then an observed outcome may not necessarily be close to where we expect it to be; if standard error is small, then observed outcomes (in the case of $\overline{x}$) are likely to be closer to where we expect they should be: near μ. Remember that in all of this discussion, the value of μ is unknown.

Some of these concerns about the accuracy of point estimators may be answered by broadening the estimation procedure from

- offering a single numerical value as the best estimate of a parameter, to
- offering a range of possible values and asserting that we are fairly sure that the parameter lies within the stated range.

The phrase 'fairly sure' may be quantified in terms of probability, but it does say something of our degree of confidence in the estimation procedure.

An interval with random endpoints

When we offer a range of possible values to cover the unknown parameter, we do so in the form of an interval. When a is less than b, the notation $[a,\ b]$ represents the set of all numbers lying between a and b inclusive. Such notation defines the interval from a to b.

Definition 10.8 *Let* $\widehat{T_1}$ *and* $\widehat{T_2}$ *be two statistics with values determined by a random sample from a population with unknown parameter* θ. *The interval* $[\widehat{T_1},\ \widehat{T_2}]$ *is a* **100(1 - α)% confidence interval for θ** *if, for any* θ *in the parameter space,*

$$P[\widehat{T_1} \le \theta \le \widehat{T_2}] = 1 - \alpha.$$

$100(1-\alpha)\%$ *is called the* **confidence level** *of the interval.*

□

Several important ideas spring from this definition. The endpoints $\widehat{T_1}$ and $\widehat{T_2}$ of the interval are random variables and therefore it makes sense to consider the probability that the *fixed* unknown parameter lies between two such random endpoints.

Further, $\widehat{T_1}$ and $\widehat{T_2}$ are statistics. According to Definition 9.2, this means that $\widehat{T_1}$ and $\widehat{T_2}$ may be completely evaluated from the information contained in the random sample. Therefore, the endpoints of the confidence interval may be determined explicitly and become numerically fixed, once the sample has been realised. An observed version of the interval $[\widehat{T_1},\ \widehat{T_2}]$ with random endpoints is represented by $[\widehat{t_1},\ \widehat{t_2}]$ with fixed endpoints. At this point the role for probability is finished and probability leaves the stage: the now fixed interval $[\widehat{t_1},\ \widehat{t_2}]$ either contains the unknown parameter θ or it does not. What therefore can be said? *We* are $100(1-\alpha)\%$ confident that the parameter θ

lies in the fixed interval $[\widehat{t_1},\ \widehat{t_2}]$. Notationally it is unfortunate that both the random endpoint interval $[\widehat{T_1},\ \widehat{T_2}]$, and the fixed observed realisation $[\widehat{t_1},\ \widehat{t_2}]$ of it are both called $100(1-\alpha)\%$ confidence intervals.

Here we recommend NEVER to use the confidence interval format

$$\widehat{t_1} \le \theta \le \widehat{t_2}.$$

Such notation conveys the wrong impression that θ is forced to lie between the constants $\widehat{t_1}$ and $\widehat{t_2}$. Instead, we will use the interval format $[\widehat{t_1},\ \widehat{t_2}]$.

Example 10.6

Cutting down to size: hose-length from a guillotine

A guillotine is set to cut lengths of hose for heating and airconditioning units to a mean length θ cm. Suppose that by monitoring the output of the guillotine it has been established that the actual length of cut hoses produced by the machine is likely to be anywhere within 1 cm of the mean setting. This implies that the errors in cutting are uniformly distributed over a 2 cm interval about θ.

If we are not told the machine setting (so that, in particular, θ is unknown) how may θ be estimated on the basis of one observation from the machine?

Worked example:

Show that $[X-0.9, X+0.9]$ is a 90% confidence interval for θ based on a single observation X of the actual cut length of a hose produced on the machine.

In this introductory example, it is not so important as to how the interval was constructed; rather, it is intended to establish that the interval has the right characteristics to be a confidence interval. Certainly the interval has the form $[\widehat{T_1},\ \widehat{T_2}]$ for $\widehat{T_1} = X - 0.9$ and $\widehat{T_2} = X + 0.9$.

In this case the sample X is trivial (of size one) and the interval end-points are statistics; their values become known once the value of X is observed. The distribution of X may be written as $X \sim U[\theta-1, \theta+1]$ with probability density function sketched below. Finally,

$$\begin{aligned}
P(\widehat{T_1} \le \theta \le \widehat{T_2}) &= P(X-0.9 \le \theta \le X+0.9) \\
&= P(-0.9 \le \theta - X \le 0.9) \\
&= P(-0.9 \le X - \theta \le 0.9) \\
&= \tfrac{1}{2}[(0.9) - (-0.9)] \\
&= 0.9,
\end{aligned}$$

since $X - \theta \sim U[-1, 1]$. The conditions for a 90% confidence interval are satisfied with $\alpha = 0.1$.

Graph of the probability density function for length of hose X

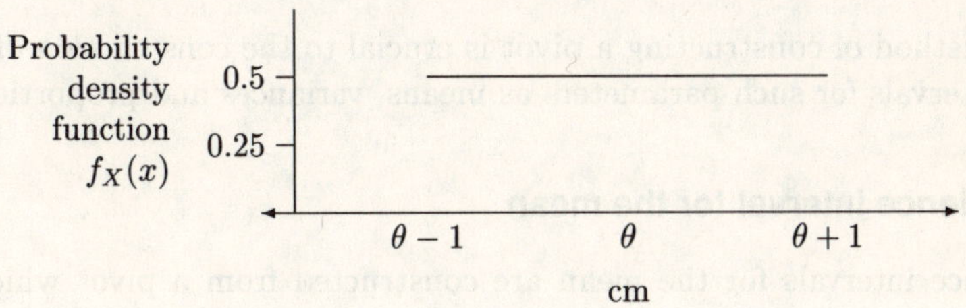

Graph of the probability density function for $Q = X - \theta$

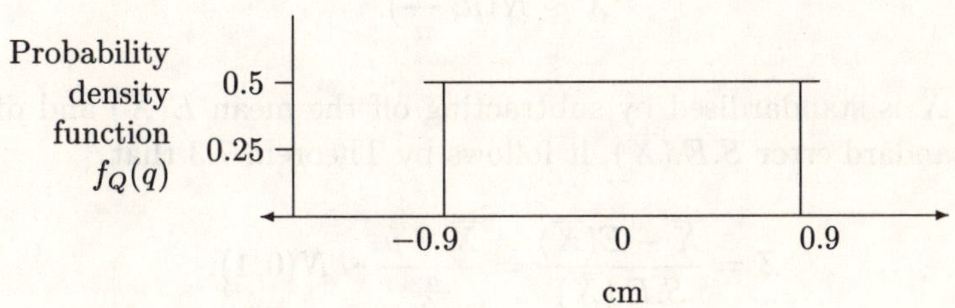

Finding pivotal quantities

The construction of a confidence interval often involves finding a function of the data and the unknown parameter that has a distribution that does not depend on the unknown parameter. In Example 10.6, the actual length X of hose cut on the guillotine had a uniform distribution ranging 1 cm about the mean setting θ. That is,

$$X \sim U[\theta - 1, \theta + 1].$$

The probability distribution for X certainly does depend on θ. However, if we define a new random variable Q by $Q = X - \theta$, then the probability distribution of Q does not depend on θ. In fact $Q \sim U[-1, 1]$. Therefore, a numerical value for θ is not required when probabilities concerning Q need to be evaluated (as in Example 10.6). Such probabilities are determined as areas under the $U[-1, 1]$ density. Q is an example of a *pivotal quantity.*

Definition 10.9 *Consider a random sample $X_1, X_2, X_3, \ldots, X_n$ taken from a population with parameter θ. If q is a function of only the data and the parameter, then the quantity Q, defined by $Q = q(X_1, X_2, X_3, \ldots, X_n, \theta)$, is called a* **pivot** *if the probability distribution for Q does not depend on θ.*

□

The method of constructing a pivot is crucial to the construction of confidence intervals for such parameters as means, variances and proportions.

A confidence interval for the mean

Confidence intervals for the mean are constructed from a pivot which is a standardised version of the sample mean. Suppose that $X_1, X_2, X_3, \ldots, X_n$ is a random sample from a normal population with mean μ and variance σ^2. Theorem 9.3 shows that $\overline{X}$, the sample mean, is also normally distributed:

$$\overline{X} \sim N(\mu, \frac{\sigma^2}{n}).$$

When $\overline{X}$ is standardised by subtracting off the mean $E(\overline{X})$ and dividing by the standard error $S.E.(\overline{X})$, it follows by Theorem 7.3 that

$$Z = \frac{\overline{X} - E(\overline{X})}{S.E.(\overline{X})} = \frac{\overline{X} - \mu}{\frac{\sigma}{\sqrt{n}}} \sim N(0, 1).$$

That is (and this is very important), the distribution of Z does not depend on μ or σ^2.

$$Z = \frac{\overline{X} - \mu}{\frac{\sigma}{\sqrt{n}}} \text{ is a pivot.}$$

A confidence interval is now easily created from the distribution for Z. From the pictured density of a $N(0, 1)$ random variable,

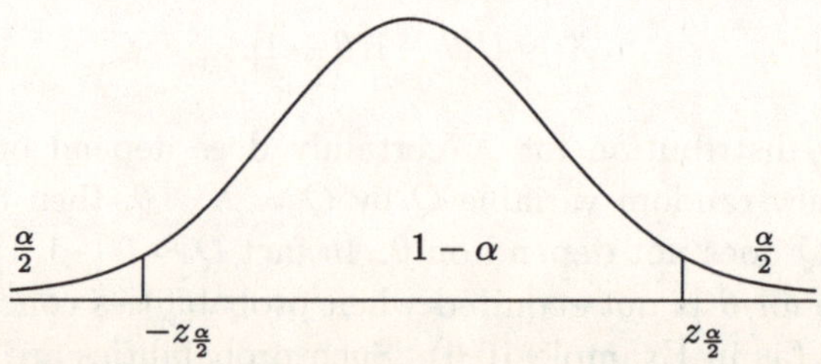

it follows that when $z_{\frac{\alpha}{2}}$ is the upper $\frac{\alpha}{2}$ point (with $\frac{\alpha}{2}$ trapped under the curve beyond that point),

$$P(-z_{\frac{\alpha}{2}} \leq Z \leq z_{\frac{\alpha}{2}}) = 1 - \alpha.$$

Substituting the pivot into this probability statement gives, with the same probability $1 - \alpha$,

$$P(-z_{\frac{\alpha}{2}} \leq \frac{\overline{X} - \mu}{\frac{\sigma}{\sqrt{n}}} \leq z_{\frac{\alpha}{2}}) = 1 - \alpha.$$

Now, unwrapping the pivot in the centre of the probability inequality leaves the mean μ alone, trapped between two known endpoints (when σ is known):

$$\begin{aligned} 1 - \alpha &= P(-z_{\frac{\alpha}{2}} \leq \frac{\overline{X} - \mu}{\frac{\sigma}{\sqrt{n}}} \leq z_{\frac{\alpha}{2}}) \\ &= P(-z_{\frac{\alpha}{2}} \frac{\sigma}{\sqrt{n}} \leq \overline{X} - \mu \leq z_{\frac{\alpha}{2}} \frac{\sigma}{\sqrt{n}}) \\ &= P(-\overline{X} - z_{\frac{\alpha}{2}} \frac{\sigma}{\sqrt{n}} \leq -\mu \leq -\overline{X} + z_{\frac{\alpha}{2}} \frac{\sigma}{\sqrt{n}}) \\ &= P(\overline{X} - z_{\frac{\alpha}{2}} \frac{\sigma}{\sqrt{n}} \leq \mu \leq \overline{X} + z_{\frac{\alpha}{2}} \frac{\sigma}{\sqrt{n}}) \\ &= P(\widehat{T_1} \leq \mu \leq \widehat{T_2}). \end{aligned}$$

This means that with probability $1 - \alpha$, the unknown mean μ is trapped between the two random endpoints

$$\widehat{T_1} = \overline{X} - z_{\frac{\alpha}{2}} \frac{\sigma}{\sqrt{n}} \quad \text{and} \quad \widehat{T_2} = \overline{X} + z_{\frac{\alpha}{2}} \frac{\sigma}{\sqrt{n}}.$$

When σ is known, $\widehat{T_1}$ and $\widehat{T_2}$ are statistics and their values provide the endpoints of the confidence interval. The process just described shows how a pivot may be used to construct a confidence interval. We will see further examples, although this construction is the prototype. In summary, the following theorem has been established:

Theorem 10.2

$$[\overline{X} - z_{\frac{\alpha}{2}} \frac{\sigma}{\sqrt{n}}, \quad \overline{X} + z_{\frac{\alpha}{2}} \frac{\sigma}{\sqrt{n}}]$$

is a $100(1 - \alpha)\%$ *confidence interval for the population mean* μ*, based on a random sample* $X_1, X_2, X_3, \ldots, X_n$ *from a* $N(\mu, \sigma^2)$ *population when* σ *is the known standard deviation of the population.*

■

The only major impediment to the usefulness of this result is that σ must be known. This will only rarely be the case—it is unlikely that the variance in a population would be explicitly known if the mean (the subject of the confidence interval construction) remained unknown. However, there are instances: The variability of output from a machine may be reasonably constant, but the machine may be set at different mean levels.

On the question of a normal population assumption, this certainly is not an impediment to usefulness, because the CLT ensures that our pivot is normally distributed (or, at least closely approximately so) for large sample sizes.

Example 10.7

Average restaurant spending

The owners of a small restaurant—it used to be called the Colusseo on Karangahape Road—wished to estimate μ, the mean of the normally distributed random variable 'dollar amount spent by each customer'.

Past records showed that, on average, customers spend different amounts according to the season; however the variability in customer spending always remains relatively constant: $\sigma = \$10$. A random sample of size 64 customers spent an average amount of \$40.60.

Worked example:

Construct a 99% confidence interval for μ.

To determine the α value for a 99% interval, solve $100(1-\alpha) = 99$, giving $\alpha = 0.01$. Therefore $\frac{\alpha}{2} = 0.005$, and, from the normal probability tables, $z_{\frac{\alpha}{2}} = 2.58$. The observed lower confidence interval endpoint is

$$\overline{x} - z_{\frac{\alpha}{2}}\frac{\sigma}{\sqrt{n}} = 40.60 - (2.58)(\frac{10}{\sqrt{64}}) = 37.375.$$

The observed upper confidence interval endpoint is

$$\overline{x} + z_{\frac{\alpha}{2}}\frac{\sigma}{\sqrt{n}} = 40.60 + (2.58)(\frac{10}{\sqrt{64}}) = 43.825.$$

Combining these endpoints gives a 99% confidence interval for μ as [37.38, 43.83] to 2 dp.

The interpretation of a confidence interval for the mean

In Example 10.7 a 99% confidence interval for average restaurant spending was determined as [37.38, 43.83]. This is one realisation of the interval

$$[\overline{X} - z_{\frac{\alpha}{2}}\frac{\sigma}{\sqrt{n}}, \quad \overline{X} + z_{\frac{\alpha}{2}}\frac{\sigma}{\sqrt{n}}]$$

with random endpoints when the observed value of $\overline{X}$ is 40.60. A different sample of customers would provide a different observed value of $\overline{X}$ simply because different samples have different sample means. The effect of these different values for $\overline{X}$ is to change the location of the centre of the interval: if we put

$$c = z_{\frac{\alpha}{2}} \frac{\sigma}{\sqrt{n}},$$

so that c is constant for specified values of n, α, and σ, then a specific interval is represented in the diagram:

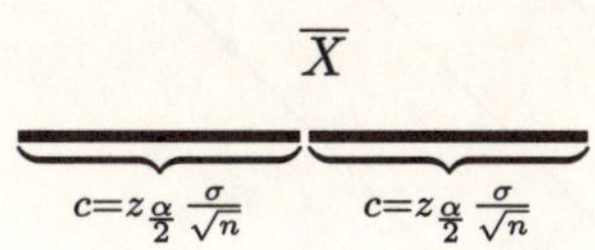

Even though the centre changes, the width

$$\widehat{T_2} - \widehat{T_1} = \left(\overline{X} + z_{\frac{\alpha}{2}} \frac{\sigma}{\sqrt{n}}\right) - \left(\overline{X} - z_{\frac{\alpha}{2}} \frac{\sigma}{\sqrt{n}}\right) = (\overline{X} + c) - (\overline{X} - c) = 2c$$

is constant. The results of repeated sampling produce confidence intervals for the unknown mean μ with constant width but with centres positioned randomly on the horizontal axis according to the sampling distribution of $\overline{X}$.

A possible geometric approach (used, for example, by Watson, Holton and Dear, 1990) is to display these constant-width intervals in a **statistic-parameter diagram**. The horizontal axis depicts possible values of the statistic $\overline{X}$ upon which the interval is based. The parameter space is shown on the vertical axis. Lines $\mu = \overline{x} - c$ and $\mu = \overline{x} + c$, expressed in terms of the constant $c = z_{\frac{\alpha}{2}} \frac{\sigma}{\sqrt{n}}$, are drawn in the diagram below relative to the (statistic, parameter) $= (\overline{x}, \mu)$ axes. They represent the extremities of a $100(1 - \alpha)\%$ confidence interval for μ, when a vertical slice (dotted line) is drawn from a realised value of $\overline{x}$ of the sample mean on the horizontal axis. The diagram shows a confidence interval $[\widehat{t_1}, \ \widehat{t_2}]$, centred at $\overline{x}$. If these intervals are removed from the vertical cross-sections in the statistic-parameter diagram and copied as solid lines, one under the other, the sampling variability becomes apparent.

A parameter-statistic diagram showing confidence intervals for μ based on $\overline{X}$

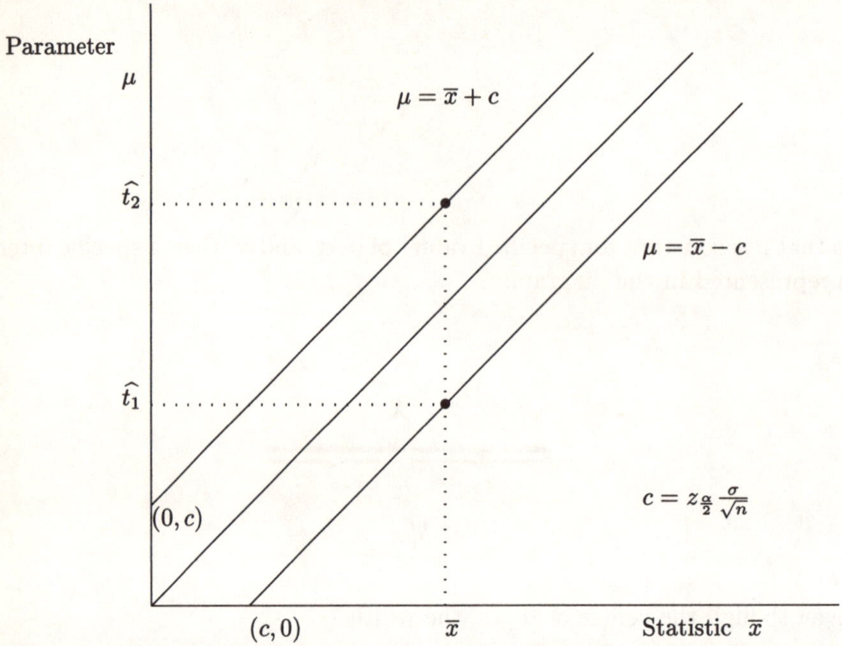

The sequence in the next diagram represents long-run repetitions of a confidence interval constructing experiment: a simulation of 50 different random samples, each of size 5, from a $N(30, 2^2)$ population. For each sample a 90% confidence interval is determined and sketched as a solid line against the scale on the horizontal axis. The population mean, positioned at the vertical dotted line, should, according to our theory, be trapped inside **approximately** 90% of the intervals in the sequence.

If we count the intervals from the top of the diagram, we find that the 14th, 24th and 25th intervals fail to trap the population mean. It would be unfortunate indeed if, when you perform your experiment/survey *once*, and obtain *one* data set, and calculate *one* confidence interval, that it turned out to be one of the three intervals listed above which does not contain μ. You will never know! The interval that you have calculated either contains μ or it does not contain μ. However, *you* may be 90% confident that your *one* 90% confidence interval contains μ. In this particular simulation, the odds have been in our favour: 47 of the 50 intervals have successfully contained the population mean.

Fifty 90% confidence intervals
for μ based on random samples from a $N(30, 2^2)$ population

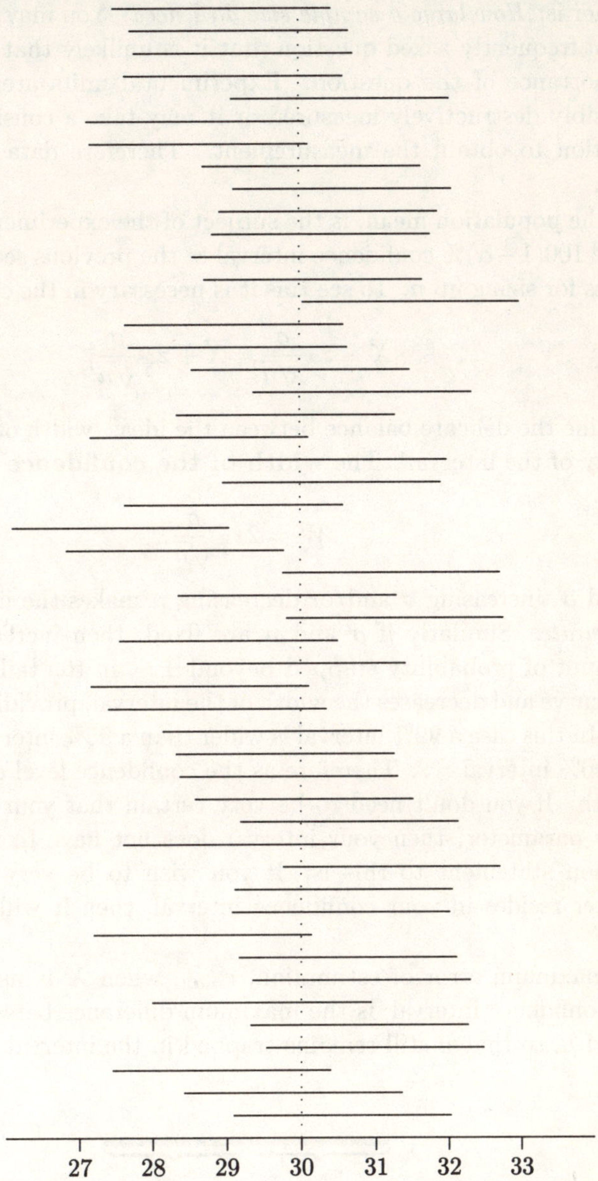

Size of n?

The most frequently asked question of the statistician by the experimental researcher is: *How large a sample size do I need?* You may doubt that this is the *most* frequently asked question, but it is unlikely that you would doubt the importance of the question. Experimental units are doubtless costly, are possibly destructively measured or it may take a considerable period of observation to obtain the measurement. Therefore data wastage is to be avoided.

If μ, the population mean, is the subject of the experimental investigation, then the $100(1-\alpha)\%$ confidence interval of the previous section provides the resources for sizing up n. To see this it is necessary in the confidence interval

$$[\overline{X} - z_{\frac{\alpha}{2}}\frac{\sigma}{\sqrt{n}}, \ \overline{X} + z_{\frac{\alpha}{2}}\frac{\sigma}{\sqrt{n}}]$$

to examine the delicate balance between the ideas 'width of the interval' and 'accuracy of the interval'. The **width of the confidence interval** is given by

$$W = 2z_{\frac{\alpha}{2}}\frac{\sigma}{\sqrt{n}}.$$

For fixed α, increasing σ and/or decreasing n makes the interval wider and *less accurate.* Similarly if σ and n are fixed, then increasing α increases the amount of probability trapped beyond $\pm z_{\frac{\alpha}{2}}$ in the tails of the standard normal curve and decreases the width of the interval, providing more accurate results. In this case a 99% interval is wider than a 95% interval which is wider than a 90% interval .... Therefore as the confidence level decreases, so does the width. If you don't need to be very certain that your interval contains the true parameter, then your interval does not have to be very wide. A companion statement to this is: If you wish to be very certain that the parameter resides in your confidence interval, then it will need to be very wide.

The maximum error of estimation, e_{max}, when $\overline{X}$ is used to estimate μ with a confidence interval, is the maximum difference between the locations of $\overline{X}$ and μ, so that μ still remains trapped in the interval.

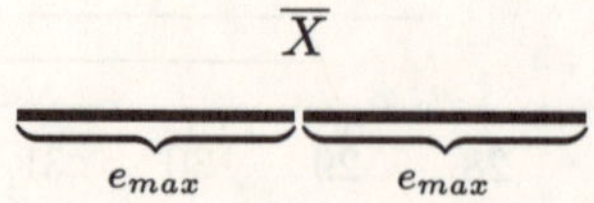

Therefore,

$$e_{max} = z_{\frac{\alpha}{2}}\frac{\sigma}{\sqrt{n}}, \tag{10.2}$$

which is exactly one half of the width, $\frac{1}{2}W$ of the interval. Equation (10.2) may be solved explicitly for the sample size n once the population standard deviation is known (or reasonably estimated) and the maximum error of estimation has been decided on. We do not solve Equation (10.2) explicitly for n; the intention is to keep away from ready-made formulas that may be 'used' with little thought. It is recommended that Equation (10.2) be derived and solved on every occasion where a sample size question needs to be answered.

Example 10.8

How many chipolatas are needed?

Airline meals must appear appetising and attractive to the general public. For this, the airline industry spends considerable research time on the latest trends in food presentation. However, because of space and payload requirements, both the volume and dry weight of a packaged meal are of particular importance.

An initial study of the weight of the sausage (chipolata) component of packaged breakfasts showed such sausages to be quite variable in weight.

Weights of 30 chipolata sausages

Scale: stems—10 g; leaves—g

2|8 represents 28 grams

1f	45
s	6777
.	8899
2*	01
t	22233
f	55
s	6
.	89
3*	000
t	23
f	
s	6
.	8
4*	
t	3

Calculations show that the standard deviation of this data set is 7.32. Suppose that for this **pilot study** 7.32 is taken as the best available point estimate of the population standard deviation σ. That is, $\widehat{\sigma} = 7.32$.

Worked example:

Determine the (approximate) minimum sample size required so as to be 99% sure that the true average weight μ of all sausages will not differ by more than 3 grams from an observed sample mean weight.

In this case $\alpha = 0.01$ so that $z_{\frac{\alpha}{2}} = 2.58$. Direct substitution into Equation (10.2) gives

$$3 = 2.58(\frac{\widehat{\sigma}}{\sqrt{n}}),$$

which, on substituting for the point estimate of the standard deviation leaves the sample size n as the only unknown:

$$3 = 2.58(\frac{7.32}{\sqrt{n}}).$$

In terms of $\sqrt{n}$,

$$\sqrt{n} = \frac{(2.58)(7.32)}{3} = 6.2952,$$

and squaring each side of this equation gives $n = 39.63$. Since as n increases, the width of the interval decreases, it follows that if the limit of the maximum error must not be broken, we must take the next largest sample size beyond 39.63; namely, a sample size of 40.

Large sample confidence intervals from the CLT

In Example 10.8, the population standard deviation was unknown, but was point estimated by using the sample standard deviation from a pilot study. The S/L plot showed that the data from this pilot study was right-skewed and therefore unlikely to have come from a population which is normally distributed. However, if n is large (as was the sample size required in Example 10.8 in order to achieve the specified experimental accuracy) the CLT operates:

$$\frac{\overline{X} - \mu}{\frac{\sigma}{\sqrt{n}}} \approx N(0,1), \qquad n \geq 25,$$

for $\overline{X}$ the mean of a random sample $X_1, X_2, X_3, \ldots, X_n$ from ANY population. If the population standard deviation σ is unknown, then, with the accuracy implied by using large sample sizes, we may point estimate it with the sample standard deviation S. The probability that the ratio $\frac{S}{\sigma}$ differs from 1, gets closer and closer to zero as n gets larger and larger. The loss on replacing σ by S becomes negligible for $n \geq 25$. The CLT then gives:

$$\frac{\overline{X} - \mu}{\frac{S}{\sqrt{n}}} = \left(\frac{\overline{X} - \mu}{\frac{\sigma}{\sqrt{n}}}\right)\underbrace{\left(\frac{\sigma}{S}\right)}_{\to 1} \approx \frac{\overline{X} - \mu}{\frac{\sigma}{\sqrt{n}}} \sim N(0,1),$$

and this suggests the following pivot:

$$\frac{\overline{X} - \mu}{\frac{S}{\sqrt{n}}} \approx N(0,1).$$

This pivot creates an *approximate* $100(1-\alpha)\%$ confidence interval in precisely the same way as described in the proof of Theorem 10.2. We take an immediate short-cut to the result:

Theorem 10.3 *When $n \geq 25$,*

$$[\overline{X} - z_{\frac{\alpha}{2}} \frac{S}{\sqrt{n}},\ \ \overline{X} + z_{\frac{\alpha}{2}} \frac{S}{\sqrt{n}}]$$

represents an APPROXIMATE $100(1-\alpha)\%$ confidence interval for the population mean μ based on a random sample $X_1, X_2, X_3, \ldots, X_n$ with sample mean $\overline{X}$ and sample standard deviation S taken from ANY population.

■

Example 10.9

Comparing intervals using big n

Continuing with the data of Example 10.8, for which $n = 30$, we can construct the following approximate confidence intervals. The lack of normality in the chipolata population (evidenced by the S/L plot in Example 10.8) is of little concern to us since n is large. Substituting into the observed interval

$$[\overline{x} - z_{\frac{\alpha}{2}} \frac{s}{\sqrt{n}},\ \ \overline{x} + z_{\frac{\alpha}{2}} \frac{s}{\sqrt{n}}]$$

from Theorem 10.1 with $\overline{x} = 24.27$, $s = 7.32$ and $n = 30$, the following intervals of different widths are easily determined. Notice how the width of the interval increases as the confidence level increases. The more certain we are about wanting to trap μ, the wider the interval has to be.

An approximate 90% confidence interval for μ:

$$[24.27 - (1.64)(\frac{7.32}{\sqrt{30}}),\ 24.27 + (1.64)(\frac{7.32}{\sqrt{30}})] = [22.08,\ 26.46].$$

An approximate 95% confidence interval for μ:

$$[24.27 - (1.96)(\frac{7.32}{\sqrt{30}}),\ 24.27 + (1.96)(\frac{7.32}{\sqrt{30}})] = [21.65,\ 26.89].$$

An approximate 99% confidence interval for μ:

$$[24.27 - (2.58)(\frac{7.32}{\sqrt{30}}),\ 24.27 + (2.58)(\frac{7.32}{\sqrt{30}})] = [20.82,\ 27.72].$$

Large sample confidence intervals for proportions

Theorem 10.3 is simply applied to Bernoulli trials. In a sequence of n Bernoulli trials with probability of success p, $X_1, X_2, X_3, \ldots, X_n$ are *i.i.d.* $Bin(1,p)$, and are observations on the population random variable X having probability mass function

x	0	1
$p_X(x)$	$1-p$	p

For this mass function, we have seen that $\mu = E(X) = p$ and $\sigma^2 = p(1-p)$. These are the population mean and variance. The total number of successes in the n Bernoulli trials is determined by $Y = \sum_{i=1}^{n} X_i \sim Bin(n,p)$. As usual, the probability of success p is estimated by the sample proportion

$$\widehat{P} = \frac{Y}{n} = \frac{\sum_{i=1}^{n} X_i}{n} = \overline{X},$$

and the population variance σ^2 is point estimated by $\widehat{P}(1-\widehat{P})$. The confidence interval for $\mu = p$ then becomes

$$[\overline{X} - z_{\frac{\alpha}{2}}\frac{S}{\sqrt{n}},\ \overline{X} + z_{\frac{\alpha}{2}}\frac{S}{\sqrt{n}}] = \left[\widehat{P} - z_{\frac{\alpha}{2}}\frac{\sqrt{\widehat{P}(1-\widehat{P})}}{\sqrt{n}},\ \widehat{P} + z_{\frac{\alpha}{2}}\frac{\sqrt{\widehat{P}(1-\widehat{P})}}{\sqrt{n}}\right].$$

An approximate large sample $100(1-\alpha)\%$ confidence interval for p is

$$\left[\widehat{P} - z_{\frac{\alpha}{2}}\sqrt{\frac{\widehat{P}(1-\widehat{P})}{n}},\ \widehat{P} + z_{\frac{\alpha}{2}}\sqrt{\frac{\widehat{P}(1-\widehat{P})}{n}}\right].$$

Example 10.10

Sample size for the political pollster

Trying to determine how voters will vote before they actually vote is the task of the political pollster. Predictions in voting patterns become quite biased on the basis of biased samples.

Suppose that we assume that on the eve of a general election, the pollster interviews voters by telephone nationwide. If the pollster wishes to estimate the proportion of voters who will vote Liberal so as to be 95% confident that the estimate is correct to within 2.5%, then n is needed so that

$$e_{max} = 1.96\sqrt{\frac{\widehat{p}(1-\widehat{p})}{n}}.$$

If there is no pilot survey to determine $\widehat{p}$ (for the professional pollster would likely have access to such pilot information), then an extremely cautious approach may be adopted: Replace $\widehat{p}(1-\widehat{p})$ by its maximum

0.25, which occurs at $\widehat{p} = \frac{1}{2}$. Inserting this conservative estimate of the population variance the pollster will find

$$0.025 = 1.96\sqrt{\frac{0.25}{n}} \quad \text{yielding} \quad n = 1537.$$

This is a lot of phone calls to make, but generally this many are made in an election forecast.

As reported in *The Weekend Australian* 11 July 1987, on the eve of the 1987 general election the Morgan Gallup Poll predicted that Labor would have the edge in a close contest with the coalition parties. Using a sample of size $n = 1542$, the nationwide poll showed 44% support for the Liberal/National coalition. Based on this poll, a 95% confidence interval for the proportion p of all Australian voters intending to vote for the coalition is

$$\left[0.44 - 1.96\sqrt{\frac{0.44(1-0.44)}{1542}},\ 0.44 + 1.96\sqrt{\frac{0.44(1-0.44)}{1542}}\right] = [0.415, 0.465].$$

This was an omen for a close election, with preferences being a major factor in the outcome. If the confidence interval had covered 0.5, the outlook for the coalition would have been much stronger. If 0.5 lay below the confidence interval, it would have been reasonable to predict victory for the coalition. They lost.

Some large sample pivots waiting to be unwrapped

The CLT provides the following results:

- X_i *i.i.d.* $Bin(1,p)$, with $\sum_{i=1}^{n} X_i \sim Bin(n,p)$,

$$1 - \alpha \approx P(-z_{\frac{\alpha}{2}} \le \frac{\frac{\sum_{i=1}^{n} X_i}{n} - p}{\sqrt{\frac{p(1-p)}{n}}} \le z_{\frac{\alpha}{2}}).$$

- X_i *i.i.d.* $Poi(\lambda)$, with $\sum_{i=1}^{n} X_i \sim Poi(n\lambda)$,

$$1 - \alpha \approx P(-z_{\frac{\alpha}{2}} \le \frac{\frac{\sum_{i=1}^{n} X_i}{n} - \lambda}{\sqrt{\frac{\lambda}{n}}} \le z_{\frac{\alpha}{2}}).$$

- X_i *i.i.d.* $exp(\beta)$, with $\sum_{i=1}^{n} X_i \sim Erlang(n,\beta) = \gamma(n,\beta)$,

$$1 - \alpha \approx P(-z_{\frac{\alpha}{2}} \le \frac{\frac{\sum_{i=1}^{n} X_i}{n} - \beta}{\sqrt{\frac{\beta^2}{n}}} \le z_{\frac{\alpha}{2}}).$$

The standardised sample means in these probability statements are each special cases of the pivot

$$\frac{\overline{X}-\mu}{\frac{\sigma}{\sqrt{n}}} \approx N(0,1), \qquad n \geq 25,$$

discussed in this chapter, so that the appropriate confidence intervals may be written down immediately in the pattern of the proof of Theorem 10.2; the algebraic details are left to the reader.

EXERCISES

10.29 A random sample of 100 Mitsubishi (car/truck/utility/...) owners from the Perth metropolitan area shows an average increase in speedometer readings in one year of 23 500 km with a standard deviation of 3900 km.

Construct a 95% confidence interval for the average distance travelled by Mitsubishi owners (in their vehicles) in one year.

10.30 The following table (*Sunday Age Football Magazine*) shows the number of career games played by Hawthorn players to the end of their 1991 premiership season in the AFL.

(a) Construct a stem-and-leaf plot for these data.

(b) Determine an approximate 95% confidence interval for the average number of career games played by AFL players assuming that the Hawthorn Club data are typical of all AFL players.

(c) Determine an approximate 95% confidence interval for the proportion of AFL players that have played over 100 career games. Assume that the data for the Hawthorn Club are typical of all AFL players.

Player	Career games
Abbot	85
Allan	31
Anderson	64
Ayres	238
Brereton	183
Collins	101
Condon	61
Cooper	22
Dear	60
Dipierdomenico	240
Dunstall	145
Gowers	22
Hall	58
Hudson	32
Jarman	24
Jencke	99
Johnston	2
Kennedy	241
Langford	194
Lawrence	51
Madigan	25
Maginness	66
McDonald	7
Mew	218
Morrissey	84
Nugent	10
Platten	137
Pritchard	103
Robran	7
Schwab	171
Tuck	426
Whittlesea	4
Wittman	87

10.31 Construct 80, 90, 95 and 99 per cent confidence intervals for the mean extrudability of silicone sealant (g $\min^{-1}$) from cartridge applicators.

205	195	230	230
229	178	220	244
188	278	220	221
262	240	226	228
207	236	240	228
178	253	260	194
285	186	307	193
200	187	208	197
212	216	214	172
200	210	250	240
205	188	185	251
200	183	198	174
202	315	150	244

Compare the intervals that you obtain with those at the same corresponding confidence levels, but based only on the first 28 data points

205	195	230	230
229	178	220	244
188	278	220	221
262	240	226	228
207	236	240	228
178	253	260	194
285	186	307	193

from the data set.

Give a discussion of how the width of confidence intervals for the mean depends on sample size and confidence level.

10.32 A PC manufacturer is investigating the performance of its product. In particular, the company wishes to classify the type of fault that it must service under the terms of its guarantee. From service records, 105 service reports were classified according to fault type.

Screen	Keyboard	Circuiting
33	46	26

Determine 95% confidence intervals for the proportion of each type of fault.

10.33 In the past, a chemical plant has produced an average of 1100 kilograms of chemicals per day. The records for the past year, based on 256 operating days, show the following mean and standard deviation respectively:

$$\begin{aligned} \overline{x} &= 1060 \text{ kg/day;} \\ s_x &= 340 \text{ kg/day.} \end{aligned}$$

(a) Give a 95% confidence interval for the mean production based on the data of the past year.

(b) Let $\overline{X}$ denote the mean daily production for a random sample of n operating days over the past year. We may well question whether the sample size $n = 256$ days was unnecessarily large for the desired levels of accuracy.

Suppose that, at least 8% of the time, we allow the absolute difference between $\overline{X}$ and the true mean daily production to exceed an error e.

Use tables of the standard normal distribution to show that

$$\frac{e\sqrt{n}}{\sigma} = 1.75$$

where σ is the population standard deviation for past daily production.

Hence determine the minimum sample size required when the error is taken as 20% of the population standard deviation.

10.34 When a random sample of size n is selected from a $N(\mu, \sigma^2)$ population, where σ is known, a $100(1-\alpha)\%$ confidence interval for μ has endpoints given by

$$\overline{x} \pm z_{\frac{\alpha}{2}} \frac{\sigma}{\sqrt{n}}.$$

(a) Which of the following increase the width of this confidence interval?

- taking a larger sample size;
- removing extreme outliers;
- increasing α.

Justify your answer in each case.

(b) 100 measurements of the diameter of steel rods used in the manufacture of bolts gave a sample mean of 10 mm and a sample standard deviation of 3 mm. Determine:

i. a point estimate of the mean diameter of all such bolts;

ii. the estimated standard error of the estimate in (i);

iii. a 98% confidence for the mean diameter of all such bolts;

iv. the sample size required to double the width of the interval in (iii).

10.35 Two independent observations, X_1 and X_2, were made on the population random variable X, distributed symmetrically about its mean μ. Show that the interval so formed with endpoints specified by X_1 and X_2 contains μ with probability 0.5.

10.36 (a) Based on a large number $X_1, X_2, \ldots, X_n$ of Bernoulli trials in which the observed outcome of each trial is 1 with probability p and 0 with probability $1-p$, show that an approximate 95% confidence interval for p is given by $[p_1,\ p_2]$, where p_1 and p_2 are solutions to the quadratic equation

$$(p - p^2)(1.96)^2 = n(\overline{X} - p)^2,$$

and $\overline{X}$ is the sample mean.

(b) Give expressions for p_1 and p_2.

(c) When the Health Department examined 250 chicken sandwiches available at lunch bars in Sydney, some 42 were found to be unsatisfactory according to at least one criterion. Determine a 90% confidence interval for the proportion of all chicken sandwiches which would fail to meet Health Department standards.

10.37 400 random numbers $X_1, X_2, \ldots, X_{400}$ are selected from an interval of width 2 but centred at an unknown point θ.

(a) Show that for the estimation of θ, the quantity

$$Q = \frac{\sum_{i=1}^{400} X_i}{400} - \theta$$

is a pivot and give the approximate distribution of Q.

(b) Construct a statistic-parameter diagram for $(\overline{x},\ \theta)$, and demonstrate how $100(1-\alpha)\%$ confidence intervals may be obtained from this diagram.

Write down an expression for such a confidence interval.

10.38 Suppose that the times between arrivals of customers at a 7-Eleven store (a 24-hour convenience store) has an exponential distribution with mean β.

If a random sample of n inter-arrival times $X_1, X_2, \ldots, X_n$ is observed (in minutes), show that

$$Q = \frac{\sum_{i=1}^{n} X_i}{\beta}$$

is a pivot.

Hence construct a 95% confidence interval for β if the sample average is 18 minutes when based on 45 observations from the late night period midnight to 3 a.m. monitored on successive weeknights.

10.4 BUILDING CONFIDENCE INTERVALS FOR σ^2 FROM A PIVOT

The Central Limit Theorem describes the sampling distribution of the mean when sample sizes are large. This has been the basis for large sample point and interval estimation of the population mean μ. We seek a similar basis for estimation of the population variance σ^2. In most circumstances, S^2, the sample variance, is the 'best' available point estimator of σ^2. We have seen that S^2 is unbiased for σ^2 (Theorem 10.1). But what is the sampling distribution of S^2?

Additive independent chi-squared distributions

PROPERTIES OF THE CHI-SQUARED DISTRIBUTION
$X \sim \chi^2_\nu = \gamma(\frac{1}{2}\nu, 2)$ (10.3)

Probability density function $f_X(x) = \frac{1}{2^{\frac{1}{2}\nu}\Gamma(\frac{1}{2}\nu)} x^{\frac{1}{2}\nu - 1} e^{-\frac{x}{2}}$

$0 < x < \infty; \quad \nu = 1, 2, 3, 4, \ldots.$

Mean	$E(X) = \nu$
Variance	$Var(X) = 2\nu$
Standard deviation	$SD(X) = \sqrt{2\nu}$
Coefficient of skewness	$\zeta_3 = \frac{2}{\sqrt{\frac{1}{2}\nu}}$
Coefficient of kurtosis	$\zeta_4 = 3 + \frac{12}{\nu}$
Moment generating function	$m_X(t) = (\frac{1}{1-2t})^{\frac{1}{2}\nu}, \quad t < \frac{1}{2}$
Cumulant generating function	$\kappa_X(t) = -\frac{1}{2}\nu \log_e(1 - 2t), \quad t < \frac{1}{2}$

The chi-squared distribution lies at the basis of the sampling distribution of S^2. It is the member of the gamma family of distributions described by Definition 7.8. If $X \sim \gamma(\frac{1}{2}\nu, 2)$, then X has a chi-squared distribution with positive parameter ν, the degrees of freedom: $\nu = 1, 2, 3, 4, \ldots$. The notation usually used for chi-squared distributions is

$$X \sim \chi^2_\nu.$$

Chi-squared distributions possess strong additive structure which depends heavily on the single assumption of normally distributed data. How? The cumulative distribution function method of Section 4.5 may be used to show that:

$$\boxed{\text{If } X \sim N(0,1), \quad \text{then} \quad X^2 \sim \chi^2_1.} \tag{10.4}$$

This indicates that a very simple example of a chi-squared random variable is the square of a standard normal random variable. The properties of the mean and variance of the gamma family dictate:

$$E(X) = (\tfrac{1}{2}\nu)(2) = \nu \quad \text{and} \quad Var(X) = (\tfrac{1}{2}\nu)(2)^2 = 2\nu,$$

and these and other properties of chi-squared distributions are collected in (10.3). Particularly important is the moment generating function of $X \sim \chi^2_\nu$:

$$m_X(t) = (\frac{1}{1-2t})^{\frac{1}{2}\nu}, \quad t < \frac{1}{2}.$$

Since the product of such mgf's

$$(\frac{1}{1-2t})^{\frac{1}{2}\nu_1}(\frac{1}{1-2t})^{\frac{1}{2}\nu_2}(\frac{1}{1-2t})^{\frac{1}{2}\nu_3}\cdots(\frac{1}{1-2t})^{\frac{1}{2}\nu_n} = (\frac{1}{1-2t})^{\frac{1}{2}(\nu_1+\nu_2+\nu_3+\ldots+\nu_n)}$$

is again a chi-squared mgf with the degrees of freedom added, then sums of independent chi-squared random variables have a chi-squared distribution with the degrees of freedom added. This is the basis of proof for the following theorem.

Theorem 10.4 *(chi-squared additions) If $X_1, X_2, X_3, \ldots, X_n$ is a random sample from a $N(0,1)$ population, and $Y_1, Y_2, Y_3, \ldots, Y_m$ is a second random sample from the same population, independent of the first sample, then, the sums of squares random variables defined by*

$$X = X_1^2 + X_2^2 + X_3^2 + \ldots + X_n^2$$

and

$$Y = Y_1^2 + Y_2^2 + Y_3^2 + \ldots + Y_m^2$$

satisfy

$$X \sim \chi^2_n, \quad Y \sim \chi^2_m \quad \textit{and} \quad X + Y \sim \chi^2_{m+n}.$$

■

The sampling distribution of S^2

This result is put to immediate use in determining the sampling distribution of S^2. However, we will need the following important characterisation of normal probability distributions. A clever proof by induction may be found in Berry and Lindgren (1990).

> When a random sample is taken from a normally distributed population, the sample mean $\overline{X}$ and the sample variance S^2 are independent random variables.

Theorem 10.5 *Suppose that S^2 is the sample variance of a random sample $X_1, X_2, X_3, \ldots, X_n$ taken from a $N(\mu, \sigma^2)$ population. Then,*

$$\frac{(n-1)S^2}{\sigma^2} \sim \chi^2_{n-1}.$$

Proof: Since

$$\sum_{i=1}^{n}(X_i - \mu)^2 = \sum_{i=1}^{n}(X_i - \overline{X})^2 + n(\overline{X} - \mu)^2,$$

by dividing each side of this equation by σ^2, the key structural result for the relationship between sample means and sample variances may be written as

$$\sum_{i=1}^{n}\left(\frac{X_i - \mu}{\sigma}\right)^2 = \frac{(n-1)}{\sigma^2}\frac{\sum_{i=1}^{n}(X_i - \overline{X})^2}{n-1} + \left(\frac{\overline{X} - \mu}{\frac{\sigma}{\sqrt{n}}}\right)^2,$$

or in summary, $A = B + C$. Since $\frac{X_i - \mu}{\sigma} \sim N(0,1)$, Theorem 10.4 shows that $A \sim \chi^2_n$. Similarly, since C is the square of a $N(0,1)$ random variable (recall Theorem 9.3), by (10.4), $C \sim \chi^2_1$. By the independence of $\overline{X}$ and S^2, $B = \frac{(n-1)S^2}{\sigma^2}$ and C are independent and finally, by the additivity of independent chi-squared distributions, $B \sim \chi^2_{n-1}$. That is,

$$\underbrace{A}_{\sim\chi^2_n} = \underbrace{B}_{\sim\chi^2_{n-1}} + \underbrace{C}_{\sim\chi^2_1}.$$

■

The results of this theorem are not sample size dependent (although a sample size of at least 2 is needed for the calculation of a sample variance). Note well that Theorem 10.5 depends on the assumption that the sample data used to determine S^2 has come from a normally distributed population. This assumption should always be checked by using a normal probability plot.

Chi-squared tables

Example 10.11

Degrees of freedom and the shape of chi-squared densities

Pictured below are probability density functions for $X_1 \sim \chi_4^2$ and $X_2 \sim \chi_8^2$ drawn on the same scale. These pdf's are centred about their means at $E(X_1) = 4$ and $E(X_2) = 8$ degrees of freedom. Since the variance of a χ^2 random variable is twice the degrees of freedom, it follows that X_2 has twice the variance of X_1. This is exhibited in the wider support for the pdf of X_2. The coefficients of skewness and kurtosis both decrease as the degrees of freedom increase.

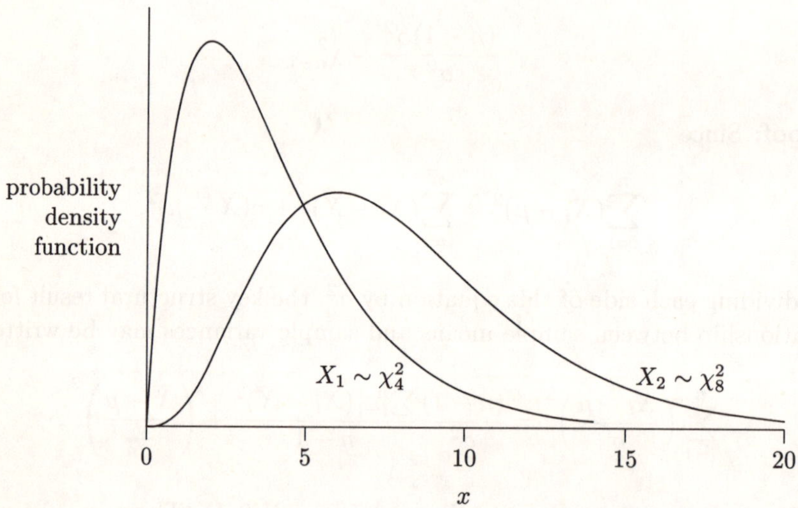

If the upper α point of a chi-squared distribution with ν degrees of freedom is denoted by $\chi^2_{\alpha,\nu}$, then

$$P(W > \chi^2_{\alpha,\nu}) = \alpha \quad \text{when} \quad W \sim \chi^2_\nu,$$

so that α is the area under a χ^2_ν density beyond the upper α point $\chi^2_{\alpha,\nu}$. The table in Appendix 5 specifies these 'upper-tail' probabilities for values $\nu = 1, 2, 3, 4, \ldots, 30$ and values of α from 0.995 (for evaluating lower-tail probabilities) to 0.005 (for evaluating upper-tail probabilities).

Example 10.12
Chi-squared percentiles
Worked example:
Determine the 90th percentile of a chi-squared distribution with 3 degrees of freedom.

$\xi_{0.9} = \chi^2_{0.1,3} = 6.251.$

A pivot for variance

The probability distribution of

$$\frac{(n-1)S^2}{\sigma^2}$$

does not depend on the unknown parameter σ^2 and so satisfies the definition of a pivot. By Theorem 10.5,

$$P\left(\chi^2_{1-\frac{\alpha}{2},n-1} \leq \frac{(n-1)S^2}{\sigma^2} \leq \chi^2_{\frac{\alpha}{2},n-1}\right) = 1-\alpha.$$

When this pivot is unwrapped, by inverting the inequality in

$$P\left(\frac{1}{\chi^2_{\frac{\alpha}{2},n-1}} \leq \frac{\sigma^2}{(n-1)S^2} \leq \frac{1}{\chi^2_{1-\frac{\alpha}{2},n-1}}\right) = 1-\alpha,$$

and then cross-multiplying by $(n-1)S^2$, it follows that

$$P\left(\frac{(n-1)S^2}{\chi^2_{\frac{\alpha}{2},n-1}} \leq \sigma^2 \leq \frac{(n-1)S^2}{\chi^2_{1-\frac{\alpha}{2},n-1}}\right) = 1-\alpha.$$

In summary:

Theorem 10.6 *When a random sample $X_1, X_2, X_3, \ldots, X_n$ is taken from a $N(\mu, \sigma^2)$ population, then a $100(1-\alpha)\%$ confidence interval for σ^2 is given by*

$$\left[\frac{(n-1)S^2}{\chi^2_{\frac{\alpha}{2},n-1}}, \frac{(n-1)S^2}{\chi^2_{1-\frac{\alpha}{2},n-1}}\right].$$

■

Example 10.13

Interval estimates for score variability

Every year in the 'Statistics for Engineering' course, each student receives an integer score out of 10 for the class tests done during the semester. Let X denote a typical score in 1991 and Y denote a typical score for 1992.

1991 scores (X)

Key: 3|x denotes a score of 3

0	
1	x
2	
3	xxx
4	xx
5	xxxx
6	xxxxxx
7	xxxx
8	xxx
9	xx
10	x

1992 scores (Y)

Key: 3|x denotes a score of 3

0	x
1	
2	xx
3	xxx
4	xxxxx
5	xxxxx
6	xxx
7	x
8	xx
9	
10	

It seems likely that the standard deviation σ in the score is approximately constant from year to year. This is borne out by the plots for the 1991 and 1992 scores: the shapes are similar, they are symmetric, and bell-shaped. Therefore, it seems reasonable to assume that the scores are normally distributed.

A 95% confidence interval for σ^2 based on the 1991 data is determined by a sample size $n = 26$. Substituting directly into the interval of Theorem 10.6 with an observed sample variance of $s_x^2 = (2.134)^2$ and $\alpha = 0.05$ gives

$$\left[\frac{(n-1)s_x^2}{\chi^2_{\frac{\alpha}{2},n-1}}, \frac{(n-1)s_x^2}{\chi^2_{1-\frac{\alpha}{2},n-1}}\right] = \left[\frac{(25)(2.134)^2}{\chi^2_{0.025,25}}, \frac{(25)(2.134)^2}{\chi^2_{0.975,25}}\right]$$
$$= \left[\frac{(25)(2.134)^2}{40.646}, \frac{(25)(2.134)^2}{13.120}\right] = [2.8,\ 8.7].$$

A 95% confidence interval for σ^2 based on the 1992 data is determined by a sample size $n = 22$:

$$\left[\frac{(n-1)s_y^2}{\chi^2_{\frac{\alpha}{2},n-1}}, \frac{(n-1)s_y^2}{\chi^2_{1-\frac{\alpha}{2},n-1}}\right] = \left[\frac{(21)(1.946)^2}{\chi^2_{0.025,21}}, \frac{(21)(1.946)^2}{\chi^2_{0.975,21}}\right]$$
$$= \left[\frac{(21)(1.946)^2}{35.478}, \frac{(21)(1.946)^2}{10.283}\right] = [2.2,\ 7.7].$$

There is similar variability apparent in each of these interval estimates. However, it appears that, on average, scores were higher in 1991: $\overline{x} = 5.9$, $\overline{y} = 4.5$.

EXERCISES

10.39 Use chi-squared tables to determine the values of $\chi^2_{0.1,10}$, $\chi^2_{0.9,10}$, $\chi^2_{0.99,10}$ and $\chi^2_{0.01,10}$.

10.40 Determine the 95th percentile of a chi-squared distribution with 25 degrees of freedom.

10.41 If $Y \sim \chi^2_{15}$, determine the value of the constant k so that:

(a) $P(Y > k) = 0.95$;

(b) $P(Y < \chi^2_{k,15}) = 0.95$;

(c) $P(Y > \chi^2_{0.03,15}) = k$.

Use a sketch diagram to illustrate each of your answers.

10.42 An author has written a book containing 500 pages. By selecting a random sample of 16 pages with replacement and counting X, the number of words per page, the author wishes to estimate σ^2, the variance in the number of words per page.

The sample data gave $\overline{x} = 180$ and $s^2 = 12$. Assuming that the number of words per page is (approximately) normally distributed, determine a 95% confidence interval for σ^2.

10.43 The data below give plasma glucose concentration (mg dL^{-1}) for 50 normal fasting people.

98	105	75	92	120	91
112	98	103	94	100	86
100	99	84	110	100	108
100	96	86	102	112	106
100	109	104	103	106	88
103	112	98	106	92	95
95	108	90	90	95	112
84	102	96	88	93	93
93	87				

(*Source*: Massod, M.F. (1977). Nonparametric percentile estimate of clinical normal ranges. *American Journal Medical Technology*, 43, 243-52.)

(a) Construct a S/L plot of the data.

(b) Construct a normal probability plot of the data.

(c) Are there any points in the normal probability plot that are possible outliers? Do these points appear as outliers in a box-and-whisker plot of the data? Explain why the data appear to be normally distributed.

(d) Construct a 95% confidence interval for the variance in fasting glucose concentration. [An approximation to $\chi^2_{\alpha,k}$, when k is large, is given in Exercise 10.51.]

10.44 Determine a 99% confidence interval for the standard deviation, σ, of test score based on the 1991 data of Example 10.13.

10.45 Explain why it is not valid to construct a confidence interval, based on the chi-squared pivot, for the variance σ^2 of a population if sample data from the population is as displayed in the S/L plot below. (Stems—units; leaves—10^{-1} units.)

0	23455677777788888899999
1	222244778999999
2	1122355556788889
3	011111233448
4	00112224799
5	0122344568
6	135555
7	34677
8	2569
9	144
10	12227
11	45
12	7
13	
14	3

10.46 A machine is producing metal pieces that are cylindrical in shape with a circular cross-section. After recent adjustments, the standard deviation in cross-section diameter, σ, is unknown.

A random sample of 9 pieces from the machine gave the following cross-section diameters:

1.01 0.97 1.03 1.04 0.99
0.98 0.99 1.01 1.03

(a) Construct a normal probability plot of these data. Do the data appear to be normally distributed?

(b) Determine a 99% confidence interval for σ^2.

(c) Determine a 95% confidence interval for σ^2.

(d) Determine a 99% confidence interval for σ.

(e) Determine a 99% confidence interval for the variance in the radius of cross-section.

10.47 Determine the probability that a random sample of 25 observations from a $N(1, 6)$ population will have a sample variance S^2:

(a) greater than 9.1;

(b) between 3.462 and 10.745.

10.48 For a random sample (of size 20 with sample variance S^2) from a normal population with mean μ and variance σ^2, determine the value of L in terms of σ so that $P(S^2 < L) = 0.05$.

10.49 Suppose that $X_1, X_2, X_3, \ldots, X_n$ are n random variables each with the same mean μ and the same variance σ^2. In terms of these n random variables, define the quantities

$$\overline{X} = \frac{\sum_{i=1}^{n} X_i}{n},$$

$$S^2 = \frac{\sum_{i=1}^{n}(X_i - \overline{X})^2}{n-1}.$$

Consider the following list of **statements**:

(a) $E(\overline{X}) = \mu$.

(b) $Var(\overline{X}) = \frac{\sigma^2}{n}$.

(c) $E(S^2) = \sigma^2$.

(d) The sampling distribution of

$$\frac{(n-1)S^2}{\sigma^2}$$

has a chi-squared distribution with $n - 1$ degrees of freedom.

Consider the following list of **assumptions**:

Assumption A: $X_1, X_2, X_3, \ldots, X_n$ are (mutually) independent.

Assumption B: $X_1, X_2, X_3, \ldots, X_n$ are normally distributed.

All the above statements are true when both Assumptions A and B hold. Some of the above statements still remain true if one (or both) of Assumptions A and B are dropped. Indicate, by completing the table below, which assumptions are necessary for each statement to hold.

	Assumption A necessary	**Assumption B necessary**
(a)		
(b)		
(c)		
(d)		

10.50 Two different independent samples (1 and 2) of respective sizes n and m have sample variances S_1^2 and S_2^2 respectively. The samples are taken from respective normal populations with the same variance σ^2 (but possibly different means).

Show that

$$\frac{(n-1)S_1^2 + (m-1)S_2^2}{\sigma^2}$$

is a pivot and determine the probability distribution of the pivot.

Use this result to construct a $100(1-\alpha)\%$ confidence interval for σ^2 based on both samples combined.

Construct a 95% confidence interval for the variance in test score using the data for both years 1991 and 1992 in Example 10.13.

10.51 Establish the following relationship between upper alpha points of the chi-squared and standard normal distributions when the degrees of freedom (k) are large:

$$\chi^2_{\alpha,k} \approx \tfrac{1}{2}[z_\alpha + \sqrt{2k-1}]^2.$$

Use the approximation to estimate the upper 10% point of a χ^2_{25} distribution.

10.5 HEAVY TAILS AND t

When $\overline{X}$ is the mean and S^2 is the variance of a random sample of size n from a population with mean μ and variance σ^2, then

$$\frac{\overline{X}-\mu}{\frac{S}{\sqrt{n}}} \sim N(0,1)$$

(approximately) **when the sample size n is large.**

Sometimes, however, it becomes necessary to attempt to estimate μ when only a small sample size is available. This situation is very likely to arise when sampling is extremely expensive, and/or destructive, or observations are simply difficult to obtain. For example, it took many years to accrue a data set on the survival time of patients after a heart transplant (Crowley and Hu, 1977).

For sample sizes less than 20, the Central Limit Theorem is not applicable; the sampling distribution of the mean has greatly increased variability, especially when the population standard deviation is unknown and needs to be estimated by the sample standard deviation S. Then, variability in both $\overline{X}$ and S contribute to increased variability in

$$\frac{\overline{X}-\mu}{\frac{S}{\sqrt{n}}}.$$

The sampling distribution remains essentially symmetric and bell-shaped, but is no longer normally distributed.

Example 10.14

Small n and normally distributed data: sampling distributions $\frac{\overline{X}-\mu}{\frac{\sigma}{\sqrt{n}}}$ and $\frac{\overline{X}-\mu}{\frac{S}{\sqrt{n}}}$ compared

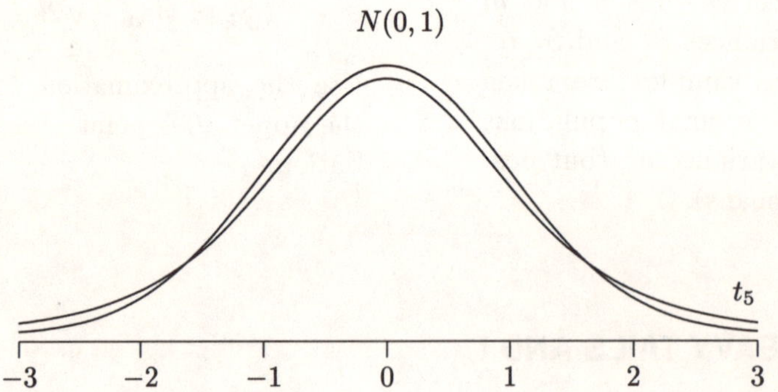

When based on a sample of size $n = 6$ from a normally distributed population, the sampling distribution of $\frac{\overline{X}-\mu}{\frac{\sigma}{\sqrt{6}}}$ is a standard normal. In the above graph this is represented by the pdf which is higher in the centre and lower in the tails.

When based on a sample of size $n = 6$ from a normally distributed population, the sampling distribution of $\frac{\overline{X}-\mu}{\frac{S}{\sqrt{6}}}$ has a t distribution with parameter value 5. (Such a probability distribution is called a t_5 distribution.) In the above graph the t_5 distribution is represented by the pdf which is lower in the centre and higher in the tails.

The structure of the t distribution

Definition 10.10 *The random variable T defined by the equation*

$$T = \frac{Z}{\sqrt{\frac{W}{k}}},$$

where $Z \sim N(0,1)$ is independent of $W \sim \chi^2_k$, is said to have a **t distribution with k degrees of freedom** *and we use the distributional notation: $T \sim t_k$.*

□

Specialised transformation methods are required to show that the probability density function for T is as given below. (See, for example, Mood, Graybill and Boes, 1974.)

PROPERTIES OF THE t DISTRIBUTION $T \sim t_k$

(10.5)

Probability density function $\quad f_T(t) = \frac{\Gamma(\frac{k+1}{2})}{\Gamma(\frac{k}{2})} \frac{1}{\sqrt{k\pi}} \frac{1}{\left(1+\frac{t^2}{k}\right)^{\frac{k+1}{2}}}$

$-\infty < t < \infty; \quad k > 0$

Mean	$E(T) = 0$
Variance	$Var(T) = \frac{k}{k-2}, \quad k > 2$
Standard deviation	$SD(T) = \sqrt{\frac{k}{k-2}}, \quad k > 2$
Coefficient of skewness	$\zeta_3 = 0$
Coefficient of kurtosis	no simple form
Moment generating function	does not exist
Cumulant generating function	does not exist

The t distribution is a 'heavy-tailed' probability distribution compared with the standard normal distribution, as may be seen in the pdf graphs of Example 10.14. This means that for a large positive constant c, more probability is trapped beyond c under a t_5 curve than under a $N(0,1)$ curve. The formula for the pdf of a t random variable

$$c_1(1 + c_2t^2)^{-c_3} \qquad (c_1,\ c_2 \text{ and } c_3 \text{ constants})$$

shows that for increasingly large values of t, the pdf is *algebraic* in nature and decreases more slowly than the exponential decline $e^{-\frac{1}{2}t^2}$ characteristic of the bell-shaped symmetry of the standard normal curve.

As the degrees of freedom of a t curve increase, the shape of the curve gradually approaches that of the standard normal curve. The pdf for t_2 is flatter than the pdf for t_3, which is flatter than the pdf for t_4, which is flatter than ..., which is flatter than the pdf for t_{24}, which is flatter than the pdf for t_{25} and

$$t_{25} \approx N(0,1).$$

Of course, the t curves which are flatter than the standard normal curve are the pdf's of random variables with larger variance than the standard normal random variable (which has variance 1).

Notation for *t* tables

If the upper α point of a t distribution with k degrees of freedom is denoted by $t_{\alpha,k}$, then

$$P(T > t_{\alpha,k}) = \alpha \quad \text{when} \quad T \sim t_k,$$

so that α is the area under a t_k density beyond $t_{\alpha,k}$. These 'upper-tail' probabilities are tabulated in Appendix 4 for values $k = 2, 3, 4, 5, \ldots, 30, \infty$. Beyond $k = 30$ degrees of freedom, standard normal probability tables replace t tables because of the impact of the Central Limit Theorem. The symmetry of the t distribution about $\mu = 0$ allows determinations of both upper- and lower-tail probabilities using the given choices of α values.

Example 10.15

Inter-90th percentile ranges: $N(0,1)$ and t_{20} curves compared

The inter-90th percentile range, as a measure of spread, is the difference between the 90th and 10th percentiles covering the whereabouts of the central 80% of the distribution: $\xi_{0.90} - \xi_{0.10}$.

For the t_{20} distribution,

$$\xi_{0.90} - \xi_{0.10} = t_{0.10,20} - (-t_{0.10,20}) = 1.325 - (-1.325) = 2.65.$$

For the $N(0,1)$ curve,

$$\xi_{0.90} - \xi_{0.10} = z_{0.9} - z_{0.1} = 1.28 - (-1.28) = 2.56.$$

This indicates a slightly higher spread for the t curve.

A pivot from the *t* distribution

The pivot $\frac{\overline{X}-\mu}{\frac{S}{\sqrt{n}}}$, on which we rely for interval estimation, has precisely the t distribution structure when the data are normally distributed.

Theorem 10.7 *When $\overline{X}$ is the sample mean and S is the sample standard deviation of a random sample of size n from a $N(\mu, \sigma^2)$ population,*

$$\frac{\overline{X} - \mu}{\frac{S}{\sqrt{n}}} \sim t_{n-1}.$$

Proof:

$$\frac{\overline{X} - \mu}{\frac{S}{\sqrt{n}}} = \frac{\frac{\overline{X} - \mu}{\frac{\sigma}{\sqrt{n}}}}{\sqrt{\frac{\frac{(n-1)S^2}{\sigma^2}}{(n-1)}}} = \frac{Z}{\sqrt{\frac{W}{n-1}}} \sim t_{n-1},$$

for

$$Z = \frac{\overline{X} - \mu}{\frac{\sigma}{\sqrt{n}}} \sim N(0, 1)$$

independently (since $\overline{X}$ and S^2 are independent) of

$$W = \frac{(n-1)S^2}{\sigma^2} \sim \chi^2_{n-1}.$$

■

In the proof of Theorem 10.7, the chi-squared distribution for W (Theorem 10.5) and the normal distribution for Z (Theorem 9.3) are constructed from normally distributed data. It must always be remembered:
For probability calculations involving t distributions to be valid, it must be assumed that the sample data are from a normally distributed population.

When the pivot for the t distribution is unwrapped according to the inequality manipulations described in the proof of Theorem 10.2 the following confidence interval for μ is obtained:

A $100(1 - \alpha)\%$ confidence interval for the population mean μ based on a random sample $X_1, X_2, X_3, \ldots, X_n$ with sample mean $\overline{X}$ and sample standard deviation S from a $N(\mu, \sigma^2)$ population is $[\overline{X} - t_{\frac{\alpha}{2}, n-1}\frac{S}{\sqrt{n}}, \ \overline{X} + t_{\frac{\alpha}{2}, n-1}\frac{S}{\sqrt{n}}]$.

Example 10.16

Emergency lock on seatbelt webbing

In a car accident, there is intense sudden pressure on the seatbelt assembly. An emergency locking retractor prevents payout of the seatbelt webbing so that front seat passengers remain constrained by their seatbelts and do not hit the dashboard.

However, for ordinary seatbelt usage, the lock is not activated when the webbing is tugged gently.

A random sample of 8 retractors from a seatbelt production line gave the following payouts (mm) in a controlled experiment where the webbing of each was subjected to the same sudden force of a simulated car accident:

18, 17, 19, 23, 19, 21, 21, 20.

Worked example:

Determine a 90% confidence interval for the mean seatbelt payout under simulated impact test conditions.

First priority is to check whether it is reasonable that the data have come from a normally distributed population. A normal probability plot is constructed for this purpose.

Normal probability plot of seatbelt payouts

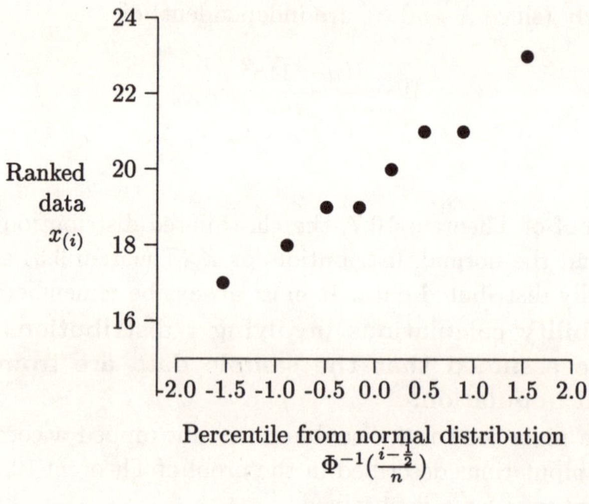

Although it is difficult to be certain from such a small sample size, the points do fall in a straight line pattern. This linear effect is convincing enough to proceed with the construction of a t interval: Calculations for this data set give a sample mean $\overline{x} = 19.75$ and a sample standard deviation of $s = 1.9086$. A 90% confidence interval for the mean seatbelt payout is

$$[\overline{x} - t_{0.05,7}\frac{s}{\sqrt{n}}, \ \overline{x} + t_{0.05,7}\frac{s}{\sqrt{n}}].$$

Substituting into this interval the observed values of the sample statistics, the interval is fixed as

$$[19.75 - 1.895\frac{1.9086}{\sqrt{8}}, 19.75 + 1.895\frac{1.9086}{\sqrt{8}}] = [18.47, \ 21.03].$$

EXERCISES

10.52 Use t tables to determine the values of $t_{0.9,5}$, $t_{0.95,7}$, $t_{0.99,12}$, $t_{0.01,10}$.

10.53 If $T \sim t_{12}$, determine the value of the constant c so that:

(a) $P(T > c) = 0.05$;

(b) $P(T < c) = 0.05$;

(c) $P(|T| > c) = 0.05$.

10.54 Seven adult dugites (venomous snakes) had measured lengths (in metres):

1.20 1.15 1.25 1.45 1.35 1.30 2.65.

Discuss the presence of an outlier in this data set.

We wish to determine the effect of outliers on confidence intervals with small sample sizes.

Answer each of the following questions, first using all 7 data points, and second, with the largest data point removed from consideration.

(a) Construct a normal probability plot of the data.

(b) Construct a B/W plot of the data.

(c) Give a point estimate of the mean length of dugites. Give the associated (estimated) standard error of estimate.

(d) Construct a 95% confidence interval for the mean length of dugites.

Comment.

10.55 The recent acceleration of the corrosion of concrete buildings is thought to be due to pollution and acidity levels in rainwater.

Five core samples drilled in randomly chosen positions from the concrete facade of a King Street building of 1960's vintage showed depth of corrosion (cm) of:

7.5 1.5 3.5 5.5 4.5.

(a) Show that the standard deviation of these data is $\sqrt{5}$.

(b) Construct a 98% confidence interval for the mean depth of corrosion of the concrete facade of this building.

(c) List the assumptions required for the construction of the interval in (b) to be valid.

10.56 A doctor recorded the following temperatures (in degrees Celsius) for five of his patients at the time of diagnosis of a new strain of influenza virus:

37.4 37.2 37.7 37.8 38.1.

(a) Construct a 98% confidence interval for the mean body temperature of people who have the virus.

(b) What assumptions about the body temperatures of people afflicted with the virus must be satisfied in order that your interval in (a) be valid?

(c) Would you support this statement?

Evidence suggests that the average temperature of people suffering from the new strain of flu virus is above normal (37.0 degrees Celsius).

How does the confidence interval in (a) help you decide?

10.57 An observed random sample $x_1, x_2, \ldots, x_{16}$, from a normal population with unknown mean and variance gave

$$\sum_{i=1}^{16} x_i = 4 \text{ and } \sum_{i=1}^{16} x_i^2 = 61.$$

Estimate the population mean and variance with both point and 98% interval estimates.

Is it possible (likely?) that the population mean is 0?

10.58 IQs are normally distributed with mean $\mu = 100$ and $\sigma = 16$.

When n randomly selected people have their IQs tested, the average sample IQ is $\overline{X}$ and the sample standard deviation IQ is S. What is the probability distribution of each of

$$\frac{\overline{X} - 100}{\frac{16}{\sqrt{n}}}$$

and

$$\frac{\overline{X} - 100}{\frac{S}{\sqrt{n}}}$$

for

(a) $n = 10$?

(b) $n = 100$?

10.59 The upper α points of a t_{20} curve and a standard normal curve are given by $t_{\alpha,20}$ and z_α respectively. Would you expect the inequality

$$t_{\alpha,20} > z_\alpha$$

to hold for all $\alpha < 0.05$?

Explain with the assistance of diagrams of the t and normal curves.

10.60 To investigate the coverage of a new 'no-drip' brand of paint, the quantity used to paint 8 metre by 2 metre panels was carefully measured. 10 such panels were painted and the quantities of paint used had a mean of 976 mL with a standard deviation of 36 mL per panel.

(a) Give a point estimate of the average quantity of paint needed to cover a square metre of panel.

(b) Give a point estimate of the variance in the average quantity of paint needed to cover a square metre of panel.

(c) Assuming that the quantity of paint, X, required to cover an 8 metre by 2 metre panel is a normally distributed random variable, determine a 95% confidence interval for $E(X)$.

Explain in words the meaning of the interval that you have constructed.

10.61 Show that a t distribution with 1 degree of freedom is a Cauchy distribution.

CHAPTER ELEVEN

One-Sample Significance Tests

In this chapter you will find

11.1 SIGNIFICANCE TESTING PROCEDURES: STATISTICIAN AS JUROR

Suppose that you regularly buy a brand of breakfast cereal in 500 gram packets. 500 g is inscribed on the side of the packet with an air of authority from the manufacturer. Undoubtedly, the manufacturer implements checks on the production line for such cereal packets to ensure quality of product and justify the label weight: these packets contain 500 g of breakfast cereal. Even though the machine that fills the packets with weight X, in grams, is as accurate as possible, we must accept that X is not a constant; that variabilities in the production process ensure that X is a random variable.

The outcomes of X, the actual net content of packets on the supermarket shelf, varies from packet to packet. X is a population random variable with unknown distribution, unknown mean and unknown variance.

It remains fair for a consumer to ask: Do the packets contain 500 g? The one just purchased? The one that was purchased last week? On average? The packet that was bought last week certainly may have had less than 500 g, but is this simply due to random fluctuation alone, or to bad setting of the filling machine? What happens if the mean μ of the probability distribution for X is centred much lower than 500 g? Then it is possible that the packets are being repeatedly underfilled in the sense that $P(X < 500)$ may exceed $\frac{1}{2}$. Suppose that normal probability plots of packet weights indicate that it is reasonable to assume that X is normally distributed. Last week, when you purchased a 500 g breakfast cereal packet and weighed the contents, you observed a value or possible outcome of the random variable X. Such an **observed outcome** will be denoted by $x_{observed}$. The two normal probability curves below show possible locations for $x_{observed}$.

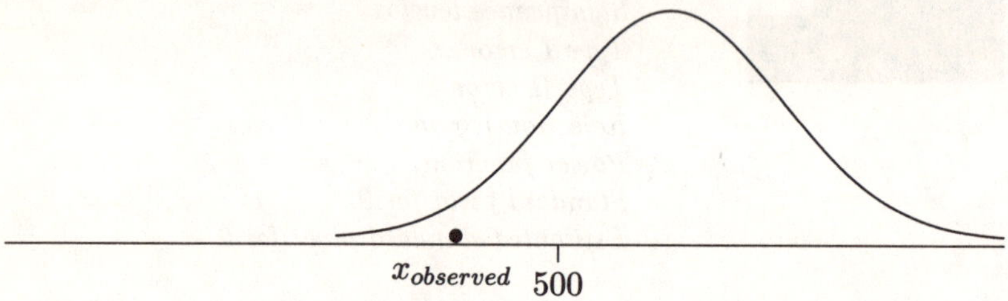

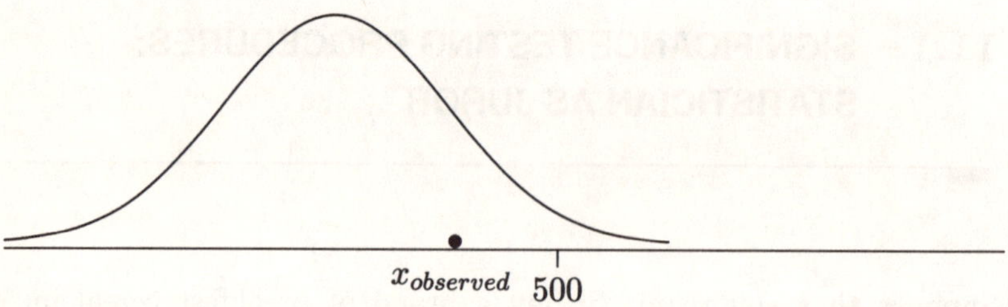

From $x_{observed}$ alone it is impossible to detect which of the two curves is the best probability model for X. In both cases, $x_{observed}$ lies below 500 g, but it is a question of where the rest of the probability distribution lies. The top diagram indicates a distribution that seems fair to consumers: only a few packets have less than the label weight (and, as luck would have it, $x_{observed}$ is the weight of such a packet). In this case our observed underweight packet is due to chance. In the bottom diagram the underweight packet observed is not due to chance but rather the fact that nearly all the packets are underweight.

Null and alternative hypotheses

A **significance test** or **hypothesis test** enables us to check, with a measure of accuracy, which of two such scenarios is the most likely. Since the normal distributions pictured have the same shape, our choice of distribution becomes a choice of centre. That is, where is $\mu = E(X)$ situated?
We must proceed on the following assumption:
ASSUMPTION:

the manufacturer is filling the packets fairly (μ= 500)

until

EVIDENCE: the data

convinces us 'beyond reasonable doubt' that the assumption is correct, or that the assumption is invalid. Statisticians are like jurors who assume that the defendant is innocent until the evidence has indicated guilt beyond a reasonable doubt. That is, we assume that the manufacturer is filling the packets fairly until the data indicate otherwise, beyond reasonable doubt.

The above **ASSUMPTION** is termed the **null hypothesis** and is denoted by the symbol H_0. It is the truth or otherwise of this hypothesis that is being challenged. In many examples, the null hypothesis represents the *status quo*, or conditions as they have been, or conditions to be tested. If, after due consideration of the experimental evidence (the data), H_0 is untenable, then an **alternative hypothesis**, H_1 is adopted. H_1 usually represents a position of change, of difference from, of reduction compared with, of improvement over, ..., the status quo. Mathematically, H_0 and H_1 are expressed in terms of the parameter (in this case μ) to be tested:

$$\begin{aligned} H_0: &\quad \mu = 500 \\ H_1: &\quad \mu < 500. \end{aligned}$$

The alternative hypothesis in the cereal packet example is phrased in terms of a poor performance on the manufacturer's part—the population mean is lower than it should be. Only population parameters are specified in the statements of the null and alternative hypotheses; **never** sample statistics. How is the evidence provided by the data used in the hypothesis test? A **test statistic** is a statistic used as a **decision rule** to determine whether H_0 is accepted; or whether H_0 is rejected in favour of H_1. The substance of a significance test is an assessment, with the aid of the test statistic, of the truth or falsity of H_0. For tests concerning population means, the obvious test statistic to use is the sample mean. The Central Limit Theorem indicates that for a large sample of size n, if σ^2 is the variance in weight of all breakfast cereal packets of the brand under consideration, then, as a close approximation, $\overline{X} \sim N(\mu, \frac{\sigma^2}{n})$, which is expressed as

$$Z = \frac{\overline{X} - \mu}{\frac{\sigma}{\sqrt{n}}} \sim N(0,1)$$

in standardised form (Theorem 9.4). In all significance testing procedures, H_0 is assumed to be true until the test statistic indicates otherwise, beyond reasonable doubt. This means that the test statistic is evaluated under the assumption that the null hypothesis is true.

Suppose that now, after setting H_0 and H_1, we observe the data: $n = 100$ packets gave $\overline{x}_{observed} = 497$ and $s_{observed} = 20$. Then, with such a large sample size, s may be appropriately used as a point estimator of σ. On substitution, the observed value of the test statistic is found to be

$$z_{observed} = \frac{\overline{x}_{observed} - \mu}{\frac{s_{observed}}{\sqrt{100}}},$$

and since this is determined under the assumption that H_0 is true,

$$z_{observed} = \frac{\overline{x}_{observed} - 500}{\frac{s_{observed}}{\sqrt{100}}} = \frac{497 - 500}{\frac{20}{\sqrt{100}}} = -1.5.$$

$z_{observed}$ should be placed on a diagram of the sampling distribution of the test statistic:

(Standardised) sampling distribution of the test statistic

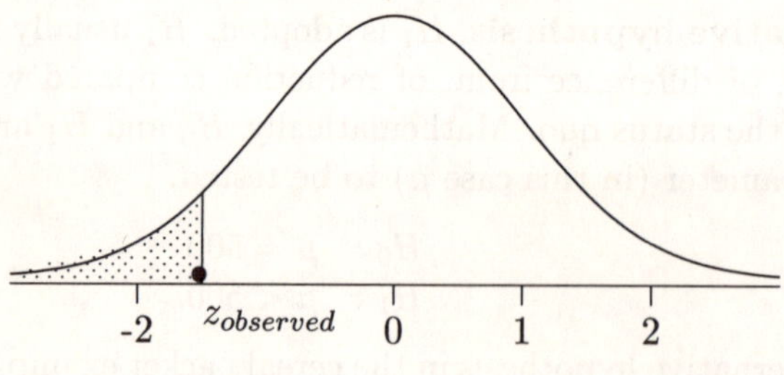

p-values

How likely is it to obtain $z_{observed}$ under the supposition that H_0 is true? Since a single fixed point in a normal (continuous) distribution has probability 0, this question is phrased more usefully as:
How likely is it to obtain $z_{observed}$, or *worse*, for the observed value of the (standardised) test statistic Z, given that H_0 is true? Such a probability is determined as follows:

$$p\text{-value} = P(Z \leq z_{observed} | H_0 \textit{ is true}) = P(Z \leq -1.5) = 0.0668$$

and is depicted by the shaded area of the lower tail in the above diagram of the standard normal curve.

This probability, called the **p-value of the test** or the **significance probability of the test**, is a measure of how likely/unlikely it is to experience the observed data if the null hypothesis is indeed true. If the p-value is large, then the observed value of the test statistic is quite likely when H_0 is true; this, in turn, leads us to favour the null hypothesis in the sense that there is not enough evidence to reject it. If the p-value is small, then it is quite unlikely to obtain a test statistic of the magnitude observed when H_0 is true; this, in turn, leads us to favour the alternative hypothesis. p-values are measured on a sliding scale between very small (near 0) and very large (near 1). The dividing line between large and small p-values is often taken as 0.05. Many researchers feel comfortable with this value; if there is less than a 5% chance of obtaining $\leq z_{observed}$ as an outcome of the test statistic, then $z_{observed}$ is indeed rare and our assessment of where the sampling distribution of the test statistic is centred needs to be updated.

When p-value< 0.05, the result of the test is said to be **statistically significant** and many researchers place the notation (*) beside the test statistic giving a significant outcome to a significance test. When p-value< 0.01, the result of the test is said to be **highly statistically significant** and many researchers place the notation (**) beside the test statistic that gives a highly significant outcome to a test. This gives

An operator's manual of p-values

p-value	Interpretation	Notation	Decision
$p < 0.01$	highly significant	$(**)$	H_0 not supported
$0.01 < p < 0.05$	significant	$(*)$	H_0 not supported
$p > 0.05$	not significant	$(n.s.)$	H_0 supported

The interpretation that we will follow for clear-cut decisions between 'reject H_0' and 'do not reject H_0' based on the significance or non-significance of the test statistic is: If

$$p\text{-value} \begin{cases} \leq \alpha, & \text{reject } H_0 \text{ in favour of } H_1; \\ > \alpha, & \text{do not reject } H_0. \end{cases}$$

Here α is most often chosen as 0.05. Under this criterion, a p-value of size 0.0668 is large (> 0.05) and we are inclined not to reject the null hypothesis. This implementation of the decision rule provides us with a conclusion to the hypothesis test: the mean weight of contents of the cereal packets is not **significantly** less than 500 g.

The tails of a test

In general terms, if θ is the parameter of interest in the probability model for the population, then three cases arise when a test statistic $\widehat{T}$ is used to test the null hypothesis that θ assumes a known value, $H_0 : \theta = \theta_0$. These cases depend on the nature of the alternative hypothesis and may be listed as

CASE 1 $H_1 : \theta < \theta_0$ (one-tailed test in the lower tail)
CASE 2 $H_1 : \theta > \theta_0$ (one-tailed test in the upper tail)
CASE 3 $H_1 : \theta \neq \theta_0$ (two-tailed test).

This indicates that whether or not a one or two-tailed test is used depends on the *direction* of the alternative hypothesis. The choice is made with full knowledge of the response being measured.

For CASE 1, if H_0 is false, there is most likely a decrease in the value of the parameter. For CASE 2, if H_0 is false, there is most likely an increase in the value of the parameter. For CASE 3, if H_0 is false, there is most likely a change in the value of the parameter but we hold no prior information as to whether the value of the parameter will have increased or decreased. To avoid bias in the test results, the choice of alternative hypothesis must be made before the data are seen.

Example 11.1

Alternative hypotheses for the car yard

Scenario 1. The manager of a Mitsubishi car dealership is considering a new bonus plan that may increase the sales volume of the dealership. Past records show that μ, the mean sales volume per sales employee, has been 4 cars per month. A sample of sales employees is to sell under the new bonus plan for a trial period. The manager is interested in implementing the new plan if the sample evidence indicates that the plan increases the mean sales volume. In this scenario, before the data are seen, H_0 and H_1 would be set as:

$$H_0: \quad \mu = 4$$
$$H_1: \quad \mu > 4.$$

Scenario 2. The manager of a Mitsubishi car dealership is considering whether the recession is having a significant effect on the sales volume of the dealership. Past records show that μ, the mean sales volume per sales employee, has been 4 cars per month. Recent data for a sample of sales employees are obtained. Is there significant evidence that the mean sales volume per employee is reduced in the current financial climate? In this scenario, before the data are seen, H_0 and H_1 would be set as:

$$H_0: \quad \mu = 4$$
$$H_1: \quad \mu < 4.$$

Scenario 3. The manager of a Mitsubishi car dealership constantly monitors the sales volume of the dealership. Past records show that μ, the mean sales volume per sales employee, has been 4 cars per month. Recent data for a sample of sales employees are obtained. Is there significant evidence that the mean sales volume per employee has recently changed? In this scenario, before the data are seen, H_0 and H_1 would be set as:

$$H_0: \quad \mu = 4$$
$$H_1: \quad \mu \neq 4.$$

After examining Example 11.1 there is a nagging question: Why does the null hypothesis always dictate that the parameter is *equal* to a specified value? Why not $\geq$ or $\leq$? The answer has two aspects. If, in Scenario 2 of Example 11.1, after seeing the data, $H_0 : \mu = 4$ is rejected in favour of $H_1 : \mu > 4$, then surely, $H_0 : \mu \leq 4$ would also be rejected in favour of the same H_1. By using the equality $\mu = 4$ in H_0, we are starting at the extreme boundary of the range of possible null hypothesis values $\mu \leq 4$. The interpretation of the inequality $\mu \leq 4$ is: past sales records show that employees have, on average, been selling *at most* 4 cars per month.

The *p*-value of the test will depend on whether or not a **one-tailed test** or a **two-tailed test** is appropriate. Since the *p*-value denotes the probability of obtaining a *more extreme value* than actually observed when H_0 is true, we may formalise the meaning of 'more extreme' in terms of the tails of the probability distribution for T. In summary:
For a one-tailed test (lower tail), *p*-value $= P(\widehat{T} \leq \widehat{t}_{observed})$.
For a one-tailed test (upper tail), *p*-value $= P(\widehat{T} \geq \widehat{t}_{observed})$.
For a two-tailed test, the *p*-value is doubled, since we are unable to tell in advance whether $\widehat{t}_{observed}$ will be observed in the top or the bottom tail of the distribution of T. When $\widehat{t}_{observed}$ is a value from the parameter space specified by H_1,

$$p\text{-value} = \tau[P(\widehat{T} \leq \widehat{t}_{observed})]^{\delta}[P(\widehat{T} \geq \widehat{t}_{observed})]^{1-\delta}, \qquad (11.1)$$

where τ is the number of tails of the test (with values $\tau = 1$ or $\tau = 2$) and

$$\delta = \begin{cases} 1 & \text{if } \widehat{t}_{observed} \text{ is in the lower tail of the probability distribution of } \widehat{T}; \\ 0 & \text{if } \widehat{t}_{observed} \text{ is in the upper tail of the probability distribution of } \widehat{T}. \end{cases}$$

Example 11.2
***p*-values for the car yard: Example 11.1 continued**
Worked example:
A random sample of 30 employees showed an average of 4.45 sales per month with a standard deviation of 1.3 sales per month. Determine the *p*-value of the tests in Scenario 1 and Scenario 3.

In this case, the general notation reduces to $\theta = \mu, \theta_0 = 4, \widehat{T} = Z$ and $\widehat{t}_{observed} = z_{observed}$. Then the observed value of the (standardised) test statistic is

$$z_{observed} = \frac{\overline{x}_{observed} - 4}{\frac{s_{observed}}{\sqrt{30}}} = \frac{4.45 - 4}{\frac{1.3}{\sqrt{30}}} = 1.896.$$

For Scenario 1, the hypotheses

$$\begin{aligned} H_0 &: \quad \mu = 4 \\ H_1 &: \quad \mu > 4 \end{aligned}$$

are tested to see if the new bonus plan has increased mean sales volume. This is a one-tailed test in the upper tail. Therefore, with $\tau = 1$ and $\delta = 0$ in (11.1),

$$\begin{aligned} p\text{-value} &= \tau[P(Z \leq z_{observed})]^{\delta}[P(Z \geq z_{observed}]^{1-\delta} \\ &= (1)P(Z \geq 1.896) = 0.0294. \end{aligned}$$

With such a small p-value the null hypothesis must be rejected in favour of the alternative hypothesis. Conclusion: There has been a significant increase in the mean sales volume (p-value $= 0.0294$).

For Scenario 3, the hypotheses

$$\begin{aligned} H_0 &: \quad \mu = 4 \\ H_1 &: \quad \mu \neq 4 \end{aligned}$$

are tested to see if the mean sales volume per employee has changed. This is a two-tailed test. Therefore, with $\tau = 2$ and $\delta = 0$, since the observed value of the test statistic is in the upper tail, in (11.1),

$$\begin{aligned} p\text{-value} &= \tau[P(Z \leq z_{observed})]^{\delta}[P(Z \geq z_{observed}]^{1-\delta} \\ &= (2)P(Z \geq 1.896) = 0.0588. \end{aligned}$$

Since the p-value exceeds the critical 'cut-off' boundary of 0.05 that is usually chosen to differentiate between small and large p-values, the null hypothesis is accepted in the sense that there is not enough evidence to reject it. Conclusion: There has not been a significant change in the mean sales volume (p-value $= 0.0588$).

This is the opposite conclusion to that reached in Scenario 1! The reason is that because of the factor of $\tau = 2$ in two-tailed tests, the p-values tend to be larger so that such tests tend to reject less often. In this sense a two-tailed test is more conservative than a one-tailed test.

Example 11.3

Impact resistance of clay pavers

In a test designed to examine the durability of clay pavers, the Pavement Research Group at RMIT designed the following test. A sharpened hammer drill impacts the surface of a test paver for a period of 1 minute. The volume of material eroded in this process, as measured by the volume of displaced liquid, becomes a measure of impact resistance of the paver. If less than 0.265 mL is eroded from the paver, then the clay paver is suitably 'hard' for heavy traffic densities: 'hard', because it is difficult to dig material from the surface of the paver.

If X denotes the volume of material (mL $\times 10^{-3}$) eroded in such a test then, for a particular batch of pavers in which $\mu = E(X)$ denotes the mean volume eroded, we wish to test whether the batch of pavers has an appropriate degree of hardness. This may be expressed as

$$H_0: \quad \mu = 265$$
$$H_1: \quad \mu < 265.$$

The observed data are:

240	255	258	267	233	229	259	273	228	234	259
226	235	223	248	238	229	233	247	244	229	239
257	256	218	228	242	248	232	215	240	243	208
201	245	242	240	228	233	224	246	253	234	223
230	249	242	239	267	258	250	261	259	262	249
265	259	262	231	256	285	252	254	264	260	263
250	263	248	258	252	254	252				

Volume of material eroded from clay pavers
Scale: Stems—10 units; leaves—units
23|5 represents 235 units

Stem	Leaves
20*	1
20.	8
21*	
21.	58
22*	334
22.	6888999
23*	01233344
23.	5899
24*	00022234
24.	56788899
25*	00222344
25.	56678889999
26*	0122334
26.	577
27*	3
27.	
28*	
28.	5

$$z_{observed} = \frac{\overline{x}_{observed} - 265}{\frac{s_{observed}}{\sqrt{73}}} = \frac{244.88 - 265}{\frac{15.82}{\sqrt{73}}} = -10.89.$$

For this result $\overline{x}_{observed} = 244.88$, (**), p-value ≈ 0.0000. This result is immediately apparent in advance if we examine a stem-and-leaf plot of the data.

If it is quite obvious from the S/L plot that this batch of pavers meets the required standard of hardness, we could indeed ask, if a larger sample mean had been recorded, would the result have been so obvious? How close would $\overline{x}$ have to be to 265 in order to be on the borderline between acceptance/rejection of H_0? We take up this search for the so-called *rejection region* again in the next section.

EXERCISES

11.1 Which of the following are valid statements for determining whether a population mean μ is significantly different from 1 when the sample mean $\overline{x}$ is to be used as the test statistic?

(a) $H_0: \mu = 1$
$H_1: \mu < 0$

(b) $H_0: \mu = 1$
$H_1: \mu > 1$

(c) $H_0: \mu = 1$
$H_1: \mu \neq 1$

(d) $H_0: \overline{x} = 1$
$H_1: \overline{x} \neq 1$

(e) $H_0: \mu = 1$
$H_1: \overline{x} \neq 1$.

11.2 After re-routing the flight path for jets on the international route between Auckland and London, Air New Zealand wishes to determine whether the mean flight time has increased.

Specify H_0 and H_1 in this situation, taking care to define all parameters used.

If n flights are monitored, what would be a useful test statistic to test the hypotheses if

(a) $n = 1$?

(b) $n = 5$?

(c) $n = 50$?

What assumptions need to be made for your choice of test statistic?

11.3 The diameter measurements of electric conduit (plastic pipe to house electric wiring), taken at 100 points along the pipe, yield a sample mean of 2.1 centimetres and a sample standard deviation of 0.3 centimetre. We wish to determine whether the true mean diameter, μ, exceeds the manufacturer's specification of 2 centimetres.

State the null and alternative hypothesis.

Is it necessary to assume that the data is normally distributed in this test?

11.4 In Exercise 11.3, what is the observed size of the (approximately) normally distributed test statistic

$$\frac{\overline{X} - \mu}{\frac{S}{\sqrt{n}}}$$

when H_0 is true? Is $z_{observed}$ significantly different from 0? Would the result of the test be described as (*), (**), or (*n.s.*)? Relate this conclusion back to deciding whether there is compliance with the manufacturer's specification.

11.5 A study of the operation of a multistorey car park showed that, in the past, average parking time was 220 minutes. Recently the car park was remodelled and parking fees increased. The management wants to know if the changes have had any effect on mean parking time.

Give the appropriate null and alternative hypotheses for a test of significance.

11.6 In a procedure designed to test the durability of exterior house paint, 100 wooden panels were painted under identical conditions. These panels were then exposed to a sequence of tests that involved extreme temperature, light and moisture conditions. At the end of the experiment, the quality of the paint surface was rated on several characteristics (such as paint loss, lustre of painted surface, cracking and peeling). Scores, X, between 0 (extremely poor) and 100 (excellent) were assigned to the 100 panels.

Since individual scores are often based on several characteristics through averaging, it is reasonable to assume that X is normally distributed.

Suppose that $X \sim N(\mu, 20^2)$. That is, the true average, μ, may vary from paint to paint, but experience in the industry has shown that most paints with similar viscosities show a similar variability in score of around 400.

(a) Determine the p-value of the test

$$H_0 : \mu = 50$$
$$H_1 : \mu > 50$$

if the 100 panels show an average sample score of 55.

(b) Draw a labelled sketch of the distribution of the test statistic used in (a). In this sketch, shade an area representing the p-value of the test

$$H_0 : \mu = 50$$
$$H_1 : \mu \neq 50$$

11.7 In a hypothesis test

$$H_0 : \mu = \mu_0$$
$$H_1 : \mu \neq \mu_0$$

for a population mean μ, suppose that the observed sample mean $\overline{x}_{observed}$ lies 2 standard errors from μ_0.

Are $\overline{x}_{observed}$ and μ_0 significantly different?

11.8 In a hypothesis test

$$H_0 : \mu = \mu_0$$
$$H_1 : \mu < \mu_0$$

for a population mean μ suppose that the observed sample mean $\overline{x}_{observed}$ lies 2 standard errors below μ_0.

Are $\overline{x}_{observed}$ and μ_0 significantly different? Compare your result with that obtained in Exercise 11.7. [If there is a choice to be had between one- and two-tailed tests, many researchers favour the cautionary nature of two-tailed tests; H_0 is rejected less often.]

11.9 If a test statistic follows a $N(0,1)$ distribution when the null hypothesis is true and the test is τ-tailed (for $\tau = 1$ or $\tau = 2$), complete the missing entries (◇) in the following table if the observed value of the test statistic is k:

k	τ	p-value
1	1	◇
−1.5	2	◇
◇	2	0.06
−1.9	1	◇
2.1	◇	0.0358
2.5	1	◇

11.10 Test the hypotheses

$$H_0 : \mu = 250$$
$$H_1 : \mu > 250$$

phrased in terms of μ, the mean volume (mL $\times 10^{-3}$) eroded from the surface of a clay paver in an impact resistance test by using the data of Example 11.3.

11.11 Imagery ratings for a large selection of abstract nouns and adjectives were determined by 50 psychology students; ratings on a scale of 1 (low imagery rating) to 7 (high imagery rating). The results included the following mean and standard deviation rating assessment:

Noun—*House*; $\overline{x} = 6.58$; $s = 0.97$.

(*Source*: Abernethy, M. (1991). Imagery ratings for 230 nouns and adjectives. *Australian Journal of Psychology*, 43, 3.)

Is the rating for *House* significantly larger than 6?

11.12 The results of a random sample of candidates in a tertiary admissions exam (TAE) designed to give a mean score μ of 50 were as follows:

TAE scores
Stems—tens; leaves—units
5.|5 represents a score of 55

3.	7
4∗	0344
4.	5677888899
5∗	12222333334
5.	5667778
6∗	012233
6.	8
7∗	

The test of interest is:

$$H_0 : \quad \mu = 50$$
$$H_1 : \quad \mu \neq 50.$$

Determine the sample mean and decide which of the following is an appropriate description to be placed beside the value of the sample mean to indicate the result of the test: (∗), (∗∗), (*n.s.*).

11.13 A steel wire manufacturer has found from past experience that an alloy filament has a mean breaking strength of 16 kg with a standard deviation of 2.5 kg. In a re-appraisal of manufacturing techniques to increase efficiency, the order of execution of certain stages in the process is changed.

A sample of 50 filaments from the new process is examined and quality tested, giving a mean breaking strength of 15.5 kg. Would you accept that the new process produces, on average, a weaker filament?

11.14 The concentration of a certain form of active agent in a liquid form of pesticide must not exceed 12 parts per million.

In preparation for wholesale distribution, the pesticide is made up in batches in which the concentration is normally distributed. 30 batches show a sample mean of 12.5 parts per million and a sample standard deviation of 2.5 parts per million.

Is the average concentration of all batches significantly larger than 12?

11.2 ERROR RATES

We have reasoned that large p-values tend to favour the null hypothesis and small p-values tend to favour the alternative hypothesis. The boundary between large and small p-values is the *significance level* of the test.

Definition 11.1 *The* **significance level**, *denoted by* α, *is defined as the largest p-value leading to the rejection of* H_0:

$$p\text{-value} \begin{cases} \leq \alpha, & \text{reject } H_0 \text{ in favour of } H_1; \\ > \alpha, & \text{do not reject } H_0. \end{cases}$$

□

The most common choice of significance level is $\alpha = 0.05$. It is a choice made by the researcher based on what is felt to be an acceptable/unacceptable sized p-value.

Type I and Type II errors

Society often demands a black and white decision to a posed question. Is H_0 accepted or is H_0 rejected? Is the mean pollution level in the Yarra river less now than it was a year ago? On average, does *Disprin* relieve headache faster than *Tylenol*?

If we are forced into a 'reject H_0' or 'do not reject H_0' conclusion, we could be making a mistake! We wish to keep the chances of arriving at the wrong conclusion to an hypothesis test as small as possible. The following chart indicates where it is possible to go wrong:

		Reality	
		H_0 true	H_0 false
Decision	Accept H_0	correct	Type II error
	Reject H_0	Type I error	correct

In a courtroom drama, where, as the jury, we test the hypotheses

$$\begin{aligned} H_0 &: \text{ defendant is innocent} \\ H_1 &: \text{ defendant is guilty,} \end{aligned}$$

a **Type I error** is committed if the jury convicts an innocent defendant. This type of error must be minimised. Morally, many people would prefer to see a small number of guilty people escape conviction, rather than see an innocent person wrongly convicted.

> A Type I error occurs if we reject H_0 when, in fact, H_0 is true.

How can we be sure that such Type I error is controlled? Simply, we maintain the significance level at a low level. If the probabilities of making Type I and Type II errors are assessed, then since the significance level, α, is the largest p-value that would lead to the rejection of H_0:

$$\text{P(Type I error)} = \text{significance level} = \alpha.$$

Therefore, by controlling the significance level of our test to about 0.05, the Type I error is kept under control. By doing this, we must be careful that the Type II errors do not run out of control.

In our previous analogy of an hypothesis test to a courtroom drama for testing

$$\begin{aligned} H_0 &: \text{ defendant is innocent} \\ H_1 &: \text{ defendant is guilty,} \end{aligned}$$

a **Type II error** is committed if the jury finds a guilty defendant innocent. That is, when the jury makes the mistake of accepting the null hypothesis when it is false.

> A Type II error occurs if we accept H_0 when, in fact, H_0 is false.

Notice that an 'innocence' verdict from a jury really means that the prosecution failed to produce enough evidence to convict the defendant. Thus **the term 'accept H_0' is best interpreted as 'fail to reject H_0'.**

Consider the following general hypotheses for testing whether an unknown parameter θ assumes a known fixed value θ_0:

$$H_0: \quad \theta = \theta_0$$
$$H_1: \quad \theta \neq \theta_0.$$

If H_0 is false, the true value of the parameter θ is some fixed value θ_1 different from θ_0. The size of the probability of making a Type II error therefore depends on which is the correct value for the parameter from among those specified by the alternative hypothesis. If the probability of a Type II error is denoted by $\beta(\theta_1)$, when $\theta = \theta_1$, then

$$\text{P(Type II error when } \theta = \theta_1 \text{)} = \beta(\theta_1) = \text{P(accept } H_0 | H_0 \text{ is false, } \theta = \theta_1\text{)}.$$

We illustrate these concepts with a simple example where the error probabilities are easily calculated.

Example 11.4

Testing the mean length of airconditioning hoses: Example 10.6 continued

Hoses are guillotined to a mean length, θ, to fit airconditioning units. Manufacturer's specifications for the current units require $\theta = 10$. However, the actual hoselength cut, X, is anywhere in a 2 cm range about θ. Based on one observation of X, test whether the lengths produced are following the manufacturer's guidelines for units requiring 10 cm hose. The hypotheses may be stated as:

$$H_0: \quad \theta = 10$$
$$H_1: \quad \theta \neq 10$$

Let us select $\alpha = 0.05$. Putting such a brake on the Type I error will have the effect of partitioning the support of X into two sets of values: those values for which H_0 is accepted and those leading to the rejection of H_0. The test statistic is X. Clearly H_0 will appear false if the observed hoselength $x_{observed}$ is a long way from 10: either above 10, or below 10. That is, if the absolute difference between $x_{observed}$ and 10 is large ... say, greater than some constant k. Such a policy could be phrased as:

$$\text{reject } H_0 \text{ if } |x_{observed} - \theta| > k.$$

The value of k may be determined from the fact that the Type I error is under control. The value of θ is determined since the null hypothesis is assumed true during the test procedure (the defendant is assumed innocent until the evidence indicates guilt beyond reasonable doubt). Then, k may be determined from α:

$$\begin{aligned} 0.05 &= \alpha \\ &= P(\text{Type I error}) \\ &= P(\text{reject } H_0 \text{ when } H_0 \text{ is true}) \\ &= P(|X-\theta| > k | H_0 \text{ is true}) \\ &= P(|X-\theta| > k | \theta = 10) \\ &= P(|X-10| > k) \text{ when } X \sim U[9,11] \\ &= (2)(\tfrac{1}{2})(1-k). \end{aligned}$$

Solving this for k gives $k = 0.95$. Therefore, the observed values of X for which H_0 is rejected are those values for which $|x_{observed} - 10| > 0.95$. The remainder of the support of X, consisting of outcomes $x_{observed}$ which satisfy

$$|x_{observed} - 10| \leq 0.95$$

may be written as the interval [9.05, 10.95]. These are the values for which H_0 is accepted. Then $\beta(10.5)$ is determined by

$$\begin{aligned} \beta(10.5) &= P(\text{accept } H_0 \mid H_0 \text{ is false and } \theta = 10.5) \\ &= P(9.05 \leq X \leq 10.95 | \theta = 10.5) \\ &= P(9.05 \leq X \leq 10.95) \text{ given that } X \sim U[9.5, 11.5] \\ &= 0.725. \end{aligned}$$

In Example 11.4, Type II errors at other values of the parameter θ are determined as follows:

$$\begin{aligned} \beta(11) &= P(\text{accept } H_0 \mid H_0 \text{ is false and } \theta = 11) \\ &= P(9.05 \leq X \leq 10.95 | \theta = 11) \\ &= P(9.05 \leq X \leq 10.95) \text{ given that } X \sim U[10, 12] \\ &= 0.475. \end{aligned}$$

$$\begin{aligned} \beta(11.5) &= P(\text{accept } H_0 \mid H_0 \text{ is false and } \theta = 11.5) \\ &= P(9.05 \leq X \leq 10.95 | \theta = 11.5) \\ &= P(9.05 \leq X \leq 10.95) \text{ given that } X \sim U[10.5, 12.5] \\ &= 0.225. \end{aligned}$$

$$\begin{aligned}\beta(11.9) &= P(\text{accept } H_0 \mid H_0 \text{ is false and } \theta = 11.9)\\ &= P(9.05 \le X \le 10.95 | \theta = 11.9)\\ &= P(9.05 \le X \le 10.95) \text{ given that } X \sim U[10.9, 12.9]\\ &= 0.025.\end{aligned}$$

From these calculations it is clear that it is likely that we make a Type II error if the true value of θ is close to the value specified by H_0. If the true value of θ is 10.5, and we hypothesise θ to be 10, then it is easy to make a mistake and come to the wrong conclusion: the Type II error is $\beta(10.5) = 0.725$. However, if the true value of θ is set apart from the value specified by H_0, then it is unlikely that we will make a Type II error: $\beta(11.9) = 0.025$. The ability of the test to differentiate between appropriate and inappropriate values of the parameter is assisted by increasing the sample size.

Rejection region

Definition 11.2 *The* **rejection region**, *RR, is the set of possible outcomes of the test statistic which lead to the rejection of the null hypothesis.* □

It is very important to understand that there is a one-to-one relationship between the size of the probability of making a Type I error and the size of the rejection region. That is, by specifying α (as 0.05, say) the rejection region becomes automatically fixed. In Example 11.4, the selection of $\alpha = 0.05$, automatically fixed the size of the rejection region as

$$RR = \{x_{observed} : |x_{observed} - 10| > 0.95\}.$$

However, the next example illustrates how a significance level of α becomes automatically fixed once a rejection region has been specified.

Example 11.5

The powers of a clairvoyant

A clairvoyant claims to have heightened powers of perception and agrees to take a simple test. Unseen by the clairvoyant, a sequence of n cards coloured simply either red or black are held up, one at a time, independently, as subjects for the clairvoyant's concentration. For each card held up the clairvoyant must indicate whether it is coloured red or black. If X is the number of cards correctly identified by colour, then in $n = 20$ trials,

$$X \sim Bin(20, p),$$

where $p = P(\text{clairvoyant correctly identifies a card colour})$.
We wish to test whether or not the clairvoyant is a fake. Without special powers, the clairvoyant can do no better than guessing, in which case

$p = \frac{1}{2}$. If the clairvoyant has special powers, then we would expect p to be larger than $\frac{1}{2}$. The hypotheses to be tested are:

$$H_0: \; p = \tfrac{1}{2}$$
$$H_1: \; p > \tfrac{1}{2}.$$

We will begin by specifying some rejection regions for the test statistic X and the α levels implied by them. The most relevant question is then: How many cards out of 20 need to be correctly identified before **you** are convinced beyond reasonable doubt that the clairvoyant has special extrasensory powers? More than 19? More than 17? More than 13? These suggestions may operate as rejection regions: If $RR = \{19, 20\}$, then

$$\begin{aligned} \alpha &= P(\text{reject } H_0 \text{ given that } H_0 \text{ is true}) \\ &= P(X = 19, 20 \mid X \sim Bin(20, \tfrac{1}{2})) \\ &= 1 - P(X \leq 18) \quad \text{for } X \sim Bin(20, \tfrac{1}{2}) \\ &= 0.00002. \end{aligned}$$

It is therefore almost impossible to make a Type I error with such a rejection region. We could afford to be a little more generous with the size of the rejection region and still keep mistakes to a minimum. If $RR = \{17, 18, 19, 20\}$, then

$$\begin{aligned} \alpha &= P(\text{reject } H_0 \text{ given that } H_0 \text{ is true}) \\ &= P(X \geq 17 \mid X \sim Bin(20, \tfrac{1}{2})) \\ &= 1 - P(X \leq 16) \quad \text{for } X \sim Bin(20, \tfrac{1}{2}) \\ &= 0.001. \end{aligned}$$

If $RR = \{14, 15, 16, 17, 18, 19, 20\}$, then

$$\begin{aligned} \alpha &= P(X \geq 14 \mid X \sim Bin(20, \tfrac{1}{2})) \\ &= 1 - P(X \leq 13) \quad \text{for } X \sim Bin(20, \tfrac{1}{2}) \\ &= 0.058. \end{aligned}$$

Essentially we have found an appropriate RR which gives a significance level of reasonable size (close to 0.05). What is the meaning of this RR? If the clairvoyant gets 14 (or more) correct out of 20 it is unlikely that this was done by guessing. From the viewpoint of p-values, an observed count of 14 correct out of 20 tries would be summarised as

$$x_{observed} = 14 \; (n.s.) \quad p = 0.058;$$

while an observed count of 15 correct out of 20 would be summarised as

$$x_{observed} = 15 \; (*) \quad p = 0.021.$$

The power function

A graph that illustrates the size of the Type I and Type II errors is that of the *power function.*

Definition 11.3 *The* **power function** *is the function $\gamma(\theta)$ of the parameter θ defined by*

$$\gamma(\theta) = P(\textit{reject } H_0 \mid \theta)$$

for all values of θ in the parameter space.

□

The following theorem shows the relationship between the Type I errors, the Type II errors and the power function.

Theorem 11.1 *For testing the null hypothesis $H_0 : \theta = \theta_0$ against an appropriate alternative:*

1. $\alpha = \gamma(\theta_0)$;
2. $\beta(\theta_1) = 1 - \gamma(\theta_1)$,

where θ_1 is a value of the parameter specified by the alternative hypothesis.

Proof: $\alpha = P(\text{reject } H_0 \mid H_0 \text{ true}) = P(\text{reject } H_0 \mid \theta = \theta_0) = \gamma(\theta_0)$. Further, $\beta(\theta_1) = P(\text{accept } H_0 \mid H_0 \text{ false and } \theta = \theta_1) = 1 - P(\text{reject } H_0 \mid \theta = \theta_1) = 1 - \gamma(\theta_1)$.

■

Theorem 11.1 indicates the kind of properties that we would like a power curve to have. Clearly, it is desirable to have both error rates as small as possible and in most hypothesis testing situations, controls are put on α. Therefore, in the notation of Theorem 11.1, we require $\gamma(\theta_0)$ to be small (0.05? 0.01?). To maintain small Type II errors *at the same time*, then it is clear that for $\beta(\theta_1)$ to be small, the requirement is equivalent to $\gamma(\theta_1)$ being close to 1 in that part of the parameter space specified by the alternative hypothesis.

Example 11.6 shows a typical sketch of the power curve for an upper-tailed test. The curve does not rise as steeply towards 1 as we would like: the larger the power, the smaller the probability of Type II errors.

In **uniformly most powerful tests** it is possible to determine an ideal rejection region so that the probability of Type I error is controlled at an acceptable level α, but that the probability of a Type II error is as small as possible at all points specified by H_1 (that is, the power curve is as large as possible at all points in this range). Such tests are discussed in more advanced texts such as Hogg and Craig (1978).

Example 11.6

Power curve for a clairvoyant: Example 11.5 continued

For the clairvoyant example we plot the points $(p, \gamma(p))$ below. The calculations are based on the choice of rejection region $RR = \{14, 15, \ldots, 20\}$ which has $\alpha = 0.058$.

p	0.5	0.6	0.7	0.8	0.9
$\gamma(p)$	0.058	0.25	0.608	0.913	0.998

The power function $\gamma(p)$ as a function of p

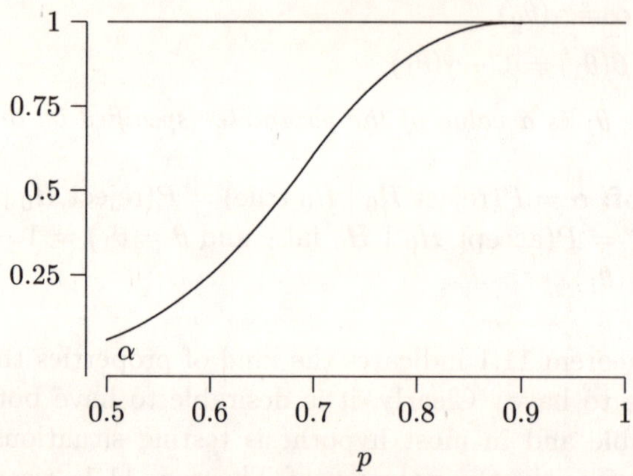

α is the height of the curve at $p = 0.05$.

EXERCISES

11.15 Which **one** of the following is correct? A Type II error is committed if you

(a) reject a false alternative hypothesis;

(b) fail to reject a false alternative hypothesis;

(c) reject a false null hypothesis;

(d) fail to reject a false null hypothesis.

11.16 Discuss whether each of the following statements is true or false:

(a) If H_0 is rejected at the 0.05 significance level, then it would also be rejected at the 0.01 significance level.

(b) If H_0 is accepted at the 0.05 significance level, then it would also be accepted at the 0.01 significance level.

(c) If H_0 is rejected at the 0.05 significance level, then the p-value of the test is greater than 0.05.

(d) If H_0 is accepted at the 0.05 significance level, then the p-value of the test is greater than 0.01.

11.17 Last year only 20% of the employees in a large organisation used public transport to commute to and from work. To determine if a recent campaign encouraging the use of public transport (and reduce congestion in carparks) has been effective, 20 randomly chosen employees are to be interviewed and the number of them currently using public transport recorded.

(a) Discuss the application of a binomial model in this problem. In particular, why would the binomial model apply only to a *very large* company?

(b) Formulate appropriate hypotheses in terms of p, the proportion of all employees currently using public transport.

(c) Determine the largest possible rejection region which still has $\alpha < 0.05$.

(d) What is the value of α for the rejection region in (c)?

11.18 Suppose now that the sample of employees is selected in Exercise 11.17.

(a) After interviewing 20 employees, it is found that 9 are currently using public transport. Determine the p-value of the test. What are your conclusions?

(b) If, in fact, $p = 0.4$, what is the probability that the test will fail to reject H_0?

11.19 In Example 11.4, the mean length, θ, of airconditioning hoses may be tested with a single observation X at significance level α with the hypotheses:

$$H_0: \ \theta = 10$$
$$H_1: \ \theta \neq 10.$$

Actual hoselength is uniformly distributed in a 2 cm width interval about θ. For the rejection region

$$\text{reject } H_0 \text{ if } |x_{observed} - \theta| > k:$$

(a) determine k if $\alpha = 0.01$;

(b) determine α if $k = 0.92$.

11.20 A coin is tossed 4 times in an experiment to check if it is unbiased: that is, if

$$p = P(\text{heads}) = \tfrac{1}{2}.$$

The hypothesis of unbiasedness is to be rejected if the number of heads observed is 0 or 4.

(a) Give the rejection region for the test.

(b) Is the test one-tailed or two-tailed?

(c) Determine the level of significance of the test.

11.21 In Exercise 11.20:

(a) determine the power of the test at $p = \frac{1}{3}$;

(b) determine the probability of a Type II error at $p = \frac{1}{3}$.

11.22 Many psychological experiments are conducted to determine whether animals exhibit a preference between two rewards. In an experiment consisting of seven independent trials with different animals, the results were:

Trial	1	2	3	4	5	6	7
Reward	1	1	2	1	2	2	2

Do the data support the null hypothesis of no reward preference?

Determine the probability of a Type II error if the probability of preference for reward 2 is 0.8.

11.23 It is thought that 40% of flight arrivals at Tullamarine Airport originated internationally. Airport officials claim that the actual percentage is less than 40%.

On a short visit to the airport, I notice that X planes, from a total of 20 observed arrivals, were international.

(a) Write down null and alternative hypotheses for this problem.

(b) Determine the largest rejection region which has $\alpha \leq 0.05$.

(c) State the significance level of the test.

(d) Complete the following table by using the symbols $(*)$, $(**)$, $(n.s.)$ to indicate the significance of $x_{observed}$.

$x_{observed}$	1	2	3	4	5
significance					

11.24 In an experiment to determine customer colour preference, a Holden car dealer places a current model black 'ute' and a white 'ute' of the same model in the showroom window.

In a busy Saturday of trading, 5 customers purchased current model utilities, with 1 customer choosing black and 4 customers choosing white.

(a) Write down the car dealer's null and alternative hypothesis.

(b) Let the random variable Y denote the number of buyers who choose black. If the values of $y_{observed}$ comprising the rejection region are $RR = \{0, 5\}$, determine the probability of a Type I error.

(c) If, in fact, 90% of buyers purchase white utilities, determine the probability of a Type II error.

(d) Explain in words the meaning of the Type I and Type II errors in the context of this problem.

11.25 A switchboard receives λ calls per minute according to a Poisson process.

(a) Determine the probability that there is at least a 1 minute gap between consecutive incoming calls.

(b) The hypotheses

$$H_0: \ \lambda = 5$$
$$H_1: \ \lambda > 5$$

are to be tested by using a single observation of X, the number of calls received in a 1 minute period as a test statistic.

It has been decided that the rejection region for $x_{observed}$ will be

$$RR = \{10, 11, 12, 13, \ldots\}.$$

i. Determine the probability of a Type I error.

ii. Determine the power function at

$$\lambda = 6, 8, 10.$$

Plot the power curve.

11.26 Suppose that $X \sim U[-b, b]$; that is, X is a uniformly distributed random variable on the interval $[-b, b]$. Based on one observation of X we wish to test

$$H_0: \ b = 1$$
$$H_1: \ b > 1$$

by using the rejection region

$$\{x : |x| > 1 - \alpha\},$$

where α is a fixed number between 0 and 1.

(a) Show that this test has significance level α.

(b) Derive an expression for the power function $\gamma(b)$ of the test. Show that $\gamma(b)$ may be expressed in the form

$$\gamma(b) = 1 - \left(\frac{1-\alpha}{b}\right), \quad \text{for } b > 1.$$

(c) Draw a sketch of $\gamma(b)$ for values of b specified by H_1. Illustrate on your sketch the relative sizes of

$$P(\text{Type I error})$$

and

$$P(\text{Type II error at } b = 2).$$

(d) Demonstrate that for fixed b, if $P(\text{Type I error})$ is decreased, then, $P(\text{Type II error})$ increases.

(e) Determine the p-value of the test if

$$x_{observed} = -0.98.$$

11.27 After extensive data collection on lift waiting time X, it is found that, when hailed from the ground floor, a lift takes 36 seconds on average to arrive.

The S/L plot of arrival times was decidedly exponential in flavour, so we assume the probability model

$$f_X(x) = \frac{1}{36} e^{-\frac{x}{36}}, \quad x > 0.$$

During an agonising wait for the lift my temper boils over. I do not believe that the average arrival time is 36 seconds; it *must* be more! I decide to test my belief based on a single observation of X.

State the hypotheses and rejection region that I should use for a significance level of 0.05. We leave $x_{observed}$ to the imagination.

11.3 TESTS FOR MEANS/PROPORTIONS/VARIANCES FROM A PIVOT

The methods for constructing confidence intervals for an unknown parameter θ relied heavily on constructing a pivot based on a point estimate $\widehat{\theta}$ of θ. For example, if the unknown parameter is the sample mean of a normal population with known standard deviation σ, then $\theta = \mu$ and $\widehat{\theta} = \overline{X}$ is the appropriate point estimator of μ having standard error

$$S.E.(\widehat{\theta}) = S.E.(\overline{X}) = \frac{\sigma}{\sqrt{n}}.$$

In confidence interval construction, this estimator is standardised into the pivot

$$\frac{\overline{X} - \mu}{\frac{\sigma}{\sqrt{n}}} \sim N(0, 1).$$

When σ is unknown and must be estimated by using the standard deviation of a large data set, then the distribution of this pivot is approximately $N(0, 1)$. In this case we use the 'hat' notation to mark the estimate: the realised value of

$$\widehat{S.E.}(\widehat{\theta}) = \frac{s}{\sqrt{n}}$$

is the (estimated) standard error of estimate. The same procedure is followed for small sample sizes; however with normally distributed data, the pivot has a t distribution.

Pivots as test statistics

Definition 11.4 *By a* **standard pivot for** θ *we mean a pivot of the form*

$$\frac{\widehat{\theta} - \theta}{S.E.(\widehat{\theta})}.$$

By an **estimated standard pivot for** θ *we mean a pivot of the form*

$$\frac{\widehat{\theta} - \theta}{\widehat{S.E.}(\widehat{\theta})},$$

where the standard error of $\widehat{\theta}$ *is replaced by its estimator.*

□

We use both types of pivot as standardised test statistics in relation to the standard normal distribution; the estimated standard pivot is used in the context of the t distribution. Some examples of these pivots are summarised below:

Pivot		**Data type**	**Distribution**
$\dfrac{\widehat{\theta}-\theta}{S.E.(\widehat{\theta})}$			
	$\dfrac{\overline{X}-\mu}{\frac{\sigma}{\sqrt{n}}}$	normal data;	$N(0,1)$
	$\dfrac{\overline{X}-p}{\sqrt{\frac{p(1-p)}{n}}}$	Bernoulli data;	approximate $N(0,1)$ for n large
	$\dfrac{\overline{X}-\lambda}{\sqrt{\frac{\lambda}{n}}}$	Poisson data;	approximate $N(0,1)$ for n large
	$\dfrac{\overline{X}-\beta}{\sqrt{\frac{\beta^2}{n}}}$	exponential data;	approximate $N(0,1)$ for n large
$\dfrac{\widehat{\theta}-\theta}{\widehat{S.E.}(\widehat{\theta})}$			
	$\dfrac{\overline{X}-\mu}{\frac{S}{\sqrt{n}}}$	any data;	approximate $N(0,1)$ for n large
	$\dfrac{\overline{X}-\mu}{\frac{S}{\sqrt{n}}}$	normal data;	t_{n-1}

By learning how to construct 'blueprint' significance tests for the standard pivot for θ and the estimated standard pivot for θ given in the first column of the above table, all the examples of these pivots, namely the test statistics listed in the second column of the table, may be used in tests by simply following the 'blueprint'. Notice that all the listed test statistics are standardised versions of the sample mean using data (in either large amounts or small amounts) from the various distributional sources listed. These 'blueprints' will be used for the two-sample tests of the next chapter.

A 10-point blueprint test for population means

The following procedure is simply followed for hypothesis tests concerning means:

1. Identify and label the parameters of interest. In particular, describe the parameter θ which is the subject of the hypothesis test.
2. Compose the null and alternative hypotheses, phrased in terms of θ.
3. Identify which of the pivots
$$\frac{\widehat{\theta}-\theta}{S.E.(\widehat{\theta})} \quad \text{or} \quad \frac{\widehat{\theta}-\theta}{\widehat{S.E.(\widehat{\theta})}}$$
is required as the (standardised) test statistic for the test. This choice will depend on whether the sample size available is small or large, and on whether the population standard deviation σ is known for use in $S.E.(\widehat{\theta})$.
4. Set the significance level. (This will control the probability of a Type I error through the size of the rejection region.)
5. Observe the data.
6. Determine the observed value of the test statistic under the assumption that the null hypothesis is true.
7. If possible, determine the p-value of the test. (t tables may give only an approximate p-value.)
8. Use the p-value to decide whether H_0 should/should not be rejected in favour of H_1.
9. Classify the test result as $(*)$, $(**)$, or $(n.s.)$.
10. If possible, for cases where the standard pivot
$$\frac{\widehat{\theta}-\theta}{S.E.(\widehat{\theta})}$$
is used for the test, determine the probability of Type II error.

Example 11.7
Long nails?
In the manufacture of HC-27 wire nails used in pneumatic guns, 90% of nails meet the manufacturer's specification of length: 75 ± 2 mm. Assume that 'length of HC-27' nail is a normally distributed random variable. Recent changes to the coating, heading and cleaning processes may have altered the overall nail length. As a check measure, 10 nails sampled from a day's output showed an average length of 75.09 mm.

Worked example:

Assuming that the variability in nail length has not changed, test whether the new process is producing nails of significantly different average length.

If, in the standard process, 90% of nails conform to the specifications, then, by symmetry the population mean length for the standard process is 75 mm. If σ denotes the population standard deviation in nail length, the upper bound of the tolerance interval 75 ± 2 must correspond to a point 1.645 standard deviations above the mean. Therefore,

$$1.645\sigma = 2 \quad \text{or} \quad \sigma = 1.216.$$

Following the hypothesis testing blueprint:

1. Let μ denote the mean length of HC-27 nails produced under the new process.
2. $$H_0: \ \mu = 75$$ $$H_1: \ \mu \neq 75$$
3. The standardised test statistic is $$Z = \frac{\widehat{\theta} - \theta}{S.E.(\widehat{\theta})} = \frac{\overline{X} - \mu}{\frac{\sigma}{\sqrt{n}}} = \frac{\overline{X} - \mu}{\frac{1.216}{\sqrt{n}}} \sim N(0,1)$$ when H_0 is true.
4. Select $\alpha = 0.05$.
5. Data: $\overline{x}_{observed} = 75.09$ is the only available summary measure based on the individual data items.
6. $$z_{observed} = \frac{\overline{x}_{observed} - 75}{\frac{1.216}{\sqrt{10}}} = \frac{75.09 - 75}{\frac{1.216}{\sqrt{10}}} = 0.234.$$
7. p-value $= 2P(Z > z_{observed}|H_0 \ true) = 2P(Z > 0.234) = 2(0.409) = 0.818$.
8. Since p-value $> \alpha$, we accept H_0 at the 0.05 level of significance. There is not enough evidence to reject the null hypothesis.
9. The test result is not significant and we summarise: $\overline{x}_{observed} = 75.09$ $(n.s.)$. Nails are not significantly different in length under the new process.
10. This is discussed in the next section.

Type II errors for tests of means

Suppose that the hypotheses

$$H_0: \ \theta = \theta_0$$
$$H_1: \ \theta > \theta_0$$

are to be tested at significance level α with test statistic $\widehat{\theta}$ which, in standardised form, is the pivot

$$\frac{\widehat{\theta} - \theta}{S.E.(\widehat{\theta})}$$

having a $N(0,1)$ distribution (either exactly so, or approximately so). Because the test is upper-tailed, the rejection region is based on $\widehat{\theta}_{observed}$ and is of the form

$$\text{reject } H_0 \text{ if } \widehat{\theta} > k,$$

where the constant k is determined as soon as α has been selected. How is the constant k determined? For significance level α,

$$\begin{aligned} \alpha &= P(\text{reject } H_0 \mid H_0 \text{ true}) \\ &= P(\widehat{\theta} > k \mid \theta = \theta_0) \\ &= P(\frac{\widehat{\theta} - \theta}{S.E.(\widehat{\theta})} > \frac{k - \theta}{S.E.(\widehat{\theta})} \mid \theta = \theta_0) \\ &= P(Z > \underbrace{\frac{k - \theta_0}{S.E.(\widehat{\theta})|_{\theta=\theta_0}}}_{=z_\alpha}), \qquad \text{where } Z \sim N(0,1). \end{aligned}$$

Then,

$$\frac{k - \theta_0}{S.E.(\widehat{\theta})|_{\theta=\theta_0}} = z_\alpha$$

implies that

$$k = \theta_0 + z_\alpha S.E.(\widehat{\theta})|_{\theta=\theta_0}.$$

Therefore, once α has been specified, k has been specified and the rejection region in terms of k has been specified.

For the Type II error at θ_1 (specified by the alternative hypothesis),

$$\begin{aligned} \beta(\theta_1) &= P(\text{accept } H_0 \mid H_0 \text{ false and } \theta = \theta_1) \\ &= P(\widehat{\theta} < k \mid \theta = \theta_1) \\ &= P(\frac{\widehat{\theta} - \theta}{S.E.(\widehat{\theta})} < \frac{k - \theta}{S.E.(\widehat{\theta})} \mid \theta = \theta_1) \\ &= P(Z < \frac{k - \theta_1}{S.E.(\widehat{\theta})|_{\theta=\theta_1}}) \qquad \text{where } Z \sim N(0,1) \\ &= \Phi\left(\frac{k - \theta_1}{S.E.(\widehat{\theta})|_{\theta=\theta_1}}\right), \end{aligned}$$

where Φ represents (as usual) the cumulative distribution function of a standard normal random variable. The same type of result is easily obtained for two-tailed and lower-tailed tests as is illustrated in the example which follows.

Example 11.8

Looking for Type II errors from the impact resistance of clay pavers: Example 11.3 continued

Suppose that $X \times 10^{-3}$ mL is gouged from the surface of a paver after a hammer drill impacts the surface for 1 minute. We also assume that experience with the impacting equipment suggests that the standard deviation of X is usually near 20. We will use this value, $\sigma = 20$, to assess the error rates. (The data of Example 11.3 showed a tighter sample standard deviation; the pavers of this data batch were uniformly hard after being subjected to a heat treatment.)

Again, let $\mu = E(X)$ denote the mean volume eroded (for the hypothetical population of all pavers of the current type under test). In terms of μ, the hypotheses may be phrased as:

$$\begin{aligned} H_0: &\quad \mu = 265 \\ H_1: &\quad \mu < 265. \end{aligned}$$

(In terms of the general notation, we are using $\theta = \mu$ and $\widehat{\theta} = \overline{X}$.) If 30 pavers are tested, how close does the sample average $\overline{X}$ have to be to 265 before the pavers are branded as 'too soft' for heavy traffic densities? It is a question of the size of α, the probability of a Type I error. From a commercial point of view, it is unwise to declare that a batch of pavers is suitably 'hard' when in fact they are not. (Such a declaration may pave the way for an expensive court case.) Thus we wish to restrict the chance of rejecting H_0 when, in fact, H_0 is true. We need a small, controlled Type I error.

Clearly H_0 is rejected if $\overline{x}_{observed}$ is too small. That is, reject H_0 if $\overline{X} < k$. Then for $\alpha = 0.05$,

$$\begin{aligned} 0.05 &= P(\text{reject } H_0 \mid H_0 \text{ true}) \\ &= P(\overline{X} < k \mid \mu = 265) \\ &= P(\frac{\overline{X} - \mu}{S.E.(\overline{X})} < \frac{k - \mu}{\frac{\sigma}{\sqrt{n}}} \mid \mu = 265) \\ &= P(Z < \underbrace{\frac{k - 265}{\frac{20}{\sqrt{30}}}}_{=-1.645}), \qquad \text{where} \quad Z \sim N(0,1). \end{aligned}$$

Therefore, since $P(Z < 1.645) = 0.05$,

$$k = 265 - 1.645(\frac{20}{\sqrt{30}}) = 259.0 \quad (1 \text{ dp}).$$

For this borderline between a batch that is 'too soft' ($\overline{x}_{observed} > 259$) and 'hard enough' ($\overline{x}_{observed} < 259$) the Type II errors may be determined by:

$$\begin{aligned} \beta(\mu_1) &= P(\text{accept } H_0 \mid H_0 \text{ false and } \mu = \mu_1) \\ &= P(\overline{X} > 259 \mid \mu = \mu_1) \\ &= P(\frac{\overline{X} - \mu_1}{\frac{\sigma}{\sqrt{n}}} > \frac{259 - \mu_1}{\frac{\sigma}{\sqrt{n}}} \mid \mu = \mu_1) \\ &= P(Z > \frac{259 - \mu_1}{\frac{20}{\sqrt{30}}}) \quad \text{where } Z \sim N(0,1) \\ &= 1 - \Phi\left(\frac{259 - \mu_1}{3.6515}\right). \end{aligned}$$

For example:

$\beta(262) = 0.791$
$\beta(260) = 0.606$
$\beta(259) = 0.500$
$\beta(257) = 0.295$
$\beta(255) = 0.138$
$\beta(250) = 0.007$

These points may be used to plot the power curve

$$\gamma(\mu) = 1 - \beta(\mu) = \Phi\left(\frac{259 - \mu_1}{3.6515}\right).$$

The relative sizes of $\alpha = 0.05$ and $\beta(257)$ comprise the tails of the two normal curves in the diagram below:

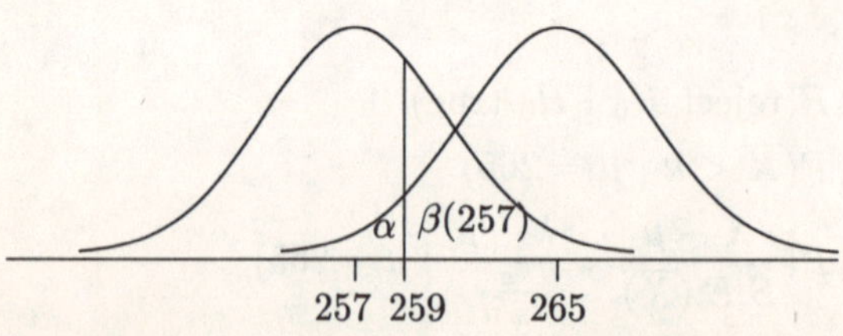

The effect of increasing the sample size

If the significance level α is fixed, then the probability of Type II error reduces (equivalently, the power curve increases) as the sample size n increases. This will occur at each point specified in the parameter space by the alternative hypothesis.

Example 11.9

More pavers put on test: Example 11.8 continued

Suppose that in Example 11.8 n pavers are put on test. This results in a sample mean $\overline{x}_{observed}$ being located relative to the rejection region:

$$\text{reject } H_0 \text{ if } \overline{X} < 265 - 1.645(\frac{20}{\sqrt{n}}).$$

That is, the exact boundary of the rejection region will depend on n. (We have seen in Example 11.8 that when $n = 30$, the boundary value is 259.) Following the same reasoning as was given in Example 11.8, the power curve is now

$$\gamma(\mu_1) = \Phi\left(-1.645 + \frac{265 - \mu_1}{\frac{20}{\sqrt{n}}}\right),$$

where Φ is the cdf for the standard normal distribution. *As a function of n*, the power curve is increasing for values of μ_1 less than the null hypothesis value of 265. This means that the probability of Type II errors $\beta(\mu_1) = 1 - \gamma(\mu_1)$ decreases with n for all μ_1 less than 265.
The following chart shows the comparative Type I and some Type II errors for two different sample sizes, $n = 30$ and $n = 100$.

	$n = 30$	$n = 100$
α	0.05	0.05
$\beta(262)$	0.791	0.558
$\beta(260)$	0.606	0.198
$\beta(259)$	0.500	0.090
$\beta(257)$	0.295	0.009
$\beta(255)$	0.138	0.000
$\beta(250)$	0.007	0.000

The Type II errors are dramatically less likely when the sample size is increased. It is important to realise during this process, the probability of a Type I error has stayed at 0.05; still well under control.

Confidence intervals and acceptance regions

The complement of the rejection region RR of a hypothesis test is the **acceptance region** consisting of all those values of the test statistic for which the null hypothesis is accepted.

The acceptance region for a two-tailed test

$$\begin{aligned} H_0 &: \quad \theta = \theta_0 \\ H_1 &: \quad \theta \neq \theta_0 \end{aligned}$$

of the population mean θ, is sketched in the statistic–parameter diagram below: In this diagram, from a point θ_0 on the parameter axis, a horizontal line is drawn intersecting the confidence interval equations in a horizontal slice (projected as 'accept H_0' on the horizontal axis).

A parameter-statistic diagram showing the acceptance region for a two-tailed test when $H_0 : \theta = \theta_0$ is true.

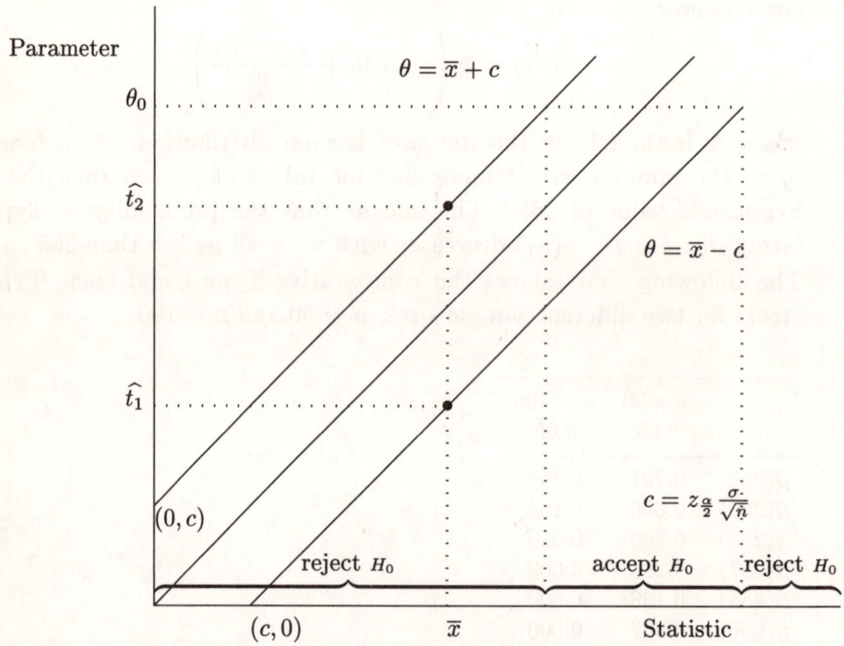

In this diagram it is clear that

$$\theta_0 \notin [\widehat{t_1}, \widehat{t_2}] \text{ exactly when } \overline{x} \notin \mathrm{A},$$

the acceptance region of the two-sided hypothesis test. That is, the confidence interval $[\widehat{t_1}, \widehat{t_2}]$ consists of all those values θ_0 for which $H_0 : \theta = \theta_0$ is accepted.

Example 11.10

Confidence interval/acceptance region: Example 11.7 continued

From the example on the length of HC-27 wire nails for pneumatic guns, we find that $X \sim N(\mu, (1.216)^2)$, where X denotes nail-length.

Because the test

$$\begin{aligned} H_0 &: \ \mu = \mu_0 \\ H_1 &: \ \mu \neq \mu_0 \end{aligned}$$

is two-tailed, for a given μ_0, the RR consists of values of $\overline{x}_{observed}$ for which

$$\left| \frac{\overline{x}_{observed} - \mu_0}{\frac{1.216}{\sqrt{10}}} \right| > 1.96.$$

This corresponds to the acceptance region

$$\left| \frac{\overline{x}_{observed} - \mu_0}{\frac{1.216}{\sqrt{10}}} \right| \leq 1.96$$

for the test. As an interval of parameter values, the acceptance region is

$$[\overline{x}_{observed} - 1.96\frac{1.216}{\sqrt{10}}, \ \overline{x}_{observed} + 1.96\frac{1.216}{\sqrt{10}}],$$

and these are the realised endpoints of a 95% confidence interval for μ.

Tests for proportions

For data from a sequence of Bernoulli trials, the test statistic $\overline{X}$ represents the proportion of successes in the sample. The standard pivot

$$\frac{\widehat{\theta} - \theta}{S.E.(\widehat{\theta})}$$

becomes

$$\frac{\overline{X} - p}{\sqrt{\frac{p(1-p)}{n}}},$$

which has an approximate standard normal distribution when n is large. Therefore the blueprint test for means may be used for

$$\begin{aligned} H_0 &: \ p = p_0 \\ H_1 &: \ p \neq p_0 \end{aligned}$$

and the obvious one-tailed tests having alternatives $H_1 : p < p_0$ or $H_1 : p > p_0$.

Example 11.11
What we throw away
The Environment Protection Agency published a breakdown of the contents of large 240 litre domestic rubbish bins used by many city councils.

Material	% by weight
Garden	20.0
Wood	0.7
Paper	23.9
Aluminium	0.6
Plastic	7.7
Other metal	0.1
Food	31.2
Rags	2.2
Inert	3.4
Steel	3.3
Glass	6.9

(*Source:* *The Sunday Age*, 24 November 1991.)

Suppose that a city council will sponsor compost tumblers for biodegradable material (so that ratepayers may recycle such material as compost) if at least 70% of garbage collected is biodegradable. (The EPA figures show a combined biodegradable content of 78%.)

Worked example:
In a random sample of 100 ratepayers, 78% of garbage disposed (collected by the council) was biodegradable. Test to see whether the compost tumblers proposal will be implemented.

1. Let p denote the proportion of biodegradable material (garden/wood /paper/food/rags) in weekly rubbish collected throughout the municipality.
2. $$\begin{aligned} H_0 &: \ p = 0.7 \\ H_1 &: \ p > 0.7. \end{aligned}$$
3. The standardised test statistic is
$$Z = \frac{\widehat{\theta} - \theta}{S.E.(\widehat{\theta})} = \frac{\overline{X} - p}{\sqrt{\frac{p(1-p)}{n}}} = \frac{\overline{X} - 0.7}{\sqrt{\frac{0.7(1-0.7)}{100}}} \sim N(0,1)$$
(approximately) when H_0 is true.

4. Select $\alpha = 0.05$.
5. Data: $\overline{x}_{observed} = \widehat{p} = 0.78$.
6.

$$z_{observed} = \frac{0.78 - 0.7}{\sqrt{\frac{0.7(1-0.7)}{100}}} = 1.75.$$

7. p-value $= P(Z > z_{observed} | H_0\ true) = P(Z > 1.75) = 0.04$.
8. Since p-value $< \alpha$, we reject H_0 at the 0.05 level of significance. The data shows support for the alternative hypothesis.
9. The test result is significant and we summarise: $\overline{x}_{observed} = 0.78$ $(*)$. Significantly more than 70% of collected garbage is biodegradable.
10. H_0 is rejected if $\overline{X} > k = 0.7 + 1.645\sqrt{\frac{0.7(1-0.7)}{100}} = 0.775$. Then, for any $p_1 > 0.7$,

$$\begin{aligned}
\beta(p_1) &= P(\text{accept } H_0 \mid H_0 \text{ false and } \ p = p_1) \\
&= P(\overline{X} < 0.775 \mid p = p_1) \\
&= P(\frac{\overline{X} - p_1}{\sqrt{\frac{p_1(1-p_1)}{100}}} < \frac{0.775 - p_1}{\sqrt{\frac{p_1(1-p_1)}{100}}} \mid p = p_1) \\
&= \Phi\left(\frac{0.775 - p_1}{\sqrt{\frac{p_1(1-p_1)}{100}}}\right).
\end{aligned}$$

Hypothesis tests from the estimated standard pivot

Hypothesis tests concerning the population mean that are based on the estimated standard pivot

$$\frac{\widehat{\theta} - \theta}{\widehat{S.E.}(\widehat{\theta})}$$

proceed with the same blueprint plan as before. The sample standard deviation S is often called upon as a point estimator of the unknown population standard deviation. This change makes little difference if n is large—the estimated standard pivot is still normally distributed; but, for small sample sizes the increased uncertainty in estimation causes the pivot to follow a t distribution.

We have seen the large-sample case in operation for 500 gram packets of cereal in the first discussions at the outset of this chapter. We will use the t distribution for the pivot

$$\frac{\overline{X} - \mu}{\frac{S}{\sqrt{n}}},$$

when $n < 25$.

Example 11.12

Testing the emergency locking on seatbelts: Example 10.16 continued

When a fastened seatbelt is given a sharp tug, a locking device prevents payout of the belt webbing. A controlled experiment was devised to simulate the sudden shock of accidents.

On average, is the payout less than 21 mm ?

Using the data of Example 10.16, the recorded payouts were:

18, 17, 19, 23, 19, 21, 21, 20.

A normal probability plot shows that the assumption of normally distributed data is reasonable. The hypothesis testing blueprint is applied as follows:

1. Let μ denote the mean length of of payout.
2. $$\begin{aligned} H_0 &: \ \mu = 21 \\ H_1 &: \ \mu < 21 \end{aligned}$$
3. The estimated standard pivot to be used as a test statistic is
$$T = \frac{\widehat{\theta} - \theta}{\widehat{S.E.(\widehat{\theta})}} = \frac{\overline{X} - 21}{\frac{S}{\sqrt{8}}} \sim t_7$$
when H_0 is true. Here $\overline{X}$ and S are the usual sample mean and sample standard deviation respectively.
4. Select $\alpha = 0.05$.
5. Data: $\overline{x}_{observed} = 19.75$. Individual data points are given as above.
6. $$t_{observed} = \frac{\overline{x}_{observed} - 21}{\frac{1.9086}{\sqrt{8}}} = \frac{19.75 - 21}{\frac{1.9086}{\sqrt{8}}} = -1.85.$$
7. $$\begin{aligned} p\text{-value} &= P(T < t_{observed} | H_0 \ true) \\ &= P(T < -1.85) \quad \text{where } T \sim t_7 \\ &= P(T > 1.85) \quad \text{(by symmetry of the } t \text{ distribution)} \\ &> P(T > 1.895) \\ &= 0.05. \end{aligned}$$

The inequality is necessary in this calculation since few upper-tail values for the t_7 distribution are tabulated. By examining more carefully adjacent values in the t_7 table we find:

$$t_{0.05,7} = 1.895, \qquad t_{0.10,7} = 1.415.$$

Therefore, since $1.415 < 1.85 < 1.895$, we may write $t_{0.10,7} < t_{p\text{-value},7} < t_{0.05,7}$, indicating that $0.05 < p$-value < 0.10. This is the easiest way to obtain bounds on the p-value from t tables. (The bounds are required so that one of the judgments (*), (**), ($n.s.$) may be made concerning the sample mean.)

8. Since p-value $> \alpha$, we do not reject H_0 at the 0.05 level of significance. The data do not provide enough evidence to reject the null hypothesis.
9. The result is not significant and we summarise: $\bar{x}_{observed} = 19.75$ ($n.s.$). The payout of the emergency locking device is not significantly less than 21 mm on average.

Significance tests for the variance σ^2

We call on Theorem 10.5 to provide a pivot that may be used as a test statistic for variance. When $X_1, X_2, X_3, \ldots, X_n$ is a random sample of size n from a $N(\mu, \sigma^2)$ population,

$$\frac{(n-1)S^2}{\sigma^2} \sim \chi^2_{n-1}.$$

For the null hypothesis $H_0 : \sigma^2 = \sigma_0^2$ against an appropriate alternative, the standardised variance test statistic

$$\frac{(n-1)S^2}{\sigma_0^2}$$

has a χ^2_{n-1} distribution when H_0 is true.

Example 11.13

Pollution on Port Phillip Bay

During the summer months Melbourne's beaches on Port Phillip Bay suffer from bacterial pollution, especially after rain; material containing *Escherichia coli* bacteria is washed directly into the bay via stormwater drains and river/canal outflows. The Environment Protection Agency monitors the bacterial count in seawater samples taken at beaches around the bay. A beach is considered polluted if the **geometric mean** of the *E-coli* bacterial count exceeds 200 per 100 mL seawater.

Recall from Chapter 2, that if the geometric mean is an appropriate measure of the centre of a data set, then the arithmetic mean becomes a suitable measure of the data on the logarithm scale. Are the bacterial counts normally distributed on the $\log_{10}$ scale?

Worked example:
Is it reasonable to assume that $\log_{10} X \sim N(\log_{10} 200, 0.001)$, where X is the *E-coli* count for 100 mL of seawater taken at Elwood beach? The probability distribution of X is then described as a log-normal distribution.

Suppose that 14 determinations of X from Elwood beach were:

150	180	170	160	220	210	220
220	205	260	190	180	200	170

We need to consider three aspects for the data transformed to the $\log_{10}$ scale: Is the transformed data normally distributed? Is the mean count of the transformed data significantly different from $\log_{10} 200$? Is the variance of the transformed count 0.001 or is it larger than this?

The transformed data appear to be normally distributed as evidenced in the following normal probability plot:

Normal probability plot of $\log_{10}$(E-coli) *bacterial counts*

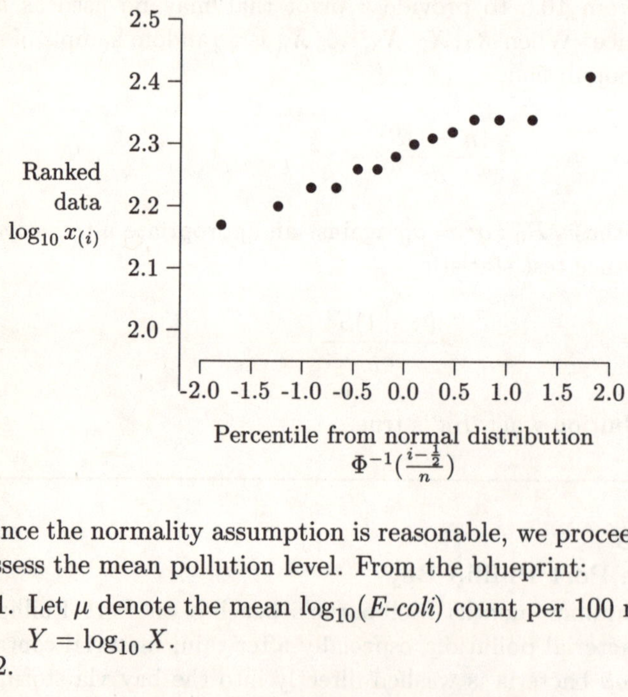

Since the normality assumption is reasonable, we proceed with a t test to assess the mean pollution level. From the blueprint:

1. Let μ denote the mean $\log_{10}$(*E-coli*) count per 100 mL seawater. Let $Y = \log_{10} X$.
2.
$$H_0 : \mu = \log_{10} 200$$
$$H_1 : \mu \neq \log_{10} 200.$$

3. The estimated standard pivot to be used as a test statistic is

$$T = \frac{\widehat{\theta} - \theta}{\widehat{S.E.(\widehat{\theta})}} = \frac{\overline{Y} - 2.3010}{\frac{S_y}{\sqrt{14}}} \sim t_{13}$$

when H_0 is true.

4. Select $\alpha = 0.05$.
5. Data: $\overline{y}_{observed} = 2.2863$, $s_{observed} = 0.0652$.
6.

$$t_{observed} = \frac{\overline{y}_{observed} - 2.3010}{\frac{0.0652}{\sqrt{14}}} = \frac{2.2863 - 2.3010}{\frac{0.0652}{\sqrt{14}}} = -0.84.$$

7. p-value > 0.1
8. Since p-value $> \alpha$, we accept H_0 at the 0.05 level of significance.
9. The test result is not significant and we summarise: $\overline{y}_{observed} = 2.2863\ (n.s.)$. On the log scale, the mean *E-coli* count per 100 mL of seawater is not significantly different from the average safe level.

The test concerning the variance may be organised within the same framework as the blueprint test for means:

1. Let σ^2 denote the variance of $\log_{10}(E\text{-}coli)$ counts per 100 mL seawater.
2.

$$H_0: \quad \sigma^2 = 0.001$$
$$H_1: \quad \sigma^2 > 0.001.$$

3. The estimated standard pivot to be used as a test statistic is

$$V = \frac{(n-1)S_y^2}{\sigma^2} = \frac{13S_y^2}{0.001}$$

when H_0 is true.
4. Select $\alpha = 0.05$. This gives an overall significance level for both tests simultaneously of at most $0.05 + 0.05 = 0.1$.
5. Data: $s_{observed} = 0.0652$.
6. $v_{observed} = \frac{(13)(0.0652)^2}{0.001} = 55.26$.
7. For $V \sim \chi^2_{13}$, p-value is $P(V > 55.26) < 0.005$.
8. Since p-value $< \alpha$, we reject H_0 at the 0.05 level of significance.
9. The test result is highly significant and we summarise: $s_{observed} = 0.0652\ (**)$. The variance in $\log_{10}(E\text{-}coli)$ counts per 100 mL of seawater is significantly larger than 0.001 per 100 mL.

EXERCISES

11.28 The data give plasma glucose concentration (mg dL^{-1}) for 50 normal fasting people.

98	105	75	92	120	91
112	98	103	94	100	86
100	99	84	110	100	108
100	96	86	102	112	106
100	109	104	103	106	88
103	112	98	106	92	95
95	108	90	90	95	112
84	102	96	88	93	93
93	87				

(*Source*: Massod, M.F. (1977). Nonparametric percentile estimate of clinical normal ranges. *American Journal Medical Technology*, 43, 243-52.)

Is the mean glucose concentration significantly different from 100 mg dL^{-1}?

11.29 'Health Masters' weights are designed for adjustable dumbbells and barbells, but are compatible with many types of gymnasium equipment. It is important that the label weight is 'very close to' the true weight; weightlifters, powerlifters and bodybuilders would not be impressed if their weights were in reality heavier than stated. Because of the manufacturing process the true weights are normally distributed.

It is of interest to determine whether weights labelled 25 kg are, on average, significantly different in weight from 25 kg.

16 weights labelled 25 kg were found to have a sample mean of 23.2 kg and a sample standard deviation of 3.2 kg.

Propose a suitable hypothesis test for this situation and determine the p-value of the test.

11.30 A thermostat switched on at the following temperatures (degrees Celsius) in 34 trials.

12.5	12.0	12.5	10.5	11.5	12.5
12.0	10.5	13.0	12.5	12.0	11.5
12.0	11.5	10.0	11.0	13.0	12.5
10.0	13.0	12.5	12.5	12.0	12.0
14.0	11.5	12.5	12.0	12.0	11.0
12.5	12.5	12.5	13.0		

(a) Are these data normally distributed?

(b) Do the data support the view that, on average, the thermostat is cutting in at a temperature higher than 12 degrees Celsius? Does your test assume that the data are normally distributed?

11.31 To test the claim that a certain device decreases average petrol consumption when fitted to a car, the following experiment was devised. Five cars were chosen at random from a rental company's pool of cars. Each car was run as far as it would go on 20 litres of petrol without the device, then the experiment was repeated with the same cars now fitted with the device. The increases in kilometres are recorded below.

3, 1, −1, 0, 2.

Was the claim was justified?

Are any distributional assumptions required?

11.32 Suppose that of people taking a certain drug, 25% complain of nausea as a side effect.

If the drug is administered to a random sample of n patients, determine the mean and standard error of the proportion of the sample experiencing nausea.

A new form of the drug is tried out on such a group and only 20% experience nausea. Determine the smallest value of n so that this result is statistically significant at the 5% level.

11.33 *The Age, 19 October 1991*, reported the following starters in the 1991 Caulfield Cup.

Horse	Weight (kg)
Sydeston	58.5
Braashee	57
Castletown	56.5
Dr Grace	56.5
Just A Dancer	56.5
Lord Revenir	54
Mantlepiece	53.5
Prince Salieri	53.5
Royal Creation	53.5
Shiva's Revenge	53.5
Bold Rancher	53
Livistona Lane	52.5
Magnolia Hall	52
Lee's Bid	51
Weekend Delight	50.5
Princess Pushy	49.5
Let's Elope	48.5
Shavano Miss	48.5
Cool Credit	51
Ivory Way	50.5

(a) Construct a normal probability plot of the weights of these horses. Explain why the weights appear to be normally distributed.

(b) Show that these 20 horses have a sample mean weight of 53 kg. Show that the standard error of the estimate is 0.648 when this sample mean is used as a point estimate of the mean weight μ of all (possible) horses that may take part in the Spring Racing Carnival (in which the Caulfield Cup is one race).

(c) Is μ significantly different from 54 kg?

11.34 On a production line, two conveyor belts feed their output into the same hopper. Items from the first conveyor belt are red and items produced on the second are blue.

It is of interest to the management to know if the production rates of red and blue items are the same. 64 items are sampled from the hopper and the colours noted; the following decision rule is adopted:

Accept the hypothesis of equal proportions of colours if between 28 and 36 (inclusive) red items are selected; reject the hypothesis of equal proportions otherwise.

(a) Find the probability of rejecting the hypothesis when it is actually correct.

(b) Find a general expression for the probability of accepting the hypothesis when it is actually false and 25% more red items are produced than blue items.

11.35 (continuation) Specify the rejection regions in terms of the number of red items sampled that will achieve approximate significance levels:

(a) $\alpha = 0.05$;

(b) $\alpha = 0.01$.

Use this information to construct approximate 95% and 99% confidence intervals for the proportion of red items manufactured on the production line.

11.36 (continuation) Suppose that we wish to test the hypothesis that a greater proportion of red items than blue items are produced.

(a) Give the null and alternative hypotheses in this case.

(b) Indicate whether a one-tailed test or a two-tailed test should be used.

(c) Give the p-value if 39 red items are observed in the sample.

11.37 A study of the operation of a multi-storey car park showed that, in the past, average parking time was 220 minutes. Recently the car park was remodelled and parking fees increased. The management wants to know if the changes have had any effect on mean parking time.

A random sample of n cars had an average parking time of 208 minutes and a standard deviation of 40 minutes.

(a) Give the appropriate null and alternative hypotheses H_0 and H_1.

(b) If $n = 50$:

i. Is there sufficient evidence to indicate a change in the mean parking time at a 0.05 significance level? Show your reasoning clearly.

ii. Does your test assume that parking times are normally distributed? (Yes/No)

(c) If $n = 5$:

i. Is there sufficient evidence to indicate a change in the mean parking time at a 0.05 significance level? Show your reasoning clearly.

ii. Does your test assume that parking times are normally distributed? (Yes/No)

(d) Which of the tests, that in (b) based on $n = 50$ or that in (c) based on $n = 5$, is the most reliable in the sense of having smaller Type II errors? Explain.

11.38 Show how to construct confidence intervals by using the acceptance regions of the hypothesis tests in Exercise 11.37.

11.39 In Example 11.11, determine $\beta(0.85)$.

11.40 A Toledo Hardness Tester is used for testing the hardness of manufactured parts by a compression test. A rotor arm impinges the test material on a flat bench-pad. For material currently under test, the hardness, X, in units, is expected to be normally distributed with unknown mean μ and standard deviation $\sigma = 11$ units. The following data are obtained in 20 trials:

198	203	187	180	204
189	188	196	192	177
211	203	205	201	204
179	201	222	198	201

(a) Construct a stem-and-leaf plot for these data.

(b) Construct a normal probability plot for these data.

(c) Assuming that the data are normally distributed, decide if it is consistent with having come from a population with standard deviation $\sigma = 11$.

(d) Assuming that the data are normally distributed:

i. perform the hypothesis test

$$H_0: \ \mu = 200$$
$$H_1: \ \mu < 200$$

at $\alpha = 0.05$ by using the population standard deviation $\sigma = 11$ in your standardised test statistic;

ii. determine values of the Type II error at $\mu =$ 190, 193, 196, 198;

iii. sketch the power curve of the test.

(e) Are you satisfied with the conclusion

$$X \sim N(200, 121)?$$

11.41 Ten determinations of the weight, X, of '7 kg' sacks of flour gave a sample variance s^2. It is known that X is normally distributed with unknown variance σ^2.

In order to judge whether the sample estimate of variance is compatible with a population variance of 3 kg^2, consider testing

$$H_0: \ \sigma^2 = 3$$
$$H_1: \ \sigma^2 < 3.$$

(a) Let p denote the p-value of the above test when the observed sample variance is $s^2_{observed} = 1$. Determine the values of k, l and m so that

$$\begin{aligned} 0.95 &= P(Y > k) \\ 0.975 &= P(Y > l) \\ 1 - p &= P(Y > m), \end{aligned}$$

where $Y \sim \chi^2_9$. Hence show that

$$0.025 < p < 0.05.$$

(b) Show that the rule: reject H_0 if $S^2 < 1.11$, leads to a significance level of $\alpha = 0.05$.

(c) Estimate, as closely as tables permit, the probability, q, of a Type II error, when in fact $\sigma^2 =$ 2.396.

(d) On a sketch of the power curve, indicate the relative positions of α and q.

11.42 To test the accuracy of a weighing device, a technician carries out the following procedure:

A standard mass (known to be 10 grams with very high precision) is weighed 10 times. The results are assumed to be 10 independent observations of the observed weight

$$X \sim N(\mu, 0.2^2).$$

(a) i. Write down the null and alternative hypotheses to test whether the device is under-recording the true weight.

ii. What are your conclusions if

$$\overline{x}_{observed} = 9.9?$$

iii. Determine $\beta(9.8)$ when $\alpha =$ 0.05.

(b) i. Write down the null and alternative hypotheses to test whether the device is over-recording the true weight.

ii. What are your conclusions if

$$\overline{x}_{observed} = 10.1?$$

iii. Determine $\beta(10.2)$ when $\alpha = 0.05$.

(c) Determine acceptance and rejection regions for testing

$$H_0: \quad \mu = 10$$
$$H_1: \quad \mu \neq 10$$

at $\alpha = 0.05$. Show that

$$\beta(10 + k) = \beta(10 - k)$$

for a positive constant k. What does this equation indicate about the shape of the power curve?

11.43 A manufacturing process produces bathroom tiles that are always exactly square but with a length of side which varies from tile to tile.

As a quality control check, large batches of tiles are accepted or rejected according to the following acceptance sampling plan: A random sample of n (> 30) tiles is selected from a batch and the length of side of the tiles measured. After determination of the mean length of side $\overline{x}$ of tiles in the batch, if

$\overline{x} > k$, accept the whole batch;
$\overline{x} \leq k$, reject the whole batch.

Determine n and k so that

$$P(\text{accepted batch} \mid \mu = 20.4 \text{ cm}) = 0.95$$
$$P(\text{accepted batch} \mid \mu = 19.6 \text{ cm}) = 0.05,$$

where μ is the mean length of side of tiles in the batch. It is known from experience that σ, the standard deviation of side length for tiles in the batch, is given by $\sigma = 2$ for any batch with a mean length of side near 20 cm.

CHAPTER TWELVE

Multi-Sample Investigations

In this chapter you will find

12.1 STRUCTURED DATA: ANALYSING DEPENDENT SAMPLES

The discussion of confidence intervals and significance tests has relied completely on results found from a single sample. The standardised sample mean was often used as a test statistic or as a pivot locating the centre of a confidence interval. The same ideas will work well for our current discussion of tests and intervals based on more than one sample. We will again use the canons of recent chapters: the Central Limit Theorem, and the (estimated) standard pivot.

With two (or more) samples we are able to discuss more sophisticated statistical questions about the populations from which the samples came: we are positioned to be able to detect **difference**, to detect changes in the same population (perhaps across time), or changes in different populations (each possibly the subject of different experimental conditions). The differences of interest may be population means, proportions, variances.

Structural relationships for two samples

An investigation that involves more than one sample must address the relationship between the samples. Statistical analysis concerning the samples will depend on this relationship.

Example 12.1
Structured diet data

Scenario 1. Suppose that 100 people (many possibly overweight) are weighed giving Data set 1. The same 100 people then try out a 'calorie reduction diet' and after 3 months are reweighed giving Data set 2. In all, 100 people are involved in the study.

Scenario 2. Suppose that 100 people (many possibly overweight) are weighed giving Data set 3. A completely independent group of 100 people after experiencing a 'calorie reduction diet' for 3 months are weighed giving Data set 4. In all, 200 people are involved in the study.

In Scenario 1, Data set 1 and Data set 2 are **dependent** because the measurements are taken on the same experimental subjects. These data sets may be used to detect if there has been a mean reduction in weight due to the effect of the diet.

In Scenario 2, Data set 3 and Data set 4 are **independent** because the measurements are taken on different and unrelated experimental subjects. These data sets may not be used to detect if there has been a mean reduction in weight due to the effect of the diet. At most they may be used to detect if two groups of people have the same mean weight.

The effect of the diet cannot be assessed by using Data sets 3 and 4. The detecting that may be accomplished with the data in Scenario 1 is very different from that which may be accomplished with the data in Scenario 2. If an experimenter wishes to detect 'diet effect', then Scenario 1 is the most desirable experimental design.

In this chapter, when considering two samples, X notation will be used for the first, and Y notation for the second:

Independent samples of sizes m and n: $X_1, X_2, X_3, \ldots, X_m$
$Y_1, Y_2, Y_3, \ldots, Y_n$

Dependent samples of size n: $\underbrace{(X_1, Y_1)}_{unit\ 1}, \underbrace{(X_2, Y_2)}_{unit\ 2}, \underbrace{(X_3, Y_3)}_{unit\ 3}, \ldots, \underbrace{(X_n, Y_n)}_{unit\ n}.$

The dependent samples have been braced together in matched pairs where 'unit' refers to experimental unit on which the data are measured. The notation for both sample and population is as follows:

Population	1	2
Population mean	μ_1	μ_2
Population variance	σ_1^2	σ_2^2
Sample size	m	n
Sample mean	$\overline{X}$	$\overline{Y}$
Sample variance	S_1^2	S_2^2

Analysing dependent samples

In this section we examine dependent samples in matched pairs which means that $m = n$: data sets $X_1, X_2, X_3, \ldots, X_n$ and $Y_1, Y_2, Y_3, \ldots, Y_n$ are dependent, but when written as matched pairs—$(X_1, Y_1), (X_2, Y_2), (X_3, Y_3), \ldots,$ (X_n, Y_n)—then each pair is independent of each other matched pair, since the experimental units on which the (X, Y) measurements are taken are randomly chosen. This holds the key to the analysis comparing the average of the X data with the average of the Y data.

If the two data values taken on a given experimental unit are **subtracted**, then the distortion effect of that experimental unit on the overall experimental variability is removed. In the diet Example 12.1, this means that observed weight change is due to the effect of the diet, not due to the fact that the experimental subject (person) is fat or skinny or has a particular disposition to weight gain. That is, the effect of 'person-type' has been removed from the experiment.

Following this policy, form the differences

$$D_i = X_i - Y_i, \quad i = 1, 2, 3, \ldots, n,$$

between the data in each matched pair. These differences are *i.i.d.* observations from a probability distribution with mean μ_D and variance σ_D^2 where by the results of Chapter 8,

$$\mu_D = E(D_i) = E(X_i - Y_i) = E(X_i) - E(Y_i) = \mu_1 - \mu_2$$

and

$$\sigma_D^2 = Var(D_i) = Var(X_i - Y_i) = Var(X_i) + Var(Y_i) - 2Cov(X_i, Y_i).$$

This simplifies to

$$\sigma_D^2 = \sigma^2 + \sigma^2 - 2(\sigma)(\sigma)\rho = 2\sigma^2(1-\rho)$$

in the case that $\sigma_1^2 = \sigma_2^2 = \sigma^2$ and ρ is the correlation coefficient between X and Y.

This process has reduced the two samples $X_1, X_2, \ldots, X_n$ and $Y_1, Y_2, \ldots, Y_n$ to a single sample $D_1, D_2, D_3, \ldots, D_n$. Using the methods of the previous chapter, this sample contains 'sufficient' information for point and/or interval estimation of μ_D and also hypothesis tests concerning μ_D. The differences D_i have average value

$$\overline{D} = \frac{\sum_{i=1}^n D_i}{n} = \frac{\sum_{i=1}^n (X_i - Y_i)}{n} = \frac{\sum_{i=1}^n X_i}{n} - \frac{\sum_{i=1}^n Y_i}{n} = \overline{X} - \overline{Y}.$$

This result says that the average sample difference is the difference between the sample averages.

Detecting mean difference

Clearly $\overline{D} = \overline{X} - \overline{Y}$ is the appropriate point estimator of $\mu_D = \mu_1 - \mu_2$. In fact, $\overline{D}$, when standardised, is the appropriate test statistic for hypothesis tests concerning μ_D. To standardise $\overline{D}$, note that

$$E(\overline{D}) = \mu_D \quad \text{and} \quad S.E.(\overline{D}) = \frac{\sigma_D}{\sqrt{n}}.$$

Without knowledge of the correlation coefficient between the X and Y data, σ_D remains unknown. Therefore the single-sample test for μ_D must be based on the **estimated standard pivot**

$$\frac{\widehat{\theta} - \theta}{S.E.(\widehat{\theta})} = \frac{\overline{D} - \mu_D}{\frac{S_D}{\sqrt{n}}},$$

where $S_D^2 = \frac{\sum_{i=1}^n (D_i - \overline{D})^2}{n-1}$ is the sample variance of the observed differences.

For any of the hypothesis testing situations:

$$\begin{array}{lll} H_0: \ \mu_D = k & H_0: \ \mu_D = k & H_0: \ \mu_D = k \\ H_1: \ \mu_D < k & H_1: \ \mu_D \neq k & H_1: \ \mu_D > k \end{array}$$

expressed in terms of a constant k, often taken as 0, we will utilise the test statistic

$$\frac{\overline{D} - \mu_D}{\frac{S_D}{\sqrt{n}}} \sim t_{n-1}. \qquad (12.1)$$

Tests concerning μ_D are called **paired comparison tests.**

Example 12.2

Comparing measurements on the same storage tanks

The inside and outside walls of storage tanks have been painted with a primer undercoat that is spray-painted on at the time of manufacture. For a given tank, it is believed that a greater quantity of paint is deposited on the outside walls than the inside walls.

For 10 randomly selected tanks, the spray-painters recorded the volume of paint applied (in litres) to both the inside and outside walls.

Tank i	**Outside wall** X_i	**Inside wall** Y_i	**Volume difference** $D_i = X_i - Y_i$
1	2.9	2.2	0.7
2	2.8	2.3	0.5
3	2.0	2.7	−0.7
4	2.7	2.4	0.3
5	2.3	2.6	−0.3
6	2.8	2.3	0.5
7	2.7	2.7	0.0
8	2.5	2.3	0.2
9	2.6	2.2	0.4
10	2.9	2.4	0.5

In this example, the tanks on which the paint measurements are taken are regarded as the experimental units.

1. Let μ_1 denote the mean volume of paint used for the outside wall of a storage tank and let μ_2 denote the mean volume of paint used for the inside wall of the same storage tank. In terms of these means, define $\mu_D = \mu_1 - \mu_2$.
2. $$H_0: \ \mu_D = 0$$ $$H_1: \ \mu_D > 0$$
3. The estimated standard pivot to be used as a test statistic is

$$T = \frac{\widehat{\theta} - \theta}{\widehat{S.E.(\widehat{\theta})}} = \frac{\overline{D} - 0}{\frac{S_D}{\sqrt{10}}} \sim t_9$$

when H_0 is true. Here $\overline{D}$ and S_D are the respective usual sample mean and sample standard deviation of the difference values $D_1, D_2, D_3, \ldots, D_{10}$.

4. Select $\alpha = 0.05$.
5. Data: $\overline{d}_{observed} = 0.210$ and $s_d = 0.431$. Individual differences are given in the table above.
6.

$$t_{observed} = \frac{0.210 - 0}{\frac{0.431}{\sqrt{10}}} = 1.54.$$

7.

$$\begin{aligned} p\text{-value} &= P(T > t_{observed} | H_0 \ true) \\ &= P(T > 1.54) \quad \text{where } T \sim t_9 \\ &> P(T > 1.833) \\ &= 0.05. \end{aligned}$$

8. Since p-value $> \alpha$, we accept H_0 at the 0.05 level of significance. The data does not provide sufficient evidence to reject the null hypothesis.
9. The test result is not significant and we summarise: $\overline{d}_{observed} = 0.210$ (*n.s.*). The mean volume of paint required to paint the outside of the storage tank is not significantly greater than the mean volume required to paint the inside of the storage tank.

Assumptions required for the paired comparison t test

Recall from Section 10.5 that use of the t distribution is restricted to data that are normally distributed. That is, the observed differences $d_i = x_i - y_i$ are observations on a random variable $D \sim N(\mu_D, \sigma_D^2)$. As usual, this normality assumption may be checked by constructing a normal probability plot for the observed differences.

A confidence interval for μ_D

By unwrapping the estimated standard pivot $\frac{\overline{D} - \mu_D}{\frac{S_D}{\sqrt{n}}}$, as in the proof of Theorem 10.2, a $100(1-\alpha)\%$ confidence interval for the mean difference μ_D may be written as

$$[\overline{D} - t_{\frac{\alpha}{2},n-1}\frac{S_D}{\sqrt{n}}, \ \overline{D} + t_{\frac{\alpha}{2},n-1}\frac{S_D}{\sqrt{n}}].$$

This has the same structure as any single small-sample confidence interval for a population mean.

Example 12.3

Ducted heating and July electricity bills

Six homes with ducted electrical central heating installed early in the winter had their electricity bills ($/day) compared for both the current July and the previous July (one year earlier).

Home	1	2	3	4	5	6
Current July ($/day) X	1.34	1.45	1.12	1.29	0.98	1.04
Previous July ($/day) Y	0.95	1.21	1.07	0.99	0.88	0.90

Worked example:

Determine a 90% confidence interval for the mean increase in electricity cost.

(Note that increases in the daily cost of electricity may be due to: increased charges from the State Electricity Commission; the installation of ducted heating which may consume more electricity than heating methods previously used in each home; greater general use of electricity in a particular home.)

The 6 independent observations on the difference random variable $D = X - Y$ are:

0.39 0.24 0.05 0.30 0.10 0.14.

The normality assumption for these data seems reasonable as is shown by the normal probability plot below.

Normal probability plot of electricity cost differences ($/day)

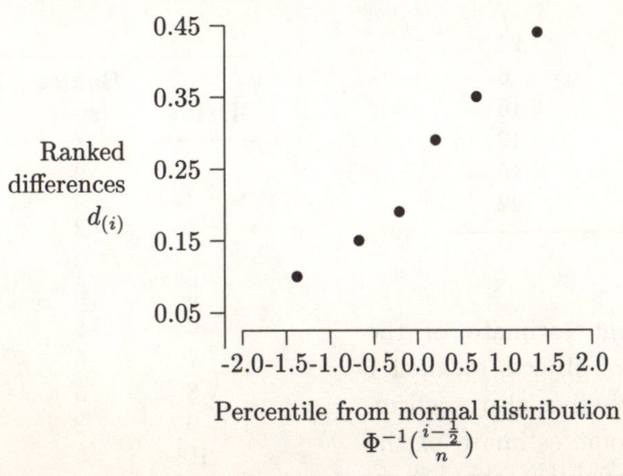

An observed 90% confidence interval for μ_D may be written as

$$[\bar{d} - t_{0.05,5}\frac{s_D}{\sqrt{6}},\ \bar{d} + t_{0.05,5}\frac{s_D}{\sqrt{6}}]$$
$$= [0.2033 - 2.015\frac{0.1294}{\sqrt{6}},\ 0.2033 + 2.015\frac{0.1294}{\sqrt{6}}] = [0.097,\ 0.310].$$

Note that since this confidence interval does not contain the point 0 we would conclude in a hypothesis testing framework that μ_D is significantly different from 0.

EXERCISES

12.1 Most would agree that once something has been memorised recollection becomes more difficult as time passes.

In a study designed to assess the effect of time on memory recall, 8 subjects viewed 30 objects on a table for 1 minute. Each subject was asked to list as many objects as could be remembered both 1 hour later and 24 hours later.

Subject	1 hour later	24 hours later
1	19	15
2	17	7
3	28	24
4	7	6
5	18	16
6	18	12
7	20	16
8	25	22

(a) Give a point estimate of the proportion of objects recalled 1 hour later. Give also a corresponding point estimate of the proportion of objects that are recalled 24 hours later.

(b) Give a point estimate of the mean difference between the number of objects recalled 1 hour and the number of objects recalled 24 hours after memorisation.

(c) Is there evidence to suggest that the mean number of objects recalled after 1 hour exceeds the mean recall after 24 hours by more than 4?

12.2 The following reaction times (measured to the nearest second) were recorded for 10 drivers both before and after having a controlled amount of alcohol.

Driver	Before (sec)	After (sec)
1	3	3
2	4	4
3	3	5
4	3	6
5	6	7
6	5	4
7	5	5
8	3	4
9	2	4
10	5	9

(a) Determine the increase in reaction time for each driver.

(b) If X denotes reaction time after alcohol consumption and Y denotes reaction time before alcohol consumption, use the data to verify that

$$\overline{D} = \overline{X} - \overline{Y}.$$

(c) Estimate the standard error of $\overline{D}$.

(d) Would you expect the standard error of $\overline{D}$ to equal the sum of the standard error of $\overline{X}$ and the standard error of $\overline{Y}$? Explain.

(e) Determine a 95% confidence interval for the mean increase in reaction time caused by consuming the given amount of alcohol.

(f) What assumptions are necessary for the construction of your confidence interval in (e)? Does it appear that these assumptions have been met?

12.3 (continuation) If the Accident Transport Commission funded 50 drivers to take part in this before/after study (instead of the 10 drivers of the previous question), and it was not reasonable to assume that reaction time is a normally distributed random variable, give the test statistic (and its distribution) that you would use to test whether mean reaction time is increased by consumption of the given amount of alcohol.

12.4 A manufacturer of 'Parka' outdoor clothing is reassessing the sizes of the rainwear that it produces with the aim of offering a more limited choice of sizes. Parkas are worn over the top of other clothing in the rain, so are generally not tight-fitting. An important size dimension is an individual's 'span' which is fingertip to fingertip width with arms outstretched sideways. Experience in the industry suggests that 'span' may be approximated by adding 4 cm to an individual's height. The manufacturer wishes to test this hypothesis and selects 12 files on customers who have been outfitted in the past. These records show:

Record	**Height (cm)**	**Span (cm)**
1	183	190
2	176	180
3	167	170
4	160	165
5	191	189
6	176	183
7	180	182
8	166	163
9	178	180
10	176	181
11	180	189
12	170	180

Let μ_D denote the mean of the random variable $D = (\text{Span}) - (\text{Height})$.

(a) Construct a table showing the values of D for each customer record. Determine the mean and the standard deviation of these values.

(b) State the null and alternative hypotheses of interest to the manufacturer.

(c) Perform the significance test at $\alpha = 0.05$.

(d) Determine a point estimate of μ_D. Give the (estimated) standard error of estimate.

(e) Give a 95% confidence interval for μ_D.

12.5 A 95% confidence interval for μ_D, the mean weight loss (kg) experienced by dieters after a 12 week low-calorie diet program, was determined from 'before' and 'after' measurements on 8 volunteers as

$$[0.002, 2.264].$$

Use this information to answer the following:

(a) If μ_B denotes mean before-diet weight and μ_A denotes the mean after-diet weight, express μ_D in terms of μ_B and μ_A.

(b) Discuss the result of the hypothesis test

$$H_0: \quad \mu_B = \mu_A$$
$$H_1: \quad \mu_B \neq \mu_A.$$

(c) What was the observed average weight loss for the 8 volunteers?

(d) What was the observed standard deviation in weight loss for the 8 volunteers?

12.6 The following data show quantitative changes in reading speed (words/min) for seven dyslexic subjects recorded with and without the assistance of coloured lenses.

Subject	With	Without
1	61	31
2	102	81
3	260	159
4	102	104
5	233	212
6	286	210
7	150	134

(*Source*: Chelva, E., Collins, D.W.K., Levy, R.L., and McLaren, T.L. (1986). Effects of exposure to coloured lenses. *Proceedings of Seminar on Advances in Learning Disabilities, May 1986.* Dislexia Research Foundation of Australia: Perth, Western Australia.)

Discuss what is revealed by these data.

12.2 VARIANCE RATIOS

For a single data set, the sample variance S^2 is a key measure of spread. It is an unbiased estimator of the population variance, σ^2.

Heuristic variance comparisons

The sample variance is often used as a comparative measure for comparing the variability in more than one data set.

Example 12.4

Variable flight times

The following flight sector times were provided by Air New Zealand (*Source: Pacific Way*, July, 1991). The original data in 'hr/min' has been expressed as hours to make calculations easier.

Auckland to:	(hr)
Adelaide	4.75
Apia	4.00
Bali	9.33
Brisbane	3.42
Cairns	5.42
Hong Kong	11.00
Honolulu	8.33
Los Angeles	11.92
Melbourne	3.83
Nadi	3.00
Nagoya	11.42
Norfolk Island	1.92
Noumea	2.92
Nuku'alofa	3.00
Papeete	4.92
Perth	7.67
Rarotonga	3.75
Singapore	10.92
Sydney	3.34
Tokyo	10.75

Christchurch to:	(hr)
Brisbane	3.50
Hobart	3.33
Melbourne	3.75
Norfolk Island	2.75
Perth	7.25
Sydney	3.33

The data shows flight times for international Air New Zealand flights out of Auckland (Data set 1) and out of Christchurch (Data set 2).

How variable is flight-length? By comparing the sample variances we find

$$\frac{s_1^2}{s_2^2} = \frac{12.03}{2.67} \approx 4.5.$$

Therefore the standard deviation of flights out of Auckland is just over twice that for flights out of Christchurch.

It would be unwise to try and force an interpretation more precise than this. These data sets really invite descriptive summary; a formal inference about Air New Zealand flights is not required. Therefore it is useful to say that flight times of international flights originating in Auckland have 'twice the standard deviation' of the Christchurch equivalent. It is even more useful to inspect the S/L plot of each data set.

Air New Zealand flight sector times
Scale: stems—hours; leaves—10^{-1} hours
|7| 3 represents 7.3 hours

Christchurch		Auckland
	1	9
7	2	9
7533	3	003478
	4	079
	5	4
	6	
2	7	6
	8	3
	9	3
	10	79
	11	049

The data sets tend to be bimodal; the observed flight times fall quite naturally into flight times for 'long haul' routes and flight times for 'short haul' routes.

Example 12.4 gives two data sets that invite only heuristic comparison of variance. If we wish to compare variances more formally, it will be necessary to have two independent data sets that are observations on normally distributed random variables.

F distributions

Data set 1 and Data set 2 show similar variability if their sample variances S_1^2 and S_2^2 are similar; this is conveniently expressed through the variance ratio:

$$\frac{S_1^2}{S_2^2} \approx 1.$$

When these data sets (1 and 2) are random samples from respective normal populations 1 and 2, then S_1^2 is an (unbiased) estimator of σ_1^2, the variance of the first population; S_2^2 is an (unbiased) estimator of σ_2^2, the variance of the second population. It would then seem reasonable that if the populations from which the data sets came were equally variable (the data sets may even be from the same population), then this will be reflected in close sample variances.

In this section we develop the ratio of independent sample variances of normal data as a statistic for testing whether the population variances have ratio 1. This so-called **variance ratio test** is built from a pivot having an *F distribution*; this probability distribution is structured as follows:

Definition 12.1 *If a random variable F is defined by the equation*

$$F = \frac{(\frac{U}{\nu_1})}{(\frac{V}{\nu_2})}$$

in terms of $U \sim \chi^2_{\nu_1}$ and $V \sim \chi^2_{\nu_2}$, then F is said to have an **F distribution with ν_1 and ν_2 degrees of freedom** *when the numerator and denominator components U and V are independent random variables and we use the distributional notation: $F \sim F_{\nu_1,\nu_2}$.*

□

F distributions are 2-parameter distributions; the parameter ν_1 represents the degrees of freedom associated with the chi-squared distribution in the numerator of F and the parameter ν_2 represents the degrees of freedom associated with the denominator of F. By changing the values of ν_1 and ν_2, a wide variety of shapes may be obtained for the probability density function of F; of the distributions discussed in this text the characteristic shape of an F distribution most closely matches that of the chi-squared distribution.

Notation for F tables

The F distribution is for positive and right-skewed data. In this sense the distributional shape differs from the positive/negative symmetry of normal or t distributions.

Tables in Appendix 6 have upper α points F_{α,ν_1,ν_2} for the F_{ν_1,ν_2} distribution for $\alpha = 0.05$ and $\alpha = 0.01$ and a wide choice of values of ν_1 and ν_2. Check that you are able to read the following directly from the tables:

$$\begin{aligned} F_{0.05,6,8} &= 3.58 \quad \text{whereas} \quad & F_{0.05,8,6} &= 4.15 \\ F_{0.01,6,8} &= 6.37 & F_{0.01,8,6} &= 8.10 \\ F_{0.01,30,20} &= 2.78 & F_{0.01,20,30} &= 2.55 \end{aligned}$$

These upper α points show that F_{ν_1,ν_2} and F_{ν_2,ν_1} distributions are not the same. The relationship between the two is easily seen, however.

By Definition 12.1, it is clear that if $F \sim F_{\nu_1,\nu_2}$, then $\frac{1}{F} \sim F_{\nu_2,\nu_1}$. Simply by inverting the random variable F, the roles of the independent ratio of chi-squared distributions are reversed, causing a reversal in the order of the degrees of freedom. The following result shows how this affects the corresponding upper α points.

Theorem 12.1 $F_{\alpha,\nu_1,\nu_2} = \frac{1}{F_{1-\alpha,\nu_2,\nu_1}}$.

Proof: Suppose that $F \sim F_{\nu_2,\nu_1}$. Then $G = \frac{1}{F} \sim F_{\nu_1,\nu_2}$. By definition,

$$1 - \alpha = P(F > F_{1-\alpha,\nu_2,\nu_1}) = P(\frac{1}{F} < \frac{1}{F_{1-\alpha,\nu_2,\nu_1}}).$$

Solving this last equation for α gives

$$\begin{aligned}\alpha &= 1 - P(\frac{1}{F} < \frac{1}{F_{1-\alpha,\nu_2,\nu_1}}) = P(\frac{1}{F} > \frac{1}{F_{1-\alpha,\nu_2,\nu_1}}) \\ &= P(G > \frac{1}{F_{1-\alpha,\nu_2,\nu_1}}) = P(G > k),\end{aligned}$$

where $k = \frac{1}{F_{1-\alpha,\nu_2,\nu_1}}$ is the upper α point of the probability distribution for G. Therefore k may also be written as $k = F_{\alpha,\nu_1,\nu_2}$. ■

Example 12.5

Changing tails in the F distribution

Theorem 12.1 shows that the relationship $F_{\alpha,\nu_1,\nu_2} = \frac{1}{F_{1-\alpha,\nu_2,\nu_1}}$ holds between the upper α and $1-\alpha$ points of respective F_{ν_1,ν_2} and F_{ν_2,ν_1} distributions.

Worked example:

Determine the value of the upper 95% point $F_{0.95,8,12}$. Does this enable us to determine the 5th percentile of a $F_{8,12}$ distribution?

$$F_{0.95,8,12} = \frac{1}{F_{1-0.95,12,8}} = \frac{1}{F_{0.05,12,8}} = \frac{1}{3.28} = 0.305 \quad (3 \text{ dp}).$$

By definition, 95% of the $F_{8,12}$ distribution lies above this value; therefore it is the point below which 5% of the distribution lies. $F_{0.95,8,12}$ is the 5th percentile of the $F_{8,12}$ distribution.

The variance ratio test

Theorem 12.2 *If S_1^2 is the sample variance of a random sample of size m from a $N(\mu_1, \sigma_1^2)$ population and S_2^2 is the sample variance of a second* **independent** *random sample of size n from a $N(\mu_2, \sigma_2^2)$ population, then*

$$\frac{\left(\frac{S_1^2}{\sigma_1^2}\right)}{\left(\frac{S_2^2}{\sigma_2^2}\right)} \sim F_{m-1,n-1}.$$

Proof: By Theorem 10.5,

$$U = \frac{(m-1)S_1^2}{\sigma_1^2} \sim \chi^2_{m-1} \quad \text{and} \quad V = \frac{(n-1)S_2^2}{\sigma_2^2} \sim \chi^2_{n-1}.$$

The sample variances S_1^2 and S_2^2 are independent random variables since they are constructed from independent random samples. Therefore, by dividing the 'numerator' and 'denominator' chi-squared distributions by their appropriate degrees of freedom, and cancelling the assumed equal population variances from both numerator and denominator,

$$\frac{S_1^2\sigma_2^2}{S_2^2\sigma_1^2} = \frac{\left(\frac{(m-1)S_1^2}{\sigma_1^2}\right)\frac{1}{(m-1)}}{\left(\frac{(n-1)S_2^2}{\sigma_2^2}\right)\frac{1}{(n-1)}} = \frac{\frac{U}{(m-1)}}{\frac{V}{(n-1)}} \sim F_{m-1,n-1}.$$

■

Theorem 12.2 shows that

$$F = \frac{\left(\frac{S_1^2}{\sigma_1^2}\right)}{\left(\frac{S_2^2}{\sigma_2^2}\right)} \tag{12.2}$$

is a pivot that may be employed for confidence intervals and hypothesis tests concerning population variance ratios.

Example 12.6

Analysing variability in salt samples

In the manufacture of chlorine, large quantities of raw salt containing impurities and various trace elements such as calcium and magnesium are dissolved in water to create a brine solution. The following data show calcium content (ppm) in salt samples from two different supply sources. It is suspected that the calcium content is more variable in raw material from the second source.

Source 1

0.034	0.112	0.104	0.084	0.074	0.127	0.087	0.097	0.032
0.108	0.088	0.097	0.097	0.097	0.097	0.099	0.091	0.099
0.114	0.132	0.070	0.075	0.083	0.099	0.076	0.112	0.111
0.123	0.096	0.112						

Source 2

0.134	0.117	0.132	0.111	0.151	0.146	0.154	0.131	0.133
0.130	0.102	0.113	0.133	0.139	0.131	0.141	0.166	0.142
0.119	0.124	0.126	0.135	0.156	0.137	0.114		

In back-to-back S/L plots, the relative variability of the two sources becomes visible:

Calcium content in salt samples (ppm)
Scale: stems—10^{-2} ppm; leaves—10^{-3} ppm
|7| 4 represents 0.074 ppm

Source 2		Source 1
	3	24
	4	
	5	
	6	
	7	0456
	8	3478
	9	1677777999
2	10	48
97431	11	12224
64	12	37
9754332110	13	2
621	14	
641	15	
6	16	

1. Let X and Y denote the calcium content from raw salt samples from Source 1 and Source 2 respectively. We assume that $X \sim N(\mu_1, \sigma_1^2)$ and $Y \sim N(\mu_2, \sigma_2^2)$. The data consist of observations on X and Y.
2.
$$H_o: \ \frac{\sigma_1^2}{\sigma_2^2} = 1$$
$$H_1: \ \frac{\sigma_1^2}{\sigma_2^2} > 1.$$
This is an upper-tail test since the raw salt from Source 1 is suspected of being more variable.
3. The estimated standard pivot to be used as a test statistic is
$$F = \frac{\left(\frac{S_1^2}{\sigma_1^2}\right)}{\left(\frac{S_2^2}{\sigma_2^2}\right)} \sim F_{29,24}.$$
Note that when H_o is true, F simplifies to the variance ratio $F = \frac{S_1^2}{S_2^2}$.
4. Select $\alpha = 0.05$.
5. Data: $s^2_{observed} = (0.02265)^2$ for the data from Source 1;
$s^2_{observed} = (0.01522)^2$ for the data from Source 2.
6. $f_{observed} = \frac{(0.02265)^2}{(0.01522)^2} = 2.215.$
7.
$$\begin{aligned} p\text{-value} &= P(F > f_{observed} | H_0 \ true) \\ &= P(F > 2.215) \quad \text{where } F \sim F_{29,24} \\ &\approx P(F > 2.215) \quad \text{where } F \sim F_{30,24} \\ &< P(F > 1.94) \quad \text{where } F \sim F_{30,24} \\ &= 0.05. \end{aligned}$$

The approximation $F_{29,24} \approx F_{30,24}$ is used in this p-value calculation because of the limited extent of the F tables.

8. Since p-value $< \alpha$, we reject H_o at the 0.05 level of significance. The data do not support the null hypothesis.
9. The test result is significant and we summarise: $f_{observed} = 2.215$ (*). There is a significant difference in variability between the calcium determinations from different sources.

The test statistic (12.2) may be used for each of the one/two-tailed tests:

$$H_o: \ \frac{\sigma_1^2}{\sigma_2^2} = 1 \qquad H_o: \ \frac{\sigma_1^2}{\sigma_2^2} = 1 \qquad H_o: \ \frac{\sigma_1^2}{\sigma_2^2} = 1$$

$$H_1: \ \frac{\sigma_1^2}{\sigma_2^2} > 1 \qquad H_1: \ \frac{\sigma_1^2}{\sigma_2^2} < 1 \qquad H_1: \ \frac{\sigma_1^2}{\sigma_2^2} \neq 1.$$

For calculating p-values and lower tail probabilities, use Theorem 12.1: the value **below** which $100\alpha\%$ of a F_{ν_2,ν_1} distribution lies is $F_{1-\alpha,\nu_2,\nu_1}$ and by Theorem 12.1 this value is determined by $\frac{1}{F_{\alpha,\nu_1,\nu_2}}$. Such tail-switching is necessary for the determination of the lower and upper endpoints of confidence intervals for the population variance ratio.

Confidence intervals for variance ratio

Exploiting the pivot $F = \frac{\left(\frac{S_1^2}{\sigma_1^2}\right)}{\left(\frac{S_2^2}{\sigma_2^2}\right)}$ from (12.2), we may write

$$P\left(F_{1-\frac{\alpha}{2},m-1,n-1} \leq \frac{\left(\frac{S_1^2}{\sigma_1^2}\right)}{\left(\frac{S_2^2}{\sigma_2^2}\right)} \leq F_{\frac{\alpha}{2},m-1,n-1}\right) = 1-\alpha.$$

The interval endpoints in this probability statement isolate $\frac{\alpha}{2}$ in the top tail of the $F_{m-1,n-1}$ distribution, $\frac{\alpha}{2}$ in the bottom tail of the same $F_{m-1,n-1}$ distribution, leaving $1-\alpha$ of the distribution between the endpoints. Now, tail switching with Theorem 12.1,

$$P\left(\frac{1}{F_{\frac{\alpha}{2},n-1,m-1}} \leq \frac{\left(\frac{S_1^2}{\sigma_1^2}\right)}{\left(\frac{S_2^2}{\sigma_2^2}\right)} \leq F_{\frac{\alpha}{2},m-1,n-1}\right) = 1-\alpha.$$

so that by unwrapping the central pivot, the population variance ratio lies isolated in the centre of the interval:

$$P\left(\frac{S_1^2}{S_2^2}\left(\frac{1}{F_{\frac{\alpha}{2},m-1,n-1}}\right) \leq \frac{\sigma_1^2}{\sigma_2^2} \leq \frac{S_1^2}{S_2^2}\left(F_{\frac{\alpha}{2},n-1,m-1}\right)\right) = 1-\alpha.$$

Therefore a $100(1-\alpha)\%$ confidence interval for the ratio of variances $\frac{\sigma_1^2}{\sigma_2^2}$ of two normal populations is given by

$$\left[\frac{S_1^2}{S_2^2}\left(\frac{1}{F_{\frac{\alpha}{2},m-1,n-1}}\right),\ \frac{S_1^2}{S_2^2}\left(F_{\frac{\alpha}{2},n-1,m-1}\right)\right].$$

For example, if $s_1^2 = 10.3$ is an observed sample variance based on 8 observations and $s_2^2 = 6.4$ is an observed sample variance based on 9 observations, then **provided both samples are independent and from normally distributed populations**, the variance ratio 90% confidence interval for $\frac{\sigma_1^2}{\sigma_2^2}$ may be expressed as

$$\begin{aligned}\left[\frac{10.3}{6.4}\left(\frac{1}{F_{0.05,8-1,9-1}}\right),\ \frac{10.3}{6.4}\left(F_{0.05,9-1,8-1}\right)\right] &= \left[\frac{10.3}{6.4}\left(\frac{1}{3.5}\right),\ \frac{10.3}{6.4}(3.73)\right]\\ &= [0.46,\ 6.00].\end{aligned}$$

A 90% confidence interval for the ratio $\frac{\sigma_1}{\sigma_2}$ of standard deviations of the two populations is found by taking the square root of the endpoints for the variance ratio confidence interval. Numerically, the interval may be expressed as $[\sqrt{0.46}, \sqrt{6.00}] = [0.68,\ 2.45]$.

Pooling sample variances

If $H_o : \frac{\sigma_1^2}{\sigma_2^2} = 1$ is accepted, we are admitting that the two independent random samples may come from respective $N(\mu_1, \sigma^2)$ and $N(\mu_2, \sigma^2)$ populations, where $\sigma^2 = \sigma_1^2 = \sigma_2^2$ is the variance common to the two populations. This invites an estimation problem: If the variances of the two populations are equal, how should this common variance be estimated? It is most efficient to use as much sample information as possible and *pool* both sample variances together to obtain an overall estimate of the common population variance. This has the effect of reducing the standard error of estimate.

Definition 12.2 *If S_1^2 is the sample variance based on a sample of size m from a $N(\mu_1, \sigma^2)$ population, and S_2^2 is the sample variance based on a sample, independent of the first, of size n from a $N(\mu_2, \sigma^2)$ population, then the* **pooled sample variance estimator**, *denoted by S_p^2, of the common population variance σ^2 is given by*

$$S_p^2 = \frac{(m-1)S_1^2 + (n-1)S_2^2}{m+n-2}.$$

□

Essentially, S_p^2 consists of a total sum of (mean corrected) squares in numerator and total degrees of freedom in denominator. That S_p^2 is indeed unbiased for σ^2 is left as an exercise. In fact, of all unbiased estimators of the form $aS_1^2 + bS_2^2$ where a and b are constants, S_p^2 has the smallest variance. In this sense S_p^2 is most efficient.

EXERCISES

12.7 Use tables to determine each of the following:

(a) $F_{0.05,6,8}$;

(b) $F_{0.05,7,9}$;

(c) $F_{0.05,20,30}$;

(d) $F_{0.01,12,10}$;

(e) $F_{0.05,10,10}$.

12.8 Use tables to determine each of the following:

(a) $F_{0.95,6,8}$;

(b) $F_{0.95,7,9}$;

(c) $F_{0.99,20,30}$;

(d) $F_{0.99,12,10}$;

(e) $F_{0.95,10,10}$.

12.9 Find the 1st and 99th percentiles of an $F_{15,20}$ distribution.

12.10 Suppose that Data set i is from a normally distributed population $N(\mu_i, \sigma_i^2)$, for $i = 1, 2$, and that the data sets are independently generated.

Scale: stems—units; leaves—10^{-1} units
|7| 3 represents 7.3 units

Data set 2		Data set 1
	1	9
54	2	1447
6	3	00001589
8777	4	012667
433200	5	24
933210	6	
77653332	7	2
99976444310	8	
863200	9	
7521	10	
220	11	
66	12	
9	13	

(a) Does the sample variance *appear* to be the same for each data set? Explain.

(b) Test whether $\sigma_1^2 < \sigma_2^2$.

(c) Determine a 95% confidence interval for σ_1^2.

12.11 In a paper by Morgan (1973), two sets of flow rates for the Antarctic Hoseason Glacier were examined through photogrammetry: Flow rates 1 were based on photographs taken 3 years apart and Flow rates 2 were based on photographs taken 5 years apart.

In order to decide which time span is best for accurately assessing flow rates through this photogrammetric survey technique, test whether results across the two time scales are equally variable ($\alpha = 0.05$).

Flow rates 1 (3 years)	Flow rates 2 (5 years)
0.73	0.72
0.76	0.74
0.75	0.74
0.77	0.72
0.73	0.72
0.75	
0.74	

(*Source*: Morgan, P.J. (1973). A photogrammetric survey of the Hoseason Glacier, Kemp Coast, Antarctica. *Journal of Glaciology*, 12, 113-120.)

12.12 The following were completion times for turning a table leg on a lathe. Two operators were each producing tables of the same design specifications. The data are in the form (min/sec).

Operator 1	Operator 2
4/40	4/15
4/12	4/05
3/30	4/00
4/15	4/15
4/00	3/50
4/40	4/00
3/20	4/25
3/25	4/15
4/05	

(a) Is the time taken to complete one leg of a table equally variable for the two operators? Test with $\alpha = 0.1$.

(b) Construct plots to determine whether the assumptions involved in the test are satisfied.

12.13 Suppose that a 95% confidence interval for $\frac{\sigma_1}{\sigma_2}$ based on two independent samples is calculated as [2.13, 6.19]. What conclusions can be made concerning the test

$$H_o: \frac{\sigma_1^2}{\sigma_2^2} = 1$$

$$H_1: \frac{\sigma_1^2}{\sigma_2^2} \neq 1?$$

12.14 Determine a 90% confidence interval for

$$\frac{\sigma_1}{\sigma_2} \quad \text{and} \quad \frac{\sigma_1^2}{\sigma_2^2}$$

for the calcium determination of salt from the two different sources in Example 12.6.

12.15 If $Z \sim N(0,1)$, determine the probability distribution of Z^2.

If $T \sim t_\nu$, determine the probability distribution of T^2.

[*Hint*: Use Definition 12.1 to show that T^2 is the ratio of two independent chi-squared distributions.]

12.16 Suppose that S_1^2 is the sample variance based on a sample of size m from a $N(\mu_1, \sigma^2)$ population, and S_2^2 is the sample variance based on a sample, independent of the first, of size n from a $N(\mu_2, \sigma^2)$ population.

(a) Show that if a and b are constants, then $aS_1^2 + bS_2^2$ is an unbiased estimator of σ^2 provided $a + b = 1$.

(b) Show that

$$S_p^2 = \frac{(m-1)S_1^2 + (n-1)S_2^2}{m+n-2}$$

is an unbiased estimator of σ^2 and find $Var(S_p^2)$.

(c) Show that in the family of all unbiased estimators of σ^2 of the form

$$aS_1^2 + bS_2^2,$$

S_p^2 has the smallest variance.

12.3 COMPARING TWO POPULATION MEANS (INDEPENDENT SAMPLES)

When faced with data from two or more samples, the first priority is to understand the relationship between the samples: the methods for constructing significance tests and confidence intervals for two *dependent* samples have been discussed in Section 12.1. In this section, consideration is given to tests and intervals for two *independent* samples. Such consideration is based on the construction of an (estimated) standard pivot.

Estimating mean difference

We review the notation for two independent random samples from respective Populations 1 and 2.
Independent samples of sizes m and n:

$$\begin{array}{ll} X_1, X_2, X_3, \ldots, X_m & \text{from Population 1;} \\ Y_1, Y_2, Y_3, \ldots, Y_n & \text{from Population 2.} \end{array}$$

Means and variances for both sample and population are:

Population	1	2
Population mean	μ_1	μ_2
Population variance	σ_1^2	σ_2^2
Sample sizes	m	n
Sample mean	$\overline{X}$	$\overline{Y}$
Sample variance	S_1^2	S_2^2

The objective is to estimate and test the nature of $\theta = \mu_1 - \mu_2$, the population mean difference, on the basis of both samples. The obvious estimation choice is $\widehat{\theta} = \overline{X} - \overline{Y}$, the difference between the sample means of the samples taken from the respective populations. The mean and variance of this sample difference are determined from **SUM 1** and **SUM 2** (5.5) and the corresponding single sample results (Theorem 9.2):

$$E(\overline{X}-\overline{Y}) = E(\overline{X}) - E(\overline{Y}) = \mu_1 - \mu_2; \tag{12.3}$$

$$Var(\overline{X}-\overline{Y}) = (1)^2 Var(\overline{X}) + (-1)^2 Var(\overline{Y}) = \frac{\sigma_1^2}{m} + \frac{\sigma_2^2}{n}. \tag{12.4}$$

Therefore, to estimate $\theta = \mu_1 - \mu_2$, we use the estimator $\widehat{\theta} = \overline{X} - \overline{Y}$, which has standard error $S.E.(\widehat{\theta}) = \sqrt{\frac{\sigma_1^2}{m} + \frac{\sigma_2^2}{n}}$ and estimated standard error of estimate,

$$\widehat{S.E.(\widehat{\theta})} = \sqrt{\frac{s_1^2}{m} + \frac{s_2^2}{n}}.$$

The independence of the two samples has no effect in (12.3), but it ensures that there are no covariance terms in (12.4). Notice that (12.3) and (12.4) hold for sample proportions which are sample averages of zeros and ones: If $X_1, X_2, X_3, \ldots, X_m$ is a random sample of size m from a $Bin(1, p_1)$ population (so that $X_1, X_2, X_3, \ldots, X_m$ represents a sequence of m Bernoulli trials with probability of success p_1) and $Y_1, Y_2, Y_3, \ldots, Y_n$ is an independent random sample of size n from a $Bin(1, p_2)$ population (so that $Y_1, Y_2, Y_3, \ldots, Y_n$ represents a sequence of n Bernoulli trials with probability of success p_2), then the sample proportion from the first population is $\widehat{P_1} = \frac{\sum_{i=1}^m X_i}{m} = \overline{X}$ and from the second, $\widehat{P_2} = \frac{\sum_{i=1}^n Y_i}{n} = \overline{Y}$. Since, by the results in Example 10.3, $E(\widehat{P_1}) = p_1$ and $Var(\widehat{P_1}) = \frac{p_1(1-p_1)}{m}$; $E(\widehat{P_2}) = p_2$ and $Var(\widehat{P_2}) = \frac{p_2(1-p_2)}{n}$, it follows that $\widehat{P_1} - \widehat{P_2}$ is an appropriate (unbiased) estimator of $p_1 - p_2$ with (estimated) standard error of estimate

$$\sqrt{\frac{\widehat{p_1}(1-\widehat{p_1})}{m} + \frac{\widehat{p_2}(1-\widehat{p_2})}{n}}.$$

Example 12.7

Comparing counts of orange/yellow Smarties in a 25 g packet

There are usually 25 small candy coated Smarties in a 25 g packet. There are 8 different colours—red/orange/yellow/green/blue/mauve/pink—on average, there should be about 3 of each colour per packet.

The number, X, of orange Smarties was counted in 27 packets (25 g) with results:

2 1 4 5 2 5 1 4 2 2 3 3 1 2
3 2 2 2 1 0 3 3 3 4 5 2 2

In an independent sample, the number, Y, of yellow Smarties was counted in 27 packets (25 g) with results:

2 6 1 3 2 2 7 1 2 5 0 4 3 2
3 4 3 0 5 3 4 4 5 3 1 1 5

Worked example:
Determine point estimates and standard errors of:

1. **the difference in average packet count for orange and yellow Smarties;**
2. **the difference in the proportion of under-coloured packets for orange and yellow Smarties (under-coloured means containing 2 or less Smarties of the given colour).**

Comparative boxplots of these data sets show reasonably symmetric distributions (for the orange Smarties data the lower quartile and median coincide at 3).

B/W plot of number of orange/yellow Smarties in 25g packets

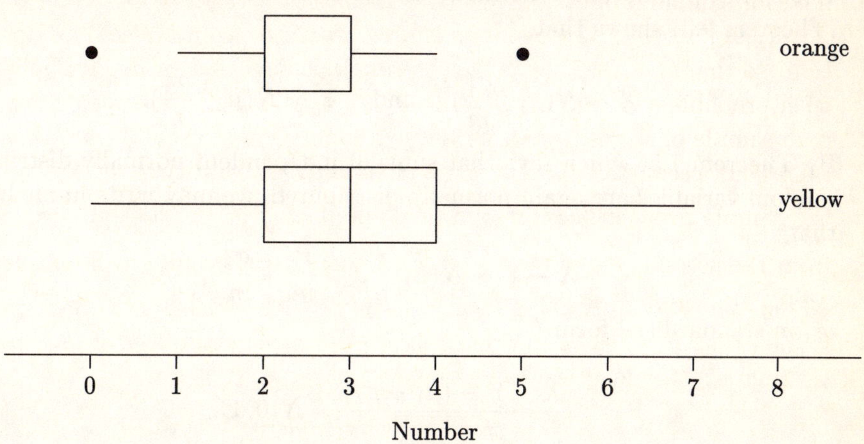

From this data we obtain:

$$\overline{x} = 2.556, \quad s_1^2 = (1.311)^2 = 1.72;$$
$$\overline{y} = 3.000, \quad s_2^2 = (1.797)^2 = 3.23.$$

Let μ_1 denote the average number of orange Smarties per packet and μ_2, the average number of yellow per packet. Then a point estimate of $\theta = \mu_1 - \mu_2$ is given by

$$\widehat{\theta} = \overline{x} - \overline{y} = 2.556 - 3.000 = -0.444,$$

with (estimated) standard error of estimate

$$\widehat{S.E.(\theta)} = \sqrt{\frac{s_1^2}{m} + \frac{s_2^2}{n}} = \sqrt{\frac{1.72}{27} + \frac{3.23}{27}} = 0.428.$$

By counting the proportion in each sample which are less than 2 colours per packet, we find that 15 packets in the first sample contained two or less orange Smarties while 11 packets in the second sample contained 2 or less yellow Smarties. The difference is

$$\widehat{p_1} - \widehat{p_2} = \frac{15}{27} - \frac{11}{27} = 0.148,$$

with (estimated) standard error of estimate

$$\sqrt{\frac{\widehat{p_1}(1-\widehat{p_1})}{m} + \frac{\widehat{p_2}(1-\widehat{p_2})}{n}} = \sqrt{\frac{\frac{15}{27}(1-\frac{15}{27})}{27} + \frac{\frac{11}{27}(1-\frac{11}{27})}{27}} = 0.134. \quad (3 \text{ dp})$$

Pivots of independent means

When each population is normally distributed, then the single sample result (Theorem 9.3) shows that

$$\overline{X} \sim N(\mu_1, \frac{\sigma_1^2}{m}) \quad \text{and} \quad \overline{Y} \sim N(\mu_2, \frac{\sigma_2^2}{n}).$$

By Theorem 7.5, which says that sums of independent normally distributed random variables are again normally distributed, we may write immediately that

$$\overline{X} - \overline{Y} \sim N(\mu_1 - \mu_2, \frac{\sigma_1^2}{m} + \frac{\sigma_2^2}{n}),$$

or, in standardised form

$$\frac{\overline{X} - \overline{Y} - (\mu_1 - \mu_2)}{\sqrt{\frac{\sigma_1^2}{m} + \frac{\sigma_2^2}{n}}} \sim N(0,1). \qquad (12.5)$$

With the identifications

$$\theta = \mu_1 - \mu_2, \quad \widehat{\theta} = \overline{X} - \overline{Y}, \quad S.E.(\widehat{\theta}) = \sqrt{\frac{\sigma_1^2}{m} + \frac{\sigma_2^2}{n}}.$$

(12.5) is an example of the standard pivot

$$\frac{\widehat{\theta} - \theta}{S.E.(\widehat{\theta})}$$

for θ. The proof of the Central Limit Theorem may be modified to show:

Theorem 12.3 *If $X_1, X_2, X_3, \ldots, X_m$ is a random sample from* **ANY** *population with mean μ_1 and variance σ_1^2 and $Y_1, Y_2, Y_3, \ldots, Y_n$ is an independent random sample from* **ANY** *population with mean μ_2 and variance σ_2^2, then the probability distribution of*

$$\frac{\overline{X} - \overline{Y} - (\mu_1 - \mu_2)}{\sqrt{\frac{\sigma_1^2}{m} + \frac{\sigma_2^2}{n}}}$$

tends to that of the standard normal probability distribution as $m \to \infty$ and $n \to \infty$.

■

Essentially, the Central Limit Theorem applies to sums of independent random variables where the variance of individual component variables in the sum contributes negligibly to the overall variability of the sum.

As in the single sample case, the standard pivot for θ and the estimated standard pivot for θ are used for confidence interval construction and as test statistics for significance tests concerning $\theta = \mu_1 - \mu_2$. We have discussed, or will discuss in this section, examples of the following pivots:

Pivot	**Data type**	**Distributions**
$\dfrac{\widehat{\theta}-\theta}{S.E.(\widehat{\theta})}$		
$\dfrac{\overline{X}-\overline{Y}-(\mu_1-\mu_2)}{\sqrt{\frac{\sigma_1^2}{m}+\frac{\sigma_2^2}{n}}}$	normal data;	$N(0,1)$
$\dfrac{\overline{X}-\overline{Y}-(\mu_1-\mu_2)}{\sqrt{\frac{\sigma_1^2}{m}+\frac{\sigma_2^2}{n}}}$	any data;	approximate $N(0,1)$ for m and n large
$\dfrac{\widehat{\theta}-\theta}{\widehat{S.E.(\widehat{\theta})}}$		
$\dfrac{\overline{X}-\overline{Y}-(\mu_1-\mu_2)}{\sqrt{\frac{S_1^2}{m}+\frac{S_2^2}{n}}}$	any data;	approximate N(0,1) for m and n large
$\dfrac{\overline{X}-\overline{Y}-(p_1-p_2)}{\sqrt{\frac{\widehat{P}_1(1-\widehat{P}_1)}{m}+\frac{\widehat{P}_2(1-\widehat{P}_2)}{n}}}$	Bernoulli data;	approximate N(0,1) for m and n large
$\dfrac{\overline{X}-\overline{Y}-(\mu_1-\mu_2)}{S_p\sqrt{\frac{1}{m}+\frac{1}{n}}}$	normal data;	t_{m+n-2}

In this table,

- S_p^2 refers to the pooled sample variance of Definition 12.2 which is a combination of both sample variances.
- $\frac{\widehat{\theta}-\theta}{\widehat{S.E.(\widehat{\theta})}}$ is the pivot which is used when population variances are unknown.

Results from the standard pivot

When $\widehat{\theta} = \overline{X} - \overline{Y}$ estimates $\theta = \mu_1 - \mu_2$, in populations with known variances, the standard pivot

$$\frac{\widehat{\theta} - \theta}{S.E.(\widehat{\theta})}$$

(introduced in Chapter 11) is either exactly normally distributed (12.5) or approximately so (Theorem 12.3). It is therefore a useful test statistic for detecting significant differences between means and since, for normal populations,

$$P\left(-z_{\frac{\alpha}{2}} \le \frac{\widehat{\theta} - \theta}{S.E.(\widehat{\theta})} \le z_{\frac{\alpha}{2}}\right) = 1 - \alpha,$$

by unwrapping the pivot from the centre of this probability statement to create random variables $\widehat{T_1}$ and $\widehat{T_2}$ such that $P(\widehat{T_1} \le \theta \le \widehat{T_2}) = 1 - \alpha$, it follows that a $100(1-\alpha)\%$ confidence interval for θ may be written as

$$\left[\widehat{\theta} - z_{\frac{\alpha}{2}}\, S.E.(\widehat{\theta}),\ \ \widehat{\theta} + z_{\frac{\alpha}{2}}\, S.E.(\widehat{\theta})\right]. \tag{12.6}$$

This structure has wide application.

Example 12.8

Are two automatic coffee machines dispensing the same sized coffees on average?

Automatic coffee machines are set to dispense 250 mL cups of coffee. To minimise the wastage of overfill, usually approximately 200 mL of liquid are dispensed, and for mean settings in this range, the variability in the dispensed volume is approximately constant at $\sigma = 20$ mL.

Suppose that the volume, X mL, dispensed by Machine 1 satisfies $X \sim N(\mu_1, 20^2)$ and the volume, Y mL, dispensed by Machine 2 satisfies $Y \sim N(\mu_2, 20^2)$.

10 observations on X gave $\overline{x} = 205$ and 15 observations of Y gave $\overline{y} = 215$.

Worked example:

1. **Are the machines, on average, dispensing the same volume of coffee?**
2. **Give a 95% confidence interval for the mean difference in the dispensed volume.**

In answer to the first of these problems, we follow the blueprint for significance tests.

1. In the above notation, μ_i denotes the mean volume of coffee dispensed by Machine i, $i = 1, 2$.

2.
$$H_o: \quad \mu_1 - \mu_2 = 0$$
$$H_1: \quad \mu_1 - \mu_2 \neq 0.$$

The test is two-tailed, since before seeing the data, it is unknown which machine dispenses the greatest volume.

3. The standard pivot to be used as a test statistic is

$$Z = \frac{\widehat{\theta} - \theta}{S.E.(\widehat{\theta})} = \frac{\overline{X} - \overline{Y} - (\mu_1 - \mu_2)}{\sqrt{\frac{\sigma_1^2}{m} + \frac{\sigma_2^2}{n}}} \sim N(0,1).$$

4. Select $\alpha = 0.05$.
5. $m = 10$ and $\overline{x}_{observed} = 205$; $n = 15$ and $\overline{y}_{observed} = 215$.
6. Under the assumption that the null hypothesis is true,

$$z_{observed} = \frac{205 - 215 - (0)}{\sqrt{\frac{20^2}{10} + \frac{20^2}{15}}} = -1.22.$$

7.
$$\begin{aligned} p\text{-value} &= 2P(Z < z_{observed} | H_0 \; true) \\ &= 2P(Z < -1.22) \quad \text{where } \; Z \sim N(0,1) \\ &= 2(0.1112) \\ &= 0.2224. \end{aligned}$$

8. Since p-value $> \alpha$, we accept H_o at the 0.05 level of significance. The data provide insufficient evidence to reject H_0.
9. The test result is not significant and we summarise: $\overline{x}_{observed} - \overline{y}_{observed} = -10$ $(n.s.)$. There is not a significant mean difference in the mean volume of coffee dispensed by the two automatic coffee machines.
10. The acceptance region of the hypothesis test is given by those observed values of of $\overline{X} - \overline{Y}$ for which

$$\left| \frac{\overline{X} - \overline{Y} - (0)}{\sqrt{\frac{20^2}{10} + \frac{20^2}{15}}} \right| \leq 1.96,$$

or $-16 \leq \overline{X} - \overline{Y} \leq 16$. Therefore, as a simple example of the evaluation of the probability of a Type II error:

$$\begin{aligned} \beta(-20) &= P(\text{accept } H_o \mid H_o \text{ is false and } \; \mu_1 - \mu_2 = -20) \\ &= P(-16 \leq \overline{X} - \overline{Y} \leq 16 \mid \mu_1 - \mu_2 = -20) \\ &= P\left(\frac{-16 - (-20)}{\sqrt{\frac{20^2}{10} + \frac{20^2}{15}}} \leq Z \leq \frac{16 - (-20)}{\sqrt{\frac{20^2}{10} + \frac{20^2}{15}}} \right) \\ &= P(0.49 \leq Z \leq 4.4) \\ &= 0.312. \end{aligned}$$

Implicit in this result is that $\beta(20) = \beta(-20)$; the power curve is symmetric for a two-tailed test. The sample sizes would each need to be increased in order to decrease the size of the probability of a Type II error. As they stand, if there is only a sample mean difference of 20 mL, we are quite likely (about 31%) to make a Type II error; such an observed sample mean difference is not significant for these sample sizes and population standard deviations of 20 mL.

Finally note that by (12.6), a 95% confidence interval for $\mu_1 - \mu_2$ may be written as

$$\begin{aligned}
&\left[\overline{x} - \overline{y} - z_{\frac{\alpha}{2}}\sqrt{\frac{\sigma_1^2}{m} + \frac{\sigma_2^2}{n}},\ \overline{x} - \overline{y} + z_{\frac{\alpha}{2}}\sqrt{\frac{\sigma_1^2}{m} + \frac{\sigma_2^2}{n}}\right] \\
&= \left[-10 - 1.96\sqrt{\frac{20^2}{10} + \frac{20^2}{15}},\ -10 + 1.96\sqrt{\frac{20^2}{10} + \frac{20^2}{15}}\right] \\
&= [-26, 6].
\end{aligned}$$

Results from the estimated standard pivot when sample sizes are large

When $\widehat{\theta} = \overline{X} - \overline{Y}$ estimates $\theta = \mu_1 - \mu_2$, in normal, binomial, ... populations with **unknown** variances, the estimated standard pivot $\frac{\widehat{\theta} - \theta}{\widehat{S.E.(\widehat{\theta})}}$ is approximately normally distributed (Theorem 12.3) when sample sizes are large (≥ 25). The unknown variances in the denominator of the pivot must be point estimated; this makes little difference to the probability distribution of the pivot for large sample sizes.

There is nothing new about the confidence interval *methods* and significance test *procedures*; they are as in the previous section. For example, following (12.6), a large-sample $100(1-\alpha)\%$ confidence interval for θ may be written as

$$\left[\widehat{\theta} - z_{\frac{\alpha}{2}}\ \widehat{S.E.(\widehat{\theta})},\ \widehat{\theta} + z_{\frac{\alpha}{2}}\ \widehat{S.E.(\widehat{\theta})}\right]. \tag{12.7}$$

(12.7) allows us to write down a confidence interval for the difference between two population proportions $p_1 - p_2$:

$$\begin{aligned}
&\left[\widehat{\theta} - z_{\frac{\alpha}{2}}\ \widehat{S.E.(\widehat{\theta})},\ \widehat{\theta} + z_{\frac{\alpha}{2}}\ \widehat{S.E.(\widehat{\theta})}\right] \\
&= \left[\widehat{p_1} - \widehat{p_2} - z_{\frac{\alpha}{2}}\sqrt{\frac{\widehat{p_1}(1-\widehat{p_1})}{m} + \frac{\widehat{p_2}(1-\widehat{p_2})}{n}},\right. \\
&\qquad\qquad \left.\widehat{p_1} - \widehat{p_2} + z_{\frac{\alpha}{2}}\sqrt{\frac{\widehat{p_1}(1-\widehat{p_1})}{m} + \frac{\widehat{p_2}(1-\widehat{p_2})}{n}}\right].
\end{aligned}$$

Example 12.9

Smarties counts: Example 12.7 continued

A 95% confidence interval for the difference between the proportion of packets containing too few 'orange' Smarties and the proportion of packets containing too few 'yellow' Smarties, based on independent searches of 27 packets for each colour may be written down immediately as:

$$\left[\frac{15}{27} - \frac{11}{27} - 1.96\sqrt{\frac{\frac{15}{27}(1-\frac{15}{27})}{27} + \frac{\frac{11}{27}(1-\frac{11}{27})}{27}}, \frac{15}{27} - \frac{11}{27} + 1.96\sqrt{\frac{\frac{15}{27}(1-\frac{15}{27})}{27} + \frac{\frac{11}{27}(1-\frac{11}{27})}{27}}\right]$$
$$= [-0.11, 0.41].$$

Since this interval crosses 0, the two sample proportions are not significantly different at a 5% significance level. It is reasonable to assume that packets with low counts of 'orange' and low counts of 'yellow' are produced in similar numbers.

Large-sample inference techniques will be illustrated with the following data set which gives the size, in hectares of burnt land, of bushfires reported to the EPA in 1980–81 and 1986–87. This is a small fraction of a much larger data set.

Example 12.10

The size of bushfires

The accompanying data give the size of bushfires reported by the Environment Protection Agency in 1980–81 and 1986–87. Only fires that burned an area of one hectare (or more) are recorded. One hectare fires are recorded in the data set as '1'; such fires occur more frequently (perhaps because they are brought under control by firefighters) than larger fires burning thousands of hectares.

The data are highly right-skewed and it seems appropriate to transform by logarithms. (The reader may wish to experiment with other power transformations.)

The data recorded in the comparative boxplots are observations on the variable log_e(firesize).

1986–87			1980–81		
Area	**Cause**	**Log(Area)**	**Area**	**Cause**	**Log(Area)**
110	1	4.70048	1	23	0.0000
1	15	0.00000	10	6	2.3026
1	21	0.00000	400	23	5.9915
3	6	1.09861	3500	16	8.1605
10	23	2.30259	2530	13	7.8360
1	7	0.00000	5200	13	8.5564
900	22	6.80239	20	20	2.9957
2	24	0.69315	60	5	4.0943
30	23	3.40120	1	24	0.0000
5	1	1.60944	10	13	2.3026
800	3	6.68461	500	7	6.2146
5	7	1.60944	40	11	3.6889
200	22	5.29832	50	1	3.9120
450	10	6.10925	120	11	4.7875
1	1	0.00000	2530	12	7.8360
1	24	0.00000	100	1	4.6052
1	23	0.00000	20	7	2.9957
21	10	3.04452	1	1	0.0000
2	6	0.69315	70	11	4.2485
99	15	4.59512	300	22	5.7038
1000	10	6.90776	540	1	6.2916
20	3	2.99573	520	22	6.2538
700	10	6.55108	1	7	0.0000
1	13	0.00000	200	20	5.2983
80	7	4.38203	200	13	5.2983
40	11	3.68888	5	10	1.6094
150	21	5.01064	1	1	0.0000
3500	1	8.16052	2	1	0.6931
200	10	5.29832	30	13	3.4012
1	1	0.00000	3	1	1.0986
5	1	1.60944	200	13	5.2983
20	1	2.99573	30	10	3.4012
2160	1	7.67786	15	24	2.7081
100	3	4.60517	200	1	5.2983
1	16	0.00000	20	1	2.9957
10	11	2.30259	1	22	0.0000
80	14	4.38203	10	13	2.3026
10	6	2.30259	1	13	0.0000
1	1	0.00000	42500	18	10.6573
30	1	3.40120	750	10	6.6201
360	13	5.88610	1330	13	7.1929
1	23	0.00000	2550	13	7.8438
82	5	4.40672	650	13	6.4770
200	22	5.29832	1	7	0.0000
2	1	0.69315	1	15	0.0000
2	11	0.69315	400	10	5.9915
400	24	5.99146	140	1	4.9416
5800	22	8.66561	74	13	4.3041
7	13	1.94591	20	1	2.9957
8	7	2.07944	200	13	5.2983
300	13	5.70378	200	11	5.2983
20	1	2.99573	1	23	0.0000
300	24	5.70378	300	23	5.7038
10	23	2.30259	1	24	0.0000
20	13	2.99573	1000	1	6.9078
10	9	2.30259	2	24	0.6931
300	10	5.70378	2	16	0.6931
2	1	0.69315	500	13	6.2146
200	6	5.29832	2	23	0.6931
10	1	2.30259			
2300	13	7.74066			
10	7	2.30259			
1	10	0.00000			
1	11	0.00000			
1	12	0.00000			
50	1	3.91202			
1	24	0.00000			
4	22	1.38629			
55	11	4.00733			
120	7	4.78749			
10	1	2.30259			
1	7	0.00000			
1	17	0.00000			
150	11	5.01064			
20	1	2.99573			

Cause Codes	
1	Lightning
2	Exhaust chainsaw
3	Exhaust other
5	Burn vehicle machine
6	Pipe cigarette match
7	Campfire BBQ
9	Burn off rail
10	Burn off grass
11	Burn off windrow
12	Train
13	Deliberately lit
14	Waste disposal domestic
15	Waste disposal industry
16	Power transmission
17	House stove flue
18	Burn building
20	Relight wildfire
21	Relight pres burn
22	Relight burn off
23	Other
24	Unknown

B/W plot of bushfire size (hectares) for 1980–81 and 1986–87

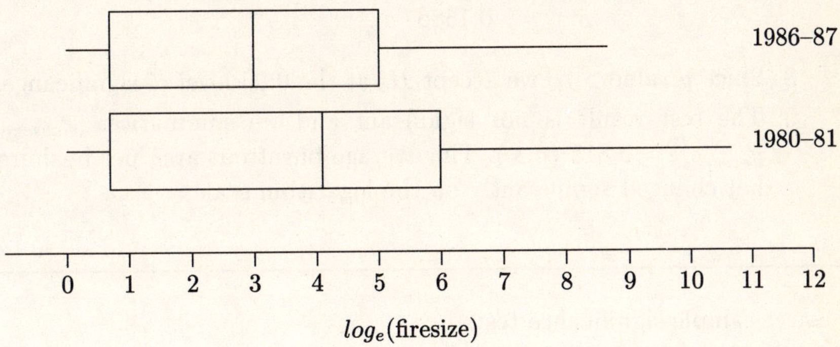

Is the average size of fires significantly different for the two time periods? That is, is the average fire size in 1987 significantly different from the previous 1981 average?

In this example of a large-sample hypothesis test for means, the data are not normally distributed, but by results from the Central Limit Theorem, the estimated standard pivot is approximately normally distributed.

1. Suppose that fire size is transformed to the log scale. Let μ_1 denote the average log_e(firesize) in 1980–81; μ_2 the average for the same variable in 1986–87.
2. $$\begin{aligned} H_o: &\ \mu_1 - \mu_2 = 0 \\ H_1: &\ \mu_1 - \mu_2 \neq 0. \end{aligned}$$

   The test is two-tailed, since before seeing the data, it is not clear whether the average bushfire size has recently increased or decreased.
3. The (approximately) normally distributed estimated standard pivot to be used as a test statistic is
$$Z = \frac{\widehat{\theta} - \theta}{\widehat{S.E.}(\widehat{\theta})} = \frac{\overline{X} - \overline{Y} - (\mu_1 - \mu_2)}{\sqrt{\frac{S_1^2}{m} + \frac{S_2^2}{n}}}.$$
4. Select $\alpha = 0.05$.
5. $m = 59$ and $\overline{x}_{observed} = 3.842$; $n = 75$ and $\overline{y}_{observed} = 3.027$.
6. Under the assumption that the null hypothesis is true,
$$z_{observed} = \frac{3.842 - 3.027 - (0)}{\sqrt{\frac{(2.791)^2}{59} + \frac{(2.477)^2}{75}}} = 1.41.$$

7.

$$\begin{aligned} p\text{-value} &= 2P(Z > z_{observed} | H_0 \ true) \\ &\approx 2P(Z > 1.41) \quad \text{where } Z \sim N(0,1) \\ &= 2(0.0793) \\ &= 0.1586. \end{aligned}$$

8. Since p-value $> \alpha$, we accept H_o at the 0.05 level of significance.
9. The test result is not significant and we summarise: $\overline{x}_{observed} - \overline{y}_{observed} = 0.815$ ($n.s.$). The average burnt out area per bushfire has not changed significantly on the logarithm scale.

Large-sample significance tests

$$H_o: \ p_1 - p_2 = 0$$
$$H_1: \ p_1 - p_2 \neq 0,$$

for comparing proportions p_1 and p_2, or any such test of H_o against a single-sided alternative, specify the same proportion $p_1 = p_2$ occurs in Populations 1 and 2 when H_o is true. The test statistic

$$\frac{\overline{X} - \overline{Y} - (p_1 - p_2)}{\sqrt{\frac{p_1(1-p_1)}{m} + \frac{p_2(1-p_2)}{n}}}$$

when H_o is true, is not completely specified; indeed, since H_o does not specify what the common population proportion **is**, but rather only that $p_1 = p_2$, this common proportion must be estimated on the basis of both samples. The effect is termed **pooling**. If $X_1, X_2, X_3, \ldots, X_m$ is a random sample of size m from a $Bin(1, p_1)$ population (so that $X_1, X_2, X_3, \ldots, X_m$ represents a sequence of m Bernoulli trials with probability of success p_1) and $Y_1, Y_2, Y_3, \ldots, Y_n$ is an independent random sample of size n from a $Bin(1, p_2)$ population (so that $Y_1, Y_2, Y_3, \ldots, Y_n$ represents a sequence of n Bernoulli trials with probability of success p_2), then the sample proportion from the first population is $\widehat{P_1} = \frac{\sum_{i=1}^{m} X_i}{m} = \overline{X}$ and from the second $\widehat{P_2} = \frac{\sum_{i=1}^{n} Y_i}{n} = \overline{Y}$. These sample proportions estimate p_1 and p_2 unbiasedly.

However, if $p_1 = p_2 = p_{pooled}$ then this common parameter p_{pooled} is estimated by combining both sequences of Bernoulli trials together (this is allowable since the two sequences of trials were independent). That is: $X_1, X_2, X_3, \ldots, X_m, Y_1, Y_2, Y_3, \ldots, Y_n$ is a sequence of $m + n$ Bernoulli trials with probability of success p_{pooled}. Then p_{pooled} is estimated by the **pooled estimator**

$$\widehat{P_{pooled}} = \frac{\text{total number of successes}}{\text{total number of trials}} = \frac{\sum_{i=1}^{m} X_i + \sum_{i=1}^{n} Y_j}{m+n}.$$

Recalling the structure of the pooled sample variance of Definition 12.2, it is convenient to write the pooled proportion estimator in the form

$$\widehat{P_{pooled}} = \frac{m\widehat{P_1} + n\widehat{P_2}}{m+n}. \tag{12.8}$$

When $p_1 = p_2$, the test statistic

$$Z = \frac{\overline{X} - \overline{Y} - (p_1 - p_2)}{\sqrt{\frac{p_1(1-p_1)}{m} + \frac{p_2(1-p_2)}{n}}}$$

is estimated by

$$Z = \frac{\overline{X} - \overline{Y} - (0)}{\sqrt{\frac{\widehat{P_{pooled}}(1-\widehat{P_{pooled}})}{m} + \frac{\widehat{P_{pooled}}(1-\widehat{P_{pooled}})}{n}}} = \frac{\overline{X} - \overline{Y} - (0)}{\sqrt{\widehat{P_{pooled}}(1 - \widehat{P_{pooled}})}\sqrt{\frac{1}{m} + \frac{1}{n}}}.$$

Example 12.11

Is there a change in the proportion of deliberately lit bush fires?

In this example we use the observations on the 'cause' variable for the bushfire data of Example 12.10, comparing the two year groups with respect to 'Cause 13'—fires that have been deliberately lit. Notice that such fires burn out large areas, since they are not reported at the time of lighting!

Worked example:

Use a test of significance to detect if there has been a change from 1980–81 to 1986–87 in the proportion of fires that have been deliberately lit?

1. Let p_1 (respectively p_2) denote the proportion of 1980–81 (respectively 1986–87) fires that are deliberately lit.
2. $$H_o: \; p_1 - p_2 = 0$$ $$H_1: \; p_1 - p_2 \neq 0.$$
3. The estimated standard pivot to be used as a test statistic is $$Z = \frac{\widehat{\theta} - \theta}{\widehat{S.E.(\widehat{\theta})}} = \frac{\overline{X} - \overline{Y} - (0)}{\sqrt{\widehat{P_{pooled}}(1 - \widehat{P_{pooled}})}\sqrt{\frac{1}{m} + \frac{1}{n}}},$$ which has an approximate standard normal distribution when H_o is true since the sample sizes are large.
4. Select $\alpha = 0.05$.

5. To extract the observed sample proportions from the data we count: '13' occurs 14 times in the $m = 59$ 'cause' records for 1980–81 so that $\widehat{p1}_{observed} = \frac{14}{59}$; '13' occurs only 6 times in the $n = 75$ 'cause' records for 1986–87 so that $\widehat{p2}_{observed} = \frac{6}{75}$.
6. When the null hypothesis is true, the common proportion of deliberately lit fires is estimated by

$$\widehat{p_{pooled}} = \frac{m\widehat{p1}_{observed} + n\widehat{p2}_{observed}}{m+n} = \frac{59(\frac{14}{59}) + 75(\frac{6}{75})}{59+75} = \frac{20}{134}.$$

This is used in the observed value of the test statistic

$$z_{observed} = \frac{\frac{14}{59} - \frac{6}{75} - (0)}{\sqrt{\frac{20}{134}(1 - \frac{20}{134})}\sqrt{\frac{1}{59} + \frac{1}{75}}} = 2.54.$$

7.

$$\begin{aligned} p\text{-value} &= 2P(Z > z_{observed} | H_0 \; true) \\ &\approx 2P(Z > 2.54) \quad \text{where } Z \sim N(0,1) \\ &= 2(0.0055) \\ &= 0.011. \end{aligned}$$

8. Since p-value $< \alpha$, we reject H_o at the 0.05 level of significance.
9. The test result is significant: $\widehat{p1}_{observed} - \widehat{p2}_{observed} = 0.24$ $(*)$. The proportion of deliberately lit fires has changed significantly.

In Example 12.11 a 95% confidence interval for $p_1 - p_2$ is not based on the pooled sample estimate, since there is no restriction that $p_1 = p_2$. As in Example 12.9, the separate population proportions are point estimated by their appropriate sample proportions and the confidence interval may be written down immediately as

$$\left[\frac{14}{59} - \frac{6}{75} - 1.96\sqrt{\frac{\frac{14}{59}(1-\frac{14}{59})}{27} + \frac{\frac{6}{75}(1-\frac{6}{75})}{27}}, \; \frac{14}{59} - \frac{6}{75} + 1.96\sqrt{\frac{\frac{14}{59}(1-\frac{14}{59})}{27} + \frac{\frac{6}{75}(1-\frac{6}{75})}{27}}\right].$$

Results from the estimated standard pivot when sample sizes are small

The Central Limit Theorem (Theorem 12.3) cannot be used for comparing population means when only small sample sizes are available. If we are able to assume that the data are normally distributed, then when $\widehat{\theta} = \overline{X} - \overline{Y}$ is used to estimate $\theta = \mu_1 - \mu_2$, the estimated standard pivot $\frac{\widehat{\theta}-\theta}{S.E.(\widehat{\theta})}$ follows a t distribution. Recall that the t distribution is flatter than the normal; the larger variance is caused by the inaccuracy resulting from estimating the unknown population variance from a small sample. In this situation of sparse information, the biggest problem is the estimation of the denominator $\widehat{S.E.}(\widehat{\theta})$ of the above estimated standard pivot (which will be used as a test statistic).

Since the samples are independent:

$$\frac{(m-1)S_1^2}{\sigma_1^2} \sim \chi^2_{m-1} \quad \text{and} \quad \frac{(n-1)S_2^2}{\sigma_2^2} \sim \chi^2_{n-1}$$

and by the additivity of chi-squared distributions (Theorem 10.4),

$$\frac{(m-1)S_1^2}{\sigma_1^2} + \frac{(n-1)S_2^2}{\sigma_2^2} \sim \chi^2_{m+n-2}.$$

With the t_k distribution structure (Definition 10.10):

$$T = \frac{Z}{\sqrt{\frac{W}{k}}} \quad \text{for} \quad Z \sim N(0,1) \quad \text{independently of} \quad W \sim \chi^2_k,$$

we identify (following the same reasoning as was used in the single-sample case, Theorem 10.7)

$$\begin{aligned} Z &= \frac{\overline{X} - \overline{Y} - (\mu_1 - \mu_2)}{\sqrt{\frac{\sigma_1^2}{m} + \frac{\sigma_2^2}{n}}}, \\ W &= \frac{(m-1)S_1^2}{\sigma_1^2} + \frac{(n-1)S_2^2}{\sigma_2^2}, \\ k &= m+n-2. \end{aligned}$$

When we substitute into $\frac{Z}{\sqrt{\frac{W}{k}}}$, we obtain the unyielding fraction

$$T = \frac{\frac{\overline{X}-\overline{Y}-(\mu_1-\mu_2)}{\sqrt{\frac{\sigma_1^2}{m}+\frac{\sigma_2^2}{n}}}}{\sqrt{\frac{\frac{(m-1)S_1^2}{\sigma_1^2}+\frac{(n-1)S_2^2}{\sigma_2^2}}{m+n-2}}}.$$

The unknown population variances still prevent this from being used as a test statistic. An appropriate way out of this predicament is to make the simplifying assumption of equal population variances. That is, assume that $\sigma_1^2 = \sigma_2^2 = \sigma^2$, where σ^2 is still unknown and common to both populations. By Definition 12.2, the pooled sample variance estimator

$$S_p^2 = \frac{(m-1)S_1^2 + (n-1)S_2^2}{m+n-2}$$

is an 'efficient' and unbiased estimator of the common population variance σ^2. Introducing σ^2 and S_p^2 into the above T structure creates the following simplification:

$$T = \frac{\dfrac{\overline{X}-\overline{Y}-(\mu_1-\mu_2)}{\sqrt{\frac{\sigma^2}{m}+\frac{\sigma^2}{n}}}}{\sqrt{\dfrac{\frac{(m-1)S_1^2}{\sigma^2}+\frac{(n-1)S_2^2}{\sigma^2}}{m+n-2}}} = \frac{\dfrac{\overline{X}-\overline{Y}-(\mu_1-\mu_2)}{\sigma\sqrt{\frac{1}{m}+\frac{1}{n}}}}{\frac{1}{\sigma}\sqrt{S_p^2}} = \frac{\overline{X}-\overline{Y}-(\mu_1-\mu_2)}{S_p\sqrt{\frac{1}{m}+\frac{1}{n}}}.$$

This has established the following theorem:

Theorem 12.4 *Suppose that $\overline{X}$ is the sample mean and S_1^2 is the sample variance of a random sample from a $N(\mu_1, \sigma^2)$ population and $\overline{Y}$ is the sample mean and S_2^2 the sample variance of an independent random sample from a $N(\mu_2, \sigma^2)$ population, then*

$$T = \frac{\overline{X}-\overline{Y}-(\mu_1-\mu_2)}{S_p\sqrt{\frac{1}{m}+\frac{1}{n}}} \sim t_{m+n-2}.$$

where S_p^2 is the pooled sample variance calculated on the basis of both samples.

■

Since the test statistic of Theorem 12.4 is widely implemented, we reinforce now the two basic assumptions necessary for this test statistic to have a t distribution.

Assumption 1:
The random samples are from normally distributed populations.
Assumption 2:
The population variances are equal: $\sigma_1^2 = \sigma_2^2 = \sigma^2$.

How might these assumptions be checked?

The normality is checked by a normal probability plot in which both data sets are plotted. Two parallel 'straight' lines indicate normality by the straightness and equal variance by the parallelism. (Recall that the slope of a linear normal probability plot is an estimate of the population standard deviation, so that if the straight lines from two plotted data sets are reasonably parallel, this provides evidence that the populations from which the data sets came have equal variance.)

Many authors recommend the use of an F test with null hypothesis H_o : $\sigma_1^2 = \sigma_2^2$ for testing whether the population variances may be assumed to be equal. Such testing is overly stringent; the t test is reasonably robust to slight flaws in Assumptions 1 and 2. The test works well for reasonably symmetric, reasonably bell-shaped, reasonably equally spread data sets. The distribution-free methods of Chapter 15 may be used when normality and/or equi-variance fail.

A $100(1-\alpha)\%$ confidence interval for $\mu_1 - \mu_2$ follows from the general structure of the estimated standard pivot:

$$\left[\overline{X} - \overline{Y} - t_{\frac{\alpha}{2},m+n-2} S_p \sqrt{\frac{1}{m} + \frac{1}{n}},\ \overline{X} - \overline{Y} + t_{\frac{\alpha}{2},m+n-2} S_p \sqrt{\frac{1}{m} + \frac{1}{n}}\right].$$

Example 12.12

The life of retreads

An experiment was designed to test whether the tyre life ($\times 10^3$ km) of retreads of car tyres has shown significant improvement under a new manufacturing process. The new process will be implemented (involving major changes to the production line) if the average life of tyres manufactured by the new process is at least 1000 km greater than the previous standard.

Cars of the same make and model were used in the experiment, and wheel balancing and alignment checks were made every 5000 km. In the experiment the tyres were randomly assigned to positions on the cars. Four 'new process tyres' were destroyed in a car accident near the outset of the experiment; had they not been destroyed, it is clear that they would have lasted longer. In this sense they constitute *censored data* and are omitted from current consideration. After this loss, 8 'new process' and 8 'standard process' data points were independently recorded.

New process:	20.4	22.6	25.0	25.2	28.4	29.3	30.0	33.2
Standard process:	15.5	18.2	19.3	20.9	21.3	23.8	26.3	27.6

Worked example:

1. **Is the average life of retreads produced under the new process at least 1000 km more than the standard process?**
2. **Give a 95% confidence interval for $E(X - Y)$, where the random variable X denotes the life of retreads from the new process and Y denotes the life of retreads from the standard process.**

Using the significance testing blueprint:

1. Let μ_1 represent the mean tyre life for 'new process' retreads and μ_2 represent the mean tyre life for 'standard process' retreads ($\times 10^3$ km).
2.

$$\begin{aligned} H_o &: \ \mu_1 - \mu_2 = 1 \\ H_1 &: \ \mu_1 - \mu_2 > 1. \end{aligned}$$

3. The estimated standard pivot to be used as a test statistic is

$$T = \frac{\overline{X} - \overline{Y} - (\mu_1 - \mu_2)}{S_p\sqrt{\frac{1}{m} + \frac{1}{n}}} \sim t_{m+n-2}.$$

4. Select $\alpha = 0.05$.
5. Normal probability plots show that these data are normally distributed with parallel straight plots for each variable indicating that it is reasonable to assume that the population standard deviations are the same.

Normal probability plot of tyre life

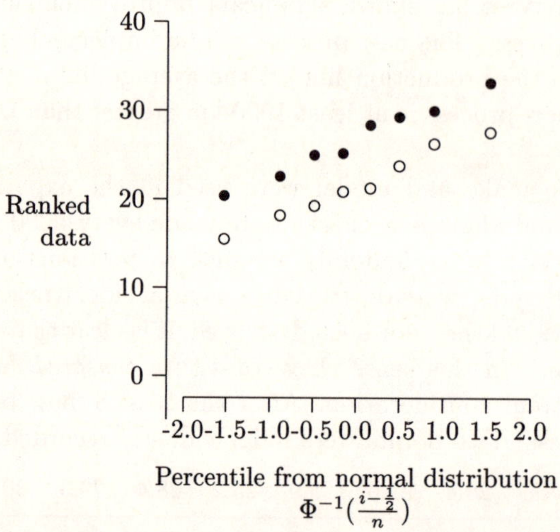

$m = n = 8$ and the sample means and variances are summarised as:

$$\overline{x} = 26.76, \quad s_1 = 4.22;$$
$$\overline{y} = 21.61, \quad s_2 = 4.10.$$

6. For a two-sample t test with small sample sizes, we assume that the population variances for retread life are equal: $\sigma_1^2 = \sigma_2^2 = \sigma^2$, where σ^2 is still unknown and common to both processes. σ^2 is point estimated by

$$s_p^2 = \frac{(m-1)s_1^2 + (n-1)s_2^2}{m+n-2} = \frac{7(4.22)^2 + 7(4.10)^2}{8+8-2} = 17.3092.$$

Under the assumption that the null hypothesis is true,

$$t_{observed} = \frac{\overline{x} - \overline{y} - (\mu_1 - \mu_2)}{s_p\sqrt{\frac{1}{m} + \frac{1}{n}}} = \frac{26.76 - 21.61 - (1)}{\sqrt{17.3092}\sqrt{\frac{1}{8} + \frac{1}{8}}} = 2.00 \ \ (2\text{ dp}).$$

7.

$$\begin{aligned} p\text{-value} &= P(T > t_{observed}|H_0 \ true) \\ &= P(T > 2.00) \quad \text{where } T \sim t_{14} \\ &< P(T > 1.76) \\ &= 0.05. \end{aligned}$$

8. Since p-value $< \alpha$, we reject H_o at the 0.05 level of significance.
9. The test result is significant and we summarise: $\overline{x}_{observed} - \overline{y}_{observed} = 5.15$ (*). The 'new process' for retreads (when compared with the 'standard process') gives significantly more than 1000 km improvement in average life.

Finally note that since $E(X-Y) = E(X) - E(Y) = \mu_1 - \mu_2$, a 95% confidence interval for $E(X-Y)$ may be written as

$$\begin{aligned} &\left[\overline{X} - \overline{Y} - t_{\frac{\alpha}{2}, m+n-2} S_p \sqrt{\tfrac{1}{m} + \tfrac{1}{n}},\ \overline{X} - \overline{Y} + t_{\frac{\alpha}{2}, m+n-2} S_p \sqrt{\tfrac{1}{m} + \tfrac{1}{n}}\right] \\ &= [5.15 - 2.145\sqrt{17.3092}\sqrt{\tfrac{1}{8} + \tfrac{1}{8}},\ 5.15 + 2.145\sqrt{17.3092}\sqrt{\tfrac{1}{8} + \tfrac{1}{8}}] \\ &= [0.69,\ 9.61]. \end{aligned}$$

Since this confidence interval contains the anticipated level of improvement 1 (thousand km), the p-value of the test in (1) is likely to be between 0.025 and 0.05.

EXERCISES

12.17 Approximately equal proportions of Melbourne and Sydney residents were questioned in a telephone survey of 1000 respondents. The following question sought attitudes towards drugs and drug use:

Should small amounts of heroin be legal?

In response, the percentages for each city, rounded to the nearest whole number were as follows:

	Melbourne	Sydney
Yes	19	24
No	75	71
Don't know	6	5

(*Source*: Saulwick, I. (1991). *The Saulwick Poll Reprints, 1989.*)

(a) Estimate the difference between the proportions of Sydney and Melbourne residents in favour of legalising small amounts of heroin.

(b) Give the (estimated) standard error of your estimate.

12.18 In Exercise 12.17, show that there is not a significant difference between the proportions of Sydney and Melbourne residents in favour of legalising small amounts of heroin.

12.19 (continuation) Use the data of Exercise 12.17 to estimate the proportion of all people in favour of legalising small amounts of heroin.

12.20 Two laboratories independently obtained the following determinations (as part of a larger data set) of the specific gravity of sulphuric acid:

Laboratory 1

1.84	1.79	1.86	1.90	1.76
1.79	1.78	1.78	1.82	1.91

Laboratory 2

1.81	1.76	1.81	1.81	1.81
1.85	1.83	1.79	1.80	1.82

(*Source*: Brookes, C.J., Betteley, I.G., and Loxston, S.M. (1966). *Mathematics and Statistics for Chemists.*)

(a) Test whether the laboratory assessments are the same on average.

(b) Are the laboratories equally variable in their determinations?

(c) If the same 10 laboratory assistants had worked in each laboratory to produce the determinations, is the method of analysis used in (a) still applicable? Explain.

Assistant	Lab 1	Lab 2
1	1.84	1.81
2	1.79	1.76
3	1.86	1.81
4	1.90	1.81
5	1.76	1.81
6	1.79	1.85
7	1.78	1.83
8	1.78	1.79
9	1.82	1.80
10	1.91	1.82

12.21 Imagery ratings for a large selection of abstract nouns and adjectives were determine by 50 Australian and 50 Canadian psychology students; ratings were constructed on a scale 1(low imagery rating) to 7 (high imagery rating). The results included the following mean and standard deviation rating assessment:

Australian:

Noun	Mean	SD
Mischief	3.12	1.98
Slave	5.14	2.04

Canadian:

Noun	Mean	SD
Mischief	4.03	1.67
Slave	6.23	0.87

(*Source*: Abernethy, M. (1991). Imagery ratings for 230 nouns and adjectives. *Australian Journal of Psychology*, 43, 4.)

Assume independence between the scores assigned by different groups of assessors.

(a) Is there a significant difference between the mean imagery ratings assigned to the noun *Slave* by Australians and Canadians?

(b) Is there a significant difference between the mean imagery ratings assigned to the nouns *Slave* and *Mischief* by Australians?

12.22 In tests to compare the durability of two brands of brick pavers, a sharpened drill impacted the surface of 10 test pavers of each brand for a period of one minute. The volume of material eroded is recorded in the S/L plot below.

Impact resistance tests
Stems—10 mL; leaves—1 mL.
|25|3 means 253 mL

Type A pavers		Type B pavers
	23	
	24	2
	25	4 8
	26	2 6 8
3	27	3 5
6 3	28	1
9 5 5 2	29	
7 3	30	
1	31	
	32	2

(a) Construct a boxplot for the Type B pavers.

(b) Test whether Type B pavers have a lower mean volume eroded than Type A pavers at a significance level of 0.05.

(c) State the assumptions required for the test in (b). Explain whether these assumptions appear to be met.

(d) Which pavers are evidently composed of the softer material? Justify your answer.

12.23 It is claimed that a new device will decrease petrol consumption when fitted to a car ignition system. To test the claim, 5 cars were chosen at random from a rental company's large pool of cars. Each car was run as far as it would go on one litre of petrol without the device and then the process was repeated with the device fitted. The distances travelled have tabular summary:

	Device fitted X (km)	Device not fitted Y (km)	Increase $X-Y$
Mean	10.0	9.0	1.0
Variance	11.8	8.2	5.0

Assuming that the data are normally distributed, construct a 95% confidence interval for μ, the mean increase in km when the device is fitted.

12.24 Year 12 students throughout Australia in a particular year are given a test.

A random sample of 40 students is chosen from the Melbourne region and an independent random sample of 50 students is chosen from the Sydney region. The observed mean score of the 40 Melbourne students is $\overline{x} = 61$ with a standard deviation $s_x = 6$. The observed mean score for the 50 Sydney students is $\overline{y} = 59$ with a standard deviation $s_y = 5$.

(a) Show that if the random variable

$$\overline{D} = \overline{X} - \overline{Y}$$

represents the difference between the means of the two samples, then the mean and variance of $\overline{D}$ are point estimated by $\overline{d} = 2$ and $s^2_{\overline{D}} = 1.4$ respectively.

(b) Write down suitable null and alternative hypotheses for testing whether the mean score, μ_x, of all Melbourne students is significantly different from the mean score, μ_y, of all Sydney students.

(c) Discuss the point estimation of

$$\mu_x - \mu_y.$$

In particular, obtain an appropriate (estimated) standard error of estimate.

(d) Use the approximately normally distributed test statistic

$$\frac{\overline{D} - (\mu_x - \mu_y)}{S_{\overline{D}}}$$

to determine the p-value of the test.

12.25 A manufacturer wishes to estimate the difference in output voltage of two different types of battery. A random sample of 10 of each battery type gives the following summary (volts):

Battery type	Mean output	Standard deviation
A	11.17	0.93
B	12.04	0.88

Determine a 95% confidence interval for the mean difference in output for the two battery types. What assumptions, concerning the probability distributions of output for each battery type, are necessary for the construction of this interval?

12.26 The following speeds (km hr^{-1}) were recorded by speed cameras for traffic on the Nepean Highway (Location 1) and crossing the Westgate Bridge (Location 2).

Nepean Highway (Location 1)	Westgate Bridge (Location 2)
65	54
81	56
67	70
59	75
48	80
75	70
82	51

Let μ_i denote the traffic speed at Location i.

(a) Determine 95% confidence intervals, based on the t distribution, for μ_1, μ_2, and $\mu_1 - \mu_2$.

(b) Suppose that the intervals determined in (a) had erroneously been based on the normal distribution. Would such intervals be wider than their counterparts in (a)?

(c) Is there evidence to suggest that at Location 1, the mean vehicle speed exceeds the speed limit of 60 km hr^{-1}? Does the same apply to Location 2?

(d) Is there a significant difference between the mean vehicle speed at the two locations?

12.27 In Exercise 12.26, suppose that more extensive records of the output from the speed cameras showed that on the Nepean Highway, 682 of 1000 recorded vehicles exceeded the speed limit, while 591 of 1000 recorded vehicles on the Westgate Bridge exceeded the limit (60 km hr^{-1} in each case).

Is there a significant difference between the *proportion* of vehicles exceeding the speed limit at the two locations?

12.28 The following firesize data is extracted from Example 12.10 and gives bushfire sizes greater than 1 hectare by two separate causes: 'lightning'—Code 1; 'deliberately lit'—Code 13. To increase symmetry, the data values have been transformed by natural logarithms.

Area	Cause	Log(Area)
360	13	5.8861
7	13	1.9459
300	13	5.7037
20	13	2.9957
2300	13	7.7406
2530	13	7.8360
5200	13	8.5564
10	13	2.3026
200	13	5.2983
30	13	3.4012
200	13	5.2983
10	13	2.3026
1330	13	7.1929
2550	13	7.8438
650	13	6.4770
74	13	4.3041
200	13	5.2983
500	13	6.2146
110	1	4.7004
5	1	1.6094
3500	1	8.1605
5	1	1.6094
20	1	2.9957
2160	1	7.6778
30	1	3.4012
2	1	0.6931
20	1	2.9957
2	1	0.6931
10	1	2.3025
50	1	3.9120
10	1	2.3025
20	1	2.9957
50	1	3.9120
100	1	4.6052
540	1	6.2916
2	1	0.6931
3	1	1.0986
200	1	5.2983
20	1	2.9957
140	1	4.9416
20	1	2.9957
1000	1	6.9078

On average, do fires that are deliberately lit burn a greater area than fires that are started by lightning strike?

12.29 In n tosses of a 'fair' die, show that the variance of the difference between the number of 'sixes' and the number of 'ones' occurring in the same sequence of tosses is $\frac{n}{3}$.

In 1000 tosses of a supposedly 'fair' die, 186 'sixes' and 100 'ones' were recorded within the same sequence. Do you doubt that the die is fair?

12.30 In this question we construct a significance test to check the claim that the addition of salt to water decreases the average time (minutes) that it takes for 1 litre of water to boil.

We will take the following as 'facts' about boiling times (inferred from previous experiments): time-to-boil is a normal random variable; in the 'without salt' case, $X \sim N(\mu, (0.3)^2)$. The corresponding distribution in the 'with salt' case is $Y \sim N(\lambda, (0.25)^2)$ with both μ and λ unknown.

(a) Derive a normally distributed test statistic based on $\overline{X} - \overline{Y}$ which can be used to test the claim that adding salt reduces boiling time on average. Here $\overline{X}$ and $\overline{Y}$ are the sample means of independent random samples of size n from the respective $N(\mu, (0.3)^2)$ and $N(\lambda, (0.25)^2)$ populations respectively.

(b) Carry out the test in (a). Make sure that you state: null and alternative hypotheses, test statistic, p-value and conclusions.

Use:

5 pots, each containing 1 litre of water (at room temperature) are heated until the water boils, giving an average boiling time

of 7.96 minutes. Independently, but under the same experimental conditions, 5 pots with one teaspoon of salt added to each gave an average boiling time of 7.76 minutes.

(c) For $\alpha = 0.05$ determine the value of the constant c so that the test has rejection region

$$\text{reject } H_0 \text{ if } \overline{X} - \overline{Y} \geq c.$$

(d) Determine the probability of a Type II error if $\mu - \lambda = 0.1$.

(e) Sketch the shape of the power curve of the test. Is the test going to be successful in detecting small discrepancies between μ and λ? Explain briefly.

12.31 Suppose that $\overline{X}$ is the sample mean and S_x^2 is the sample variance of a random sample of size n_x taken from a $N(\mu_x, \sigma^2)$ population.

Suppose that $\overline{Y}$ is the sample mean and S_y^2 is the sample variance of an independent random sample of size n_y taken from a $N(\mu_y, k\sigma^2)$ population. k is a known positive constant. σ^2 is an unknown parameter.

(a) Show that the estimator S^2 of σ^2 defined by

$$S^2 = \frac{k(n_x - 1)S_x^2 + (n_y - 1)S_y^2}{k(n_x + n_y - 2)}$$

is an unbiased estimator of σ^2.

(b) By considering the distributions of

$$\overline{X} - \overline{Y}$$

and

$$\frac{(n_x + n_y - 2)S^2}{\sigma^2}$$

(which you may assume to be independent), show that the statistic T, defined by

$$T = \frac{\overline{X} - \overline{Y} - (\mu_x - \mu_y)}{S\sqrt{\frac{1}{n_x} + \frac{k}{n_y}}},$$

has a $t_{n_x+n_y-2}$ distribution.

(c) Hence determine a 95% confidence interval for $\mu_x - \mu_y$ if 2, 4, 3, 2 is an observed random sample from a $N(\mu_x, \sigma^2)$ population and 1, 1, 4, 6, 3 is an independent observed random sample from a $N(\mu_y, 5\sigma^2)$ population.

12.4 COMPARING K POPULATION MEANS (ANOVA)

The concluding remarks of the last section addressed one of the most widely used statistical tests: the two-sample t test. In this t test, two population means μ_1 and μ_2 are compared by using the test statistic

$$T = \frac{\overline{X} - \overline{Y} - (\mu_1 - \mu_2)}{S_p\sqrt{\frac{1}{m} + \frac{1}{n}}} \sim t_{m+n-2},$$

where

$$S_p^2 = \frac{(m-1)S_1^2 + (n-1)S_2^2}{m+n-2}$$

is the pooled sample variance calculated on the basis of both samples.

In this section, this test is generalised so that the means of $k > 2$ populations may be compared. If μ_i denotes the mean of population i for each $i = 1, 2, \ldots, k$, the focus shifts to the hypothesis test

$$H_o : \mu_1 = \mu_2 = \mu_3 = \ldots = \mu_k$$

against an alternative hypothesis that at least two of the population means differ. The test is performed on the collection of samples taken independently, one from each of the k populations. Surprisingly, such a comparison of population means is termed **analysis of variance**, or **ANOVA** for short. This springs from the notion that it is dangerous to compare averages unless variability about those averages is fully understood. When there are k samples at our disposal to use in the population mean comparisons, variability occurs both within, and between, each sample.

When we set up our own experiments, we hope to use this sample/population structure. Careful consideration of how the populations are organised and how the samples are gathered may lead to increased accuracy of experimental results. This is a worthy goal.

Treatments

In the language of experimental design the different populations are referred to as **treatments**; this calls to mind experiments where the experimental units are divided into several groups and different sets of experimental conditions are administered to each group. The experimental units may be randomly assigned to these groups so that each group constitutes a realised random sample from a treatment population. For example, a group of cancer patients may be subdivided into three groups where Group 1 receives Drug 1 (and is a random sample from Treatment 1), Group 2 receives Drug 2 (and is a random sample from Treatment 2) and Group 3 receives the placebo, or no treatment (and is a random sample from Treatment 3). If we use an 'x' to denote the observed measurement (perhaps blood pressure) on the experimental unit, in this case a patient, and assign 10 patients to each treatment, then we obtain the structure:

Treatment	1 Drug 1	2 Drug 2	3 Placebo
Patient 1	x	x	x
Patient 2	x	x	x
Patient 3	x	x	x
Patient 4	x	x	x
⋮	⋮	⋮	⋮
Patient 10	x	x	x

This shows an arrangement of three independent samples of size 10. If Drug 1 is prohibitively expensive, we may be forced to opt for fewer patients to represent Treatment 1. The design

Treatment	1 Drug 1	2 Drug 2	3 Placebo
Patient 1	x	x	x
Patient 2	x	x	x
Patient 3	x	x	x
Patient 4	x	x	x
Patient 5	x	x	x
Patient 6	x	x	x
Patient 7		x	x
Patient 8		x	x
Patient 9		x	x
Patient 10		x	x
Patient 11		x	x
Patient 12		x	
Patient 13		x	

shows $n_1 = 6$ patients assigned to Treatment 1, $n_2 = 13$ patients assigned to Treatment 2 and $n_3 = 11$ patients assigned to Treatment 3. The aim of this designed experiment would be to test whether the mean blood pressure is the same for each treatment. We have used an 'x' to denote a blood pressure measurement, however applications are quite general. In other circumstances, the three treatments described above may represent different fertilisers applied to experimental units (plots of land) in preparation for 'crop yield' measurements 'x'.

The above tables show that with $k > 2$ treatments and random samples within each treatment, the complete data set is two-dimensional. We begin by concentrating on the data structure, because it is on the *structure* of the data set that the hypothesis test for comparing k means is based.

In the one-dimensional case, we may extract a single observed random sample $x_1, x_2, x_3, \ldots, x_n$ from a $N(\mu, \sigma^2)$ population. According to Definition 7.13, when $\overline{x}$ is used to estimate μ, the ith residual is given by

$$e_i = x_i - \overline{x} \qquad (i = 1, 2, 3, \ldots n),$$

and measures how far the ith data point lies from the sample mean. We may write

$$x_i = \overline{x} + (x_i - \overline{x}), \tag{12.9}$$

or $x_i = \overline{x} + e_i$. This forces the e_i to satisfy

$$\sum_{i=1}^{n} e_i = \sum_{i=1}^{n} (x_i - \overline{x}) = 0 \quad \text{and} \quad \sum_{i=1}^{n} e_i^2 = \sum_{i=1}^{n} (x_i - \overline{x})^2 = (n-1)s^2,$$

where s^2 is the observed sample variance. This way of viewing our data suggests that $X_i = \mu + \epsilon_i$, where ϵ_i are *i.i.d.* $N(0, \sigma^2)$, are the hypothetical random variables which generate the observed residuals.

Data arrays and dot notation

Now we apply the same considerations to the two-dimensional case. The two-dimensional structured list of data points

Treatment	**1**	**2**	…	k
	Sample 1	**Sample 2**	…	**Sample** k
	x_{11}	x_{12}	…	x_{1k}
	x_{21}	x_{22}	…	x_{2k}
	x_{31}	x_{32}	…	x_{3k}
	⋮	⋮		⋮
			…	$x_{n_k k}$
		⋮		
	$x_{n_1 1}$			
		$x_{n_2 2}$		

is called a **data array** for an observed random sample of size n_i from Treatment i, where i varies from 1 to k. Often the headings in the array are omitted and we write

$$\begin{bmatrix} x_{11} & x_{12} & \cdots & x_{1k} \\ x_{21} & x_{22} & \cdots & x_{2k} \\ x_{31} & x_{32} & \cdots & x_{3k} \\ \vdots & \vdots & & \vdots \\ & & \cdots & x_{n_k k} \\ x_{n_1 1} & \vdots & & \\ & x_{n_2 2} & & \end{bmatrix}$$

or even

$$[x_{ij}]$$

as a useful abbreviation for the entire array. In this notation, the entry x_{ij} in the data array denotes the ith member of the observed sample taken from population j, or, the ith observation on the jth treatment. (In all double subscript notation, the subscript representing the row position precedes the subscript representing the column position.) Notice that there is no requirement that the sample sizes be equal; that is, the columns in a data array may have differing numbers of entries. With the usual lower/upper case notation to represent data/variable, x_{ij} is the observed value of random variable X_{ij}. In a data array, dot notation is used to indicate column totals and the total of all data in the array. The subscript '.' in a row or column position indicates that the row or column has been summed across.

$$x_{.j} = \sum_{i=1}^{n_j} x_{ij} \quad \text{and} \quad \overline{x_{.j}} = \frac{\sum_{i=1}^{n_j} x_{ij}}{n_j}.$$

$$x_{..} = \sum_{j=1}^{k} \sum_{i=1}^{n_j} x_{ij} \quad \text{and} \quad \overline{x_{..}} = \frac{\sum_{j=1}^{k} \sum_{i=1}^{n_j} x_{ij}}{n} \quad \text{for} \quad n = \sum_{j=1}^{k} n_j.$$

The notation $x_{i.}$ is similarly defined when all the rows in the array have the same number of entries. The dot notation is most convenient for analysing variability both within each sample and in the sample means as we move across the array. In a generalisation of Equation (12.9) to take account of both the row and column structure, we write

$$x_{ij} = \overline{x_{..}} + (\overline{x_{.j}} - \overline{x_{..}}) + (x_{ij} - \overline{x_{.j}}),$$

where

x_{ij}	denotes the data point in row i and column j,
$\overline{x_{..}}$	denotes the **grand sample mean** of all data in the array,
$\overline{x_{.j}} - \overline{x_{..}}$	denotes the **sample treatment effect**,
$x_{ij} - \overline{x_{.j}}$	denotes the **observed residual.**

Equation (12.10) shows how to split the data array into the sum of three separate arrays: the **grand mean array**, the **treatment effect array** and the **residual array**. This array structure is summarised as

$$[x_{ij}] = [\overline{x_{..}}] + [\overline{x_{.j}} - \overline{x_{..}}] + [x_{ij} - \overline{x_{.j}}]. \tag{12.10}$$

Example 12.13
Constructing data arrays from experimental data

Guttman and Wilks (1965) give the following data when four thermometers T_1, T_2, T_3 and T_4 were used to make determinations of the melting point of hydroquinine in degrees Celsius.

T_1	T_2	T_3	T_4
174.0	173.0	171.5	173.5
173.0	172.0	171.0	171.0
173.5		173.0	
173.0			

We wish to examine how the data points in this array vary according to their row and column positions.

The four populations or treatments are clearly labelled, so that $k = 4$. The experimental results give random samples of differing sizes for each treatment: $n_1 = 4$, $n_2 = 2$, $n_3 = 3$ and $n_4 = 2$.

Let x_{ij} denote the entry in row i and column j. The column averages (or sample means) are

$$\overline{x_{.1}} = \frac{\sum_{i=1}^{n_1} x_{i1}}{n_1} = \frac{174.0+173.0+173.5+173.0}{4} = 173.375$$
$$\overline{x_{.2}} = \frac{\sum_{i=1}^{n_2} x_{i2}}{n_2} = \frac{173.0+172.0}{2} = 172.500$$
$$\overline{x_{.3}} = \frac{\sum_{i=1}^{n_3} x_{i3}}{n_3} = \frac{171.5+171.0+173.0}{3} = 171.833$$
$$\overline{x_{.4}} = \frac{\sum_{i=1}^{n_4} x_{i4}}{n_4} = \frac{173.5+171.0}{2} = 172.250$$

when determined to 3 decimal places. The average of all 11 data points is $\overline{x_{..}} = 172.591$ (3 dp). This gives observed treatment effects

$$\overline{x_{.1}} - \overline{x_{..}} = 173.375 - 172.591 = 0.784$$
$$\overline{x_{.2}} - \overline{x_{..}} = 172.500 - 172.591 = -0.091$$
$$\overline{x_{.3}} - \overline{x_{..}} = 171.833 - 172.591 = -0.758$$
$$\overline{x_{.4}} - \overline{x_{..}} = 172.250 - 172.591 = -0.341.$$

In array form (12.10), $[x_{ij}] = [\overline{x_{..}}] + [\overline{x_{.j}} - \overline{x_{..}}] + [x_{ij} - \overline{x_{.j}}]$, becomes

$$\begin{bmatrix} 174.0 & 173.0 & 171.5 & 173.5 \\ 173.0 & 172.0 & 171.0 & 171.0 \\ 173.5 & & 173.0 & \\ 173.0 & & & \end{bmatrix}$$

$$= \begin{bmatrix} 172.591 & 172.591 & 172.591 & 172.591 \\ 172.591 & 172.591 & 172.591 & 172.591 \\ 172.591 & & 172.591 & \\ 172.591 & & & \end{bmatrix} + \begin{bmatrix} 0.784 & -0.091 & -0.758 & -0.341 \\ 0.784 & -0.091 & -0.758 & -0.341 \\ 0.784 & & -0.758 & \\ 0.784 & & & \end{bmatrix}$$

$$+ \begin{bmatrix} 0.625 & 0.500 & -0.333 & 1.250 \\ -0.375 & -0.500 & -0.833 & -1.250 \\ 0.125 & & 1.167 & \\ -0.375 & & & \end{bmatrix}$$

That is:

data array = grand mean array + sample treatment effect array + observed residual array.

There are some interesting points to notice about these arrays: all entries in the treatment effect array should sum to zero; all entries in the residual array should sum to zero. (Slight inaccuracies are incurred by rounding.) Note that the individual treatment effects for the four treatments should not sum to zero.

Model for comparing population means

The data array break-up, x_{ij} = grand mean + treatment effect + residual, expressed succinctly in Equation (12.10) suggests a normal probability model of the form

$$X_{ij} = \mu + \delta_j + \epsilon_{ij} \text{ where } \epsilon_{ij} \text{ are } i.i.d.\ N(0, \sigma^2). \tag{12.11}$$

The hypothetical random variables ϵ_{ij} generate the observed residuals; the parameter μ may be interpreted as an 'overall' population mean; parameter δ_j denotes the **population treatment effect** of the jth treatment. To ensure that such effects cancel when considering the array as a whole, the treatment effects must satisfy the constraint $\sum_{j=1}^{k} n_j \delta_j = 0$. In terms of this notation, the mean μ_j of treatment j may be expressed as $\mu_j = \mu + \delta_j$.

Definition 12.3 *By a* **one-way ANOVA model** *we mean the probability model description (12.11). This probability model is* **fitted** *to an observed data array* $[x_{ij}]$ *by constructing the array* $[\widehat{x_{ij}}]$ *of* **fitted values** *given by*

$$[\widehat{x_{ij}}] = [\overline{x_{..}}] + [\overline{x_{.j}} - \overline{x_{..}}].$$

□

Notice that the fitted values array simplifies as $[\widehat{x_{ij}}] = [\overline{x_{..}}] + [\overline{x_{.j}} - \overline{x_{..}}] = [\overline{x_{.j}}]$, and that the elements, e_{ij}, of the observed residual array are indeed the differences between the observed and fitted values:

$$[e_{ij}] = [x_{ij} - \widehat{x_{ij}}] = [x_{ij} - \overline{x_{.j}}].$$

To see this in the numerical context of Example 12.13, the model

$$X_{ij} = \mu + \delta_j + \epsilon_{ij} \quad \text{where for each } i \text{ and } j, \quad \epsilon_{ij} \text{ are } i.i.d. \ N(0, \sigma^2)$$

when fitted to the data array

$$\begin{bmatrix} 174.0 & 173.0 & 171.5 & 173.5 \\ 173.0 & 172.0 & 171.0 & 171.0 \\ 173.5 & & 173.0 & \\ 173.0 & & & \end{bmatrix}$$

produces the grand mean array and treatment effect array

$$\begin{bmatrix} 172.591 & 172.591 & 172.591 & 172.591 \\ 172.591 & 172.591 & 172.591 & 172.591 \\ 172.591 & & 172.591 & \\ 172.591 & & & \end{bmatrix} \quad \begin{bmatrix} 0.784 & -0.091 & -0.758 & -0.341 \\ 0.784 & -0.091 & -0.758 & -0.341 \\ 0.784 & & -0.758 & \\ 0.784 & & & \end{bmatrix}$$

respectively. These add together to give the array of fitted values:

$$[\widehat{x_{ij}}] = \begin{bmatrix} 173.375 & 172.5 & 171.833 & 172.25 \\ 173.375 & 172.5 & 171.833 & 172.25 \\ 173.375 & & 171.833 & \\ 173.375 & & & \end{bmatrix}$$

When the probability model fits the data well, the observed data points x_{ij} will be 'close' to the fitted data points $\widehat{x_{ij}}$ thus producing small observed residuals e_{ij}.

Implicit in the ANOVA model is

Assumption 1: Normally distributed data.

Assumption 2: Independent samples from each of the treatments.

Assumption 3: σ^2 is the population variance common to each treatment.

We shall see that the normality assumption is necessary for the construction of a successful test statistic for $H_o : \mu_1 = \mu_2 = \ldots = \mu_k$ or, equivalently, $H_o : \delta_1 = \delta_2 = \ldots = \delta_k = 0$, against an alternative that at least two of the treatment effects differ. When H_o is rejected, so that at least some treatment effects are present, point estimates of the parameters δ_j and μ_j are taken directly from the fitted data arrays:

$$\widehat{\delta_j} = \overline{x_{.j}} - \overline{x_{..}}$$
$$\widehat{\mu_j} = \overline{x_{.j}}$$

ANOVA

Consider (12.10) rearranged in the form

$$x_{ij} - \overline{x_{..}} = (\overline{x_{.j}} - \overline{x_{..}}) + (x_{ij} - \overline{x_{.j}}).$$

When each side of this equation is squared and all terms summed across every array entry (that is for all values of i and for all values of j), all cross-product terms on the right-side of the resulting equation vanish and we obtain the remarkable result

$$\sum_{j=1}^{k}\sum_{i=1}^{n_j}(x_{ij} - \overline{x_{..}})^2 = \sum_{j=1}^{k}\sum_{i=1}^{n_j}(\overline{x_{.j}} - \overline{x_{..}})^2 + \sum_{j=1}^{k}\sum_{i=1}^{n_j}(x_{ij} - \overline{x_{.j}})^2 \qquad (12.12)$$

which leads to the title ANOVA (analysis of variance). It is worth considering each term in this formula separately. The left-side sum of squares is the **total sum of squares**, for it measures the total variability of all array entries relative to the array mean. It is conveniently abbreviated SS TOTAL:

$$\text{SS TOTAL} = \sum_{j=1}^{k}\sum_{i=1}^{n_j}(x_{ij} - \overline{x_{..}})^2.$$

Next

$$\text{SS TREATMENT} = \sum_{j=1}^{k}\sum_{i=1}^{n_j}(\overline{x_{.j}} - \overline{x_{..}})^2 = \sum_{j=1}^{k} n_j(\overline{x_{.j}} - \overline{x_{..}})^2,$$

the so-called **treatment sum of squares** reflects variation between the samples by comparing how the individual sample means relate to the grand mean through adding the squared deviations. Suppose that s_j^2 is the sample variance of the jth sample. The last formula on the right-side of Equation (12.12) is the **residual sum of squares**

$$\text{SS RESIDUALS} = \sum_{j=1}^{k}\sum_{i=1}^{n_j}(x_{ij} - \overline{x_{.j}})^2$$

and gives an overall measure of how well the model (12.11) fits the array. (For this reason many authors use the unfortunate title 'error sum of squares'.) If s_j^2 is the sample variance of the jth sample, then SS RESIDUALS may be re-expressed as

$$\text{SS RESIDUALS} = \sum_{j=1}^{k}(n_j - 1)s_j^2.$$

In summary: SS TOTAL = SS TREATMENT + SS RESIDUALS.

When $H_o : \mu_1 = \mu_2 = \mu_3 = \ldots = \mu_k$ is true, according to Theorem 10.5, the normality and independence assumptions of the model (12.11) give

$$\frac{\text{SS TOTAL}}{\sigma^2} \sim \chi^2_{n-1}, \quad \frac{\text{SS TREATMENT}}{\sigma^2} \sim \chi^2_{k-1}, \quad \text{and} \quad \frac{\text{SS RESIDUALS}}{\sigma^2} \sim \chi^2_{n-k},$$

the distributions being independent. ($n = \sum_{j=1}^{k} n_j$.) The degrees of freedom associated with these chi-squared distributions are additive: abbreviating 'degrees of freedom' as d of f, we may write

$$\text{d of f SS TOTAL} = \text{d of f SS TREATMENT} + \text{d of f SS RESIDUALS},$$

since $n - 1 = (k - 1) + (n - k)$. Further, since the mean of a chi-squared distribution is simply the degrees of freedom,

$$E(\frac{\text{SS RESIDUALS}}{\sigma^2}) = n - k, \quad \text{and} \quad E(\frac{\text{SS TREATMENTS}}{\sigma^2}) = k - 1,$$

which provides a choice of two unbiased mean squared estimators of σ^2:

$$\text{MS RESIDUALS} = \frac{\text{SS RESIDUALS}}{n - k},$$

$$\text{MS TREATMENTS} = \frac{\text{SS TREATMENTS}}{k - 1}.$$

Therefore we anticipate that the ratio $\frac{\text{MS TREATMENTS}}{\text{MS RESIDUALS}}$ should be close to 1 in value. How close? Measure with the F distribution of Definition 12.1: when H_o is true,

$$F = \frac{\text{MS TREATMENTS}}{\text{MS RESIDUALS}} \sim F_{k-1,n-k}.$$

F is the ANOVA test statistic.

What happens when H_o is not true? Essentially, by the same types of arguments which showed sample variances to be unbiased, it can be shown that still $E(\text{MS RESIDUALS}) = \sigma^2$, but the non-zero treatment effects emerge in the expected mean treatment sum of squares.

$$E(\text{MS TREATMENTS}) = \sigma^2 + \frac{\sum_{i=1}^{k} n_j \delta_j^2}{k - 1}.$$

The ratio of $E(\text{MS TREATMENTS})$ to $E(\text{MS RESIDUALS})$ is therefore larger than 1, and this is an indication that the test statistic F is likely to be larger than 1 when H_o is false. That is, H_o is a one-sided test. Heuristically, if the variability between samples is much larger than the variability within each sample, so that the F ratio is much larger than 1, it is unlikely that the treatment means are the same.

The information required for this hypothesis test is gathered together in a table, called an **ANOVA table**. This table has the added advantage that it shows how the total variability in the data array is partitioned with a corresponding partition for the degrees of freedom.

Source	Degrees of freedom	Sums of squares	Mean squares	F
Treatments	$k-1$	SS TREATMENTS	MS TREATMENTS	$\frac{\text{MS TREATMENTS}}{\text{MS RESIDUALS}}$
Residuals	$n-k$	SS RESIDUALS	MS RESIDUALS	
Total	$n-1$	SS TOTAL		

Since Sample i has $n_i - 1$ degrees of freedom (being one less than the sample size), the total degrees of freedom associated with the residuals is the sum of the degrees of freedom associated with each sample:

$$\sum_{i=1}^{k}(n_i - 1) = \sum_{i=1}^{k} n_i - k = n - k,$$

as is suggested in the ANOVA table. When calculating an ANOVA table by hand (rather than by reading a computer output), it is easiest to first determine SS RESIDUAL from the sample variances s_j^2, $j = 1, 2, 3, \ldots, k$:

$$\text{SS RESIDUALS} = \sum_{j=1}^{k}(n_j - 1)s_j^2.$$

Next determine SS TOTAL from the hand calculation formula

$$\text{SS TOTAL} = \sum_{j=1}^{k} x_{ij}^2 - n\overline{x}_{..}^{\,2},$$

and finally, obtain SS TREATMENTS by subtraction. If required, a hand calculation formula for SS TREATMENTS is

$$\text{SS TREATMENTS} = \sum_{j=1}^{k} \frac{x_{.j}^2}{n_j} - n\overline{x}_{..}^{\,2}.$$

Example 12.14

Comparing batch means

Samples of 25 kg label weight packets were selected from 5 batches of packets of PVC powder. The discrepancy between contents weight and label weight was noted ($\times 10^{-2}$kg). The batches were produced in time order on a production line. However to apply the ANOVA procedures the 5 samples must be assumed independent, normally distributed with a common variance. The data was recorded as:

Batch 1	**Batch 2**	**Batch 3**	**Batch 4**	**Batch 5**
0	6	7	−25	1
−15	−28	−13	−45	10
8	5	−13	17	−45
36	5	−40	1	−15
7	−19	15	−14	−3
27	−10	−15	1	−28
10	25	17	−5	30
15	−14	−38	−50	11
15	10	−5	10	−5
24	0	−2	−35	−50
29	15	−37	10	9

The first step in a formal ANOVA procedure is to graph the data to assess whether the required assumptions are likely to hold. Comparative boxplots drawn against the same scale are ideal for this purpose.

B/W plot of fill discrepancy for 25 kg bags PVC powder ($\times 10^{-2}$*kg*)

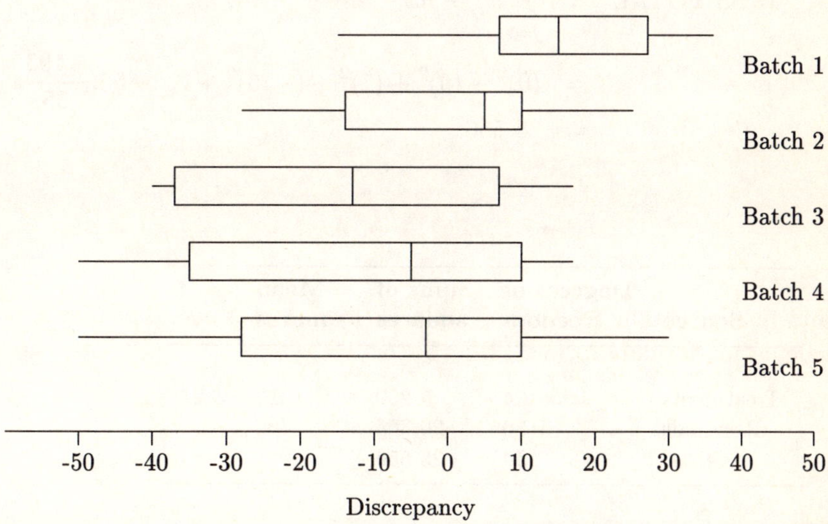

These plots highlight Batch 1 as centred visibly higher than the other batches. The samples tend to be left-skewed. Proceeding to an ANOVA for all 5 batches we obtain the following values for the summary statistics—sample total, sample mean, sample standard deviation:

	Batch 1	Batch 2	Batch 3	Batch 4	Batch 5
j	1	2	3	4	5
n_j	11	11	11	11	11
$x_{.j}$	156	−5	−124	−135	−85
$\overline{x_{.j}}$	14.18	−0.45	−11.27	−12.27	−7.73
s_j	14.55	15.73	20.42	23.30	24.73

$$\begin{aligned}\text{SS RESIDUALS} &= \sum_{j=1}^{k}(n_j-1)s_j^2 \\ &= (10)(14.55)^2+(10)(15.73)^2+(10)(20.42)^2 \\ &\quad +(10)(23.30)^2+(10)(24.73)^2 \\ &= 20\ 306.\end{aligned}$$

$$\begin{aligned}\text{SS TOTAL} &= \sum_{j=1}^{k} x_{ij}^2 - n\overline{x_{..}}^2 \\ &= (0)^2+(6)^2+(7)^2+(-25)^2+\ldots-55(\frac{-193}{55})^2 \\ &= 25\ 556.\end{aligned}$$

Source	Degrees of freedom	Sums of squares	Mean squares	F
Treatments	4	5 250	1312	3.23
Residuals	50	20 306	406	
Total	54	25 556		

Since $F_{0.05,4,50} = 2.57$, the p-value of the test $H_o : \mu_1 = \mu_2 = \mu_3 = \mu_4 = \mu_5$ (where μ_i is the mean discrepancy for Batch i) exceeds 0.05. Hence we reject H_o at this significance level and conclude that the means of at least two of the batches are significantly different. A glance at the boxplots for the separate samples indicates that Batch 1 is likely to be the cause of the rejection of H_o.

ANOVA assumptions

The F test performed from the ANOVA table remains approximately valid for moderately large samples from distinctly non-normal populations. Further, the constancy of σ^2 across each treatment may be relaxed without destroying the usefulness of the F test. In Example 12.14, the samples from Batches 1 and 2 appear to have much smaller standard deviations than the samples from the remaining batches, but the pertinent question is: Has the dispensing equipment filling the packets really changed the *accuracy* of fill from batch to batch? The independence of the k samples representing the treatments remains an important consideration.

Estimating model parameters

$$\text{MS RESIDUALS} = \frac{\sum_{j=1}^{k}(n_j - 1)S_j^2}{n-k},$$

which may be read directly from the ANOVA table, is the best estimator of the unknown error variance σ^2 common to all treatments. When $k = 2$, MS RESIDUALS is identical to the formula

$$S_p^2 = \frac{(n_1 - 1)S_1^2 + (n_2 - 1)S_2^2}{n_1 + n_2 - 2}$$

for the two-sample t test of the last section. The treatment means μ_j are point estimated by the observed sample means from each treatment, $\widehat{\mu_j} = \overline{X_{.j}}$ and interval estimated with confidence intervals

$$\left[\overline{X_{.j}} - t_{\frac{\alpha}{2},n-k}\sqrt{\frac{\text{MS RESIDUALS}}{n_j}},\ \overline{X_{.j}} + t_{\frac{\alpha}{2},n-k}\sqrt{\frac{\text{MS RESIDUALS}}{n_j}}\right],$$

which have width based on MS RESIDUALS, and the degrees of freedom $n - k$ associated with this quantity. Many computer packages produce such confidence intervals automatically with ANOVA output.

When H_o is rejected, so that there exist at least two treatment means that are significantly different, then for assessing the difference between pairs $\mu_i - \mu_j$, we use the point estimate $\overline{X_{.i}} - \overline{X_{.j}}$ and the interval estimate

$$\Bigg[\overline{X_{.i}} - \overline{X_{.j}} - t_{\frac{\alpha}{2},n-k}\sqrt{\text{MS RESIDUALS}(\frac{1}{n_i} + \frac{1}{n_j})},$$
$$\overline{X_{.i}} - \overline{X_{.j}} + t_{\frac{\alpha}{2},n-k}\sqrt{\text{MS RESIDUALS}(\frac{1}{n_i} + \frac{1}{n_j})}\Bigg].$$

Notice that if all $c = \binom{k}{2}$ pairs of treatment means are to be **simultaneously** compared in this way, then the overall level of confidence is no longer

$100(1-\alpha)$: indeed, for these c confidence intervals, if R_i is the event that the ith interval fails to cover the required parameter, so that $P(R_i) = \alpha$, then the **Bonferroni Inequality** suggests the probability, α^*, that at least one of the c intervals does not cover the required parameter is

$$\begin{aligned} \alpha^* &= P(R_1 \cup R_2 \cup \ldots \cup R_c) \\ &\leq P(R_1) + P(R_2) + \ldots + P(R_c) \\ &= \alpha + \alpha + \ldots + \alpha \\ &= c\alpha. \end{aligned}$$

Therefore the simultaneous confidence that all c intervals cover their respective mean differences $\mu_i - \mu_j$, is **at least** $100(1 - c\alpha)\%$. For example, if $\alpha = 0.05$ and $k = 3$, the simultaneous confidence is only 85%.

EXERCISES

12.32 In an experiment to compare 3 treatments, 4 experimental subjects were randomly assigned to each treatment. Suppose that the first sample treatment effect is -8 and the second sample treatment effect is 10.

(a) Determine the value of the sample treatment effect for Treatment 3.

(b) If $\overline{x}_{..} = 10$, write down the grand mean array.

(c) From (a) determine the array of sample treatment effects.

(d) Suppose that the residual array is as follows:

$$\begin{bmatrix} 0 & -1 & -2 \\ 0 & 1 & 0 \\ -1 & 0 & 1 \\ 1 & 0 & 1 \end{bmatrix}$$

Determine the original data.

12.33 An experiment is designed to examine the effect of different types of surface coatings of television tubes on their conductivity.

The televisions to be tested were randomly assigned to the treatments. The following is a partially completed ANOVA for the experiment.

Source	d of f	Sums of squares	Mean squares	F
Treatments	3	1 134	378	-
Residuals	-	-	-	
Total	19	1 342		

(a) Complete the ANOVA table.

(b) Is there a significant difference in average conductivity between the different coatings?

(c) If columns denote separate treatments, which of the following is a possible structure for the data array which may have been obtained in this experiment?

(A)

x	x	x	x
x	x	x	x
x	x	x	x
x	x	x	
x		x	
x		x	

(B)

x	x	x
x	x	x
x	x	x
x	x	x
x	x	x
x	x	x
	x	x

(C)

x	x	x	x
x	x	x	x
x	x	x	x
	x	x	
	x	x	
	x	x	
	x		
	x		

(D)

x	x	x
x	x	x
x	x	x
x	x	x
x	x	x
x	x	
x	x	

12.34 Consider the melting point data of Example 12.13:

T_1	T_2	T_3	T_4
174.0	173.0	171.5	173.5
173.0	172.0	171.0	171.0
173.5		173.0	
173.0			

Produce a residual array $[e_{ij}]$ for this data when the ANOVA model is fitted.

Now consider the same data set, but with a recording entry error where 1000 has mistakenly been added to the data entry x_{11}.

T_1	T_2	T_3	T_4
1174.0	173.0	171.5	173.5
173.0	172.0	171.0	171.0
173.5		173.0	
173.0			

Produce a residual array $[e_{ij}]$ for this 'error' data when the ANOVA model is fitted.

Are we able to detect the error by an examination of the residuals? Explain briefly.

12.35 Determine the array $[\widehat{x_{ij}}]$ of fitted values when a one-way ANOVA model is applied to the following data which results when 100 is subtracted off all entries in the data array of Example 12.13.

T_1	T_2	T_3	T_4
74.0	73.0	71.5	73.5
73.0	72.0	71.0	71.0
73.5		73.0	
73.0			

12.36 Consider the batches of PVC powder in Example 12.14. Test whether the means of Batches 3, 4 and 5 are significantly different. Summarise your conclusions as (∗), (∗∗) or (*n.s.*)

(*Comment*: Since strategies for hypothesis tests are decided on before the data are seen—so as to avoid bias—imagine for the purpose of this question that Batches 3, 4 and 5 were the only batches for which data were obtained.)

12.37 Suppose that three different methods were used to determine the amount of oxygen dissolved in lake water. A water sample with a known dissolved oxygen concentration (10.80 mg L^{-1}) was analysed by the three methods, giving the following results.

Method 1	Method 2	Method 3
10.96	10.88	11.13
10.77	10.75	10.99
10.90	10.80	10.98
10.69	10.81	11.02
10.87	10.70	11.02
10.60	10.82	11.08

(a) Construct B/W plots for the three groups. Explain what is revealed by the plots.

(b) Show that there is a significant difference in the mean determination recorded by the three methods. Record your results in an ANOVA table.

12.38 An accelerated life test experiment (where experimental conditions cause the time scale to be compressed) compared the durability of three types of exterior housepaint. 10 wooden panels were coated with an oil-based paint (Paint 1); 10 different wooden panels were painted with a latex paint (Paint 2); a further 10 panels were painted with enamel (Paint 3).

For each $i = 1, 2, 3$, let the random variable X_i denote the durability score (compiled from qualitative and quantitative characteristics) for Paint i.

The following data were collected on these random variables:

Score for Paint 1 X_1	Score for Paint 2 X_2	Score for Paint 3 X_3
20	13	31
35	37	86
44	45	80
90	83	54
43	40	63
58	62	69
52	50	54
43	50	99
70	64	54
75	78	46

Assume that $X_i \sim N(\mu_i, \sigma^2)$, $i = 1, 2, 3$, so that μ_1, μ_2 and μ_3 may be different but the three paints have the same variability in score σ^2.

(a) Do the data support the assumption that $\sigma = 20$ for different paints? Explain with the help of box-and-whisker plots drawn on the same scale

(b) Construct an ANOVA table for this experiment.

(c) Show how to use the ANOVA table to test

$$H_o : \mu_1 = \mu_2 = \mu_3$$

against an alternative that at least two of the means differ.

(d) Give a point estimate for σ based on the three samples.

(e) Determine a 95% confidence interval for $\mu_1 - \mu_3$.

12.39 Construct the residual array $[e_{ij}]$ for the data of Exercise 12.37.

(a) Identify the largest residual(s). What percentages do they contribute to the overall size of SS RESIDUALS?

(b) State the sum of squares of all the entries in this residual array.

(c) If the e_{ij} were regarded as data points to which an ANOVA could be fitted, write down the resulting ANOVA table.

[*Hint*: Use the table of Exercise 12.37.]

12.40 The following results are raw weight loss test data (grams) from a series of laboratory abrasion tests on four materials.

Material G	Material H
35.3	41.4
86.7	18.9
72.4	44.8
74.6	34.1
85.1	37.7
79.3	40.0
80.5	27.8
69.7	25.7

Material Y1	Material Y2
13.4	5.7
19.9	5.2
18.0	6.4
18.6	4.5
17.9	11.4
15.4	11.1
17.1	10.7
15.8	5.0

(*Source*: Sare, I.R. and Constantine, A.G. (1991). Design and Analysis of Jaw Crusher Gouging and Abrasion Tests. *Journal of Testing and Evaluation*, 19, 115-122.)

Construct four B/W plots to compare the abrasion test results for the four materials after the data have been transformed by logarithms. (This improves symmetry so that ANOVA techniques may be applied.)

Test whether there is a significant difference between the means on the log scale.

12.41 Suppose that an ANOVA results in the following array of residuals:

$$\begin{bmatrix} 0 & -1 & -2 \\ 0 & 1 & 0 \\ -1 & 0 & 1 \\ 1 & 0 & 1 \end{bmatrix}$$

Does this residual array satisfy the equation

$$\sum_{i=1}^{n_j}(x_{ij} - \overline{x_{.j}}) = 0?$$

Explain briefly. (The equation is expressed in the usual notation where x_{ij} denotes the ith data point in the jth sample.)

12.42 Write down an ANOVA table for $k = 2$ treatment groups. Use the following notation: Sample 1 has size n_1, mean $\overline{X}$ and variance S_1^2 and is from the first treatment which has mean μ_1 and variance σ^2; Sample 2 has size n_2, mean $\overline{Y}$ and variance S_2^2 and is from the second treatment which has mean μ_2 and variance σ^2.

If F denotes the F statistic in the ANOVA table, and T denotes the two-sample t test statistic, show that

$$F = T^2.$$

12.43 When each side of the equation

$$x_{ij} - \overline{x_{..}} = (\overline{x_{.j}} - \overline{x_{..}}) + (x_{ij} - \overline{x_{.j}})$$

is squared and summed across all values of i and all values of j, show that the resulting cross-product terms

$$\sum_{j=1}^{k}\sum_{i=1}^{n_j}(\overline{x_{.j}} - \overline{x_{..}})(x_{ij} - \overline{x_{.j}})$$

are zero.

[*Hint*: $\sum_{i=1}^{n}(a_i - \overline{a}) = 0$ for any constants $a_1, a_2, a_3, \ldots, a_n$.]

CHAPTER THIRTEEN

Categorical Data

In this chapter you will find

13.1 CHECKING THE FIT OF PROBABILITY MODELS

The 'goodness-of-fit' of a probability model to a data set (of observations on a continuous variable) was discussed in Section 7.5 through the superposition of a probability density function over a relative frequency histogram. Hanging rootograms provide a pictorial means of assessing adequacy of fit of the model to the data.

In this section we begin by fitting probability models to frequency data which occur in categories. Such data are termed **categorical data**. A chi-squared test involving standardised residuals is developed to formally test whether the proposed model adequately fits the data.

The multinomial model

We want our goodness-of-fit test to be widely applicable, so we endeavour to make few restrictions:

1. There are k categories (which we will call BINS).
2. Observed data must fall into one and only one BIN.

Our measurement is recorded on only the most basic scale; in Chapter 1, it was called the **nominal scale**. In fact scales (ordinal, interval and ratio) all possess the basic qualities of the nominal scale that we seek: that is, that there are BINS. We wish to record frequencies of occurrence in each BIN.

Example 13.1
BINS
200 temperatures are recorded on different scales of measurement. On a **nominal scale**

BIN 1	cold	67
BIN 2	hot	133
Total		200

the BINS are simply two categories. On an **ordinal scale**

BIN 1	very cold	25
BIN 2	cold	42
BIN 3	hot	71
BIN 4	very hot	62
Total		200

there is an inherent ordering between the BINS. The data are qualitative and relate to levels of a quantitative temperature variable. On **interval** or **ratio scales**

BIN 1	$[-20, 0)$	30
BIN 2	$[0, 20)$	39
BIN 3	$[20, 40)$	41
BIN 4	$[40, 60)$	57
BIN 5	$[60, 80)$	33
Total		200

the data are quantitative, being observations on the continuous variable 'temperature'. In this last representation, the BINS are intervals in a frequency distribution.

The question of most concern is: Are these observed (relative) frequencies of occurrence compatible with hypothesised probabilities of data being found in a particular BIN?

In Section 8.2, the great multinomial experiment consisted of throwing n balls into k BINS: on record is the number X_i landing in BIN number i.

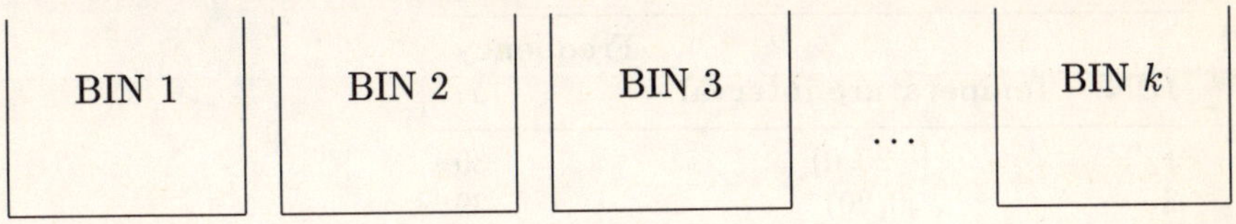

If we label $p_i = P(\text{ball lands in BIN } i)$, then, under the rules of no spillage (each data point is uniquely assigned to a BIN), so that all the balls land in *some* BIN, it follows that

$$P(X_1 = x_1, X_2 = x_2, \ldots, X_x = x_k) = \frac{n!}{x_1!x_2!\ldots x_k!} p_1^{x_1} p_2^{x_2} \ldots p_k^{x_k} \qquad (13.1)$$

represents the probability that x_1 balls land in BIN 1, x_2 balls land in BIN 2, $\ldots x_{k-1}$ balls land in BIN $(k-1)$ and $x_k = n - (x_1 + x_2 + \ldots + x_{k-1})$ balls land in BIN k. That is

$$(X_1, X_2, X_3, \ldots, X_k) \sim \mathit{Multinomial}(n, p_1, p_2, p_3, \cdots, p_{k-1}),$$

where $\sum_{i=1}^{k} X_i = n$ and $\sum_{i=1}^{k} p_i = 1$. We will also use the following information which was established in Section 8.2 and becomes important to us now:

> Marginally, the observed count X_i in BIN i is binomial: $X_i \sim Bin(n, p_i)$with $E(X_i) = np_i$, $Var(X_i) = np_i(1 - p_i)$, for $i = 1, 2, 3, \ldots, k$.

How can we actually use this multinomial model?

If the probabilities $p_1, p_2, p_3, \cdots, p_k$ are known, then the observed BIN counts x_i may be compared with the BIN counts, $\widehat{x}_i$, fitted by the probability model—these are the BIN counts that we would expect to get if the multinomial model is correct; the fitted values are the known quantities $E(X_i) = np_i$. If observed and fitted values are 'close', then the model fits well. The ith **observed residual** e_i is defined as

$$e_i = x_i - \widehat{x}_i = \text{observed BIN i count} - \text{fitted BIN i count} = x_i - np_i.$$

Example 13.2

Determining fitted values from a probability model

Consider the following hypothesised probability model for temperatures, Y, in degrees Celsius. Suppose that we expect temperature measurements to be uniformly distributed in the interval $[-20, 80)$ of width 100 degrees. That is, $Y \sim U[-20, 80]$. Is this in conflict with the observed data encountered in Example 13.1?

BIN	Temperature interval	Frequency x_i
1	$[-20, 0)$	30
2	$[0, 20)$	39
3	$[20, 40)$	41
4	$[40, 60)$	57
5	$[60, 80)$	33
Total		200

Assuming that the model $Y \sim U$[-20,80] is correct, the probability of an observation occurring in the first interval (or BIN 1) $[-20, 0)$, of width 20, is $(\frac{1}{100})(20) = \frac{1}{5}$. That is, $p_1 = \frac{1}{5}$. Then the number that we would expect to see in the first interval (BIN 1) is $np_1 = (200)(\frac{1}{5}) = 40$. Therefore $\widehat{x_1} = 40$. Similarly:

$$p_2 = P(0 < X < 20) = \frac{1}{5}; \quad \widehat{x_2} = np_2 = (200)(\frac{1}{5}) = 40\ldots$$

In fact, since the intervals are of equal width, $p_1 = p_2 = \ldots p_5 = \frac{1}{5}$. Values of $\widehat{x}_i$ are often placed in parentheses alongside the observed values x_i in the frequency distribution:

BIN	Temperature interval	Observed x_i	Fitted $(\widehat{x}_i)$	Residual e_i
1	$[-20, 0)$	30	(40)	−10
2	$[0, 20)$	39	(40)	−1
3	$[20, 40)$	41	(40)	1
4	$[40, 60)$	57	(40)	17
5	$[60, 80)$	33	(40)	−7
Total		200	(200)	0

The simple formation of differences is not a sufficient overall measure for assessing a good fit: the difference of 1 found in an 'observed' – 'fitted' residual difference 1000−999 seems (as a percentage of the size of the numbers involved) not as large as the difference of 1 from the residual difference 1.999−0.999. This is overcome by standardising the residuals.

Definition 13.1 *If observed residuals* $e_1, e_2, e_3, \ldots, e_k$ *are such that* e_i *is an observation on a random variable with mean* μ_i *and variance* σ_i^2*, then the* **observed standardised residuals** *are given by*

$$e_i^* = \frac{e_i - \mu_i}{\sigma_i}, \quad i = 1, 2, 3, \ldots k.$$

□

Since such residuals have mean 0 and variance 1, intuitively it seems reasonable to consider a standardised residual as 'large' if it exceeds 2 (standard deviations) in absolute value. In the current application,

$$E_i = X_i - np_i$$

have mean $\mu_i = 0$ and $\sigma_i^2 = Var(E_i) = np_i(1-p_i) \approx np_i$. The approximation seems reasonable if n is large and p_i small so that there is an expected BIN count in each BIN of at least 5. Then the standardised residual e_i^* may be written approximately as

$$e_i^* = \frac{x_i - np_i}{\sqrt{np_i}}, \quad i = 1, 2, 3, \ldots k.$$

An overall measure of fit then consists of examining whether the sum of squares of these standardised residuals is too large. This measure is the goodness-of-fit test statistic

$$Q = \sum_{i=1}^{k} (e_i^*)^2 = \sum_{i=1}^{k} (\frac{x_i - np_i}{\sqrt{np_i}})^2 = \sum_{i=1}^{k} \frac{(x_i - np_i)^2}{np_i}.$$

In casual notation Q is often written as

$$Q = \sum_{\text{all BINS}} \frac{(\text{observed BIN count} - \text{fitted BIN count})^2}{\text{fitted BIN count}}. \tag{13.2}$$

The probability distribution of Q is given by the following theorem.

Theorem 13.1 *If the frequency counts* X_i *in BIN* i *satisfy, for each* i, $E(X_i) = np_i \geq 5$, *then when* n *is large, the probability distribution of* Q *is approximately chi-squared:*

$$Q = \sum_{i=1}^{k} \frac{(X_i - np_i)^2}{np_i} \sim \chi^2_{k-1}.$$

Proof: For illustration, the proof is given here for $k = 2$ categories only: the relationships $X_1 + X_2 = n$ and $p_1 + p_2 = 1$ imply that

$$\begin{aligned} Q &= \sum_{i=1}^{k} \frac{(X_i - np_i)^2}{np_i} \\ &= \frac{(X_1 - np_1)^2}{np_1} + \frac{(X_2 - np_2)^2}{np_2} \\ &= \frac{(X_1 - np_1)^2}{np_1} + \frac{[(n - X_1) - n(1 - p_1)]^2}{n(1 - p_1)} \\ &= \frac{(X_1 - np_1)^2}{np_1(1 - p_1)} \\ &= (\frac{X_1 - np_1}{\sqrt{np_1(1 - p_1)}})^2 \\ &= Z^2, \quad \text{where } Z \sim N(0,1) \text{ approximately.} \end{aligned}$$

(It is at this last line that we use the fact that a $Bin(n, p_1)$ random variable is approximately normally distributed for large sample sizes.) Finally, by (10.4), the square of a standard normal random variable has a chi-squared distribution with 1 degree of freedom.

■

Theorem 13.1 is the test statistic to check whether hypothesised probabilities p_{io}, for $i = 1, 2, 3 \ldots, k$, are appropriate for BIN 1, BIN 2, ..., BIN k. The hypothesis test is one-tailed (in the upper tail) since when the fitted probabilities p_{io} are at odds with the observed BIN counts x_i, the standardised residuals, and hence the sum of squares of such residuals, is large. That is, we reject

$$H_0 : p_1 = p_{1o}, p_2 = p_{2o}, \ldots p_k = p_{ko}, \quad \text{where } \sum_{i=1}^{k} p_{io} = 1,$$

in favour of an alternative hypothesis that the BIN probabilities are not as specified, if

$$q_{observed} > \chi^2_{\alpha,k-1}.$$

The degrees of freedom, ν, required for this hypothesis test are easily remembered by drawing a schematic diagram of all the BINS and flagging those BINS whose counts must be known in order that the counts of all BINS be known **assuming a fixed total count of all BINS combined**. For example, for 6 BINS

x	x	x	x	x	

five are flagged, giving $\nu = 6 - 1 = 5$: once the totals of five of the BINS have been freely chosen, the totals of all six BINS are fixed—since they must add to the fixed total. Such reasoning makes the term **degrees of freedom** easier to understand.

Example 13.3

Quarterly hospital births

Suppose that the number of births at a hospital in four consecutive quarterly periods were recorded as:

Quarter	Jan-Mar	Apr-June	July-Sept	Oct-Dec
Number of births	110	57	53	80

The hospital board claims that twice as many babies are born in the Jan-Mar quarter as are born in any one of the other three quarters. The question of interest is: Do the data strongly contradict the stated conjecture?

Moulding the blueprint used for hypothesis tests of means in Section 11.3 to a goodness-of-fit testing framework, we proceed as follows:

1. In a goodness-of-fit test, the four quarters are represented by four BINS, where we denote p_i as the probability that a birth occurs in quarter (or BIN) number i.
2. The null hypothesis may be phrased as $H_0 : p_1 = \frac{2}{5}, p_2 = \frac{1}{5}, p_3 = \frac{1}{5}, p_4 = \frac{1}{5}$, against an alternative hypothesis that these BIN probabilities are not correct.
3. $Q \sim \chi_3^2$ when H_0 is true.
4. Take $\alpha = 0.05$.
5. When H_0 is true, of the 300 observed births for the period, $p_{1_0} n = (\frac{2}{5})(300) = 120$, are fitted to BIN 1; $(\frac{1}{5})(300) = 60$ to each of the other three BINS. Then:

BIN Quarter	1 Jan-Mar	2 Apr-June	3 July-Sept	4 Oct-Dec
Number of births	110 (120)	57 (60)	53 (60)	80 (60)

$$
\begin{aligned}
q_{observed} &= \sum_{\text{all BINS}} \frac{(\text{observed BIN count} - \text{fitted BIN count})^2}{\text{fitted BIN count}} \\
&= \frac{(110-120)^2}{120} + \frac{(57-60)^2}{60} + \frac{(53-60)^2}{60} + \frac{(80-60)^2}{60} \\
&= 8.47.
\end{aligned}
$$

6. Since $\chi^2_{0.05,3} = 7.815$, and $\chi^2_{0.025,3} = 9.348$, it follows that for this one-tailed test
$$0.025 < p\text{-value} < 0.05.$$
7. Since p-value $< \alpha$ we reject H_0 at the 0.05 significance level. The data are in conflict with the claim.
8. Summary: $q_{observed} = 8.47$ (*).

Many authors relax the condition that the fitted BIN count should be at least 5 for all BINS; that is $\widehat{x}_i = np_{io} \geq 5, \;\; i = 1, 2, 3, \ldots, k$. Often it appears reasonable that these fitted BIN counts may fall as low as 1, provided that the corresponding standardised residual, e_i^*, does not contribute overwhelmingly to $q_{observed}$.

BINS with a fitted count that is too small should be combined with nearby BINS (with an overall loss of degrees of freedom) until the resulting total fitted BIN count is suitably large.

Tests of composite hypotheses

The hypothesis tested in Example 13.3, $H_0 : p_1 = \frac{2}{5}, p_2 = \frac{1}{5}, p_3 = \frac{1}{5}, p_4 = \frac{1}{5}$, completely specifies the probability $p_i = P(\text{observation occurs in BIN } i)$ for each i. Such a null hypothesis is called a **simple hypothesis**. Fitting and testing of the fit may proceed even when these multinomial probabilities are not completely specified. H_0 is termed a **composite hypothesis** when the BIN probabilities are a function of an unknown parameter.

Consider now the population of all families comprising four children. Suppose that from this population, 1000 families are surveyed and the number, X, of boys in each family recorded. When the data are collated it is found that X_i have i boys (and comprise BIN $(i+1)$), $i = 0, 1, 2, 3, 4$. Is a binomial model with probability p of a male birth, suitable for these data? If so, with which value of p?

From Example 6.1, the **masculinity ratio**, defined as the number of male births per 100 female births is currently at 105. Therefore, a reasonable estimate of p is $\frac{105}{105+100} = 0.512195$.

Case 1: **Simple null hypothesis**:
$H_0 : X \sim Bin(4, 0.512195)$

Case 2: **Compound null hypothesis**:
$H_0 : X \sim Bin(4, p)$

Case 1 is simple since the 5 BIN probabilities are specified:

$$P(X = i) = \binom{4}{i}(0.512195)^i(1 - 0.512195)^{4-i}, \quad i = 0, 1, 2, 3, 4.$$

i	0	1	2	3	4
$p_{i+1} = P(X = i)$	0.0566	0.2378	0.3746	0.2622	0.0688

A goodness-of-fit test proceeds by comparing observed and fitted values from the probability model as before.

However, Case 2 is compound since the 5 BIN probabilities are expressed only in terms of the unknown parameter p.

i	0	1	2	3	4
$p_{i+1} = P(X = i)$	$(1-p)^4$	$4p(1-p)^3$	$6p^2(1-p)^2$	$4p^3(1-p)$	p^4

Now it is the binomial nature that is being tested, not simply the value of a specific parameter. Before proceeding p must be estimated from the data through the process of maximum likelihood. When, as in this case, only the observed BIN counts x_i are available as data, the estimation procedure proceeds on the BINS and not the original data. Since the multinomial mass function is

$$P(X_1 = x_1, X_2 = x_2, \ldots, X_x = x_k) = \frac{n!}{x_1!x_2!\ldots x_k!} p_1^{x_1} p_2^{x_2} \ldots p_k^{x_k},$$

the multinomial likelihood for the families of four children is, by Definition 10.5, in terms of a constant $(const)$ not depending on p:

$$l(p) = (const) p_1^{x_1} p_2^{x_2} p_3^{x_3} p_4^{x_4} p_5^{x_5}; \quad \sum_{i=1}^{5} p_i = 1, \quad \sum_{i=1}^{5} x_i = 1000,$$

for the binomial probabilities $p_{i+1} = P(X = i) = \binom{4}{i} p^i (1-p)^{4-i}$, $i = 0, 1, 2, 3, 4$. Therefore,

$$\begin{aligned} l(p) &= (const)[(1-p)^4]^{x_1}[4p(1-p)^3]^{x_2}[6p^2(1-p)^2]^{x_3}[4p^3(1-p)]^{x_4}[p^4]^{x_5} \\ &= (const)(1-p)^{4x_1+3x_2+2x_3+x_4} p^{x_2+2x_3+3x_4+4x_5}. \end{aligned}$$

Maximising l as a function of p (in fact it is easier, and gives the same result, to maximise $\log l$ as a function of p) gives the seemingly complicated formula

$$\begin{aligned} \widehat{p} &= \frac{x_2 + 2x_3 + 3x_4 + 4x_5}{4x_1 + 4x_2 + 4x_3 + 4x_4 + 4x_5} = \frac{x_2 + 2x_3 + 3x_4 + 4x_5}{4(x_1 + x_2 + x_3 + x_4 + x_5)} \\ &= \frac{\text{total number of boys in sample}}{\text{total number of children in the sample}} \end{aligned}$$

These simplifications mean that the maximum likelihood estimator is

$$\widehat{p} = \text{observed proportion of boys in the entire sample,}$$

and this is the solution which would have been obtained for the original data set of 4000 points (one thousand families of four children each) consisting of 0's (girls) and 1's (for boys). Often things don't work out so nicely, and the likelihood equation is difficult to solve for the maximum likelihood estimator of the unknown parameter. This difficulty is apparent in Example 13.4. After determining the maximum likelihood estimator $\widehat{p}$ of p, the model $H_0 : X \sim Bin(4, \widehat{p})$ may be fitted as before, but with a consequent loss of one degree of freedom due to the estimation of p from the data.

Theorem 13.2 *Suppose that for a composite null hypothesis expressed in terms of ω unknown parameters, the fitted BIN counts to k BINS are given by $\widehat{x}_i = n\widehat{p}_i$ where, for each $i = 1, 2, \ldots, k$, $\widehat{p}_i$ is the maximum likelihood estimator of p_i based on the observed BIN frequencies $X_1, X_2, X_3, \ldots, X_k$. Then, under certain regularity conditions on the BIN probabilities, when n is large, approximately,*

$$Q = \sum_{i=1}^{k} \frac{(X_i - n\widehat{p}_i)^2}{n\widehat{p}_i} \sim \chi^2_{k-\omega-1}.$$

■

The proof of goodness-of-fit theorems of this kind may be found in advanced texts on mathematical statistics such as Kendall and Stuart (1967). In summary: If ω parameters require estimation from the BIN counts data, ω degrees of freedom are lost but an approximate chi-squared distribution still holds for the distribution of Q, namely $Q \sim \chi^2_{k-\omega-1}$. If the maximum likelihood procedure is based on the full original data set, then the limiting distribution of Q is bounded between a chi-squared distribution with $k - \omega - 1$ degrees of freedom and a chi-squared distribution with $k - 1$ degrees of freedom. Using the full data set is a more efficient way of estimating the BIN probabilities than simply using the BIN totals. It is then wise to reject the null hypothesis if $q_{observed} > \chi^2_{\alpha,k-1}$ and accept the null hypothesis if $q_{observed} < \chi^2_{\alpha,k-\omega-1}$. If $\chi^2_{k-\omega-1} < q_{observed} < \chi^2_{\alpha,k-1}$, the result of the test is doubtful. This is, however, a narrow band when only one or two parameters require estimation.

Example 13.4

Poisson counts of kina per square metre

Kina is the Maori name for a variety of sea urchin (or sea egg) found in New Zealand waters. Prized as a great food delicacy, *kina* may be gathered at low tide on some of the more isolated rocky coastlines on the South Island of New Zealand.

Suppose that at low tide at Resolution Bay, Marlborough Sounds, the following counts of *kina* per square metre were recorded:

Number of *kina*	0	1	2	3	4	5 or more
Square metre units (frequency)	22	56	52	38	20	12

Worked example:
Fit a Poisson distribution to these data.
Suppose that X, the number of *kina* observed in a square metre follow a Poisson distribution with rate λ. That is, $H_0 : X \sim Poi(\lambda)$ against an alternative that the data are not Poisson. For 6 BINS (representing *kina* counts 0,1,2,3,4,$\geq$ 5) with probability p_i that an observation is found in BIN i for $i = 1, 2, 3, 4, 5, 6$, it follows that when H_0 is true, $p_{i+1} = \frac{\lambda^i e^{-\lambda}}{i!}$, for $i = 0, 1, 2, 3, 4$, and $p_6 = 1 - \sum_{i=1}^{5} p_i$.
The parameter λ must be estimated by maximum likelihood before the model may be fitted. The multinomial likelihood is, by Equation (13.1),

$$l(\lambda) = (const) p_1^{x_1} p_2^{x_2} p_3^{x_3} p_4^{x_4} p_5^{x_5} p_6^{x_6},$$

where $x_1 = 22, x_2 = 56, x_3 = 52, x_4 = 38, x_5 = 20$ and $x_6 = 12$. Substitution of probabilities p_i specified by the model and data x_i gives

$$l(\lambda) = \left(\frac{\lambda^0 e^{-\lambda}}{0!}\right)^{22} \left(\frac{\lambda^1 e^{-\lambda}}{1!}\right)^{56} \left(\frac{\lambda^2 e^{-\lambda}}{2!}\right)^{52} \left(\frac{\lambda^3 e^{-\lambda}}{3!}\right)^{38} \left(\frac{\lambda^4 e^{-\lambda}}{4!}\right)^{20} \left(1 - \sum_{i=0}^{4} \frac{\lambda^i e^{-\lambda}}{i!}\right)^{12}$$

and this function has no algebraic solution for the maximum value $\widehat{\lambda}$. (l may be maximised numerically.)
However, in the binomial case we noted that the $\widehat{p}$ corresponded to the sample mean, so we can reasonably assume that if a Poisson distribution is to fit the data well, then the sample mean and the fitted mean based on all 200 observations should coincide.
Fortunately, for these particular BINS, the identity of the original data is not completely lost. If we approximate BIN 6 as $X = 5$ rather than $X \geq 5$, then we may take

$$\widehat{\lambda} = \frac{\text{total kina count}}{\text{total area}} = \frac{(0)(22) + (1)(56) + \ldots + (4)(20) + (5)(12)}{200} = 2.07$$

as the maximum likelihood estimator of λ based on the full sample, rather than simply the BIN counts. We may now attempt to fit a Poisson model.

1. For a Poisson goodness-of-fit test the six BINS are labelled as above.
2. The null hypothesis may be phrased as $H_0 : X \sim Poi(\lambda)$ against an alternative hypothesis that the Poisson model is inappropriate.
3. Since one parameter λ requires estimation from the full data set before all BIN probabilities are completely specified by H_0, it follows that H_0 is rejected only if $q_{observed} > \chi^2_{\alpha,k-1} = \chi^2_{\alpha,5}$ and accepted if $q_{observed} < \chi^2_{\alpha,k-\omega-1} = \chi^2_{\alpha,6-1-1} = \chi^2_{\alpha,4}$.
4. Take $\alpha = 0.05$.

5. When H_0 is true, and λ is estimated by $\widehat{\lambda} = 2.07$, so that

$$\widehat{p_{i+1}} = \frac{2.07^i e^{-2.07}}{i!}, \quad \text{for } i = 0, 1, 2, 3, 4, \text{ and } \widehat{p_6} = 1 - \sum_{i=1}^{5} \widehat{p_i}.$$

The fitted values ($\widehat{x_i} = n\widehat{p_i}$) and residuals ($e_i = x_i - \widehat{x_i}$) are found in the following table:

BIN	*kina*	x_i	$\widehat{p_i}$	$\widehat{x_i}$	e_i
1	0	22	0.12618	25.24	-3.24
2	1	56	0.26120	52.24	3.76
3	2	52	0.27035	54.07	-2.07
4	3	38	0.18654	37.31	-0.69
5	4	20	0.09653	19.31	0.69
6	≥ 5	12	0.05920	11.84	0.16

None of the residuals seems excessive. When the fitted values are rounded to the nearest whole number and placed in the original data set we obtain

Kina	0	1	2	3	4	5 or more
Frequency	22 (25)	56 (52)	52 (54)	38 (37)	20 (19)	12 (12)

$$\begin{aligned} q_{observed} &= \sum_{\text{all BINS}} \frac{(\text{observed BIN count} - \text{fitted BIN count})^2}{\text{fitted BIN count}} \\ &= \frac{(22-25)^2}{25} + \frac{(56-52)^2}{52} + \frac{(52-54)^2}{54} + \frac{(38-37)^2}{37} \\ &\quad + \frac{(20-19)^2}{19} + \frac{(12-11)^2}{11} \\ &= 0.82. \end{aligned}$$

6. Since $\chi^2_{0.95,4} = 0.711$, p-value ≈ 0.95.
7. The large p-value indicates that we fail to reject the null hypothesis that the distribution of *kina* follows a Poisson distribution.
8. Summary: $q_{observed} = 0.82$ $(n.s.)$.

EXERCISES

13.1 For the temperature data recorded on a nominal scale under experimental conditions test the hypothesis that cold temperatures occur 25% of the time.

BIN 1	cold	67
BIN 2	hot	133
Total		200

Use two methods:

Method 1: A test for proportions using an (approximately) normally distributed test statistic;

Method 2: A chi-squared goodness-of-fit test based on $k = 2$ BINS.

[The proof of Theorem 13.1 should indicate why the test statistics used in these two methods are equivalent.]

13.2 The following temperature data were recorded on an ordinal scale under experimental conditions.

BIN 1	very cold	65
BIN 2	cold	40
BIN 3	hot	33
BIN 4	very hot	62
Total		200

(a) Test the hypothesis that

$$p_1 = p_2 = p_3 = p_4.$$

(b) Construct the residuals and standardised residuals when a probability model in which all temperature classifications are equally likely is fitted to the data.

(c) What information do these residuals show?

(d) Now fit the following probability model to the data: extreme temperature conditions (either very cold or very hot) are equally likely, and moderate temperature conditions (cold or hot) are equally likely with extremely hot conditions twice as likely as hot conditions.

13.3 For the following temperature data

BIN 1	$[-20, 0)$	29
BIN 2	$[0, 20)$	26
BIN 3	$[20, 40)$	47
BIN 4	$[40, 60)$	53
BIN 5	$[60, 80)$	45
Total		200

recorded on a ratio scale under experimental conditions, test these hypotheses:

(a) All five temperature groupings are equally likely.

(b) The four temperature groupings

$$[-20, 20),\ [20, 40),\ [40, 60),\ [60, 80)$$

are equally likely.

(c) The temperature lies below 20 degrees one-quarter of the time.

13.4 A survey of elderly nursing home residents in Western Australia by Finley (1985) resulted in the following data on hearing aid usage.

Hearing aid usage	n	%
Mostly	18	60
Sometimes	4	13
Never	9	27
Total	31	100

Test whether the categories

Mostly/Sometimes/Never occur in the ratio 5 : 1 : 3.

13.5 In a survey of runners in the 7th Los Angeles Marathon, 309 entrants were asked:

Does it require physical or mental endurance to run a marathon?

213 responded that it required mental endurance; the remaining 96 felt the physical challenge. (*Source: The Sunday Age*, 1 March 1992.)

Use a chi-squared test to determine whether two thirds of marathon runners believe that it requires mental endurance (as opposed to physical endurance) to run a marathon.

13.6 The distribution of the number of females in 160 litters each consisting of 4 mice was as follows:

Number of females	Number of litters
0	12
1	37
2	55
3	47
4	9

(a) Assuming that there is a 50:50 chance for each mouse to be male or female, what is a reasonable probability model for the random variable X denoting the number of females in litters of size 4?

(b) Test whether the model in (a) adequately describes these data.

(c) Determine the residuals, standardised residuals and fitted values when the model is fitted. What information do these quantities convey?

13.7 Champagne glasses are produced in packs of 6. Suppose that defects in the glasses occur independently during manufacture, so that it is reasonable to assume that the number of defective glasses per pack follows a binomial distribution.

Number of Defective glasses	Number of Packs
0	50
1	37
2	10
3	2
4	1
5	0
6	0

(a) Does a $Bin(6, 0.1)$ distribution fit these data? Explain why or why not.

(b) Does a $Poi(0.6)$ distribution fit these data? Explain why or why not.

13.8 The number of cosmic particles striking an apparatus in forty consecutive periods of one minute were recorded as:

0	0	0	1	0	2	1	0	1	1
3	3	4	1	2	0	2	0	3	1
1	0	1	2	0	2	1	0	1	2
3	1	0	0	2	1	0	3	1	2

Suppose that these particles are arriving randomly at a constant overall rate. Then it is reasonable to suspect that the random variable X denoting 'the number of particles arriving in a one minute period' may follow a Poisson distribution.

(a) Construct a frequency distribution for these data.

(b) Determine the sample mean.

(c) Fit a Poisson probability model to these data using a maximum likelihood procedure.

(d) Use an appropriate chi-squared test to ascertain how well the model fits these data.

13.9 If the frequency counts X_i in BIN i satisfy, for each i, $E(X_i) = np_i$, determine the value of $E(Q)$ when

$$Q = \sum_{i=1}^{k} \frac{(X_i - np_i)^2}{np_i}.$$

Is this the result that you would have expected given the approximate distribution for Q when n is large?

13.2 CONTINGENCY TABLES

The concept of BINS has been used to indicate groups, categories, intervals, ... We imagine such BINS arranged in a row so as to give one-dimensional categorical data. In Section 13.1, the suitability of a probability model for these BINS was assessed by comparing observed BIN counts with those counts that would be fitted under the probability model. The assessment was performed with a chi-squared test statistic.

In this section, our categorical data is two-dimensional; we continue using BINS to hold the observed frequency counts, and we continue using the chi-squared test statistic to assess goodness-of-fit.

Definition 13.2 *An* $r \times c$ **contingency table** *is an array* $[x_{ij}]$, $1 \leq i \leq r$, $1 \leq j \leq c$ *of frequency data in* r *rows of equal length and* c *columns of equal length.*

□

Example 13.5

Car pooling

The following data on car pooling was the result of a large 1981 survey of employees of a firm in Melbourne's central business district. The survey was taken at a time of great interest in car pooling; a time of rising petrol prices and decreasing patronage of public transport.

Key: N—Never
S—Strike days only
A—Alternate week days only
W—All week days

The results for car drivers appear in the following 2×4 contingency table:

	Car pooling			
	N	S	A	W
Drivers				
Male	11	52	26	68
Female	2	6	1	9

The results for employees who do not drive themselves to work also comprise a 2×4 contingency table.

	Car pooling			
	N	S	A	W
Non-drivers				
Male	15	51	1	24
Female	9	13	0	7

(*Source*: Raper, R.L. and Kumar, S. (1981). Car pooling—A feasibility study in an Australian environment. *Scientific Management of Transport Systems.*)

Contingency tables summarise the changing frequencies across different categories (BINS). The researchers noted in this survey that there were many more male drivers than male non-drivers, whereas there were more female non-drivers than female drivers. Drivers seem more interested in car pooling on weekdays than non-drivers.

(Note that if these two tables were combined, a three-dimensional table $[x_{ijk}]$ is created, where each respondent may be classified according to gender/driving/pooling preference in a $2 \times 2 \times 4$ contingency table. See Everitt (1977).)

Contingency tables differ from the type of array used for comparing several samples in ANOVA (Section 12.4) as follows: The data of contingency tables consists of whole numbers only (corresponding to frequencies of occurrence of an event of interest) rather than interval scale measurements; there are no gaps in a contingency table since the columns are of equal length so that a frequency measurement lies in each (row i, column j) position.

$$\begin{bmatrix} x_{11} & x_{12} & \cdots & x_{1c} \\ x_{21} & x_{22} & \cdots & x_{2c} \\ x_{31} & x_{32} & \cdots & x_{3c} \\ \vdots & \vdots & & \vdots \\ x_{r1} & x_{r2} & \cdots & x_{rc} \end{bmatrix}$$

Two cases are of interest to us: contingency tables with fixed column totals and contingency tables with fixed grand total. Although the chi-squared statistic method of analysis is the same for each type of table, the nature of the hypotheses being tested differs for the two cases.

Contingency tables with fixed column totals

Suppose that from each of c populations, samples (of possibly differing sizes) are selected independently and **each sample** classified as frequencies occupying r different BINS.

The sample sizes are fixed in advance; what is of interest is how the samples fall into, or are distributed across, the different BINS. Is this distribution the same for each of the populations? The chi-squared test that addresses this question is called a test of **homogeneity of *c* populations** with respect to BIN classifications.

Population	1	2	...	c
		Samples		
BIN 1	x_{11}	x_{12}	...	x_{1c}
BIN 2	x_{21}	x_{22}	...	x_{2c}
BIN 3	x_{31}	x_{32}	...	x_{3c}
$\vdots$	$\vdots$	$\vdots$		$\vdots$
BIN r	x_{r1}	x_{r2}	...	x_{rc}
Fixed column totals	$x_{.1}$	$x_{.2}$	...	$x_{.r}$

If p_{ij} denotes the classification probability for the members sampled from the jth population, then

$$p_{ij} = P(\text{member of population } j \text{ is classified in BIN } i), \quad 1 \le i \le r, \ 1 \le j \le c.$$

The test of homogeneity may be phrased in terms of H_0: the probability of

BIN i classification is the same for each population. That is,

$$H_0 : p_{i1} = p_{i2} = p_{i3} = \ldots = p_{ic}, \quad i = 1, 2, 3, \ldots r.$$

Contingency tables with fixed grand total

When a fixed sample size is drawn from a population and two categorical variables measured (on a nominal or ordinal scale) on each sampling unit, then frequencies of occurrence may be observed in any of rc different BINS. Instead of listing these BINS in a row, as was done in Section 13.1, we list the BINS as cells in a contingency table; the two-dimensional nature of the table reflects the effect of one of the measured variables on the other.

		Variable A			
		Level 1	**Level 2**	...	**Level c**
		BIN_{11}	BIN_{12}		BIN_{1c}
	Level 1	x_{11}	x_{12}	...	x_{1c}
		BIN_{21}	BIN_{22}		BIN_{2c}
	Level 2	x_{21}	x_{22}	...	x_{2c}
Variable B		BIN_{31}	BIN_{32}		BIN_{3c}
	Level 3	x_{31}	x_{32}	...	x_{3c}
		⋮	⋮		⋮
		BIN_{r1}	BIN_{r2}		BIN_{rc}
	Level r	x_{r1}	x_{r2}	...	x_{rc}

In this array, $\sum_{i=1}^{r}\sum_{j=1}^{c} x_{ij} = n$ is fixed as the total number sampled. As before, if we put

$$p_{ij} = P(\text{observation is classified in BIN}_{ij}), \quad 1 \le i \le r, \quad 1 \le j \le c,$$

then the null hypothesis of interest about Variables A and B is an assertion of independence; that is,

$$H_0 : \text{The levels A are independent of the levels of B}$$

against an alternative that the independence fails. The test procedure outlined below is termed a **chi-squared test of independence**. When independence fails (and the null hypothesis is rejected) the two Variables A and B are said to **interact**. In terms of the dot notation applied to the BIN probabilities p_{ij},

$$p_{i.} = \sum_{j=1}^{c} p_{ij} = P(\text{observation is classified in row } i), \quad 1 \le i \le r,$$

$$p_{.j} = \sum_{i=1}^{r} p_{ij} = P(\text{observation is classified in column } j), \quad 1 \le j \le c.$$

Then,

$$\begin{aligned} p_{ij} &= P(\text{observation is classified in BIN}_{ij}) \\ &= P(\text{observation is classified in row } i \text{ AND column } j) \\ &= P(\text{observation is in row } i)P(\text{observation is in column } j) \\ &= p_{i.}p_{.j}, \quad 1 \leq i \leq r, \quad 1 \leq j \leq c. \end{aligned}$$

The null hypothesis may then be expressed as $H_0 : p_{ij} = p_{i.}p_{.j}, \quad 1 \leq i \leq r,\ 1 \leq j \leq c$, against an alternative that the equality fails for at least one BIN. (Of course this process may be simply extended to three or more dimensions by recording the results of three categorical variables on each experimental unit.)

Example 13.6

Homogeneity and independence

If 50 Labor voters, 50 National voters and 50 Liberal voters are classified by (approve/do not approve/no comment) for a new tax reform package, the results may form the 3 × 3 Table A with column totals fixed 50 in advance—before interviewing commences. If however, simply 150 (Labor/Liberal/National) voters are sampled, it is unlikely that equal numbers of Labor, National and Liberal supporters would be gathered by the survey. If sampling was conducted randomly, the sample proportions of party support would be indicative of party support in the population being sampled; see Table B.

Table A

	Labor supporter	Liberal supporter	National supporter
Approve	31	11	3
Do not approve	12	36	43
No comment	7	3	4
Total	50	50	50

Table B

	Labor supporter	Liberal supporter	National supporter
Approve	67	12	3
Do not approve	19	29	6
No comment	2	11	1
Total	88	52	10

In Table A a **test of homogeneity** would examine whether each of the parties is evenly divided in support or lack of support of the tax reform proposals.

In Table B, a **test of independence** would examine whether political party support is independent of tax reform support.

Both tests would involve fitted values determined by the same formula (13.3) below; both tests use the same chi-squared statistic expressed in terms of these fitted values, and even though the tests for Table A and Table B differ in type,

$$\text{degrees of freedom Table A} = \text{degrees of freedom Table B.}$$

Determining the test statistic for a test of independence

When the probability model asserted under the null hypothesis is fitted to each entry in the contingency table, the fitted values $\widehat{x_{ij}}$ are compared to the observed values x_{ij} through the (approximately) standardised residuals

$$e_{ij}^* = \frac{x_{ij} - \widehat{x_{ij}}}{\sqrt{\widehat{x_{ij}}}}$$

The main question before the chi-squared statistic can be calculated is: What are the values $\widehat{x_{ij}}$ fitted under the assumption that H_0 is true? In the test for independence the BIN counts are again marginally binomial. Accordingly, estimated mean BIN count is sample size times estimated probability:

$$\begin{aligned}
\widehat{x_{ij}} &= n\widehat{p_{ij}} \\
&= n\widehat{p_{i.}}\widehat{p_{.j}} \\
&= n(\text{estimated probability observation in row } i) \\
&\qquad (\text{estimated probability observation in column } j) \\
&= n(\text{observed proportion in row } i)(\text{observed proportion in column } j) \\
&= n(\frac{x_{i.}}{n})(\frac{x_{.j}}{n}) \\
&= \frac{(x_{i.})(x_{.j})}{n},
\end{aligned}$$

so that

$$\text{fitted BIN}_{ij}\text{ count} = \widehat{x_{ij}} = \frac{(\text{row } i \text{ total})(\text{column } j \text{ total})}{\text{grand total}}. \tag{13.3}$$

Then the chi-squared statistic Q may be expressed as

$$Q = \sum_{\text{all BINS}} \frac{(\text{observed BIN count} - \text{fitted BIN count})^2}{\text{fitted BIN count}} = \sum_{all\ i,j} \frac{(x_{ij} - \widehat{x_{ij}})^2}{\widehat{x_{ij}}}.$$

By Theorem 13.2, approximately, $Q \sim \chi^2_{k-\omega-1}$, where k is the number of BINS (in this case, $k = rc$) and ω is the number of parameters that require estimation under H_0. According to their definitions, and the fact that they must sum to 1, $r-1$ of the r $p_{i.}$ quantities, and $c-1$ of the c $p_{.j}$ quantities, must be estimated when H_0 is true. In total this is, $\omega = (r-1)+(c-1) = r+c-2$. Therefore the degrees of freedom associated with Q are $k-\omega-1 = rc-(r+c-2)-1 = (r-1)(c-1)$. Therefore, approximately, $Q \sim \chi^2_{(r-1)(c-1)}$ when H_0 is true. As was the case in Section 13.1 when the BINS were arranged in a row, the degrees of freedom for a particular contingency table are easily remembered by drawing a schematic diagram of all BINS and flagging those BINS requiring their counts to be known in order that the counts of all BINS be known **assuming fixed marginal counts for the contingency table**. For example, a contingency table with 6 rows and 8 columns contains $(6-1)(8-1) = 35$ x's and consequently 35 degrees of freedom.

x	x	x	x	x	x	x	
x	x	x	x	x	x	x	
x	x	x	x	x	x	x	
x	x	x	x	x	x	x	
x	x	x	x	x	x	x	

Example 13.7

Is hearing ability related to psychiatric disorder in elderly nursing home residents?

Finley (1985) gives data on the hearing ability of 100 elderly nursing home residents. Hearing ability is classified by Pure Tone Average (PTA), which is a hearing test record of the minimum decibel volume required for a subject to hear pure tone. PTA was classified as Good (0–25dB), Medium (26–40dB), Poor (41+dB) against a 0–1 diagnosis record of non-psychiatric disorders (0), psychiatric disorders (1).

	Pure Tone Average (PTA)			
	Good	**Medium**	**Poor**	**Totals**
Diagnosis 0	16	17	39	72
Diagnosis 1	4	9	15	28
Total	20	26	54	100

Worked example:
Test whether hearing ability (PTA score) is independent of psychiatric diagnosis.
This is a standard chi-squared test of independence on the 6 BINS:

16	17	39
4	9	15

1. Let the above array of BINS be denoted by $[x_{ij}]$, $i = 1, 2$; $j = 1, 2, 3$. In this notation, $[x_{ij}]$ denotes the observed frequency for, and p_{ij} denotes the probability that an observation appears in BIN_{ij}.
2. The null hypothesis may be phrased as $H_0 : p_{ij} = p_{i.}p_{.j}, \quad i = 1, 2, \; j = 1, 2, 3$. That is, H_0: PTA score is independent of psychiatric diagnosis, against an alternative hypothesis that the 'diagnosis' and 'score' are dependent.
3. The appropriate degrees of freedom are flagged in the BINS below:

x	x	

   Once entries 'x' are known (assuming known marginal totals), then all BIN entries are known. Formally, the degrees of freedom calculation is $(r - 1)(c - 1) = (2 - 1)(3 - 1) = 2$. Therefore, when H_0 is true $Q \sim \chi^2_2$ approximately.
4. Take $\alpha = 0.05$.
5. When H_0 is true, the fitted BIN frequencies $\widehat{x_{ij}}$ are, by (13.3), $\widehat{x_{ij}} = \frac{(\text{row } i \text{ total})(\text{column } j \text{ total})}{\text{grand total}}$. Therefore, we fit

$$\widehat{x_{11}} = \tfrac{(20)(72)}{100} = 14.40 \;\text{ to BIN}_{11}$$
$$\widehat{x_{12}} = \tfrac{(26)(72)}{100} = 18.72 \;\text{ to BIN}_{12}$$
$$\cdots$$

   Collectively, the fitted values are bracketed in the array

16 (14.40)	17 (18.72)	39 (38.88)
4 (5.60)	9 (7.28)	15 (15.12)

   and the discrepancies between observed and fitted values create no unusually large residuals.

$$\begin{aligned}
q_{observed} &= \sum_{\text{all BINS}} \frac{(\text{observed BIN count} - \text{fitted BIN count})^2}{\text{fitted BIN count}} \\
&= \frac{(16-14.40)^2}{14.40} + \frac{(17-18.72)^2}{18.72} + \frac{(39-38.88)^2}{38.88} \\
&\quad + \frac{(4-5.60)^2}{5.60} + \frac{(9-7.28)^2}{7.28} + \frac{(15-15.12)^2}{15.12} \\
&= 1.20.
\end{aligned}$$

6. Since $\chi^2_{0.05,2} = 5.992$ and $q_{observed} = 1.20$, it follows that the p-value exceeds the stated significance level and so H_0 is not rejected: 'PTA' and 'Diagnosis' appear statistically independent at a 0.05 level of significance.
7. Summary: $q_{observed} = 1.20$ ($n.s.$).

Determining the test statistic for a test of homogeneity

In the chi-squared test for homogeneity we suppose that from each of c populations, samples (of possibly differing sizes) are selected independently and **each sample** classified as frequencies occupying r different BINS. The sample sizes are fixed in advance and form the contingency table $[x_{ij}]$ to which independent probability models are fitted under the null hypothesis

$$H_0 : p_{i1} = p_{i2} = p_{i3} = \ldots = p_{ic}, \quad i = 1, 2, 3, \ldots r.$$

As in the test for independence the test statistic is (see exercises)

$$\begin{aligned} Q &= \sum_{\text{all BINS}} \frac{(\text{observed BIN count} - \text{fitted BIN count})^2}{\text{fitted BIN count}} \\ &= \sum_{all\ ij} \frac{(x_{ij} - \widehat{x_{ij}})^2}{\widehat{x_{ij}}} \sim \chi^2_{(r-1)(c-1)}, \end{aligned}$$

where

$$\text{fitted BIN}_{ij} \text{ count} = \widehat{x_{ij}} = \frac{(\text{row } i \text{ total})(\text{column } j \text{ total})}{\text{grand total}}.$$

> The chi-squared test for independence and the chi-squared test for homogeneity both use the test statistic $Q \sim \chi^2_{(r-1)(c-1)}$ under H_0.

We will discuss a special case ($c = 2$) in the next section.

A test for homogeneity: pattern recognition

Image processing translates a photographic image into a tonal grid. When each tone is represented by a digit, the processed image collectively forms an array. An image that has been segmented into a basic **binary array** $[u_{ij}]$, where for all i and j, $u_{ij} = 0$ or $u_{ij} = 1$ consists of a sequence of zeros and ones. Even so, the shape and form of the original remains recognisable (?).

A sheep, or is it a polar bear, may look like:

```
00000000000000000000000000000000000000000000
00000000000000010010011110101000000000000000
00000000000010000110111111111101000000000000
00000000001010000111111111111111110000000000
00000000001010000111111111111111111000000000
00000000000000000000101111111111111000000000
00000000000000000000011000000011000000000000
00000000000000000000011000000011000000000000
00000000000000000000011100001111000000000000
```

In a binary array, the '1' s are called **shape pixels** and may be coloured black to make the binary image stand out more clearly especially if the '0's are left blank. (See Example 13.8.)

Definition 13.3 *A shape pixel at position* (i, j) *in the array is termed a* **k-neighbourhood pixel**, $k = 1, 2, 3, 4$, *if it has at most* k *nearest neighbours in the array that are also shape pixels.*

□

If • denotes a shape pixel then the diagrams show examples of the $k = 4$ cases that arise for shape pixels neighbouring •. In sequence, • is respectively classified as a 1-neighbourhood, 2-neighbourhood, 3-neighbourhood and finally a 4-neighbourhood shape pixel.

```
000000     001000     001000     001000
00 • 100   00 • 100   01 • 100   01 • 100
000000     000000     000000     001000
```

For example 4-neighbourhood pixels are found in solidly '1' areas of an image, so that collectively they measure image size. On the other hand, 1-neighbourhood pixels occur at isolated points (outliers) and we may choose to omit them from consideration. We will compare shapes (as binary arrays) by comparing their neighbourhood pixel counts. Consider Image 1 as the reference template to be compared with a test image (Image 2).

Suppose that X_{ij} denotes the number of i-neighbourhood shape pixels from the reference template ($j = 1$) and test shape ($j = 2$). Then if there are n reference template shape pixels and m (possibly different from n) shape pixels from the test pattern, we may use a chi-squared test statistic to test the hypothesis

H_0: reference template and test shape are the same
H_1: reference template and test shape differ,

or

H_0: $p_{i1} = p_{i2}$, $\quad i = 1, 2, 3, 4$
H_1: reference template and test shape differ.

The following 8 BINS may be used:

	Population	1 Reference template	2 Test sample
1-neighbourhood pixels	BIN 1	x_{11}	x_{12}
2-neighbourhood pixels	BIN 2	x_{21}	x_{22}
3-neighbourhood pixels	BIN 3	x_{31}	x_{32}
4-neighbourhood pixels	BIN 4	x_{41}	x_{42}
	Fixed column totals	$x_{.1} = n$	$x_{.2} = m$

$$Q = \sum_{\text{all BINS}} \frac{(\text{observed BIN count} - \text{fitted BIN count})^2}{\text{fitted BIN count}}$$

may be determined once BIN counts have been fitted assuming H_0 true. When H_0 is true and $p_{i1} = p_{i2} = p_i$, the fitted number of i-neighbourhood pixels may be taken as $\frac{(\text{row } i \text{ total})(\text{column } j \text{ total})}{\text{grand total}}$ which simplifies to $n\widehat{p_i}$ and $m\widehat{p_i}$ for reference and test shapes respectively, where

$$\widehat{p_i} = \frac{X_{i1} + X_{i2}}{n+m}.$$

Then,

$$\begin{aligned} Q &= \sum_{i=1}^{4} \frac{(X_{i1} - n\widehat{p_i})^2}{n\widehat{p_i}} + \sum_{i=1}^{4} \frac{(X_{i2} - m\widehat{p_i})^2}{m\widehat{p_i}} \\ &= \sum_{i=1}^{4} \frac{(mX_{i1} - nX_{i2})^2}{nm(X_{i1} + X_{i2})} \end{aligned}$$

has approximately a chi-squared distribution with $(4-1)(2-1) = 3$ degrees of freedom.

Example 13.8

Spotting the difference between aircraft images

Ryall and Sandor (1989) apply a chi-squared test for homogeneity to the problem of discerning a recognisable shape—in their paper *Statistical Pattern Matching*, Pattern Recognition Letters 9, 163–168, the authors engage in aircraft detection (at the Aeronautical Research Laboratories in Melbourne). They use a comparison between a basic reference template produced manually by 'contour tracking' of an aircraft image and a test shape of an aircraft image extracted by image segmentation.

The aircraft images compared in Ryall and Sandor (1989) have 1 (black) and 0 (white) so that the black entries in the array so formed comprise the **shape pixels**. One template/test pattern discussed by the authors is:

Template pattern *Test pattern*

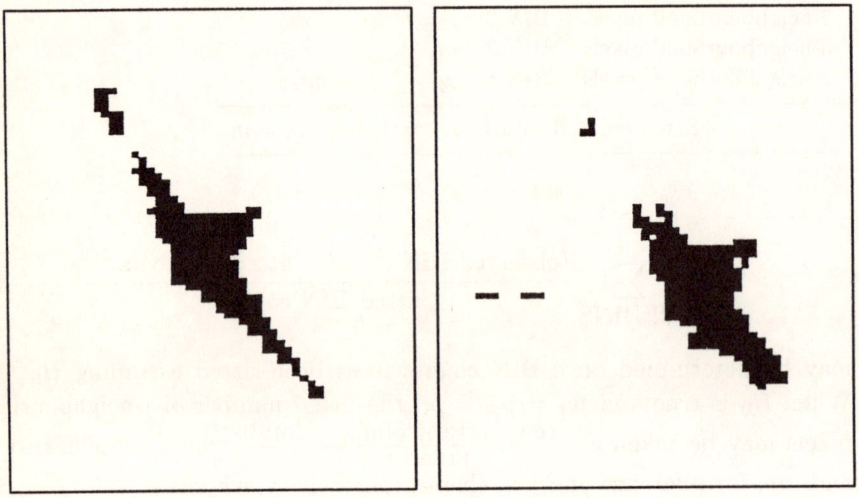

Technical considerations: The aircraft image was recorded at 64×64 resolution spanning $\frac{1}{16}$ of a standard 625 line CCIR television frame at 64 levels of greyscale. An image processing technique assigned the image to a binary scale (test pattern).
How effective is the technique?
The experiment (which did not consider 1-neighbourhood shape pixels) resulted in $q_{observed} = 0.74$, against a $\chi^2_{0.05,3} = 7.81$, clearly leading to agreement between template and test image. The authors used a slightly different technique for estimating the fitted i-neighbourhood pixels to ensure that a template image and a sample image which differ by a scale factor are classified as different by the test. Results were compared for five other aircraft image test/template pairs.

EXERCISES

13.10 Determine the missing data entries in the contingency table:

					Total
	54	48		41	190
	43				170
			38	43	
Total	117	114		126	491

13.11 For the contingency table:

				Total
	22	27		60
	13			58
			20	52
	21	20		60
Total	68	102	60	230

(a) Determine the values of r and c that classify this table as an $r \times c$ contingency table.

(b) Determine the values that are fitted under the assumption of independence of row and column classification.

13.12 The car pooling data of Example 13.5 shows employee response to the question:

Would you participate in car pooling if it became available?

Key: N—Never
S—Strike days only
A—Alternate week days only
W—All week days

	Car pooling			
	N	S	A	W
Non-drivers				
Male	15	51	1	24
Female	9	13	0	7

Obtain the fitted values for this contingency table. In particular, indicate why the categories A and W should be combined for a test of independence.

Is the response to the car pooling question gender-based?

13.13 *Year Book Australia (1985)* records the following numbers of births both from within marriages (nuptial births) and from outside marriages (ex-nuptial births). (The frequencies are rounded to the nearest thousand births.)

Year	'78	'79	'80	'81	'82
Nuptial	197	195	195	203	205
Ex-nuptial	25	26	28	31	33

Assuming that the numbers of births across different years are independent (and this may not be the case because of economic circumstances), has the proportion of births 'out of wedlock' changed significantly over this period?

13.14 For the 2×2 contingency table

	Attribute A	
Attribute	a	b
B	c	d

(a) Show that when Attribute A is independent of Attribute B, then the absolute difference between the observed BIN counts and the fitted BIN counts is

$$\frac{|D|}{a+b+c+d}, \quad D = ad - bc,$$

for all BINS. (Here | | denotes absolute value.)

(b) Use (a) to conclude that the chi-squared test statistic Q, for a 2×2 contingency table may be expressed in the form

$$Q = \frac{D^2(a+b+c+d)}{(a+c)(b+d)(a+b)(c+d)}.$$

(c) Suppose that 900 responses to the question

Should Australia become a republic?

were:

Response	Gender Men	Women
In favour	170	122
Against	280	328

What conclusion may be drawn from these data?

13.15 In a large factory, products are produced on a machine press by various press operators. An experiment was designed to test the manufacturer's claim that 'quality of product' (high/medium/low) is independent of the 'operator' handling the press.

Two skilled operators ran the same press for consecutive four-hour shifts in the same eight-hour day. The following bar charts were tendered to management to display frequencies, f, of products produced in the various quality categories.

Operator 1

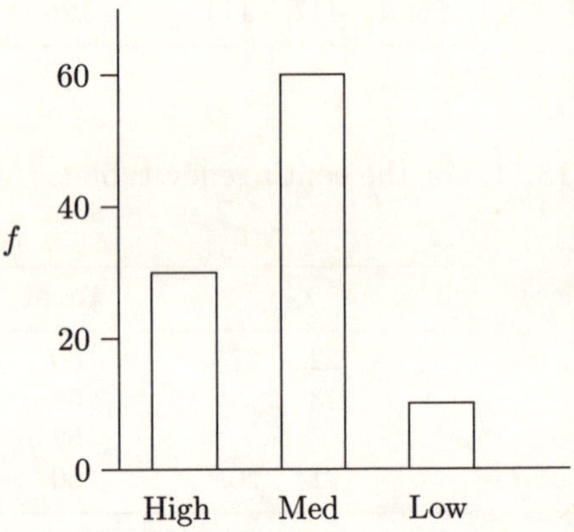

Operator 2

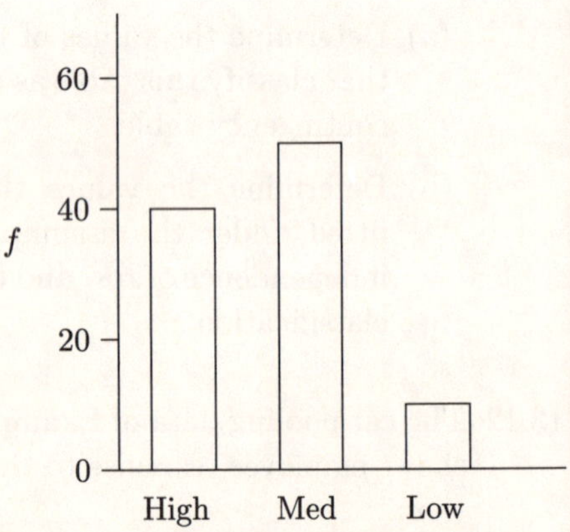

(a) Do these bar charts provide visual evidence to support the manufacturer's claim? Explain.

(b) By constructing a contingency table for the data, test the manufacturer's claim at a 0.05 significance level.

(c) Manufacturing guidelines specify that high/medium/low quality products should be produced in the ratio 2 : 3 : 1. Do the results of this experiment conform to these guidelines?

13.16 Wilks (1991) discusses the problem of trained and certified underwater divers who had not dived since gaining their certificates. Data, experience and training problems for 95 such divers (55 men and 40 women) were collected from the files of the National Association of Underwater Divers.

The sample consisted of the following occupational groups and these are classified by gender as percentages.

Occupation	**Male** %	**Female** %
Professional	27	28
Managerial/White collar	14	18
Manual	32	13
Tertiary students	15	18
Other	12	23

(*Source*: Diving dropouts: The Australian experience. *Australian Journal of Science and Medicine in Sport*, 23, 17-20.)

Is there evidence that occupation is independent of gender for 'dropout divers'?

13.17 Consider the 2 × 2 contingency table in the usual notation with fixed column totals:

	Sample 1	2
Success	x_{11}	x_{12}
Failure	x_{21}	x_{22}
Totals	$x_{.1}$	$x_{.2}$

(a) Show that the test statistic

$$Z = \frac{\left(\frac{X_{11}}{X_{.1}} - \frac{X_{12}}{X_{.2}}\right)}{\sqrt{\frac{X_{1.}}{X_{..}}\frac{X_{2.}}{X_{..}}\left(\frac{1}{X_{.1}} + \frac{1}{X_{.2}}\right)}}$$

may be used to test whether the proportion of successes is the same for Populations 1 and 2.

(b) Show that $Z^2 = Q$, where Q is the usual chi-squared test statistic for testing the homogeneity of two populations against success/failure categories.

[This result shows that the test for the difference between two independent sample proportions is equivalent to the chi-squared test for homogeneity.]

(c) Suppose that 450 men and 450 women were asked to respond to the question

Should Australia become a republic?

The responses were:

Response	**Gender** Men	Women
In favour	170	122
Against	280	328
Totals	450	450

What conclusion may be drawn from these data?

Estimate the standard error of the estimated difference between the proportion of men and women in favour of republic status.

13.18 Referring to Example 13.8, suppose that the test statistic

$$Q = \sum_{i=1}^{4} \frac{(mX_{i1} - nX_{i2})^2}{nm(X_{i1} + X_{i2})}$$

is used for comparing test and template aircraft images. Suppose that one image is a scaled version of the other; that is, for some scale factor s,

$$X_{i1} = sX_{i2}$$

for each $i = 1, 2, 3, 4$. Show that this leads to $Q = 0$.

However, show that if the fitted values for i-neighbourhood pixels are given by

$$\frac{X_{i1} + X_{i2}}{2},$$

then

$$Q = \sum_{i=1}^{4} X_{i2} \frac{(s-1)^2}{s+1}.$$

How does this assist in discrimination between template and test? (Ryall and Sandor, 1989.)

13.19 In a test for homogeneity based on the $r \times c$ contingency table data $[x_{ij}]$, show that the values $\widehat{x_{ij}}$ fitted when the null hypothesis is true are given by

$$\widehat{x_{ij}} = \frac{x_{i.}x_{.j}}{x_{..}}.$$

[*Hint*: Generalise the case $c = 2$ discussed in the text.]

CHAPTER FOURTEEN

Regression and Correlation

In this chapter you will find

14.1 SCATTERPLOTS AND THE METHOD OF LEAST SQUARES

The frequencies observed in a contingency table often represent the outcomes at different levels of a pair (X, Y) of qualitative variables. If a gender $\times$ eyecolour table classifies observed counts for: gender levels—female (1), male (0); eyecolour levels—blue (1), non-blue (0); then an observed data point $(1, 1)$ represents a blue-eyed woman. Measurement on a nominal or ordinal scale is a process of recording labels like 1 and 0 in the previous example.

In this chapter we move to the representation of bivariate data (x, y), where y is measured at a higher level than group membership as the outcome of a **response random variable** Y. From an experimenter's point of view, it is these responses that hold our interest. There is a suggestion, however, that the values of the response that we are able to observe are dependent in some way on an independent variable, X. (X may, in fact, denote group membership.) Do values of Y depend on outcomes of X? If so, in what way? How strongly? We may structure the data-gathering aspects of our experiment to help answer these questions. We fix values of the variable X at pre-determined points specified by us. Not only that, but it is assumed that we are able to control, measure and fix these values of X **without error**. Then, we conduct the experiment at these fixed x values and gather data for the response Y. If X does hold a key to our understanding of Y, then this will be revealed in data-gathered in this way. For this reason, the variable X is called the **explanatory variable**. (Some authors use the terms (independent variable, dependent variable) for (X, Y); however this does not seem to stress the role of X and Y in the data-gathering process.)

For example, an agricultural researcher may be interested in the rate of growth of a corn plant from germination to maturity. Of interest is the 'height of plant' so this is taken as the response variable Y. This is a random variable because we do not know its value before we go to the field and measure it. Then we ask: What factors affect the height of the plant as it grows? This is a search for appropriate explanatory variables; obvious important ones are 'age of plant', 'hours of sunlight', 'soil acidity', ... If we had to choose only one, it would likely be 'time' or 'age of plant'. Therefore, $(X, Y) =$ ('time','height'). Both are ratio scale quantitative variables. For values x of X to be measured without error, we fix an origin on the time scale (germination date) and then measure the height y of the plant at fixed pre-determined values of X; for example, Day 1, Day 3, Day 5, Day 7, ...

Values of the explanatory variable are assumed to be measured

without error. The experiment will yield **bivariate data**

$$(x_1, y_1),\ (x_2, y_2),\ (x_3, y_3), \ldots, (x_n, y_n),$$

when n observations are made on the pair (X, Y). There may be more than one y value for each pre-determined x value.

Plotting the data

Before we embark on a probabilistic formulation on the relationship between the values of X and Y, we concentrate in this section on the information that is conveyed in the arrangement of the data.

Of highest priority is a plot of the data showing the values of the response variable at each value of x. Why is this such a high priority?

1. To help reveal any surprise mis-recordings of the response variable, any atypical values, outliers; these data points will lie separate from the other data points. Their presence is an interesting aspect of the data, but we must try to be sure that they are not transcribing errors from the original experimental outcomes.
2. The overall shape of the bivariate data set becomes immediately apparent; are points clustered together or are they spread apart? Are they in positions that we would have anticipated that they should be?
3. Is there a trend in the data set? Is it a straight line trend? Curvilinear trend?

Definition 14.1 *A* **scatterplot** *is a plot of the data values* $(x_i, y_i), i = 1, 2, 3 \ldots, n$, *on* x-y *axes, showing the values,* y_i, *of the response on the vertical axis and the fixed values,* x_i, *of the explanatory variable on the horizontal axis.*

□

Example 14.1

If cement is well cured, will it more easily withstand compression?

Does the drying time (or curing time) of cement have an effect on the final compressive strength? Does this change for various soil/cement compositions that may be required in road construction? The response of interest here is the compressive strength of a soil/cement mix; an explanatory variable is drying time.

A civil engineer expects the compressive strength to increase linearly with time. Accordingly, compressive strength readings are taken (with mix date as time 0) after 2 days, after 7 days and then after every week on laboratory 8% cement specimen with the same initial mixing date. Suppose that the following data were obtained:

Curing period (days)	2	7	14	21	28
Compressive strength (mPa)	2.53	3.12	3.49	3.95	4.36

Little information is immediately seen in this data summary. However, even with such a small amount of data, the scatterplot is revealing. There is a trend in the data that appears to be linear, not quadratic, cubic or exponential. Simply, the points (as was suspected) appear to follow a straight line.

Scatterplot of compressive strength of cured 8% cement mix

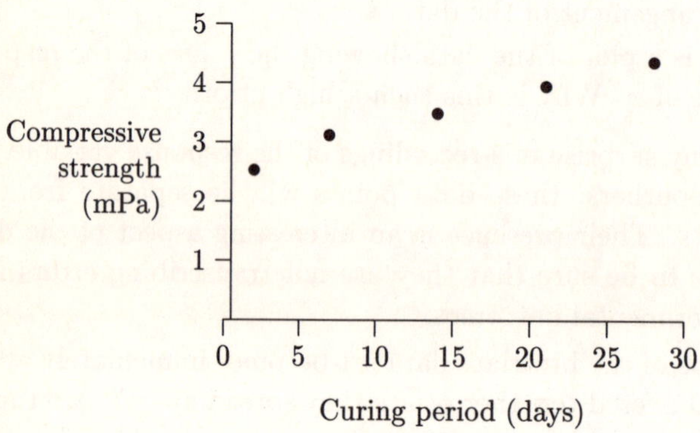

When we look at the scatterplot in Example 14.1, there is a feeling that a linear relationship exists between the X and Y variables, if only we knew what it was! On x-y axes, such a linear relationship may be described by the straight line function

$$\mu(x) = \beta_0 + \beta_1 x, \tag{14.1}$$

where β_0 and β_1 are unknown constant parameters and μ is the **regression function**, seen before in Section 8.4. The equation $y = \mu(x)$ is that of a deterministic straight line with unknown **slope** β_1 and unknown **intercept** β_0. Since slope and intercept are unknown, we cannot draw the graph. Certainly, it would be useful to obtain, and draw on the graph, the straight line that **best fits the data** in order to estimate $\mu(x)$. Even if the scatterplot shows some other trend in the relationship between X and Y, we will discuss later in this section how the data may be transformed towards linearity.

The method of least squares

It is indeed a difficult task to determine the line which 'best fits the data'. The difficulty is not with the mathematical calculation, but with the concept of 'best': Is a 'best' line one which goes through the middle of the scatterplot? Through the middle of the data? Should it take account of outliers? Is it robust to the presence of outliers? (By **robust** in this context we mean 'unduly affected by'.)

We take an approach of seeking a line describing the general trend of all data present in the scatterplot. That is, unwanted data points of influence have been noted and already a decision made as to whether they should be discarded or kept. (See Section 14.3.) Therefore the scatterplot is ready for outfitting with a line.

We take the equation of the line that we fit as $y = \widehat{\beta_0} + \widehat{\beta_1}x$, where $\widehat{\beta_0}$ and $\widehat{\beta_1}$ estimate respectively the unknown values of β_0 and β_1 in (14.1). Since this will obviously be an estimate of the 'true' straight line (14.1), it is appropriate to use the 'hat' notation that has been consistently used to denote estimates. The fitting is achieved by the **method of least squares** which we now describe. Algebraically, the calculations are more manageable if we begin by centring the x's by subtracting off the mean $\overline{x}$ from each x value.

The proposal is to fit a line of the form $y = a + b(x - \overline{x})$ to the data (x_1, y_1), $(x_2, y_2), \ldots, (x_n, y_n)$. To do this, we need the 'best fitting' values $\widehat{a}$ of a and $\widehat{b}$ of b; then, in terms of the previous parametrisation $y = \beta_0 + \beta_1 x$, the required estimates $\widehat{\beta_0}$ and $\widehat{\beta_1}$ of β_0 and β_1 are obtained from $\widehat{a}$ and $\widehat{b}$ via

$$\begin{aligned} \widehat{\beta_0} &= \widehat{a} - \widehat{b}\,\overline{x} \\ \widehat{\beta_1} &= \widehat{b}. \end{aligned}$$

At the point $x = x_i$, let $y = \widehat{y_i}$ denote the corresponding value of y on the line $y = a + b(x - \overline{x})$ that we seek. That is, $\widehat{y_i} = a + b(x_i - \overline{x})$, $i = 1, 2, 3, \ldots, n$. At the point $x = x_i$, the difference between the observed response y_i and the fitted response $\widehat{y_i}$ is

$$y_i - \widehat{y_i}, \quad i = 1, 2, 3, \ldots, n.$$

These discrepancies are drawn in the scatterplot below.

Discrepancies between observed and fitted values

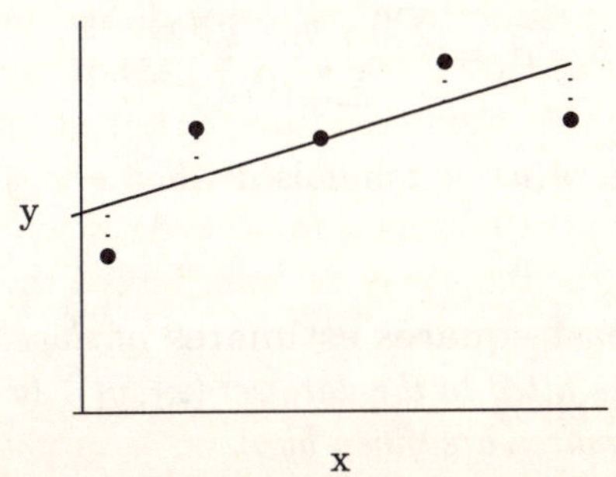

Some of the values $y_i - \widehat{y_i}$ are positive (when the observed data point lies above the line); negative (when the observed data point lies below the line); zero (when the observed data point lies on the line). If we used their sum as a measure of goodness-of-fit, the positive discrepancies would cancel with the negative discrepancies to give 0 every time! Instead, the differences are squared before adding.

Method of least squares: Choose the slope and intercept of the line so that the sum of squared discrepancies $\sum_{i=1}^{n}(y_i - \widehat{y_i})^2$ is as small as possible.

By making the sum of squared discrepancies as small as possible, it will ensure that the line of best fit deviates as little as possible from the observed data points.

$$\begin{aligned}
\sum_{i=1}^{n}(y_i - \widehat{y_i})^2 &= \sum_{i=1}^{n}[y_i - \{a + b(x_i - \overline{x})\}]^2 \\
&= \sum_{i=1}^{n}[(y_i - a) - b(x_i - \overline{x})]^2 \\
&= \sum_{i=1}^{n}(y_i - a)^2 - 2b\sum_{i=1}^{n} y_i(x_i - \overline{x}) + b^2\sum_{i=1}^{n}(x_i - \overline{x})^2 \\
&= \underbrace{[\sum_{i=1}^{n}(y_i - a)^2]}_{A(a)} + \underbrace{[-2b\sum_{i=1}^{n}(y_i - \overline{y})(x_i - \overline{x}) + b^2\sum_{i=1}^{n}(x_i - \overline{x})^2]}_{B(b)}.
\end{aligned}$$

The bracketing of the sum of squares into two pieces (one involving a alone and the other involving b alone) indicates that the joint minimisation over all (a, b) pairs may be done separately: first the minimising value of a may be found, and then the minimising value of b . The second quantity, $B(b)$, is a quadratic function of b, so that it is readily seen to have a minimum (either by calculus, or properties of quadratics) at

$$b = \widehat{b} = \widehat{\beta_1} = \frac{\sum_{i=1}^{n}(y_i - \overline{y})(x_i - \overline{x})}{\sum_{i=1}^{n}(x_i - \overline{x})^2}.$$

The first sum of squares, $A(a)$, is minimised when $a = \overline{y}$. Therefore $\widehat{a} = \overline{y}$ and $\widehat{\beta_0} = \overline{y} - \widehat{\beta_1}\overline{x}$.

Definition 14.2 *The* **least squares estimates** *of slope, β_1, and intercept, β_0, when a straight line is fitted to the data set (x_1, y_1), $(x_2, y_2), \ldots, (x_n, y_n)$, by the method of least squares are given by*

$$\begin{aligned}
\widehat{\beta_1} &= \frac{\sum_{i=1}^{n}(y_i - \overline{y})(x_i - \overline{x})}{\sum_{i=1}^{n}(x_i - \overline{x})^2} \\
\widehat{\beta_0} &= \overline{y} - \widehat{\beta_1}\overline{x}.
\end{aligned}$$

The fitted line $y = \widehat{\beta_0} + \widehat{\beta_1}x$ is called the **least squares regression line of y on x**.

□

Simple algebraic manipulations involving summations suggest that

$$\widehat{\beta_1} = \frac{\sum_{i=1}^{n} x_i y_i - n\overline{x}\,\overline{y}}{\sum_{i=1}^{n} x_i^2 - n\overline{x}^2}, \qquad \widehat{\beta_0} = \overline{y} - \widehat{\beta_1}\overline{x}. \tag{14.2}$$

These formulae are particularly useful when calculations need to be done by hand-held calculator. Many computer packages such as MINITAB or SAS perform such calculations and provide detailed output of the results. Once $\widehat{\beta_0}$ and $\widehat{\beta_1}$ have been obtained, the least squares regression line $y = \widehat{\beta_0} + \widehat{\beta_1}x$ may be plotted on the scatterplot. In the diagram below the least squares line is fitted to the scatterplot of Example 14.1.

Scatterplot of compressive strength of cured 8% cement mix

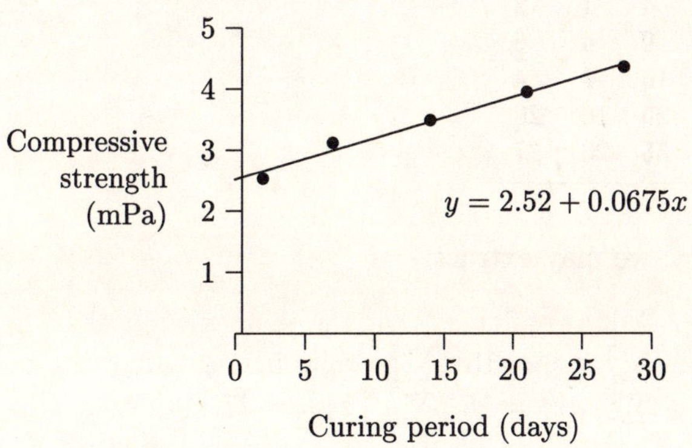

Once the line has been fitted, we must consider how *badly* it fits the scatterplot. The *fitted values* and the *residuals* are two important familiar quantities that are used in this assessment of lack of fit.

Definition 14.3 *When $y = \widehat{\beta_0} + \widehat{\beta_1}x$ is the least squares regression line of y on x, then the* **fitted value** *at $x = x_i$ is given by*

$$\widehat{y_i} = \widehat{\beta_0} + \widehat{\beta_1}x_i$$

and the ith **residual**, *e_i is determined by*

$$e_i = y_i - \widehat{y_i}.$$

□

Example 14.2

A simple fit

A sample of 5 observations of the population change, Y, of a bacterial population at different temperatures, x, in °C, are recorded below:

Population change	y	(*units*)	−1	−1	2	1	4
Temperature	x	(°C)	1	2	3	4	5

Hand calculation of the least squares regression line requires the following data manipulation and summary:

Variable	x	y	x^2	y^2	xy
	1	−1	1	1	−1
	2	−1	4	1	−2
	3	2	9	4	6
	4	1	16	1	4
	5	4	25	16	20
Totals	15	5	55	23	27

From this summary, we may extract:

$$\begin{array}{ll} \sum_{i=1}^{5} x_i = 15 & \sum_{i=1}^{5} y_i = 5 \\ \sum_{i=1}^{5} x_i^2 = 55 & \sum_{i=1}^{5} x_i y_i = 27. \end{array}$$

Therefore

$$\widehat{\beta_1} = \frac{\sum_{i=1}^{n} x_i y_i - n\overline{x}\,\overline{y}}{\sum_{i=1}^{n} x_i^2 - n\overline{x}^2} = \frac{27 - (5)(\frac{15}{5})(\frac{5}{5})}{55 - (5)(\frac{15}{5})^2} = 1.2$$

and

$$\widehat{\beta_0} = \overline{y} - \widehat{\beta_1}\overline{x} = (\frac{5}{5}) - 1.2(\frac{15}{5}) = -2.6.$$

The least squares regression equation of 'population change' on 'temperature' is given by

$$y = -2.6 + 1.2x.$$

Scatterplot of bacterial population change against temperature

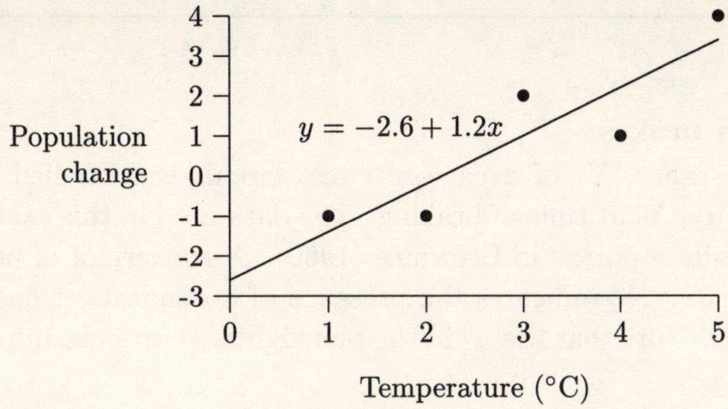

How can we interpret the slope of this least squares regression line?

We can predict the effect on the population change when the temperature is increased by 1°C. This is simply the slope of the least squares regression line.

At what temperature would you predict that there would be zero population change? Simply solve $0 = -2.6 + 1.2x$ for x to give a temperature of 2.167 degrees Celsius.

The residuals are found from the fitted values by

$$e_i = y_i - \widehat{y_i} = y_i - \widehat{\beta_0} - \widehat{\beta_1}x_i, \; i = 1, 2, 3, 4, 5.$$

These are determined as

$$\begin{aligned}
\widehat{y_1} &= -2.6 + 1.2(1) = -1.4 & e_1 &= -1 - (-1.4) = 0.4 \\
\widehat{y_2} &= -2.6 + 1.2(2) = -0.2 & e_2 &= -1 - (-0.2) = -0.8 \\
\widehat{y_3} &= -2.6 + 1.2(3) = 1.0 & e_3 &= 2 - (1) = 1.0 \\
\widehat{y_4} &= -2.6 + 1.2(4) = 2.2 & e_4 &= 1 - (2.2) = -1.2 \\
\widehat{y_5} &= -2.6 + 1.2(5) = 3.4 & e_5 &= 4 - (-3.4) = 0.6.
\end{aligned}$$

Straightening out the plot

The scatterplot of Example 14.1 showed an apparently strong linear trend in the sense that the plotted data lie close to the fitted line. In Example 14.2, the linear effect is less noticeable as the data are more scattered about the line. Even so, in both cases, the scatterplots show a linear trend.

Some scatterplots show a curvilinear trend when the data are plotted. By using a power transformation, it may be possible to bring the data 'into line' and apply linear regression techniques to find least squares estimators. The results may then be inverse-transformed back into the framework of the original scatterplot.

Example 14.3
Putting on the brakes
The stopping distance, Y, of a car on a test circuit was studied as a function of velocity, X, at time of braking. The data used in this example are based on results reported in Brownlee (1960). A scatterplot of points (x_i, y_i), $i = 1, 2, 3 \ldots, 8$, indicates the presence of a quadratic trend (or even exponential). Note that the '2' in the plot signifies two superimposed data points.

Scatterplot of stopping distance of a car

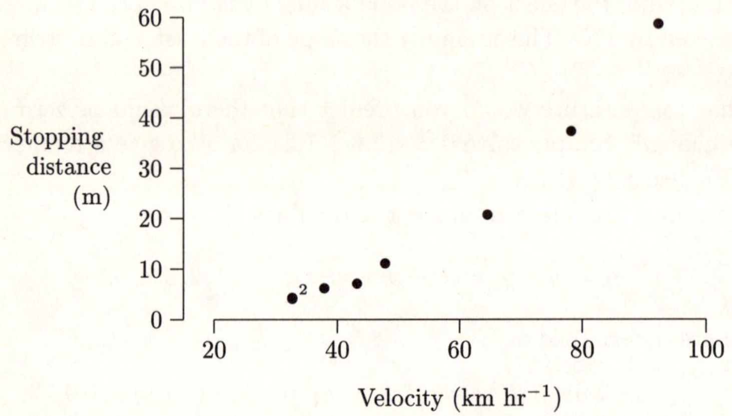

Scatterplot of square root of stopping distance of a car

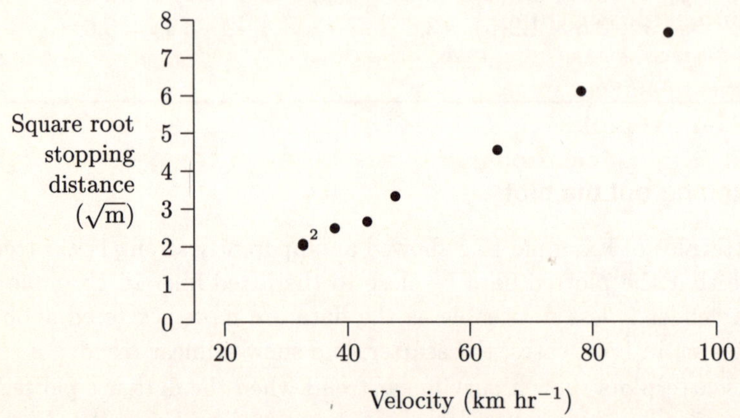

We can use the ladder of power transformations to choose a transformation that brings the large stopping distances vertically down the scatterplot to follow a linear trend. The logarithm and square root transformation could be tried; both have the effect of reducing long tails.

In the second scatterplot, the straight line effect is achieved through the square root transformation. If we set $y^* = \sqrt{y}$, then the least squares regression line of y^* on x fitted to the second scatterplot is

$$y^* = -1.11 + 0.0927x.$$

If an estimate of stopping distance is required for a given value of x (for example $x = 50$), then

$$y = (y^*)^2 = [-1.11 + (0.0927)(50)]^2 = 12.4$$

represents a stopping distance of 12.4 metres.

The transformation of a scatterplot of points (x_i, y_i), $i = 1, 2, 3 \ldots, n$, by power transformations f_p of the x coordinate and f_q of the y coordinate creates a transformed scatterplot

$$(f_p(x_i), f_q(y_i)),\ i = 1, 2, 3 \ldots, n.$$

If this results in a linear trend, then least squares estimates $\widehat{\beta_0}$ and $\widehat{\beta_1}$ may be found for the β_0 and β_1 values in $f_q(y) = \beta_0 + \beta_1 f_p(x)$. Then y may be estimated for a specified value of $x = x_o$ (within the range of the data available for x) by

$$y = f_q^{-1}[\widehat{\beta_0} + \widehat{\beta_1} f_p(x_o)].$$

[Here f_q^{-1} denotes the inverse function of f_q.]

Example 14.4

Undoing a logarithm

Suppose that logarithmic transformations of x and y values in a scatterplot create a linear trend.

Worked example:

What kind of relationship exists between the original x and y data?

Assume that a least squares regression line of $\log_e y$ on $\log_e x$ is of the form $\log_e y = \widehat{\beta_0} + \widehat{\beta_1} \log_e x$. In this case the ladder of power transformations is used with $p = 0$ and the power transformation is written as $f_0(w) = \log_e w$. The inverse function in this case is the exponential: $f_0^{-1}(w) = e^w$. Therefore, taking the exponential of each side in the least squares regression line, we find

$$y = e^{\widehat{\beta_0} + \widehat{\beta_1} \log_e x} = e^{\widehat{\beta_0}} e^{\widehat{\beta_1} \log_e x} = e^{\widehat{\beta_0}} (e^{\log_e x})^{\widehat{\beta_1}} = e^{\widehat{\beta_0}} (x)^{\widehat{\beta_1}},$$

since, by definition of the logarithm, $e^{\log_e x} = x$. The estimating equation for y is $y = e^{\widehat{\beta_0}} x^{\widehat{\beta_1}}$, so that on fitting the linear model, the least squares estimates $\widehat{\beta_0}$ and $\widehat{\beta_1}$ provide estimates $\widehat{\alpha}$ and $\widehat{\beta}$ of α and β in the **power model** $y = \alpha x^{\beta}$ through $\widehat{\alpha} = e^{\widehat{\beta_0}}$, and $\widehat{\beta} = \widehat{\beta_1}$.

EXERCISES

14.1 Suppose that the following results were obtained in a survey investigating medical expenses for Australian families.

Six families of different sizes were asked to state their medical expenses incurred over a six-week period. The data for X, representing 'number of children in family' and Y, representing 'family medical expenses ($)' were:

x_i	y_i
3	70
2	60
1	50
3	80
5	100
4	90

(a) Identify the response and the explanatory variable in this example.

(b) Draw a scatterplot of the data.

(c) Do increases in Y tend to accompany increases in X?

(d) When a line is fitted to the scatterplot, would you expect the slope to be positive?

(e) Copy and complete the following display:

x_i	y_i	x_i^2	y_i^2	x_iy_i
3	70	9	4900	210
2	60		3600	120
1	50	1		50
3	80	9	6400	
5	100	25	10000	
4	90	16	8100	

(f) What was the average number of children for families in the study?

(g) What was the average medical expenditure for families in this study?

(h) Show that the least squares regression line of y on x has equation $y = 36 + 13x$.

14.2 In order to determine the effect of a new fertiliser on yield of strawberry plants, a nursery proprietor grew plots of strawberry plants under identical conditions, except that the fertiliser dose levels were varied for each plot.

The following results were obtained:

Fertiliser dose (100g/plot)	Yield (berries/plant)
0	1.5
1	1.8
2	3.1
3	3.6
4	5.4

(a) In what sense have the levels of the explanatory variable been fixed by the nursery proprietor?

(b) Construct a scatterplot for these data.

(c) Determine the least squares regression line of 'yield' on 'dose'.

(d) Estimate the effect on 'yield' of a unit increase 'dose'.

(e) Using the least squares regression line, determine the fitted yield values at each dose level.

14.3 For the compressive strength of concrete data in Example 14.1, derive the least squares regression line of y on x. Retain as many decimal places as possible in your solution. Demonstrate that the residuals and fitted values are as given in the following table (exact numerical values are difficult to achieve because of round-off error).

x_i	y_i	e_i	$\widehat{y_i}$
2	2.53	−0.123028	2.65303
7	3.12	0.129483	2.99052
14	3.49	0.026999	3.46300
21	3.95	0.014515	3.93548
28	4.36	−0.047969	4.40797

14.4 When a least squares line is fitted to the data (x_1, y_1), $(x_2, y_2), \ldots, (x_n, y_n)$, for each i, values $\widehat{y_i}$ are fitted at x_i. Show that the sample mean of $\widehat{y_1}, \widehat{y_2}, \ldots, \widehat{y_n}$ is the same as the sample mean of $y_1, y_2, \ldots, y_n$.

14.5 In the spread of parasites in animal populations, the basic infection pressure, h, is defined as the mean rate at which a susceptible host would acquire parasites. A model in a population which lacks immunity is

$$m = ht$$

showing a linear increase in mean parasite count m with time t.

Roberts, Lawson and Gemmell (1986) give the mean larvae count of *Echinococcus granulosus* (hydatids) in female sheep of different ages, in New Zealand in 1958.

Mean larvae (m)	Age sheep t (years)
2.5	3
4.5	4
6.2	5
7.4	6

(*Source*: Roberts, M.G., Lawson, J.R. and Gemmell, M.A. (1986). Epidemiology and the control of hydatidosis and cysticercosis. *Pacific Statistical Congress, 1985*, 460-3.)

(a) Construct a scatterplot for these data.

(b) Use the data to provide a least squares estimate of the infection pressure parameter h in the presence of an initial age parameter t_0 representing the time from birth until initial infections are acquired through grazing.

14.6 The following data are the amounts, Y, of unconverted substance (in milligrams) from 6 similar chemical reactions after x minutes:

y	23.5	16.9	17.5	14.0	9.8	8.9
x	1	2	2	3	5	6

(a) Construct a scatterplot for these data.

(b) Show that the regression equation

$$Y = \gamma\delta^x \qquad \ldots(*)$$

can be put in the linear form $W = \alpha + \beta x$ by taking logarithms.

(c) Determine the least squares estimates of γ and δ in $(*)$. The following calculations may be of assistance:

$$\begin{aligned} \textstyle\sum_{i=1}^{6} x_i \log_e y_i &= 46.98 \\ \textstyle\sum_{i=1}^{6} (\log_e y_i)^2 &= 43.11 \\ \textstyle\sum_{i=1}^{6} \log_e y_i &= 15.95 \\ \textstyle\sum_{i=1}^{6} x_i &= 19 \\ \textstyle\sum_{i=1}^{6} x_i^2 &= 79 \end{aligned}$$

(d) Estimate the amount of unconverted substance in this type of reaction after 4 minutes have elapsed.

14.7 Show that

$$A(a) = \sum_{i=1}^{n} (y_i - a)^2$$

is minimised for $a = \overline{y}$.

How may this result be used when fitting an equation $y = \beta_0$, where β_0 is a constant, to a scatterplot?

14.8 Barr (1986) gives data on New Zealand Geothermal Fields used for the generation of electricity. The temperature, extent, permeability are factors influencing output from a field.

The temperature of a field is determined by discovery drilling. The data below show the maximum megawatts per well and temperature for several field locations.

Location	**Temperature** °C	**Maximum output** *(MW)*
Wairakei	260	10
Kawerau	290	25
Broadlands	270	13.5
Mokai	290	25
Rotokawa	300	10
Ngawha	225	8

(*Source*: Barr, H. (1986). Uncertainty and other impediments to geothermal power. *Pacific Statistical Congress, 1985*, 375-8.)

(a) Theory predicts that maximum output depends on the cube of the temperature. Construct a scatterplot relating the cube root of maximum output and temperature.
Explain your choice of response variable in this scatterplot.

(b) Fit a least squares line to the plot in (a) and comment on whether the observed data support the theory.

14.9 Show that

$$\sum_{i=1}^{n} (y_i - a)(x_i - \overline{x}) = \sum_{i=1}^{n} y_i(x_i - \overline{x})$$

for any constant a.

14.10 (a) Show that when the line $y = \beta x$, which passes through the origin, is fitted to the scatterplot

$$(x_i, y_i),\ i = 1, 2, 3, \ldots, n,$$

the least squares estimate of β is

$$\widehat{\beta} = \frac{\sum_{i=1}^{n} x_i y_i}{\sum_{i=1}^{n} x_i^2}.$$

(b) The following data give observed values for the coded temperature X and percentage impurity Y in an industrial quality control experiment.

Coded temperature x	**Impurities** %
0	1
1	3
2	5
3	9
4	12
4	11
5	14
7	22
8	25

The coded temperature $x = 0$ corresponds to 200 degrees Celsius, while the coded temperature $x = 8$ corresponds to 240 degrees Celsius.

i. Construct a scatterplot for these data.

ii. Fit a least squares line through the origin.

iii. If the coded temperature X is a linear transformation of the temperature T in degrees Celsius used in the experiment, write down the linear equation that relates X and T.

iv. Obtain the equation of the least squares regression line of y against t which is constrained to pass through $t = 200$, representing the minimum temperature for the process.

14.11 If e_i, $i = 1, 2, 3, \ldots, n$, are the observed residuals when a least squares line is fitted to a scatterplot, show that

$$\sum_{i=1}^{n} e_i = 0.$$

Use this result to show that the point $(\overline{x}, \overline{y})$ lies on the least squares regression line

$$y = \widehat{\beta_0} + \widehat{\beta_1} x.$$

14.12 Suppose that e_i, $i = 1, 2, 3, \ldots, n$, are the observed residuals when a least squares line is fitted to a scatterplot of points

$$(x_i, y_i),\ i = 1, 2, 3, \ldots, n.$$

(a) Show that

$$\sum_{i=1}^{n} x_i e_i = 0.$$

(b) Use (a) to establish that

$$\sum_{i=1}^{n} (x_i - \overline{x})(e_i - \overline{e}) = 0.$$

[The expression on the left-side of this last equation is a 'sample covariance' measure of the strength of the relationship between $x_1, x_2, x_3, \ldots, x_n$ and $e_1, e_2, e_3, \ldots, e_n$.]

14.2 THE LINEAR REGRESSION MODEL

The study of regression is primarily a study of scatterplots. We construct scatterplots for data from a pair of variables where we suspect that there may be a functional relationship between the variables. Regression is the study of that relationship. Linear regression is the study of straight line relationships. In Section 14.1 we moved from carefully examining scatterplots exhibiting a linear trend to uncovering (and trying to estimate by least squares) a **deterministic linear equation** to describe that trend. But all the points do not fall on the deterministic linear equation. The line and its equation only summarise the trend. The line is like a 'signal' and the scatter about the line, 'noise'.

Our aim is to construct a probability model with a fixed deterministic linear component for the signal and a random scatter component to describe the noise. This type of probability model has been used in our study of comparing several samples in ANOVA. [See (12.11).]

The **simple linear regression model** is as follows:

$$Y_i = \beta_0 + \beta_1 x_i + \epsilon_i, \quad \text{where the} \quad \epsilon_i \text{ are } i.i.d.\ N(0, \sigma^2) \text{ for each } i = 1, 2, 3, \ldots, n. \tag{14.3}$$

The terms in the model are explained as follows:

1. β_0 and β_1 are fixed, unknown parameters.
2. x_i is the ith observation of the explanatory variable X. Since these are often pre-selected by the experimenter, they are **fixed, known** constants which are measured **without error**.
3. Y_i is the **response variable** at $X = x_i$.
4. The **regression function** is linear in x: $\mu(x_i) = E(Y|X = x_i) = \beta_0 + \beta_1 x_i$.
5. The errors ϵ_i are independent and identically normally distributed with mean 0 and variance σ^2. This error variance represents the 'noise' component of random scatter about the line.

By writing the model as

$$Y_i = \underbrace{\mu(x_i)}_{\text{deterministic component}} + \underbrace{\epsilon_i}_{\text{random component}}$$

the signal and noise become apparent. It is also apparent that it is not Y that is linear in x, but rather the deterministic part $\mu(x) = E(Y|x)$ which is chosen to be linear in x. (In curvilinear regressions, the regression function $\mu(x)$ may be modelled as a quadratic—in this text, however, we consider only the linear model.) By point 5 above, the spread of values about the line is constant for all values of x. That is, the scatter of Y values in a vertical direction is assumed to be the same for all choices of x. The errors are chosen to follow independent normal distributions so that significance tests based on normal distribution assumptions (such as the t test) may proceed.

What is the distribution of the ith response?

What does the model postulate for the probability distribution of the ith response Y_i?

If ϵ_i is normally distributed, then $const + \epsilon_i$ is normally distributed (by the methods of Section 4.5), ensuring that $Y_i = const + \epsilon_i$ is normally distributed. To determine the mean and variance of Y_i,

$$E(Y_i) = E(\beta_0 + \beta_1 x_i + \epsilon_i) = E(\beta_0 + \beta_1 x_i) + E(\epsilon_i) = \beta_0 + \beta_1 x_i + 0$$

and

$$Var(Y_i) = Var(\beta_0 + \beta_1 x_i + \epsilon_i) = Var(\beta_0 + \beta_1 x_i) + Var(\epsilon_i) = 0 + \sigma^2.$$

Combining all three results gives $Y_i \sim N(\beta_0 + \beta_1 x_i,\ \sigma^2)$. Therefore **the mean responses follow the true regression line.**

A cautionary note is relevant here: If the responses in a scatterplot are transformed, the error distribution is transformed as well, and may no longer satisfy the mean zero, constant variance demands of the linear model.

Example 14.5

Mazda price distributions for the last 20 years

The following data is of the retail price of 124 Mazda cars. (Source: *The Melbourne Age*, 8 February 1992.) The listing is by year without regard to make or model.

If we take the explanatory variable X to be the age of the car (years), then when 1991 is taken as $x = 0$, the data supply the fixed values of the explanatory variable as $x = 0, 1, 2, 3, \ldots, 20$. Above each of these fixed x values in the scatterplot is the probability distribution corresponding to 'Mazda price', the response variable Y.

Year	Price ($)	Year	Price ($)	Year	Price ($)	Year	Price ($)
79	2950	82	5999	81	5500	86	10500
82	5900	81	5990	82	11500	84	8000
83	2999	88	12500	79	2600	79	4600
88	11950	85	14000	81	5100	83	4975
82	6100	90	20500	81	4000	87	18950
90	26500	79	4250	77	4500	85	10000
82	6850	80	4500	78	2600	90	30000
81	5490	85	7750	75	2400	82	4000
86	12999	71	950	80	4900	87	7650
82	9900	79	4300	87	14950	76	2695
89	19900	74	2300	83	8900	86	12490
80	3990	88	9490	85	13490	89	36900
73	800	84	8900	84	5499	88	18000
81	3900	80	3000	84	12999	89	27900
79	7000	83	6500	88	21500	90	22950
80	2750	79	8000	79	4200	91	23900
85	11000	73	1450	82	4200	86	11900
87	9850	87	14150	84	10500	84	9800
84	7500	77	1950	84	10800	89	20980
89	26500	76	1800	84	8600	74	5000
73	1990	88	18000	83	8950	82	10980
84	7000	88	16350	86	7900	80	2800
78	3500	78	1690	79	3700	90	24500
85	6990	81	3490	91	25600	82	10995
90	39990	89	26000	84	11000	79	7879
87	29990	79	4000	86	8700	82	6990
90	29990	81	5490	80	3800	77	1498
89	25990	86	7900	89	10950	88	17900
88	21990	82	10500	83	6450	86	9990
88	19990	76	799	86	9450	91	13990
86	6000	81	4200	76	2950	89	9990

The scatterplot of Y against X shows a decreasing slope as x increases. Either a logarithm or square root transformation seems appropriate for straightening out the plot; $Y^* = \log_{10}(Y)$ against X is displayed in the second scatterplot. The trend is strongly linear, suggesting that on a logarithmic scale, Mazda cars decrease in value at a constant rate irrespective of model. The linear model to be fitted here is

$$Y_i^* = \beta_0 + \beta_1 x_i + \epsilon_i, \quad i = 1, 2, 3, \ldots, 20.$$

For each of the values of x_i, observations on the distribution of Y_i^* are displayed as vertical columns of black dots; it is the means of these distributions that follow the regression line.

Retail price of Mazdas

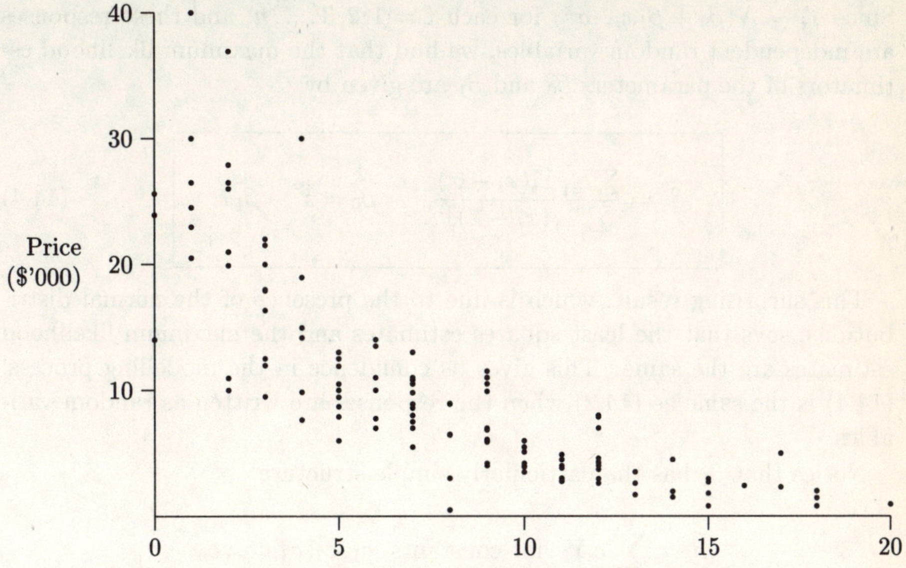

Retail price of Mazdas

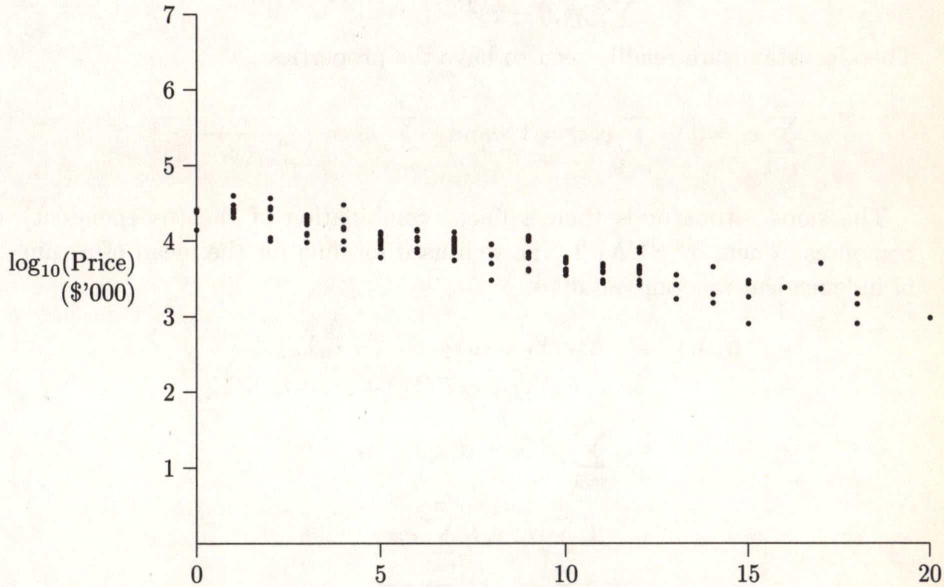

Probability distribution for $\widehat{\beta_1}$

Since $Y_i \sim N(\beta_0 + \beta_1 x_i,\ \sigma^2)$ for each $i = 1, 2, 3 \ldots, n$, and these responses are independent random variables, we find that the maximum likelihood estimators of the parameters β_0 and β_1 are given by

$$\boxed{\widehat{\beta_1} = \frac{\sum_{i=1}^n Y_i(x_i - \overline{x})}{\sum_{i=1}^n (x_i - \overline{x})^2}, \qquad \widehat{\beta_0} = \overline{Y} - \widehat{\beta_1}\overline{x}.} \tag{14.4}$$

This surprising result, which is due to the presence of the normal distributions, says that the least squares estimates and the maximum likelihood estimates are the same. This gives us confidence in the modelling process! (14.4) is the same as (14.2) when the responses are written as random variables.

Notice that $\widehat{\beta_1}$ has the particularly simple structure

$$\widehat{\beta_1} = \sum_{i=1}^n c_i Y_i \ \text{ for constants } \ c_1, c_2, c_3, \ldots, c_n,$$

where

$$c_i = \frac{x_i - \overline{x}}{\sum_{i=1}^n (x_i - \overline{x})^2}, \quad i = 1, 2, 3 \ldots, n.$$

These constants are readily seen to have the properties:

$$\sum_{i=1}^n c_i = 0, \quad \sum_{i=1}^n c_i x_i = 1 \quad \text{and} \quad \sum_{i=1}^n c_i^2 = \frac{1}{\sum_{i=1}^n (x_i - \overline{x})^2}.$$

The slope estimator is then a linear combination of the (independent) responses. Then, by **SUM 1**, the well-used formula for the mean of a sum of independent random variables,

$$\begin{aligned}
E(\widehat{\beta_1}) &= E(c_1Y_1 + c_2Y_2 + \ldots + c_nY_n) \\
&= c_1E(Y_1) + c_2E(Y_2) + \ldots + c_nE(Y_n) \\
&= \sum_{i=1}^n c_i(\beta_0 + \beta_1 x_i) \\
&= \beta_0 \underbrace{\sum_{i=1}^n c_i}_{=0} + \beta_1 \underbrace{\sum_{i=1}^n c_i x_i}_{=1} \\
&= \beta_1.
\end{aligned}$$

By **SUM 2**,

$$\begin{aligned}
Var(\widehat{\beta_1}) &= Var(c_1Y_1 + c_2Y_2 + \ldots + c_nY_n) \\
&= c_1^2 Var(Y_1) + c_2^2 Var(Y_2) + \ldots + c_n^2 Var(Y_n)
\end{aligned}$$

$$= \sum_{i=1}^{n} c_i^2 \sigma^2$$

$$= \frac{\sigma^2}{\sum_{i=1}^{n}(x_i - \overline{x})^2}.$$

Finally, $\widehat{\beta_1}$ is normally distributed since a linear combination of independent normal random variables is again normally distributed (Theorem 7.5). These arguments establish the following theorem.

Theorem 14.1 *The least squares slope estimator* $\widehat{\beta_1}$ *has sampling distribution*

$$\widehat{\beta_1} \sim N(\beta_1, \frac{\sigma^2}{\sum_{i=1}^{n}(x_i - \overline{x})^2}).$$

Estimating variability about the line

Using a normal probability model, the ith response Y_i is modelled as $Y_i \sim N(\beta_0 + \beta_1 x_i,\ \sigma^2)$. The only unknown parameter which remains to be estimated is σ^2 governing the amount of random scatter about the true regression line. According to the method of least squares, the amount of scatter about the line is measured by the sum of squared deviations

$$\sum_{i=1}^{n}(Y_i - \widehat{Y_i})^2 = \sum_{i=1}^{n} E_i^2,$$

where random variables $\widehat{Y_i}$ denotes the **fitted value** and $E_i = Y_i - \widehat{Y_i}$ is the ith *residual* at x_i. ($e_1, e_2, e_3, \ldots, e_n$ are the observed values of $E_1, E_2, E_3, \ldots, E_n$.) With slope and intercept $\widehat{\beta_0}$ and $\widehat{\beta_1}$, (compared to other choices of slope and intercept of a fitted line) this sum of squared deviations is as small as possible.

Definition 14.4 *The* **residual sum of squares** *is given by*

$$\text{SS RESIDUALS} = \sum_{i=1}^{n}(Y_i - \widehat{Y_i})^2,$$

when $\widehat{Y_i} = \widehat{\beta_0} + \widehat{\beta_1} x_i$ *are the responses fitted by the method of least squares.* □

Many computer packages produce SS RESIDUALS as automatic output, but for hand calculation, once the slope estimator $\widehat{\beta_1}$ and intercept estimator $\widehat{\beta_0}$ have been determined, the equivalent formula

$$\text{SS RESIDUALS} = \sum_{i=1}^{n} Y_i^2 - \widehat{\beta_0}\sum_{i=1}^{n} Y_i - \widehat{\beta_1}\sum_{i=1}^{n} x_i Y_i$$

provides for ease of calculation, provided rounding is done only at the end of the calculation.

The maximum likelihood estimator of σ^2 is a constant multiple of SS RESIDUALS. If we use the criterion of unbiasedness, then $E(\text{SS RESIDUALS}) = (n-2)\sigma^2$. More advanced statistical texts such as Hogg and Craig (1978) may be consulted for a proof that

$$\frac{\text{SS RESIDUALS}}{\sigma^2} \sim \chi^2_{n-2}.$$

An unbiased estimate of the variability, σ^2, about the line is

$$\widehat{\sigma^2} = \frac{\text{SS RESIDUALS}}{n-2} = \frac{\sum_{i=1}^n Y_i^2 - \widehat{\beta_0}\sum_{i=1}^n Y_i - \widehat{\beta_1}\sum_{i=1}^n x_i Y_i}{n-2}. \tag{14.5}$$

The $(n-2)$ appearing in the denominator of $\widehat{\sigma^2}$ should be no surprise. We will look more closely at the distribution of the residuals in the next section. However, we note here that only $n-2$ of the quantities $e_1, e_2, e_3, \ldots, e_n$ may be freely chosen. There are *two* restrictions on the observed residuals:

Restriction 1: $\sum_{i=1}^n e_i = 0.$
Restriction 2: $\sum_{i=1}^n x_i e_i = 0.$

Example 14.6

Example 14.3 continued: variability about the braking line

A scatterplot of 'stopping distance', Y, against 'velocity at time of braking' X, was shown for 8 data points in Example 14.3. The tabled values below are based on that data and illustrate the numerical calculation of $\widehat{\sigma^2}$, the estimate of the variability about the line.

x	y	$y^* = \sqrt{y}$	xy^*	
32.8	4.1	2.02485	66.415	
32.8	4.3	2.07364	68.016	
37.9	6.2	2.48998	94.370	
43.2	7.1	2.66458	115.110	
47.8	11.1	3.33167	159.254	
64.5	20.7	4.54973	293.457	
78.2	37.3	6.10737	477.597	
92.3	58.6	7.65506	706.562	
	149.4	30.8969	1980.78	Totals

The required calculation proceeds by retaining as many decimal places as possible in all the intermediate steps, as the formula for $\widehat{\sigma^2}$ is notoriously sensitive to round-off error. Therefore we use $\widehat{\beta_0} = -1.1125$ and $\widehat{\beta_1} = 0.092659$ determined from columns 1 and 3 of the above table. Then

$$\begin{aligned}
\widehat{\sigma^2} &= \frac{\sum_{i=1}^{n}(Y_i^*)^2 - \widehat{\beta_0}\sum_{i=1}^{n} Y_i^* - \widehat{\beta_1}\sum_{i=1}^{n} x_i Y_i^*}{n-2} \\
&= \frac{\sum_{i=1}^{n} Y_i - \widehat{\beta_0}\sum_{i=1}^{n}\sqrt{Y_i} - \widehat{\beta_1}\sum_{i=1}^{n} x_i\sqrt{Y_i}}{n-2} \\
&= \frac{149.4 - (-1.1125)(30.8969) - (0.092659)(1980.78)}{6} \\
&= 0.03928455.
\end{aligned}$$

This gives $\widehat{\sigma} = 0.1982$. (The same calculation done with the MINITAB package produced 0.1985.)

An estimated standard pivot for assessing slope

The standardised slope estimator

$$\frac{\widehat{\beta_1} - \beta_1}{\sqrt{\frac{\sigma^2}{\sum_{i=1}^{n}(x_i - \overline{x})^2}}}$$

is of the form

$$\frac{\widehat{\theta} - \theta}{S.E.(\widehat{\theta})}$$

and has a standard normal distribution (Theorem 14.1) not depending on the parameter $\theta = \beta_1$. It is therefore a pivot. To be of use, however, the nuisance parameter σ^2 must be estimated by $\widehat{\sigma^2}$, according to (14.5). This gives the estimated standard pivot structure

$$\frac{\widehat{\theta} - \theta}{\widehat{S.E.(\widehat{\theta})}}$$

which follows a t distribution.

Theorem 14.2 *If $\widehat{\beta_1}$ represents the least squares estimator of slope in the true regression line, then*

$$T = \frac{\widehat{\beta_1} - \beta_1}{\sqrt{\frac{\widehat{\sigma^2}}{\sum_{i=1}^{n}(x_i - \overline{x})^2}}} \sim t_{n-2}.$$

Proof: The proof follows the pattern of Theorem 10.7 which established a t distribution for the sample mean when the population variance is unknown. The current scenario has $\widehat{\beta_1}$ filling the role of a sample mean and the unknown σ^2 in the role of the unknown population variance.

■

Comment: $\widehat{\beta_1} = \sum_{i=1}^{n} c_i Y_i$ is 'like' a sample mean provided that the quantities

$$c_i = \frac{x_i - \overline{x}}{\sum_{i=1}^{n}(x_i - \overline{x})^2}, \quad i = 1, 2, 3 \ldots, n,$$

behave 'like' fractions $\frac{1}{n}$. This occurs when $\sum_{i=1}^{n}(x_i - \overline{x})^2$ behaves like the sample size n and none of the discrepancies $x_i - \overline{x}$ contributes enormously to the magnitude of the summation $\sum_{i=1}^{n}(x_i - \overline{x})^2$. This helps to ensure that $\widehat{\beta_1}$ is normally distributed for large sample sizes even when the responses Y_i are not normally distributed.

According to Theorem 14.2, T is appropriate to use as a test statistic for detecting whether there is slope in the regression line: it is often crucial to be able to test the hypothesis $H_0 : \beta_1 = 0$ against a one- or two-sided alternative hypothesis that a positive/negative slope is present. If H_0 is true then we waste our time performing a linear regression. For in this case, the explanatory variable X is of no assistance in describing the behaviour of the response.

Consider:

$$\begin{aligned} H_0 : &\quad \beta_1 = 0 \\ H_1 : &\quad \beta_1 \neq 0. \end{aligned}$$

Choosing between H_0 and H_1 is equivalent to choosing between the linear regression models:

$$\begin{aligned} &\text{Model 1:} \quad Y_i = \beta_0 + \beta_1 x_i + \epsilon_i, \quad \epsilon_i \ \textit{i.i.d.}\ N(0, \sigma^2). \\ &\text{Model 2:} \quad Y_i = \beta_0 + \epsilon_i, \quad \epsilon_i \ \textit{i.i.d.}\ N(0, \sigma^2). \end{aligned}$$

If H_0 is true, Model 2 is an appropriate model for the responses. If H_1 is true, Model 1 is an appropriate model for the responses Y_i.

Of course, if we use the method of least squares to fit Model 2 to a scatterplot of y against x (and the data would likely be quite scattered in the plot with no visual linear trend), then we attempt to fit a line of the form $y = a$ to the scatterplot. This is a line of zero slope. Method of least squares: choose the value a of intercept, so that the sum of squared discrepancies

$$\sum_{i=1}^{n}(y_i - a)^2$$

is minimum. Since

$$\sum_{i=1}^{n}(y_i - a)^2 = \sum_{i=1}^{n}(y_i - \overline{y})^2 + n(\overline{y} - a)^2,$$

it follows that the left-side is minimised when $a = \overline{y}$. That is, the least squares estimator of β_0 under Model 2 is

$$\widehat{\beta_0} = \overline{Y}.$$

This means that in order to 'predict' a response when X offers no useful assistance as an explanatory variable, we simply take the average response.

The acceptance region of a two-sided hypothesis test for β_1 corresponds to a confidence interval for β_1. If we unwind the pivot T in Theorem 14.2, then we may detect whether 0 lies inside the $100(1-\alpha)\%$ confidence interval for β_1:

$$\left[\widehat{\beta_1} - t_{\frac{\alpha}{2},n-2}\frac{\widehat{\sigma}}{\sqrt{\sum_{i=1}^{n}(x_i-\overline{x})^2}},\ \widehat{\beta_1} + t_{\frac{\alpha}{2},n-2}\frac{\widehat{\sigma}}{\sqrt{\sum_{i=1}^{n}(x_i-\overline{x})^2}}\right].$$

Example 14.7

Example 14.5 continued: slope detection with a confidence interval

When Y is the Mazda retail price and X is the declared age of the car, then the scatterplot of Example 14.5 shows a dramatic linear trend with negative slope for $\log_{10} Y$ against X.
The fitted least squares line of $\log_{10} Y$ against X has equation

$$y^* = \log_{10} y = 4.425477 - 0.070747x.$$

Since the slope is small and negative, it is important to detect whether, in fact, it is present. The scatterplot makes us suspect that the slope is significantly different from 0, but is this visual effect simply due to the relative scales used on the x and $\log_{10} y$ axes?
Is the slope a *trompe-l'oeil* effect?
A simple 95% confidence interval for $\widehat{\beta_1}$ answers this: If the interval contains 0 then the slope is not significantly different from 0. Since, by calculation, the standard deviation of the x values is $s_x = \sqrt{\frac{\sum_{i=1}^{n}(x_i-\overline{x})^2}{n-1}} = 4.648$, it follows that with $n = 124$, $\sum_{i=1}^{n}(x_i-\overline{x})^2 = (4.648)^2(123)$. Further, $t_{0.025,122} \approx z_{0.025} = 1.96$ and $\widehat{\sigma} = 0.1594$. Substituting this information into the confidence interval, we obtain

$$\begin{aligned}&[-0.070747 - 1.96\tfrac{0.1594}{\sqrt{(4.648)^2(123)}}, -0.070747 + 1.96\tfrac{0.1594}{\sqrt{(4.648)^2(123)}}]\\ &= [-0.077, -0.065],\end{aligned}$$

which, significantly, does not cross 0.

An ANOVA table for detecting zero slope

When the data $(x_i, y_i), i = 1, 2, 3 \ldots, n$, are plotted in the scatterplot, there is variability in the points in a vertical direction. This may simply be random scatter (Model 2), but the variability in response in a vertical direction may be due to a scatterplot observation that the responses are following a linear trend—in which case we would use Model 1 to model the responses. In a sense, the points are less variable because they are near a line. How are these different kinds of variability measured?

Definition 14.5 *The* **total sum of squares**, *denoted by SS TOTAL, is given by*

$$\text{SS TOTAL} = \sum_{i=1}^{n}(Y_i - \overline{Y})^2.$$

The **regression sum of squares**, *denoted by SS REGRESSION, is given by*

$$\text{SS REGRESSION} = \sum_{i=1}^{n}(\widehat{Y}_i - \overline{Y})^2.$$

□

The quantities SS TOTAL, SS REGRESSION and SS RESIDUALS are reminiscent of ANOVA notations. In fact, together, these sums of squares form the basis of an ANOVA table.

Begin with the equality,

$$(Y_i - \overline{Y}) = (\widehat{Y}_i - \overline{Y}) + (Y_i - \widehat{Y}_i).$$

By squaring each side and adding the result over all points in the scatterplot,

$$\sum_{i=1}^{n}(Y_i - \overline{Y})^2 = \sum_{i=1}^{n}(\widehat{Y}_i - \overline{Y})^2 + \sum_{i=1}^{n}(Y_i - \widehat{Y}_i)^2 + 2\sum_{i=1}^{n}(\widehat{Y}_i - \overline{Y})(Y_i - \widehat{Y}_i).$$

As in the case of ANOVA, the cross-product term $\sum_{i=1}^{n}(\widehat{Y}_i - \overline{Y})(Y_i - \widehat{Y}_i)$ vanishes. (The interested manipulator should consult the exercises at the end of this section.) This leaves us with an equation fundamental to all of regression

$$\text{SS TOTAL} = \text{SS REGRESSION} + \text{SS RESIDUALS} \tag{14.6}$$

The probability distribution for SS REGRESSION is simply obtained: Since $(\overline{x}, \overline{y})$ lies on the least squares regression line, this centre of gravity of the scatterplot satisfies $\overline{Y} = \widehat{\beta_0} + \widehat{\beta_1}\overline{x}$. Using this in the formula for SS REGRESSION,

$$\begin{aligned}\text{SS REGRESSION} &= \sum_{i=1}^{n}(\widehat{Y}_i - \overline{Y})^2 \\ &= \sum_{i=1}^{n}[\widehat{\beta_0} + \widehat{\beta_1}x_i - (\widehat{\beta_0} + \widehat{\beta_1}\overline{x})]^2 \\ &= \sum_{i=1}^{n}[\widehat{\beta_1}x_i - \widehat{\beta_1}\overline{x}]^2 \\ &= \widehat{\beta_1}^2\sum_{i=1}^{n}(x_i - \overline{x})^2.\end{aligned}$$

Distributionally, when $H_0 : \beta_1 = 0$ holds,

$$\frac{\text{SS REGRESSION}}{\sigma^2} = \frac{\widehat{\beta_1}^2\sum_{i=1}^{n}(x_i-\overline{x})^2}{\sigma^2} = \left[\frac{\widehat{\beta_1}-0}{\sqrt{\frac{\sigma^2}{\sum_{i=1}^{n}(x_i-\overline{x})^2}}}\right]^2 = Z^2,$$

where $Z \sim N(0,1)$. Recalling that the square of a $N(0,1)$ random variable has a chi-squared distribution with 1 degree of freedom, we write

$$\frac{\text{SS REGRESSION}}{\sigma^2} \sim \chi_1^2.$$

This means that SS REGRESSION has 1 degree of freedom. Further, since it may be shown that

$$\frac{\text{SS RESIDUALS}}{\sigma^2} \sim \chi_{n-2}^2$$

independently of SS REGRESSION, we have

$$\frac{\text{SS TOTAL}}{\sigma^2} = \frac{\text{SS REGRESSION}}{\sigma^2} + \frac{\text{SS RESIDUALS}}{\sigma^2} \sim \chi_1^2 + \chi_{n-2}^2 = \chi_{n-1}^2.$$

Therefore, both the degrees of freedom and the sums of squares in the following ANOVA table, add.

Source	Degrees of freedom	Sums of squares	Mean squares	F
Regression	1	SS REGRESSION	MS REGRESSION	$\frac{\text{MS REGRESSION}}{\text{MS RESIDUALS}}$
Residuals	$n-2$	SS RESIDUALS	MS RESIDUALS	
Total	$n-1$	SS TOTAL		

When H_0 is true,

$$F = \frac{\text{MS REGRESSION}}{\text{MS RESIDUALS}} = \frac{\frac{\text{SS REGRESSION}}{(1)\sigma^2}}{\frac{\text{SS RESIDUALS}}{(n-2)\sigma^2}} = \frac{\left[\frac{\widehat{\beta_1}-\beta_1}{\sqrt{\frac{\sigma^2}{\sum_{i=1}^n (x_i-\overline{x})^2}}}\right]^2}{\frac{\widehat{\sigma^2}}{\sigma^2}}.$$

This establishes that $T^2 \sim F_{1,n-2}$ so that the F statistic in the ANOVA table and the T statistic in Theorem 14.2 are equivalent test statistics for testing whether or not a slope is present in a scatterplot. We may use either. However, it is very important to be able to understand the structure of the ANOVA table, since standard computer packages adopt this format for their regression output.

The ANOVA output for the Mazda data of Example 14.5 is

Source	Degrees of freedom	Sums of squares	Mean squares	F
Regression	1	13.307	13.307	532.28
Residuals	122	3.102	0.025	
Total	123	16.409		

The F ratio, which should be close to 1 if $H_0 : \beta_1 = 0$ were true, is highly significant: $f_{observed} = 532.28$ (**).

Estimating the mean response at $x = x_{fixed}$

One of the most useful aspects of the least squares line is to use it to estimate mean response for a particular value of the explanatory variable of interest to us—denote this value by $x = x_{fixed}$. Then, a point estimator of the mean response $\theta = \mu(x_{fixed}) = E(Y|X = x_{fixed})$ is given by

$$\widehat{\theta} = \widehat{\beta_0} + \widehat{\beta_1} x_{fixed}.$$

This estimator is simply a linear combination of the mean responses. To see this, write

$$\widehat{\beta_0} + \widehat{\beta_1} x_{fixed} = \sum_{i=1}^n \left[\frac{1}{n} + \frac{(x_{fixed} - \overline{x})(x_i - \overline{x})}{\sum_{i=1}^n (x_i - \overline{x})^2}\right] Y_i = \sum_{i=1}^n k_i Y_i,$$

where the coefficients $k_1, k_2, k_3, \ldots, k_n$ are constant. Following the proof of Theorem 14.1, with $\widehat{\theta} = \sum_{i=1}^n k_i Y_i$ replacing $\widehat{\beta_1} = \sum_{i=1}^n c_i Y_i$ we obtain

$$\widehat{\theta} = \widehat{\beta_0} + \widehat{\beta_1} x_{fixed} \sim N\left(\beta_0 + \beta_1 x_{fixed},\ \sigma^2 \left[\frac{1}{n} + \frac{(x_{fixed} - \overline{x})^2}{\sum_{i=1}^n (x_i - \overline{x})^2}\right]\right). \quad (14.7)$$

The variance of the distribution of $\widehat{\theta}$ is point estimated by replacing σ in the variance term of (14.7) by $\widehat{\sigma}$, in the same way that sample standard deviations are used to point estimate population standard deviations. Also the part of the variance expression that involves $x_1, x_2, x_3, \ldots, x_n$ is small when x_{fixed} is near $\overline{x}$. This means that the mean response is more accurately estimated at x values at the centre of the data set.

According to our knowledge of confidence intervals we may use the standard structure: A $100(1-\alpha)\%$ confidence interval for the mean response $\theta = \mu(x_{fixed}) = E(Y|X = x_{fixed})$, at the value $x = x_{fixed}$ of the explanatory variable, is given by

$$\left[\widehat{\theta} - t_{\frac{\alpha}{2},n-2}\widehat{S.E.(\widehat{\theta})},\ \widehat{\theta} + t_{\frac{\alpha}{2},n-2}\widehat{S.E.(\widehat{\theta})}\right],$$

for

$$\widehat{S.E.(\widehat{\theta})} = \widehat{\sigma}\sqrt{\left[\frac{1}{n} + \frac{(x_{fixed} - \overline{x})^2}{\sum_{i=1}^{n}(x_i - \overline{x})^2}\right]}.$$

From this, the distribution of, the test statistic for, the confidence interval for, ..., the intercept parameter follow immediately by the substitution $x_{fixed} = 0$. In particular, for example,

$$\widehat{\beta_0} \sim N\left(\beta_0, \sigma^2\left[\frac{1}{n} + \frac{\overline{x}^2}{\sum_{i=1}^{n}(x_i-\overline{x})^2}\right]\right).$$

Extrapolation

In general there is a **danger in extrapolation** when the mean response is sought for values x_{fixed} of the explanatory variable. We should question the mean response only across an interval of values of the explanatory variable with which we are familiar—values that have been used in the experimental settings. Any linearity that may be apparent in the scatterplot simply may *not* hold outside the support of $x_1, x_2, x_3, \ldots, x_n$; that is, outside interval $[x_{(1)}, x_{(n)}]$.

EXERCISES

14.13 A steel mill wishes to know how much fixed diameter steel rods elongate when a known tensile force is applied. The company engineers perform an experiment with the following results:

Force (units)	Elongation (mm)
1	2.1
2	3.9
3	5.2
4	6.0
5	8.3

(a) Determine the residual sum of squares when a least squares line of 'elongation' on 'tensile force' is fitted to these data.

(b) Construct a 95% confidence interval for the slope of the regression line of 'elongation' on 'tensile force'.

14.14 The following data from Salvia (1990) shows particle size, Y, (10^{-6}m) after grinding small ceramic chips for T minutes, where fixed values t of T were selected as 5 minutes, 10 minutes, 15 minutes, ...

Grinding time	Particle size
5	40
5	36
10	29
10	35
15	22
15	21
20	16
20	20

(a) Before calculations are performed, explain why you would expect the least squares line to have negative slope.

(b) Test $H_0 : \beta_1 = -1$ against an alternative hypothesis $H_1 : \beta_1 < -1$.

14.15 In Example 14.5, the recorded data for the price of Mazda cars show 'year', which we label as T, with values '71, '72, '73, ... However the scatterplot drawn in the Example is based on the explanatory variable, X, the 'age of the car' in years with $t = 91$ (representing year '91) taken as $x = 0$.

(a) Write down the linear transformation which relates X and T.

(b) Given that the fitted least squares line of $\log_{10} Y$ against X has equation

$$y^* = \log_{10} y = 4.425477 - 0.070747x,$$

determine the least squares line of $\log_{10} Y$ against T.

(c) In general, what is the effect on the slope of a least squares line when the explanatory variable X is transformed to a new explanatory variable T by $T = a + bX$?

14.16 The computer printout of the linear regression ANOVA table

Source	d of f	Sums of squares	Mean squares
Regression	1	376.41	376.41
Residuals	6	2.51	0.42
Total	7	378.92	

may be used to assess the adequacy of the model

$$Y_i = \beta_0 + \beta_1 x_i + \epsilon_i, \quad i = 1, 2, 3, \ldots, n.$$

(a) Determine n.

(b) Test whether β_1 should be taken as 0.

(c) Estimate $\sigma^2 = Var(\epsilon_i)$.

14.17 For the 'grind' data of Exercise 14.14:

(a) determine a point estimate of

$$E(Y|X = 12)$$

and give the estimated standard error of your estimate;

(b) explain why the same methods (as you used in (a)) would not be reliable in determining a point estimate of $E(Y|X = 25)$ from these data.

14.18 For the data of Example 14.5, determine a 95% confidence interval for the mean $\log_{10}$ price of a 5-year old car.

Convert this confidence interval to an approximate 95% confidence interval for the mean price of a 5-year old car.

14.19 The following information was extracted from a computer output when a simple linear regression was fitted to 7 data points.

Estimate $\widehat{\theta}$	Standard error of estimate $\widehat{S.E.(\theta)}$	t ratio $\frac{\widehat{\theta}}{\widehat{S.E.(\theta)}}$
$\widehat{\beta_1} = -0.12857$	0.07249	−1.77
$\widehat{\sigma} = 0.08577$		

Use this information to answer the following questions. In this problem you will need to examine the relationships between different regression equations.

(a) Write down a point estimate of the slope of the least squares line.

(b) Give a point estimate of the variance, σ^2, about the least squares line.

(c) Test the hypothesis

$$H_0 : \beta_1 = 0$$
$$H_1 : \beta_1 < 0.$$

(d) Complete the missing entries in the ANOVA table:

Source	d of f	Sums of squares	Mean Squares
Regression			
Residuals			
Total	6	0.05982	

14.20 Show that $\widehat{\beta_0}$ is a linear combination of normal random variables.

14.21 Using the computer output of Exercise 14.19, test the hypothesis

$$H_0 : \beta_1 = -0.2$$
$$H_1 : \beta_1 > -0.2.$$

14.22 Determine the values of constants c_i and d_i so that

$$\widehat{\beta_0} = \sum_{i=1}^{n} d_i Y_i \text{ and } \widehat{\beta_1} = \sum_{i=1}^{n} c_i Y_i.$$

Examine how these constants behave. That is, derive values for:

(a) $\sum_{i=1}^{n} c_i$

(b) $\sum_{i=1}^{n} c_i^2$

(c) $\sum_{i=1}^{n} d_i$

(d) $\sum_{i=1}^{n} d_i^2$

(e) $\sum_{i=1}^{n} c_i d_i$.

14.23 If, for each $i = 1, 2, \ldots, n$, $\widehat{Y_i}$ denotes the fitted value at x_i when a least squares line is fitted to the scatterplot, does

$$E(\widehat{Y_i}) = E(Y_i)$$

for each $i = 1, 2, 3, \ldots n$? Explain.

14.24 Following the notation of Exercise 14.22, show that

$$Cov(\widehat{\beta_0}, \widehat{\beta_1}) = \sigma^2 \sum_{i=1}^{n} c_i d_i.$$

Use this to establish a formula for the correlation between $\widehat{\beta_0}$ and $\widehat{\beta_1}$. In particular, show that $\widehat{\beta_0}$ and $\widehat{\beta_1}$ are uncorrelated when $\overline{x} = 0$; that is, when the x values have been mean corrected.

14.25 If, for each $i = 1, 2, \ldots, n$, $\widehat{Y_i}$ denotes the fitted value at x_i when a least squares line is fitted to the scatterplot, show that

$$\sum_{i=1}^{n} (\widehat{Y_i} - \overline{Y})(Y_i - \widehat{Y_i}) = 0.$$

[*Hint:* Expand the left-side of this equation with $\widehat{Y_i} = \widehat{\beta_0} + \widehat{\beta_1} x_i$ and use the fact that

$\sum_{i=1}^{n} e_i = 0 = \sum_{i=1}^{n} x_i e_i$.]

14.26 If

$$Y_i \sim N(\beta_0 + \beta_1 x_i, \sigma^2),$$

and $Y_1, Y_2, Y_3, \ldots, Y_n$ are independent, determine the maximum likelihood estimates of β_0, β_1 and σ^2.

14.27 For the regression model (14.3) involving the usual assumptions:

(a) Give the mean and variance of the slope estimator $\widehat{\beta_1}$.

(b) By taking the expected value in each side of the equation

$$\text{SS REGRESSION} = \widehat{\beta_1}^2 \sum_{i=1}^{n} (x_i - \overline{x})^2$$

show that E(SS REGRESSION) may be written in the form

$$\sigma^2 + \beta_1^2 \sum_{i=1}^{n} (x_i - \overline{x})^2.$$

(c) Use the above result to argue that when $H_0 : \beta_1 = 0$ is false, the F statistic in the ANOVA table is likely to be larger than 1. That is, the ANOVA F test is one-tailed.

14.3 MESSAGES FROM THE RESIDUALS

We have emphasised the role that residuals play when a probability model is fitted to data. Residuals represent the discrepancy between the model and fitted values. Various forms of residuals have been discussed: rootogram residuals when densities are fitted to histograms, standardised residuals in contingency tables, and even residuals as left-overs when an ANOVA model is applied to treatments.

Residuals in the regression model

When a least squares line is fitted to a scatterplot, the **residuals** are given by $E_i = Y_i - \widehat{Y_i}$, in terms of fitted values $\widehat{Y_i} = \widehat{\beta_0} + \widehat{\beta_1} x_i$, $i = 1, 2, 3, \ldots, n$, where $\widehat{\beta_0}$ and $\widehat{\beta_1}$ are the least squares estimators of β_0 and β_1 in the linear model

$$Y_i = \beta_0 + \beta_1 x_i + \epsilon_i, \quad \text{where } \epsilon_i \text{ are } i.i.d.\ N(0, \sigma^2).$$

The observed residuals are key indicators of goodness-of-fit of the model. Our aim is to extract as much information as possible from residual plots and look for warning signs that the model assumptions are being violated—for if model assumptions fail, regression predictions based on the assumptions become unreliable.

Probability distribution of the residuals

To determine the distribution of residual E_i, consider first the **fitted values**

$$\widehat{Y_i} = \widehat{\beta_0} + \widehat{\beta_1} x_i = \sum_{j=1}^{n} h_{ij} Y_j,$$

for

$$h_{ij} = \frac{1}{n} + \frac{(x_i - \overline{x})(x_j - \overline{x})}{\sum_{i=1}^{n} (x_i - \overline{x})^2}. \tag{14.8}$$

This shows them to be a linear combination of independent responses. The constant coefficients in the linear combination give important information about how the x values are clustered in the scatterplot. This information is called *leverage.*

Definition 14.6 *The ith* **leverage** *in the scatterplot* $(x_i, y_i), i = 1, 2, 3, \ldots, n$ *is given by*

$$h_{ii} = \frac{1}{n} + \frac{(x_i - \overline{x})^2}{\sum_{i=1}^{n} (x_i - \overline{x})^2}.$$

□

The $n \times n$ array $[h_{ij}]$, with h_{ij} defined at (14.8), is called a **hat matrix** and contains the leverages as its diagonal entries.

Returning to the fitted values and following the proof of Theorem 14.1,

$$Var(\widehat{Y_i}) = \sum_{j=1}^{n} h_{ij}^2 \sigma^2 = \sigma^2 \left(\frac{1}{n} + \frac{(x_i - \overline{x})^2}{\sum_{i=1}^{n} (x_i - \overline{x})^2} \right) = \sigma^2 h_{ii},$$

and this expression is in fact the same as (14.7) with $x_{fixed} = x_i$.

Since the fitted values are linear combinations of the responses, the ith fitted value is correlated with the ith observed response:

$$\begin{aligned} Cov(Y_i, \widehat{Y}_i) &= Cov(Y_i, \sum_{j=1}^{n} h_{ij} Y_j) \\ &= h_{ii} Var(Y_i) \quad \text{since } Y_1, Y_2, Y_3, \ldots, Y_n \text{ are independent} \\ &= h_{ii}\sigma^2. \end{aligned}$$

Therefore, according to **SUM 3** concerning the variance of a correlated sum of random variables,

$$\begin{aligned} Var(E_i) &= Var(Y_i - \widehat{Y}_i) \\ &= Var(Y_i) + Var(\widehat{Y}_i) - 2Cov(Y_i, \widehat{Y}_i) \\ &= \sigma^2 + \sigma^2 h_{ii} - 2h_{ii}\sigma^2 \\ &= \sigma^2(1 - h_{ii}). \end{aligned}$$

Similar arguments are used to show that $Cov(E_i, E_j) = -h_{ij}\sigma^2$. This establishes the following theorem.

Theorem 14.3 *When a linear model is fitted to a scatterplot by the method of least squares the observed residuals* $E_1, E_2, E_3, \ldots, E_n$ *are normally distributed with*

$$E(E_i) = 0, \quad Var(E_i) = (1 - h_{ii})\sigma^2, \quad \textit{and} \quad Cov(E_i, E_j) = -h_{ij}\sigma^2.$$

■

The interpretations of Theorem 14.3 are important—provided that the probability model is not mis-specified—the observed residuals have mean 0, are correlated, and have differing variances which depend directly on leverage size.

Even so we can use the observed residuals to check the validity of the probability model specification ϵ_i are $i.i.d.N(0, \sigma^2)$. We must have a clear mental picture of the difference between the quantities ϵ_i and E_i. By definition,

$$\epsilon_i = Y_i - \beta_0 - \beta_1 x_i,$$

while E_i denotes the difference between the observed and fitted response at x_i:

$$E_i = Y_i - \widehat{\beta_0} - \widehat{\beta_1} x_i.$$

Since $\widehat{\beta_0}$ is an estimator of β_0 and $\widehat{\beta_1}$ is an estimator of β_1, the observed residual E_i is an estimator of the model residual ϵ_i.

Residual plots

Diagnostic plot 1: Residuals against the explanatory variable. This is a plot of the observed points

$$(x_i,\ e_i), i = 1, 2, 3, \ldots, n.$$

It is used to check whether the residuals ϵ_i in the regression model:

1. have mean $E(\epsilon_i) = 0$;

   Success The plotted points are randomly scattered about the x axis line $e = 0$. As a quick rule of thumb, since we use $\widehat{\sigma}$ as an estimator of σ, the random scatter should occur within a band no more than $2\,\widehat{\sigma}$ either side of 0.

   Residual plot showing random scatter about x axis

   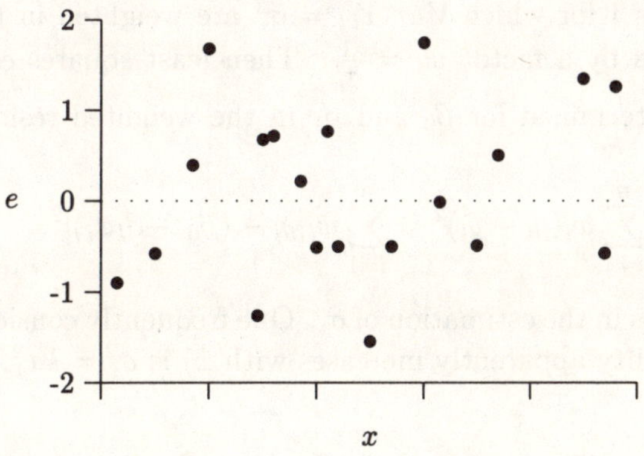

   Failure The plotted points are not evenly scattered about $e = 0$.

   Remedy If the residuals are evenly scattered about a horizontal line other than $e = 0$, it is likely that there has been a miscalculation of the intercept β_0. This is very possible with hand calculation of $y = \widehat{\beta_0} + \widehat{\beta_1} x$ but virtually impossible when using regression output from a computer package.

2. satisfy $Var(\epsilon_i) = \sigma^2$ for each i;

   Success The observed residuals lie in a uniform band over the entire range of x values.

   Failure The plot is trumpet-shaped so that, for example, there is greater variability in the observed residuals for large values of x than there is for small values of x.

Residual plot showing random scatter about x axis

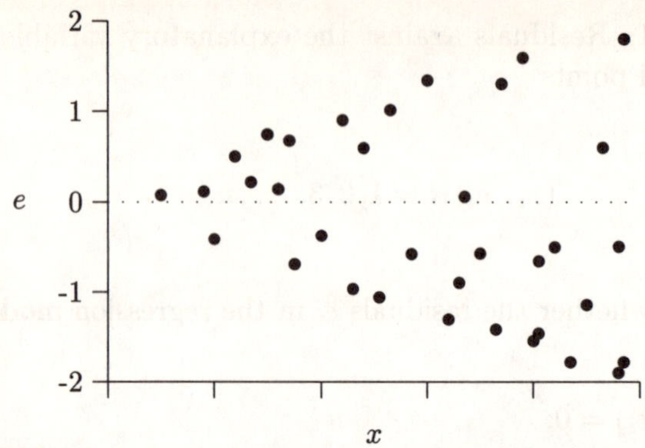

Remedy A variance stabilising transformation may be applied to the Y data (such as the square root transformation), the regression line recomputed and the residuals recalculated. A possible disadvantage of this is that the experimental errors become transformed as well. The technique of **weighted least squares** may be used here. See, for example, more advanced texts such as Myers (1990). In weighted least squares, points i for which $Var(Y_i) = \sigma_i^2$ are weighted in the residual sum of squares by a factor $w_i = \frac{1}{\sigma_i^2}$. Then least squares estimates $\widehat{\beta_0}$ and $\widehat{\beta_1}$ are determined for β_0 and β_1 in the weighted residual sum of squares

$$\sum_{i=1}^{n} w_i(y_i - \widehat{y_i})^2 = \sum_{i=1}^{n} w_i[y_i - (\beta_0 + \beta_1 x_i)]^2.$$

The difficulty is in the estimation of σ_i. One frequently considered choice (where variability apparently increases with x) is $\sigma_i^2 = kx_i^2$.

3. are *i.i.d.*;

Success No curvilinear effect from the plotted points.

Failure Plotted points follow a fluctuating curve (for example, a sine curve showing a 'seasonal' plot of points). This may occur when the subscript i is indicating time order, so that the explanatory variable values $x_1, x_2, x_3, \ldots, x_n$ in time sequence.

Remedy Simple linear regression is only appropriate for estimating the 'trend' in such plots after seasonal fluctuations have been removed by **time series analysis** methods.

In Example 14.8, when a straight line is fitted to the scatterplot having points lying along a curve, warning signals are generated from the resulting residual plot since it exhibits a curved pattern. When the plot is straightened out by transformation of the responses, the residual plot shows much more random scatter than before. Notice that, by the calculation in Example 14.6, $\widehat{\sigma} = 0.1992$. All the residuals in the scatterplot lie within $2\widehat{\sigma}$ of 0.

The residual plots of $(e_i, x_i), i = 1, 2, 3, \ldots, n$, do not take into account that the random variables $E_1, E_2, E_3, \ldots, E_n$ which yield the observed values $e_1, e_2, e_3, \ldots, e_n$ are *not* equally variable. By Theorem 14.3,

$$Var(E_i) = (1 - h_{ii})\sigma^2.$$

This quantity is appropriately estimated by

$$\widehat{Var(E_i)} = (1 - h_{ii})\widehat{\sigma^2} \quad \text{for} \quad \widehat{\sigma^2} = \frac{\text{SS RESIDUALS}}{n-2} = \text{MS RESIDUALS}.$$

Definition 14.7 *The quantities*

$$\frac{e_i}{\sqrt{(1 - h_{ii})\sigma^2}}$$

are called **standardised residuals**. *When σ is estimated by $\widehat{\sigma}$ the standardised residuals are estimated by*

$$e_i^* = \frac{e_i}{\sqrt{(1 - h_{ii})\widehat{\sigma^2}}}$$

and $e_1^*, e_2^*, e_3^*, \ldots, e_n^*$, are called the **studentised residuals**.

□

The term 'studentised' is applied to these residuals in reference to the way that the population standard deviation σ is estimated by the sample standard deviation s in Student's t test.

Example 14.8

Residual plots demonstrating a correct and a mis-specified model

The scatterplot

Scatterplot of stopping distance of a car

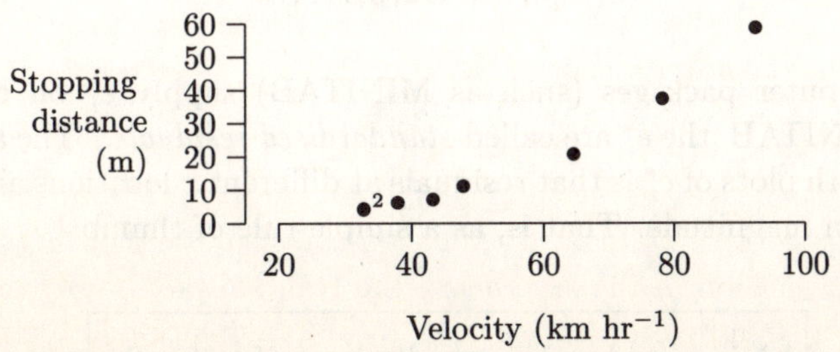

which we first encountered in Example 14.3, has residual plot

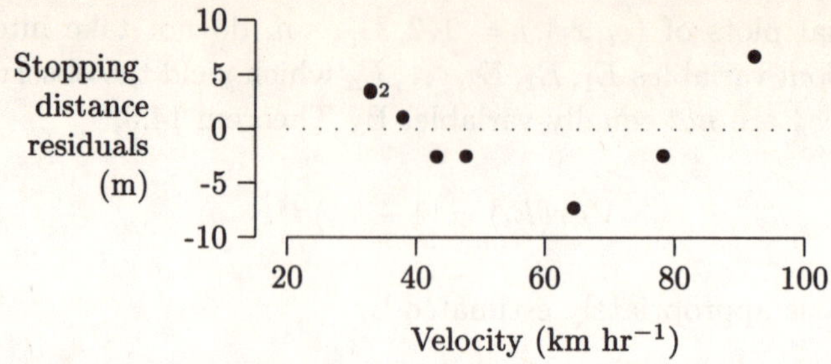

while the scatterplot that has been straightened with a square root transformation

Scatterplot of square root of stopping distance of a car

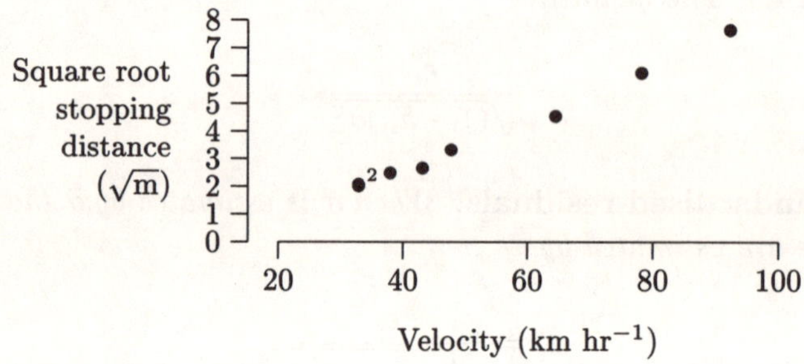

has residual plot

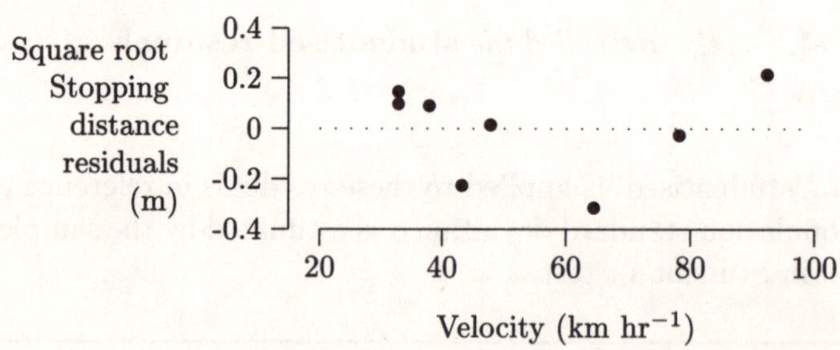

Diagnostic plot 2: Studentised residuals against the explanatory variable. The plotted points are

$$(x_i, e_i^*), i = 1, 2, 3, \ldots, n.$$

Many computer packages (such as MINITAB) supply e_i^* on command. (Note: In MINITAB, the e_i^* are called *standardised residuals.*) The advantage of working with plots of e_i^* is that residuals at different x locations are directly comparable in magnitude. That is, as a simple rule of thumb:

e_i^* is considered unusually large if $|e_i^*| > 2$.

Other diagnostic plots

Diagnostic plot 3: Fitted values against observed values. The fitted values at the observations $x_1, x_2, x_3, \ldots, x_n$ of the explanatory variable are $\widehat{y_i} = \widehat{\beta_0} + \widehat{\beta_1} x_i$. The fitted against observed plot consists of the plotted points

$$(y_i, \widehat{y_i}), i = 1, 2, 3, \ldots, n.$$

Success The plotted points lie close to a 45 degree line demonstrating that observed and fitted ith response are a similar pair for each $i = 1, 2, 3, \ldots, n$.
Failure Points consistently deviate from line; the predictive power of $\widehat{y_i}$ is called into question.
Remedy The remedies here are essentially the same as for **Diagnostic plot 1**. Transformations may be tried on either/both the X and Y variable. A non-linear model may also be appropriate: for example, a quadratic term may be included in the regression function.
Diagnostic plot 4: A normal probability plot of the studentised residuals. (See Section 7.4.) This is the standard method for checking the normality condition in ϵ_i are $i.i.d.N(0, \sigma^2)$. The normal probability plot should be fairly linear when there are a large number of normally distributed residuals being plotted.

Locating points of influence

The residuals are not the only means by which the goodness-of-fit is assessed. The leverages also provide important information about the scatterplot. From Definition 14.6 the ith leverage is

$$h_{ii} = \frac{1}{n} + \frac{(x_i - \overline{x})^2}{\sum_{i=1}^{n}(x_i - \overline{x})^2}.$$

The leverages depend only on the values of the explanatory variables. Clearly, $\sum_{i=1}^{n} h_{ii} = 2$, and the h_{ii} are all positive. This means, on average, the leverages are of size $\frac{2}{n}$ when a least squares line is fitted to a scatterplot.

Examination of the leverage formula indicates that h_{kk} will be unusually large if $x_k - \overline{x}$ is large relative to the other discrepancies $x_1 - \overline{x}, x_2 - \overline{x}, \ldots, x_n - \overline{x}$. This will happen if x_k is far removed (at either end of the scatterplot) from $\overline{x}$.

h_{kk} is considered unusually large if $h_{kk} > \frac{4}{n}$.

That is, h_{kk} is classified as 'unusually large' if it is more than double the average size of these quantities. Points with unusually large leverage may be exerting undue **influence** on the fit of the regression, in the sense that the fitted line is tipped towards them. Such points may be difficult to detect in residual plots since, if the fitted line is tipped towards them, and passes very near them, they will exhibit *small* residuals. A dangerous point is shown in Plot A below. In Plot B, however, the point of high leverage is not having an effect on the least squares estimators $\widehat{\beta_0}$ and $\widehat{\beta_1}$ since it lies in the path of the least squares line anyway.

Points of high influence should not suffer automatic exclusion as outliers: a possible policy is to refit the regression with the point in question removed, observe the effect on the residuals, and classify the point for exclusion from the scatterplot only if its absence has caused a considerable change in $\widehat{\beta_1}_{observed}$, the slope of the fitted line.

Plot A: a point of high leverage influencing the fit

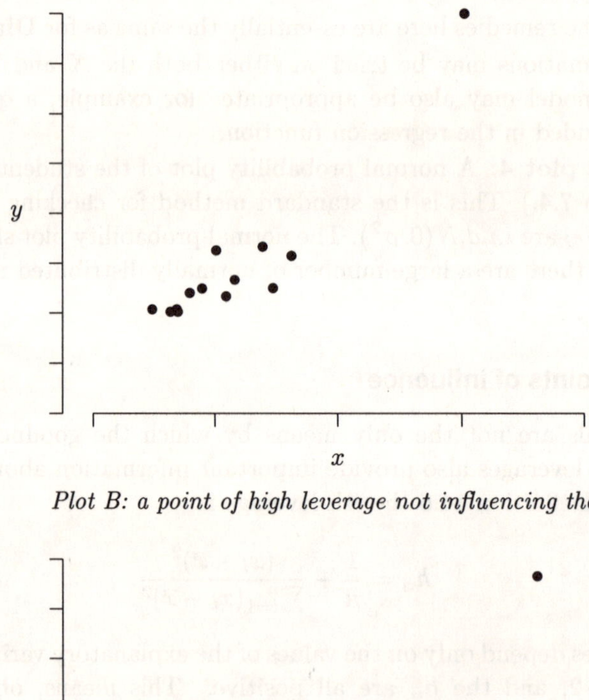

Plot B: a point of high leverage not influencing the fit

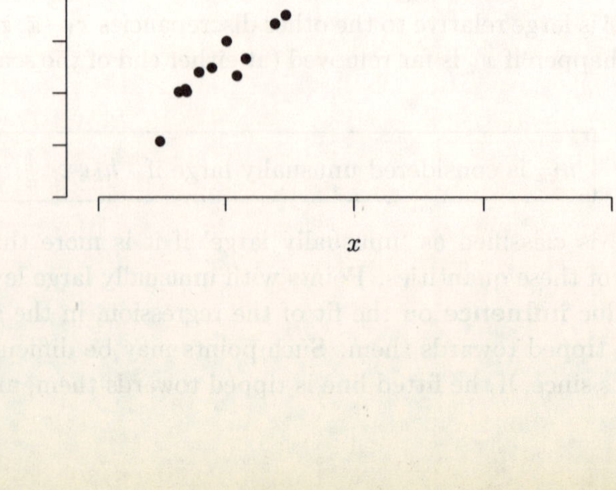

Example 14.9

Points of influence in flow rate versus stream depth

Ryan, Joiner and Ryan (1985) present data on the flow rate of stream water with observations part of an overall environmental impact study. They define flow rate as 'the total amount of water that flows past a given point in a fixed amount of time'. In this data set the highest flow rates are occurring at the greatest measured depths. Climatic (flood/drought) conditions are omitted from discussion.

The following table presents the data as well as the results of fitting a straight line to this data set. The scatterplot shows slight curvature in the plotted points, so $\sqrt{y}$ against x provides a superior fit (as the original authors point out). We plot here the least squares line $y = -3.98 + 13.8x$ so that residual plots, influential points and normal probability plot are an accumulation of evidence of a bad fit: the message here is that a straight line should not be fitted to this data set.

i	x	y	e_i^*	$\widehat{y}_i$	h_{ii}	e_i
1	0.34	0.636	−0.15203	0.72130	0.135523	−0.085304
2	0.29	0.319	0.52987	0.02962	0.181018	0.289377
3	0.28	0.734	1.55385	−0.10871	0.192336	0.842713
4	0.42	1.327	−0.87568	1.82799	0.101198	−0.500994
5	0.29	0.487	0.83749	0.02962	0.181018	0.457377
6	0.41	0.924	−1.33955	1.68966	0.102900	−0.765658
7	0.76	7.350	1.88741	6.53143	0.483503	0.818572
8	0.73	5.890	−0.49070	6.11642	0.415372	−0.226420
9	0.46	1.979	−0.70347	2.38134	0.101790	−0.402340
10	0.40	1.124	−0.74864	1.55132	0.105341	−0.427322

A comparison of the columns for e_i and e_i^* shows that the residuals e_i are standardised by dividing by different standard deviations $\widehat{\sigma}\sqrt{1 - h_{ii}}$ and these are approximately the same when n is large and the leverages h_{ii} are close to 0.

Notice in the scatterplot that the two largest observations exert undue influence on the position of the line, which is drawn up between them: their leverages are $h_{77} = 0.48$ and $h_{88} = 0.42$ to 2 dp. These values exceed the fraction $\frac{4}{n} = \frac{4}{10} = 0.4$ which would be considered potentially large for a data set of size $n = 10$. These points of influence accompany small studentised residuals $e_7^* = 1.89$ and $e_8^* = -0.49$, in the sense that both have absolute value less than 2.

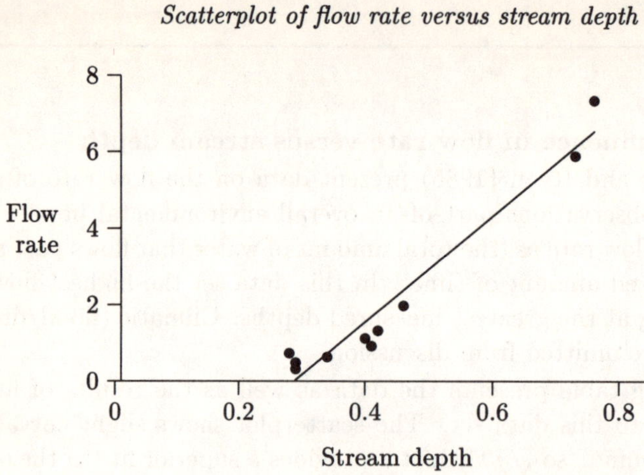

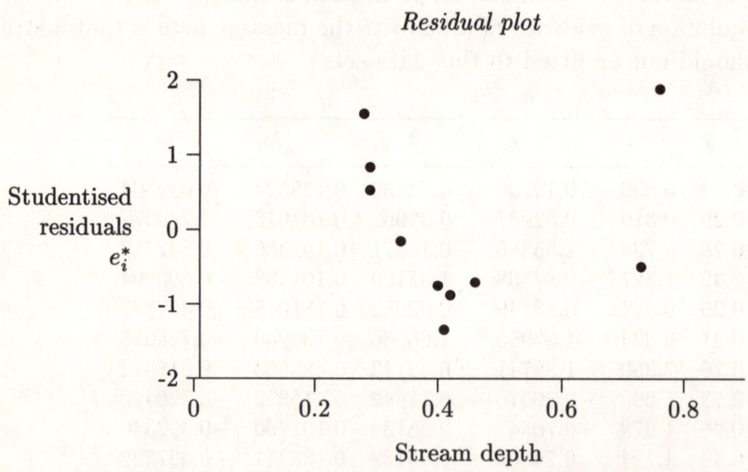

The studentised residual plot shows an undesirable arc-shaped pattern.

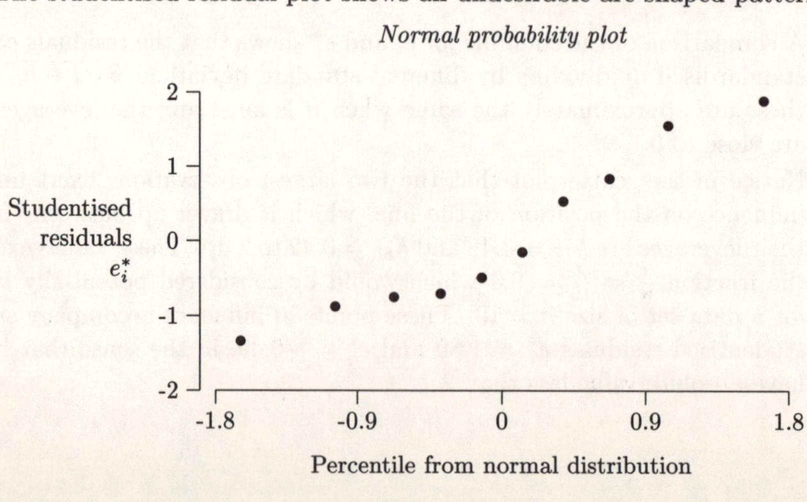

Since the plot does not follow a straight line, there is evidently a lack of normality in the residuals.

Diagnostic plot of observed versus fitted flow rate

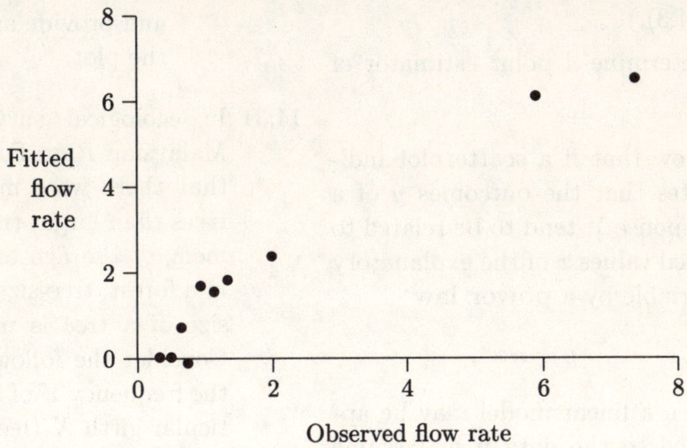

EXERCISES

14.28 Would you expect the point labelled *A* in the scatterplot below to

(a) be a point of high influence?

(b) show a large residual?

when a least squares line is fitted to the plot.

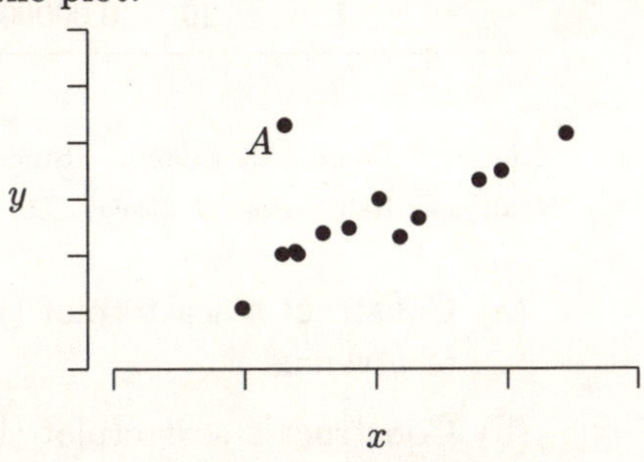

14.29 An experiment was conducted to determine the ethylene content Y of lettuce seeds (in nL/g dry weight) as a function of exposure time, X (in minutes).

The following residuals were observed when a least squares line was fitted to data collected at the given exposure times.

Exposure time x_i	Residual e_i
2	0.135
10	−0.004
20	−0.006
30	−0.051
40	−0.148
50	0.033
60	−0.069
70	0.011
80	0.047
90	0.060
100	0.001

(a) Construct plots to assess whether the assumption that the residuals in the linear model

(14.3) follow a $N(0, \sigma^2)$ distribution.

(This assumption is necessary for hypothesis testing based on (14.3).)

(b) Determine a point estimator of σ^2.

14.30 (a) Show that if a scatterplot indicates that the outcomes y of a response Y tend to be related to fixed values x of the explanatory variable by a **power law**

$$y = \alpha x^{\beta},$$

then a linear model may be applied to the data if we assume that the errors ϵ_i act **multiplicatively** in the recording of Y. That is, show that the model

$$Y_i = \alpha x_i^{\beta} \epsilon_i, \ i = 1, 2, 3, \ldots, n$$

may be linearised.

Show that the linear model will require $\log_e(\epsilon_i)$ to be *i.i.d.* $N(0, \sigma^2)$.

(b) The pressure P of a gas corresponding to various volumes V was recorded as:

V cm^3	**P** kg.wt cm^{-2}
50	64.6
60	51.4
70	40.4
90	25.7
100	7.9

The gas law relating P and V is given by the equation

$$P\,V^{\gamma} = \kappa,$$

where κ and γ are constants.

Determine least squares estimates of κ and γ.

Construct a normal probability plot of the observed residuals and provide an interpretation of the plot.

14.31 In ecological survey work in the Malaysian Rain Forest, it was found that there were many more smaller trees than larger trees. In any assessment of the age and timber volume in a forest, tree-size is important: the size of a tree is measured by *girth.* Consider the following data showing the frequency Y of trees having a particular girth X (feet).

Frequency y_i	**Girth** x_i	$\log_{10}(y_i)$
1100	3	3.04139
600	4	2.77815
340	5	2.53148
210	6	2.32222
120	7	2.07918
80	8	1.90309
60	9	1.77815
40	10	1.60206
20	11	1.30103
15	12	1.17609
2	13	0.30103
1	20	0.00000

(*Source*: Poore, M. (1968). Studies in a Malaysian Rain Forest, *J. Ecology*, 56, 143-96.)

(a) Construct a scatterplot (Plot 1) of y versus x.

(b) Construct a scatterplot (Plot 2) of $y^* = \log_{10}(y)$ versus x.

(c) Explain why Plot 2 is more effective than Plot 1 for the application of linear regression methods.

(d) Estimate the number of trees having a 10 foot girth.

(e) How is the least squares line affected if the girth measurements are converted to metres? Does the slope estimate change? Does the intercept estimate change?

(f) Construct residual plots when a least squares line is fitted to Plot 2. Are there any points of high leverage? Are there any points with large studentised residuals?

14.32 (Continuation) Suppose that a power transformation (with $p \neq 0$) is applied to the responses (tree frequencies) in the rainforest data of the previous question.

(a) Construct a scatterplot (Plot 3) of $\sqrt{y}$ versus x.

(b) Construct a scatterplot (Plot 4) of $\sqrt[5]{y}$ versus x.

(c) Do either of these plots show a linear trend?

(d) Construct residual plots when a least squares line is fitted to Plots 3 and 4. Are there any points of high leverage? Are there any points with large studentised residuals?

(e) Are the residuals from these plots normally distributed?

(f) Which of the powers $p = 1, 0.5, 0.2, 0$, seems to be the most appropriate for power transforming the responses to linearity?

14.33 Ryan, Joiner and Ryan (1985) indicate that

$$a = \sqrt{b - c},$$

where

$$\begin{aligned} a &= \text{(standard deviation of residual)} \\ b &= \text{MS RESIDUALS} \\ c &= \text{(standard deviation of fit)}^2. \end{aligned}$$

This is the formula which MINITAB uses to estimate

$$\sqrt{Var(E_i)}.$$

Show that the MINITAB formula and

$$\sqrt{\widehat{Var(E_i)}} = \widehat{\sigma}\sqrt{1 - h_{ii}}$$

are in complete agreement.

14.4 INTERPRETING SAMPLE CORRELATION

Regression is the study of the *nature* of the relationship between the response Y and explanatory variable X. In the first sections of this chapter we have examined linear regression, where the regression function, or mean response at a fixed value $X = x$, is modelled linearly with x: that is, $\mu(x) = \beta_0 + \beta_1 x$. Using least squares methods, we may fit such a line to a scatterplot. Consider cases where this is successful; the fitted line is a good summary of trend in the scatterplot. Even so, the scatterplot remains as it was: full of scatter. (It is highly unlikely that the points in the scatterplot fall exactly on the least squares line.) The linear relationship appears visually stronger in plots where the scatterplot points are close to the fitted line, compared to plots of diffuse scatter well away from the line.

Correlation is a measure of the *strength* of the linear relationship between response and explanatory variable.

Determining sample correlations

Many formulas for two data sets (with the same number of data points) involve the *sample covariance*:

Definition 14.8 *The* **sample covariance,** s_{ab}, *between two data sets* $a_1, a_2, a_3, \ldots, a_n$ *and* $b_1, b_2, \ldots, b_n$ *is given by*

$$s_{ab} = \frac{\sum_{i=1}^{n}(a_i - \overline{a})(b_i - \overline{b})}{n-1} = \frac{\sum_{i=1}^{n} a_i b_i - n\overline{a}\overline{b}}{n-1},$$

for $\overline{a}$ *and* $\overline{b}$ *the sample means of data sets* $a_1, a_2, a_3, \ldots, a_n$ *and* $b_1, b_2, b_3, \ldots, b_n$ *respectively.*

□

Note that $s_{xx} = s_x^2$, the sample variance of $x_1, x_2, \ldots, x_n$.

Following the convention of using upper case letters for random variables, we use s_{XY} to denote the sample covariance between the sequences $X_1, X_2, \ldots, X_n$ and $Y_1, Y_2, Y_3, \ldots, Y_n$ of random variables. s_{XY} provides a measure of association between the variables X and Y: observed outcomes s_{xy} will be positive when large values of X tend to accompany large values of Y, and small values of X accompany small values of Y. s_{xy} will be negative when extreme values of one variable are observed with values of the opposite extreme of the second variable.

However, s_{XY} has no in-built scale by which we can compare one pair of variables (X, Y) with another. For the n data points $(x_1, y_1), (x_2, y_2), \ldots, (x_n, y_n)$, we may approach this problem of scale by considering

$$s_{uv} \quad \text{for } u_i = \frac{x_i - \overline{x}}{s_x} \quad \text{and } v_i = \frac{y_i - \overline{y}}{s_y}.$$

Here, s_x and s_y denote the sample standard deviations of the x data and the y data respectively. If we standardise the data $x_1, x_2, x_3, \ldots, x_n$ and the data $y_1, y_2, y_3, \ldots, y_n$ before considering sample covariance, we obtain (for verification, see the exercise section)

$$s_{uv} = \frac{s_{xy}}{\sqrt{s_{xx}}\sqrt{s_{yy}}}.$$

This becomes a measure by which we can compare the outcomes of one pair of variables with the outcomes of another pair of variables. This is a measure of correlation.

Definition 14.9 *The* **sample correlation coefficient**, $r_{x,y}$, *for the n data points*

$$(x_1, y_1), (x_2, y_2), (x_3, y_3), \ldots, (x_n, y_n),$$

is defined by

$$r_{x,y} = \frac{s_{xy}}{\sqrt{s_{xx}}\sqrt{s_{yy}}}.$$

□

When there is no ambiguity as to which data sets are used to determine a sample correlation, it is convenient to write $r_{x,y} = r$. When sample correlations require hand calculation, it is usually easier to use the equivalent formula

$$r_{x,y} = \frac{\sum_{i=1}^{n} x_i y_i - n\bar{x}\,\bar{y}}{\sqrt{\sum_{i=1}^{n} x_i^2 - n\bar{x}^2}\sqrt{\sum_{i=1}^{n} y_i^2 - n\bar{y}^2}}. \tag{14.9}$$

Example 14.10

Are curing time and compressive strength highly correlated?

The data of Example 14.1 consisted of 5 observations on the pair

$$(X, Y) = (\text{curing time, compressive strength})$$

for samples of cement/soil mix. The points in a scatterplot appear to lie close to the least squares line $y = 2.52 + 0.0675x$.

Worked example:

Determine the correlation coefficient for this data.

Calculations are based on the following display of the data.

i	x_i	x_i^2	y_i	y_i^2	$x_i y_i$	
1	2	4	2.53	6.4009	5.06	
2	7	49	3.12	9.7344	21.84	
3	14	196	3.49	12.1801	48.86	
4	21	441	3.95	15.6025	82.95	
5	28	784	4.36	19.0096	122.08	
	72	1474	17.45	62.9275	280.79	Totals

Then,

$$
\begin{aligned}
r_{x,y} &= \frac{\sum_{i=1}^n x_i y_i - n\overline{x}\,\overline{y}}{\sqrt{\sum_{i=1}^n x_i^2 - n\overline{x}^2}\sqrt{\sum_{i=1}^n y_i^2 - n\overline{y}^2}} \\
&= \frac{280.79 - 5(\frac{72}{5})(\frac{17.45}{5})}{\sqrt{1474 - 5(\frac{72}{5})^2}\sqrt{62.975 - 5(\frac{17.45}{5})^2}} \\
&= 0.98 \quad (2 \text{ dp}).
\end{aligned}
$$

The calculation is numerically simple. The interpretation of the answer is of interest to us.

Relationship to $\widehat{\beta_1}$

In terms of the sample covariance, we may write from (14.4),

$$\widehat{\beta_1} = \frac{\sum_{i=1}^n x_i y_i - n\overline{x}\,\overline{y}}{\sum_{i=1}^n x_i^2 - n\overline{x}^2} = \frac{s_{xy}}{s_{xx}}.$$

From this it follows that

$$r_{x,y} = \frac{s_{xy}}{\sqrt{s_{xx}}\sqrt{s_{yy}}} = \widehat{\beta_1}\sqrt{\frac{s_{xx}}{s_{yy}}}$$

and on squaring each side of this equation,

$$r_{x,y}^2 = \frac{\widehat{\beta_1}^2 s_{xx}}{s_{yy}}.$$

As preparation for the ANOVA table in Section 14.2, the total sum of squares was partitioned as

$$\text{SS TOTAL} = \text{SS REGRESSION} + \text{SS RESIDUAL},$$

where, in the current notation, and format of observed data points,

$$\text{SS TOTAL} = (n-1)s_{yy} \quad \text{and} \quad \text{SS REGRESSION} = (n-1)\widehat{\beta_1}^2 s_{xx}.$$

Using these results, we have that

$$r_{x,y}^2 = \frac{\text{SS REGRESSION}}{\text{SS TOTAL}} \quad \text{and} \quad 1 - r_{x,y}^2 = \frac{\text{SS TOTAL} - \text{SS REGRESSION}}{\text{SS TOTAL}}.$$

This establishes the following theorem, which provides us with an appropriate interpretation of the sample correlation coefficient.

Theorem 14.4 *If a least squares regression line is fitted to the points*

$$(x_i, y_i), i = 1, 2, 3, \ldots, n,$$

the square of the sample correlation coefficient satisfies

$$r_{x,y}^2 = \frac{SS\ REGRESSION}{SS\ TOTAL} \quad and \quad 1 - r_{x,y}^2 = \frac{SS\ RESIDUALS}{SS\ TOTAL}.$$

■

The following **correlation properties** spring directly from Theorem 14.4. Their proofs are simple, but provide an enhanced understanding of sample correlation for a data set

$$(x_1, y_1), (x_2, y_2), (x_3, y_3), \ldots, (x_n, y_n).$$

Sample correlation property 1: $|r_{x,y}| \leq 1$.
This follows by Theorem 14.4 since $1 - r_{x,y}^2 \geq 0$.
Sample correlation property 2: $r_{x,y}^2 = 1$ when all data in the scatterplot lies on the least squares line.
By Theorem 14.4, $1 - r_{x,y}^2 = 0$ precisely when SS RESIDUALS $= 0$, which occurs when all observed residuals are 0.
Sample correlation property 3: $r_{x,y} = 0$ when there is no linear relationship between the points in the scatterplot.
Again by Theorem 14.4, $r_{x,y}^2 = 0$ precisely when SS REGRESSION $= 0$, which occurs when $\widehat{\beta_1} = 0$ and the fitted least squares line has slope 0.
Sample correlation property 4: $r_{x,y}^2$ measures the proportion of variability in the responses (the Y values) that results from the linear fit with x. (This is Theorem 14.4 put into words.)
Property (4) allows the size of correlation coefficients to be compared. Properties (1) to (3) indicate that correlations lie in the interval $[-1, 1]$, however (4) indicates that the true measure of the strength of a linear relationship is $r_{x,y}^2$, not $r_{x,y}$.
By this we mean that a correlation of $r_{x,y} = 0.9$ (and likewise $r_{x,y} = -0.9$) is reasonably strong, since, by (4), $r_{x,y}^2 = (0.9)^2 = 0.81$ so that 81% of the variability in the y values results from the straight line fit with x.
Sample correlation property 5: The sign of $r_{x,y}$ is the same as the sign of $\widehat{\beta_1}$. This follows immediately from

$$r_{x,y} = \widehat{\beta_1}\sqrt{\frac{s_{xx}}{s_{yy}}}.$$

Example 14.11
Earthquake patterns

The following data are part of a larger data set from Worsley (1986) showing earthquakes having magnitude exceeding 5 on the Richter Scale recorded off the coast of New Zealand between 1950 and 1975.

Year	Magnitude	Year	Magnitude
1951	5.2	1960	5.2
1952	5.2	1961	5.3
1953	5.1	1964	5.1
1954	5.0	1965	6.0
1956	5.0	1966	5.0
1957	5.0	1967	5.1
1959	6.0	1968	5.5
		1972	5.2

(*Source*: Worsley, K.J. (1986). Confidence regions and tests for a change-point in a sequence of exponential random variables. *Pacific Statistical Congress, 1985.*)

The scatterplot of 'magnitude' Y against 'year' X is shown below with overlaid least squares line $y = -16.0 + 0.0109x$. $r^2_{x,y} = 0.046$ represents a weak correlation between 'magnitude' and 'year'; only 4.6% of the observed variability in Y is accounted for by the straight line fit with X. The least squares line has a non-significant slope.

Scatterplot of high-magnitude earthquakes against year

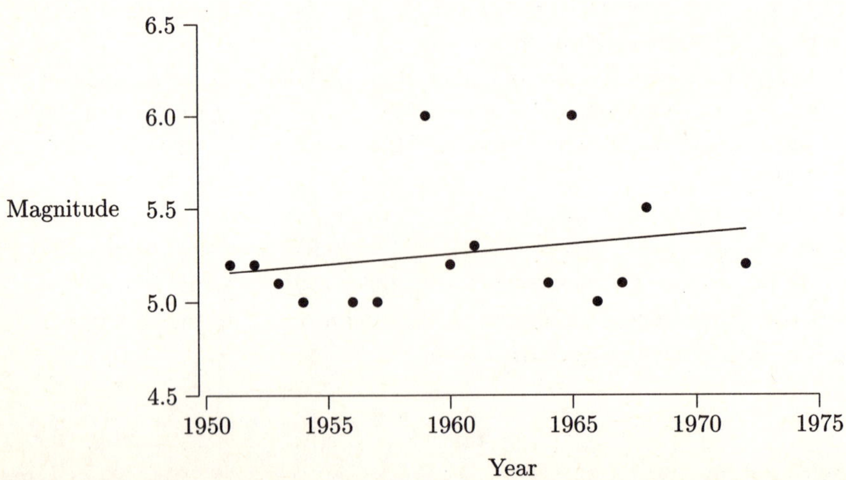

This is an interesting data set in the sense that there is a sine-wave pattern occurring with time, a so-called **time series**, along a non-significant trend. This is punctuated by two data points corresponding to extreme earthquakes. These points are not outliers in the sense that they represent real earthquake behaviour and should be included in any prediction mechanism based on the data.

Correlation for residuals

Correlation may be used to detect if the observed residuals are correlated with one another when the explanatory variable is indexed by time. For this, the points $(e_2, e_1), (e_3, e_2), (e_4, e_3), \ldots, (e_n, e_{n-1})$ are plotted. That is, the vertical axis supports the residuals and the horizontal axis supports the **lag-1 residuals** with time index of each residual reduced by one unit.

residuals	e_2	e_3	e_4	$\ldots$	e_n
lag-1 residuals	e_1	e_2	e_3	$\ldots$	e_{n-1}

If the size of one residual naturally carries over to the size of the next, this 'carry-over' will manifest itself as correlation in the residual by lag-1 residual plot. In this way we can sense whether the ϵ_i from the linear model are *i.i.d.* How big is the correlation, at least approximately? Since, on average, residuals have mean 0, then for large sample sizes the correlation is approximately

$$\frac{\sum_{i=2}^{n} e_{i-1}e_i - (n-1)(0)(0)}{\sqrt{\sum_{i=2}^{n} e_i^2 - (n-1)(0)^2}\sqrt{\sum_{i=1}^{n-1} e_i^2 - (n-1)(0)^2}} \approx \frac{\sum_{i=2}^{n} e_{i-1}e_i}{\sum_{i=1}^{n} e_i^2}.$$

Definition 14.10 *The* **lag-1 autocorrelation** *of the residuals is given by*

$$r_{auto} = \frac{\sum_{i=2}^{n} e_{i-1}e_i}{\sum_{i=1}^{n} e_i^2}.$$

□

The statistic $DW = 2(1 - r_{auto})$ called the **Durbin-Watson statistic** appears on the output from many computer packages and assesses whether the residuals are autocorrelated (or correlated with themselves lag-1). Some packages base the calculation of DW on the studentised residuals, however, when n is large, $e_i^* \approx e_i/\widehat{\sigma}$ and since correlation is scale invariant, the value of r_{auto} is not affected by the scale change. If $r_{auto} \approx 0$, then DW is close to 2, a value corresponding to no lag-1 autocorrelation. Since observed values of DW must lie in the interval $[0, 4]$, autocorrelation in the observed residuals corresponds to values of r_{auto} near $+1$ or -1; in translation, to values of DW near 0 or 4.

$DW = 2.66$ for the residual by lag-1 residual plot below for the earthquake data. Although the sample size is small, the indications are that the residuals are uncorrelated; the picture would change substantially if the points with unusually large studentised residuals were removed. However, as mentioned in Example 14.11, this is not an option that we should consider.

Scatterplot residuals against lagged residuals for earthquake data

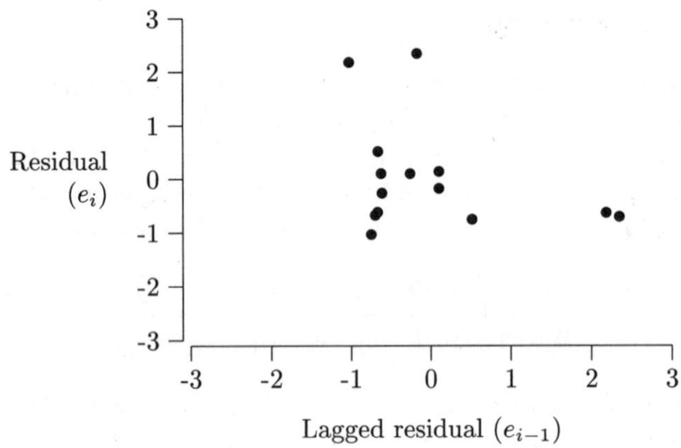

What does $r_{X,Y}$ estimate?

$R = r_{X,Y}$ is a point estimator of $\rho_{X,Y} = \rho$, the population correlation between random variables X and Y defined by

$$\rho_{X,Y} = \frac{Cov(X,Y)}{\sqrt{Var(X)}\sqrt{Var(Y)}}$$

in Definition 8.8. (When regarding sample correlation as a random variable, upper case notation is appropriate.)

Example 14.11 is essentially different from other examples that we have considered; in this example, no control was exerted over the explanatory variable (in this case, year). If we had the predictive power of knowing when high-magnitude earthquakes were going to occur then .... In some years no large earthquakes occur, while in other years multiple quakes occur.

If we consider *both* X and Y to be random, and assume that (X, Y) has a *bivariate normal distribution*, then by the results of Section 8.3,

$$(X,Y) \sim N(\mu_X, \mu_Y, \sigma_X^2, \sigma_Y^2, \rho),$$

where ρ represents the population correlation between X and Y. In terms of the standardised variables

$$U = \frac{X - \mu_X}{\sigma_X} \quad \text{and} \quad Z = \frac{Y - \mu_Y}{\sigma_Y},$$

the conditional distribution of Z given $U = u$ is $N(\rho u, 1-\rho^2)$. In terms of the original variables X and Y, the conditional distribution of Y given $X = x$ is

$$N(\underbrace{[\mu_Y - \rho(\frac{\sigma_Y}{\sigma_X})\mu_X]}_{\beta_0} + \underbrace{[\rho\frac{\sigma_Y}{\sigma_X}]}_{\beta_1}x,\ (1-\rho^2)\sigma_Y^2).$$

Clearly shown here is the linear regression function $\mu(x) = \beta_0 + \beta_1 x$ and, of particular interest for this section, the correlation coefficient in

$$\beta_1 = \rho\frac{\sigma_Y}{\sigma_X}. \tag{14.10}$$

The conclusion here is as follows: If (X, Y) is assumed to be bivariate normal, then the marginal and conditional variables are normally distributed; in particular, $E(Y|X = x) = \beta_0 + \beta_1 x$ with $Var(Y|X = x)$ constant and therefore the conditions for the linear model are satisfied.

From this bivariate normal distribution, it follows by Equation 14.10 that the test $H_0 : \beta_1 = 0$ is equivalent to the test $H_0 : \rho = 0$ against a non-zero, positive, or negative alternative. The test statistic is the 'test for slope' test statistic, T, where, when H_0 is true and $\beta_1 = 0$,

$$T = \frac{\widehat{\beta_1} - 0}{\sqrt{\frac{\widehat{\sigma^2}}{\sum_{i=1}^n (x_i - \overline{x})^2}}} = \frac{R\sqrt{\frac{s_{YY}}{s_{XX}}}}{\sqrt{\frac{(1-R^2)s_{YY}}{(n-2)s_{XX}}}} = \frac{R\sqrt{n-2}}{\sqrt{1-R^2}}.$$

When H_0 is true, T has a t distribution with $n - 2$ degrees of freedom.

Example 14.12

A test on the correlation coefficient for earthquake patterns

In the earthquake pattern data of Example 14.11, we may pursue the hypothesis test for zero correlation as follows:

Let ρ denote the population correlation between 'high magnitude earthquakes' and 'year'. For testing

$$\begin{aligned} H_0 &: \ \rho = 0 \\ H_1 &: \ \rho \neq 0 \end{aligned}$$

we find that the observed value of the test statistic $T = \frac{R\sqrt{n-2}}{\sqrt{1-R^2}}$ is $t_{observed} = \frac{\sqrt{0.046}\sqrt{15-2}}{\sqrt{1-0.046}} = 0.79$ which has a p-value exceeding 0.1. Therefore there is not significant evidence of correlation between these variables: $t = 0.79$, $(n.s.)$.

EXERCISES

14.34 Barrett et al. (1984) give data for the mineral content of sedimentary core samples taken at or below the seabed at Granite Harbour, southern Victoria Land in the Antarctica.

Part of the reported data gives parts per million (ppm) of nickel and cobalt, the observed pairs being taken on the same core sample.

Nickel X (ppm)	**Cobalt** Y (ppm)
13	8
14	10
16	10
13	10
12	10
18	9
24	9
18	10
12	12
18	13
10	9
8	9
14	10
6	8
10	8
12	15
12	15
10	8
7	10
4	9
7	9
7	7
12	10

(*Source*: Barrett, P.J. et al. (1984). Texture, mineralogy and composition of four sediment cores from Granite Harbour and New Harbour, southern Victoria Land, Antarctica. *New Zealand Journal of Geology and Geophysics*, 27, 477-85.)

(a) Construct B/W plots for the observed values of X and Y separately.

(b) Construct a scatterplot for these data.

(c) Based on the visual impression created by the scatterplot, try to 'guess', or rather anticipate, the value of the sample correlation between the observed values of X and Y.

(d) Determine the value of $r_{X,Y}$. Comment on whether high nickel content tends to accompany high cobalt content in core samples of this region.

14.35 Would you expect the following variables to be highly correlated?

$(X, Y) =$ ('Age of husband', 'Age of wife')

Would you expect the correlation to be positive? Explain why.

Determine the correlation between X and Y based on the following data:

X	Y
55	51
21	24
28	26
33	33
19	17
64	61
40	38
59	56
34	36
50	44
51	51

14.36 Given data pairs

$$(x_1, y_1), (x_2, y_2), (x_3, y_3), \ldots, (x_n, y_n),$$

show that $s_{uv} = r_{x,y}$, for

$$u_i = \frac{x_i - \overline{x}}{s_x} \quad \text{and} \quad v_i = \frac{y_i - \overline{y}}{s_y}.$$

Explain in words what this result means.

14.37 Thirty observations were recorded for the variables X, denoting counts of freeway vehicles per kilometre and Y, the average speed (km hr^{-1}) of vehicles in the traffic flow on the South Eastern Freeway during the Easter holiday weekend. Which one of the following is the most likely value of $r_{X,Y}$?

$-1.00 \quad -0.89 \quad -0.21 \quad 0.21 \quad 0.89 \quad 1.00.$

Justify your choice.

14.38 Show that the square of the correlation coefficient $r^2_{X,Y}$ is unaffected by linear transformations of X and Y. That is,

$$r^2_{aX+b,cY+d} = r^2_{X,Y}.$$

Does $r_{aX+b,cY+d} = r_{X,Y}$? Explain.

14.39 Determine the sample correlation between the observed residuals $e_1, e_2, e_3, \ldots, e_n$ and the values $x_1, x_2, x_3, \ldots, x_n$ of the explanatory variable. That is, find $r_{e,x}$.

14.40 Suppose that $y_1, y_2, y_3, \ldots, y_n$, are the observed responses and $\widehat{y}_1, \widehat{y}_2, \widehat{y}_3, \ldots, \widehat{y}_n$ are the fitted values when least squares line is fitted to the scatterplot data $(x_1, y_1), (x_2, y_2), \ldots, (x_n, y_n)$.

Discuss whether the following statement is true or false:

$$r^2_{x,y} = r^2_{x,\widehat{y}}.$$

14.41 The following data show quantitative changes in reading speed (words/min) and 'age' for 7 dyslexic subjects recorded with and without the assistance of coloured lenses.

Subject	With	Without	Age
1	61	31	11
2	102	81	21
3	260	159	44
4	102	104	11
5	233	212	49
6	286	210	19
7	150	134	15

(*Source*: Chelva, E., Collins, D.W.K., Levy, R.L., and McLaren, T.L. (1986). Effects of exposure to coloured lenses. *Proceedings of Seminar on Advances in Learning Disabilities, May 1986.* Dyslexia Research Foundation of Australia: Perth, Western Australia.)

Is the *improvement* in reading speed (when coloured lenses are worn) highly correlated with age?

Calculate a correlation coefficient to justify your answer.

14.42 Suppose that random variables X and Y satisfy

$$\left(\frac{X-1}{3}, \frac{Y+2}{2}\right) \sim N(0,0,1,1,0.5).$$

Write down the conditional normal distribution for Y given $X = x$.

14.43 Are uncorrelated normally distributed random variables independent? Is the converse true? Explain.

14.44 Suppose that the period of heat treatment X (in seconds) and the depth of hardening Y (in cm) of an enamel underlay follow a bivariate normal distribution where

$$\left(\frac{X-20}{3}, \frac{Y-0.2}{0.09}\right) \sim N(0,0,1,1,0.9).$$

(a) Determine the regression function $\mu(x)$.

(b) Determine the correlation coefficient between

$$\frac{X - 25}{3} \quad \text{and} \quad \frac{Y - 0.4}{0.09}.$$

(c) Determine the variance of Y given that $X = 30$. Does the size of this variance depend on the correlation between X and Y?

CHAPTER FIFTEEN

Distribution-Free Methods

In this chapter you will find

15.1 SIGN TEST: DISCARD THE DATA BUT KEEP THE SIGNS

We have developed probability models for the random variables which generate our data: $X \sim Poi(\lambda)$, $X \sim Bin(n,p)$, $X \sim N(\mu, \sigma^2)$ ... We have developed methods of assessing the goodness-of-fit of a probability model to a data set: chi-squared tests for detecting binomial and Poisson data, probability plots for tracking Weibull, exponential, normal data.

When we become convinced that the model works well for our data (residual plots are well behaved and random, rootograms are not over-spiked, probability plots are straight) it is time to sit back and fold our arms. With a probability distribution at our disposal, we can sail directly into the population at large, we are ready to point estimate, seek significance, create confidence intervals. With this probability distribution at our disposal, we will know how accurate our inferences will be. Thus, for example, if we accept that a small amount of data is normally distributed, we can create a $100(1-\alpha)\%$ confidence interval for the population mean μ based on a pivot having a t distribution.

The 'if' is highlighted because it is the crunch point or a branching point in our decision process—with flawed premises, we head toward statistical misadventure—with well-founded premises, we can be confident of our final conclusions.

For example, with a t test: if we pretend that our data are normally distributed when they are not, our conclusions based on the normality assumption could be badly misguided. The endpoints of confidence intervals may be in the wrong place and unavoidably carry a larger α value than we claim. Therefore, we must always test data for normality, when normality is required by the inference procedure that we wish to use (for example, in t tests, t confidence intervals, F tests, ...).

What if a test for normality fails on our data? In this chapter we answer: think again, there are alternatives to procedures which assume a *known* distributional shape to the data. These are the **distribution-free** procedures. Some texts use the term **non-parametric**, but this seems counter-productive to our efforts: after all, there are still parameters present with estimates to be inferred from our data.

What is meant by distribution-free? It does not mean that there is no probability distribution for the population random variable; the data still follow a probability distribution, it is just that we are not prepared to state what that probability distribution is. The following notation is basic to distribution-free analysis of a single sample:

Population random variable X is continuous and has • unknown cumulative distribution function F and • unknown median $\xi_{0.5}$.

In contrast, an equivalent normality assumption would be: X has cdf $F(x) = \Phi(\frac{x-\mu}{\sigma})$, where $\Phi(z) = \int_{-\infty}^{z} \frac{1}{\sqrt{2\pi}} e^{-\frac{1}{2}t^2}\,dt$ and because of the symmetry of this distribution, $\xi_{0.5} = \mu$. The notation for the population median emphasises the equivalent name—the 50th percentile.

Always remember that distribution-free procedures come with a sacrifice to the data. After all, if one concedes that there is no suitable distributional shape to data, then the actual sizes of the data in the data set start to lose meaning. The parts of the data set that we can rescue are: the **sign** $(+/-)$ of each data point and the **relative ranks** of each data point. The parts of the data set that we throw away are the magnitudes of the data points.

Assessments through '+' and '–'

We now outline the implementation of the **sign test**; this is the most basic of all the distribution-free tests concerning the 'centre' of the population.

Since we are most likely to operate the sign test on skewed data sets (for which a normality assumption is untenable), the population median $\xi_{0.5}$ is the best measure of the 'centre' of the population. However, in what follows in the testing procedures, $\xi_{0.5}$ may be replaced by μ if the population is thought to be symmetric. Recall: The population median $\xi_{0.5}$ is the unique (since the population is assumed continuous) solution to the equation $F(x) = \frac{1}{2}$; that is, $F(\xi_{0.5}) = \frac{1}{2}$.

Then, by definition of the median, $P(X > \xi_{0.5}) = P(X < \xi_{0.5}) = 0.5$. (The probability that the continuous random variable X equals $\xi_{0.5}$ *exactly* is 0.) In the procedure for the sign test, all data points in the data set are replaced by '+' or '−' signs, depending upon whether they lie above or below the value of the population median specified by the null hypothesis. If a data point falls *exactly* at the value of the population median specified by the null hypothesis, then the data point is thrown away completely; it carries no evidence of positivity or negativity relative to the hypothesised median. Of course, the sample size is reduced by the number of such points. It is evident that a lot of information is being discarded in this process.

For example, if the null hypothesis of interest is $H_0 : \xi_{0.5} = 0$ against an appropriate alternative, then the data set with 49 points on the left in the S/L plot below, is transformed to the signed data set on the right with '+' signs representing data above the hypothesised median (in this case 0) and '−' signs below 0. There are two 0.0 values on this data set which apparently have no sign. The test procedure is then based on the remaining $49 - 2 = 47$ data points. In reality, it is possible that the experimenter would have more knowledge of a 0.0 reading; at least the experimenter is the most qualified person to be able to assess whether 0.0 should really be regarded as positive or negative.

Discard the data but keep the signs
Scale: stems—units; leaves—10^{-1} units
3|2 represents 3.2 units

Stem	Leaves	Stem	Signs
−4	3	−4	−
−3	6	−3	−
−2	1 2 3 3 8 9 9 9	−2	− − − − − − − −
−1	3 4 4 4 5 6 7 7 7	−1	− − − − − − − − −
0	0 0 2 4 5 5 6 7 8 9	0	0 0 + + + + + + + +
1	0 0 1 2	1	+ + + +
2	1 6 7 7 8 8 8 9	2	+ + + + + + + +
3	2 6	3	+ +
4	8	4	+
5		5	
6	2	6	+
7		7	
8	3 4 6	8	+ + +
9	8	9	+

The process of creating a sign for each data point may be easily measured by a simple counting random variable.

Definition 15.1 *The* **sign indicator function,** *defined on a data set $w_1, w_2, w_3, \ldots, w_n$ of non-zero values, is given by*

$$sign(w_i) = \begin{cases} 1 & \textit{if } w_i > 0; \\ 0 & \textit{if } w_i < 0. \end{cases}$$

□

To test $H_0 : \xi_{0.5} = \xi^0$, simply add the '+' signs. When H_0 is true, and the specified median is in the correct location, this will reflect in approximately equal numbers in the sample $X_1, X_2, X_3, \ldots, X_n$ lying above/below the median. Therefore if

$$Y_i = sign(X_i - \xi^0) = \begin{cases} 1 & \text{if } X_i > \xi^0; \\ 0 & \text{if } X_i < \xi^0, \end{cases}$$

then $Y_1, Y_2, Y_3, \ldots, Y_n$ is a random sample from a $Bin(1, \frac{1}{2})$ or Bernoulli distribution. (Suddenly a distribution emerges from within the structure of distribution-free tests!) Since the sum of independent Bernoulli distributions has a binomial distribution, $Y = \sum_{i=1}^{n} Y_i \sim Bin(n, \frac{1}{2})$. From this, the usual significance tests concerning p may be used: in each case Y is the test statistic.

The Sign Test:
Null hypothesis $H_0 : \xi_{0.5} = \xi^0$
Test statistic: $Y = \sum_{i=1}^{n} sign(X_i - \xi^0).$

Example 15.1
Petrol yield from crude oil
As part of an evaluation program for determining new supply sources, an oil company makes regular determinations of crude oil quality, based on the percentage yield of petrol. The company's current benchmark median yield is 20% for their current supplier. An analysis of crude oil samples from a prospective supplier gave the following percentage yields:

18.1, 35.6, 20.6, 20.4, 22.5.

The company does not wish to assume that 'yield' is normally distributed.
Worked example:
Use a sign test to determine whether the prospective supply source would provide a median yield exceeding 20%.
The sign test procedure is as follows:

1. Let $\xi_{0.5}$ denote the median percentage petrol yield from crude oil samples from the prospective supplier.
2. $H_0 : \xi_{0.5} = 20$ versus $H_1 : \xi_{0.5} > 20$. Here the alternative hypothesis is designed to detect a preference for the new supplier.
3. Test statistic: $Y = \sum_{i=1}^{5} sign(X_i - 20)$ is expressed in terms of the sample X_1, X_2, X_3, X_4, X_5.
4. α should be as close as possible, but not exceeding 0.05.
5. Data: $x_1 = 18.1,\ x_2 = 35.6,\ x_3 = 20.6,\ x_4 = 20.4,\ x_5 = 22.5.$
6. When H_0 is true we determine the observed value of the test statistic as follows:

$x_i - 20$	$sign(x_i - 20)$
−1.9	0
15.6	1
0.6	1
0.4	1
2.5	1

$$y_{observed} = \textstyle\sum_{i=1}^{5} sign(x_i - 20) = 0 + 1 + 1 + 1 + 1 = 4.$$

7. The p-value may be determined from binomial probability tables or by direct calculation:

$$\begin{aligned} p\text{-value} &= P(Y \geq 4 | H_0 \text{ is true}) \\ &= P(Y \geq 4 | Y \sim Bin(5, \tfrac{1}{2})) \\ &= P(Y = 4) + P(Y = 5) \quad \text{for} \quad Y \sim Bin(5, \tfrac{1}{2}) \\ &= \tbinom{5}{4}(\tfrac{1}{2})^5 + \tbinom{5}{5}(\tfrac{1}{2})^5 \\ &= 0.1875. \end{aligned}$$

8. Hence since p-value exceeds α, we fail to reject H_0 and in summary: $y_{observed} = 4$ $(n.s.)$ The median percentage yield is not significantly different for the new supply source.

Notice that to obtain a significant result with a sample size as small as this it would be necessary to have $y_{observed} = 5$, for then p-value $= (\frac{1}{2})^5 = 0.03125 < \alpha$.

Therefore an acceptance region is of the form: 'Accept H_0 if $Y \leq 4$'. Based on such an acceptance region, the Type II errors are large; for example, $\beta(0.9) = P(Y \leq 4 | Y \sim Bin(5, 0.9)) = 0.41$. Much more data are needed to control β for fixed α.

Large-sample sign test

Binomial tables do not operate for large sample sizes. Instead we proceed as follows: If we set $Y_i = sign(X_i - \xi_0)$, then, when H_0 is true, the Y_i are an *i.i.d.* random sample from a $Bin(1, \frac{1}{2})$ population with mean $\mu = \frac{1}{2}$ and variance $\sigma^2 = (1)(\frac{1}{2})(1 - \frac{1}{2}) = 0.25$. By the Central Limit Theorem,

$$Z = \frac{\overline{Y} - \mu}{\frac{\sigma}{\sqrt{n}}} = \frac{\frac{Y}{n} - \frac{1}{2}}{\frac{\frac{1}{2}}{\sqrt{n}}} = \frac{Y - \frac{n}{2}}{\sqrt{\frac{n}{4}}} \sim N(0,1).$$

Hypothesis tests are then large-sample tests for whether a binomial parameter is equal to $\frac{1}{2}$. (See Section 11.3.)

Example 15.2

Research funding in Australia

The Australian Research Council (ARC) and National Health and Medical Research Council (NHMRC) grants for 1992 are listed (in census form) in the S/L plot below, for the 39 universities and colleges in Australia.

Total ARC + NHMRC grants, Australian Universities, 1992.

Scale: stems—millions (\$); leaves—10^{-1} millions (\$)

5|4 represents 5.4 million dollars.

Stem	Leaves	Stem	Signs
0	0 0 0 0 0 0 0 0 0 0 1 2 2 3 6 6 8 8	−0	− − − − − − − − − − − − − − − − − −
1	0 1 2 5 8 9	0	+ + + + + +
2	3 5	1	+ +
3		2	
4		3	
5	2 2 4 7	4	+ + + +
6	8	5	+
7		6	
8	7	7	+
9		8	
10		9	
11	1 5	10	+ +
12		11	
13	5	12	+
14		13	
15	5	14	+
16		15	
17		16	
18		17	
19	2	18	+
20	4 †	19	+
21		20	
22	1 *	21	+

† denotes Sydney University; * denotes Melbourne University.

(*Source*: *Australian Campus Review Weekly*, 16–22 January 1992.)

This is a census of university funding rather than a sample; however interpreted as data from a hypothetical population, X, of 'possible' research grant funding now, or in the past, or in the future, in Australia or overseas—we may question:

Do the data indicate a median funding support of \$100 000 ?
To test whether the median of this theoretical population may be reasonably taken as \$100 000 we construct the right-side S/L plot above, showing the signs of $x_i - 100\ 000$.
Consider $H_0 : \xi_{0.5} = 100\ 000$ against $H_1 : \xi_{0.5} > 100\ 000$ the standardised sign test statistic gives

$$z_{observed} = \frac{y_{observed} - \frac{n}{2}}{\sqrt{\frac{n}{4}}} = \frac{21 - \frac{39}{2}}{\sqrt{\frac{39}{4}}} = 0.48,$$

which carries a p-value of 0.32, providing little evidence to reject H_0. This seems to be a long way short of the 'grantmanship' displayed by the institutions receiving the most support.

EXERCISES

15.1 Is the median uric acid level 6 mg /100 mL for people on an experimental diet?

15 people trying the diet had the following measured uric acid levels:

5.9	5.5	6.1	6.1	6.1
5.9	5.9	5.9	5.7	5.8
5.6	5.7	6.3	6.4	4.1

15.2 An aeromagnetic survey of an outback area, when analysed, revealed the following measurements of percentage of the trace element zinc in the soil of independent survey sectors:

23.4	16.8	21.9	12.2
4.6	15.9	11.8	16.9
12.2	28.6	13.3	8.6
13.4	20.1	19.3	19.0

If further ground research and investigation is warranted only if the median percentage exceeds 20% (otherwise the aerial survey moves to adjacent sectors), use the sign test to provide an initial indication of the course of action.

15.3 The following ranked data result from 9 determinations of the heat of sublimation of platinum (kcal/mole):

134.00	134.50	134.70	134.80	134.88
134.96	135.08	135.30	136.00	

(a) Construct a normal probability plot for these data.

(b) Explain why it is unreasonable to assume that these data are taken from a normally distributed population.

(c) What evidence is contained in the normal probability plot in (a) to suggest that the data have come from a population which is symmetrically distributed?

(d) The sign test is to be used to test

$$H_0 : \mu = 135$$
$$H_1 : \mu < 135$$

where μ is the population mean.

i. Explain how, in using the sign test, the data set is reduced to the sequence of symbols

$- - - - - - + + +$

Is the ordering of signs crucial to the outcome of the test?

ii. How likely is it to get 3 or fewer + signs when H_0 is true?

iii. Give the p-value of the test.

15.4 The following list of 20 'prices realised' resulted from auction realisations of Melbourne properties on the weekend of 17 August 1991. These data appeared in Example 7.11.

520 000	90 000	120 000
(U) 88 000	147 000	(U) 137 000
(U) 60 000	170 000	165 000
130 000	133 000	232 000
(U) 87 500	337 000	120 000
119 000	158 500	211 500
(U) 137 000	200 000	

(*Source: The Sunday Age*, 18 August 1991.)

(U) signifies a price realised for a 'flat', 'apartment' or 'unit'.

(a) Use the sign test to determine whether the median price for 'units' exceeds \$85 000.

(b) Use the sign test to determine whether the median price for property exceeds \$130 000.

15.5 The following is a 5-point number summary for 28 laboratory data on the measured ionisation potential of some second period elements in the periodic table of elements (from lithium to fluorine).

28	**Ionisation potential (eV)**	
	12.4	
	7.3	15.0
	5.4	17.4

Use the sign test to test the hypotheses

$$H_0 : \xi_{0.5} = 15$$
$$H_1 : \xi_{0.5} < 15.$$

15.6 If X is a continuous random variable for which $P(X > 0) = 0.3$, determine the mean and variance of $Y = sign(X)$.

15.2 THE SIGNED-RANK TEST

If we are willing to impose the extra condition of symmetry on the nature of the population from which our data has been sampled, then we do not have to discard so much data information as we did in the sign test. In the **Wilcoxon signed-rank test** we are able to retain not only the sign of each data point (the sign test procedure), but also the relative rankings of the magnitudes of the data points—and this is why the terms 'sign' and 'rank' appear in the name of the test.

Using ranks

Recall from Chapter 1 that if distinct observations $w_1, w_2, w_3, \ldots, w_n$ are ranked from smallest to largest, in the ordered data set $w_{(1)} < w_{(2)} < \ldots < w_{(n)}$ then the rank of $w_{(i)}$ is i. Therefore, the function '*rank*' is defined by

$$rank(w_j) = i \text{ if } w_j = w_{(i)}.$$

Often data sets have several points the same. This results in **tied data.** How do we calculate ranks when the data are tied? Simply push the data points apart a little and pretend that instead of being the same, they were, in fact, minutely different. Then assign the ranks to each and average the results of these rankings to obtain the rank of the tie.

Example 15.3

Ranking wave size at Bells Beach

Bells Beach, near Torquay, is one of the great surfing beaches of Victoria. This has often been the stage for surfing championships. Suppose that a one week period resulted in the following wave-height data (in metres). These are outcomes of random variables X_i = wave-height on Day i.

1, 1.5, 1.5, 1.5, 2, 2.5, 1.

The smallest waves for the period were 1 metre, so these would occupy ranks 1 and 2 (if, as data points, x_1 and x_7 were slightly different). Therefore, $x_1 = 1$ and $x_7 = 1$ are assigned rank $\frac{1+2}{2} = 1.5$ each. Similarly, $x_2 = x_3 = x_4 = 1.5$ are each assigned the rank $\frac{3+4+5}{3} = 4$. We summarise the averaging process in the following table:

x_i	1	1.5	1.5	1.5	2	2.5	1
$rank(x_i)$	1.5	4	4	4	6	7	1.5

This method of averaging tied ranks has the advantage that the sum of the ranks of the data set $x_1, x_2, x_3, \ldots, x_n$ is *always*

$$1 + 2 + 3 + 4 + \ldots + n = \frac{n(n+1)}{2}.$$

Creating a test statistic

The following are the basic assumptions of the signed-rank test:

> The probability distribution of the continuous population random variable X is
> - symmetric
> - has population mean μ and median $\xi_{0.5}$ with $\mu = \xi_{0.5}$.

Suppose that to test the null hypothesis $H_0 : \mu = \mu_0$ we have a random sample $X_1, X_2, X_3, \ldots, X_n$ from the population. The Wilcoxon signed-rank test statistic is determined by ranking the observed absolute differences $|x_i - \mu_0|$ from smallest to largest and then adding the ranks of only those x_i which exceed μ_0. For a reasonably large amount of data, the ranking is easily done in a S/L plot with ordered leaves (which may be annotated to retain the sign of a data point).

When the null hypothesis is true, and the population is symmetric, a data point of large magnitude is just as likely to have come from the top tail of the distribution (with a positive value of $x_i - \mu_0$) as it is to have come from the bottom tail of the distribution (with a negative value of $x_i - \mu_0$). In fact, for a given ranking of $|x_i - \mu_0|$, the data point x_i has a 50:50 chance of representing either tail of the distribution.

Therefore we can assess the likely truth of H_0 by seeing if there are too many high or too many low ranks associated with the (for example) top tail of the data set. The test statistic, S^+, counts up the ranks of $|X_i - \mu_0|$ for which $X_i - \mu_0 > 0$. That is

> **The Wilcoxon Signed-Rank Test**
> Null hypothesis: $H_0 : \mu = \mu_0$
> Test statistic: $S^+ = \sum_{X_j - \mu_0 > 0} rank(|X_j - \mu_0|)$. (15.1)

Of course, if we define

$$S^- = \sum_{X_j - \mu_0 < 0} rank(|X_j - \mu_0|),$$

as the signed-rank for the negative ranks then

$$S^+ + S^- = \sum_{i=1}^{n} i = \frac{n(n+1)}{2},$$

since all the ranks, even when some are averaged, always add to the same total. This means that S^+ can be as small as 0 and as large as $\frac{n(n+1)}{2}$.

The probability distribution of S^+ when $n = 5$

It is easy to give the null distribution of S^+ for small values of n like $n = 5$. If we rank the (distinct) absolute values of the observed differences $|d_j| = |x_j - \mu_0|$, from smallest to largest, then when H_0 is true, for a fixed value of $rank(|d_j|)$, the sign of the difference d_j is **equally likely** to be '+' as '−'. This provides the 2^5 possibilities, which we could call **signed ranks**, listed below.

$rank(d_j)$					s^+	$rank(d_j)$					s^+
1	2	3	4	5		1	2	3	4	5	
+	+	+	+	+	15	−	+	+	+	+	14
+	+	+	+	−	10	−	+	+	+	−	9
+	+	+	−	+	11	−	+	+	−	+	10
+	+	+	−	−	6	−	+	+	−	−	5
+	+	−	+	+	12	−	+	−	+	+	11
+	+	−	+	−	7	−	+	−	+	−	6
+	+	−	−	+	8	−	+	−	−	+	7
+	+	−	−	−	3	−	+	−	−	−	2
+	−	+	+	+	13	−	−	+	+	+	12
+	−	+	+	−	8	−	−	+	+	−	7
+	−	+	−	+	9	−	−	+	−	+	8
+	−	+	−	−	4	−	−	+	−	−	3
+	−	−	+	+	10	−	−	−	+	+	9
+	−	−	+	−	5	−	−	−	+	−	4
+	−	−	−	+	6	−	−	−	−	+	5
+	−	−	−	−	1	−	−	−	−	−	0

For example, the calculation in the third line of the left-side table is

$$s^+ = \sum_{d_i>0} rank(|d_i|) = 1 + 2 + 3 + 5 = 11.$$

The probability distribution for S^+ arises simply because each of the 32 sequences are equally likely. Thus, for example, there are three occurrences where s^+ equals 5. Therefore, $P(S^+ = 5) = \frac{3}{32}$.

s	0	1	2	3	4	5	6	7	8	9	10	11	12	13	14	15
$P(S^+ = s)$	$\frac{1}{32}$	$\frac{1}{32}$	$\frac{1}{32}$	$\frac{2}{32}$	$\frac{2}{32}$	$\frac{3}{32}$	$\frac{3}{32}$	$\frac{3}{32}$	$\frac{3}{32}$	$\frac{3}{32}$	$\frac{3}{32}$	$\frac{2}{32}$	$\frac{2}{32}$	$\frac{1}{32}$	$\frac{1}{32}$	$\frac{1}{32}$

Probability mass function of the Wilcoxon signed-rank statistic S^+ ($n = 5$)

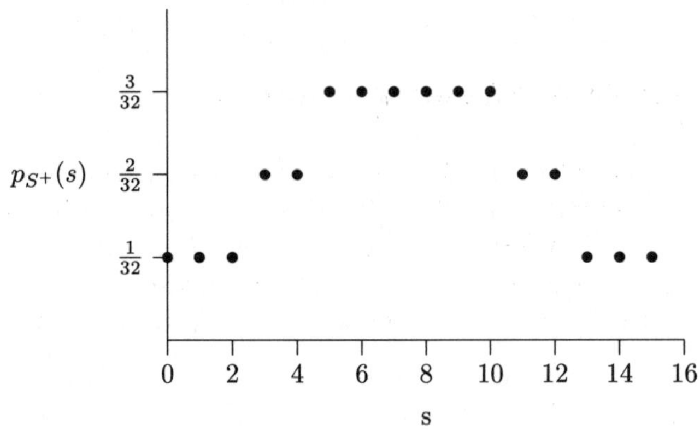

The graph of the probability mass function shows that the probability distribution is symmetric about its centre. To locate the centre based on a sample size n, note that

$$0 \leq s^+_{observed} \leq \frac{n(n+1)}{2},$$

so that the centre lies in the middle of this interval at $\frac{n(n+1)}{4}$. This means that probabilities for the upper and lower tail of the distribution are linked by

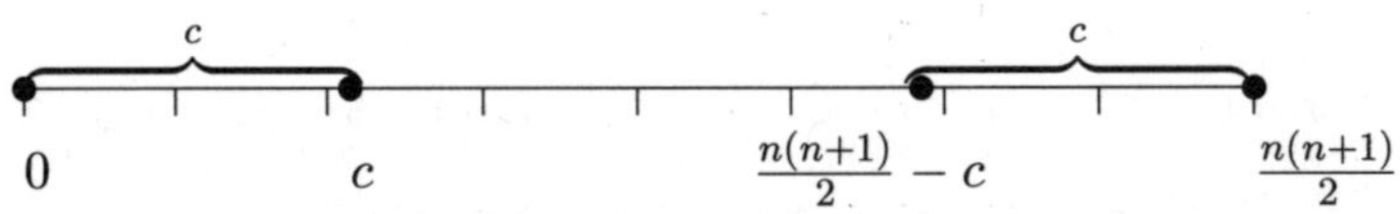

$$P(S^+ \leq c) = P(S^+ \geq \frac{n(n+1)}{2} - c). \tag{15.2}$$

From the listing of the probability distribution of S^+, determined under the assumption that H_0 is true, a table of p-values is simply made for $n = 5$. Starting from the highest possible values for S^+, for these are the ones which will cause us to reject H_0, it is easy to form $P(S^+ \geq c)$ for various values of c.

$$\begin{aligned} P(S^+ \geq 15) &= \tfrac{1}{32} = 0.03125 \\ P(S^+ \geq 14) &= \tfrac{2}{32} = 0.0625 \\ P(S^+ \geq 13) &= \tfrac{3}{32} = 0.09375 \\ P(S^+ \geq 12) &= \tfrac{5}{32} = 0.15625 \\ &\vdots \end{aligned}$$

Notice (as is usually the case for discrete distributions) there is no *exact* upper 5% point. In the above calculation, 6.25% lies at 14 or above, while only 3.125% lies at 15 or above. This means that it is easier to operate significance tests through p-values than to fix a significance level at 5% and not be able to achieve it.

Since the distribution is symmetric, obviously, the lower tail probabilities are identical to these p-values. Using (15.2) repeatedly:

$$\begin{aligned} P(S^+ \leq 0) &= \tfrac{1}{32} = 0.03125 \\ P(S^+ \leq 1) &= \tfrac{2}{32} = 0.0625 \\ P(S^+ \leq 2) &= \tfrac{3}{32} = 0.09375 \\ P(S^+ \leq 3) &= \tfrac{5}{32} = 0.15625 \\ &\vdots \end{aligned}$$

This allows us to prepare for confidence intervals and two-tailed tests.

Example 15.4

Completion times for electrical drafting tasks

A mining company produces 'in progress' details of budgeted time and actual time taken for construction tasks currently in operation. The following times were the actual hours recorded for tankhouse cabling layout tasks which had a budgeted timeframe of 15 hours.

12.50, 15.25, 14.25, 16.75, 11.75.

Worked example:

Use a distribution-free test to determine whether the mean layout time is 15 hours as budgeted.

We shorten the hypothesis testing blueprint to the following steps:

1. Let μ denote the mean layout time for tasks of this nature.
2. $H_0 : \mu = 15$ against an alternative $H_1 : \mu \neq 15$.
3. Test statistic: Wilcoxon signed-rank statistic $S^+ = \sum_{D_i>0} rank(|D_i|)$, for the discrepancies $D_i = X_i - 15$.
4. Significance level should be as close as possible, but not exceeding 0.05.
5. Data: $x_1 = 12.50$, $x_2 = 15.25$, $x_3 = 14.25$, $x_4 = 16.75$, $x_5 = 11.75$.
6. When H_0 is true we determine the observed value of the test statistic as follows:

$$\begin{aligned} |d_1| &= |x_1 - 15| = |12.50 - 15| = 2.50 \\ |d_2| &= |x_2 - 15| = |15.25 - 15| = 0.25 \\ |d_3| &= |x_3 - 15| = |14.25 - 15| = 0.75 \\ |d_4| &= |x_4 - 15| = |16.75 - 15| = 1.75 \\ |d_5| &= |x_5 - 15| = |11.75 - 15| = 3.25. \end{aligned}$$

Ranked absolute differences	$rank(\|d_i\|)$	$sign(d_i)$
0.25	1	1
0.75	2	0
1.75	3	1
2.50	4	0
3.25	5	0

The sum of the ranks associated with positive discrepancies is $s^+_{observed} = 1 + 3 = 4$.

7. By (15.2) which links upper and lower tail probabilities,

$$\begin{aligned} p\text{-value} &= 2P(S^+ \leq 4) = 2P(S^+ \geq \tfrac{5(5+1)}{2} - 4) = 2P(S^+ \geq 11) \\ &= 2(\tfrac{7}{32}) = 0.4375. \end{aligned}$$

8. Hence since p-value exceeds α and in summary: $s^+_{observed} = 4$ $(n.s.)$. The mean layout time is not significantly different from the budgeted time of 15 hours.

Appendix 7 gives p-values for the null distribution of S^+ for sample sizes up to 20; for larger sample sizes we exploit the properties of a standard pivot which is reasonably approximated by a standard normal probability distribution.

Theorem 15.1 *The null distribution of S^+ (or the probability distribution of S^+ when $H_0 : \mu = \mu_0$ is true) which is based on a random sample of size n satisfies*

$$E(S^+) = \frac{n(n+1)}{4}, \quad Var(S^+) = \frac{1}{24}n(n+1)(2n+1).$$

Proof: According to (15.1),

$$\begin{aligned} S^+ &= \sum_{X_i - \mu_0 > 0} rank(|X_i - \mu_0|) \\ &= \sum_{i=1}^{n} sign(X_i - \mu_0) rank(|X_i - \mu_0|) \\ &= \sum_{i=1}^{n} i\ sign(D_i) \quad \text{for} \quad D_i = X_i - \mu_0. \end{aligned}$$

For a given $rank(|D_i|)$, the discrepancy D_i has a 50:50 chance of being '+' or '−'. This means that we may write

$$S^+ = \sum_{i=1}^{n} i\, Y_i,$$

where $Y_i \sim Bin(1, \frac{1}{2})$ have identical means $E(Y_i) = \frac{1}{2}$ and variances $Var(Y_i) = (1)(\frac{1}{2})(1 - \frac{1}{2}) = \frac{1}{4}$. By **SUM 1**:

$$E(S^+) = E(\sum_{i=1}^{n} iY_i) = \sum_{i=1}^{n} iE(Y_i) = \tfrac{1}{2}\sum_{i=1}^{n} i = \tfrac{1}{2}\frac{n(n+1)}{2} = \frac{n(n+1)}{4}.$$

(This was the centre of the distribution which we determined earlier by symmetry.) By **SUM 2**:

$$Var(S^+) = Var(\sum_{i=1}^{n} iY_i) = \sum_{i=1}^{n} i^2 Var(Y_i) = \frac{1}{4}\sum_{i=1}^{n} i^2 = \frac{1}{24}n(n+1)(2n+1).$$

■

When S^+ is standardised we may use a normal approximation:

$$\frac{S^+ - E(S^+)}{\sqrt{Var(S^+)}} = \frac{S^+ - \frac{n(n+1)}{4}}{\sqrt{\frac{1}{24}n(n+1)(2n+1)}} \sim N(0,1).$$

Accounting for tied observations

This results in the ranks being averaged as discussed in Example 15.3.

Example 15.5

Testing for an average meal weight

We re-introduce the data of Example 1.6 showing the combined weights (g) of the ham/crepe/sauce/tomato/onion components from a sample of in-flight meals served by Ansett Airlines.

189.1	188.4	205.2	188.4	176.3
199.1	188.1	181.9	189.8	182.5
198.1	197.9	197.4	198.3	211.6
186.5	188.2	198.1	202.4	203.6
177.1	171.9	172.3	183.6	183.7
191.6	204.2	204.5	183.2	

The S/L plot is symmetric:

The weight of airline meals
Scale: stems—10 g; leaves—g
17. | 6 represents 176 grams

17*	1 2
17.	6 7
18*	1 2 3 3 3
18.	6 8 8 8 8 9 9
19*	1
19.	7 7 8 8 8 9
20*	2 3 4 4
20.	5
21*	1

Worked example:

Apply a distribution-free test to determine whether the average meal weight exceeds 190 g.

1. Let μ denote the mean weight of an airline meal.
2. $H_0 : \mu = 190$ against an alternative $H_1 : \mu > 190$.
3. Test statistic: Since n is large we can use the normalised version of the Wilcoxon signed-rank statistic $S^+ = \sum_{D_i>0} rank(|D_i|)$, for the discrepancies $D_i = X_i - 190$.
4. α should be as close as possible, but not exceeding 0.05.
5. Data: Denote the 29 listed data points as realisations $x_1, x_2, \ldots .x_{29}$.
6. When H_0 is true we determine the observed value of the test statistic by ranking the $|d_i|$ from smallest to largest, and including only those ranks corresponding to a positive discrepancy in the determination of $s^+_{observed}$.

Ranked absolute differences	$rank(\|d_j\|)$	$sign(d_j)$	**Ranked absolute differences**	$rank(\|d_j\|)$	$sign(d_j)$
0.2	1	0	8.1	16	0
0.9	2	0	8.1	16	1
1.6	4	0	8.1	16	1
1.6	4	1	8.3	18	1
1.6	4	0	9.1	19	1
1.8	6	0	12.4	20	1
1.9	7	0	12.9	21	0
3.5	8	0	13.6	22	1
6.3	9	0	13.7	23	0
6.4	10	0	14.2	24	1
6.8	11	0	14.5	25	1
7.4	12	1	15.2	26	1
7.5	13	0	17.7	27	0
7.9	14	1	18.1	28	0
			21.6	29	1

$$s^+_{observed} = 4+12+14+16+16+18+19+20+22+24+25+26+29 = 245$$

and with $n = 29$,

$$z_{observed} = \frac{s^+_{observed} - \frac{n(n+1)}{4}}{\sqrt{\frac{1}{24}n(n+1)(2n+1)}} = \frac{245 - \frac{29(29+1)}{4}}{\sqrt{\frac{1}{24}29(29+1)(58+1)}} = 0.59.$$

7. p-value $= P(Z > 0.59) = 1 - 0.7224 \approx 0.28$.
8. Hence, since p-value exceeds α, this provides little evidence to reject H_0 and in summary: $z_{observed} = 0.59$ ($n.s.$). The mean weight of airline meals is not significantly greater than 190 g.

(Could we have made this assessment by looking at the S/L plot? Could we have assumed the data normally distributed and applied a t test?)

Signed-rank test for paired data

The Wilcoxon signed-rank test applies to data from a continuous symmetrically distributed variable that need not be normally distributed. When the underlying population is normally distributed, the t test is a superior performer, in the sense that for fixed significance levels α, the Type II errors are smallest compared to other competing tests.

For very heavy-tailed symmetric populations, that are non-normal, the Wilcoxon signed-rank procedure is recommended. For example, the mean of a Cauchy distribution is not defined, but the median is; therefore the signed-rank test is a good detector of the centre of Cauchy-type distributions.

The continuous/symmetry conditions are often met when considering paired data, since, by the paired comparison test of Section 12.1, the analysis of the paired data (X_i, Y_i), $i = 1, 2, 3, \ldots, n$ begins by subtracting paired observations (to destroy their dependency on one another as observations on the same experimental unit) and this creates a single data set

$$X_i - Y_i, \; i = i, 2, 3, \ldots, n,$$

to which the Wilcoxon signed-rank methods may be applied for hypothesis tests concerning the true mean difference μ_D.

EXERCISES

15.7 Acidity of rainwater causes the facades of concrete buildings to corrode.

$n = 5$ core samples drilled in randomly chosen positions from the concrete facade of a Collins Street building of 1960's vintage showed depth of corrosion (in cm) of

7.7, 1.5, 3.5, 5.5, 4.5.

(a) Use a t test to determine whether the depth of corrosion of this building is significantly greater than 1.6 cm.

(b) Test the hypothesis in (a) by using signed-rank procedures.

15.8 Apply the Wilcoxon signed-rank test to the Bells Beach wave size data of Example 15.3. Test whether the average wave size at Bells Beach exceeds 1.75 metres.

15.9 The following measurements of wind velocity (km hr^{-1}) were monitored during a period of summer thunderstorms.

65	32	37	41	56
43	66	74	138	21
51	43	47	58	49
31	30	48	52	107

(a) Construct a S/L plot for these data.

(b) Do these data satisfy the assumptions of the Wilcoxon signed-rank test?

(c) Test whether the median wind velocity significantly exceeds 35 km hr^{-1}.

15.10 The hypothesis $H_0 : \mu = 0$ is to be tested against a two-sided alternative.

8 data points produced an observed value of $s^+_{observed} = 4$. Determine, for a sample of size 8,

$$P(S^+ \leq 4|H_0 \text{ is true}).$$

Hence determine the p-value of the test.

15.11 Devore (1991) reports the accompanying data on lead concentration in samples gathered during eight different summer rainfalls in Christchurch, New Zealand.

17.0, 21.4, 30.6, 5.0, 12.2, 11.8, 17.3, 18.8.

Assuming that the lead content distribution is symmetric, use the Wilcoxon signed-rank test to determine whether the average lead content is significantly different from 10 μ g L^{-1}.

(*Source*: J. L. Devore, (1991). *Probability and Statistics for Engineering and the Sciences.* Brooks/Cole: California. From: The Lead Content and Acidity of Christchurch Precipitation, *New Zealand Journal of Science*, 1980.)

15.12 The residuals

0.135	−0.004	−0.006
−0.051	−0.148	0.033
−0.069	0.011	0.047
0.060	0.001	0.009

resulted when a least squares model was fitted to a scatterplot.

Assuming only that the distribution of the residuals is symmetric, test whether the distribution has mean zero.

15.13 An upper-tailed Wilcoxon signed-rank test is applied to 50 data points. What values of S^+ lead to the rejection of the null hypothesis at $\alpha = 0.05$?

15.14 In the citation of published articles, the *immediacy index* is a measure of how quickly an 'average article' in a journal is cited during the first year of publication. This is a measure of the profile of the journal: are the journal's articles accessed and then quoted?

The data show the immediacy indices for two journals by year.

Year	Online	Online Review
1979	0.667	0.190
1980	0.808	0.154
1981	1.174	0.042
1982	0.257	0.043
1983	0.150	0.217
1984	0.362	0.000
1985	0.380	0.056
1986	0.448	0.000

(*Source*: Sievert, M.C. (1990). Ten years of the literature of online searching: An analysis of Online and Online Review. *Journal for the American Society for Information Science*, 41, 560-8.)

It is unlikely that such indices are normally distributed. Test whether there is a significant difference between the mean immediacy indices for the two journals.

15.3 THE RANK-SUM TEST

To compare the means μ_1 and μ_2 of two populations using independent random samples, one drawn from each population, the two-sample t test is the standard test procedure. However, the t test assumes that the population random variables for the two populations are $N(\mu_1, \sigma^2)$ and $N(\mu_2, \sigma^2)$ respectively, where σ^2 is the common population variance.

When we do not wish to assume a distributional shape for the data, the **Wilcoxon rank-sum test** is applicable. In particular, it is not necessary to assume that the populations are normally distributed.

Creating a test statistic

The following are the basic assumptions for the rank-sum test:

> The probability distribution for the two populations
> - have the same shape
> - are centred at (possibly) different means μ_1 and μ_2 respectively.

For testing $H_0 : \mu_1 - \mu_2 = 0$ against an appropriate single- or double-sided alternative we proceed as follows.
From Population 1, with mean μ_1, extract a random sample $X_1, X_2, \ldots, X_m$.
From Population 2, having mean μ_2, extract a random sample $Y_1, Y_2, \ldots, Y_n$.
To facilitate using probability tables for the test statistic now being developed, we ensure that $m \leq n$; that is, the X data is the label for the data set with the smallest sample size.

Begin by collecting both samples together and ranking all $m+n$ data points from smallest to largest. If μ_1 is very different from μ_2, then the samples extracted from the populations will show very little overlap; that is, there will be very few data points that could have come from either population. In fact, if μ_1 is located far below μ_2, then the ranks assigned to the X data will be all the smallest ranks. If we define the random variable W to be the sum of the ranks of the X values in the combined sample, then the situation just described, with the X data lying below all the Y data, means that ranks 1,2,3...,m (assuming no ties) are assigned the X values. These are the smallest possible assigned ranks. Therefore, $w_{minimum}$, the smallest possible value for the sum of the X ranks is

$$w_{minimum} = 1 + 2 + 3 + 4 + \ldots + m = \frac{m(m+1)}{2}.$$

Of course, the observed value of W is largest in the situation where μ_1 lies far above μ_2, and all the X data are larger than all the Y data, and has the largest ranks $(n+1)$, $(n+2)$, $(n+3), \ldots, (n+m)$ assigned. These are the largest ranks for the combined sample. Therefore, $w_{maximum}$, the largest possible observed value of the random variable W is

$$\begin{aligned} w_{maximum} &= (n+1)+(n+2)+(n+3)+\ldots+(n+m) \\ &= \underbrace{(n+n+n+\ldots+n)}_{m\ terms}+(1+2+3+\ldots+m) \\ &= mn+\tfrac{1}{2}m(m+1) \\ &= \tfrac{1}{2}m(m+2n+1). \end{aligned}$$

Usually, significance testing decisions are not so clearcut, so that when the samples concerned are reflecting close proximity of population means, there will be considerable overlap and mixing of the X and Y data. This will provide a range of ranks assigned to the X values so that, in general,

$$\frac{m(m+1)}{2} \leq w_{observed} \leq \tfrac{1}{2}m(m+2n+1).$$

The Wilcoxon Rank-Sum Test
Null hypothesis: $H_0 : \mu_1 = \mu_2$
Test statistic: $W = \sum_{i=1}^{m} R_i$,
where R_i = rank assigned to X_i in the combined sample. (15.3)

When H_0 is true, the probability distribution for W is symmetric in the interval $[w_{minimum}, w_{maximum}]$. Upper α points are given in Appendix 8 for $3 \leq m \leq 8$ and $3 \leq n \leq 8$. These provide p-values for upper-tailed tests. The symmetry of the distribution of W may be exploited to find p-values in the lower tail: for a constant c in the interval $[w_{minimum}, w_{maximum}]$,

$$\begin{aligned} P(W \leq c) &= P(W \geq w_{maximum} - (c - w_{minimum})) \\ &= P(W \geq w_{maximum} + w_{minimum} - c) \\ &= P(W \geq \tfrac{1}{2}m(m+2n+1) + \tfrac{1}{2}m(m+1) - c). \end{aligned}$$

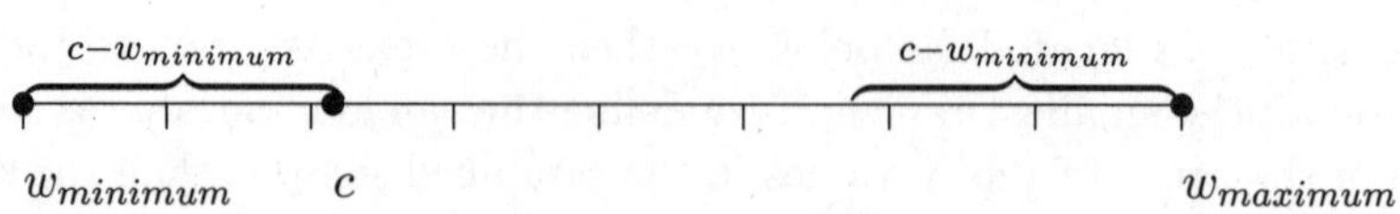

That is,

$$P(W \leq c) = P(W \geq m(m+n+1) - c). \tag{15.4}$$

Example 15.6
Probability of a more extreme sequence of ranks
Suppose that 6 observed X data points and 8 observed Y data points were ranked, in order from smallest to largest as

$$x\ x\ x\ y\ y\ x\ y\ x\ y\ y\ y\ x\ y\ y\ .$$

If we place the ranks under this sequence

combined sample	x	x	x	y	y	x	y	x	y	y	y	x	y	y
rank	1	2	3	4	5	6	7	8	9	10	11	12	13	14

then $w_{observed} = 1 + 2 + 3 + 6 + 8 + 12 = 32$.
The probability of a more extreme value than was actually observed is the probability of an x rank which is smaller than 32. According to (15.4),

$$P(W \leq 32) = P(W \geq 6(6 + 8 + 1) - 32) = P(W \geq 58) = 0.054.$$

Example 15.6 is essentially an exercise in reading the probability tables for the rank-sum test. The next example shows how to use a back-to-back S/L plot with ordered leaves to assist in the ranking of the combined data set.

Example 15.7
The die is cast
A metal-stamping die must be replaced at frequent intervals on a production line. Lifetimes of local and imported replacements (selected at random from either source) have been recorded (in days) as follows:

Local	64	22	16	48	8		
Imported	61	70	17	36	90	29	73

Since the cost of the dies from either source is approximately the same, the decision on whether to use the local or the imported replacements is to be taken on their reliability. The maintenance engineer replacing the dies wishes to know if the imported replacement dies have longer lifetimes, on average, than do the locally produced replacement dies.
Worked example:
Explain why a t test is inappropriate for these data and analyse the data using an appropriate distribution-free test.

If we begin by constructing a back-to-back S/L plot, then this assists, not only in detecting distributional shape, but also in speeding the ranking process for the entire data set.

Time to replacement for metal stamping die (days)
Scale: stems–tens; leaves—units
1|6 represents 16 days

Imported		Local
	0	8
7	1	6
9	2	2
6	3	
	4	8
	5	
1	6	4
3 0	7	
	8	
0	9	

Imported		Local
	0	$\boxed{1}$
$\boxed{3}$	1	$\boxed{2}$
$\boxed{5}$	2	$\boxed{4}$
$\boxed{6}$	3	
	4	$\boxed{7}$
	5	
$\boxed{8}$	6	$\boxed{9}$
$\boxed{11}$ $\boxed{10}$	7	
	8	
$\boxed{12}$	9	

The S/L plots show data that are quite dispersed; probably from a distribution that has more probability in the upper tails than the normal distribution. (Reliability distributions are often modelled by the Weibull distribution.)

For this lack-of-normality reason, a t test is inappropriate for these data. However, it is anticipated that the probability distributions for locally and imported dies could easily be of the same type, with the same dispersed shape, but simply shifted in location. This makes a Wilcoxon rank-sum test appropriate.

1. Let μ_1 denote the mean lifetimes of the local dies (X data with $m = 5$) and μ_2 denote the mean lifetimes of the imported dies (Y data with $n = 7$).
2. $H_0 : \mu_1 = \mu_2$ against an alternative $H_1 : \mu_1 < \mu_2$.
3. Test statistic: $W = \sum_{i=1}^{12} R_i$ where the observed values of R_i are given in the right-side S/L plot above.
4. α should be as close as possible, but not exceeding 0.05.
5. Data: Given appropriately in the above left-side S/L plot.
6. When H_0 is true we determine the observed value of the test statistic by adding the observed ranks which appear in the boxes for the local dies:
   $w_{observed} = r_1 + r_2 + r_3 + r_4 + r_5 = \boxed{1} + \boxed{2} + \boxed{4} + \boxed{7} + \boxed{9} = 23.$
7.

$$\begin{aligned} p\text{-value} &= P(W \leq w_{observed} | H_0 \text{ is true}) \\ &= P(W \leq 23) \\ &= P(W \geq 5(5+7+1) - 23) \end{aligned}$$

$$
\begin{aligned}
&= P(W \geq 42) \\
&\geq P(W \geq 43) \\
&= 0.053.
\end{aligned}
$$

8. Hence, since p-value exceeds α, $w_{observed} = 23$ $(n.s.)$ The mean lifetime of the imported replacement dies is not significantly different from the mean lifetimes of locally produced replacements.

A standard pivot for m, n large

Appendix 8 gives the p-values for the null distribution of W for sample sizes up to $m = 8$, $n = 8$. For sample sizes larger than this, the standard pivot created by standardising W has an approximate standard normal distribution.

Theorem 15.2 *When $H_0 : \mu_1 - \mu_2 = 0$ is true, the probability distribution of W satisfies*

$$E(W) = \frac{m(m+n+1)}{2}, \quad Var(W) = \frac{mn(m+n+1)}{12}.$$

Proof: We begin with W as a sum of ranks:

$$W = \sum_{i=1}^{m} R_i$$

where, since we assume H_0 true, the ranks are equally likely to be any of the integers $1, 2, 3, 4, \ldots, m+n$. That is, for each i, R_i has a discrete uniform distribution on these integers. This ensures that (see exercises)

$$\mu = E(R_i) = \frac{m+n+1}{2}, \quad \text{and} \quad \sigma^2 = Var(R_i) = \frac{(m+n)^2 - 1}{12}. \qquad (15.5)$$

Therefore $R_1, R_2, R_3, \ldots, R_m$ is a without-replacement sample of size m from a finite population of ranks of size $m + n$. By Theorem 9.6, the mean and variance of the without-replacement sample mean $\overline{R} = \frac{W}{m}$ are given by

$$E(\overline{R}) = \mu \quad \text{and} \quad Var(\overline{R}) = \tfrac{\sigma^2}{m}(\tfrac{N-m}{N-1}),$$

where $N = m + n$ is the population size. The result follows by inserting the population mean and variance from (15.5) into

$$E(W) = mE(\overline{R}) = m\mu \quad \text{and} \quad Var(W) = m^2 Var(\overline{R}) = m\sigma^2(\tfrac{N-m}{N-1}).$$

■

The results of Theorem 15.2, together with a Central Limit Theorem (for sums of weakly correlated random variables) give a standard normal random variable approximation

$$Z = \frac{W - E(W)}{\sqrt{Var(W)}} = \frac{W - \frac{m(m+n+1)}{2}}{\sqrt{\frac{mn(m+n+1)}{12}}} \sim N(0,1).$$

Example 15.8

How long does it take to fill your car with petrol?

As part of a survey into forecourt design and accessibility in self-service petrol stations, the following 'activity times' (seconds) were recorded for the time per customer a petrol pump is in operation.

Hawthorn service station	49	100	161	65	34	101	157	61	55	99
Windsor service station	114	63	117	57	103	72	53	47	102	51

Is the mean activity time the same for each station? The S/L plots below show non-normal distributions of similar shape, so that a two-sample t test is inapplicable.

Service station activity times (seconds)
Scale: stems—tens; leaves—units
11|4 represents 114 seconds

Hawthorn		Windsor
4	3	
9	4	7
5	5	1 3 7
5 1	6	3
	7	2
	8	
9	9	
1 0	10	2 3
	11	4 7
	12	
	13	
	14	
7	15	
1	16	

Hawthorn		Windsor
[1]	3	
[3]	4	[2]
[6]	5	[4] [5] [7]
[10] [8]	6	[9]
	7	[11]
	8	
[12]	9	
[14] [13]	10	[15] [16]
	11	[17] [18]
	12	
	13	
	14	
[19]	15	
[20]	16	

The following is a condensed version of the significance testing format. If we (arbitrarily) label the Hawthorn service station as Population 1 providing activity data $x_1, x_2, x_3, \ldots, x_{10}$, and the Windsor service station as Population 2, then for testing $H_0 : \mu_1 - \mu_2 = 0$ against a two-sided alternative,

$$w_{observed} = \boxed{1} + \boxed{3} + \boxed{6} + \boxed{8} + \boxed{10} + \boxed{12} + \boxed{13} + \boxed{14} + \boxed{19} + \boxed{20} = 106$$

and

$$z_{observed} = \frac{106 - \frac{10(10+10+1)}{2}}{\sqrt{\frac{100(10+10+1)}{12}}} = 0.08.$$

Hence there is not a significant difference between the mean activity times at the two stations: p-value $= 2P(Z > 0.08) = 0.93$.

A comparison with the two-sample t test

The two-sample t test is the best test to use according to the criterion of minimising Type II errors for a fixed Type I error when the sampled populations are $N(\mu_1, \sigma^2)$ and $N(\mu_2, \sigma^2)$ respectively. The Wilcoxon rank-sum test also holds in these situations (where the t test is more efficient) but is superior to the t test in situations of non-normality, especially for very heavy-tailed distributions.

EXERCISES

15.15 Use the data from Example 15.7 to test whether the mean lifetime of locally produced die replacements is 1 month.

15.16 In a sawmill two breast-bench saws are set to cut lengths of timber into 2 metre battens.

The output from the saws was sampled independently, with the following results for batten length measured to the nearest centimetre.

Machine 1	Machine 2
2.10	1.99
1.98	2.00
1.90	1.93
2.05	2.11
2.08	2.10
	2.13

(a) Use a distribution-free test to determine if there is a difference in the average length of battens produced by the two machines.

(b) Use a distribution-free test to determine if the mean length of battens produced on Machine 2 is significantly different from 2 metres.

15.17 In a biology laboratory that is updating its weighing equipment, the new weights are calibrated against the old: ten molluscs are weighed on both scales to see how the distributions of weights compare.

The results are as follows:

Mollusc	Weight on old balance	Weight on new balance
1	6.14	6.13
2	5.90	5.88
3	7.15	7.14
4	8.86	8.87
5	4.99	4.99
6	6.74	6.72
7	7.81	7.78
8	8.15	8.12
9	6.26	6.22
10	6.97	6.93

Use a distribution-free procedure to test whether there is any difference between the mean weight of molluscs as measured by the two balances.

Explain your reasons for the choice of test you have made.

15.18 Suppose that Sample 1 is selected from a population with mean μ_1 and, independently, Sample 2 is selected from a population with mean μ_2.

Sample 1		Sample 2
	0	1
	1	2
	2	9
6	3	1 2 3
8	4	7
5	5	
1	6	
5	7	
	8	
	9	
	10	
	11	
	12	
2	13	

(a) Explain whether the two-sample t test or the Wilcoxon rank-sum test is the most appropriate for these data when testing $H_0 : \mu_1 = \mu_2$ against a two-sided alternative. (Use a scale: stems—tens; leaves—units.)

(b) Perform both the tests. Do they produce the same conclusions?

(c) Suppose that the data value 132 was misrecorded and should really have been 56. How does this affect each of the testing procedures? Explain briefly.

15.19 The following list of 20 'prices realised' resulted from auction realisations of Melbourne properties on the weekend of 17 August 1991. These data appeared in Example 7.11.

520 000	90 000	120 000
119 000	(U) 87 500	(U) 88 000
147 000	(U) 137 000	158 500
337 000	(U) 60 000	170 000
165 000	211 500	120 000
130 000	133 000	232 000
(U) 137 000	200 000	

(*Source: The Sunday Age*, 18 August 1991.)

(U) signifies a price realised for a 'flat', 'apartment' or 'unit'. The other realised prices listed refer to 'house' prices.

Assuming that the 'unit' and 'house' price distributions have the same shape, use the Wilcoxon rank-sum test to determine whether the median price for 'houses' exceeds the median price for 'units' by at least \$50 000.

15.20 Construct the probability distribution for W when the null hypothesis of equal population means is true for a sample of size $m = 3$ of x data and $n = 4$ of y data.

You could begin by listing the possible rank orderings for the combined sample: this is represented by the number of distinct sequences of 3 x symbols and 4 y symbols.

$$\begin{array}{l} xxxyyyy \\ xxyxyyy \\ xxyyxyy \\ xyxyxyy \\ \vdots \end{array}$$

How many such sequences are there? Are all sequences equally likely?

15.21 If the random variable R, representing 'rank', is uniformly distributed on the integers

$$1, 2, 3, \ldots, m+n,$$

show that

$$\mu = E(R) = \frac{m+n+1}{2}$$

and

$$\sigma^2 = Var(R) = \frac{(m+n)^2 - 1}{12}.$$

Appendices

APPENDIX ONE

Cumulative Binomial Probabilities $B(x; n, p) = P(X \le x)$ for $X \sim \text{Bin}(n, p)$

a. $n = 5$

x	0.01	0.05	0.10	0.20	0.25	0.30	0.40	0.50	0.60	0.70	0.75	0.80	0.90	0.95	0.99
0	.951	.774	.590	.328	.237	.168	.078	.031	.010	.002	.001	.000	.000	.000	.000
1	.999	.977	.919	.737	.633	.528	.337	.188	.087	.031	.016	.007	.000	.000	.000
2	1.000	.999	.991	.942	.896	.837	.683	.500	.317	.163	.104	.058	.009	.001	.000
3	1.000	1.000	1.000	.993	.984	.969	.913	.812	.663	.472	.367	.263	.081	.023	.001
4	1.000	1.000	1.000	1.000	.999	.998	.990	.969	.922	.832	.763	.672	.410	.226	.049

b. $n = 10$

x	0.01	0.05	0.10	0.20	0.25	0.30	0.40	0.50	0.60	0.70	0.75	0.80	0.90	0.95	0.99
0	.904	.599	.349	.107	.056	.028	.006	.001	.000	.000	.000	.000	.000	.000	.000
1	.996	.914	.736	.376	.244	.149	.046	.011	.002	.000	.000	.000	.000	.000	.000
2	1.000	.988	.930	.678	.526	.383	.167	.055	.012	.002	.000	.000	.000	.000	.000
3	1.000	.999	.987	.879	.776	.650	.382	.172	.055	.011	.004	.001	.000	.000	.000
4	1.000	1.000	.998	.967	.922	.850	.633	.377	.166	.047	.020	.006	.000	.000	.000
5	1.000	1.000	1.000	.994	.980	.953	.834	.623	.367	.150	.078	.033	.002	.000	.000
6	1.000	1.000	1.000	.999	.996	.989	.945	.828	.618	.350	.224	.121	.013	.001	.000
7	1.000	1.000	1.000	1.000	1.000	.998	.988	.945	.833	.617	.474	.322	.070	.012	.000
8	1.000	1.000	1.000	1.000	1.000	1.000	.998	.989	.954	.851	.756	.624	.264	.086	.004
9	1.000	1.000	1.000	1.000	1.000	1.000	1.000	.999	.994	.972	.944	.893	.651	.401	.096

c. $n = 15$

x	0.01	0.05	0.10	0.20	0.25	0.30	0.40	0.50	0.60	0.70	0.75	0.80	0.90	0.95	0.99
0	.860	.463	.206	.035	.013	.005	.000	.000	.000	.000	.000	.000	.000	.000	.000
1	.990	.829	.549	.167	.080	.035	.005	.000	.000	.000	.000	.000	.000	.000	.000
2	1.000	.964	.816	.398	.236	.127	.027	.004	.000	.000	.000	.000	.000	.000	.000
3	1.000	.995	.944	.648	.461	.297	.091	.018	.002	.000	.000	.000	.000	.000	.000
4	1.000	.999	.987	.836	.686	.515	.217	.059	.009	.001	.000	.000	.000	.000	.000
5	1.000	1.000	.998	.939	.852	.722	.403	.151	.034	.004	.001	.000	.000	.000	.000
6	1.000	1.000	1.000	.982	.943	.869	.610	.304	.095	.015	.004	.001	.000	.000	.000
7	1.000	1.000	1.000	.996	.983	.950	.787	.500	.213	.050	.017	.004	.000	.000	.000
8	1.000	1.000	1.000	.999	.996	.985	.905	.696	.390	.131	.057	.018	.000	.000	.000
9	1.000	1.000	1.000	1.000	.999	.996	.966	.849	.597	.278	.148	.061	.002	.000	.000
10	1.000	1.000	1.000	1.000	1.000	.999	.991	.941	.783	.485	.314	.164	.013	.001	.000
11	1.000	1.000	1.000	1.000	1.000	1.000	.998	.982	.909	.703	.539	.352	.056	.005	.000
12	1.000	1.000	1.000	1.000	1.000	1.000	1.000	.996	.973	.873	.764	.602	.184	.036	.000
13	1.000	1.000	1.000	1.000	1.000	1.000	1.000	1.000	.995	.965	.920	.833	.451	.171	.010
14	1.000	1.000	1.000	1.000	1.000	1.000	1.000	1.000	1.000	.995	.987	.965	.794	.537	.140

d. $n = 20$

x	0.01	0.05	0.10	0.20	0.25	0.30	0.40	0.50	0.60	0.70	0.75	0.80	0.90	0.95	0.99
0	.818	.358	.122	.012	.003	.001	.000	.000	.000	.000	.000	.000	.000	.000	.000
1	.983	.736	.392	.069	.024	.008	.001	.000	.000	.000	.000	.000	.000	.000	.000
2	.999	.925	.677	.206	.091	.035	.004	.000	.000	.000	.000	.000	.000	.000	.000
3	1.000	.984	.867	.411	.225	.107	.016	.001	.000	.000	.000	.000	.000	.000	.000
4	1.000	.997	.957	.630	.415	.238	.051	.006	.000	.000	.000	.000	.000	.000	.000

(continued)

Cumulative Binomial Probabilities (cont.)

d. $n = 20$ *(continued)*

x \ p	0.01	0.05	0.10	0.20	0.25	0.30	0.40	0.50	0.60	0.70	0.75	0.80	0.90	0.95	0.99
5	1.000	1.000	.989	.804	.617	.416	.126	.021	.002	.000	.000	.000	.000	.000	.000
6	1.000	1.000	.998	.913	.786	.608	.250	.058	.006	.000	.000	.000	.000	.000	.000
7	1.000	1.000	1.000	.968	.898	.772	.416	.132	.021	.001	.000	.000	.000	.000	.000
8	1.000	1.000	1.000	.990	.959	.887	.596	.252	.057	.005	.001	.000	.000	.000	.000
9	1.000	1.000	1.000	.997	.986	.952	.755	.412	.128	.017	.004	.001	.000	.000	.000
10	1.000	1.000	1.000	.999	.996	.983	.872	.588	.245	.048	.014	.003	.000	.000	.000
11	1.000	1.000	1.000	1.000	.999	.995	.943	.748	.404	.113	.041	.010	.000	.000	.000
12	1.000	1.000	1.000	1.000	1.000	.999	.979	.868	.584	.228	.102	.032	.000	.000	.000
13	1.000	1.000	1.000	1.000	1.000	1.000	.994	.942	.750	.392	.214	.087	.002	.000	.000
14	1.000	1.000	1.000	1.000	1.000	1.000	.998	.979	.874	.584	.383	.196	.011	.000	.000
15	1.000	1.000	1.000	1.000	1.000	1.000	1.000	.994	.949	.762	.585	.370	.043	.003	.000
16	1.000	1.000	1.000	1.000	1.000	1.000	1.000	.999	.984	.893	.775	.589	.133	.016	.000
17	1.000	1.000	1.000	1.000	1.000	1.000	1.000	1.000	.996	.965	.909	.794	.323	.075	.001
18	1.000	1.000	1.000	1.000	1.000	1.000	1.000	1.000	.999	.992	.976	.931	.608	.264	.017
19	1.000	1.000	1.000	1.000	1.000	1.000	1.000	1.000	1.000	.999	.997	.988	.878	.642	.182

e. $n = 25$

x \ p	0.01	0.05	0.10	0.20	0.25	0.30	0.40	0.50	0.60	0.70	0.75	0.80	0.90	0.95	0.99
0	.778	.277	.072	.004	.001	.000	.000	.000	.000	.000	.000	.000	.000	.000	.000
1	.974	.642	.271	.027	.007	.002	.000	.000	.000	.000	.000	.000	.000	.000	.000
2	.998	.873	.537	.098	.032	.009	.000	.000	.000	.000	.000	.000	.000	.000	.000
3	1.000	.966	.764	.234	.096	.033	.002	.000	.000	.000	.000	.000	.000	.000	.000
4	1.000	.993	.902	.421	.214	.090	.009	.000	.000	.000	.000	.000	.000	.000	.000
5	1.000	.999	.967	.617	.378	.193	.029	.002	.000	.000	.000	.000	.000	.000	.000
6	1.000	1.000	.991	.780	.561	.341	.074	.007	.000	.000	.000	.000	.000	.000	.000
7	1.000	1.000	.998	.891	.727	.512	.154	.022	.001	.000	.000	.000	.000	.000	.000
8	1.000	1.000	1.000	.953	.851	.677	.274	.054	.004	.000	.000	.000	.000	.000	.000
9	1.000	1.000	1.000	.983	.929	.811	.425	.115	013	.000	.000	.000	.000	.000	.000
10	1.000	1.000	1.000	.994	.970	.902	.586	.212	.034	.002	.000	.000	.000	.000	.000
11	1.000	1.000	1.000	.998	.980	.956	.732	.345	.078	.006	.001	.000	.000	.000	.000
12	1.000	1.000	1.000	1.000	.997	.983	.846	.500	.154	.017	.003	.000	.000	.000	.000
13	1.000	1.000	1.000	1.000	.999	.994	.922	.655	.268	.044	.020	.002	.000	.000	.000
14	1.000	1.000	1.000	1.000	1.000	.998	.966	.788	.414	.098	.030	.006	.000	.000	.000
15	1.000	1.000	1.000	1.000	1.000	1.000	.987	.885	.575	.189	.071	.017	.000	.000	.000
16	1.000	1.000	1.000	1.000	1.000	1.000	.996	.946	.726	.323	.149	.047	.000	.000	.000
17	1.000	1.000	1.000	1.000	1.000	1.000	.999	.978	.846	.488	.273	.109	.002	.000	.000
18	1.000	1.000	1.000	1.000	1.000	1.000	1.000	.993	.926	.659	.439	.220	.009	.000	.000
19	1.000	1.000	1.000	1.000	1.000	1.000	1.000	.998	.971	.807	.622	.383	.033	.001	.000
20	1.000	1.000	1.000	1.000	1.000	1.000	1.000	1.000	.991	.910	.786	.579	.098	.007	.000
21	1.000	1.000	1.000	1.000	1.000	1.000	1.000	1.000	.998	.967	.904	.766	.236	.034	.000
22	1.000	1.000	1.000	1.000	1.000	1.000	1.000	1.000	1.000	.991	.968	.902	.463	.127	.002
23	1.000	1.000	1.000	1.000	1.000	1.000	1.000	1.000	1.000	.998	.993	.973	.729	.358	.026
24	1.000	1.000	1.000	1.000	1.000	1.000	1.000	1.000	1.000	1.000	.999	.996	.928	.723	.222

Source: Adapted from *Statistics for Management*, by Lincoln L. Chao. Copyright © 1980 by Wadsworth, Inc. Reprinted by permission.

APPENDIX TWO

Cumulative Poisson Probabilities $P(X \leq x)$ for $X \sim Poi(\lambda)$

		λ									
		.1	.2	.3	.4	.5	.6	.7	.8	.9	1.0
	0	.905	.819	.741	.670	.607	.549	.497	.449	.407	.368
	1	.995	.982	.963	.938	.910	.878	.844	.809	.772	.736
	2	1.000	.999	.996	.992	.986	.977	.966	.953	.937	.920
x	3		1.000	1.000	.999	.998	.997	.994	.991	.945	.981
	4				1.000	1.000	1.000	.999	.999	.989	.996
	5							1.000	1.000	.998	.999
	6									1.000	1.000

		λ										
		2.0	3.0	4.0	5.0	6.0	7.0	8.0	9.0	10.0	15.0	20.0
	0	.135	.050	.018	.007	.002	.001	.000	.000	.000	.000	.000
	1	.406	.199	.092	.040	.017	.007	.003	.001	.000	.000	.000
	2	.677	.423	.238	.125	.062	.030	.014	.006	.003	.000	.000
	3	.857	.647	.433	.265	.151	.082	.042	.021	.010	.000	.000
	4	.947	.815	.629	.440	.285	.173	.100	.055	.029	.001	.000
	5	.983	.916	.785	.616	.446	.301	.191	.116	.067	.003	.000
	6	.995	.966	.889	.762	.606	.456	.313	.207	.130	.008	.000
	7	.999	.988	.949	.867	.744	.599	.453	.324	.220	.018	.001
	8	1.000	.996	.979	.932	.847	.729	.593	.456	.333	.037	.002
	9		.999	.992	.968	.916	.830	.717	.587	.458	.070	.005
	10		1.000	.997	.986	.957	.901	.816	.706	.583	.118	.011
	11			.999	.995	.980	.947	.888	.803	.697	.185	.021
	12			1.000	.998	.991	.973	.936	.876	.792	.268	.039
	13				.999	.996	.987	.966	.926	.864	.363	.066
	14				1.000	.999	.994	.983	.959	.917	.466	.105
	15					.999	.998	.992	.978	.951	.568	.157
	16					1.000	.999	.996	.989	.973	.664	.221
	17						1.000	.998	.995	.986	.749	.297
x	18							1.000	.999	.993	.819	.381
	19								1.000	.997	.875	.470
	20									.998	.917	.559
	21									.999	.947	.644
	22									1.000	.967	.721
	23										.981	.787
	24										.989	.843
	25										.994	.888
	26										.997	.922
	27										.998	.948
	28										.999	.966
	29										1.000	.978
	30											.987
	31											.992
	32											.995
	33											.997
	34											.999
	35											.999
	36											1.000

Source: Lincoln L. Chao, *Statistics: Methods and Analysis*, (2nd ed.), New York: McGraw-Hill Book Company,

APPENDIX THREE

Cumulative Standard Normal Probabilities

$$\Phi(z) = P(Z \leq z)$$

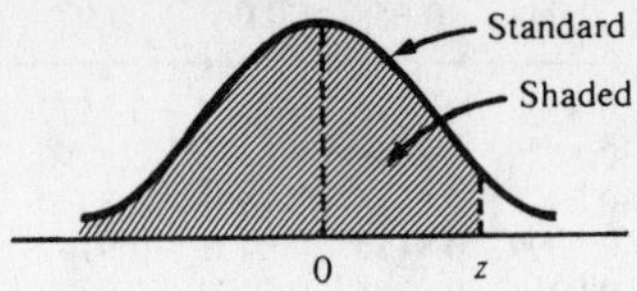

z	0.00	0.01	0.02	0.03	0.04	0.05	0.06	0.07	0.08	0.09
−3.4	0.0003	0.0003	0.0003	0.0003	0.0003	0.0003	0.0003	0.0003	0.0003	0.0002
−3.3	0.0005	0.0005	0.0005	0.0004	0.0004	0.0004	0.0004	0.0004	0.0004	0.0003
−3.2	0.0007	0.0007	0.0006	0.0006	0.0006	0.0006	0.0006	0.0005	0.0005	0.0005
−3.1	0.0010	0.0009	0.0009	0.0009	0.0008	0.0008	0.0008	0.0008	0.0007	0.0007
−3.0	0.0013	0.0013	0.0013	0.0012	0.0012	0.0011	0.0011	0.0011	0.0010	0.0010
−2.9	0.0019	0.0018	0.0017	0.0017	0.0016	0.0016	0.0015	0.0015	0.0014	0.0014
−2.8	0.0026	0.0025	0.0024	0.0023	0.0023	0.0022	0.0021	0.0021	0.0020	0.0019
−2.7	0.0035	0.0034	0.0033	0.0032	0.0031	0.0030	0.0029	0.0028	0.0027	0.0026
−2.6	0.0047	0.0045	0.0044	0.0043	0.0041	0.0040	0.0039	0.0038	0.0037	0.0036
−2.5	0.0062	0.0060	0.0059	0.0057	0.0055	0.0054	0.0052	0.0051	0.0049	0.0048
−2.4	0.0082	0.0080	0.0078	0.0075	0.0073	0.0071	0.0069	0.0068	0.0066	0.0064
−2.3	0.0107	0.0104	0.0102	0.0099	0.0096	0.0094	0.0091	0.0089	0.0087	0.0084
−2.2	0.0139	0.0136	0.0132	0.0129	0.0125	0.0122	0.0119	0.0116	0.0113	0.0110
−2.1	0.0179	0.0174	0.0170	0.0166	0.0162	0.0158	0.0154	0.0150	0.0146	0.0143
−2.0	0.0228	0.0222	0.0217	0.0212	0.0207	0.0202	0.0197	0.0192	0.0188	0.0183
−1.9	0.0287	0.0281	0.0274	0.0268	0.0262	0.0256	0.0250	0.0244	0.0239	0.0233
−1.8	0.0359	0.0352	0.0344	0.0336	0.0329	0.0322	0.0314	0.0307	0.0301	0.0294
−1.7	0.0446	0.0436	0.0427	0.0418	0.0409	0.0401	0.0392	0.0384	0.0375	0.0367
−1.6	0.0548	0.0537	0.0526	0.0516	0.0505	0.0495	0.0485	0.0475	0.0465	0.0455
−1.5	0.0668	0.0655	0.0643	0.0630	0.0618	0.0606	0.0594	0.0582	0.0571	0.0559
−1.4	0.0808	0.0793	0.0778	0.0764	0.0749	0.0735	0.0722	0.0708	0.0694	0.0681
−1.3	0.0968	0.0951	0.0934	0.0918	0.0901	0.0885	0.0869	0.0853	0.0838	0.0823
−1.2	0.1151	0.1131	0.1112	0.1093	0.1075	0.1056	0.1038	0.1020	0.1003	0.0985
−1.1	0.1357	0.1335	0.1314	0.1292	0.1271	0.1251	0.1230	0.1210	0.1190	0.1170
−1.0	0.1587	0.1562	0.1539	0.1515	0.1492	0.1469	0.1446	0.1423	0.1401	0.1379
−0.9	0.1841	0.1814	0.1788	0.1762	0.1736	0.1711	0.1685	0.1660	0.1635	0.1611
−0.8	0.2119	0.2090	0.2061	0.2033	0.2005	0.1977	0.1949	0.1922	0.1894	0.1867
−0.7	0.2420	0.2389	0.2358	0.2327	0.2296	0.2266	0.2236	0.2206	0.2177	0.2148
−0.6	0.2743	0.2709	0.2676	0.2643	0.2611	0.2578	0.2546	0.2514	0.2483	0.2451
−0.5	0.3085	0.3050	0.3015	0.2981	0.2946	0.2912	0.2877	0.2843	0.2810	0.2776
−0.4	0.3446	0.3409	0.3372	0.3336	0.3300	0.3264	0.3228	0.3192	0.3156	0.3121
−0.3	0.3821	0.3783	0.3745	0.3707	0.3669	0.3632	0.3594	0.3557	0.3520	0.3483
−0.2	0.4207	0.4168	0.4129	0.4090	0.4052	0.4013	0.3974	0.3936	0.3897	0.3859
−0.1	0.4602	0.4562	0.4522	0.4483	0.4443	0.4404	0.4364	0.4325	0.4286	0.4247
−0.0	0.5000	0.4960	0.4920	0.4880	0.4840	0.4801	0.4761	0.4721	0.4681	0.4641
0.0	0.5000	0.5040	0.5080	0.5120	0.5160	0.5199	0.5239	0.5279	0.5319	0.5359
0.1	0.5398	0.5438	0.5478	0.5517	0.5557	0.5596	0.5636	0.5675	0.5714	0.5753
0.2	0.5793	0.5832	0.5871	0.5910	0.5948	0.5987	0.6026	0.6064	0.6103	0.6141
0.3	0.6179	0.6217	0.6255	0.6293	0.6331	0.6368	0.6406	0.6443	0.6480	0.6517
0.4	0.6554	0.6591	0.6628	0.6664	0.6700	0.6736	0.6772	0.6808	0.6844	0.6879
0.5	0.6915	0.6950	0.6985	0.7019	0.7054	0.7088	0.7123	0.7157	0.7190	0.7224
0.6	0.7257	0.7291	0.7324	0.7357	0.7389	0.7422	0.7454	0.7486	0.7517	0.7549
0.7	0.7580	0.7611	0.7642	0.7673	0.7704	0.7734	0.7764	0.7794	0.7823	0.7852
0.8	0.7881	0.7910	0.7939	0.7967	0.7995	0.8023	0.8051	0.8078	0.8106	0.8133
0.9	0.8159	0.8186	0.8212	0.8238	0.8264	0.8289	0.8315	0.8340	0.8365	0.8389

(continued)

Cumulative Standard Normal Probabilities (cont.) $\Phi(z) = P(Z \leq z)$

z	0.00	0.01	0.02	0.03	0.04	0.05	0.06	0.07	0.08	0.09
1.0	0.8413	0.8438	0.8461	0.8485	0.8508	0.8531	0.8554	0.8577	0.8599	0.8621
1.1	0.8643	0.8665	0.8686	0.8708	0.8729	0.8749	0.8770	0.8790	0.8810	0.8830
1.2	0.8849	0.8869	0.8888	0.8907	0.8925	0.8944	0.8962	0.8980	0.8997	0.9015
1.3	0.9032	0.9049	0.9066	0.9082	0.9099	0.9115	0.9131	0.9147	0.9162	0.9177
1.4	0.9192	0.9207	0.9222	0.9236	0.9251	0.9265	0.9278	0.9292	0.9306	0.9319
1.5	0.9332	0.9345	0.9357	0.9370	0.9382	0.9394	0.9406	0.9418	0.9429	0.9441
1.6	0.9452	0.9463	0.9474	0.9484	0.9495	0.9505	0.9515	0.9525	0.9535	0.9545
1.7	0.9554	0.9564	0.9573	0.9582	0.9591	0.9599	0.9608	0.9616	0.9625	0.9633
1.8	0.9641	0.9649	0.9656	0.9664	0.9671	0.9678	0.9686	0.9693	0.9699	0.9706
1.9	0.9713	0.9719	0.9726	0.9732	0.9738	0.9744	0.9750	0.9756	0.9761	0.9767
2.0	0.9772	0.9778	0.9783	0.9788	0.9793	0.9798	0.9803	0.9808	0.9812	0.9817
2.1	0.9821	0.9826	0.9830	0.9834	0.9838	0.9842	0.9846	0.9850	0.9854	0.9857
2.2	0.9861	0.9864	0.9868	0.9871	0.9875	0.9878	0.9881	0.9884	0.9887	0.9890
2.3	0.9893	0.9896	0.9898	0.9901	0.9904	0.9906	0.9909	0.9911	0.9913	0.9916
2.4	0.9918	0.9920	0.9922	0.9925	0.9927	0.9929	0.9931	0.9932	0.9934	0.9936
2.5	0.9938	0.9940	0.9941	0.9943	0.9945	0.9946	0.9948	0.9949	0.9951	0.9952
2.6	0.9953	0.9955	0.9956	0.9957	0.9959	0.9960	0.9961	0.9962	0.9963	0.9964
2.7	0.9965	0.9966	0.9967	0.9968	0.9969	0.9970	0.9971	0.9972	0.9973	0.9974
2.8	0.9974	0.9975	0.9976	0.9977	0.9977	0.9978	0.9979	0.9979	0.9980	0.9981
2.9	0.9981	0.9982	0.9982	0.9983	0.9984	0.9984	0.9985	0.9985	0.9986	0.9986
3.0	0.9987	0.9987	0.9987	0.9988	0.9988	0.9989	0.9989	0.9989	0.9990	0.9990
3.1	0.9990	0.9991	0.9991	0.9991	0.9992	0.9992	0.9992	0.9992	0.9993	0.9993
3.2	0.9993	0.9993	0.9994	0.9994	0.9994	0.9994	0.9994	0.9995	0.9995	0.9995
3.3	0.9995	0.9995	0.9995	0.9996	0.9996	0.9996	0.9996	0.9996	0.9996	0.9997
3.4	0.9997	0.9997	0.9997	0.9997	0.9997	0.9997	0.9997	0.9997	0.9997	0.9998

APPENDIX FOUR

Upper Alpha Points for the t Distribution $P(T > t_{\alpha,\nu}) = \alpha$ for $T \sim t_\nu$

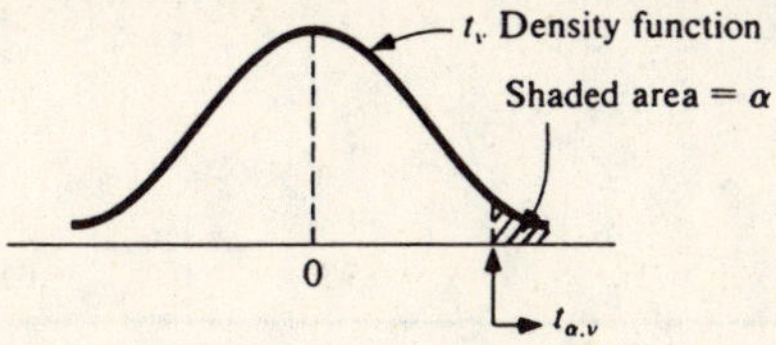

ν \ α	.10	.05	.025	.01	.005	.001	.0005
1	3.078	6.314	12.706	31.821	63.657	318.31	636.62
2	1.886	2.920	4.303	6.965	9.925	22.326	31.598
3	1.638	2.353	3.182	4.541	5.841	10.213	12.924
4	1.533	2.132	2.776	3.747	4.604	7.173	8.610
5	1.476	2.015	2.571	3.365	4.032	5.893	6.869
6	1.440	1.943	2.447	3.143	3.707	5.208	5.959
7	1.415	1.895	2.365	2.998	3.499	4.785	5.408
8	1.397	1.860	2.306	2.896	3.355	4.501	5.041
9	1.383	1.833	2.262	2.821	3.250	4.297	4.781
10	1.372	1.812	2.228	2.764	3.169	4.144	4.587
11	1.363	1.796	2.201	2.718	3.106	4.025	4.437
12	1.356	1.782	2.179	2.681	3.055	3.930	4.318
13	1.350	1.771	2.160	2.650	3.012	3.852	4.221
14	1.345	1.761	2.145	2.624	2.977	3.787	4.140
15	1.341	1.753	2.131	2.602	2.947	3.733	4.073
16	1.337	1.746	2.120	2.583	2.921	3.686	4.015
17	1.333	1.740	2.110	2.567	2.898	3.646	3.965
18	1.330	1.734	2.101	2.552	2.878	3.610	3.922
19	1.328	1.729	2.093	2.539	2.861	3.579	3.883
20	1.325	1.725	2.086	2.528	2.845	3.552	3.850
21	1.323	1.721	2.080	2.518	2.831	3.527	3.819
22	1.321	1.717	2.074	2.508	2.819	3.505	3.792
23	1.319	1.714	2.069	2.500	2.807	3.485	3.767
24	1.318	1.711	2.064	2.492	2.797	3.467	3.745
25	1.316	1.708	2.060	2.485	2.787	3.450	3.725
26	1.315	1.706	2.056	2.479	2.779	3.435	3.707
27	1.314	1.703	2.052	2.473	2.771	3.421	3.690
28	1.313	1.701	2.048	2.467	2.763	3.408	3.674
29	1.311	1.699	2.045	2.462	2.756	3.396	3.659
30	1.310	1.697	2.042	2.457	2.750	3.385	3.646
40	1.303	1.684	2.021	2.423	2.704	3.307	3.551
60	1.296	1.671	2.000	2.390	2.660	3.232	3.460
120	1.289	1.658	1.980	2.358	2.617	3.160	3.373
∞	1.282	1.645	1.960	2.326	2.576	3.090	3.291

Source: This table is reproduced with the kind permission of the Trustees of Biometrika from E. S. Pearson and H. O. Hartley (eds.), *The Biometrika Tables for Statisticians*, vol. 1, 3rd ed., *Biometrika*, 1966.

APPENDIX FIVE

Upper Alpha Points for the Chi-Squared Distribution $P(X > \chi^2_{\alpha,\nu}) = \alpha$ for χ^2_ν

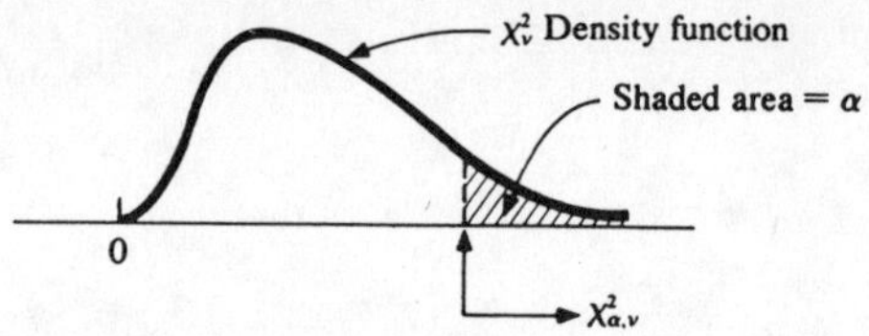

ν \ α	.995	.99	.975	.95	.90	.10	.05	.025	.01	.005
1	0.000	0.000	0.001	0.004	0.016	2.706	3.843	5.025	6.637	7.882
2	0.010	0.020	0.051	0.103	0.211	4.605	5.992	7.378	9.210	10.597
3	0.072	0.115	0.216	0.352	0.584	6.251	7.815	9.348	11.344	12.837
4	0.207	0.297	0.484	0.711	1.064	7.779	9.488	11.143	13.277	14.860
5	0.412	0.554	0.831	1.145	1.610	9.236	11.070	12.832	15.085	16.748
6	0.676	0.872	1.237	1.635	2.204	10.645	12.592	14.440	16.812	18.548
7	0.989	1.239	1.690	2.167	2.833	12.017	14.067	16.012	18.474	20.276
8	1.344	1.646	2.180	2.733	3.490	13.362	15.507	17.534	20.090	21.954
9	1.735	2.088	2.700	3.325	4.168	14.684	16.919	19.022	21.665	23.587
10	2.156	2.558	3.247	3.940	4.865	15.987	18.307	20.483	23.209	25.188
11	2.603	3.053	3.816	4.575	5.578	17.275	19.675	21.920	24.724	26.755
12	3.074	3.571	4.404	5.226	6.304	18.549	21.026	23.337	26.217	28.300
13	3.565	4.107	5.009	5.892	7.041	19.812	22.362	24.735	27.687	29.817
14	4.075	4.660	5.629	6.571	7.790	21.064	23.685	26.119	29.141	31.319
15	4.600	5.229	6.262	7.261	8.547	22.307	24.996	27.488	30.577	32.799
16	5.142	5.812	6.908	7.962	9.312	23.542	26.296	28.845	32.000	34.267
17	5.697	6.407	7.564	8.682	10.085	24.769	27.587	30.190	33.408	35.716
18	6.265	7.015	8.231	9.390	10.865	25.989	28.869	31.526	34.805	37.156
19	6.843	7.632	8.906	10.117	11.651	27.203	30.143	32.852	36.190	38.580
20	7.434	8.260	9.591	10.851	12.443	28.412	31.410	34.170	37.566	39.997
21	8.033	8.897	10.283	11.591	13.240	29.615	32.670	35.478	38.930	41.399
22	8.643	9.542	10.982	12.338	14.042	30.813	33.924	36.781	40.289	42.796
23	9.260	10.195	11.688	13.090	14.848	32.007	35.172	38.075	41.637	44.179
24	9.886	10.856	12.401	13.848	15.659	33.196	36.415	39.364	42.980	45.558
25	10.519	11.523	13.120	14.611	16.473	34.381	37.652	40.646	44.313	46.925
26	11.160	12.198	13.844	15.379	17.292	35.563	38.885	41.923	45.642	48.290
27	11.807	12.878	14.573	16.151	18.114	36.741	40.113	43.194	46.962	49.642
28	12.461	13.565	15.308	16.928	18.939	37.916	41.337	44.461	48.278	50.993
29	13.120	14.256	16.147	17.708	19.768	39.087	42.557	45.772	49.586	52.333
30	13.787	14.954	16.791	18.493	20.599	40.256	43.773	46.979	50.892	53.672
31	14.457	15.655	17.538	19.280	21.433	41.422	44.985	48.231	52.190	55.000
32	15.134	16.362	18.291	20.072	22.271	42.585	46.194	49.480	53.486	56.328
33	15.814	17.073	19.046	20.866	23.110	43.745	47.400	50.724	54.774	57.646
34	16.501	17.789	19.806	21.664	23.952	44.903	48.602	51.966	56.061	58.964
35	17.191	18.508	20.569	22.465	24.796	46.059	49.802	53.203	57.340	60.272
36	17.887	19.233	21.336	23.269	25.643	47.212	50.998	54.437	58.619	61.581
37	18.584	19.960	22.105	24.075	26.492	48.363	52.192	55.667	59.891	62.880
38	19.289	20.691	22.878	24.884	27.343	49.513	53.384	56.896	61.162	64.181
39	19.994	21.425	23.654	25.695	28.196	50.660	54.572	58.119	62.426	65.473
40	20.706	22.164	24.433	26.509	29.050	51.805	55.758	59.342	63.691	66.766

For $\nu > 40$, $\chi^2_{\alpha,\nu} \doteq \nu\left(1 - \frac{2}{9\nu} + z_\alpha \sqrt{\frac{2}{9\nu}}\right)^3$

Source: This table is reproduced with the kind permission of the Trustees of Biometrika from E. S. Pearson and H. O. Hartley (eds.), *The Biometrika Tables for Statisticians*, vol. 1, 3rd ed., *Biometrika*, 1966.

APPENDIX SIX

Upper 0.05 Points for the F Distribution $F_{0.05, \nu_1, \nu_2}$

$\alpha = .05$

ν_2 \ ν_1	1	2	3	4	5	6	7	8	9	10	12	15	20	24	30	40	60	120	∞
1	161.4	199.5	215.7	224.6	230.2	234.0	236.8	238.9	240.5	241.9	243.9	245.9	248.0	249.1	250.1	251.1	252.2	253.3	254.3
2	18.51	19.00	19.16	19.25	19.30	19.33	19.35	19.37	19.38	19.40	19.41	19.43	19.45	19.45	19.46	19.47	19.48	19.49	19.50
3	10.13	9.55	9.28	9.12	9.01	8.94	8.89	8.85	8.81	8.79	8.74	8.70	8.66	8.64	8.62	8.59	8.57	8.55	8.53
4	7.71	6.94	6.59	6.39	6.26	6.16	6.09	6.04	6.00	5.96	5.91	5.86	5.80	5.77	5.75	5.72	5.69	5.66	5.63
5	6.61	5.79	5.41	5.19	5.05	4.95	4.88	4.82	4.77	4.74	4.68	4.62	4.56	4.53	4.50	4.46	4.43	4.40	4.36
6	5.99	5.14	4.76	4.53	4.39	4.28	4.21	4.15	4.10	4.06	4.00	3.94	3.87	3.84	3.81	3.77	3.74	3.70	3.67
7	5.59	4.74	4.35	4.12	3.97	3.87	3.79	3.73	3.68	3.64	3.57	3.51	3.44	3.41	3.38	3.34	3.30	3.27	3.23
8	5.32	4.46	4.07	3.84	3.69	3.58	3.50	3.44	3.39	3.35	3.28	3.22	3.15	3.12	3.08	3.04	3.01	2.97	2.93
9	5.12	4.26	3.86	3.63	3.48	3.37	3.29	3.23	3.18	3.14	3.07	3.01	2.94	2.90	2.86	2.83	2.79	2.75	2.71
10	4.96	4.10	3.71	3.48	3.33	3.22	3.14	3.07	3.02	2.98	2.91	2.85	2.77	2.74	2.70	2.66	2.62	2.58	2.54
11	4.84	3.98	3.59	3.36	3.20	3.09	3.01	2.95	2.90	2.85	2.79	2.72	2.65	2.61	2.57	2.53	2.49	2.45	2.40
12	4.75	3.89	3.49	3.26	3.11	3.00	2.91	2.85	2.80	2.75	2.69	2.62	2.54	2.51	2.47	2.43	2.38	2.34	2.30
13	4.67	3.81	3.41	3.18	3.03	2.92	2.83	2.77	2.71	2.67	2.60	2.53	2.46	2.42	2.38	2.34	2.30	2.25	2.21
14	4.60	3.74	3.34	3.11	2.96	2.85	2.76	2.70	2.65	2.60	2.53	2.46	2.39	2.35	2.31	2.27	2.22	2.18	2.13
15	4.54	3.68	3.29	3.06	2.90	2.79	2.71	2.64	2.59	2.54	2.48	2.40	2.33	2.29	2.25	2.20	2.16	2.11	2.07
16	4.49	3.63	3.24	3.01	2.85	2.74	2.66	2.59	2.54	2.49	2.42	2.35	2.28	2.24	2.19	2.15	2.11	2.06	2.01
17	4.45	3.59	3.20	2.96	2.81	2.70	2.61	2.55	2.49	2.45	2.38	2.31	2.23	2.19	2.15	2.10	2.06	2.01	1.96
18	4.41	3.55	3.16	2.93	2.77	2.66	2.58	2.51	2.46	2.41	2.34	2.27	2.19	2.15	2.11	2.06	2.02	1.97	1.92
19	4.38	3.52	3.13	2.90	2.74	2.63	2.54	2.48	2.42	2.38	2.31	2.23	2.16	2.11	2.07	2.03	1.98	1.93	1.88
20	4.35	3.49	3.10	2.87	2.71	2.60	2.51	2.45	2.39	2.35	2.28	2.20	2.12	2.08	2.04	1.99	1.95	1.90	1.84
21	4.32	3.47	3.07	2.84	2.68	2.57	2.49	2.42	2.37	2.32	2.25	2.18	2.10	2.05	2.01	1.96	1.92	1.87	1.81
22	4.30	3.44	3.05	2.82	2.66	2.55	2.46	2.40	2.34	2.30	2.23	2.15	2.07	2.03	1.98	1.94	1.89	1.84	1.78
23	4.28	3.42	3.03	2.80	2.64	2.53	2.44	2.37	2.32	2.27	2.20	2.13	2.05	2.01	1.96	1.91	1.86	1.81	1.76
24	4.26	3.40	3.01	2.78	2.62	2.51	2.42	2.36	2.30	2.25	2.18	2.11	2.03	1.98	1.94	1.89	1.84	1.79	1.73
25	4.24	3.39	2.99	2.76	2.60	2.49	2.40	2.34	2.28	2.24	2.16	2.09	2.01	1.96	1.92	1.87	1.82	1.77	1.71
26	4.23	3.37	2.98	2.74	2.59	2.47	2.39	2.32	2.27	2.22	2.15	2.07	1.99	1.95	1.90	1.85	1.80	1.75	1.69
27	4.21	3.35	2.96	2.73	2.57	2.46	2.37	2.31	2.25	2.20	2.13	2.06	1.97	1.93	1.88	1.84	1.79	1.73	1.67
28	4.20	3.34	2.95	2.71	2.56	2.45	2.36	2.29	2.24	2.19	2.12	2.04	1.96	1.91	1.87	1.82	1.77	1.71	1.65
29	4.18	3.33	2.93	2.70	2.55	2.43	2.35	2.28	2.22	2.18	2.10	2.03	1.94	1.90	1.85	1.81	1.75	1.70	1.64
30	4.17	3.32	2.92	2.69	2.53	2.42	2.33	2.27	2.21	2.16	2.09	2.01	1.93	1.89	1.84	1.79	1.74	1.68	1.62
40	4.08	3.23	2.84	2.61	2.45	2.34	2.25	2.18	2.12	2.08	2.00	1.92	1.84	1.79	1.74	1.69	1.64	1.58	1.51
60	4.00	3.15	2.76	2.53	2.37	2.25	2.17	2.10	2.04	1.99	1.92	1.84	1.75	1.70	1.65	1.59	1.53	1.47	1.39
120	3.92	3.07	2.68	2.45	2.29	2.17	2.09	2.02	1.96	1.91	1.83	1.75	1.66	1.61	1.55	1.50	1.43	1.35	1.25
∞	3.84	3.00	2.60	2.37	2.21	2.10	2.01	1.94	1.88	1.83	1.75	1.67	1.57	1.52	1.46	1.39	1.32	1.22	1.00

(continued)

Upper 0.01 Points for the F Distribution

$F_{0.01, \nu_1, \nu_2}$

$\alpha = .01$

ν_2 \ ν_1	1	2	3	4	5	6	7	8	9	10	12	15	20	24	30	40	60	120	∞
1	4052	4999.5	5403	5625	5764	5859	5928	5981	6022	6056	6106	6157	6209	6235	6261	6287	6313	6339	6366
2	98.50	99.00	99.17	99.25	99.30	99.33	99.36	99.37	99.39	99.40	99.42	99.43	99.45	99.46	99.47	99.47	99.48	99.49	99.50
3	34.12	30.82	29.46	28.71	28.24	27.91	27.67	27.49	27.35	27.23	27.05	26.87	26.69	26.60	26.50	26.41	26.32	26.22	26.13
4	21.20	18.00	16.69	15.98	15.52	15.21	14.98	14.80	14.66	14.55	14.37	14.20	14.02	13.93	13.84	13.75	13.65	13.56	13.46
5	16.26	13.27	12.06	11.39	10.97	10.67	10.46	10.29	10.16	10.05	9.89	9.72	9.55	9.47	9.38	9.29	9.20	9.11	9.02
6	13.75	10.92	9.78	9.15	8.75	8.47	8.26	8.10	7.98	7.87	7.72	7.56	7.40	7.31	7.23	7.14	7.06	6.97	6.88
7	12.25	9.55	8.45	7.85	7.46	7.19	6.99	6.84	6.72	6.62	6.47	6.31	6.16	6.07	5.99	5.91	5.82	5.74	5.65
8	11.26	8.65	7.59	7.01	6.63	6.37	6.18	6.03	5.91	5.81	5.67	5.52	5.36	5.28	5.20	5.12	5.03	4.95	4.86
9	10.56	8.02	6.99	6.42	6.06	5.80	5.61	5.47	5.35	5.26	5.11	4.96	4.81	4.73	4.65	4.57	4.48	4.40	4.31
10	10.04	7.56	6.55	5.99	5.64	5.39	5.20	5.06	4.94	4.85	4.71	4.56	4.41	4.33	4.25	4.17	4.08	4.00	3.91
11	9.65	7.21	6.22	5.67	5.32	5.07	4.89	4.74	4.63	4.54	4.40	4.25	4.10	4.02	3.94	3.86	3.78	3.69	3.60
12	9.33	6.93	5.95	5.41	5.06	4.82	4.64	4.50	4.39	4.30	4.16	4.01	3.86	3.78	3.70	3.62	3.54	3.45	3.36
13	9.07	6.70	5.74	5.21	4.86	4.62	4.41	4.30	4.19	4.10	3.96	3.82	3.66	3.59	3.51	3.43	3.34	3.25	3
14	8.86	6.51	5.56	5.04	4.69	4.46	4.28	4.14	4.03	3.94	3.80	3.66	3.51	3.43	3.35	3.27	3.18	3.09	3
15	8.68	6.36	5.42	4.89	4.56	4.32	4.14	4.00	3.89	3.80	3.67	3.52	3.37	3.29	3.21	3.13	3.05	2.96	2
16	8.53	6.23	5.29	4.77	4.44	4.20	4.03	3.89	3.78	3.69	3.55	3.41	3.26	3.18	3.10	3.02	2.93	2.84	2
17	8.40	6.11	5.18	4.67	4.34	4.10	3.93	3.79	3.68	3.59	3.46	3.31	3.16	3.08	3.00	2.92	2.83	2.75	2
18	8.29	6.01	5.09	4.58	4.25	4.01	3.84	3.71	3.60	3.51	3.37	3.23	3.08	3.00	2.92	2.84	2.75	2.66	[illegible]
19	8.18	5.93	5.01	4.50	4.17	3.94	3.77	3.63	3.52	3.43	3.30	3.15	3.00	2.92	2.84	2.76	2.67	2.58	[illegible]
20	8.10	5.85	4.94	4.43	4.10	3.87	3.70	3.56	3.46	3.37	3.23	3.09	2.94	2.86	2.78	2.69	2.61	2.52	2.42
21	8.02	5.78	4.87	4.37	4.04	3.81	3.64	3.51	3.40	3.31	3.17	3.03	2.88	2.80	2.72	2.64	2.55	2.46	2.36
22	7.95	5.72	4.82	4.31	3.99	3.76	3.59	3.45	3.35	3.26	3.12	2.98	2.83	2.75	2.67	2.58	2.50	2.40	2.31
23	7.88	5.66	4.76	4.26	3.94	3.71	3.54	3.41	3.30	3.21	3.07	2.93	2.78	2.70	2.62	2.54	2.45	2.35	2.26
24	7.82	5.61	4.72	4.22	3.90	3.67	3.50	3.36	3.26	3.17	3.03	2.89	2.74	2.66	2.58	2.49	2.40	2.31	2.21
25	7.77	5.57	4.68	4.18	3.85	3.63	3.46	3.32	3.22	3.13	2.99	2.85	2.70	2.62	2.54	2.45	2.36	2.27	2.17
26	7.72	5.53	4.64	4.14	3.82	3.59	3.42	3.29	3.18	3.09	2.96	2.81	2.66	2.58	2.50	2.42	2.33	2.23	2.13
27	7.68	5.49	4.60	4.11	3.78	3.56	3.39	3.26	3.15	3.06	2.93	2.78	2.63	2.55	2.47	2.38	2.29	2.20	2.10
28	7.64	5.45	4.57	4.07	3.75	3.53	3.36	3.23	3.12	3.03	2.90	2.75	2.60	2.52	2.44	2.35	2.26	2.17	2.06
29	7.60	5.42	4.54	4.04	3.73	3.50	3.33	3.20	3.09	3.00	2.87	2.73	2.57	2.49	2.41	2.33	2.23	2.14	2.03
30	7.56	5.39	4.51	4.02	3.70	3.47	3.30	3.17	3.07	2.98	2.84	2.70	2.55	2.47	2.39	2.30	2.21	2.11	2.01
40	7.31	5.18	4.31	3.83	3.51	3.29	3.12	2.99	2.89	2.80	2.66	2.52	2.37	2.29	2.20	2.11	2.02	1.92	1.80
60	7.08	4.98	4.13	3.65	3.34	3.12	2.95	2.82	2.72	2.63	2.50	2.35	2.20	2.12	2.03	1.94	1.84	1.73	1.60
120	6.85	4.79	3.95	3.48	3.17	2.96	2.79	2.66	2.56	2.47	2.34	2.19	2.03	1.95	1.86	1.76	1.66	1.53	1.38
∞	6.63	4.61	3.78	3.32	3.02	2.80	2.64	2.51	2.41	2.32	2.18	2.04	1.88	1.79	1.70	1.59	1.47	1.32	1.00

Source: This table is reproduced with the kind permission of the Trustees of Biometrika from E. S. Pearson and H. O. Hartley (eds.), *The Biometrika Tables for Statisticians*, vol. 1, 3rd ed., *Biometrika*, 1966.

APPENDIX SEVEN

Upper-tail probabilities for the Wilcoxon Signed-Rank Test

$P_0(S^+ \geq c_1) = P(S^+ \geq c_1 \text{ when } H_0 \text{ is true})$

n	c_1	$P_0(S^+ \geq c_1)$	n	c_1	$P_0(S^+ \geq c_1)$
3	6	.125		78	.011
4	9	.125		79	.009
	10	.062		81	.005
5	13	.094	14	73	.108
	14	.062		74	.097
	15	.031		79	.052
6	17	.109		84	.025
	19	.047		89	.010
	20	.031		92	.005
	21	.016	15	83	.104
7	22	.109		84	.094
	24	.055		89	.053
	26	.023		90	.047
	28	.008		95	.024
8	28	.098		100	.011
	30	.055		101	.009
	32	.027		104	.005
	34	.012	16	93	.106
	35	.008		94	.096
	36	.004		100	.052
9	34	.102		106	.025
	37	.049		112	.011
	39	.027		113	.009
	42	.010		116	.005
	44	.004	17	104	.103
10	41	.097		105	.095
	44	.053		112	.049
	47	.024		118	.025
	50	.010		125	.010
	52	.005		129	.005
11	48	.103	18	116	.098
	52	.051		124	.049
	55	.027		131	.024
	59	.009		138	.010
	61	.005		143	.005
12	56	.102	19	128	.098
	60	.055		136	.052
	61	.046		137	.048
	64	.026		144	.025
	68	.010		152	.010
	71	.005		157	.005
13	64	.108	20	140	.101
	65	.095		150	.049
	69	.055		158	.024
	70	.047		167	.010
	74	.024		172	.005

Source: Adapted from W. J. Dixon and F. J. Massey, Jr., *Introduction to Statistical Analysis*, (3rd ed.), New York: McGraw-Hill Book Company, Copyright © 1969. Reprinted by permission.

APPENDIX EIGHT

Upper-tail probabilities for the Wilcoxon Rank-Sum Test

$P_0(W \geq c) = P(W \geq c \text{ when } H_0 \text{ is true})$

m	n	c	$P_0(W \geq c)$
3	3	15	.05
	4	17	.057
		18	.029
	5	20	.036
		21	.018
	6	22	.048
		23	.024
		24	.012
	7	24	.058
		26	.017
		27	.008
	8	27	.042
		28	.024
		29	.012
		30	.006
4	4	24	.057
		25	.029
		26	.014
	5	27	.056
		28	.032
		29	.016
		30	.008
	6	30	.057
		32	.019
		33	.010
		34	.005
	7	33	.055
		35	.021
		36	.012
		37	.006
	8	36	.055
		38	.024
		40	.008
		41	.004
5	5	36	.048
		37	.028
		39	.008
		40	.004
	6	40	.041
		41	.026
		43	.009
		44	.004
	7	43	.053
		45	.024
		47	.009
		48	.005
	8	47	.047
		49	.023
		51	.009
		52	.005
6	6	50	.047
		52	.021
		54	.008
		55	.004
	7	54	.051
		56	.026
		58	.011
		60	.004
	8	58	.054
		61	.021
		63	.01
		65	.004
7	7	66	.049
		68	.027
		71	.009
		72	.006
	8	71	.047
		73	.027
		76	.01
		78	.005
8	8	84	.052
		87	.025
		90	.01
		92	.005

Source: Adapted from W. J. Dixon and F. J. Massey, Jr., *Introduction to Statistical Analysis,* (3rd ed.), New York: McGraw-Hill Book Company, Copyright © 1969. Reprinted by permission.

REFERENCES

1. Abernethy, M. (1991). Imagery ratings for 230 nouns and adjectives. *The Australian Journal of Psychology,* 43, 155-57.

2. Adams, G. and Watson, R. (1989). The analysis of failure time data. *The Australian Journal of Statistics,* 31, 3.

3. Arnold, S. F. (1990). *Mathematical Statistics.* Prentice Hall: New Jersey.

4. Barr, H. (1986). Uncertainty and other impediments to geothermal power. *Pacific Statistical Congress, Proceedings of the Pacific Statistical Congress 1985.* I.S. Francis, B.F.J. Manly and F.C. Lam (eds). North-Holland: Amsterdam.

5. Barrett, P.J. (1984). Texture, mineralogy and composition of four sediment cores from Granite Harbour and New Harbour, southern Victoria Land, Antarctica. *New Zealand Journal of Geology and Geophysics,* 27, 477-85.

6. Berry, D.A. and Lindgren, B.W. (1990). *Statistics—Theory and Methods.* Brooks/Cole: California.

7. Box, G.E.P. and Cox, D.R. (1964). An analysis of transformations (with discussion). *Journal of the Royal Statistical Society (B),* 26, 211-52.

8. Brick, M.J., Michael, J.R. and Morganstein, D. (1989). Using Statistical Thinking to Solve Maintenance Problems. *Quality Progress*, May.

9. Broad, B.J. (1986). Shaft plumbing techniques. *The Australian Surveyor*, 33, 4, 292-304.

10. Brookes, I.G., Betteley, I.G. and Loxston, S.M. (1966). *Mathematics and Statistics for Chemists.* Wiley: London.

11. Brown, B.M. and Maritz, J.S. (1982). Distribution-free methods in regression, *Australian Journal of Statistics*, 24, 3, 318-31.

12. Brownlee, K.A. (1960). *Statistical Theory and Methodology in Science and Engineering.* Wiley: New York.

13. Cameron, R.J. (1985). *Year Book Australia 1985.* Australian Bureau of Statistics: Canberra.

14. Castles, I. (1990). *Year Book Australia 1990.* Australian Bureau of Statistics: Canberra.

15. Chatfield, C. (1983). *Statistics for Technology—A Course in Applied Statistics*, 3rd edn. Chapman and Hall: London.

16. Chelva, E., Collins, D.W.K., Levy, R.L., and McLaren, T.L. (1986). Effects of exposure to coloured lenses. *Proceedings of Seminar on Advances in Learning Disabilities, May.* Dyslexia Research Foundation of Australia Inc: Perth.

17. Cheung, B. and Elford, M.T. (1990). The drift velocity of electrons in water vapour at low values of E/N. *Australian Journal of Physics*, 43, 755-63.

18. Crowley, J. and Hu, M. (1977). The covariance analysis of heart transplant data. *J. Am. Statist. Assoc.* 72, 27-36.

19. Devore, J.L. (1991). *Probability and Statistics for Engineering and the Sciences,* 3rd edn. Brooks/Cole: California.

20. Dixon, W.J. and Kronmal, R.A. (1965). The choice of origin and scale for graphs. *Journal of the Association of Computing Machinery*, 12, 259-61.

21. Everitt, B.S. (1977). *The Analysis of Contingency Tables.* Chapman and Hall: London.

22. Feller, W. (1968). *An Introduction to Probability Theory and Its Applications, Volume 1*, 3rd edn. Wiley: New York.

23. Finley, V.E. (1985). 'Hearing Handicap and Social Functioning in the Aged Nursing Home Population,' M. Appl. Sci. Thesis, Curtin University of Technology.

24. Freund, J.E. (1971). *Mathematical Statistics*, 2nd edn. Prentice Hall: New Jersey.

25. Guttman, I. and Wilks, S.S. (1965). *Introductory Engineering Statistics.* Wiley: New York

26. Hamilton, L.C. (1990). *Modern Data Analysis: A First Course in Applied Statistics.* Brookes-Cole: California.

27. Hoaglin, D.C., Mosteller, F. and Tukey, J. W. (1983). *Understanding Robust and Exploratory Data Analysis.* Wiley: New York.

28. Hogg, R. V. and Craig, A. T. (1978). *Introduction to Mathematical Statistics*, 4th edn. Macmillam: New York.

29. Hogg, R. V. and Ledolter, J. (1987). *Engineering Statistics.* Macmillan: New York.

30. Kendall, M.G. and Stuart, A. (1967). *The Advanced Theory of Statistics, Vol. 2, Inference and Relationship.* Griffin: London.

31. Khazanie, R. (1979). *Elementary Statistics in a World of Applications.* Goodyear: Santa Monica.

32. Lapin, L.L. (1990). *Probability and Statistics for Modern Engineering,* 2nd edn. PSW-Kent: Boston.

33. Larsen, R.J. and Marx, M.L. (1981). *An Introduction to Mathematical Statistics and its Applications.* Prentice Hall: New Jersey.

34. Leitch, E.C. and Feenan, J.P. (1989). Quartz-kaolinite rocks from northeastern New South Wales: Distal products of a pluton-driven hydrothermal system. *Australian Journal of Earth Sciences,* 36, 275-81.

35. McConnell, J. (1986). *The Seven Tools of TQC.* Delaware: New South Wales.

36. Massod, M.F. (1977). Nonparametric percentile estimate of clinical normal ranges. *Am. J. Medical Technology,* 43, 243-52.

37. Mauri, P. (1989). A computer system for the analysis of integrated circuit reliability. *Quality and Reliability Engineering International,* 5, 273-82.

38. Mood, A.M., Graybill, F.A. and Boes, D.C. (1974). *Introduction to the Theory of Statistics*, 3rd edn. McGraw-Hill: New York.

39. Morgan, P.A. (1988). *Air Quality in Victoria.* EPA Scientific Services.

40. Morgan, P.J. (1973). A photogrammetric survey of Hoseason Glacier, Kemp Coast, Antarctica. *Journal of Glaciology*, 12, 113-20.

41. Moupfouma, F., Martin, L., Spanjaard, N. and Hughes, K. (1990). Rainfall rate characteristics for microwave systems in tropical and equitorial areas. *International Journal of Satellite Communications*, 8, 151-61.

42. Myers, R.H. (1990). *Classical and Modern Regression with Applications*, 2nd edn. PWS-Kent: Boston.

43. Paton, D.C. and Collins, B.G. (1989). Bills and tongues of nectar-feeding birds: A review of morphology, function and performance, with intercontinental comparisons. *Australian Journal of Ecology*, 14, 473-506.

44. Pearson, K. (ed.) (1922). *Tables of the Incomplete* Γ*-function.* Cambridge University Press: London.

45. Pearton, S.J., Ren, F., Katz, A., Fullowan, T.R., Abernathy, C.R., Hobson, W.S. and Kopf, R.F. (1992). Rapid Isothermal Processing for Fabrication of GaAs-Based Electronic Devices. *IEEE Transactions On Electron Devices*, 39, 154-58.

46. Poore, M. (1968). Studies in a Malaysian Rain Forest. *Journal of Ecology*, 56, 143-96.

47. Raper, R.L. and Kumar, S.(1981). Car pooling—a feasibility study in an Australian environment. In *Scientific Management of Transport Systems.* N.K. Jaiswal (ed). North-Holland: Amsterdam.

48. Roberts, M.G., Lawson, J.R. and Gemmell, M.A. (1986). Epidemiology and control of hydatidosis and cysticercosis. *Pacific Statistical Congress, Proceedings of the Pacific Statistical Congress,1985.* I.S. Francis, B.F.J. Manly and F.C. Lam (eds). North-Holland: Amsterdam.

49. Ryall, G.R. and Sandor, J. (1989). Statistical pattern matching. In *Pattern Recognition Letters*, 9, 163-68. North Holland: Amsterdam.

50. Ryan, B.F., Joiner, B.L. and Ryan, T.A. (1985). *Minitab Handbook*, 2nd edn. PWS-Kent: Boston.

51. Salvia, A.A. (1990). *Introduction to Statistics.* Saunders College Publishing: Philadelphia.

52. Sare, I.R. and Constantine, A.G. (1991). Design and Analysis of Jaw Crusher Gouging Abrasion Tests. *Journal of Testing and Evaluation*, 19, 115-22.

53. Saulwick, I. (1991). *The Saulwick Poll Reprints, 1989.* David Syme and Co. Ltd: Melbourne.

54. Sievert, M.C. (1990). Ten years of the literature of Online searching: An analysis of Online and Online Review. *Journal of the American Society for Information Science*, 41, 560–68.

55. Snedecor, G.W. and Cochran, W. G. (1967). *Statistical Methods,* 6th edn. The Iowa State University Press: Iowa.

56. Staudte, R.G. (1990). *Seeing, Through Statistics.* Prentice Hall: Sydney.

57. Thomas, G.B. and Finney, R.L. (1984). *Calculus and Analytic Geometry,* 6th edn. Addison-Wesley: Massachusetts.

58. Trivedi, K.S. (1982). *Probability and Statistics with Reliability, Queuing, and Computer Science Applications.* Prentice Hall: New Jersey.

59. Tukey, J.W. (1977). *Exploratory Data Analysis.* Addison-Wesley: Massachusetts.

60. Velleman, P.F. and Hoaglin, D.C. (1981). *Applications, Basics, and Computing of Exploratory Data Analysis.* Duxbury: Massachusetts.

61. Verzijlbergen, J.F., Vermeersch, P., Laarman, G. and Ascoop, C. (1991). Myocardial perfusion imaging with dipyridamole and exercise Thallium-201. *Journal of Nuclear Medicine*, 32, 2071-8.

62. Watson, R., Holton, D. and Dear, R. (1990). *Reasoning and Data.* Pilot Edition, Mathematics Today: Senior Series. McGraw-Hill: Sydney.

63. Weibull, W. (1951). A Statistical Distribution of Wide Applicability. *Journal of Applied Mechanics*, 18, 293-97.

64. Wilks, J. (1991). Diving dropouts: the Australian experience. *The Australian Journal of Science and Medicine in Sport*, 23, 17-20.

65. Woodroofe, M. (1975). *Probability With Applications.* McGraw-Hill: New York.

66. Worsley, K.J. (1986). Confidence regions and tests for a change-point in a sequence of exponential random variables. *Pacific Statistical Congress, Proceedings of the Pacific Statistical Congress, 1985,* I.S. Francis, B.F.J. Manly and F.C. Lam (eds). North-Holland: Amsterdam.

ANSWERS

Chapter 1

1.1 Year of birth—interval; Blood type—nominal; Religion—nominal; Blood pressure—ordinal.

1.3 Discrete. $\frac{D}{ND}$ may assume only the values 9, 4, 2.5, 2.333, 1, 0.667, 0.429, 0.25, 0.111.

1.5 3.5, 1.2, 1.0.

1.7 22 000 in each case.

1.9 (a)i. 365.2422; ii. 365.2 (b) Total number of days in 1991-94 is (3)(365) + 366 = 1461. Four solar years comprise (4)(365.242199) = 1460.968796. Absolute difference is 3.1204×10^{-2} days.

1.11 (a) 99 500 (b) 99 250 (c) 99 240 (d) 99 237.

1.13 (a) $-0.054 - (-0.144) = 0.09$
(b)

Failure analysis profile
Scale: stems—10^{-2}mm; leaves—10^{-3}mm
−14 | 4 represents −0.144 units

−14	4
−13	1 0 0
−12	8 5 5 3 1
−11	9 9 8 7 6 5 1 1 0
−10	9 9 8 7 7 6 6 5 5 5 5 5 3 2 0
−9	9 9 6 1 1 0 0
−8	9 9 8 7 6 5 3 3 2 1 1 1
−7	9 9 9 8 6 5 5 5 4 3 2 2 1
−6	7 6 6 5 3 1 1 1
−5	9 5 4

(c) yes.

1.15 The S/L plot is as follows:

\$NZ100 in terms of \$A
Scale: stems—10^{-2} \$A; leaves—10^{-3} \$A
74 | 5 represents 0.745\$A

74.	7
75*	0444
75.	55666888
76*	12334
76.	558
77*	0
77.	9
78*	01

The skewness is reversed.

1.17 Data:

2.0	2.1	2.2	2.2	2.3
2.3	2.5	2.5	2.5	2.6
2.7	2.7	2.8	2.9	3.1
3.2	3.3	3.5		

Same plot.

1.19 The S/L plot is as follows:

Days onset vaginal cancer rats
Scale: stems—10^2days; leaves—10 days
1*f* | 4 represents 140–149 days

1f	4 5
s	6
.	9
2*	0 0†
t	3 3 3 3 3 3
f	4
s	6
.	8 9 9
3*	
t	2
f	4†

† *Censored*

1.21 The S/L plot is as follows:

Rainfall acidity (pH)
Scale: stems—units; leaves—10^{-1} units
4∗ | 0 represents 4.00–4.09

3.	9
4*	0 1 1 1
t	2 2 2 2 3 3 3 3
f	4 4 4 4 4 5 5 5 5
s	6 6 6 6 7
.	8
5*	0
t	3
f	5
s	6 6 7
.	9

No. The distribution appears to be centred in two locations rather than one; however, most results clustered about the acidic pH 4.5.

1.23 (a) Source 1: 150 days; Source 2: 20 days. (b) Yes. Key indicates range values for 3 ∗ |2. (c) Source 1. Most long-lasting. (d) 40 days.

1.25 (a) $C = \frac{5}{9}(F - 32) = -\frac{160}{9} + \frac{5}{9}F$ (b) 37.8.

1.27 $Y = 1.01X - (0.01)(1.01X) = 0.9999X$. Cheaper to use the prompt repayment credit card facility.

1.29 $Y = (45)(X) + (36)(97 - X) = 3492 + 9X$.

$P = Y - (97)(20) = Y - 1940$. Therefore, $P = 1552 + 9X$.

1.31 (b) $y = 1 - x^{-1}$ (d) $y = x^2$.

1.33 $y = (const)_1 + (const)_2 \log_e x$ for $(const)_1 = 0$ and

$(const)_2 = \frac{1}{\log_e a}$.

1.35 The exponential series is $e^r = 1 + r + \frac{r^2}{2} + \dots$ and $x^p = e^r$ when $r = p \log_e x$. Therefore, for $r = p \log_e x$,

$$\frac{e^r - 1}{p} = \frac{1 + r + \frac{r^2}{2} + \dots - 1}{p}$$

$= \log_e x +$ *small terms in powers of* p. In the limit, as $p \to 0$, only $\log_e x$ remains on the right-side.

1.37 median 5; lower quartile 3; upper quartile 10.

1.39 S/L plot shows data in two separate groups. Sample median falls between the groups, unrepresentative of either.

1.41 $x_{(5)} = 2494$; rank median = 3.5.

1.43 8.3, 25.0, 41.7, 58.3, 75.0, 91.7.

1.45 (a)

Yield strength (MPa)
Scale: stems—10 units; leaves—units
40 | 6 represents 406 MPa

40	6
41	0 1 3 7 9
42	0 5 6 7 8
43	1 2 2 3 4 5 5 9
44	0 1 2 6
45	0

(b) 2 (c) 95.8%.

1.47 Median = 90×10^{-3}

Contact resistance (ohms.mm $\times 10^{-3}$)
Scale: stems—100 units; leaves—10 units
1 ∗ |0 represents 100 – 109($\times 10^{-3}$) ohms.mm

0f	5
s	77
.	8888888899999999
1*	0001
t	3
f	5
s	
.	
2*	
t	
f	
s	
.	8

1.49 $np + \frac{1}{2} = 24.5$. Therefore $x_{0.8} = \frac{x_{(24)}+x_{(25)}}{2} = \frac{129+131}{2} = 130$.

1.51

Capacity	Up rank	Down rank	Depth
30 000	4	6	4
27 000	3	7	3
96 000	6	4	4
22 000	2	8	2
3 000	1	9	1
200 000	7	3	3
40 000	5	5	5
287 000	8	2	2
1 068 000	9	1	1

1.53 Yes. Add 11th data point which is identical to the median of the original 10 data points. Although rankings change, the numerical values of the median and quartiles do not.

1.55 (a) Depths provide ranks from each end of the data set. $x_{0.25} = x_{(6)} = 10.25$, so that $x_{0.75}$ is the 6th data point from the top, namely, $x_{0.75} = x_{(16)} = 10.33$. $x_{0.5} = 10.29.q_s = 0.08$; lower fence $= 10.25 - (1.5)(0.08) = 10.13$; upper fence $= 10.33 + (1.5)(0.08) = 10.45$. (b) No.

1.57 $x_{0.25} = 10.25; x_{0.5} = 10.29; x_{0.75} = 10.33$. Lower whisker extends down to 10.22 (the smallest data point inside the lower inner fence). Upper whisker extends up to 10.38 (the largest data point inside the upper inner fence).

1.59 $x_{0.25} = 0.9840; x_{0.5} = 1.0005; x_{0.75} = 1.0180$. Lower whisker extends down to the smallest data point. Upper whisker extends up to the largest data point.

1.61 **Ozone** $x_{0.25} = 0.2; x_{0.5} = 1.0; x_{0.75} = 1.7$. Lower whisker extends down to -2.0. Upper whisker extends up to 3.9. Outliers: 4.0, 4.4, 5.4.

Control $x_{0.25} = 1.9; x_{0.5} = 2.2; x_{0.75} = 2.6$. Lower whisker extends down to 1.3. Upper whisker extends up to 2.9. Outliers: -0.4, 3.8, 4.1.

1.63 Interquartile range; range.

1.65

$n = 24$	Data label	
	31.5	
6.5	19.5	37
1	6	50

Second data set (y values) obtained from first (x values) by the transformation $y = -0.01x - 14$. All points in the 5-point number summary are transformed in this way.

Chapter 2

2.1 (a) Time has been discretised into years. (b) 5 January 1976. (c)

Year	Relative frequency
1979	0.152
1980	0.158
1981	0.166
1982	0.177
1983	0.175
1984	0.173

2.3 (a) State/territory where traffic death recorded; levels—states/territories. (b) Road deaths have decreased between 1978 and 1983. (c) 2755. (d) The proportion of deaths in each state changes by 1983.

2.5 (a)

	Relative frequency
Speeding	0.62
Not speeding	0.38

(b) Less speeding in heavy traffic.

2.7 (a) Frequencies in order listed from top; A: 3, 10, 12, 16, 5, 4, 4, 6; B: 5, 20, 18, 7, 9, 1. (b)

A: Highest 6–under 8. B: Highest 3–under 6.

A: Lowest 0–under 2. B: Lowest 15–under 18.

2.9 (a) Data may be uniquely assigned to intervals. (b) Frequency 40.

Concentration	Frequency
0.05–2.05	10
2.05–4.05	20
4.05–6.05	40
6.05–8.05	20
8.05–12.05	10

(c) 10%.

2.11 Evidence of low frequencies between 45 and 50 disappears.

Interval	Relative frequency
25–35	0.14
35–45	0.28
45–55	0.22
55–65	0.28
65–75	0.08

2.13 (b) Yes. One possible answer: Histogram for first offenders is right-skewed, whereas repeat offenders have more left-skewed histogram which is centred at higher levels.

2.15 (a) $Y = X^{0.5}$. (b) $Y = X^{0.5}$.

2.17 13.

2.19 9100.

2.21 (a) 30.08 gives an approximate interquartile range of 30.08×10^4 cm s^{-1}. (b) Approximation only.

2.23 (c) 0.88.

2.25 0.

2.27 (a) $\overline{x} = 0.033$; $s_x^2 = 1.780$. (b) Sample mean very close to 0. (c) 2 of 30 lie outside interval. This is 6.7%, much less than the maximum 25% (for any distribution) allowed by Chebychev's Inequality.

2.29 \$19.14; 162.

2.31 $\overline{x} = 4$; $s_x = 3$.

2.33 (a) 10 ones. (b) 5 zeros; 5 ones. (c) 1 zero; 9 ones (must have more ones than zeros). (d) 5 zeros; 5 ones.

2.35 $(x_i - \overline{x})^2 = (\frac{x_1-x_2}{2})^2$, $i = 1, 2$. These two quantities add to give the stated result.

2.37 (a) $\overline{x^2}$ is the average of squares. $\overline{x}^2$ is the square of an average. (c) Data points identical.

2.39 Cross-product terms cancel: $\sum_{i=1}^n \sum_{j=1}^n (x_i - x_j)^2 = \sum_{i=1}^n \sum_{j=1}^n (x_i - \overline{x})^2 + \sum_{i=1}^n \sum_{j=1}^n (x_j - \overline{x})^2 = n\sum_{i=1}^n (x_i - \overline{x})^2 + n\sum_{j=1}^n (x_j - \overline{x})^2 = 2n(n-1)s_x^2$.

2.41 (a) Each template has area $\frac{\sqrt{3}}{4}x_i^2$. These average to an amount $\overline{A}$ as given. (b) $\overline{A} = \frac{\sqrt{3}}{4}(\overline{x}^2 + \frac{n-1}{n}s_x^2)$.

2.43 Square each side of the inequality $|x_i - \overline{x}| \geq r s_x$ and add across all ℓ terms. Divide each side of the resulting inequality (as stated in the text) by $nr^2 s_x^2$.

2.45 (a) April. (b) June. (c) April. (d) June. (e) June. (f) June. (g) April.

2.47 (a) The cubes of the data have opposite signs and so cancel. (b) $m_3 = 0$ as in (a) since the addition of the constant 0.1 to all the data points does not alter the skewness. (c) 28.

2.49 (a) B/W plot slightly right skewed, so $\overline{x}$ slightly larger than $\widetilde{x} = 6.5$. (b) $\frac{2}{1.35} = 1.48$.

2.51 2×10^{-4}.

2.53 1.18.

2.55 (a)

25 300 000
25 400 000
25 300 000
25 900 000
25 300 000
25 000 000

(b) y data 3, 4, 3, 9, 3, 0 has $\overline{y} = 3.67$ and $s_y^2 = 8.67$. Then $\overline{x} = 25\,000\,000 + 100\,000\overline{y}$ and $s_x^2 = 10^{10} s_y^2$.

2.57 75% by Chebychev's Inequality.

2.59 For an x km circuit, $v_i = \frac{x}{t_i}$, $i = 1, 2, \ldots, n$. Total time $= \sum_{i=1}^{n} \frac{x}{v_i}$. Total distance $= nx$. Average speed from total distance by total time.

2.61 Geometric mean is $10^{5.5}$. Harmonic mean is 90.

Chapter 3

3.1 Experiment: sample 4 cables from lot. Each either breaks B or does not break N. There are $2^4 = 16$ points in the sample space. They are sequences such as $BBNB\ldots$. Only the sequence $NNNN$ leads to acceptance.

3.3 (a) i.

$$\begin{aligned} A &= \{bbg, bgg, gbg, ggg\} \\ B &= \{bgg, gbg, ggb\} \\ C &= \{bbb, bbg, bgg\} \\ D &= \{bbb, bbg, bgb, bgg, gbb, gbg, ggb\} \end{aligned}$$

ii. $P(A) = 0.5$; $P(C \cap D) = 0.375$; $P(A \cup B) = 0.625$.

3.5 Death rates in each age category are smaller for women than for men. Therefore, regardless of the relative mix of these age categories in the general population, women have higher survival chances at each age.

3.7 To use the classical definition, assume that the lift is equally likely to be on any of the 12 floors (excluding the 8 th). $P(\textit{Going up}) = \frac{7}{12}$; $P(\textit{Going down}) = \frac{5}{12}$.

3.9 (a) $(A_1 \cap \overline{A_2} \cap \overline{A_3}) \cup (\overline{A_1} \cap A_2 \cap \overline{A_3}) \cup (\overline{A_1} \cap \overline{A_2} \cap A_3)$. (b) No. The A_i may overlap and not cover $\mathcal{S}$.

3.11 (b) $P(A) = 0.75$; $P(B) = 0.35$. (c) $A \cup B = \mathcal{S}$ which has probability 1. (d) $P(\overline{A}) = 0.25$. (e) Yes; $P(\overline{A} \cup \overline{B}) = 0.9$. (f) i. 0.65; ii. 0.25; iii. 0.1.

3.13 $\frac{9}{13}$.

3.15 (a) 0.2. (b) 0.8. (c) 0.1. (d) 0.7.

3.17 True; true; true; true.

3.19 (a) Points in the union of A and B which do not lie in the intersection of A and B. (e) $\frac{5}{26}$.

3.21 (a) $a+b-c$. (b) $0 \le a+b-c \le 1$. Further, c must be positive and smaller than each of a and b.

3.23 120; 20.

3.25 24.

3.27 3780.

3.29 $P(Seated\ together) = \frac{(2)(15)13!}{15!} = \frac{1}{7}$.

3.31 (a) Provided that the remaining cards in the hand are not a 10 (four of a kind) or a pair (full house); 0.021. (b) Applying classical law, $P(Quickdraw\ wins) = \frac{1}{3}$.

3.33 0.0014.

3.35 $\binom{n}{n-r} = \frac{n!}{[n-(n-r)]!(n-r)!}$. The triangle is symmetric.

3.37 (c) $\binom{n}{r}$. (d) Sum over r all events with r outcomes. (e) Put $x = 1$.

3.39 (a) 0.8. (b) 0.878. (c) 0.531.

3.41 0.0018.

3.43 0.65.

3.45 (a) In the order listed: $P(T|C) = \frac{1}{2}$; $P(T|\overline{C}) = \frac{1}{2}$; $P(\overline{T}|C) = \frac{1}{2}$; $P(\overline{T}|\overline{C}) = \frac{1}{2}$; $P(C|T) = \frac{1}{3}$; $P(C|\overline{T}) = \frac{1}{3}$; $P(\overline{C}|T) = \frac{2}{3}$; $P(\overline{C}|\overline{T}) = \frac{2}{3}$.

3.47 (b)i. $P(Live) = \frac{5}{6}$. ii. $P(Box\ 1|Live) = \frac{1}{50}$.

3.49 (a) 0.3; 0.1; 1; 0.27. (b) i. $P(W|B) = 0.5$.

3.51 (a) 0.25. (b) 0.25.

3.53 $n \ge 14$.

3.55 Only if either one, or both, of the events are empty.

3.57 0.8829; 0.8968.

3.59 (b) 0.936.

3.61 $P(A) = P(A \cap B) + P(A \cap \overline{B})$ and $P(A \cap B) = P(A)P(B)$. Therefore, $P(A \cap \overline{B}) = P(A) - P(A)P(B)$, which factorises on the right-side.

Chapter 4

4.1 (a) 0.1. (b) 0.2. (c) 0.7. (d) 3. (e) 0; 0.7; 0.7; 0.9; 1.

4.3 (a) 0.07 (b) 0.15. (c) 0.3. (d) $(0.7)^5$ (e)

y	100	99	98	97	96	≤ 95
$p_Y(y)$	0.7	0.1	0.05	0.05	0.03	0.07

4.5 $P(X = 0) = 0.4$; $P(X = 1) = 0.5$; $P(X = 2) = 0.1$. $P(X \ge 1) = 0.6$.

4.7 (a) $F_Y(y) = 1 - \frac{1}{y^2}$, $y > 1$ and $F_Y(y) = 0$ for $y \leq 1$. (b) 0.0486. (c) 0.4444. The probability that the train is more than 3 minutes late given that it is more than 2 minutes late.

4.9 $p_X(x) = pq^{x-1}$, for $x = 1, 2, 3, \ldots$, $q = 1 - p$. By summing a geometric series, $P(X > k) = q^k$. Then, $P(X > a + b|X > a) = \frac{q^{a+b}}{q^a} = q^b$.

4.11 (a) 0.2387.(b) $f_T(t) = \frac{1}{10}e^{-\frac{t}{10}}, t > 0$.

4.13 (a) 8.9×10^7 (2 sf) (b) 0.417. (c) 4007 hours.

4.15 (a) 0.25. (b) 0.32.

4.17 (a) Change variable in integrand. (b) $I = \int_{-\infty}^{c} f_X(x)dx$. Then $2I = 1$, so that $I = \frac{1}{2}$. Hence $c = \xi_{0.5}$.

4.19 (a) $c = 0.1$. (c) 1.2.

4.21 $c = 1$ and $\mu_X = \frac{4}{3}$.

4.23 (a) $P(H) = 0.25$; $P(T) = 0.25$; $P(O) = 0.5$. In the given sequences, 5 odds have not occurred in a row. (b) $P(Win) = P(H) + P(OH) + \ldots + P(OOOOH) = \frac{31}{64}$. (c)

x	5	−5
$p_X(x)$	$\frac{31}{64}$	$\frac{33}{64}$

(d) $E(X) = -0.156$. A long-run loss of 16 cents per game for the player. A long-run gain of 16 cents per game for the casino.

4.25 0.25.

4.27 $b = \xi_{0.5}$.

4.29 $b = E(X)$.

4.31 (a) $E(X) = 12$; $Var(X) = 48$. (b) 0.577. (c) 0.6928.

4.33 $m_T(t) = (1 - t)^{-1}$. Coefficient of t^r in this expansion is 1. Coefficient of t^r in the expansion of $E(e^{tX})$ is $\frac{E(X^r)}{r!}$. Result follows by equating coefficients.

4.35 (a) $\frac{n+1}{2}$. (b) $\frac{n^2-1}{12}$. (c) $\frac{7}{2}$; $\frac{35}{12}$. (d) 0. (e) i. 10^6. ii. 500 000.5; No. iii. 10^6.iv. $\frac{10^{12}-1}{12}$.

4.37 $f_Z(z)$ has mean 0, variance 1, skewness 0, and kurtosis 3. $f_X(x)$ has mean 0, variance 0.05, skewness 0, and kurtosis 2.144. (The parabolic density is flatter in the centre than the normal curve.)

4.39 $E(C) = 550$. $Var(C) = 26100$. $E(C^2) = 328\,600$.

4.41 $f_Y(y) = \frac{1}{\sqrt{2\pi y}}e^{-\frac{y}{2}}$, for $y \geq 0$.

4.43 (a) $U[0, 1]$. (b) Yes. (c) $[0.5, 1]$. (d) $\frac{2}{3}$. (e) No. (f) $F_Y(y) = \frac{y}{1+y}$, for $y > 0$. (g) $f_Y(y) = \frac{1}{(1+y)^2}$, for $y > 0$. (h) $f_L(l) = \frac{e^l}{(1+e^l)^2}$, for $-\infty < l < \infty$.

4.45 (a) 0.578. (b) \$37 798.

Chapter 5

5.1 (a) 0.15. (b) 0.6. (c) 0.25. (d) $1 - p_{X,Y}(0, 0) = 1 - 0.1 = 0.9$.

5.3 (a)

$p_{X,Y}(x,y)$	y	-1	0	1	2
x					
-2		0	0.25	0	0
-1		0	0	0.25	0
0		0	0	0	0.25
1		0.25	0	0	0

(b) Marginal distribution of Y:

y	-1	0	1	2
$p_Y(y)$	0.25	0.25	0.25	0.25

Marginal distribution of X:

x	-2	-1	0	1
$p_X(x)$	0.25	0.25	0.25	0.25

(c) No. X and Y assume different values.

5.5 No. The tabled probabilities do not add to 1.

5.7 $0.4179 + 0.0815 + * + 0 + 0 = 0.4994$.

5.9 (a) $\frac{2}{36}$ below diagonal; $\frac{1}{36}$ on diagonal; 0 above diagonal. (b) $\frac{5}{6}$ (c)

x	1	2	3	4	5	6
$p_X(x)$	$\frac{1}{36}$	$\frac{3}{36}$	$\frac{5}{36}$	$\frac{7}{36}$	$\frac{9}{36}$	$\frac{11}{36}$

(d)4.47. (e) 2.53. (f) Yes.

5.11 $f_R(r) = 2r$, for $0 < r < 1$; $E(R) = \frac{2}{3}$.

5.13 Exponential densities with $\beta = 1$.

5.15 0.5.

5.17 No. $p_{X,Y}(0,1) \neq p_X(0)p_Y(1)$.

5.19 No. pmf does not factorise.

5.21 (a)$f_{X,Y}(x,y) = 1$ for $0 \leq x \leq 1$, and $0 \leq y \leq 1$. (b) $\frac{7}{16}$. (c) Meet if absolute difference between their arrival times X and Y is less than $\frac{1}{4}$ with probability as in (b).

5.23 Product of densities for X_1 and X_2.

5.25 (a), (b) Exponential distribution with mean $\frac{2000}{3}$. (c) $e^{-\frac{3}{2}}$.

5.27 Y denotes lifetime. (a) $F_Y(y) = (1 - e^{-\frac{y}{\beta}})^2$.

(b) $F_Y(y) = (1 - e^{-\frac{y}{\beta_1}})(1 - e^{-\frac{y}{\beta_2}})$.

(c) $F_Y(y) = 1 - [1 - (1 - e^{-\frac{y}{\beta}})^2]^2$.

5.29 $\frac{4}{5}$.

5.31 $(1 - e^{-2})^{20} = 0.055$.

5.33 (a) Values t and $p_T(t)$ are given by

4	5	6	7	8	9	10
0.08	0.18	0.29	0.19	0.08	0.13	0.05

(b) $E(X) = 2.3; E(Y) = 4.3; E(T) = 6.6$.

5.35 $P(Y > X) = 0.8$.

5.37 (a) mean = 7; variance = $\frac{35}{6}$. (b) mean = 0; variance = $\frac{35}{6}$. (c) mean = 3.5; variance = $\frac{35}{24}$.

5.39 (a) sum of units times cost per unit (plus fixed cost). (b) $E(C) = 1950; Var(C) = 3756.25$.

5.41 $Var(C) = \frac{9}{4}[p^2 + (1-p)^2]$ which is minimised when $p = 0.5$.

5.43 (a) Time taken = $T^* = \frac{X_1}{v}$. $E(T^*) = \frac{5}{2v}; Var(T^*) = \frac{25}{12v^2}$. (b) In hours, $T_{total} = 2T^* + T_{attend}$, so that $E(T_{total}) = \frac{1}{3}$ and $Var(T_{total}) = \frac{1}{27}$.

5.45 X_i is weight on Balance i: $E(X_i) = \mu$, $Var(X_i) = i\sigma^2$. Strategy 1 is X_1 with variance σ^2. Strategy 2 is $\frac{X_2+X_3}{2}$ with variance $\frac{5\sigma^2}{4}$. Strategy 1 more accurate.

Chapter 6

6.1 (a) 0.064. (b) 0.022. (c) 0.909. (d) 0.783. (e) 0.091. (f) 0.217.

6.3 0.966.

6.5 (a) 0.651. (b) 0.736. (c) 0.387.

6.7 X denotes total number of speed cameras catching me. $X \sim Bin(4, 0.7)$. $P(X \geq 1) = 0.9919$.

6.9 (a) 10 fixed trials; 2 outcomes each; constant prob. side-effect person to person; patients independent.

(b)$P(X = x) = \binom{10}{x}(0.2)^x(0.8)^{10-x}$; $x = 0, 1, 2, \ldots, 10$. (c) 0.201. (d)i. 0.376; ii. 0.503; iii. 0.624.

6.11 (a) 9. (b) 0.07. (c) 0.012.

6.13 The respective values of $h(p)$ for $p = 0.1, 0.2, \ldots, 0.7$ are 0.919, 0.737, 0.529, 0.337, 0.188, 0.087, 0.031.

6.15 $E(X) = 30$; $Var(X) = 12$.

6.17 Solution to $1 - (0.999)^n \geq 0.99$ is $n \geq 4603$.

6.19 (a) $p \geq 0$ implies that $p^x(1-p)^{n-x} \geq 0$. (b) $(p+r)^n = \sum_{x=0}^{n} \binom{n}{x} p^x r^{n-x}$. (c) left-side is $[p + (1-p)]^n = 1^n = 1$.

6.21 (a) $\frac{n!}{(n-r)!r!} = \frac{n!}{(n-r)![n-(n-r)]!}$. (b) symmetric. (c) $\binom{n}{r}(\frac{1}{2})^r(\frac{1}{2})^{n-r} = \binom{n}{n-r}(\frac{1}{2})^{n-r}(\frac{1}{2})^{n-(n-r)}$ by (a).

6.23 $p(1-p) \geq \frac{1}{4}$ with equality occurring when $p = \frac{1}{2}$.

6.25 (a) $R_i \sim Bin(n_i, \frac{1}{i})$ has mean $n_i(\frac{1}{i})$ and variance $n_i(\frac{1}{i})(1 - \frac{1}{i})$. (b) M_i = mark for n_i questions $= (1)(R_i + (-\frac{1}{2})(N_i - R_i)$. (c) $E(M_3) = 0$. (d) $E(M_4) < 0$; $P(Mark < 25) = P(M_2 < 7) = P(R_2 \leq 7) = 0.945$.

6.27 Suppose $X \sim Bin(n, p)$; $Y \sim Bin(n-1, p)$. $P(1st\ trial\ success|X = k) = \frac{[p][p_Y(k-1)]}{p_X(k)} = \frac{k}{n}$.

6.29 $\frac{(\frac{1}{2})^n}{1-(\frac{1}{2})^{m+n}}$.

6.31 0.815; 0.647; 0.647; 0.168.

6.33 (a) $\frac{3^3 e^{-3}}{3!} = 0.224$. (b) $(\frac{1^1 e^{-1}}{1!})^3 = 0.050$.

6.35 (a) X denotes number machine failures in day. $X \sim Poi(3)$. $P(X > 3) = 0.353$. (b) $P(Y = y) = 0.647$ if $y = 0$; $P(Y = y) = P(X = 3 + y)$ if $y = 1, 2, 3, \ldots$.

6.37 λ denotes mean defect rate per plate. R denotes number defects. (a) $P(R = r) = \frac{\lambda^r e^{-\lambda}}{r!}$; $r = 0, 1, 2, 3, \ldots$. (b) $P(R = 0) = e^{-\lambda}$. (c) $e^{-\lambda} = 0.9$, giving $\lambda = 0.1054$ (4 dp). Then $P(R = 2) = \frac{\lambda^2 e^{-\lambda}}{2!} = 0.005$ (0.5%).

6.39 (a) 0.632. (b) 0.264. (c) 1; 1. (d) $P(X \geq 1) + P(Y \geq 1) - P(X \geq 1)P(Y \geq 1) = 0.950$. (e) No. Left and right turning traffic may have the same accident. (f) Y.

6.41 $\lambda + \lambda^2 = 12$.

6.43 15; $P(X \geq 15) = 0.041$.

6.45 $\lambda = 0.25$ flaws per 5 metre piece. X denotes total no flawless pieces (out of 100) $\sim Bin(100, e^{-0.25})$. $E(X) = 77.88$. $E(cost) = 5E(X) = 389.4$ dollars.

6.47 (a) Profit $= P = kt - (X_t^2 + X_t)$. (b) $\frac{k-2}{2}$. (c) $t = k - 2$.

6.49 (a) $C = 1X_1 + 2X_2$. (b) $Var(C) = 5\lambda = \frac{5}{3}E(C)$.

6.51 $\kappa_Z(t) = \lambda(e^t - 2 + e^{-t})$.

6.53 (a) X_t discrete; Y continuous. (b) 6. (c) e^{-2t}. (d) As events $\{X_t = 0\} \equiv \{Y > t\}$ (e) $F_Y(y) = 1 - e^{-2y}$ for $y > 0$.

6.55 (a) $p_X(x) = \frac{1}{20}(\frac{19}{20})^{x-1}$, $x = 1, 2, 3, \ldots$ (b) $p_Y(y) = \binom{y-1}{2-1}p^2(1-p)^{y-2}$ for $p = \frac{1}{20}$ and $y = 2, 3, 4, \ldots$.

6.57 $p_X(x) = p(1-p)^x$, for $x = 0, 1, 2, 3, \ldots$.

6.59 (a) $2, 3, 4, \ldots$ (b) $T = T_1 + T_2$ where T_1 is the number of tosses till first tail inclusive; T_2 is the number of tosses until first head inclusive. (c) $E(T) = \frac{1}{1-p} + \frac{1}{p}$ (d) $Var(T) = \frac{p}{(1-p)^2} + \frac{(1-p)}{p^2}$.

6.61 $1 + \frac{5}{4} + \frac{5}{3} + \frac{5}{2} + \frac{5}{1} = 11.42$. All animals equally likely.

6.63 $\frac{1}{2}$.

6.65 $\left[\frac{pe^t}{1-qe^t}\right]^r$ for $q = 1 - p$.

6.67 (a) $W = Y + r \sim Neg\ Bin(r, p)$. (b) $m_W(t) = e^{rt}m_Y(t)$ with $m_W(t)$ as in the previous question.

6.69 $Neg\ Bin(\sum_{i=1}^n r_i, p)$.

Chapter 7

7.1 (a) 0.8413. (b) 0.8413. (c) 0.

7.3 (a) 0.3085. (b) 0.2417. (c) 0.9772. (d) 0.5. (e) 0.6915.

7.5 (a) -0.74. (b) 0.44. (c) 2.22. (d) 1.18.

7.7 0.45.

7.9 0.253.

7.11 0.1587.

7.13 9.2%.

7.15 (a) V = volume dispensed $\sim N(200, 20^2)$. $P(V > 250) = P(Z > \frac{250-200}{20}) = P(Z > 2.5) = 0.0062$. (b) $X \sim Bin(1000, 0.0062)$; $E(X) = (1000)(0.0062) = 6.2$, $Var(X) = (1000)(0.0062)(1 - 0.0062) = 6.16$. Approximately, X has a $Poi(6.2)$ distribution. (c) $Y = (0.25)(1000 - X) - 0.15X$. $E(Y) = 250 - (0.40)(6.2) = 247.52$; $Var(Y) = (-0.4)^2(6.1615) = 0.986$.

7.17 (a) $A - B \sim N(0.06, 0.0004)$. (b) $P(A - B > 0) = P(Z > \frac{0-0.06}{0.02}) = 0.9987$.

7.19 (a) Tasman flight time $= X \sim N(180, 10^2)$. Akd/Well flight time $= Y \sim N(50, 5^2)$. $E(Stopover) = 240 - E(X) = 60$. (b) $P(Y > k) = 0.05$ gives $\frac{k-50}{5} = 1.65$, so that $k = 58.25$. Arrive Wellington approx 2 minutes to 5.(c) No. (d) $240 + Y \sim N(290, 5^2)$.

7.21 (a) $P(M - F < 0) = P(Z < \frac{0-(178-165)}{\sqrt{41+59}}) = P(Z < -1.3) = 0.0968$. (b) Same probability. M and F heights independent; M and F selected randomly.

7.23 $N \sim Bin(1000, P(W > 0.19))$. Since $P(W > 0.19) = 0.001$, N has an approximate Poisson distribution with mean $(1000)(0.001) = 1$. $P(N < 2) = P(N = 0) + P(N = 1) \approx \frac{1^0 e^{-1}}{0!} + \frac{1^1 e^{-1}}{1!} = \frac{2}{e}$.

7.25 For W: mean = 1; median = 1; variance = 2; standard deviation $= \sqrt{2}$; upper quartile $= 1 + (0.675)\sqrt{2} = 1.95$; pdf $f_W(w) = \frac{1}{2\sqrt{\pi}} e^{-\frac{(w-1)^2}{4}}$.

7.27 1; 4.

7.29 $\int_0^\infty e^{-x^2} dx = \sqrt{2} \int_0^\infty e^{-\frac{1}{2}z^2} dz$
$= \frac{\sqrt{2}}{2} \sqrt{2\pi} \int_{-\infty}^\infty \frac{1}{\sqrt{2\pi}} e^{-\frac{1}{2}z^2} dz = \sqrt{\pi}$.

7.31 0.2387.

7.33 (a) $exp(6)$. (b) $e^{-\frac{20}{6}}$. (c) $6 \log_e 2$.

7.35 (a) $exp(5)$; mean 5; variance 25. (b) $Erlang(3, 20)$; mean 60; variance 1200. (c) $Poi(10)$.

7.37 (a) 0.134. (b) $1 - \frac{0.99}{e} = 0.636$. (c) $f_T(t) = 0.01$ if $t = 0$ and $f_T(t) = 0.0099 e^{-0.01t}$ if $t > 0$.

7.39 99 hours.

7.41 $Y \sim exp[(\frac{1}{300} + \frac{1}{400} + \frac{1}{500})^{-1}] = exp(127.7)$.
$Y \sim exp[(\frac{1}{\beta_1} + \frac{1}{\beta_2} + \ldots + \frac{1}{\beta_m})^{-1}]$.

7.43 (a) $E(X) = 6$. (b) 0.816. (c) $E(\sqrt{X}) = \frac{2\Gamma(2)}{\Gamma(1.5)} = \frac{4}{\sqrt{\pi}}$.

7.45 $R \sim \gamma(1 - b, \frac{1}{u}) = \gamma(0.2, 50)$. $E(R) = 10$ and $Var(R) = 500$.

7.47 (a) $\alpha = 2$; $\beta = 10$. (b) 0.1. (c) i. 0.264; ii. 0.330; iii. 0.406.

7.49 The momemt generating function for $Y = \sum_{i=1}^n X_i$ is given by
$(\frac{1}{1-\beta t})^{\alpha_1} (\frac{1}{1-\beta t})^{\alpha_2} \ldots (\frac{1}{1-\beta t})^{\alpha_n} = (\frac{1}{1-\beta t})^{\sum_{i=1}^n \alpha_i}$
and is of the gamma type.

7.51 (a) Solving $S(t) = 0.05$ for t gives $t = 10(\log_e 0.05)^2 = 89.7$. (b) 0.95. (c) 0.95. (d) $(0.95)^{20} = 0.358$. (e) $P(Success) = P(Censored) = 0.05$. Y denotes number of censored from 20 motors. $Y \sim Bin(20, 0.05)$. $P(Y = 2) = 0.189$.

7.53 Integrate each side of $h(t) = \frac{f(t)}{S(t)} = -\frac{d}{dt}[\log_e S(t)]$.

7.55 (a) $h_s(y) = 2h_n(y)$. (b) $exp[-\int_a^b h_s(t)dt]$
$= exp[-\int_a^b 2h_n(t)dt] = (exp[-\int_a^b h_n(t)dt])^2$.

7.57 Data not normally distributed: $x_{0.25} = 15.65$; $x_{0.5} = 16.3$; $x_{0.75} = 18.15$; outliers 26.2, 27.5, 48.0.

7.59 $\frac{1}{2}$.

7.61 (a) $\overline{x}$ close to $\widetilde{x} = 6$. (b) $f(x) \geq 0$ and $\int_0^{12} f(x)dx = 1$. (c) Isosceles triangle standing on $[0, 12]$. (d) lower quartile from f is $3\sqrt{2} = 4.24$. $x_{0.25} = 4.8$.

7.63 (a) $F^{-1}(x) = \beta \sqrt[\alpha]{-\log_e(1-x)}$. (b) Take logarithms of each side in (a).

7.65 (a) $F_X(x) = 1 - e^{-\frac{1}{4}x^2}$. (b) $P(a < X < b) = e^{-\frac{1}{4}a^2} - e^{-\frac{1}{4}b^2}$. Then $P(0 < X < 0.5) = 0.061$. (c) Use (b) for each interval: rounded to 1 dp expect

6.1 16.1 20.9 20.2 15.8 10.4 5.9 2.9.

7.67 $f(p) = \sin^{-1}\sqrt{p}$ satisfies $f'(p) = \frac{1}{\sqrt{1-(\sqrt{p})^2}}(\frac{1}{2\sqrt{p}})$ and therefore the stated equation. f is variance stabilising for binomials .

Chapter 8

8.1 Conditional distribution:

y	5	6
$p_{Y\|50}(y)$	$\frac{2}{3}$	$\frac{1}{3}$

8.3 (a) $E(X) = 1$. (b) $P(X > Y) = 0.25$. (c) $E(X|Y = 0) = 1.25$. (d) $E(Y|X = 1) = 1.5$.

8.5 Joint mass function:

$p_{X_1,X_2}(x_1, x_2)$	x_2	0	1
x_1			
0		$\frac{56}{90}$	$\frac{16}{90}$
1		$\frac{16}{90}$	$\frac{2}{90}$

X_1 and X_2 are dependent.

8.7 (a)i. $f_{Y|x}(y|x) = \frac{1}{2\sqrt{1-x^2}}$ for $-\sqrt{1-x^2} < y < \sqrt{1-x^2}$ and $-1 < x < 1$; ii. 0; iii. 0.
(b) $Y|x \sim U[-\sqrt{1-x^2}, \sqrt{1-x^2}]$. The distribution of $Y|x$ lies uniformly on a vertical slice through x about mean 0 corresponding to the x axis.

8.9 Clearly, $q(x) \geq 0$ for all x. That the values of $q(x)$ sum to 1 is seen by summing over all x values:
$q(x) = P(X = x|Y = y) = \frac{P(X=x|Y=y)P(Y=y)}{P(Y=y)} = \frac{p_{X,Y}(x,y)}{p_Y(y)}$.

8.11 X denotes number of words scanned. Y is 1 if 'operation' is in list and 0 otherwise. $E(X) = E(E(X|Y)) = (\frac{10^6+1}{2})(\frac{1}{2})+(10^6)(\frac{1}{2}) = 750000.25$

8.13 $p_Y(y) = \sum_{x=y}^{\infty} \binom{x}{y} p^y(1-p)^{x-y}(\frac{\lambda^x e^{-\lambda}}{x!})$

$$= \frac{(\lambda p)^y e^{-\lambda p}}{y!} \underbrace{\sum_{w=0}^{\infty} \frac{[\lambda(1-p)]^w e^{-\lambda(1-p)}}{w!}}_{=1}.$$

8.15 (a) increases.(b) $2E(X) - 1$.

8.17 $\frac{2}{27}$.

8.19 $E(X + Y) = -1$; $Var(X + Y) = Var(X) + Var(Y) + 2Cov(X, Y) = 4.33$.

8.21 $Cov(X, n - X) = -Cov(X, X) = -np(1 - p)$.

8.23 $Cov(X-Y, X-Y) = Cov(X, X) - Cov(X, Y) - Cov(Y, X) + Cov(Y, Y)$ and $Cov(W, W) = Var(W)$; **SUM 3** with $n = 2$, $a_1 = 1$ and $a_2 = -1$.

8.25 $Cov(c, X) = E(cX) - E(c)E(X) = cE(X) - cE(X) = 0$.

8.25 $Cov(c, X) = E(cX) - E(c)E(X) = cE(X) - cE(X) = 0$.

8.27 (a) larger values of X are associated with smaller values of Y. (b) -0.6. (c) Same. (d)

Variable	Mean	Variance
X	3.4	1.84
Y	2.2	2.16
T	5.6	2.80
C	368	736

8.29 (a) Negative covariance—opposite price movements. (b) Positive covariance—similar price movements. (c) $\sqrt{1750} = 41.83$. (d) $\sqrt{3200} = 56.57$.

8.31 (b) $c = \frac{Cov(X_1,X_2)}{\sigma_2^2}$. (c) Mean is $\mu_2 - \frac{Cov(X_1,X_2)}{\sigma_1^2}\mu_1$. Variance is $\sigma_2^2[1 - \rho^2]$ for $\rho = \frac{Cov(X_1,X_2)}{\sigma_1\sigma_2}$.

8.33 (a) $(X_1, X_2, \ldots, X_6) \sim Multinomial(10, \frac{1}{7}, \frac{1}{7}, \frac{1}{7}, \frac{1}{7}, \frac{1}{7})$. (b) $\frac{10!}{5!5!}(\frac{1}{7})^5(\frac{2}{7})^5 = 2.85 \times 10^{-5}$. (c) $Cov(X_1, X_6) = -\frac{20}{49}$.

8.35 -0.164.

8.37 -0.111.

8.39 (a) $E(T) = 5$; $Var(T) = 23.5$. (b) $Cov(T, X) = 12.7$. (c) 0.828.

8.41 (a) 1.5. (b) 0.75. (c) 7; 86.

8.43 $N(\frac{x-5}{3}, 3)$.

8.45 $Var(Z|U = u) = 1 - \rho^2$. Therefore, $Var(\frac{Y-\mu_Y}{\sigma_Y}|X = x) = 1 - \rho^2$, implying that $Var(Y|X = x) = \sigma_Y^2(1 - \rho^2)$.

Chapter 9

9.1 (a) $E(X) = 0.9$; $Var(X) = 0.09$. (b) $\frac{0.3}{\sqrt{2}} = 0.212$. (c) $\frac{0.3}{\sqrt{100}} = 0.030$. (d) Decreases. (e) Stays same.

9.3 178%.

9.5 The discrete uniform distribution on $1, 2, \ldots, 100$ has variance $\sigma^2 = \frac{100^2-1}{12}$. Then $Var(\overline{X}) = \frac{\sigma^2}{100} = 8.33$.

9.7 (a) 0.5. (b) Area under pdf beyond 0.75 is 0.125. (d) 0.75.

9.9 X has pmf:

x	α_1	α_2	$\ldots$	α_n
$p_X(x)$	$\frac{1}{n}$	$\frac{1}{n}$	$\ldots$	$\frac{1}{n}$

$E(X) = \sum_{i=1}^n(\frac{1}{n})\alpha_i = \overline{\alpha}$. $Var(X) = E(X^2) - E(X)^2 = \sum_{i=1}^n(\frac{1}{n})\alpha_i^2 - (\overline{\alpha})^2$ equivalent to the stated result.

9.11 (a) 100. (b) 10.

9.13 (a) $E(\overline{X}) = 98$; $S.E.(\overline{X}) = 1$. (b) CLT. (c) 0.0228.

9.15 (a) $a = 0.5$; $b = \frac{1}{400}$. (b) Total $= T = 100\overline{X}$. $P(T > 59.8) = P(\overline{X} > 0.598) = 0.025$.

9.17 $E(X) = 9.5$; $Var(X) = 22.25$. Four 10 week terms consist of 200 days at an average of $\overline{X}$ km per day. $P(200\overline{X} > 2000) = 0.0668$.

9.19 (a) $F_X(x) = P(X \le x) = P(\log_e X \le \log_e x) = P(Z \le \frac{\log_e x - 5}{\sqrt{2}})$ where $Z \sim N(0,1)$. (b) Solve $F_X(t) = 0.01$ for t to give $t = 5.58$ years. (c) 0.0024.

9.21 0.8413.

9.23 0.0294.

9.25 0.016 without continuity correction. 0.037 with continuity correction.

9.27 Using continuity correction, for $Y \sim Bin(1000, \frac{1}{3})$, and $Z \sim N(0,1)$, $P(Y = 340) \approx P(0.414 < Z < 0.481) = 0.025$.

9.29 (a) Mean error in length. (b) 0.8413.

9.31 0.5.

9.33 (a) 225 792 840. (c) No. (d) $\binom{31}{11} = \frac{12}{32}\binom{32}{12}$.

9.35 $Cov(X,Y) = -\frac{\sigma^2}{N-1}$. Result follows since $\rho_{X,Y} = \frac{Cov(X,Y)}{\sigma_X \sigma_y}$ and $\sigma_X = \sigma_Y = \sigma$.

9.37 Case 1: $X \sim Hypergeom(20, 18, 10)$. $P(X = x) = \frac{\binom{18}{x}\binom{2}{10-x}}{\binom{20}{10}}$ for $8 \le x \le 10$.

Case 2: $X \sim Hypergeom(20, 2, 10)$. $P(X = x) = \frac{\binom{2}{x}\binom{18}{10-x}}{\binom{20}{10}}$ for $0 \le x \le 2$.

9.39 (a) 1000. (b) $Y \sim Bin(10000, 0.1)$. (c) $P(Y \ge 1100) = P(Z \ge 3.3) = 0.0004$. (d) No. 1100 unlikely to occur by chance if digits truly random.

9.41 Simplify the given variance formula for the stratified sample mean by replacing $\frac{n_1}{N_1}$ and $\frac{n_2}{N_2}$ by the sampling fraction f; also $\frac{w_1}{n_1}$, $\frac{w_2}{n_2}$ by $\frac{1}{n}$.

9.43 $\frac{(0.1)300^2+(0.9)200^2}{50}(1 - \frac{5}{24}) = 712.5$.

9.45 Sampling frame is {Row 1, Row 2, ..., Row 24}. Use random number tables to select 6 rows from the 24. Determine the observed sample mean $\overline{x}$. This has estimated variance $\frac{s^2}{6}(1-f)$, where s^2 is the observed sample variance of the number of millable trees in a row and $f = \frac{1}{4}$ is the sampling fraction.

9.47 (a) $n_1 = 12$, $n_2 = 40.5 \approx 41$, $n_3 = 22.5 \approx 22$. (b) Ignoring sampling fractions and common factor σ^2,
Method 1 variance $\frac{(2)(0.16)+(1)(0.54)+(1)(0.3)}{75}$ is smaller than Method 2 variance $\frac{(2)(0.3333)+(1)(0.3333)+(1)(0.3333)}{75}$.

9.49 (a) $Var(\widehat{T}) = N^2 Var(\overline{X}_{strat})$. (b) Stratum totals. (c) $\overline{x_1} = 18.75$; $\overline{x_2} = 5.00$. Therefore, $\widehat{t_1} = 16(18.75) = 300$ and $\widehat{t_2} = 8(5.00) = 40$.

Chapter 10

10.1 $\overline{x} = 0.608$ estimates μ, the true mean heat output. The standard error of estimate is 0.060.

10.3 Estimated total spending is 390 000 dollars. The standard error is estimated by 20 000 dollars.

10.5 (a) S/L plot:

Electrical resistance (ohms)
Scale: stems—10^{-1} ohms; leaves—10^{-2} ohms
28|5 represents 2.85 ohms

28.	5 6 8
29*	0 0 0 1 1 1 2 2 2 3 3 3 4 4
29.	5 5 5 5 5 5 6 8 8 8 8 9
30*	0 0 0 3 3 4 4 4
30.	5 5 5 7 9 9
31*	0 0 2 2 2
31.	7 8

(b) 2.990. (c) Sample variance is 0.0066. (d) 0.0115. (e) Range is 0.33. One-quarter of this gives $\widehat{\sigma} = 0.0825$.

10.7 (a) $E(\widehat{\lambda}) = E(\frac{1}{n}\sum_{i=1}^{n} X_i) = \frac{1}{n}\sum_{i=1}^{n} E(X_i) = \frac{1}{n}\sum_{i=1}^{n} \lambda = \lambda$. (b) $S.E.(\widehat{\lambda}) = \sqrt{\frac{\lambda}{n}}$. (c) $\widehat{\lambda} = 4.6$. $S.E.(\widehat{\lambda}) = \sqrt{\frac{4.6}{15}} = 0.55$.

10.9 (a) Yes. (b) $\widehat{T_1}$ has smallest standard error.

10.11 Because of unbiasedness, $M.S.E.(\widehat{T}) = Var(\widehat{T}) = (const)[\frac{p^2}{16} + (1-p)^2]$. This is minimised for $p = \frac{16}{17}$.

10.13 $Var(\overline{X}) = \frac{\sigma^2}{n} = E(\overline{X}^2) - \mu^2$. Therefore, $E(\overline{X}^2) = \frac{\sigma^2}{n} + \mu^2$ and so $\overline{X}^2$ is biased for μ^2.

10.15 (a) Coefficient of kurtosis is 3. (c) $\frac{n-1}{n+1}$.

10.17 $\prod_{i=1}^{n} e^{-\frac{1}{2}(x_i-\mu)^2} = e^{-\frac{1}{2}\sum_{i=1}^{n} x_i^2 + \sum_{i=1}^{n} x_i\mu - \frac{1}{2}n\mu^2} = (const)e^{\mu\sum_{i=1}^{n} x_i - \frac{1}{2}n\mu^2}$. Maximising value (and method of moments estimate) is $\widehat{\mu} = \overline{x}$.

10.19 $\widehat{\sigma} = \sqrt{\frac{1}{10}\sum_{i=1}^{10} x_i^2} = 0.0932$.

10.21 (a) 1. (b) $\frac{1}{2}$.

10.23 (a) $E(\delta_i) = (1)P(X_i > c) + (0)P(X_i \leq c) = \theta$. Then $E(\widehat{\theta}) = \frac{1}{n}\sum_{i=1}^{n} E(\delta_i) = \frac{1}{n}\sum_{i=1}^{n} \theta = \theta$. (b) larger than c. (c) $\frac{\theta(1-\theta)}{n}$. (d) δ_i has a Bernoulli distribution with parameter θ. (e) $c\theta^{\sum_{i=1}^{n}\delta_i}(1-\theta)^{n-\sum_{i=1}^{n}\delta_i}$.

10.25 $X \sim Poi(\lambda)$, where X denotes the number of faults in 100 metres of cloth. Required MLE is $e^{-3\overline{x}} = 0.097$.

10.27 (a) $\theta = \widehat{\theta} = \frac{\sum_{i=1}^{n}\delta_i}{n}$ is the value of θ which maximises the likelihood. (b) MLE of c is $\widehat{c} = F^{-1}(1-\widehat{\theta})$.

10.29 $[22\,735.6, 24\,264.4]$.

10.31 For $n = 52$,

80% CI is $[212.39, 224.61]$
90% CI is $[210.62, 226.38]$
95% CI is $[209.05, 227.95]$
99% CI is $[205.91, 231.09]$.

For $n = 28$,

80% CI is $[219.13, 235.23]$
90% CI is $[216.74, 237.62]$
95% CI is $[214.60, 239.75]$
99% CI is $[210.20, 244.16]$.

As confidence level decreases so does the width. Smaller n makes for wider intervals.

10.33 (a) $[1018.4, 1101.7]$. (b) The standardised error $\frac{e}{\sigma/\sqrt{n}}$ is $z_{0.04} = 1.75$ standard deviations away from 0. With $e = \frac{1}{5}\sigma$, solution is $n = 77$.

10.35 P(μ between X_1 and X_2) $= 2P(X_1 < \mu < X_2) = 2P(X_1 < \mu)P(X_2 > \mu) = (2)(\frac{1}{2})(\frac{1}{2}) = \frac{1}{2}$.

10.37 (a) $X_i \sim U[\theta-1, \theta+1]$. Therefore, $Y_i = X_i - \theta \sim U[-1, 1]$ and this distribution does not depend on θ. $Q = \overline{Y} \sim N(0, \frac{1}{1200})$ approximately. (b) $[\overline{X} - z_{\frac{\alpha}{2}}\sqrt{\frac{1}{1200}}, \overline{X} + z_{\frac{\alpha}{2}}\sqrt{\frac{1}{1200}}]$.

10.39 15.987; 4.865; 2.558; 23.209.

10.41 7.261; 0.05; 0.03.

10.43 (a)S/L plot:

Glucose plasma data (mg dL^{-1})
Scale: stems—10 units; leaves—units
10|5 represents 105 mg dL^{-1}

7	5
8	44
8	66788
9	001223334
9	555668889
10	00000223334
10	5666889
11	02222
11	
12	0

(c) No outliers. (d) $s = 9.08$; $[57.94, 129.82]$.

10.45 Data not normally distributed.

10.47 (a) 0.05. (b) 0.94.

10.49 (a) No; no. (b) Yes; no. (c) Yes; no. (d) Yes; yes.

10.51 A chi-squared distribution with k degrees of freedom has mean k and variance $2k$. Then for large k, $\frac{W-k}{\sqrt{2k}} \approx N(0,1)$ for $W \sim \chi^2_k$. Let $Y = \sqrt{2W} - \sqrt{2k-1}$. Then $P(Y \le y) = P(\frac{W-k}{\sqrt{2k}} < \frac{\frac{1}{2}y^2 - \frac{1}{2} + y\sqrt{2k-1}}{\sqrt{2k}})$ $\approx \Phi(\frac{\frac{1}{2}y^2 - \frac{1}{2} + y\sqrt{2k-1}}{\sqrt{2k}}) \to \Phi(y)$ as $k \to \infty$.

$\alpha = 0.1$ and $k = 25$ give $z_{0.1} = 1.28$ and $\chi^2_{0.1,25} \approx 34.28$.

10.53 (a) 1.782. (b) -1.782. (c) 2.179.

10.55 (b) $[0.75, 8.25]$. (c) Data are normally distributed.

10.57 $\bar{x} = 0.25$. A 98% CI for μ is $[-1.05, 1.55]$. The population mean is likely to be 0 since this value lies in the CI; $s^2 = 4$. A 98% CI for σ^2 is $[1.96, 11.47]$.

10.59 Yes. t_{20} curve has more probability in its tails (than the standard normal curve) so that its upper α points are higher than for the standard normal curve.

10.61 Put $k = 1$ in the formula for the pdf of $T \sim t_k$ and a *Cauchy*(1) density is obtained.

Chapter 11

11.1 Only (b) and (c) are appropriate.

11.3 Hypotheses are:

$$H_0: \quad \mu = 2$$
$$H_1: \quad \mu > 2.$$

n is large so no need to assume normal data.

11.5 μ denotes mean parking time. Hypotheses are:

$$H_0: \quad \mu = 220$$
$$H_1: \quad \mu \neq 220.$$

11.7 $z_{observed} = 2$. p-value $= 2P(Z > 2) = 0.046 < 0.05$, so result just near borderline between significance and non-significance.

11.9 Entries from top to bottom of table: 0.1587; 0.1336; ±1.88; 0.0287; 2; 0.0062.

11.11 In terms of μ, the mean imagery rating for *house*, hypotheses are:

$$H_0: \quad \mu = 6$$
$$H_1: \quad \mu > 6.$$

$z_{observed} = 4.22$ (**) which is highly significant.

11.13 In terms of μ, the mean breaking strength, hypotheses are:

$$H_0: \quad \mu = 16$$
$$H_1: \quad \mu < 16.$$

$z_{observed} = -1.41$ ($n.s.$); $p - value = 0.079$.

11.15 (d).

11.17 (a) Bernoulli trials only if number of employees, N, is exceedingly large compared with $n = 20$, otherwise sampling is without replacement from a finite population and independence fails. (b)

$$H_0: \quad p = 0.2$$
$$H_1: \quad p > 0.2.$$

(c) Reject when $Y \geq 8$, where Y denotes total number public transport commuters in sample. (d) $P(Y \geq 8) = 0.032$.

11.19 (a) 0.99. (b) 0.08.

11.21 (a) $(\frac{2}{3})^4 + (\frac{1}{3})^4 = 0.2099$. (b) 0.790. (We seem very likely to accept H_0 when it is false!)

11.23 (a)

$$H_0: \quad p = 0.4$$
$$H_1: \quad p < 0.4.$$

(b) $RR = \{0, 1, 2, 3\}$. (c) 0.016. (d) (**), (**), (*), (*n.s.*), (*n.s.*) for $x_{observed} = 1, 2, 3, 4, 5$ respectively.

11.25 (a) $e^{-\lambda}$. (b) i. For $X \sim Poi(5)$, $\alpha = P(X \geq 10) = 0.032$. ii. 0.084; 0.283; 0.542.

11.27 Hypotheses:

$$H_0: \quad \mu = 36$$
$$H_1: \quad \mu > 36.$$

where μ is the true mean waiting time.

Reject H_0 if $X > k$ where $P(X > k) = 0.05$, so that $k = -36 \log_e 0.05 \approx 108$.

11.29 Hypotheses:

$$H_0: \quad \mu = 25$$
$$H_1: \quad \mu \neq 25.$$

$t_{observed} = -2.25$. $0.02 < p-\text{value} < 0.05$.

11.31 Hypotheses:

$$H_0: \quad \mu_d = 0$$
$$H_1: \quad \mu_d > 0,$$

where μ_d denotes mean increase in km travelled on 20 L petrol with device fitted. $t_{observed} = 1.414(n.s.)$. Fail to reject H_0. Claim not justified. Require normally distributed differences.

11.33 (a) linear plot. (b) Use $S.E.(\overline{X}) = \frac{s}{\sqrt{n}}$. (c) $t_{observed} = -1.54$; (*n.s.*). $0.1 < p\text{-value} < 0.2$.

11.35 (a) Reject for $X \geq 40$ or $X \leq 24$. 95% CI is $[\frac{25}{64}, \frac{39}{64}]$. (b) Reject for $X \geq 43$ or $X \leq 21$. 99% CI is $[\frac{22}{64}, \frac{42}{64}]$.

Take confidence intervals as providing acceptance regions for the hypothesis tests.

11.37 (a) Hypotheses:

$$H_0: \quad \mu = 220$$
$$H_1: \quad \mu \neq 220.$$

where μ denotes mean parking time. (b)i. $z_{observed} = -2.12$ (*). ii. No. (c)i. $t_{observed} = -0.67$ (*n.s.*). ii.Yes. (d) Test in (b) with larger n.

11.39 0.0179.

11.41 (a) $k = 3.325; l = 2.700; m = 3$. (b) $P(S^2 < 1.11|\sigma^2 = 3) = 0.05$. (c) $P(S^2 > 1.11|\sigma^2 = 2.396) = 0.90$.

11.43 $k = 20$; $n = 68$.

Chapter 12

12.1 (a) 1 hr: 0.63; 24 hr: 0.49. (b) 4.25. (c) No: $t_{observed} = 0.26$ (*n.s.*).

12.3 $Z = \frac{\overline{D} - \mu_D}{\frac{S_D}{\sqrt{50}}} \approx N(0, 1)$.

12.5 (a) $\mu_D = \mu_B - \mu_A$. (b) Reject H_0 at $\alpha = 0.05$. (c) 1.133. (d) $2.262 = (2)(t_{0.025,7})\frac{s_D}{\sqrt{8}}$ has solution $s_D = 1.353$.

12.7 (a) 3.58. (b) 3.29. (c) 1.93. (d) 4.71. (e) 2.98.

12.9 99th percentile: 3.09. 1st percentile: 0.297.

12.11 $f_{observed} = \frac{s_1^2}{s_2^2} = 1.87$ (*n.s.*).

12.13 Reject H_0 at the 0.05 level since the CI does not contain 1.

12.15 $Z^2 \sim \chi_1^2$. If $W \sim \chi_\nu^2$ independently of Z, then $T = \frac{Z}{\sqrt{\frac{W}{\nu}}} \sim t_\nu$. Therefore, $T^2 = \frac{(\frac{Z^2}{1})}{(\frac{W}{\nu})} \sim F_{1,\nu}$.

12.17 (a) $\widehat{P}(Sydney) - \widehat{P}(Melbourne) = 0.05$. (b) 0.026.

12.19 0.215.

12.21 (a) Yes. $z_{observed} = 3.48$, (**). (b) Yes. $z_{observed} = 5.02$, (**).

12.23 Data paired, NOT independent. $[\overline{D} - t_{0.025,4}\frac{s_D}{\sqrt{5}}, \overline{D} + t_{0.025,4}\frac{s_D}{\sqrt{5}}] = [-1.776, 3.776]$.

12.25 $s_p = 0.905345$; 95% CI is $[-1.72, -0.02]$. Battery types are normally distributed with common variance.

12.27 Yes. $z_{observed} = 4.23$, (**).

12.29 $Var(X_1) = Var(X_6) = (n)(\frac{1}{6})(\frac{5}{6})$. $Cov(X_1, X_6) = -(n)(\frac{1}{6})(\frac{1}{6})$. This gives $Var(X_1 - X_6) = \frac{n}{3}$. Observed difference is 86. $S.E.(diff) = \sqrt{\frac{1000}{3}} = 18.3$. The difference appears significant.

12.31 (c) $s = 0.952$. 95% CI is $[-2.77, 2.27]$.

12.33 (a)

Source	Deg of freedom	Sums of squares	Mean squares	F
Treatments	3	1 134	378	29.1
Residuals	16	208	13	
Total	19	1 342		

(b) Yes. $f_{observed} = 29.1$, (**). (c) (C).

12.35 The array of fitted values is:

$$\begin{bmatrix} 73.375 & 72.500 & 71.833 & 72.25 \\ 73.375 & 72.500 & 71.833 & 72.25 \\ 73.375 & & 71.833 & \\ 73.375 & & & \end{bmatrix}$$

12.37 (b)

Source	Deg of freedom	Sums of squares	Mean squares	F
Treatments	2	0.23208	0.1160	13.48
Residuals	15	0.12915	0.0086	
Total	17	0.36123		

12.39 (a) -0.198; 30%. (b) SS RESIDUALS = 0.12915. (c)

Source	Deg of freedom	Sums of squares	Mean squares	F
Treatments	2	0	0	0
Residuals	15	0.12915	0.0086	
Total	17	0.12915		

12.41 Equation indicates that column sums in residual array should be 0. This is true because the columns have been mean corrected.

12.43 $\sum_{j=1}^{k}\sum_{i=1}^{n_j}(\overline{x}_{.j} - \overline{x}_{..})(x_{ij} - \overline{x}_{.j})$

$$= \sum_{j=1}^{k}(\overline{x}_{.j} - \overline{x}_{..})\underbrace{\sum_{i=1}^{n_j}(x_{ij} - \overline{x}_{.j})}_{=0}.$$

Chapter 13

13.1 $z_{observed} = 2.77$, (**); $q_{observed} = 7.7$, (**).

13.3 (a) $q_{observed} = 14$, (**). (b) $q_{observed} = 1.36$, (*n.s.*). (c) $q_{observed} = 0.67$, (*n.s.*).

13.5 $q_{observed} = 0.71$, (*n.s.*).

13.7 (a) The 3 categories $x = 0$, $x = 1$, $x \geq 2$ have probabilities 0.5314, 0.3543, 0.1143 respectively under H_0 after pooling. This gives $q_{observed} = 0.47$, (*n.s.*). Good fit. (b) Poisson model gives similar results for same categories; $q_{observed} = 1$, (*n.s.*). Good fit. But allows for more defects than six.

13.9 Taking the expected value of each term with $E(X_i - np_i)^2 = Var(X_i)$ we find on substitution that

$$E(Q) = \sum_{i=1}^{k}(1 - p_i) = k - \sum_{i=1}^{k} p_i = k - 1.$$

13.11 (a) 4×3. (b) Fitted values are:

17.7	26.6	15.6
17.1	25.7	15.1
15.4	23.1	13.6
17.7	26.6	15.6

13.13 Test for homogeneity across years in 2×5 table has $q_{observed} \approx 1$, (*n.s.*).

13.15 (a) Yes. Distributions have similar shape. (b) $q_{observed} = 2.34$, (*n.s.*). (c) No. $q_{observed} = 6.48$, (*).

13.17 (a) Test for difference between binomial proportions with pooled estimator of the common probability of success

$$\widehat{p} = \frac{X_{11} + X_{12}}{X_{.1} + X_{.2}} = \frac{X_{1.}}{X_{..}}.$$

(c) The allocation of the responses is gender-dependent. $z_{observed} = 3.4$, (**). Estimated standard error (without pooling) is 0.031.

Chapter 14

14.1 (a) Response - y; explanatory variable - x. (c) Yes. (d) Yes. (e) 4; 2500; 240; 500; 360. (f) 3. (g) 75 dollars.

14.3 $y = 2.51803 + 0.067498x$. For example, $e_1 = 2.53 - [2.51803 + (0.067498)(2)] = -0.123026 \approx -0.123$.

14.5 (b) $m = h(t - t_0)$ estimated by $m = 1.640(t - 1.3598)$.

14.7 $A(a) = \sum_{i=1}^{n}(y_i - \overline{y})^2 + n(\overline{y} - a)^2$. Right-side is smallest when $a = \overline{y}$. When fitting $y = \beta_0$, predicted value is $\widehat{\beta_0} = \overline{y}$.

14.9
$$\sum_{i=1}^{n}(y_i - a)(x_i - \overline{x}) = \sum_{i=1}^{n} y_i(x_i - \overline{x}) - a\underbrace{\sum_{i=1}^{n}(x_i - \overline{x})}_{=0}.$$

14.11 $\sum_{i=1}^{n} e_i = \sum_{i=1}^{n}(y_i - \widehat{y_i}) = n\overline{y} - n\overline{y} = 0.$

14.13 (a) SS RESIDUALS = 0.475. (b) [1.05, 1.85].

14.15 (a) $X = 91 - T$. (b) $y^* = -2.0125 + 0.070747t$. (c) New intercept $\widehat{\beta_0} - \widehat{\beta_1}\frac{a}{b}$. New slope $\frac{\widehat{\beta_1}}{b}$.

14.17 (a). $\widehat{\theta} = 28.08$; $S.E.(\widehat{\theta}) = 1.03$. (b) Outside x range.

14.21 $t_{observed} = 0.985$, $(n.s.)$.

14.23 $E(\widehat{Y_i}) = E(\widehat{\beta_0} + \widehat{\beta_1}x_i) = E(\widehat{\beta_0}) + x_i E(\widehat{\beta_1}) = \beta_0 + \beta_1 x_i = E(Y_i)$.

14.25 Left-side $= \sum_{i=1}^{n}(\widehat{Y_i} - \overline{Y})e_i = \widehat{\beta_1}\sum_{i=1}^{n}(x_i - \overline{x})e_i = 0$.

14.27 (a) $E(\widehat{\beta_1}) = \beta_1 . Var(\widehat{\beta_1}) = \frac{\sigma^2}{\sum_{i=1}^{n}(x_i-\overline{x})^2}$. (b) Take expected value of each side of the equation, substitute $E(\widehat{\beta_1}^2) = Var(\widehat{\beta_1}) + [E(\widehat{\beta_1})]^2$, and use (a). (c) Ratio with E(MS RESIDUALS) $= \sigma^2$ will be $1 + \epsilon$ where ϵ is small and positive.

14.29 (b) 0.00606.

14.31 (c) More linear. (d) $\log_{10} y = 3.46 - 0.192X$. At $x = 10$, $y = 34.67 \approx 35$. (e) Scale (slope) changes. Intercept does not. (f) Points $(13, 0.3010)$, $(20, 0)$ have large residuals. $(20, 0)$ has high influence.

14.33 Estimated variance of residual $= \widehat{\sigma^2}(1 - h_{ii}) = \widehat{\sigma^2} - \widehat{\sigma^2}h_{ii} =$ MS RESIDUALS – estimated variance of fit.

14.35 Marriages tend to be between people of similar age. $r_{X,Y} = 0.988$.

14.37 More cars per kilometre means that they drive more slowly; -0.898.

14.39 $r_{e,x} = 0$ since $\sum_{i=1}^{n}(x_i - \overline{x})(e_i - \overline{e}) = 0$.

14.41 Very weakly correlated, $r = 0.451$. The improvement is not dictated by youth. (*Caution*: The sample size is really too small for sound experimental conclusions.)

14.43 Yes. Yes. The joint density factorises if $\rho = 0$.

Chapter 15

15.1 $y_{observed} = 5$, $(n.s.)$.

15.3 (c) S shape. (d)i. Subtract 135 from all data; retain only sign. Order irrelevant. ii. 0.254. iii. 0.254.

15.5 15 lies at the upper quartile, so that $(0.25)(28) = 7 = y_{observed}$ + signs lie above 15. (Assume no data point of 15, although such discrepancies are negligible for large sample sizes.) $z_{observed} = -2.6$, $(**)$.

15.7 (a) $t_{observed} = 2.85$, $(*)$. (b) $s^+_{observed} = 14$, $(n.s.)$.

15.9 (b) Test for median wind velocity satisfies assumptions. (c) $z_{observed} = 3.19$, $(**)$.

15.11 $s^+_{observed} = 33$, $(*)$.

15.13 $s^+ \geq 807.9$.

15.15 Subtract 30 days from each observation. $s^+_{observed} = 8$, $(n.s.)$.

15.17 Subtract Old - New. Single sample signed-rank test. $n = 9$, as 0 value discarded. $s^+_{observed} = 43$, $(*)$

15.19 Add 50 000 to the U data. Then the large-sample rank-sum test gives $z_{observed} = -0.39$, $(n.s.)$.

15.17 Subtract Old - New. Single sample signed-rank test. $n = 9$, as 0 value discarded. $s^{+}_{observed} = 43$, (*)

15.19 Add 50 000 to the U data. Then the large-sample rank-sum test gives $z_{observed} = -0.39$, (*n.s.*).

15.21 Let $k = m + n$. Then $E(R) = \sum_{i=1}^{k} i(\frac{1}{k}) = \frac{1}{k}\frac{k(k+1)}{2}$. Similarly, $E(R^2) = \sum_{i=1}^{k} i^2(\frac{1}{k}) = \frac{1}{k}\frac{k(k+1)(2k+1)}{6}$. Then $Var(R) = E(R^2) - E(R)^2 = \frac{k^2-1}{12}$.

Index